Hormones and Aggressive Behavior

Hormones and Aggressive Behavior

Edited by

Bruce B. Svare
State University of New York at Albany
Albany, New York

PLENUM PRESS • NEW YORK AND LONDON

Library of Congress Cataloging in Publication Data

Main entry under title:

Hormones and aggressive behavior.

Includes bibliographical references and index.

1. Aggressiveness (Psychology)—Physiological aspects. 2. Hormones. 3. Psychology, Comparative. I. Svare, Bruce B., 1949- . [DNLM: 1. Aggression—Physiology. 2. Hormones—Physiology. WK 102 H8118]

BF575.A3H66 1983 156'5 83-4020

ISBN 0-306-41055-9

A Division of Plenum Publishing Corporation
233 Spring Street, New York, N.Y. 10013

Printed in the United States of America

Contributors

DAVID B. ADAMS, Department of Psychology, Wesleyan University, Middletown, Connecticut

WILLIAM J. BELL, Department of Entomology, University of Kansas, Lawrence, Kansas

DAVID BENTON, Department of Psychology, University College of Swansea, Swansea, Wales, United Kingdom

IRWIN S. BERNSTEIN, Department of Psychology, University of Georgia, Athens, Georgia, and Yerkes Regional Primate Research Center, Emory University, Atlanta, Georgia

MARIE-FRANCE BOUISSOU, Station de Physiologie de la Reproduction, I.N.R.A., Nouzilly, France

PAUL F. BRAIN, Department of Zoology, University College of Swansea, Swansea, Wales, United Kingdom

MICHAEL D. BREED, Department of Environmental, Population, and Organismic Biology, University of Colorado, Boulder, Colorado

ROBERT L. CONNER, Department of Psychology, Bowling Green State University, Bowling Green, Ohio

ANTHONY P. CONSTANTINO, Department of Psychology, Bowling Green State University, Bowling Green, Ohio

DAVID CREWS, Department of Zoology, University of Texas, Austin, Texas

JOSEPH F. DEBOLD, Department of Psychology, Tufts University, Medford, Massachusetts

OWEN R. FLOODY, Department of Psychology, Bucknell University, Lewisburg, Pennsylvania

RONALD GANDELMAN, Department of Psychology, Rutgers University, New Brunswick, New Jersey

THOMAS P. GORDON, Yerkes Regional Primate Research Center, Emory University, Atlanta, Georgia

NEIL GREENBERG, Department of Zoology, University of Tennessee, Knoxville, Tennessee

CHERYL F. HARDING, Department of Psychology, Hunter College, New York, New York and Department of Ornithology, American Museum of Natural History, New York, New York

DAVID W. INGERSOLL, Department of Psychology, Fordham University, Rose Hill Campus, Bronx, New York

CHING-TSE LEE, Department of Psychology, Brooklyn College of the City University of New York, Brooklyn, New York

ALAN I. LESHNER, National Science Foundation, Washington, D.C.

WILLIAM G. LUTTGE, Department of Neuroscience and Center for Neurobiological Sciences, University of Florida College of Medicine, Gainesville, Florida

MARTHA A. MANN, Department of Psychology, State University of New York at Albany, Albany, New York

STEPHEN C. MAXSON, Department of Biobehavioral Sciences, University of Connecticut, Storrs, Connecticut

ALLAN MAZUR, Department of Sociology, Syracuse University, Syracuse, New York

KLAUS A. MICZEK, Department of Psychology, Tufts University, Medford, Massachusetts

ROBERT M. ROSE, University of Texas Medical School Branch at Galveston, Galveston, Texas

GEORGE C. SCHEUCH, Department of Psychology, Bowling Green State University, Bowling Green, Ohio

PAUL SHRENKER, Department of Biobehavioral Sciences, University of Connecticut, Storrs, Connecticut

BRUCE B. SVARE, Department of Psychology, State University of New York at Albany, Albany, New York

LYNNE C. VIGUE, Department of Biobehavioral Sciences, University of Connecticut, Storrs, Connecticut

TRUDY A. VILLARS, Department of Psychology, Wheaton College, Norton, Massachusetts

FREDERICK S. VOM SAAL, Division of Biological Sciences and Department of Psychology, University of Missouri, Columbia, Missouri

PAULINE YAHR, Department of Psychobiology, School of Biological Sciences, University of California, Irvine, California

Preface

This volume is an overview of research examining the relationship between hormones and aggressive behavior. The last 15 years have witnessed a tremendous growth of knowledge in this area, yet reviews written by specialists are virtually nonexistent. This work is an attempt to provide a comprehensive and cohesive synthesis of this literature. Chapters 1–7 provide an analysis of hormonal influences on the major forms of aggressive behavior, including intermale, interfemale, shock-induced, maternal, territorial, and predatory aggression. The focus of Chapters 8–12 is an examination of the mechanisms through which hormones might act to produce changes in agonistic responding. Genetic, developmental, neural, and biochemical influences are considered. It is well known that environment, social context, and experience modulate the effects of hormones on behavior. Thus, Chapters 13–15 are designed to review the literature concerning hormone–pheromone interactions, hormonal responses to competition, and the influence of social context on the endocrine system and aggressive behavior. Frequently, the principles advanced by behavioral endocrinologists are based on research in one species, the rodent. To provide a more comparative perspective and to examine specifically the generality of those principles generated for rodents, Chapters 16–22 examine hormone–aggression relationships in a variety of species, including fish, birds, amphibians, reptiles, infrahuman primates, humans, ungulates, and insects. This volume should be useful to both beginning and advanced researchers in animal behavior, behavioral endocrinology, physiological psychology, neuroendocrinology, zoology, physiology, and psychiatry.

I am grateful to the contributors of the volume and the staff of Plenum Press for their advice and encouragement during the preparation of this work. I am also indebted to the National Science Foundation (Grant BNS80-08546), the National Institute of Aging (Grant AGO1319), the Research Foundation of the State University of New York, and the Harry Frank Guggenheim Foundation for their support during the preparation of this volume. Special thanks are extended to Martha Mann for her help in translating Chapter 19. Finally, my children, John and Mark, and my wife, Maryalice, provided me with encouragement and patience during the preparation of this volume; I dedicate this book to them.

BRUCE B. SVARE

Albany, New York

Contents

I

HORMONAL DETERMINANTS OF AGGRESSIVE BEHAVIOR

1

Pituitary-Gonadal Influences on Social Aggression

Paul F. Brain

INTRODUCTION

It is obviously very difficult to define the precise involvement of a complex physiological system in a rather nebulous concept dealt with, for example, in Brain (1981a). Consequently, some initial care will be directed toward defining the precise area of this review. This account will deliberately look at the phenomena of hormone–aggression correlations in a wide range of species, but this is not meant to imply the existence of genes for aggression. It is likely that similar strategies will lead to analogous hormone–aggression relationships even in diverse species. One should emphasize (as in Brain, Haug, & Kamis, in press) that different models of aggression generate radically different views of the relationships between hormones and this behavior even within the same species. Some attempt will also be made to relate the material to clinical investigations.

SOCIAL AGGRESSION

Barnett (1975) described *aggression* as "a word with many meanings and a source of much confusion." One must consequently take care to define precisely what we mean by the use of the term *social aggression.* It is obvious that aggression is *not* a unitary concept, and this fact has been underlined by attempts to provide classifications (e.g., Brain, 1977, 1978, 1979a; Moyer, 1968) of conflict behaviors. Note also that Blanchard and his colleagues (Blanchard & Blanchard, 1977; Blanchard, Fukunaga, Blanchard, & Kelley, 1975; Blanchard, Blanchard, & Takahashi, 1977; Blanchard, Blanchard, Takahashi, & Kelley, 1977; Blanchard, Blanchard, Takahashi, & Takahashi, 1977) have advocated the distinction between defensive and aggressive reactions (both may involve attack but the pat-

---

PAUL F. BRAIN • Department of Zoology, University College of Swansea, Swansea SA2 8PP, Wales, United Kingdom. The author thanks the Medical Research Council for grant G 975/58/N, which supported much of his own research in this area.

terns of such responses are different). A similar dichotomy between defensive and offensive attack has been proposed by Brain (1981b).

All attack behavior is a consequence of interactions among (a) situational factors, (b) experiental influences, and (c) internal variables.

While there are difficulties associated with establishing the presence of some criteria, many workers (e.g., Archer, 1976; Brain, 1977) feel that for an act to be unequivocably described as "aggression," one must demonstrate that the action (a) involves actual or potential physical (or psychological?) damage to another organism, (b) is perceived as intentional rather than accidental, and (c) is accompanied by signs of emotional arousal. Note that it is virtually impossible to establish these three criteria in encounters involving infrahuman organisms. Social aggression (it will become apparent later) comes closer to fulfilling the criteria than do many others models of attack.

Moyer in his 1968 classification proposed two categories of aggression, which will be combined for the purposes of this review: *Intermale fighting,* defined by Brain (1977) as "fighting or threat displayed between intact males of the same species which have not become habituated to each other" (p. 7). Brain (1977) noted that this activity is not always the exclusive province of the male since it can also be elicited in masculinized female rodents and females of such species as the Golden hamster. The response often serves the function of establishing dominance hierarchies in social groups. *Territorial aggression,* characteristic of animals (usually males) that defend particular exclusive areas (larger than a nesting site and smaller than a home range) from intrusion by like-sexed conspecifics.

A third type, namely, *mate-selection-related aggression* (Brain, 1977), which is fighting behavior to determine which males have mating access to females (as a consequence, the behavior often shows a circannual variation) is also included.

Brain (1979a) advocated combining the above categories as *social aggression* for five reasons:

1. It is often difficult to distinguish the categories clearly (e.g., intermale fighting may lead to territorality). The lack of a clear distinction between territorial and other "male" forms of aggression has been emphasized by Crabtree and Moyer 1977), who advocated dropping the term *territorial aggression.* It is also of interest to note that the forms of fighting induced in the mouse by individual housing, reproductive experience, and dominance within a group are virtually identical (Al-Maliki, 1980; Childs, 1979).

2. All categories generally involved exclusively intraspecific competition for a substrate that directly influences reproductive success—that is, aggression is used as a reproductive strategy to obtain, respectively, social status, an exclusive breeding area, or a mate. Consequently, the different categories may merely reflect the social organization of the particular species employed.

3. All these forms of attack are generally (but not exclusively) the province of the male and these responses are much influenced by pituitary-gonadal function.

4. All these reactions tend to utilize "ritualized" forms of attack and threat display, reducing the chances of serious tissue damage to the combatants (see Blanchard & Blanchard, 1977; Brain, 1981b). Certainly, particularly vulnerable

regions of the opponent's body tend to be avoided in biting attack, in contrast to reflexive fighting, predatory aggression, etc.

5. All three phenomena may show circannual variations (which can be related to cycles of pituitary-gonadal activity).

One should note that the term *social aggression* is not particularly apposite since *all* forms of aggression are, to a degree, social. Use of the adjective *social* is, however, meant to emphasize (a) the intraspecific nature of these responses, and (b) the role of these reactions in the normal social organization of the species. It is also worth commenting that all behavior seen in a social aggression context would not fall into this category. Fighting of this type can range from encounters between evenly matched opponents to biting attacks by dominants against clearly submissive social subordinates or intruder animals. As Blanchard and Blanchard (1977) have pointed out, any biting by the attacked animal in the latter type of response is more accurately termed a defensive rather than an offensive response. One should always be aware of the importance of the behavior of "losers" as well as winners (see Leshner, 1980). This account will limit itself to offensive responses.

The Pituitary-Gonadal System

A brief description of the vertebrate pituitary-gonadal system is necessary before one can hope to understand the role(s) of the hormones associated with this system. A highly simplified diagram is provided in Figure 1 (and a more detailed account is available in Brain, 1979b).

The most pertinent features to note are

1. The functioning of this system is an intimate interaction among hypothalamic releasing factors, pituitary gonadotrophins (luteinizing hormone—LH—and follicle stimulating hormone—FSH), and sex steroids (from the gonads and the adrenal cortices).
2. Androgens and estrogens are *not* exclusively the province of the male and the female, respectively.
3. Endocrine systems do not function in isolation, so environmental factors, neural input, and other endocrine systems are likely to modulate pituitary-gonadal function.
4. Hormones operate in (usually negative) feedback systems, so disturbance of one factor is likely to have wide-ranging repercussions.

ATTEMPTS TO RELATE PITUITARY-GONADAL HORMONES TO SOCIAL AGGRESSION

Studies that attempt to relate hormones to aggression may be very diverse (see Table I). For example, the variation in hormone levels may be produced by endogenous or exogenous manipulation. In some cases, the precise levels of circulating material is accurately determined; in other cases, it is merely presumed.

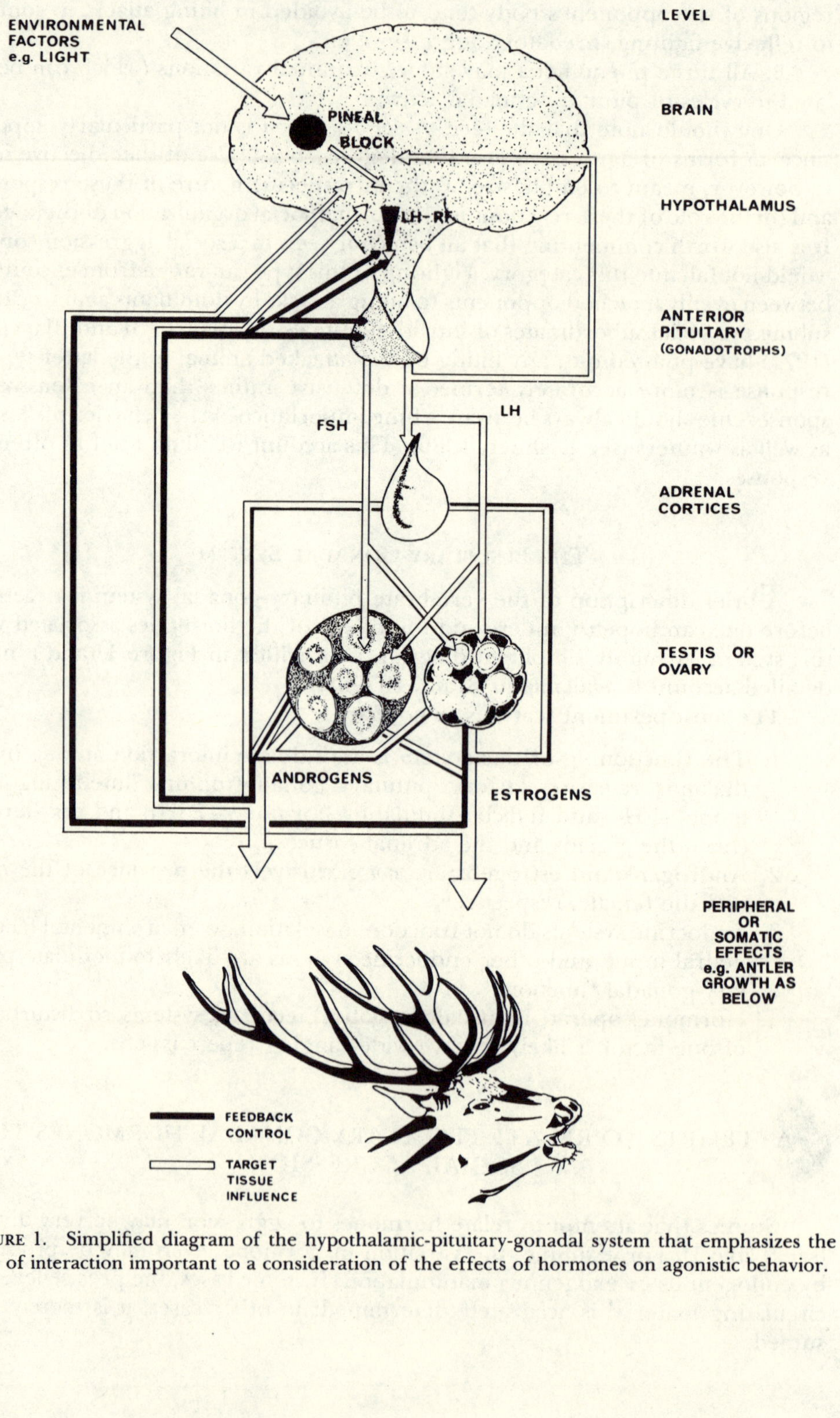

FIGURE 1. Simplified diagram of the hypothalamic-pituitary-gonadal system that emphasizes the sites of interaction important to a consideration of the effects of hormones on agonistic behavior.

TABLE I
Types of Attempts to Establish Hormone–Aggression Correlations

Approach	Method
Endogenous	1. Studying changes within circadian, circannual, estrous cycles, etc.
	2. Investigating hormonal and behavioral changes with age
	3. Utilizing genetic variability
Exogenous	1. Surgical ablation
	2. Systemic replacement
	3. Neural implantation
	4. Correlating neural uptake with behavioral changes

Variables likely to alter the directions of hormone–social aggression correlations include

1. Use of different hormones that may differ in purity.
2. Dosage of hormone—it is not always the case that linear dose–response curves apply here.
3. Route of application (e.g., intraneural, intramuscular, subcutaneous, etc.).
4. Form of the material—long acting preparations often have actions different from those of their shorter-surviving counterparts.
5. Duration of application.
6. Species or even strain of animal employed.
7. Temporal relationships between treatments and testing—this is especially important where the possibility of experiental effects is introduced by repeated testing.
8. Housing or seasonal influences—selection of a particular housing condition or time of year may be essential for showing actions of hormones in particular species.
9. Use of different tests—these may generate very different conclusions, especially when (see below) no attempt is made to distinguish between motivational and somatic influences of these factors.

Pituitary-gonadal hormones (particularly androgens) have been implicated in the modulation of fighting in vertebrates in a number of ways. These include (a) early influences on the potential for subsequent aggressive behavior (this area is covered by Vom Saal, (this volume), and (b) adult influences on social aggression.

ADULT INFLUENCES ON SOCIAL AGGRESSION

Hormones can have very diverse effects on the organisms that are exposed to them. It is often difficult to assess precise influences of hormones on aggression in adults, but it is likely that it is worth distinguishing two basic subdivisions of action, covered in the following two sections.

Peripheral or Somatic Influences on Fighting

Peripheral influences of sex steroids on fighting are mediated through actions on somatic structures rather than by changes in CNS activity. The term *peripheral influences* encompasses all nonmotivational (see below) actions on attack behavior. These include hormonal influences on morphology (e.g., on antler growth, cock comb size, body size, etc.), changes in the sensitivity of sense organs, secretion of "pheromones," control of marking behavior, etc. For example, the olfactory sense is the primary means of social communication in rodents (congenitally blind mice will spontaneously fight, but anosmic individuals rarely show this behavior). Consequently, because sex steroids stimulate production of largely urinary olfactory cues that alter the incidence and intensity of fighting and threat in rodents (Brain & Evans, 1974a,b; Mackintosh & Grant, 1966; Mugford & Nowell, 1970), these actions may account for some documented hormone–aggression correlations evident in the literature.

Motivational Effect on Fighting

Another type of hormonal involvement in fighting is through a presumed CNS effect that changes "motivation" for attack behavior (i.e., aggression). Motivational changes may be evident in studies of the effects of postneonatal castration on reproductive and aggressive behaviors in rats and mice.

Circumstantial lines of evidence for sex steroid effects on motivation being mediated by the CNS (Brain, 1977) include the following facts:

1. Some steroid treatments cause morphological/receptor changes in neural structures.
2. Sex steroids are sometimes more behaviorally effective when the material is placed in particular neural loci.
3. Neural regions commonly accumulate specific behaviorally active steroids.
4. Steroids may alter neuronal activity in particular regions of the CNS.
5. Hormonal changes may be correlated with alterations in localized neurotransmitters. Changes in these biogenic amines have, in some cases, been related to behavior.

Note that not all the above actions have been demonstrated with respect to social aggression. Hormones can consequently be regarded as acting on social aggression by altering internal (motivational) and situational factors (by altering signaling or perception of signaling between conspecifics). They may also be implicated in experiental effects (Brain, 1980; Leshner, 1980).

The following account will be used to (a) provide a brief synopsis of demonstrated motivational and somatic hormone–aggression correlations in a wide range of species, (b) assess detailed considerations raised by our very extensive studies of the effects of sex steroids on the motivation for social fighting in laboratory mice, and (c) critically assess studies that have attempted to relate pituitary-gonadal function to aggression in humans.

A PHYLOGENETIC SUMMARY OF THE ROLE OF THE PITUITARY-GONADAL SYSTEM IN SOCIAL AGGRESSION

Brain (1977, 1978, 1979a) has provided wide-ranging reviews of the direct and indirect influences of hormones on various forms of aggression in a wide range of species. It should be noted that it is often difficult, because of the widely differing methodologies, to evaluate accurately the diverse studies in this area, but a number of trends are evident.

Direct (Motivational) Influences

Gonadotropins

There are a number of studies suggesting that males of seasonally breeding species (mainly birds and fish) may show an independent influence of LH on fighting potential. It has been argued that this relationship is likely when it is important that fighting *preceeds* mate selection. The small number of studies available provide little support for the views that (a) LH directly alters social fighting in laboratory rodents, and (b) FSH and prolactin have motivational influences on social aggression.

Sex Steroids

Effects of Postneonatal Gonadectomy. Removal of the major source of sex steroids generally suppresses the incidence and intensity of social aggression in fish, lizards, rodents, dogs, cats, deer, sheep, and (perhaps) man (see Brain, 1981a). It should be noted that the response to this operation is very dependent on (a) the species, (b) the animal's prior experience, and (c) (in some species) the time of the year. Limited studies on a number of infrahuman primates have not generally shown this operation to have much of an influence on this form of fighting.

Effects of Therapy with Aromatizable Androgens. Treatment of (generally) castrated males with testosterone or androstenedione (i.e., hormones that can be converted to estrogenic metabolites) generally leads to a restoration/induction of social aggression. Such actions have been shown in fish, lizards, birds, rodents, infrahuman primates, and man. One should note that many of the studies make it difficult to distinguish between motivational and somatic actions.

Effects of Therapy with Estrogens. Estrogen (e.g., estradiol, estrone, etc.) administration has been shown to suppress social aggression in intact fish, chickens, and man. Conversely, this class of hormones will restore fighting in castrated male mice, rats, and sheep. This is one line of support for the aromatization hypothesis (see below).

Effects of Therapy with Nonaromatizable Androgens. These compounds (which cannot be converted to estrogens) generally have less influence on social aggression than do the two preceeding categories. Certain examples, however, (notably

5α-dihydrotestosterone) will restore the motivation for this activity in castrated mice, rats, hamsters, sheep, and rhesus monkeys. This type of finding must be taken as evidence against the aromatization hypothesis.

Effects of Therapy with Androgen–Estrogen Intermediates. The limited studies with these transitional materials suggests that they will restore aggressive motivation in castrated mice and rats. The fact that they are less potent than testosterone or estradiol may be taken to suggest that (a) aromatization is not obligatory in steroid influences on the brain, or (b) the low stability of these materials in biological systems renders them less effective.

Effects of Therapy with Progesterone. Progesterone has variable effects on the motivation for social aggression. This steroid suppresses fighting in intact or castrated/testosterone-replaced mice. It may restore fighting in castrated male hamsters and augment the activity in gonadectomized female hamsters and intact rhesus monkeys. Some of the changes in females may, however, be related to altered receptivity.

Effects of Antihormone Preparations. This area has been revolutionized by the recently developed compounds that modify sex steroid activity/metabolism. For example, the anti-androgens (cyproterone acetate, flutamide, etc.) are said to reduce social aggression in intact sticklebacks and man, but have little influence on the motivation for this activity in mice and gerbils. Anti-estrogens (e.g., MER-25) reduce fighting in castrated/testosterone-replaced mice. It is particularly interesting to note that aromatase inhibitors (compounds that depress the conversion of androgens to estrogens) block testosterone but not estradiol-maintained social aggression in castrated mice. Provera (a progesterone derivative) has also been reported to produce some amelioration of clinical hyperaggressiveness.

Indirect (Somatic) Influences

Hormones may produce peripheral changes in physiology and/or anatomy changing the likelihood of observing fighting behavior in a wide range of species. These effects are not explained in terms of hormonal influences on the CNS and hence on motivation. Types of influence that are cited here include (a) changes in somatic growth—body size, antler growth in deer, etc.; (b) alterations in the production and secretion of aggression-modifying odors ("pheromones"); (c) induction of territorial marking behavior; (d) acquisition of fear-motivated responses that may not be conducive to social aggression; and (e) variations in body coloration, plumage, bird song, etc. The involvement of many indirect hormonal influences on social aggression is discussed by Brain (1977, 1978, 1979a).

Gonadotropins

These have not been clearly implicated in indirect effects.

Sex Steroids

Effects of Therapy with Androgens. Similar actions are generally produced by aromatizable and nonaromatizable androgens. These compounds generally stimulate social aggression. For example, androgens have been reported to augment body size in rodents and birds. The production of changes in color and skin derivatives related to social aggression (e.g., antlers) have been correlated with these steroids in fish, birds, deer, etc. Sound production may be modified by androgens in birds, rodents, and man. The production of aggression-enhancing pheromones has been reported to be related to androgen secretion in tortoises, rodents, dogs, deer, pigs, and prosimians. Territorial marking is also increased by these hormones in rodents and lagomorphs, and the detection of odors by rodents and man can be altered by these compounds. Persistence (the likelihood of continuing a specific behavior) is also modified by androgens in birds and rodents.

Effects of Therapy with Estrogens. Altered levels of these hormones have been associated with altered sexual motivation in rodents and primates. Changes in this activity can obviously modify social aggression. Estrogens also alter the production of odors, in rodents and primates that change social aggression—generally, individuals carrying the odor are less likely to be attacked, whereas other individuals may be *more* likely to receive social aggression. Both the ability to perform territorial marking and to detect odors are modified by estrogens in a variety of rodent species.

Effects of Therapy with Progesterone. This steroid has been implicated in modifying sexual receptiveness and pheromone production in rodent species.

STUDIES ON THE ROLE OF SEX STEROIDS IN MODIFYING THE MOTIVATION FOR SOCIAL AGGRESSION IN LABORATORY MICE

In this laboratory, a large number of studies on the abilities of sex steroids to *maintain* motivation for social aggression in castrated TO strain mice, initially rendered aggressive by breeding experience, have been carried out (Bowden, 1979; Brain, 1979c; Brain & Bowden, 1979). It seems worth reiterating some of this material in order to look at the intricacies of hormone–aggression correlations in our best-known species. Note that

(a) Males were housed with their mates until 10–15 days after the birth of the first litter.

(b) They were subsequently removed, bilaterally castrated, and allocated to individual cages.

(c) Subjects "rested" for three days and then received 17 consecutive daily intramuscular injections of an oil-based (80% arachis oil, 20% ethyl oleate) hormone or control material.

(d) On days 15, 16, and 17 of treatment, mice received 10-minute encounters in their cleaned home-cages with nonaggressive "standard opponents" (rendered docile by repeated defeat or by the induction of peripheral anosmia). Two observers were generally employed in each experiment, each observer scoring subjects at least once and good observer agreement being maintained throughout testing. The measures obtained in these present studies were more detailed than those employed in most earlier investigations and included

1. Proportions of animals fighting in at least two out of three encounters.
2. The latency of attack—the interval (in seconds) from the introduction of the standard opponent to the first biting attack on it by the test animal. Mice failing to attack were allotted a maximal score of +600 seconds.
3. The accumulated attacking time (AAT)—the total number of attacking bouts directed toward the "standard opponent," that is, the number of times discrete biting attacks, separated by periods of threat or exploration, were observed.

Measures 2 and 3 were each summated over the three tests for individual mice. This appears a reasonable simplification, because factor analysis of data obtained in this type of test (Goldsmith, 1977) has indicated that the factor structure is similar over the three days.

(e) Weights of sex accessory glands were subsequently determined.

(f) Mice generally received only *one* dose of *one* hormone.

The accumulated data on the efficacies of a wide range of steroids (expressed as μmols per day) required to maintain the measure of fighting at the level seen in sham-castrated/control-injected individuals are presented in Tables II, III, and IV. Effects of these treatments on ventral prostate weights (expressed as

TABLE II

Effect of Sex Steroids and Related Compounds on a Number of Subjects Fighting in at Least Two Out of Three Tests[a]

Substance	Daily dosage[b]
Diethyl stilbestrol (DES)	3.730×10^{-4}
Estradiol benzoate (EB)	1.328×10^{-3}
Estradiol-17β (E2)	1.835×10^{-3}
Estrone (E1)	1.849×10^{-3}
Estriol (E3)	1.700×10^{-2}
Testosterone (T)	8.700×10^{-2}
2α,7α-Dimethyl-androst-5-en-3β,17β-diol (RTI)	1.572×10^{-1}
5α-Dihydrotestosterone (5α-DHT)	1.720×10^{-1}
Androstenedione (A)	1.740×10^{-1}
Cholesterol (C)	1.290×10^{-1}[c]
5α,19-Hydroxydihydrotestosterone (19OHDHT)	1.284[c]
19-Hydroxytestosterone (19OHT)	1.642[c]
Androsterone (AND)	3.442[c]
5β-Dihydrotestosterone (5βDHT)	6.887[c]

[a] Modified from Brain (1978).

[b] Expressed in μmols/day required to maintain fighting level at that seen in sham-castrates/control-injected animals.

[c] Highest dose tested.

TABLE III
Effects of Sex Steroids and Related Compounds on Summated Latency to First Attack Over the Three Tests[a]

Substance	Daily dosage[b]
Estrone (E1)	1.849×10^{-3}
Diethyl stilbestrol (DES)	1.863×10^{-3}
Estradiol benzoate (EB)	2.656×10^{-3}
2α,7α-Dimenthyl-androst-5-en-3β,17β-diol (RTI)	1.572×10^{-1}
5α-Dihydrotestosterone (5α-DHT)	1.720×10^{-1}
Testosterone (T)	1.730×10^{-1}
Androstenedione (A)	1.740×10^{-1}
Estriol (E3)	1.700×10^{-1}[c]
Estradiol-17β (E2)	1.800×10^{-2}[c]
Cholesterol (C)	1.290×10^{-1}[c]
5α,19-Hydroxydihydrotestosterone (19OHDHT)	1.284[c]
19-Hydroxytestosterone (19-OHT)	1.642[c]
Androsterone (AND)	3.442[c]
5β-Dihydrotestosterone (5βDHT)	6.887[c]

[a] Modified from Brain (1978).
[b] Expressed in μmols/day required to maintain fighting level at that seen in sham-castrates/control-injected animals.
[c] Highest dose tested.

TABLE IV
Effects of Sex Steroids and Related Compounds on Summated Accumulated Attacking Times (Time Spent Biting the Opponent over the Three Tests)[a]

Substance	Daily dosage[b]
Esterone (E1)	1.849×10^{-3}
Diethyl stilbestrol (DES)	1.863×10^{-3}
Estradiol benzoate (EB)	2.656×10^{-3}
Testosterone (T)	8.700×10^{-2}
2α,7α-Dimethyl-androst-5-en-3β,17β-diol (RTI)	1.572×10^{-1}
5α-Dihydrotestosterone (5α-DHT)	1.720×10^{-1}
Androstenedione (A)	1.740×10^{-1}
19-Hydroxytestosterone (19MOHT)	1.642
Estriol (E3)	1.700×10^{-2}[c]
Estradiol-17β (E2)	1.800×10^{-2}[c]
Cholesterol (C)	1.290×10^{-1}[c]
5α,19-Hydroxydihydrotestosterone (19OHDHT)	1.284[c]
Androsterone (AND)	3.442[c]
5β-Dihydrotestosterone (5β-DHT)	6.887[c]

[a] Modified from Brain (1978).
[b] Expressed in μmols/day required to maintain fighting level at that seen in sham-castrates/control-injected animals.
[c] Highest dose tested.

TABLE V
Effects of Sex Steroids and Related Compounds on Ventral Prostate Weights[a]

Substance	Daily dosage[b]
Testosterone (T)	1.720×10^{-2}
5α-Dihydrotestosterone (5α-DHT)	1.720×10^{-1}
Androstenedione (A)	1.740×10^{-1}
Estradiol benzoate (EB)	2.000×10^{-3}[c]
Estriol (E3)	1.700×10^{-2}[c]
Estradiol-17β (E2)	1.800×10^{-2}[c]
Estrone (E1)	1.800×10^{-2}[c]
Diethyl stilbestrol (DES)	1.800×10^{-2}[c]
2α,7α-Dimethyl-androst-5-en-3β,17β-diol (RTI)	1.572×10^{-1}[c]
Cholesterol (C)	1.290×10^{-1}[c]
5α,19-Hydroxydihydrotestosterone 19OHDHT	1.284[c]
19-Hydroxytestosterone (19MOHT)	1.642[c]
Androsterone (AND)	3.442[c]
5β-Dihydrotestosterone (5β-DHT)	6.887[c]

[a] Modified from Brain (1978).
[b] Expressed in μmols/day required to restore weights of castrated mice halfway toward intact weight.
[c] Highest dose tested.

daily dose required to restore weights of castrated mice halfway toward the intact condition) are given in Table V. Similar studies, in which the effects of a variety of compounds on fighting in castrates treated daily with 25 μg of testosterone or 1 μg of estradiol, are presented in Table VI.

Finally, because it seemed important to establish that the fighting induced by androgen and estrogen treatment was identically motivated, a videotape (VTR) analysis of bite targets was performed. Thirty-six male isolates (four weeks of age) were bilaterally castrated, left for six weeks, and then allocated to categories given 17 daily intramuscular injections of (a) oil control, (b) testosterone propionate (25 μg per day), or (c) estradiol benzoate (1 μg per day). Twelve sham-castrated controls were also included.

The VTR analysis (described in Childs, 1979) involved a "frame-by-frame" determination of the bite targets used by the above mice in 10-minute encounters with anosmic opponents. Tests took place under white light in the subject's home cage and separate determinations of bites to the head, ventral surface, flanks, back, and rump were made. Mean percentages of bites to particular regions *in encounters where attack occurred* are present in Figure 2.

BASIC FINDINGS

The inherent difficulties of deducing the exact way in which steroids influence social aggression in mice arise from two features of this type of study:

1. Methodological variation in housing conditions, type of animal employed, mode of hormone application, type of behavioral test situation, etc.

TABLE VI

Effects of Various Compounds on Fighting Behavior in Steroid-Treated Castrates[a]

Compound	Maintenance steroid[b]	Behavioral action	Peripheral action	Reference
Cyproterone acetate	T	No effect	Lowers	Brain and Bowden (1978)
	E2	No effect	Increases	
MER-25	T	Weak suppression	Low doses increase	Brain and Bowden (1979)
	E2	Lowers	Lowers	Bowden (1979)
Antiaromatase	T	Lowers	No effect	Bowden and Brain (1978)
	E2	No effect	No effect	
Progesterone	T	Lowers	Increases	Brain and Bowden (1978)
	E2	No effect	Effect depends on dose	

[a] Modified from Brain (1980).
[b] T, testosterone, 25 μg/day; E2, estradiol, 1 μg/day.

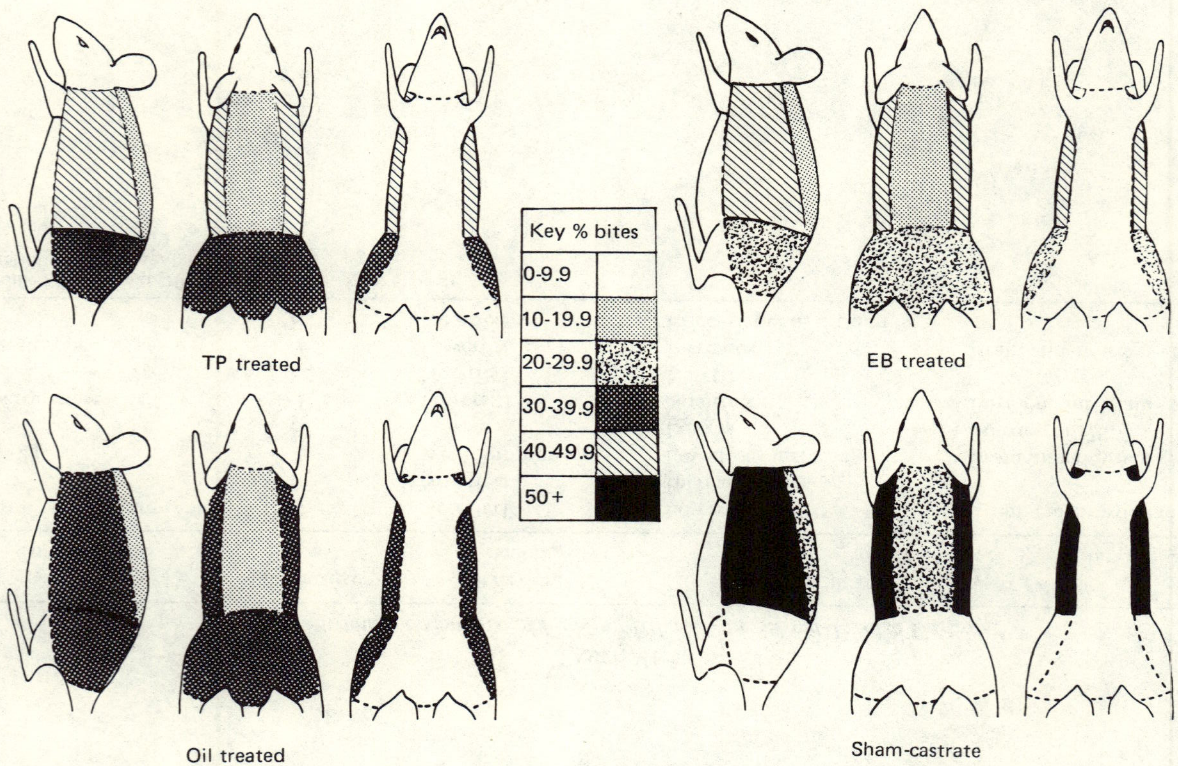

FIGURE 2. Diagram showing percentage of total bites *in attacking animals* to various regions of anosmic standard opponents in castrates treated with testosterone propionate (TP), estradiol benzoate (EB), or oil. The pattern used by sham-castrates is included for comparison. Note that picture in TP- and EB-treated animals is similar to that seen in sham-castrates, with the exception of the low incidence of rump-biting in the last-mentioned category. This may reflect the greater induction of fleeing responses in the opponent by the highly aggressive hormone-treated animals. The small number of oil-treated animals showing attack reveal patterns not too dissimilar from those seen in the others.

2. The fact that experience in prior behavioral tests often confounds the results. Test animals are often given more than one dose of material and subjected to repeated behavioral testing.

The data obtained in the present group of experiments confirm and extend the results of the earlier limited investigations by Brain and Poole (1976), and comparable studies by Edwards and Burge (1971), Luttge (1972), Luttge and Hall (1973a), and Luttge, Hall, and Wallis (1974). These experiments also (a) investigate a number of additional possibilities, generally neglected in these earlier studies, and (b) largely overcome the two areas of difficulty cited earlier.

The ranges of behavioral measures and organ-weight changes determined in this series of experiments largely serve as checks against individual variability. Most were not strictly independent measures, but they provided both confirmation of the effects and checks on observer reliability. In general, a steroid was regarded as being behaviorally effective if it augmented a majority of indicators of social "aggression," and was regarded as being effective in a somatic sense if it increased the weights of the ventral prostate in castrates.

It is worth remarking that chronic isolation was not employed in this study and that the doses used here were much closer to the physiological range than those utilized in earlier investigations. Because doses cited do not take account of loss of material by seepage, conversion of steroid, clearance of hormone, retention in injected tissues, etc., they are likely to be several orders of magnitude higher than the concentrations actually reaching the neural substrate.

Although the behavioral actions reported here are dramatic and at the upper end of the range reported by the majority of workers, fighting intensity rarely approached that of untreated, intact "aggressive" mice. It could be that the stress of castration and injection reduces levels of fighting or that subtle combinations of hormones, rather than single materials, would be more effective. Another possibility is that this means of hormone application does not adequately mimic endogenous release of sex steroids.

Looking at the hormonal treatments tested on an individual basis, the fact that even chromatographically pure 5α-DHT maintained some fighting in castrated mice suggests that aromatization of the "A" ring is not a prequisite for this behavioral influence. However, other supposedly nonaromatizable androgens (5β-DHT, AND, 19 OHDHT) were generally low in behavioral potency. A major advantage of these 5α- and 5β-reduced compounds is that they are more likely than the aromatizable androgens to have their actions on behavior and sex accessories in the form in which they were injected. Levels of fighting generated by 5α-DHT were generally lower than those produced by smaller doses of T or estrogen. It is suggested that one's interpretation of the behavioral effectiveness of nonaromatizable androgens may be greatly influenced by one's choice of androgen (among other factors). One should not base theoretical edifices on *one* example of *a* particular dose of *one* steroid. In general, nonaromatizable androgens are *not* as effective as investigated estrogens or aromatizable compounds.

All estrogens employed in these studies had dramatic stimulatory influences on aggression in castrated mice at relatively low (submicrogram) doses. Activity was not dependent on the steroidal nature of the estrogen, since DES was the most potent compound tested. Thus (in spite of the above finding with 5α-DHT), it does seem that aromatization of the "A" ring (or an equivalent structure) increases the behavioral potency of sex hormones in this situation. The finding that EB was more potent than E2 replicates a large number of studies that also suggest that esters are more active than free hormones.

All the estrogens (and the ester of E2) used here showed little sign of behavioral dose-responsiveness. It seems likely that there is a very sharp cutoff between an effective and a noneffective dose of estrogen (0.075 μg per day of E2 did not maintain fighting, whereas 0.1 μg per day did, in this study). Certainly it does not seem easy to produce a moderate degree of fighting using estrogens. It may consequently be more instructive to consider the minimal weight of compound that must be injected in order to maintain fighting at a particular level in castrates (see Tables II–IV) than to use the typical dose–response curves.

The aromatizable androgens (T or A) had potent actions on social aggression by castrated "aggressive" mice. Testosterone seemed more behaviorally effective than A.

The two 19-hydroxylated compounds provide evidence that 19-hydroxylation reduces behavioral potency of steroid hormones. It is possible that the weak behavioral effects of 19 OHT are a consequence of incomplete conversion to E2. 19 OHDHT may not be altered in this fashion, but we have already seen that 5α-reduction (and lack of potential for aromatization) does not prevent behavioral activity of all compounds (5α-DHT is behaviorally potent).

Recent studies suggest that the conversion of androgen to estrogen is very small in biological systems (e.g., Naftolin & Ryan, 1975). Although 19 OHT has some behavioral activity, it was less potent than either T or E2. This compound has been used with varying success in a variety of replacement therapies: Booth (1976), for example, reports that 19 OHT successfully "androgenized" female rats—a neonatal effect. It could be that this compound is relatively unstable or easily inactivated (see earlier comment on purity).

A strange recurring finding was that certain compounds showed lower behavioral effectiveness at higher doses. This is difficult to explain, but if aromatization (or some other binding/conversion process in the brain) is involved in the behavioral actions, it may be possible to "swamp" the limited receptor sites with nonpotent or only weakly potent material when using very high doses. This is, of course, a highly speculative suggestion at present, but it certainly could be the case that some workers would find more behavioral activity by employing lower rather than higher doses of steroid. Indeed, the range of investigated doses seems of crucial importance to this type of study.

It is possible to compare the relative potencies of materials used in these studies. It should be noted that the estrogens used in this study caused, in a number of cases, small further decrements in the weights of the sex accessories of castrated mice. Their suppressive actions would, of course, have been much more striking in intact animals.

A feature that should be noted is the apparent absence of an obvious relationship between a hormone's behavioral and somatic potencies. This finding supports the contention that the behavioral test employed here measures differences in aggression in treated animals rather than having part of its action by changing production of aggression-modifying olfactory cues (Mugford & Nowell, 1970). Fighting in this type of aggression test does not appear to be significantly influenced by olfactory cues received by the submissive "standard opponent." One may expect androgen- and estrogen-treated test animals to differ in the cues they produce, but aromatizable androgens and estrogens both seem to stimulate intense attack and threat directed toward the opponent. Pheromone production may perhaps be regarded as a "somatic" or "peripheral" action of androgens on murine fighting behavior, which is analogous to the effects of these compounds on penile sensitivity in the experiments on hormonal maintenance of ejaculatory behavior in rats (reviewed by Brain, 1979b). The pheromonal influences in this type of behavioral test appear to be unidirectional—that is, they influence the attack of the test animal on the opponent.

The sex-steroid-blocking data support the view that estrogens rather than androgens may be important in maintaining the motivation for social aggression in the laboratory mouse. Certainly, the fact that aromatase inhibitors block testosterone- but not estradiol-maintained fighting in castrates is particularly striking.

The VTR analysis confirms that the patterns of attack employed by both testosterone- and estradiol-maintained castrates are remarkably similar and comparable to those of sham-castrates. The criterion of avoiding the particularly vulnerable head and ventral surface regions has been utilized as an index of social aggression in this species (Brain, 1981b).

Comments on the Aromatization Hypothesis

It has often been maintained that the effects of castration are direct consequences of the suppression of androgens. Some recent studies have, however, indicated that androgens may exert their motivational effects after neural conversion to estrogenic metabolites. This is termed the *aromatization hypothesis.* The lines of evidence for this view have been listed by Brain (1979b). Early studies were largely concerned with ejaculatory behavior in the rat.

Attempts to assess the effects of estrogens on rank-related fighting in laboratory mice initially generated conflicting reports (reviewed by Brain, 1977). Important lines of evidence in this body of work that relate to the role of aromatization in murine aggressiveness include the following observations:

1. Estrogens and aromatizable androgens are the most effective steroids investigated, in terms of their abilities to maintain aggressive motivation in castrates.
2. An aromatase-inhibiting compound effectively blocked T- but not E2-maintained fighting in castrates.
3. Cyproterone acetate did *not* block T-maintained aggression in castrates.
4. MER-25 reduced T-maintained fighting in these animals.

5. The behavioral activity of 5α-DHT and the weak action of MER 25 on T-maintained fighting are not necessarily fatal to the aromatization hypothesis. 5α-DHT is (because it is cheap, readily available, and a major metabolite of T) the most commonly used 5α-reduced androgen. One cannot conclude, however, that *all* 5-reduced steroids are behaviorally effective, because three other "nonaromatizable" androgens (5β-DHT, AND, and 19 OHDHT) *were* ineffective. It *does* mean that dependence on one "nonaromatizable" androgen is likely to generate erroneous conclusions. 5α-DHT's intrinsic behavioral effectiveness is evident in its maintenance of male sexual behavior in mice (Luttge & Hall, 1973b), guinea pigs (Alsum & Goy, 1974), rabbits (Beyer & Rivaud, 1973), and rhesus monkeys (Phoenix, 1974).

It has been suggested that progesterone's (P) inhibition of neural T uptake is due to competition between P and T for active 5α-reductase sites (Voigt & Hsia, 1973). It is also suggestive that hypothalamic DHT receptors have been demonstrated in male mice, hamsters, guinea pigs, rabbits, and cats (Kato, 1975), and it is known that T is hypothalamically converted to DHT (Denef, Magnus, & McEwen, 1973; Juneja, Visconti, Motta, & Martini, 1976; Kniewald, Massa, & Martini, 1971). One may comment however, that this localization and conversion may be related to the negative feedback inhibition of gonadotrophin release rather than to behavioral activation. It is not possible, however, to dismiss 5α-DHT as being behaviorally ineffective.

An alternative suggestion is that 5α-DHT's behavioral actions may be consequent on metabolic conversion. Heyns and de Moor (1974) found that this steroid could be readily metabolized by erythrocytes to 3β-androstanediol in male rats. This metabolite, which has a moderate affinity for anterior hypothalamic cytosol estradiol receptors, *can* activate mounting behavior in castrated rats (Baum & Vreeburg, 1976).

Use of Lloyd Morgan's canon suggests however, that one *should* interpret the present 5α-DHT data as confirming that aromatization does *not* provide an entire description of how testosterone exerts its motivational actions.

There are numerous obstacles to a convincing demonstration that aromatization is involved in T's physiological actions on masculine behavioral motivation. Interpretations are influenced by

1. Compounds employed. For example, we have noted that erroneous conclusions may be generated by using one hormone to represent a class of compounds.
2. Hormonal dosage. Many studies use pharmacological rather than physiological doses. Interpretation then becomes difficult because elevated doses may produce atypical actions.
3. Duration of treatment.
4. Species (or even strain of animal) utilized. For example, some species may aromatize "nonaromatizable" androgens.
5. Test situation. Some tests reflect effects of hormones on peripheral structures—for example, they reflect altered pheromone production, etc.

6. Prior experience. This may be evident when animals are (a) repeatedly tested, (b) used in more than one experiment, (c) subjected to different doses throughout the test period, (d) fought against other treatment categories, or (e) prevented from developing the complete behavioral pattern before termination of testing.
7. Behavior assessed.
8. Initial behavioral baseline. For example, high baselines do not facilitate demonstration of stimulatory actions of hormones.

Studies employing relatively few subjects, using limited ranges and doses of compounds and ambivalent testing methodologies, must be regarded with suspicion. These defects are surprisingly common. Attempts were made to minimize these problems by using

1. A wide variety of (both naturally occurring and synthetic) sex hormones, antihormones, and inhibitors. Extended dose ranges for each preparation (at least three doses for each compound) have also been employed.
2. Subjects in which social aggression cannot be dismissed as a consequence of "social deprivation."
3. Behavioral tests in which fighting is unilateral and is likely to reflect central (motivational) rather than peripheral actions.

General Discussion

Conversion of T to estrogenic metabolites may be important before the former compound can masculinize neonatal rats and mice or change adult motivation for "T-mediated" behaviors in some species. It appears improbable that this is the only way in which T can change motivation, in all species.

The two modest (but important) conclusions follow.

First, one *must* consider the possibility of metabolic conversion in the actions of endogenously or exogenously changed hormones. Testicular androgens, for example, may be converted in nonneural tissues to other neutral (Bruchovsky & Wilson, 1968; Wilson & Lasnitski, 1971) or phenolic (Grodin, Suteri, & MacDonald, 1973; West, Damast, Sarro, & Pearson, 1956) metabolites. Some products not only retain androgenicity but also exhibit two- or three-fold increases in activity in some systems (Dorfman & Shipley, 1956; Liao & Fang, 1969) or in some instances different biological properties (Mann, Rowson, Baronos, & Karagianindis, 1971). 5α-Reductases may be of importance in both central and peripheral targets.

Second, aromatization of T is important but not obligatory in controlling the motivation for intermale fighting in male TO strain mice. Naftolin and Ryan (1975) have, for example, reviewed evidence that the CNS can convert androgens to estrogens.

Testosterone seems to be a precursor for a variety of other steroids. Perhaps it exerts its physiological and behavioral actions only after metabolic conversion? It seems unlikely that a single hormonal metabolite will control aggressiveness and libido in all vertebrate species. Metabolic conversions may increase the variety of candidates for behavioral control. Particular pathways have differing importance in different species and they may enable organisms to influence two or more targets relatively independently. For example, at its simplest:

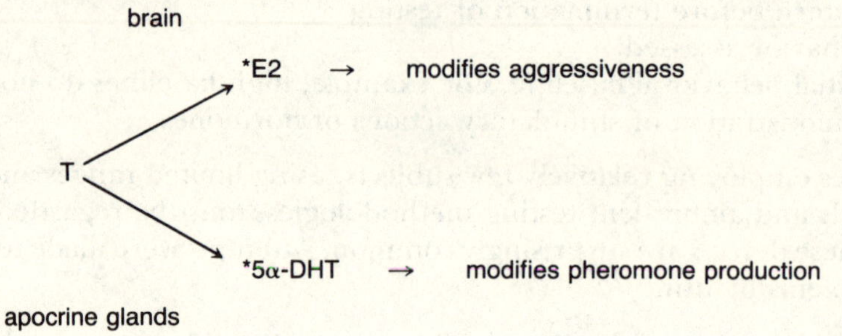

Suffice it to say that the hormone–behavior interactions involved in the modulation of social aggression in the mouse are a good deal more complex than has been suggested in the past. The same is likely to be true of other species, including man (see Mazur, this volume).

CONCLUSIONS

1. The category "social aggression" forms a logical grouping of conflict behaviors.
2. Circulating levels of LH, sex steroids, or sex steroid metabolites can often be related to changes in social aggression in a wide range of species.
3. The changes can be produced by early programming effects, motivational actions, and/or somatic influences.
4. Metabolic conversions of hormones may be important in these influences in some species.

ACKNOWLEDGMENTS

Professor J. S. Ryland is thanked for facilities in Swansea, and the animal care of Messrs. R. Guntrip and J. Young is gratefully acknowledged. The competent assistance of Dr. N. J. Bowden is much appreciated.

REFERENCES

Al-Maliki, S. *Studies on the effects of experimental manipulations on a variety of aggressive behaviors in laboratory albino mice.* Unpublished doctoral dissertation, University of Wales, 1980.

Alsum, P., & Goy, R. W. Actions of esters of testosterone, dihydrotestosterone, or estradiol on sexual behavior in castrated male guinea pigs. *Hormones and Behavior,* 1974, *5,* 207–217.

Archer, J. The organization of aggression and fear in vertebrates. In P. P. G. Bateson & P. Klopfer (Eds.), *Perspectives in ethology.* New York: Plenum Press, 1976.

Barnett, S. A. *The rat: A study in behaviour* (Rev. ed.). Chicago and London: The University of Chicago Press, 1975.

Baum, M. J., & Vreeburg, J. T. M. Differential effects of the anti-estrogen, MER-25, and of three 5α-reduced androgens on mounting and lordosis behavior in the rat. *Hormones and Behavior,* 1976, *7,* 87–104.

Beyer, C., & Rivaud, N. Differential effects of testosterone and dihydrotestosterone on the sexual behavior of prepuberally castrated male rabbits. *Hormones and Behavior,* 1973, *4,* 175–180.

Blanchard, R. J., & Blanchard, D. C. Aggressive behavior in the rat. *Behavioral Biology,* 1977, *21,* 197–224.

Blanchard, R. J., Fukunaga, K., Blanchard, D. C., & Kelley, M. J. Conspecific aggression in the laboratory rat. *Journal of Comparative and Physiological Psychology,* 1975, *89,* 1204–1209.

Blanchard, R. J., Blanchard, D. C., & Takahashi, L. K. Reflexive fighting in the albino rat: Aggressive or defensive behavior? *Aggressive Behavior,* 1977, *3,* 145–155.

Blanchard, R. J., Blanchard, D. C., Takahashi, T., & Kelley, M. J. Attack and defensive behavior in the albino rat. *Animal Behaviour,* 1977, *25,* 622–634.

Blanchard, D. C., Blanchard, R. J., Takahashi, L. K., & Takahashi, T. Septal lesions and aggressive behavior. *Behavioral Biology,* 1977, *21,* 157–161.

Booth, J. E. Effects of 19-hydroxylated androgens on sexual differentiation in the neonatal female rat. *Journal of Endocrinology,* 1976, *70,* 319–320.

Bowden, N. J. Studies on the manner in which sex steroids influence aggressiveness in *Mus musculus* L. Unpublished doctoral dissertation, University of Wales, 1979.

Brain, P. F. *Annual Research Reviews* (Vol. 1): *Hormones and aggression.* Montreal: Eden Press, 1977.

Brain, P. F. *Annual Research Reviews* (Vol. 2): *Hormones and aggression.* Montreal: Eden Press, 1978.

Brain, P. F. *Annual Research Reviews* (Vol. 3): *Hormones, drugs and aggression.* Montreal: Eden Press, 1979. (a)

Brain, P. F. Effects of the hormones of the pituitary-gonadal axis on behavior. In K. Brown & S. J. Cooper (Eds.), *Chemical influences on behavior.* New York: Academic Press, 1979. (b)

Brain, P. F. Steroidal influences on aggressiveness. In J. Obiols, C. Ballús, F. Gonzalez Monclús, & J. Pujol (Eds.), *Biological psychiatry today.* Amsterdam: Elsevier/North-Holland, 1979. (c)

Brain, P. F. Adaptive aspects of hormonal correlates of attack and defence: A study in ethobiology. In P. S. McConnell, G. J. Boer, H. J. Romijn, N. E. Van de Poll, & M. A. Corner (Eds.), *Adaptive capabilities of the nervous system: Progress in brain research* (Vol. 53). Amsterdam: Elsevier/North-Holland, 1980.

Brain, P. F. Hormones and aggression in infra-human vertebrates. In P. F. Brain & D. Benton (Eds.), *The biology of aggression.* Alphen aan den Rijn, Netherlands: Sijthoff and Noordhoff, 1981. (a)

Brain, P. F. Differentiating types of attack and defense in rodents. In P. F. Brain & D. Benton (Eds.), *A multidisciplinary approach to aggression research.* Amsterdam: Elsevier/North-Holland, 1981. (b)

Brain, P. F., & Bowden, N. J. Sex steroid control of intermale fighting in albino laboratory mice. In W. B. Essman & L. Valzelli (Eds.), *Current developments in psychopharmacology* (Vol. 5). New York: Spectrum Publications, 1979.

Brain, P. F., & Evans, C. M. Influences of two naturally occurring androgens on the attack directed by "trained fighter" TO strain mice towards castrated mice of three different strains. *IRCS Medical Science,* 1974, *2,* 1672. (a)

Brain, P. F., & Evans, C. M. Effects of androgens on the "attackability" of gonadectomized mice by TO trained fighter individuals: Confirmatory experiments. *IRCS Medical Science,* 1974, *2,* 1730. (b)

Brain, P. F., & Poole, A. E. The role of endocrines in isolation-induced intermale fighting in albino laboratory mice: II. Sex steroid influences in aggressive mice. *Aggressive Behavior,* 1976, *2,* 55–76.

Brain, P. F., Haug, M., & Kamis, A. Hormones and different tests for 'aggression' with particular reference to the effects of testosterone metabolites. In J. Balthazart, *Hormones and behaviour in higher vertebrates.* Berlin: Springer-Verlag, in press.

Bruchovsky, N., & Wilson, J. D. The conversion of testosterone to 5α-androstan-17β-ol-3-one by rat prostate in vivo and in vitro. *Journal of Biological Chemistry,* 1968, *243,* 2012–2021.

Childs, G. *Videotape analyses of social behaviour in laboratory mice.* Unpublished masters dissertation, University of Wales, 1979.

Crabtree, J. M., & Moyer, K. E. *Bibliography of aggressive behavior: A reader's guide to the research literature.* New York: Alan R. Liss, 1977.

Denef, C., Magnus, C., & McEwen, B. S. Sex differences and hormonal control of testosterone metabolism in rat pituitary and brain. *Journal of Endocrinology,* 1973, *39,* 605–621.

Dorfman, R. K., & Shipley, R. A. In R. I. Dorfman & R. A. Shipley (Eds.), *Androgens: Biochemistry, physiology and clinical significance.* New York: Wiley, 1956.

Edwards, D. A., & Burge, K. G. Estrogenic arousal of aggressive behavior and masculine sexual behavior in male and female mice. *Hormones and Behavior,* 1971, *2,* 239–245.

Goldsmith, J. F. *Behavioural and adrenocortical consequences of isolation in rodents.* Unpublished doctoral dissertation, University of Wales, 1977.

Grodin, J. M., Suteri, P. K., & MacDonald, P. C. Source of estrogen production in postmenopausal women. *Journal of Clinical Endocrinology Metabolism,* 1973, *36,* 207–214.

Heyns, W., & deMoor, P. A 3(17)β-hydroxy-steroid dehydrogenase in rat erythrocytes: Conversion of 5α-dihydrotestosterone into 5α-androstane-3β,17β-diol and purification of the enzyme by affinity chromatography. *Biochemica et Biophysica Acta,* 1974, *358,* 1–13.

Juneja, H., Visconti, F., Motta, M., & Martini, L. Effect of testosterone and its 5α-reduced metabolites on LH secretion. *Fifth International Endocrinology Congress* (Hamburg), *Abstract* 283, 1976.

Kato, J. The role of hypothalamic and hypophyseal 5α-dihydrotestosterone, estradiol and progesterone receptors in the mechanism of feedback action. *Journal of Steroid Biochemistry,* 1975, *6,* 979–987.

Kniewald, Z., Massa, R., & Martini, L. Conversion of testosterone into 5α-androstan-17β-ol-3-one at the anterior pituitary and hypothalamic level. In V. H. T. James & L. Martin (Eds.), *Proceedings of the Third International Congress on Hormonal Steroids* (Hamburg). Amsterdam: Excerpta Medica, 1971.

Leshner, A. I. The interaction of learning and neuroendocrine factors in determining behavioral adaptions to aggression. In P. S. McConnell, G. J. Boer, H. J. Romijn, N. E. Van de Poll, & M. A. Corner (Eds.), *Recent progress in brain research* (Vol. 53). Amsterdam: Elsevier/North-Holland, 1980.

Liao, S., & Fang, S. Receptor-proteins for androgens and the mode of action of androgens on gene transcription in ventral prostate. *Vitamins and Hormones,* 1969, *27,* 17–90.

Luttge, W. G. Activation and inhibition of isolation induced intermale fighting behavior in castrate male CD-1 mice treated with steroidal hormones. *Hormones and Behavior,* 1972, *3,* 71–81.

Luttge, W. G., & Hall, N. R. Androgen-induced agonistic behavior in castrate male Swiss-Webster mice: Comparison of four naturally occurring androgens. *Behavioral Biology,* 1973, *8,* 725–732. (a)

Luttge, W. G., & Hall, N. R. Differential effectiveness of testosterone and its metabolites in the induction of male sexual behavior in two strains of albino mice. *Hormones and Behavior,* 1973, *4,* 31–43. (b)

Luttge, W. G., Hall, N. R., & Wallis, C. J. Studies on the neuroendocrine, somatic and behavioural effectiveness of testosterone and its 5α-reduced metabolites in Swiss-Webster mice. *Physiology and Behavior,* 1974, *13,* 553–561.

Mackintosh, J. H., & Grant, E. C. The effect of olfactory stimuli on the agonistic behaviour of laboratory mice. *Zeitschrift für Tierpsychologie,* 1966, *23,* 584–587.

Mann, T., Rowson, L. E. A., Baronos, S., & Karagianindis, A. The androgenic potency of dihydrotestosterone and testosterone in the bull, and the effect of growth hormone on the bovine seminal vesicle. *Journal of Endocrinology,* 1971, *51,* 707–718.

Moyer, K. E. Kinds of aggression and their physiological basis. *Communications in Behavioral Biology,* 1968, *2,* 65–68.

Mugford, R. A., & Nowell, N. W. Pheromones and their effects on aggression in mice. *Nature,* 1970, *226,* 967–968.

Naftolin, F., & Ryan, K. J. The metabolism of androgens in central neuro-endocrine tissues. *Journal of Steroid Biochemistry,* 1975, *6,* 993–997.

Phoenix, C. H. Effects of dihydrotestosterone on sexual behavior of castrated male rhesus monkeys. *Physiology and Behavior,* 1974, *12,* 1045–1055.

Voigt, W., & Hsia, S. L. The antiandrogenic action of 4-androstan-3-one-17β-carboxylic acid and its methyl ester on hamster flank organ. *Endocrinology,* 1973, *92,* 1216–1222.

West, C. D., Damast, B. L., Sarro, S. D., & Pearson, O. H. Conversion of testosterone to estrogens in castrated, adrenalectomized human females. *Journal of Biological Chemistry,* 1956, *218,* 409–418.

Wilson, J. D., & Lasnitski, I. Dihydrotestosterone formation in fetal tissues of rabbit and rat. *Endocrinology,* 1971, *89,* 659–668.

2

Pituitary-Adrenocortical Effects on Intermale Agonistic Behavior

Alan I. Leshner

Until the mid-1960s, it was widely believed that the gonadal hormones, particularly the androgens, were the only hormones involved in the control of aggressive behavior. This belief was based on the rather full literature establishing that these hormones are important to aggression, and the effects of no other subsystem had received experimental attention. In the past 15 years, however, it has been established that there are indeed many other kinds of hormonal effects on aggressive behavior, and many of those are discussed in other chapters of this book. This chapter is concerned with the effects of the hormones of the pituitary-adrenocortical axis, ACTH and corticosterone.

In addition to altering our view of the hormonal control of aggression, the results of studies conducted within the past 15 years have also caused a reevaluation of our view of the broader class of behavior of which aggression is a subset—what has been called "agonistic behavior." Agonistic behavior refers to the entire range of behaviors that can occur in competitive situations. These include aggressive, defensive, and submissive situations. These traditionally were believed to exist along a single continuum, with aggressive behavior at one end, defensive behavior in the middle, and submissive behavior at the other end. That would make aggressiveness the opposite of submissiveness, and nonaggressiveness synonymous with submissiveness. What has been learned, instead, is that each of these behaviors can be altered by hormonal manipulations in quite different and independent ways and, therefore, that they are independent behavioral traits that only usually vary together. That is, some hormones seem to affect only aggression, whereas other affect only submission, etc. (e.g., Adams, 1979; Leshner, 1981; Leshner, Korn, Mixon, Rosenthal, & Besser, 1980). The focus of this chapter is on aggression, but, when appropriate, I shall include in this discussion hormonal effects on other classes of agonistic behavior.

Before proceeding to a discussion of the literature on pituitary-adrenocortical effects on aggressive responding, I should mention two limits

ALAN I. LESHNER • National Science Foundation, Washington, D.C. 20550.

imposed on this chapter. First, this discussion will consider only what has been called "intermale" or "spontaneous" aggression. It is now well established that more than one type of aggression can be observed. At the least, it seems clear that hormones affect the aggressive behavior seen under different circumstances in different ways. Intermale or spontaneous aggression refers to the attack behavior seen when two unfamiliar conspecifics meet, and it is only that context for aggression that will be discussed here. The effects of the pituitary-adrenal hormones on aggressive responses observed in other contexts is discussed in other chapters.

Second, the focus of this discussion will primarily be on the intermale aggression of mice. These have been the most frequent subjects of studies on pituitary-adrenocortical effects on aggression, and the effects of ACTH and corticosterone on the aggressive responses of other species will be mentioned in other chapters. I shall, however, make brief mention of the effects of these hormones on the aggression of other species when particularly appropriate.

ESTABLISHING A ROLE FOR ACTH AND/OR CORTICOSTERONE IN AGGRESSION

The First Hints at a Relationship

The first studies suggesting some role for the pituitary-adrenocortical hormones in the control of aggressive responding were not intended to test that relationship directly. Probably the first such study was that of Sigg and his colleagues (Sigg, Day, & Colombo, 1966), in which it was found that adrenalectomized animals require a longer period of isolation before they display aggressive responses than do intact control animals. This study really was designed to explore in more detail the depressive effects of hypophysectomy on aggressiveness, and one of the endocrine systems affected by this operation is the pituitary-adrenocortical axis. Therefore, it made sense to examine the effects of adrenalectomy as well as those of a variety of other endocrine manipulations.

A second suggestive study was conducted by Lagerspetz, Tirri, and Lagerspetz (1968), who were examining the neurochemical and endocrinological characteristics of mice selectively bred for aggressiveness and nonaggressiveness. In the course of that study, they examined the weight of the endocrine glands of those mice, and found that the more aggressive strain had heavier adrenal glands than did the less aggressive strain.

A third early experiment was conducted by Kostowski, Rewerski, and Piechocki (1970). In the course of screening a wide range of steroid hormones for their effects on aggressive behavior, these investigators examined the effects of three days of treatment with an intermediate dosage of cortisol on the aggressive behavior of mice. They found this treatment to increase aggressiveness, and they attributed this effect to an excitation of the central nervous system. A similar stimulatory effect of glucocorticoid treatment on aggressiveness was reported early on by Banarjee (1971).

All of these studies suggested some kind of relationship between the pituitary-adrenocortical hormones and aggression, and they were the basis for later direct tests. However, at the same time that these highly suggestive studies were published, two other reports argued that pituitary-adrenal manipulations did not affect aggression. Bronson and Desjardins (1971) reported that exogenous ACTH treatment did not affect aggressive responding in their mice, and Burge and Edwards (1971) reported that adrenalectomy had no effect on intermale aggressiveness, neither in aggression-experienced nor in naive mice. Although these negative findings were initially ignored by those interested in this relationship, they do raise questions about the generality of pituitary-adrenocortical effects and will be discussed again in later sections of this chapter.

The Identification That a Relationship Exists

The first direct test of the role of the pituitary-adrenal hormones in aggressiveness showing dramatic effects was published in 1971 by Brain, Nowell, and Wouters. In that study, these investigators examined the effects of long-term treatment with ACTH, adrenalectomy, and treatment with the synthetic glucocorticoid dexamethasone on aggression. They found that both ACTH treatment and adrenalectomy decrease aggressiveness, and that dexamethasone increases aggressiveness. At about the same time, Harding and Leshner (1972) published a study originally designed to test the role of the adrenals in the nonaggressiveness of group-housed mice, and in that study, they found a marked decrease in aggressiveness in adrenalectomized, isolated mice. These two reports then formed the basis for a rather intensive effort designed at analyzing the relationship between pituitary-adrenocortical hormones and aggression that still goes on.

THE EFFECTS OF LONG-TERM CHANGES IN PITUITARY-ADRENOCORTICAL HORMONE LEVELS

To summarize the findings discussed so far in this chapter, the initial studies showed that (1) hypophysectomy reduces aggressiveness; (2) adrenalectomy reduces aggressiveness; (3) long-term treatment with ACTH reduces aggressiveness; (4) dexamethasone, which both is a synthetic glucocorticoid and reduces endogenous ACTH levels, increases aggressiveness; and (5) short-term increases in adrenal corticoid levels increases aggressiveness. The next group of studies investigated the effects of long-term alterations in pituitary-adrenal hormone levels on intermale aggressive behavior.

All of those findings just summarized point superficially to a biphasic or bimodal relationship between pituitary-adrenal hormone levels and levels of aggressiveness. Lowering the levels of these hormones, by hypophysectomy, reduces aggressiveness, and injecting ACTH, which increases the levels of both ACTH and corticosterone, leads to a similar suppression of aggressive responding. However, adrenalectomy, which leads to increases in ACTH levels but decreases in corticosterone levels, also decreases aggressiveness, and recognition of this effect led to studies to determine which hormone, ACTH or corticosterone, is the critical pituitary-adrenal hormone in this relationship.

At the same time that this question was posed, great strides had been made in identifying the ways in which the pituitary-adrenal hormones affect avoidance responding, and studies had identified an extra-adrenal effect of this peptide on avoidance behavior (DeWied, 1969, 1976). Therefore, the first tactic in asking whether increasing ACTH levels decreases aggressiveness directly or through its effects on corticosterone levels was similar to those used when studying avoidance behavior. Two variations of a single approach were used. One was to take advantage of the existence of ACTH analogs that have no effect on adrenal function, and the second was to manipulate ACTH levels while experimentally holding corticosterone levels constant. Both techniques result in elevated ACTH levels with unchanged corticosterone levels.

Before discussing those studies, it should be mentioned here that some of the earlier studies bear on this problem of separating ACTH and corticosterone effects. Recall that adrenalectomy was found to decrease aggressiveness, and note Candland and Leshner's (1974) finding that corticosterone treatment restores the aggressiveness of adrenalectomized mice. Adrenalectomy leads to a condition in which ACTH levels are high and corticosteroid levels are low, and replacement therapy with corticosterone decreases ACTH levels while it restores corticosterone levels in the adrenalectomized animal. Therefore, it could be said that elevated ACTH levels—whether corticosterone levels are also high, as in the case of ACTH treatment, or are low, as in the case of adrenalectomy—always lead to decreased aggressiveness.

In using the first approach, Brain (1972) injected mice with the full ACTH molecule and with the analog $ACTH_{4-10}$. He found that the analog had the same suppressive effect as the parent molecule, which suggested that ACTH was in fact affecting aggressiveness directly. However, in the same study, Brain noticed that, although $ACTH_{4-10}$ has no tropic actions in the rat, this analog did lead to a slight increase in corticosterone levels in the mouse. Therefore, although his findings were suggestive, they were by no means conclusive.

On the other hand, Leshner and his colleagues (Leshner, Walker, Johnson, Kelling, Kreisler, & Svare, 1973) used the second approach, that of manipulating ACTH levels while holding corticosterone levels constant. To accomplish that test, they compared the effects of injected ACTH on the aggressiveness of intact mice with these effects in mice with controlled corticosterone levels. Controlling corticosterone levels was accomplished by adrenalectomizing the animals and then treating them with a fixed dosage of corticosterone—a dosage sufficient to maintain the aggressiveness of adrenalectomized mice. In this way, Leshner and his colleagues could compare treatments with ACTH leading to increases in ACTH but not in corticosterone levels with the effects of similar treatments leading to increases in the levels of both hormones. If ACTH were acting directly, then both groups should have been affected. If, however, ACTH were acting through its effects on corticosterone, then only the intact animals should have been affected.

In fact, ACTH was effective in both cases: ACTH treatment decreased the aggressiveness of both intact mice and mice with controlled corticosterone levels. This finding, then, was consistent with Brain's (1972) earlier finding suggesting that ACTH affects aggressiveness directly.

Having seen that the effects of increasing pituitary-adrenal hormone levels is an ACTH effect, one would then ask about the suppressive effects of decreasing the levels of these hormones. The only technique used to answer this question has been to examine the effects of decreasing ACTH levels while keeping corticosterone levels high. Sadly, the obvious study examining the abilities of ACTH and/or corticosterone to restore the aggressiveness of hypophysectomized mice (with concomittant increases in testosterone levels to ensure adequate supplies of that hormone) has not been done.

Candland and Leshner (1974) conducted a study in which mice were adrenalectomized and then treated with varying dosages of corticosterone, from a very low to a very high dosage. They found that high dosages of this steroid decrease aggressiveness. Since that treatment also would decrease ACTH levels, it seems likely that those manipulations leading to decreased levels of both hormones decrease aggressiveness because of the reduction in ACTH, rather than corticosterone, titers. Although this study is highly suggestive, however, it is by no means conclusive, and therefore, additional studies will be needed to resolve the question of which hormone is critical when both ACTH and corticosterone levels are reduced.

Some Questions of Generality

In spite of the apparent power of pituitary-adrenal effects just discussed, it should be mentioned before going further that these effects are observed in only some species of animals (e.g., Landau, 1975, for a positive effect in hamsters; Hull, l'Hommedieu, Kastaniotis, & Franz, 1979, for negative effects in gerbils), and not even in all strains of mice nor under all conditions of aggression testing in mice. As mentioned before, Bronson and Desjardins (1971) and Burge and Edwards (1971) had found no effect on pituitary-adrenal manipulations on aggressiveness in their animals. Furthermore, some later tests with mice (Simon & Gandelman, 1978) also found no effects of prolonged corticosterone treatment, which decreases ACTH levels. The only explanation that as yet seems appropriate is that these studies with negative findings had used quite different methods of testing and hormone administration than had Brain's and Leshner's groups, and that they also used different strains of animals. That explanation is not very comforting to this author, but in a more systematic study of strain and testing technique differences, Brain did find these effects to be rather sensitive to such differences (Brain & Poole, 1974). At the least, these negative findings must lead to reservations about the power of these effects, which certainly are not as powerful as those of pituitary-gonadal manipulations, which pass tests of generality very well (see chapter by Brain).

Another way to ask about generality is to ask whether these effects observed in studies using extreme manipulations of pituitary-adrenal hormone levels hold for the case of the intact animal, which is not subject to such extreme variations in the levels of these hormones. One study by Politch and Leshner (1977) did find a significant negative correlation between resting corticosterone levels and levels of aggressiveness. This finding suggests that within the intermediate range of pituitary-adrenal hormone levels (intermediate as compared with the experimental studies discussed before), the greater the endogenous level of ACTH is, the lower the level of aggressiveness. This relationship is identical to that predicted from the experimental studies.

Mechanism of Long-Term Action of ACTH on Aggressiveness

The next obvious question to ask concerns the mechanism by which long-term treatment with ACTH decreases aggressiveness. Rephrased, one might ask whether ACTH affects aggressiveness directly or whether it alters this behavior through its effects on another endocrine system or on a related behavior. Since ACTH can have a suppressive effect on androgen secretion, and since the withdrawal of androgens has such a dramatic depressive effect on aggressive responding, it was first asked whether ACTH might be decreasing aggressiveness because it also decreases androgen levels. Leshner *et al.* (1973) asked this question by comparing the effects of ACTH treatment in intact mice, who could respond with reduced androgen levels, with the effects of ACTH on mice with controlled testosterone levels, who could not respond with reductions in androgen titers. ACTH decreased aggressiveness in both groups of animals. Thus, increasing pituitary-adrenocortical activity could not be decreasing aggressiveness by reducing androgen levels.

The second possibility was that ACTH decreases aggressiveness through its effects on a related behavior. Since many studies have suggested that ACTH can have direct, enhancing effects on avoidance responding, which has been said to be fear-mediated, the most likely indirect mechanism of action for ACTH on aggression would be through this peptide's effects on fearfulness. It was thought that fearful animals would more likely be submissive than aggressive, and, therefore, that high levels of fearfulness would be incompatible with aggressiveness (e.g., Brain, 1971; Leshner *et al.*, 1973).

Leshner and his colleagues tested that possibility directly. A fear–mediation hypothesis would suggest that ACTH-treated animals would be nonaggressive and would avoid contact with a dangerous opponent. Therefore, they examined the effects of pituitary-adrenocortical hormones on the tendency to avoid attack (e.g., Leshner, Moyer, & Walker, 1975; Moyer & Leshner, 1976). This was accomplished by using a standard passive avoidance procedure, but substituting attack by a trained fighter mouse for the more commonly used aversive stimulus of electric shock.

In the first study, Leshner *et al.* (1975) found that increasing ACTH levels in intact animals increases the tendency to avoid attack and suggested that their findings support the fear-mediation hypothesis. However, when they proceeded to separate the effects of ACTH and corticosterone on this avoidance response, their findings were inconsistent with that hypothesis. Whereas ACTH had been shown to have an extra-adrenal effect on aggressiveness, the effects of this peptide on avoidance-of-attack were totally dependent on its effects on corticosterone levels (Moyer & Leshner, 1976). The same was found true in studies of submissiveness (e.g., Leshner & Politch, 1979). Therefore, since ACTH treatment could decrease aggressiveness without increasing fearfulness (when corticosterone levels were not also raised), it is unlikely that ACTH would be acting through an effect on fearfulness.

These data rule out the one mediational hypothesis proposed to account for ACTH's effects on aggressive responding, that of a mediation by fearfulness. No other interpretations have been suggested yet, but, to this author, it seems most likely that ACTH is somehow altering the state of the neural circuits that control aggressive responding to external stimuli. I have suggested elsewhere (Leshner, 1975, 1978) that ACTH treatment might decrease aggressiveness by altering the brain and, thereby, the animal's perception of aggression-eliciting stimuli, but this possibility has not been explored further. (Note: At the least, there is little question that ACTH could affect aggressive behavior by acting on the brain. This issue is discussed in other chapters in this book.)

THE EFFECTS OF SHORT-TERM CHANGES IN PITUITARY-ADRENOCORTICAL HORMONE LEVELS

The studies just discussed all dealt with the effects of long-term alterations in pituitary-adrenal hormone levels on levels of aggressiveness. However, recall that Kostowski *et al.* (1970) had found that relatively short-term treatments (specifically, three days of treatment) with cortisol increased aggressiveness. This relationship is the opposite of that observed in the long-term case, and, if true, would suggest that there is a biphasic relationship between pituitary-adrenal hormone levels and aggressiveness. In fact, Matte (1979) recently compared short-term and long-term treatment with ACTH and found just that. Treatment with ACTH for periods of one to four days increased aggressiveness, whereas longer-term treatments reduced aggressiveness. That short-term treatment with ACTH or with corticosterone increases aggressiveness has by now been demonstrated many times (see reviews in Brain & Poole, 1974; Leshner, 1978).

Separation of the Effects of ACTH and Corticosterone

After this stimulatory effect of short-term increases in pituitary-adrenal hormone levels was established, investigators turned to the question of whether ACTH or corticosterone was the critical hormone in this effect. Brain and Evans (1977) used the analog approach to ask that question; they treated mice with either the almost complete ACTH molecule or with analogs with varying degrees of effect on the adrenal cortex. In these studies they found that $ACTH_{1-24}$, a form that would also increase corticosterone levels, was the only form of ACTH that led to marked increases in aggressiveness. However, they did find that $ACTH_{1-10}$, which had only a minimal effect on corticosterone levels, also increased aggressiveness slightly. Therefore, they suggested that the effect of short-term ACTH treatment on aggressiveness is primarily corticosterone mediated, although there may also be a slight direct stimulatory effect of this peptide hormone as well.

We also conducted a study separating ACTH and corticosterone effects in the short-term case (Leshner & Walker, unpublished observations). In that study, we compared the effects of three days of ACTH treatment on the aggressiveness of intact mice with the effects of this same treatment in mice with controlled corticosterone levels (see discussion above of the technique for controlling corticosterone levels). Our finding supported the basic finding of Brain and Evans's studies, by showing that short-term ACTH treatment increased the aggressiveness of intact mice only; mice who could not respond to the peptide treatment with increased corticosterone levels were unaffected. These two sets of studies, combined with the observation that short-term corticoid treatment alone also increases aggressiveness, suggest that corticosterone, and not ACTH (at least not very importantly), is the critical pituitary-adrenocortical hormone in the short-term case.

Corticosterone Mediation of the Effects of Shock Treatments on Later Aggressiveness

No discussion of pituitary-adrenocortical effects on aggressiveness, particularly the effects of short-term treatments, would be complete without discussion of the elegant work of Heller (1977, 1978, 1979) on the effects of prior shock treatments on the display of aggressive behavior. In 1977, Heller reported that if mice are exposed to foot-shocks between 30 and 60 minutes before testing for aggressiveness, they exhibit increased aggression in that test situation. Furthermore, he found that castration had no effect on this postshock fighting, but that adrenalectomy or adrenalectomy combined with castration decreased the postshock increase in aggressiveness. This set of findings suggested that postshock fighting is somehow related to pituitary-adrenal hormone levels.

He next (Heller, 1978) asked whether corticosterone treatment could restore the depressed postshock fighting of adrenalectomized animals, and he found that it could. But corticosterone treatment alone did not restore this behavior in adrenalectomized and castrated animals. Therefore, some androgens must be present for the postshock enhancement of aggressiveness.

In the earlier study (Heller, 1977), however, castration alone did not alter the effect of shock; why then should testosterone treatment be necessary in mice that are both adrenalectomized and castrated? The most reasonable answer is that some androgens must be present for the effect to be observed, and in the castrated animal, androgens from the andrenal provide that necessary amount of androgen stimulation. But when animals are both adrenalectomized and castrated, there are no androgens present; therefore, exogenous androgen must be provided.

In that second paper, Heller (1978) also reported that preventing corticoid changes during the shock treatment (through controlling corticosterone levels as discussed above and in Leshner *et al.,* 1973) prevented the shock-induced facilitation of later aggressiveness. This lent even further support to the notion of critical pituitary-adrenal involvement in this postshock stimulation of fighting.

In the third paper, Heller (1979) reported that single ACTH injections without shock treatment produced the same increase in aggressiveness as did prior shock. Furthermore, preventing corticosterone responses to either ACTH or shock treatment diminished the shock-induced effect. This set of findings, then, suggests that shock treatments increase later aggressiveness because they produce acute increases in corticosterone levels. This interpretation is entirely consistent with the findings just discussed (in the previous section) and lends further credence to them. Therefore, it by now seems well established, through a variety of kinds of tests, that acute increases in corticosterone levels can increase aggressiveness.

Mechanism of Short-Term Action of Corticoids on Aggressiveness

Having established that short-term increases in corticosterone levels increase aggressiveness, the next question might concern the way in which that effect is exerted. That is, do corticoids increase aggressiveness directly, or do they act through their effects on another endocrine subsystem or through their effects on other behaviors?

Evans and Brain (1976) asked about possible mediation through ACTH's or corticosterone's effects on the testes. In their study, they used the technique described above (see also Leshner *et al.,* 1973) of comparing the effects of ACTH treatment on the aggressiveness of intact mice with those of controlled levels, in this case, of testosterone. They found that short-term ACTH treatment increased the aggressiveness of both groups of animals, suggesting that the pituitary-adrenal hormones do not affect aggressiveness through their effects on the testes.

Assuming then that the pituitary-adrenal hormones somehow affect aggressiveness independently of their effects on other endocrine glands, one might next ask about mediation through effects on other behaviors. An important finding enters into this discussion. That is, that corticosterone treatment also increases submissiveness. My colleagues and I (e.g., Leshner, 1981; Leshner & Politch, 1979; Leshner *et al.,* 1980) have conducted a series of studies concerned with the effects of the pituitary-adrenal hormones on submissive responding, and we have found that corticosterone, whether administered acutely or chronically, increases this behavior. Therefore, it appears that short-term increases in pituitary-adrenocortical hormone levels, particularly corticosterone levels, can increase both aggressiveness and submissiveness. Although, as stated above, aggressiveness and submissiveness are not necessarily opposites, they almost never occur simultaneously; they do appear to be mutually exclusive behaviors in a given situation. Thus, we must ask how corticosterone could be increasing the readiness to exhibit both of these normally exclusive behaviors.

The most likely explanation lies in Bohus's (1975) suggestion of how corticoids might affect avoidance responding in shock-mediated situations. He suggested that corticosterone alters avoidance responding by altering the motivational value of critical environmental stimuli, perhaps through an effect on general arousal. I have suggested elsewhere (Leshner, 1981) that a similar interpretation is appropriate to account for corticosterone's stimulatory effects on both aggressiveness and submissiveness, and I reiterate that opinion here. It could well be that relatively acute increases in corticoid levels increase general arousal (see also Weiss, McEwen, Silva, & Kalkut, 1969), and, thereby, that the perceived intensity of incoming stimuli is increased. In this way, an aggression-eliciting stimulus would be perceived as more aggression-provoking, and a submission-eliciting stimulus would be perceived as more threatening, and therefore, more submission-provoking. Since we all study agonistic behavior using relatively passive stimulus animals for studies of aggression and highly aggressive stimulus animals for studies of submission, it is likely that corticosterone increases both aggression and submission because it increases the test animals's arousal and, therefore, the perceived intensity of the relevant behavior-provoking stimulus. Since the behavior-provoking stimulus, then, is more intense, one would observe a more intense response.

CONCLUDING REMARKS

The case of pituitary-adrenocortical effects on aggressive responding provides a very complex and unusual example of hormone–behavior interactions. First, it is clear that these effects are not observed in all species, not even in all strains of mice. In addition, these effects are not observed under all methods of testing, but when they are observed, they appear very powerful.

Second, there is the biphasic nature of the effects of these hormones. As Matte (1979) dramatically showed, short-term increases in pituitary-adrenal hormone levels lead to increased aggressiveness, whereas long-term increases in the levels of these hormones lead to decreased aggressiveness.

Third, these short-term and long-term effects depend on different pituitary-adrenal hormones. The long-term effect appears to be one of direct ACTH action. On the other hand, the short-term effect appears to be due to corticosterone. Whether they will be found to have common mediating mechanisms or not remains to be determined.

REFERENCES

Adams, D. B. Brain mechanisms for offense, defense and submission. *Behavioral and Brain Sciences,* 1979, *2,* 201–241.

Banarjee, U. The influence of some hormones and drugs on isolation-induced aggression in male mice. *Communications in Behavioral Biology,* 1971, *6,* 163–170.

Bohus, B. The hippocampus and the pituitary-adrenal system hormones. In R. L. Isaacson & K. H. Pribram (Eds.), *The hippocampus* (Vol. I). New York: Plenum Press, 1975.

Brain, P. F. Possible role of the pituitary-adrenocortical axis in aggressive behaviour. *Nature,* 1971, *233,* 489.

Brain, P. F. Study on the effect of the 4-10 ACTH fraction on isolation-induced intermale fighting behaviour in the albino mouse. *Neuroendocrinology,* 1972, *10,* 371–376.

Brain, P. F., & Evans, A. E. Acute influences of some ACTH-related peptides on fighting and adrenocortical activity in male laboratory mice. *Pharmacology, Biochemistry and Behavior,* 1977, *7,* 425–433.

Brain, P. F., & Poole, A. E. The role of endocrines in isolation-induced intermale fighting in albino laboratory mice: 1. Pituitary adrenocortical influences. *Aggressive Behavior,* 1974, *1,* 39–69.

Brain, P. F., Nowell, N. W., & Wouters, A. Some relationships between adrenal function and the effectiveness of a period of isolation in inducing intermale aggression in albino mice. *Physiology and Behavior,* 1971, *6,* 27–29.

Bronson, F. H., & Desjardins, C. Steroid hormones and aggressive behavior in mammals. In B. E. Eleftheriou & J. P. Scott (Eds.), *Physiology of aggression and defeat.* New York: Plenum Press, 1971.

Burge, K. G., & Edwards, D. A. The adrenal gland and the pre- and post-castrational aggressive behavior of male mice. *Physiology and Behavior,* 1971, *7,* 885–888.

Candland, D. K., & Leshner, A. I. A model of agonistic behavior: Endocrine and autonomic correlates. In L. V. DiCara (Ed.), *Limbic and autonomic nervous system research.* New York: Plenum Press, 1974.

DeWied, D. Effects of peptide hormones on behavior. In W. F. Ganong & L. Martini (Eds.), *Frontiers in neuroendocrinology.* New York: Oxford University Press, 1969.

DeWied, D. Pituitary adrenal system hormones and behaviour. *Symposium on developments in endocrinology.* Oss, Netherlands: Oganon, 1976.

Evans, A. E., & Brain, P. F. Acute influences on gelatin preparation of corticotrophin and corticotrophin-related peptides in "TO" strain mice. *Journal of Endocrinology,* 1976, *69,* 7P–8P.

Harding, C. F., & Leshner, A. I. The effects of adrenalectomy on the aggressiveness of differently housed mice. *Physiology and Behavior,* 1972, *8,* 437–440.

Heller, K. E. Temporal relationships between shock treatment and intermale fighting behaviour in laboratory mice. Effects of modifying testicular and adrenal functions. *Behavioural Processes,* 1977, *2,* 337–347.

Heller, K. E. Role of corticosterone in the control of post-shock fighting behaviour in male laboratory mice. *Behavioural Processes,* 1978, *3,* 211–222.

Heller, K. E. An attempt to separate the roles of corticosterone and ACTH in the control of post-shock fighting behaviour in male laboratory mice. *Behavioural Processes,* 1979, *4,* 231–238.

Hull, E. M., l'Hommedieu, G., Kastaniotis, C., & Franz, J. R. Pituitary-adrenal hormones do not influence bulbectomy induced behavioral changes. *Physiology and Behavior,* 1979, *22,* 417–421.

Kostowski, W., Rewerski, W., & Piechocki, T. Effects of some steroids on aggressive behaviour in mice and rats. *Neuroendocrinology,* 1970, *6,* 311–318.

Lagerspetz, K. Y. H., Tirri, R., & Lagerspetz, K. M. J. Neurochemical and endocrinological studies of mice selectively bred for aggressiveness. *Scandinavian Journal of Psychology,* 1968, *9,* 157–160.

Landau, I. T. Effects of adrenalectomy on rhythmic and non-rhythmic aggressive behavior in the male golden hamster. *Physiology and Behavior,* 1975, *14,* 775–780.

Leshner, A. I. A model of hormones and agonistic behavior. *Physiology and Behavior,* 1975, *15,* 225–235.

Leshner, A. I. *An introduction to behavioral endocrinology.* New York: Oxford University Press, 1978.

Leshner, A. I. The role of hormones in the control of submissiveness. In P. F. Brain & D. Benton (Eds.), *A multidisciplinary approach to aggression research.* Amsterdam: Elsevier, 1981.

Leshner, A. I., & Politch, J. A. Hormonal control of submissiveness in mice: Irrelevance of the androgens and relevance of the pituitary-adrenal hormones. *Physiology and Behavior,* 1979, *22,* 531–534.

Leshner, A. I., Walker, W. A., Johnson, A. E., Kelling, J. S., Kreisler, S. J., & Svare, B. B. Pituitary adrenocortical activity and intermale aggressiveness in isolated mice. *Physiology and Behavior,* 1973, *11,* 705–711.

Leshner, A. I., Moyer, J. A., & Walker, W. A. Pituitary-adrenocortical activity and avoidance-of-attack in mice. *Physiology and Behavior,* 1975, *15,* 689–693.

Leshner, A. I., Korn, S. J., Mixon, J. F., Rosenthal, C., & Besser, A. K. Effects of corticosterone on submissiveness in mice: Some temporal and theoretical considerations. *Physiology and Behavior,* 1980, *24,* 283–288.

Matte, A. C. Biphasic and dissociated effects of ACTH on motor activity, aggression and "emotionality" in mice. *Psychoneuroendocrinology,* 1979, *4,* 21–27.

Moyer, J. A., & Leshner, A. I. Pituitary-adrenal effects on avoidance-of-attack in mice: Separation of the effects of ACTH and corticosterone. *Physiology and Behavior,* 1976, *17,* 297–301.

Politch, J. A., & Leshner, A. I. Relationship between plasma corticosterone levels and levels of aggressiveness in mice. *Physiology and Behavior,* 1977, *19,* 775–780.

Sigg, E. B., Day, C., & Colombo, C. Endocrine factors in isolation-induced aggressiveness in rodents. *Endocrinology,* 1966, *78,* 679–684.

Simon, N. G., & Gandelman, R. Influence of corticosterone on the development and display of androgen-dependent aggressive behavior in mice. *Physiology and Behavior,* 1978, *20,* 391–396.

Weiss, J. M., McEwen, B. S., Silva, M. T. A., & Kalkut, M. F. Pituitary-adrenal influences on fear responding. *Science,* 1969, *163,* 197–199.

3

Hormones and Aggression in Female Mammals

Owen R. Floody

Aggressive behavior commonly is thought of as the prerogative of males. However, females of a variety of species are aggressive and can compete for rank or other resources with important biological consequences.

This chapter will review studies of aggression in female mammals. My primary goals are, first, to assess the generality and functional significance of aggressive behavior by females. Second, I will try to evaluate the extent to which female aggression can be, and normally is, controlled by sex hormones.

In working toward these goals, I will emphasize studies of the effects of endogenous cycles in hormone levels and of exogenous hormones administered in adulthood on overt aggressive behavior directed by females at adult or subadult conspecifics. Some consideration also will be given to studies describing competition for food or water rewards or using other indirect measures of aggression. However, little or no attention will be devoted to predation or to several classes of aggressive behavior that can involve females but that are discussed in other chapters of this book, specifically maternal aggression (Svare & Mann), infanticide (Gandelman), and shock-induced fighting (Conner). Similarly, the changes in female aggression that can be caused by perinatal hormone manipulations are discussed briefly below, but are summarized fully in the chapter by vom Saal.

MALE-LIKE AGGRESSION IN FEMALES: EFFECTS OF PERINATAL HORMONE TREATMENTS

Previous reviews have stressed that males typically are more aggressive than and dominant to females: "In all mammalian species, from mouse to man, the male is the more aggressive sex" (Moyer, 1974, p. 335). This view is an unfor-

---

OWEN R. FLOODY • Department of Psychology, Bucknell University, Lewisburg, Pennsylvania 17837. This chapter was written while the author was a Visiting Research Fellow in the Department of Biology at Princeton University. This was made possible by the hospitality of Dr. Robert D. Lisk and by support provided by a postdoctoral fellowship (08070) and grant (33191) from the National Institute of Mental Health.

tunate exaggeration (see next section). Nevertheless, it has had important influences on the ways in which females have been used in studies of the hormonal control of aggression. First, it has helped to focus attention on species and situations that illustrate the "typical" sexual dimorphism in aggression, species such as the house mouse *(Mus musculus)*, in which males but not females readily attack adult conspecifics (e.g., Bronson & Desjardins, 1968). Second, it has encouraged the use of females as imperfect males in studies of the hormonal events controlling the development of "male-like" aggression in these species and situations. The resulting, relatively large body of work has provided important support for the hypothesis that the hormonal priming normally experienced early in life by males can permanently "organize" an initially neutral neural substrate so that aggressiveness during exposure of the adult to androgens is greatly facilitated (see chapter by vom Saal).

As applied to females, studies of perinatal hormone effects on male-like aggression describe ways in which hormones *can* act to modulate female aggression. However, they generally tell much less about the manner in which hormones act in species and situations in which females normally are aggressive. Nevertheless, recent studies have described differences in female aggression that do seem attributable to perinatal androgen exposure, that is, to natural events similar to some of the experimental procedures favored by developmental studies of male-like aggression. Specifically, vom Saal and his colleagues have found that female fetuses that develop *in utero* between two males are more aggressive as adults than females developing between other females (chapter by vom Saal). This suggests that the partial androgenization of female mammals is a much more routine event than previously realized. In turn, individual differences in degree of masculinization could lead to individual differences in female aggression, with potentially important implications for processes of sexual selection among females and the regulation of animal populations by aggressive females (also see later sections of this chapter and the chapters by vom Saal and Christian).

SEXUAL DIMORPHISMS IN AGGRESSION

Are Males Always the More Aggressive Sex?

Previous reviews often have considered the greater aggressiveness of males to be so obvious as to require little or no documentation (see classic review by Collias, 1944; also see Archer, 1976, and Gray, 1971, for relatively thorough recent reviews).

Studies that provide relatively direct comparisons between levels of aggression or dominance in intact males and females and that support the "typical" sexual dimorphism in aggression are summarized in Table I. Some possible counterexamples to the view that males always dominate females in forms of aggression common to the two sexes are identified in Table II (also see review of human aggression by Frodi, Macaulay, and Thome, 1977).

Comparison of Tables I and II makes one point quite forcefully: Viewed within or across species, it simply is not the case that females are always less aggressive than or submissive to males. To the contrary, superficial consideration of these summaries suggests that the reverse is nearer the truth. Unfortunately, this impression may stem, in large part, from limitations caused by (a) tendencies to avoid belaboring "obvious" relations (elevated aggression in males) while continuing to report relatively novel findings (elevated aggression in females) and (b) a genuine scarcity of detailed studies dealing explicitly with female aggression or with sexual differences in aggression. Additional studies clearly are needed, and one hopes that these will receive some impetus from this review, as well as from other chapters treating vertebrate orders or classes that tend to be underrepresented in experimental studies of hormones controlling aggression (see chapters by Bouissou, Greenberg and Crews, Harding, and Villars). For maximal impact, however, these studies will have to consider the issues and problems discussed below.

Different Measures Yield Different Results

The use of different measures of aggression or dominance sometimes can lead to very different conclusions regarding sexual differences in aggression. For example, in an early study that compared levels of aggression and dominance in male and female rats *(Rattus norvegicus),* Seward (1945a) found that males were more likely to form clear dominance–subordinance relations, suggesting that females are "less concerned with the problem of status" (p. 185). This difference has been used subsequently to argue that males are, in general, more aggressive than females (e.g., Archer, 1976). This interpretation, however, places disproportionate emphasis on one measure that showed a sex difference, ignoring several others that did not (frequency of aggressive acts, incidence of fights, and the percentage of fights that were decisive). In addition, it overlooks Seward's later (1945b) inability to detect stable dominance orders in pairs or groups of male or female rats.

Rowell's (1971) study of dominance relations in a captive group of vervet monkeys *(Cercopithecus aethiops)* also illustrates the discrepancies that can arise among different measures of aggression. Here, the hierarchy based on the relative frequencies of retreat suggested that the one adult male in the group was similar in dominance to the two top-ranking females. In turn, this obviously suggests that vervets are not sexually dimorphic in aggression or dominance. This analysis, however, was complicated by other results. For one, at least some classes of agonistic behavior (chases and threats) tended to be directed up, not down, the dominance hierarchy, suggesting that quite a different hierarchy could have been generated on the basis of different criteria. In fact, the male could have been ranked either second, fifth, or seventh in the group of seven based just on different measures derived from frequencies of approach–retreat interaction. In addition, the focal role played by the male in several nonagonistic behaviors (grooming, sitting, and nose-to-mouth greetings) suggests that he could even have been assigned the top rank in the group had dominance been defined more broadly to include relative access to affiliative interactions.

TABLE I

Studies Showing Elevated Aggression or Dominance in Males[a]

Subject	Source	Type of study	Finding
Human, *Homo sapiens*	Maccoby & Jacklin, 1974	Review	While crime records are unequivocal in showing higher levels of violence by adult males, studies of children in a variety of situations and cultures probably are contaminated less by sociological factors, yet concur in showing higher levels of aggression and dominance in boys.
Celebes black ape, *Cynopithecus niger*	Bernstein, 1970	Determination of status hierarchy in a captive group	The top-ranking individual was a male.
Sooty mangabey, *Cercocebus atys*	Bernstein, 1970	Determination of status hierarchy in a captive group	The top-ranking individual was a male.
Vervet monkey, *Cercopithecus aethiops*	Struhsaker, 1967; Rowell, 1971	Field study; Determination of status hierarchy in a captive group	Each male dominates all females in the group; Rowell disagrees, but the one male in her group showed a low rate of participation in status interactions and was a focal point of the group in other respects.
Crab-eating macaque, *Macaca fascicularis*	Bernstein, 1970	Determination of status hierarchy in a captive group	The top-ranking individual was a male.

Rhesus monkey, *M. mulatta*	Carpenter, 1942; Michael & Zumpe, 1970	Study of free-ranging groups on Cayo Santiago; staged encounters between captive males and females	Male always "completely dominant" over female in field; aggressive episodes in lab typically were initiated by male.
Pigtail macaque, *M. nemestrina*	Bernstein, 1970	Determination of status hierarchy in a captive group	The top-ranking individual was a male.
Baboon, *Papio cynocephalus ursinus* or *P. ursinus*	Saayman, 1971; Seyfarth, 1976	Field studies	Frequency of male–male aggression greater than that of female–female aggression.
Gelada baboon, *Theropithecus gelada*	Bernstein, 1970	Determination of status hierarchy in a captive group	The top-ranking individual was a male.
Laboratory mouse, *Mus musculus*	Gray, Whitsett, & Ziesenis, 1978; Suchowsky, Pegrassi, & Bonsignori, 1971	Staged encounters with smaller adult or juvenile mouse	The likelihood of attack by male was greater than that by female.
Laboratory rat, *Rattus norvegicus*	Barr, Gibbons, & Moyer, 1976; Lester & Cheses, 1968	Staged encounters between same-sex, food-deprived individuals	Males exceeded females in frequency of most aggressive acts, including the frequency of relatively violent behaviors. Lester and Cheses argue for greater female aggression, but on basis of less detailed measure (percentage fighting in series of round-robin encounters).

[a] Secondary accounts suggest that males also are more aggressive than females in Japanese macaques *(M. fuscata)* and chimpanzees *(Pan troglodytes)* (Gray, 1971). Studies summarized here and in Tables II and III were cited in previous reviews or were identified in computerized searches of the literature on aggression in females and on the hormonal control of aggression. The resulting sample probably is biased toward studies using laboratory or seminaturalistic techniques and dealing with primates or rodents.

TABLE II

Species in which Females Show Types of Aggression Similar to Those of Males and at Similar or Greater Intensity[a]

Species	Source	Type of study	Finding
Gibbon, *Hylobates lar*	Carpenter, 1964	Field study	Mated male and female approximately equal in "pugnacity" and dominance.
Sykes' monkey, *Cercopithecus mitis (albogularis) kolbi*	Rowell, 1971	Observations of approach–retreat interactions in captive group	Single adult male and top-ranking female approximately equal in dominance. Aggressive interactions initiated more frequently by females.
Marmoset, *Callithrix jacchus*	Abbott & Hearn, 1978, 1979	Young animals observed in peer groups with similar numbers of males and females	No consistent relation between inter- and intrasexual ranks. Most agonistic behaviors not sexually dimorphic.
Ring-tailed lemur, *Lemur catta*	Jolly, 1966	Field study	Females in one troop were similar to males in the frequency of mild agonistic acts, but showed fewer violent behaviors. Nevertheless, females were dominant over males.
Wolf, *Canis lupus*	Zimen, 1976	Study of captive groups in large enclosure	Most instances of high-intensity aggression involved two females.
European rabbit, *Oryctolagus cuniculus*	Mykytowycz & Hesterman, 1975	Staged encounters between individuals of wild or domestic strain and same or opposite sex	Males showed greater tendency to use some high-intensity aggressive acts ("ripping with hind claws"). However, females generally were at least as aggressive in same or opposite sex territorial encounters.
European hamster, *Cricetus cricetus*	Eibl-Eibesfeldt, 1953	Observations of captive animals under seminaturalistic conditions	Nonestrous female solitary and intolerant of conspecifics, including males.
Collared lemming, *Dicrostonyx groenlandicus*	Banks, Mankovich, & Huck, 1979	Review of previous field and laboratory studies by authors	Females readily dominate males in most social contexts.

Southern flying squirrel, *Glaucomys volans*	Madden, 1974	Field study	Females territorial, males not. Most aggressive encounters initiated and won by females.
Brown lemming, *Lemmus trimucronatus*	Banks *et al.*, 1979	Review of previous field and laboratory studies by authors	Females readily dominate males in most social contexts.
Yellow-bellied marmot, *Marmota flaviventris*	Downhower & Armitage, 1971	Field study	Of eight instances of a particular type of aggressive interaction, one was male–male and seven involved just adult females.
Mongolian gerbil, *Meriones unguiculatus*	Swanson, 1974; Yahr & Coquelin, 1980	Staged encounters between animals of same or opposite sex	Female was more aggressive than male toward a male intruder. Although similar levels of fighting were seen in male–male and female–female pairs, females were more likely to initiate and win fights with males in heterosexual pairings.
Golden hamster, *Mesocricetus auratus*	Brain, 1972; Payne & Swanson, 1970; Tiefer, 1970	Staged encounters with same or opposite-sex pairs, sometimes using young or estrous female as constant stimulus	Female dominates male in direct encounters. Female equals or exceeds male in likelihood of attacking stimulus animals of different types.
Woodmouse, *Peromyscus leucopus*	Metzgar, 1971	Comparisons of home-range exclusiveness in field study	Females exclude conspecifics from their home ranges more effectively than do males. Implies comparable or greater territorial aggression by females.
Laboratory mouse, *Mus musculus*	Hedricks & Daniels, 1981	Staged encounters in female's home cage	Resident females more aggressive and less submissive than strange male intruder.
Laboratory rat, *Rattus norvegicus*	Seward, 1945a	Brief round-robin encounters between same-sex pairs	Females comparable to males on variety of measures of the likelihood

(Continued)

TABLE II (*Continued*)

Species	Source	Type of study	Finding
			and intensity of aggression, though clear dominance–subordination relations were less likely in females.
Black rat, *R. rattus*	Ewer, 1971	Observations of partially provisioned, but free-ranging groups	Females were "much more aggressive" than males at a feeding station, though females would not attack alpha male of group.
Northeastern grey squirrel, *Sciurus carolinensis pennsylvanicus*	Thompson, 1978	Field study	Females dominant in approximately half of encounters with males of similar age.
Tree squirrels, *Tamiasciurus douglasii* and *T. hudsonicus*	Smith, 1968	Field study	No sex difference in mode of territorial defense
Short-tailed shrew, *Blarina brevicauda Say*	Pearson, 1944	Staged encounters	Female fights vigorously with male except when receptive.

[a] Secondary accounts suggest that females are at least as aggressive as males in chinchillas (*Chinchilla laniger;* Marques & Valenstein, 1977).

The studies by Seward and Rowell indicate that analyses of sexual differences in aggressive behavior or status depend on the measures used to define aggression and dominance operationally. Many of these differences, however, may just reflect the complexity of animal social relations based on aggression. Such complexity can create problems in the short run, but ultimately it should facilitate the application to subtle human relations of results from studies of other animals. In this context, the approaches to the study of aggression taken by Seward and Rowell are commendable. Specifically, each of these studies examined differences in levels of several distinct aggressive acts. Such an approach increases the chances of uncovering complexity (or disorder) where it exists and is far preferable to one in which an unrealistically simple or orderly picture is provided by single measures, or composite indices, of aggression.

What Should We Measure?

Aggressive Behavior?

Aggressive behavior represents the product of a complex chain of events that includes physiological, emotional, and cognitive components. Sexual differences in aggressive behavior can be introduced at any stage; however, not all of these may signal differences in aggressive motivation (Frodi, 1978). This point impacts most strongly on the evaluation of sexual dimorphisms in human aggression, where apparent motivational differences actually may stem from differences in (a) responsiveness to specific types of provocation or aggression-eliciting cues, (b) degree of empathy with the opponent or victim, (c) tendencies to inhibit aggressive performance regardless of motivational state, or (d) cognitive strategies used to deal with anger or impulses to behave aggressively (Frodi, 1978; Frodi *et al.*, 1977). In other animals, similar ambiguities can arise from the occurrence of "competing responses," which are mutually exclusive with the performance of aggressive behavior (Floody & Pfaff, 1977), or from a tendency on the part of one sex to withhold attacks directed toward opposite-sex conspecifics (e.g., Marques & Valenstein, 1977; Payne & Swanson, 1970). Such inhibitions probably are important contributors to an apparently widespread trend for levels of fighting between members of the same sex (intrasexual aggression) to exceed those in heterosexual pairs (intersexual aggression; Carpenter, 1964; Marques & Valenstein, 1977; Payne & Swanson, 1970; Robinson, 1979; Tenaza, 1975; Tiefer, 1970; Waser & Wiley, 1980; but also see Frodi *et al.*, 1977; Mykytowycz & Hesterman, 1975; Swanson, 1974). Nevertheless, inhibitions of intersexual aggression, like the possibility of competing responses, make it difficult to interpret levels of fighting seen in the male–female encounters that superficially seem to provide the most direct test for sexual differences in aggression or dominance.

Dominance?

Both aggression and dominance clearly deserve study regardless of their mutual relations in a particular species. The distinction between the two may be of minimal importance where dominance is established and maintained by threats and fighting ("aggressive dominance"). However, the distinction may be crucial in understanding social relations in other contexts (see chapter by Mazur). For example, human children show sexual differences in aggressive play, other aggressive behaviors, and in aggressive dominance; but it is not thought that these differences reflect differences in generalized (cross-situational) dominance or that they have binding effects on future social relations (Maccoby & Jacklin, 1974). In other animals, too, the variety of dominance orders that can be determined using different measures of aggression suggests that generalized dominance commonly is sacrificed for more specialized social roles (e.g., Fischer & Nadler, 1977; Jolly, 1966; Rowell, 1971; Seyfarth, 1976; also see later sections of this chapter).

Or Something Else?

Even in species in which generalized dominance may exist, simple descriptions of sexual differences in overall levels of dominance or aggression can convey an incomplete and misleading view of social relations. For example, in a study of the processes involved in the formation and integration of groups in rhesus monkeys *(Macaca mulatta),* Bernstein, Gordon, and Rose (1974) found that a group of females could dominate a newly introduced group of males. Furthermore, this outcome seemed attributable not to sexual differences in *levels* of aggression or dominance (in fact see Table I), but to the greater *stability* of the dominance hierarchy of the all-female group. This interpretation has helped to draw recent attention to sexual differences in the stability of dominance relations as a subtle but potentially important characteristic of animal social organization (e.g., Chapais & Schulman, 1980).

Aggression and Dominance Are Situation-Dependent

The results of studies of sexual differences in aggression depend not just on the measures used to compare fighting in males and females but also on the situation used to provoke aggression. Female house mice present an intriguing example because of their very extensive use as the "passive" sex in a clearly dimorphic species (above and Table I). Indeed, female mice conform to this model by fighting infrequently in the type of situation typically used to test aggression in males, that is, during pairwise encounters with adult conspecifics in neutral arenas (e.g., Fredericson, 1952). In contrast, females are as aggressive as males when confronted with immature opponents (Edwards, 1969; Gray, Whitsett, & Ziesenis, 1978) and are considerably more aggressive than males when tested in their home cages with strange males (Hedricks & Daniels, 1981) or when housed in unisexual groups and presented with lactating intruders (Haug & Mandel, 1978). Thus, even individuals from the prototypical dimorphic species can show either type of sexual dimorphism in aggression, or none at all, depending on the test situation (also see summaries of aggression by rats in Tables I–III).

TABLE III

Studies Suggesting That Females of Some Species "Specialize" in Particular Types of Aggression[a]

Species	Source	Type of study	Finding
Rhesus monkey, *Macaca mulatta*	Bernstein, Gordon, & Rose, 1974	Observation of behaviors occurring during the formation of captive groups	Females repeatedly led intergroup attacks.
Gelada baboon, *Theropithecus gelada*	Dunbar & Dunbar, 1975	Field study	Females were involved more frequently than males in group defense in intergroup encounters.
European rabbit, *Oryctolagus cuniculus*	Mykytowycz & Dudzinski, 1972	Staged encounters between resident male–female pairs and juveniles of different parentage	Females were more aggressive than males toward strange juvenile conspecifics.
House mouse, *Mus musculus*	Haug & Brain, 1979	Small groups of intact female or male mice presented with lactating female intruders	Females showed higher incidence and frequency of attacks.
Deer mouse *Peromyscus maniculatus*	Ayer & Whitsett, 1980; Whitsett, Gray, & Bediz, 1979	Adults in heterosexual pairs or asymptotic groups presented with juvenile intruder of either sex	Females attacked whereas males did not; however, this difference only was found in heterosexual groups. Unisexual groups showed no sexual difference in attack.
Laboratory rat, *Rattus norvegicus*	DeBold & Miczek, 1981	Staged intrusions into small colonies of captive animals	Female, but not male, residents vigorously attacked a variety of types of nongroup intruders.
Black rat, *R. rattus*	Ewer, 1971	Observations of partially provisioned, but free-ranging, groups	Likelihood of attack on nongroup intruder greater for female than for male residents.

[a] Instances of "maternal" and "intermale" aggression are not considered here since they are sex-specific by definition.

The situational differences described above for mice, together with the different levels of aggression commonly seen in intra- and intersexual encounters (see previous section), suggest ways in which females might specialize in forms of aggression that differ from those that occupy conspecific males. The data summarized in Table III suggest other species and other ways in which the sexes may specialize in aggression. Specifically, these studies suggest that females commonly are at least as aggressive as males in interactions with juveniles (also see Smith, 1968; Thompson, 1978) and in encounters between an established social group and one or more intruder (cf. Waser & Wiley, 1980).

Together, instances of intrasexual, intergroup, and adult–juvenile fighting suggest ways in which aggression by females can differ from that by males in terms of the classes of conspecifics "accepted" as opponents or targets. In turn, this sort of difference is suggestive of functional specializations in aggression that, when better understood, may provide useful ways of breaking down animal aggressive behaviors into meaningful types or classes for subsequent behavioral and physiological analyses (cf. Archer, 1976; Brain, 1977; Moyer, 1976).

Conclusions

The data summarized above and in Tables I–III suggest several conclusions. First, female mammals are not always subordinate to male conspecifics, even in situations in which individuals of the two sexes compete directly. On the other hand, direct competition between adult males and females may be relatively infrequent. Intrasexual competition probably is more frequent, and here it is not uncommon for levels of fighting between females to equal or exceed those by males. In addition, members of each sex may perform aggressive behaviors that differ more drastically in target selection and that possibly represent the use of aggression in specialized, sex-specific social roles. In turn, role differentiation with respect to aggression may vary with other species differences in spacing and social organization (Collias, 1944; Ralls, 1976; Waser & Wiley, 1980).

POSSIBLE FUNCTIONS OF THE AGGRESSIVE BEHAVIORS NORMALLY PERFORMED BY FEMALES

Data summarized in Table III and in the preceding section of this chapter suggest that aggressive behaviors by females function in the dispersal of young and in the defense of individual territories or group integrity. Nevertheless, the most interesting role attributed to female aggression probably is its use to suppress behavioral or hormonal preparations for reproduction, and ultimately reproduction itself, in competing female conspecifics.

In solitary species, these effects can occur as products of territorial systems of social organization. For example, mature female bank voles *(Clethrionomys glareolus)* during the breeding season distribute themselves uniformly over the available habitat, a distribution presumably maintained by territorial defense (Bujalska, 1973). If the population is depleted of older females, sexual maturity and reproductive activity seem to be accelerated in females born later the same year. This suggests that reproduction by younger females previously had been suppressed by the mature animals, perhaps due to differential access to territories (differential success at territorial defense) or to stress associated more directly with female–female aggressive interactions.

The opportunities for controlling the reproductive success of subordinates probably are considerably more numerous and varied in social canids and primates than in rodents and other small mammals. In wolves *(Canis lupus),* for example, females compete among themselves for status, and subordinate females suffer several disadvantages in reproduction (Zimen, 1976). Specifically, low-ranking females are (a) less likely to copulate and produce young, (b) less successful in raising any young that they do produce, and (c) less secure in their group membership, sometimes being forced to emigrate from the pack altogether as a result of attacks instigated by higher-ranking females.

Like wolves, subordinate females in some primate species are engaged by males in sexual behavior at disproportionately low rates (Keverne, 1979). However, this difference does not occur in other species (Abbott & Hearn, 1978; Dunbar & Dunbar, 1977; Rowell, 1963) and may generally be less important than more direct effects on female reproductive capacity. For example, Rowell (1970) reported that the stress of female–female aggression in baboons (*Papio* spp.) can delay the next sexual swelling, thus effecting at least a temporary decrease in fertility. Similarly, Drickamer (1974) summarizes evidence suggesting that low-ranking female rhesus monkeys are less likely to have young at all and tend to produce young late in the birth season, as a result, suffering unusually high rates of infant mortality. Furthermore, any daughters that survive tend to be low-ranking themselves (Chapais & Schulman, 1980) and to suffer the reproductive disadvantages mentioned above together with a delay in the age at which they first give birth (Drickamer, 1974).

More recent work has extended to other species effects such as those described by Rowell and Drickamer. In addition, this work has begun to explore in greater detail the mechanisms responsible for the changes in reproductive success with rank. For example, Dunbar and Dunbar (1977) found that the fertility of female gelada baboons *(Theropithecus gelada)* was directly related to rank. However, high- and low-ranking females did not differ in frequency of copulation, suggesting that differences in fertility and order of conception stemmed from effects of female–female aggression on the frequency of anovulatory cycles or spontaneous abortions. The first of these possibilities also is consistent with the increased susceptibility to attack experienced by subordinate females when in estrus and about to ovulate. On the other hand, overt aggression may not be essential to maintain such differences in fertility, since similar effects were seen in groups of young marmosets *(Callithrix jacchus)* that fought only rarely after their first few days of contact (Abbott & Hearn, 1978). Subordinate female members of these groups copulated but never became pregnant. Consistently low levels of progesterone suggested that these females did not ovulate while subject to the social stresses associated with low rank. However, these changes are reversible, since all of the females that had been subordinate and anovulatory while in the group were found to ovulate or become pregnant within two months of their removal from the group.

Correlational data such as those summarized above clearly implicate hormonal changes in the decreased reproductive potential shown by subordinate female primates. However, the most detailed description of these changes has been provided by Keverne and his colleagues (Bowman, Dilley, & Keverne, 1978; Keverne, 1979) on the basis of their studies of social groups of talapoin monkeys *(Miopithecus talapoin)*. The females used in these studies were ovariectomized and maintained on estradiol implants, so that it was not possible to determine effects of social rank on fertility or levels of ovarian hormones. In addition, initial differences in rank obviously cannot be explained in terms of different levels of ovarian hormones. Nevertheless, dominant and subordinate females were compared with respect to levels of sexual and aggressive behaviors received from males; rates of female–female aggression; baseline levels of plasma cortisol, luteinizing hormone (LH), and prolactin; and changes in LH levels in response to an exogenous pulse of estradiol (treatment that simulates the normal trigger for both an LH surge and ovulation). The results showed first that, over a relatively long observation period, subordinate females received fewer mounts and ejaculations from males, were attacked and displaced more frequently by males and other females, and had elevated levels of both cortisol and prolactin compared with dominant individuals (Figure 1). Furthermore, when challenged with a sudden increase in estradiol level, dominant females showed an obvious LH surge whereas subordinate females did not (Figure 2). These results suggest strongly that intact females would have differed in capacity to ovulate, and thus in fertility, as a function of rank. In addition, the results of additional experiments suggest that the absence of an estradiol-induced LH surge is due specifically to the elevated prolactin levels shown by subordinate females. In particular, treatment of subordinate animals with bromocriptine, a dopamine agonist that depresses prolactin levels, restores the ability to respond to an estradiol challenge with an LH surge (Figure 2). Conversely, treatment of a dominant female with haloperidol, a dopamine antagonist that increases prolactin levels, prevents any reliable change in LH levels in response to supplemental estradiol treatment (Figure 2).

The results summarized here and in Figures 1 and 2 do not provide a complete description of the physiological consequences of female–female competition in primates. At the behavioral level, they fail to distinguish between the consequences of aggression directed toward subordinate females by higher-ranking females versus males. At the physiological level, they do not specify the relationships among changes in prolactin levels, activity in dopaminergic neurotransmitter systems, and hormonal responses to estradiol challenge. For example, some data implicate dopamine relatively directly in the changes described in Figure 2, suggesting that the concurrent changes in prolactin are not critical. Additional work will be needed to clarify the roles of dopamine and prolactin, and to explore the possibility that responses to dopaminergic agents are mediated by some hormone other than prolactin. Nevertheless, these studies are of great value in helping to specify the critical step in reproduction at which dominant

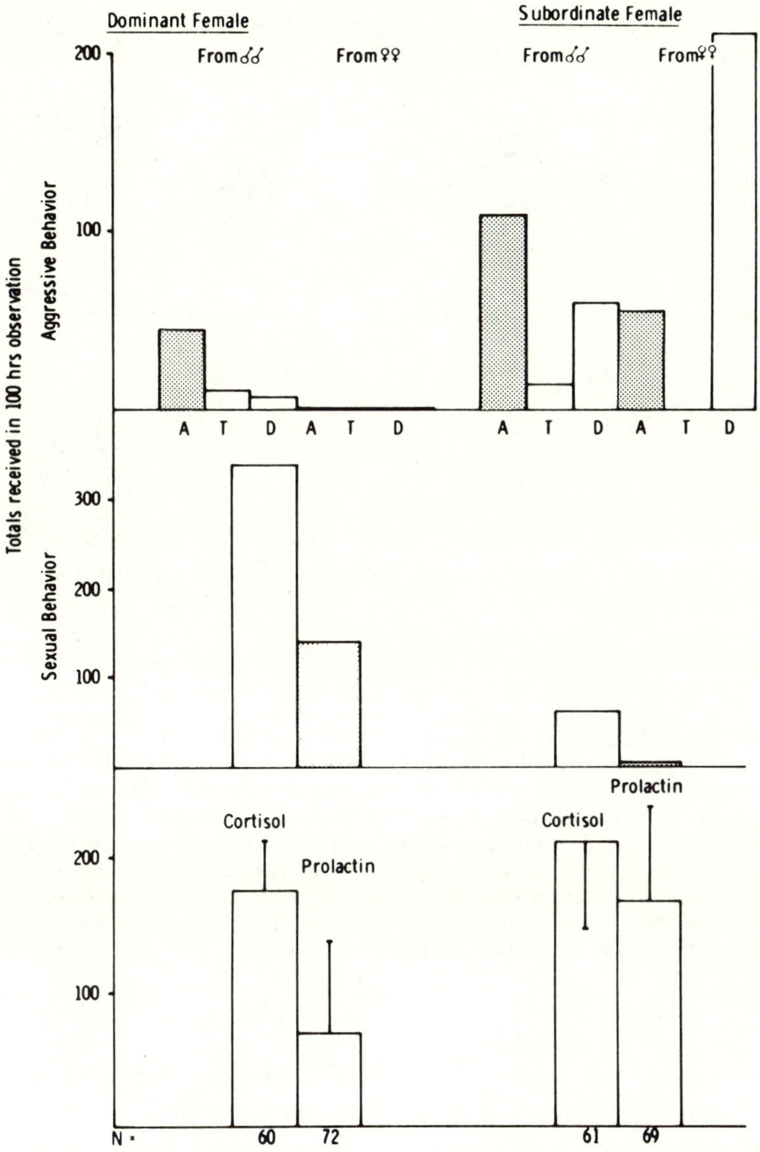

FIGURE 1. The upper two panels show the total numbers of aggressive and sexual acts received by dominant and subordinate female talapoin monkeys during five three-week periods of estrogen treatment. For aggressive behaviors: A = attack (shaded); T = threat; D = displace. For sexual behavior, open columns indicate mounts and shaded columns indicate ejaculations. The lowest panel summarizes plasma concentrations (ng/ml) of cortisol and prolactin. (From "Sexual and Aggressive Behaviour in Social Groups of Talapoin Monkeys" by E. B. Keverne, *CIBA Foundation Symposium,* 1979, *62,* 271–286. Copyright 1979 by Excerpta Medica. Reprinted by permission.)

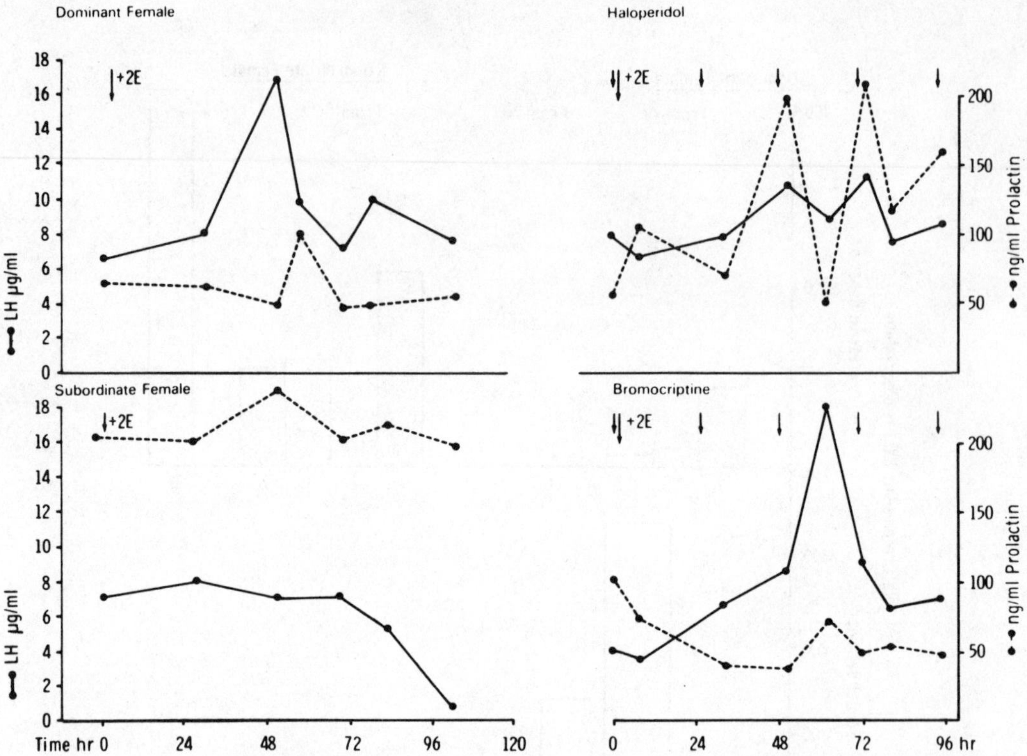

FIGURE 2. Effects of drug treatments that suppress or elevate prolactin levels (bromocriptine and haloperidol, respectively) on the LH surge that normally follows treatment of female talapoin monkeys with estrogen. Dominant females received haloperidol and are described in the upper panels; subordinate females received bromocriptine and are described in the lower panels. Times of drug treatment are indicated on the right-hand panels by arrows; +2E indicates administration of the additional estrogen used to test for the capacity to show an LH surge. Solid and dashed lines indicate plasma concentrations of LH and prolactin, respectively. (From "Suppression of Oestrogen-Induced LH Surges by Social Subordination in Talapoin Monkeys" by L.A. Bowman, S.R. Dilley, and E.B. Keverne, *Nature*, 1978, *275*, 56–58. Copyright 1978 by Macmillan Journals Limited. Reprinted by permission.)

and subordinate females diverge and, thus, in helping to direct future studies of the reproductive consequences and biological functions of female–female aggression.

CHANGES IN FEMALE AGGRESSION CORRELATED WITH ENDOGENOUS CHANGES IN REPRODUCTIVE STATE

Evidence reviewed in the preceding sections shows that female mammals are considerably more aggressive than was previously thought. This means that studies of aggressive behaviors by females are not only possible but necessary to describe completely the behavioral and physiological mechanisms controlling animal aggression. This section will begin to explore the hormonal control of female aggression by describing correlations between levels of fighting or hostility and hormone-dependent cycles in reproductive functions.

Changes in Aggression Correlated with the Rodent Estrous Cycle

Table IV summarizes the changes in aggressive behavior observed in female rodents at different phases of the estrous cycle. Female collared lemmings *(Dicrostonyx groenlandicus)* show no consistent changes in aggression across the cycle. However, other species do show differences, at least in comparisons of estrous and nonestrous females. These changes are especially clear in the golden hamster *(Mesocricetus auratus)*. Here, estrous females are much less aggressive than nonestrous individuals toward conspecifics of either sex. A similar trend has been reported in two of three studies of the laboratory rat. In addition, Hyde and Sawyer (1977) have described elevated aggression in nonestrous house mice, but the difference depends on the nonestrous phase considered. Specifically, estrous and diestrous females show similar levels of aggression, but both are less aggressive than females in proestrus or metestrus.

Regular fluctuations in aggression within the nonestrous phase of the female cycle have been described in the house mouse and woodmouse *(Peromyscus leucopus)*, but not in the rat (Table IV). In each of the first two species, levels of aggression toward other females are lower in diestrus than during other nonestrous phases.

Nonestrous hamsters also have been reported to show significant changes in aggression, but conclusions as to the existence and nature of these changes vary completely across studies. In fact, every relationship possible between the levels of aggression shown by hamsters in pro- and diestrus has been described in different studies or on the basis of different measures (Table IV). These discrepant results could stem from methodological differences or from more subtle interactions between test conditions and hormonal state (Floody & Pfaff, 1977; Grelk, Papson, Cole, & Rowe, 1974). They do not, however, argue strongly for consistent cycles in the aggressiveness of nonestrous hamsters.

Caveats

Contradictory descriptions of relationships between reproductive state and aggression in hamsters emphasize the need for care in designing studies of aggressive behavior and other social interactions. Important tasks for these designs are to distinguish between (a) effects of hormonal state and changes due to experience and (b) direct and indirect effects of hormonal state.

The aggressive behavior of rodents can change rapidly with experience (Floody & Pfaff, 1977). As a result, studies of the hormonal modulation of aggression must be designed to control for experiential changes in both the test subject and its opponent. Animals can be (a) used just once, (b) assigned to independent groups for repeated testing, (c) run in counterbalanced orders

TABLE IV

Changes in the Aggressiveness of Female Rodents over the Estrous Cycle

Subject	Source	Opponent	Result[b,c]	Controls for[a]: Experiential changes in: Subject	Controls for[a]: Experiential changes in: Opponent	Controls for[a]: Effect on stimulus quality	Comment
Collared lemming, *Dicrostonyx groenlandicus*	Hasler & Banks, 1976	Male	Females in E and D showed similar levels of aggression.	1	2	2	Females of this species are induced ovulators.
Golden hamster, *Mesocricetus auratus*	Ciaccio, Lisk, & Reuter, 1979	Male	Fighting varied: D > P > E				Some variations in pattern seen in different experiments.
	Floody & Pfaff, 1977	Female	Aggression varied: D = P > E	3	3		Same trend seen in smaller sample of female–male encounters
	Payne & Swanson, 1970	Male	Aggression varied: D = P > E; however, female won more encounters in D than in P.		2 (At least over nonestrous phase)	2	
	Perkins, Perkins, & Bunnell, 1979	Male	Aggression varied: M > E.				Report available just in abstract form. Some laboratories probably would refer to M as D in the hamster.
	Wise, 1974	Male	Levels of "fight or pounce" varied: P > D > E	1	2		Other measures showed different (or no) trends. Males did seem to

						show different levels of interest in females, even across nonestrous phase.
House mouse, *Mus musculus*	Hyde & Sawyer, 1977	Female	Aggression varied: P = M > E = D	1		Some of subjects were from a strain selected for high levels of aggression.
Woodmouse, *Peromyscus leucopus*	Gleason, Michael, & Christian, 1979	Female	Duration of aggression varied: P > D	1	1	
Laboratory rat, *Rattus norvegicus*	Barr, Gibbons, & Moyer, 1976	Female	No change across cycle	3 (random order)	3	
	Hardy, 1972	Male	Female's rejection of mount attempts by male varied: nonestrous phases > E.			Emphasis on role of female in interactions initiated by opponent.
	Seward, 1945a	Female	Estrous female subordinate to nonestrous female in five encounters.	3 (random order)	3	Five encounters is a very small fraction of those that should have had at least one female in E. Suggests that performance of lordosis, not E itself, associated with "subordinance."

[a] Codes: 1 = Use of animals just in one condition (subject) or one encounter (either animal). 2 = Verify lack of significant change in behavior over repeated tests (either animal) or in tests with subjects from different groups (opponent); the behaviors of greatest interest were instances of social contact or investigation by the opponent and instances of aggression by either animal. 3 = Use counterbalanced order of testing, at least for first series of tests.

[b] Abbreviations: E, estrus (period of sexual receptivity or "heat"); M, metestrus; D, diestrus; P, proestrus.

[c] Less detailed reports suggest that estrous squirrels (*Tamiasciurus* sp.) show depressed levels of female–male aggression and that proestrous guinea pigs *(Cavia porcellus)* are unusually aggressive toward males (Smith, 1968; Young, Dempsey, & Meyers, 1935).

through series of tests under different treatments, or (d) simply tested *post hoc* for changes in behavior due to experience. Unfortunately, a large fraction of the studies summarized in Table IV fail to report the use of any of these precautions (see table). Because of this and the small number of studies available, it is impossible to be certain if the aggressive behaviors of female rodents in general show reliable fluctuations over the estrous cycle. Furthermore, the second caveat listed above suggests that it can be difficult to interpret even those changes in aggression that are most reliable.

As discussed in a previous section, effects of reproductive state on aggressive behavior do not necessarily indicate direct effects of hormones on aggression. Indirect effects are possible, including effects initiated by changes in the stimulus properties of the test animal and mediated by altered responses to these stimuli on the part of the opponent.

To varying degrees, the studies summarized in Table IV have focused on aggressive interactions known to have been initiated by the test females. This emphasis helps to implicate direct effects of hormonal state on aggression, but it does not eliminate the possibility of indirect effects. To deal more effectively with this problem, stimulus animals can be used that (a) have shown stable patterns of response in the past (Hasler & Banks, 1976), (b) receive concurrent treatment designed to minimize or delay changes due to aggression (Ciaccio, Lisk, & Reuter, 1979), or (c) are deficient in the ability to detect at least some classes of stimuli, due to olfactory bulbectomy. However, even studies that take one of these approaches also must verify empirically that any differences in behavior shown by the opponents of animals from different experimental groups represent reactions to and not antecedents of aggressive behavior by the test subjects.

Very few of the studies summarized in Table IV have dealt adequately with the possibility that cyclic changes in aggressive behavior are mediated by changes in stimulus quality (see table). Although unfortunate, this lapse is understandable. Very few clear examples of cyclic changes in rodent aggression have been described, and without such effects, it is difficult to entertain questions of mechanism. Nevertheless, it is clear that additional work is needed to clarify some of the results summarized in Table IV, assess their generality, and determine the extent to which they represent direct effects on aggression.

Conclusions

Despite these caveats, several speculative conclusions are suggested by the data summarized in Table IV. First, females that are induced ovulators and can remain in estrus for long periods (e.g., the collared lemming) might experience unacceptable risks by showing depressed levels of aggressive behavior throughout estrus. In such species, estrous and nonestrous females may be similar in aggressiveness, with transient changes on the parts of estrous females requiring persistent mating attempts by males. Second, the data summarized in Table IV suggest that spontaneous ovulators that are highly aggressive when nonestrus (e.g., the golden hamster) show significant decreases in aggression during estrus.

Less aggressive females (e.g., mice and rats) might experience similar changes, but low baseline rates of aggressive behavior would make these changes difficult to detect, especially in intersexual encounters. Finally, data from the house mouse and woodmouse suggest that moderately aggressive females of at least some species show depressed levels of intrasexual aggression near the middle of the nonestrous phase, with aggressiveness increasing as estrus approaches. Such a pattern might function to permit females approaching estrus to dissuade other females from competing for potential mates or from harassing the mating pair.

Changes in Aggression during the Menstrual Cycle in Nonhuman Primates

Studies of Chimpanzees

Food competition tasks involving male–female and female–female pairs of chimpanzees *(Pan troglodytes)* have been used to study changes in dominance relations with the menstrual cycle. Most pairs showed no such change (Crawford, 1940; Yerkes, 1940, 1941). Instead, polar dominance relations, presumably based on individual differences in aggression, overrode any changes that might have occurred with the female cycle. Nevertheless, the access of some subordinate females to food did fluctuate reliably with their reproductive states. Specifically, these females received more food than usual at midcycle, near the time of ovulation, and during the exhibition of maximal swelling by the external genitalia (Crawford, 1940; Nowlis, 1942; Yerkes, 1940). These changes may have reflected some increase in assertiveness or aggression by the subordinates. However, indirect effects mediated by changes in the dominant individual's tolerance of competition from the subordinate seem to have been more important factors (Crawford, 1940; Yerkes, 1941). In fact, Yerkes (1939) suggests that dominant females are more prone to *yield* access to food during their own periods of genital swelling. In turn, this suggests that the middle of the menstrual cycle is associated with decreased, not increased, aggressiveness.

Studies of Rhesus Monkeys

The frequency with which female rhesus monkeys initiate aggressive interactions tends to increase over the first half of the menstrual cycle, peaking shortly before, or during, estrus: "Increased aggressiveness and wounds are obviously correlated with estrus in females" (Carpenter, 1941, p. 120; also see Mallow, 1979; Rowell, 1963). To some degree, the increased involvement of estrous females in aggressive encounters probably reflects increases in general activity and proximity to other animals, especially aggressive males (Carpenter, 1942; Michael & Zumpe, 1970). However, these indirect effects do not seem to provide a complete explanation for the changes in female aggressive behavior seen at midcycle. For example, many of the fights observed by Mallow (1979) were initiated by estrous females and were not associated with sexual interactions. In addition, the attacks initiated by estrous females are not targeted randomly, as might be expected if they were due just to increases in general activity. Instead, these attacks tend to be directed selectively toward subordinate females and males (Carpenter, 1942; Mallow, 1979). In fact, where no such appropriate target exists, it has been impossible to detect changes in female aggression over the menstrual cycle (Michael & Zumpe, 1970; also see Rowell, 1963, 1967).

The susceptibility of macaque females to aggression also has been examined for fluctuations with the menstrual cycle. However, peak levels of aggression toward and wounding of females have been described both at midcycle (Carpenter, 1942, see quote above) and at or near menstruation (Rowell, 1963; Sassenrath, Rowell, & Hendrickx, 1973). These contradictory results may be attributable, at least in part, to differences in the sex of the attacker in different studies or situations. Most of the attacks observed by Carpenter (1942) on females in or approaching estrus were initiated by males and seemed to result from the increased proximity to males sought by estrous females: "She [the estrous female] actively approaches males and must overcome their usual resistance to close association, hence she becomes an object of attacks by them" (p. 136). In contrast, reports of elevated levels of wounding near menstruation have been based on groups composed mainly or entirely of females (Rowell, 1963; Sassenrath *et al.*, 1973). As should be clear from material discussed in previous sections, the causes and functions of group interactions leading to aggression and wounding could differ greatly in these types of encounters.

Studies of Other Species

Studies of baboons, vervet monkeys, and Sykes' monkeys *(Cercopithecus mitis [albogularis] kolbi)* have not detected reliable changes in female-initiated aggression over the menstrual cycle (Rowell, 1967, 1971; Seyfarth, 1976). However, Saayman (1971) has reported that free-ranging female chacma baboons *(Papio ursinus)* do show consistent cycles, tending to attack other group members more frequently during the perimenstrual phase than at midcycle. Female baboons (*Papio* spp.) also seem to be most susceptible to attack during the perimenstrual part of the cycle (Rowell, 1970; Saayman, 1971). In contrast, subordinate female gelada baboons *(Theropithecus gelada)* are attacked most often by dominant females when in estrus (Dunbar & Dunbar, 1977).

Conclusions

These data do not present a consistent picture of changes in female-initiated aggression with the primate menstrual cycle. They are quite consistent, however, in suggesting that the levels of aggressive behavior and dominance shown by female primates are affected more by social interactions and relationships than by the hormonal changes associated with the menstrual cycle. For example, differences in social rank clearly have very profound effects on the behavior of subordinate animals. Furthermore, these effects persist despite cyclic hormonal changes (Rowell, 1972).

Even where hormonal effects are not negated by the existing social order, social controls on aggression by females can interact in important ways to help shape the results of changes in reproductive state. For example, Bernstein (1970) has described a form of interaction between hormonal and social controls in the increased rank sometimes experienced by estrous females due to the formation of temporary consortships with higher-ranking males. Similarly, Carpenter (1942) has emphasized that the changes in female aggression seen near estrus and menstruation occur in a social context involving more complex shifts in patterns of social affiliation by the cycling females.

These analyses suggest that future attempts to relate aggression by females to endogenous hormonal cycles should consider the possibility that hormonal effects are subtle and can be modulated importantly by the social milieu. In particular, these studies should more consistently (a) focus on spontaneous interactions in natural or seminatural settings; (b) describe the role played by a female in aggressive interactions; (c) distinguish among encounters with opponents differing in age, sex, and status; and (d) describe the aggressive interactions themselves in more detail, allowing for the possibility that changes correlated with the menstrual cycle are expressed in forms of aggression that fall short of complete reversals in dominance.

Although not always well differentiated from changes in aggression itself, the extent to which females are wounded or become the targets of less destructive aggressive episodes also seems to fluctuate with the menstrual cycle. For instance, the increased tolerance of competition for food shown by dominant chimpanzees toward estrous females (Crawford, 1940; Yerkes, 1941) suggests that swollen females should be relatively immune from attack in more natural settings as well. In contrast, female rhesus macaques are attacked most frequently by males as they approach estrus but seem to receive peak levels of aggression from other females near menstruation (Carpenter, 1942; Rowell, 1963; Sassenrath *et al.*, 1973). However, both trends may reflect more general changes in the social relationships of females in these reproductive states (Carpenter, 1942).

Female baboons and gelada baboons show increased susceptibility to aggression near menstruation and at estrus, respectively (Dunbar & Dunbar, 1977; Rowell, 1970; Saayman, 1971). Nevertheless, the aggression received by the females in each of these species seems to decrease fertility temporarily, in effect helping to confine reproductive activity within the group to the most dominant females (Dunbar & Dunbar, 1977; Rowell, 1970). If these effects indicate the biological function of this form of aggression (see previous section), the differences in the timing of female–female antagonism may have been required to parallel species differences in the times of maximal sensitivity to effects of stress on female reproductive potential.

Aggression, Mood, and Menstrual Cycles in Human Females

Many women report experiencing a variety of unpleasant physical and emotional changes shortly before or during menstruation (Dalton, 1969). This "premenstrual syndrome" can include increases in irritability and, perhaps, aggression. Table V summarizes previous estimates of the prevalence of premenstrual

TABLE V
Prevalence of Menstrual or Premenstrual Irritability

Source	Technique	Estimate
Coppen & Kessel, 1963	Questionnaires sent to 463 women selected randomly from records of 10 physicians.	Severe premenstrual irritability in 11%; moderate in 21%.
Greene & Dalton, 1953	Histories of 84 patients treated for premenstrual syndrome.	Premenstrual or menstrual attacks included irritability in 6%.
Moos, 1968	Questionnaires sent to 839 wives of graduate students.	Menstrual or premenstrual irritability severe in 8%–13%; moderate in 39%–41%.
Rees, 1953	Histories of 30 patients treated for severe premenstrual tension.	Premenstrual tension included irritability in 100%.
Sutherland & Stewart, 1965	Questionnaires sent to 150 registered nurses or university students.	Increased irritability regular occurrence in 59%; occasional or regular in 69%.

or menstrual irritability. The extreme variability of these estimates (6%–100%) seems attributable, in large part, to the highly select population described by Rees (1953). With this exception, the studies summarized in Table V suggest that 6%–13% of women experience severe cases of perimenstrual irritability while 20%–60% regularly show perimenstrual irritability of at least moderate intensity.

The studies summarized in Table V suggest that the irritability and aggressiveness of human females fluctuate regularly with the menstrual cycle. Studies that have tested this possibility more thoroughly, by comparing levels of irritability, hostility, or aggression at different phases of the menstrual cycle, are summarized in Table VI. Studies of overt aggression, primarily by institutionalized women (A), and of mood fluctuations in more diverse populations (B) concur in suggesting that women experience increases in aggressiveness near the time of menstruation, decreases in aggressiveness near the time of ovulation, or both of these types of change. The only clear exceptions to this generalization are an early study using just five subjects (Gottschalk, Kaplan, Gesser, & Winget, 1962) and a more recent study of subjects that had been screened rigorously for emotional stability, perhaps selecting a population with unusually mild fluctuations in mood (Persky, 1974).

Mood changes correlated with the menstrual cycle commonly are more prominent in group data or in retrospective summaries of menstrual symptoms than in daily records from the cycles of individual women (Parlee, 1973; Tonks, 1975). These differences suggest that some form of averaging, over individuals

TABLE VI

Changes in Aggression, Irritability, or Hostility during the Human Menstrual Cycle

Source	Measure	Result	Comment
A. Studies of overt behavior			
Dalton, 1961	Crimes by newly convicted prisoners	Incidence elevated during premenstrual and menstrual phases of cycle.	Most crimes were thefts or prostitution.
	Disorderly conduct by prisoners	Peak during menstruation; incidence very low at midcycle.	
	Minor rule infractions by school-girls	Incidence peaks during menstruation; low at midcycle.	
Dalton, 1975	Incidence of child-abuse	Suggestion of increase during premenstrual and menstrual phases.	
Gregory, 1957	Violent behavior by hospitalized psychotics	Incidence higher during last 10 days of cycle (including premenstrual phase) than during first 10 days. Intermediate level of violence seen at midcycle.	
Morton, Additon, Addison, Hunt, & Sullivan, 1953	Violent crimes by newly convicted prisoners	62% committed during the week preceding menstruation.	
Schonberg, Costanzo, & Carpenter, 1976	Intensity of shocks administered by females to males in simulated learning task	Levels of punishment administered early in task by premenstrual group exceeded those by menstrual and intermenstrual groups.	

(Continued)

TABLE VI *(Continued)*

Source	Measure	Result	Comment
B. Studies using self-report techniques or psychological tests			
Coppen & Kessel, 1963	Questionnaire	Women reporting irritability as part of menstrual symptoms suffer peak levels of irritability before menstruation.	Length of premenstrual period not defined.
Gottschalk, Kaplan, Gesser, & Winget, 1962	Analysis of five-minute samples of spontaneous speech by each of five women	Some women showed significant changes in outward-directed hostility; however, these were not consistent across subjects.	Unconventional definition of cycle phases may have obscured changes premenstrually and at ovulation.
Ivey & Bardwick, 1968	Gottschalk speech-analysis technique	Trend toward greater hostility during premenstrual phase than during periovulatory period.	No data or statistical analysis presented.
Moos, 1968	Menstrual Distress Questionnaire	Menstrual and premenstrual levels of irritability similar, and each significantly higher than intermenstrual level.	
Moos, Kopell, Melges, Yalom, Lunde, Clayton, & Hamburg, 1969	Moos Menstrual Distress Questionnaire or Nowlis Mood Adjective Check List	Aggression less at midcycle (days 7–14) than at other cycle phases. Within-subject changes were not correlated significantly across cycles.	Subjects selected for high or low levels of premenstrual tension.

Paige, 1971	Gottschalk speech-analysis technique	Hostility low at midcycle (day 16) and very high at premenstrual phase (day 26) compared to other points in cycle (days 4 and 10).	
Persky, 1974	Hostility and aggression scales of Buss-Durkee Hostility Inventory; hostility and aggression scales of Zuckerman's Multiple Affect Adjective Check List; clinical psychiatric evaluation of hostility/aggression	No differences between aggression or hostility levels at early follicular (days 1–4), ovulatory (14–18), and late luteal (24–28) phases of cycle.	Subjects rigorously screened for emotional stability.
Silbergeld, Brast, & Nobel, 1971	Aggression measured with Mood Adjective Check List (MACL) and in unstructured interview (IRT); Irritability measured with Moos Menstrual Distress Questionnaire (MDQ); Hostility measured with Gottschalk speech-analysis technique (FAT)	IRT aggression, MDQ irritability, and FAT hostility showed significant changes across cycle. For each, low levels of aggressiveness consistently were seen at midcycle; relatively high levels were seen in the premenstrual phase.	Correlations among different measures of aggression, irritability, and hostility generally were low. Also, the very large number of measures and comparisons used here might have yielded some "significant" differences by chance (Parlee, 1973).

or cycles, is required for the detection of reliable cyclic changes in aggression. In turn, this suggests that the mood shifts that do accompany the menstrual cycle are themselves susceptible to modulation by social influences analogous to those operating in other primate species (Frodi, 1978; Frodi *et al.*, 1977; but also see Paige, 1971; Ruble, 1977).

Seasonal Changes in Female Aggression

Seasonal changes in aggression by males have been used as one form of evidence supporting the ability of androgens to stimulate fighting (Floody & Pfaff, 1974). However, very little attention has been given to possible seasonal changes in the aggressive behavior of females. Nevertheless, seasonal fluctuations occur, at least in patterns of territorial defense by female bank voles (Bujalska, 1973) and of group defense by female rhesus monkeys (Bernstein, Rose, & Gordon, 1977). Furthermore, these changes can have interesting and potentially important effects on aspects of social organization. For example, Bernstein *et al.* (1977) introduced male rhesus monkeys into female groups in the breeding or nonbreeding season. In each case, the males were greeted with agonistic responses immediately upon their introduction into the group. However, the character of interactions in the breeding season shifted rapidly, with males initiating sexual interactions and receiving positive social responses within the first hour of contact. Within 10 days, each male had attained alpha status within the group's dominance hierarchy. In contrast, the same males, when introduced into female groups during the nonbreeding season, provoked higher initial levels of aggression and received much lower rates of positive social interaction throughout the encounter. No male achieved alpha rank within a nonbreeding group. In fact, the infrequent interactions between males and nonbreeding females suggested that the males were effectively ostracized from the groups and had no status at all. These results, then, suggest that seasonal changes in the aggressive behavior of female primates could help to determine the feasibility and timing of intergroup movements by males. In turn, changes in female aggression could be important elements helping to shape the genetic makeup of primate groups (also see Bernstein *et al.*, 1974).

Noncyclic Changes in Female Aggression: Effects of Pregnancy and Lactation

Female baboons initiate and receive fewer aggressive behaviors when pregnant or lactating than when cycling (Rowell, 1967, 1972; Saayman, 1971; Seyfarth, 1976). These results suggest that pregnancy and lactation can cause decreases in both the aggressiveness and the aggression-eliciting properties of females. This effect, however, seems to be much less common than instances of unchanged or elevated aggression during these reproductive states. Thus, female mice (Gandelman, 1972; Noirot, Goyens, & Buhot, 1975), voles (*Microtus agrestis;* Clarke, 1956) and hamsters (Wise, 1974; but also see Marques & Valenstein, 1977) have been reported to show elevated levels of aggressive behavior during both pregnancy and lactation. In addition, female deer mice (*Peromyscus maniculatus bairdii;* Ayer & Whitsett, 1980) and rats (DeBold & Miczek, 1981; Erskine, Barfield & Goldman, 1978; Price & Belanger, 1977) have been reported to show increased aggression when lactating but not when pregnant. Possibly in contrast, female rhesus monkeys have been reported to show elevated aggressive behavior during pregnancy (Michael & Zumpe, 1970); however, these animals have not been studied adequately for changes correlated with lactation.

Conclusions

The preceding sections summarize instances in which the levels of aggressive behavior shown by intact female mammals fluctuate regularly with changes in reproductive state. These fluctuations can be subtle and clearly do not provide a complete description of all changes in aggression seen within or across females. To the contrary, it is easy to find examples in which individual differences in aggressive motivation or differences in aggressive behavior attributable to rank or other social factors overshadow changes correlated with reproductive state. Nevertheless, the regularity of these changes, over time and across individuals, could lead to important effects on reproduction and other aspects of social dynamics.

The apparent dependence of female aggression on reproductive state obviously suggests the involvement of endogenous sex hormones. Not surprisingly, many of the studies reviewed above have concluded that ovarian hormones control aggression. Few, however, have attempted to specify the relevant hormones more precisely.

Table VII summarizes the relatively small number of studies in which changes in aggression across estrous cycles, menstrual cycles, pregnancy, or lactation have been related to particular gonadal hormones or patterns of hormone secretion. Experimental studies evaluating these and other hypotheses are summarized in the next section.

EXPERIMENTAL STUDIES OF THE HORMONAL MECHANISMS CONTROLLING AGGRESSION IN FEMALES

Ovariectomy and treatment with exogenous gonadal or pituitary hormones have been used in experimental attempts to confirm the involvement of hormones in the control of female aggression and to specify the hormones responsible for changes in aggressive mood and behavior.

Studies of House Mice

Ovariectomy has little or no effect on the levels of aggression shown by female mice toward adult opponents (Barkley & Goldman, 1978; Tollman &

TABLE VII

Proposed Hormone–Aggression Relationships Based on Changes in Aggression with Estrous Cycles, Menstrual Cycles, Pregnancy, or Lactation

Species	Source	Aggression increased by	Aggression decreased by
Human, *Homo sapiens*	Greene[a] & Dalton, 1953; Rees, 1953	High estrogen levels or high ratio of estrogen to progesterone[a]	Progesterone[a]
Chimpanzee, *Pan troglodytes*	Crawford, 1940	Estrogen	Progesterone
Pigtail macaque, *Macaca nemestrina*	Erwin & Anderson, 1975	Androgens	
Golden hamster, *Mesocricetus auratus*	Vandenbergh, 1971		Combination of estrogen and progesterone
	Wise, 1974	Prolactin	
	Ciaccio, Lisk, & Reuter, 1979		Estrogen
House mouse, *Mus musculus*	Gandelman, 1972	Prolactin	
	Hedricks & Daniels, 1981	Testosterone	
	Noirot, Goyens, & Buhot, 1975	Progesterone	Estrogen
	Svare & Gandelman, 1975	Corticosterone (or some other adrenal corticoid)	
Woodmouse, *Peromyscus leucopus*	Monder, Yasukawa, & Christian, 1980	Prolactin	

[a] For other, less popular theories, see reviews by Moyer (1976), Parlee (1973), Thiessen (1976), and Tonks (1975).

King, 1956). However, these negative results might be due, in part, to the very low baseline levels of aggression shown by intact females in these encounters. In other situations, ovariectomy can affect female aggression. Specifically, intact females inflict more wounds on juvenile intruders when the test females are housed in groups instead of in isolation: Ovariectomized females show no such increase and, as a result, are considerably less aggressive than intact females when housed in groups and tested with juveniles (Gray *et al.*, 1978). These results implicate ovarian hormones in the control of female–juvenile aggression and suggest that gonadal hormones could be of broader significance in the control of aggressive behavior by female mice.

Effects of Androgens

Many reports have described effects of androgens on the aggressive behavior of female mice. Most of these have focused on the effects of neonatal treatments on "male-like" aggression in females and are described in earlier sections of this chapter and in the chapter by vom Saal. Despite this emphasis, these studies have provided a relatively clear picture of the effects of adult androgen treatment on fighting between females and other adults.

Adult females typically do not show changes in aggressive behavior during androgen treatment (e.g., Bronson & Desjardins, 1970; Edwards, 1968; Tollman & King, 1956). However, chronic treatments recently have been shown to stimulate aggression in intact and ovariectomized females (Barkley & Goldman, 1977, 1978; Simon & Gandelman, 1978; Suchowsky, Pegrassi, & Bonsignori, 1971; Svare, Davis, & Gandelman, 1974; vom Saal, Gandelman, & Svare, 1976; vom Saal, Svare, & Gandelman, 1976). These chronic treatments do not seem to produce permanent "organizational" changes in brain systems for aggression, though they can temporarily facilitate aggressive responses to subsequent, more modest androgen therapies (Barkley & Goldman, 1977; vom Saal, Gandelman, & Svare, 1976). In this way, the effects of androgen treatment in adulthood differ importantly from the more permanent changes in hormone sensitivity produced by the administration of androgens to neonatal (Bronson & Desjardins, 1968; Edwards, 1968) or subadult (Edwards, 1970) females.

The effects of androgens on aggression by adult female mice can differ dramatically depending on the age, sex, and reproductive state of the opponent. The studies summarized above show that androgens *can* enhance female aggression toward adult opponents, but these studies also emphasize the need for chronic treatments to overcome an apparently low level of androgen sensitivity. In contrast, androgen treatment readily stimulates attacks by females on smaller, juvenile opponents (Edwards, 1969; Gray *et al.*, 1978). In addition, androgens can *inhibit* attacks by group-housed females on lactating intruders (Brain & Haug, 1979; Haug & Brain, 1979). This effect is consistent with the low levels of aggression shown by intact males toward lactating females. However, the sensitivity of this response and the possibility of its modulation by endogenous androgens in females remain open questions. The involvement of endogenous androgens in the control of female aggression toward juveniles is better supported by results showing that (a) the increased levels of aggression that accompany group housing are blocked by ovariectomy, and (b) increased aggression by ovariectomized, grouped females can be reinstated by testosterone propionate, but not by estradiol benzoate or progesterone (Gray *et al.*, 1978). These results suggest that at least one aspect of population regulation in mice normally involves androgen-dependent aggressive behavior by females.

Effects of Estrogens and Progestins

Estradiol has little or no effect on the aggressive responses shown by adult females toward juvenile or lactating intruders (Brain & Haug, 1979; Gray *et al.*, 1978; Haug & Brain, 1979). Similarly, developmentally normal females fail to show increased aggression toward mature male opponents during chronic estrogen treatment in adulthood, though the same treatment is very effective in females primed neonatally with testosterone or estradiol (Barkley & Goldman, 1978; Edwards & Burge, 1971; Simon & Gandelman, 1978).

These results suggest that changes in female aggression correlated with the estrous cycle (Hyde & Sawyer, 1977) are not attributable to fluctuations in estrogen levels. However, changes in aggression associated with other reproductive states still could reflect effects of estrogens. For example, together with progesterone, estrogen can help to stimulate nipple development and contribute to the high levels of maternal aggression shown by lactating females (see chapter by Svare and Mann). In contrast, estrogen alone acts to depress maternal aggression, possibly accounting for the delay shown by female mice between parturition and initiation of the intense aggression normally associated with lactation (Svare & Gandelman, 1975). These results also emphasize that in aggression, as in aspects of feminine sexual behavior, the effects of estrogen alone can differ dramatically from those of combined treatment with estrogen and progesterone.

As described above, progesterone can help to stimulate maternal aggression indirectly, through its effects on nipple development (Svare & Gandelman, 1976). Unlike estrogen, however, progesterone apparently has little or no capacity to influence maternal aggression after its initial appearance (Svare & Gandelman, 1975). Studies of nonmaternal aggression also have failed to show significant effects of progesterone treatment on the aggressive behaviors of females confronted with adult males (Bronson & Desjardins, 1968) or juveniles (Gray *et al.*, 1978). In fact, the stimulatory effects of androgens on female–juvenile aggression are unaffected by progesterone treatment, in spite of progesterone's ability to antagonize androgen-induced aggression by males and masculinized females (Erpino, 1975; Gray *et al.*, 1978). These results emphasize limits on the ability of progesterone to influence spontaneous aggression by female mice. In turn, they question the suggestion that fluctuations in progesterone levels might cause the regular changes in aggression observed over the estrous cycle (Hyde & Sawyer, 1977).

Effects of Prolactin

Correlational evidence suggests that the maternal aggression shown by lactating mice depends on elevated levels of prolactin (Gandelman, 1972). This interpretation is supported by the ability of the prolactin-inhibitor ergocornine to decrease aggression in lactating females (Politch & Herrenkohl, 1980). However, studies of other strains of mice have questioned the need for prolactin by showing that (a) virgin females primed with estrogen and progesterone show elevated aggression, but not lactation, if exposed to suckling stimulation (Svare & Gandelman, 1976); (b) progesterone treatment of parturient females inhibits lactation, but not maternal aggression (Svare & Gandelman, 1975); and, (c) suckling stimulation is not required for the maintenance of elevated levels of aggression by parturient females (Svare & Gandelman, 1973). These data suggest that an increase in prolactin level is not a necessary condition for the initiation or maintenance of maternal aggression. In addition, results showing that prolactin itself can increase aggression only in conjunction with exposure to infants suggest that elevated prolactin levels are not a sufficient stimulus for the display of maternal aggression (Politch & Herrenkohl, 1980). Although prolactin still may be important in the control of maternal aggression in mice, its role clearly is a complex one, involving interactions with tactile stimulation and gonadal hormones and begging for clarification by additional research in a variety of strains (see chapter by Svare & Mann).

Studies of Hamsters

The golden hamster has stood out for some time as a species in which females are unusually aggressive (e.g., Kislak & Beach, 1955). As a result, hamsters have been used extensively in studies of the hormonal mechanisms controlling aggression in normal females. These studies, however, have uncovered conflicting results at nearly every turn (Table VIII). For example, most recent studies suggest that ovariectomy does not produce reliable changes in hamster aggression. Several studies disagree, but few of these provide persuasive evidence for a difference between the behaviors of intact and ovariectomized females. Kislak and Beach (1955) describe a reliable effect on just one of ten measures whereas the reliability of the difference described by Tiefer (1970) was not assessed at all. More consistent effects are described by Payne and Swanson (1971, 1972). These authors argue that ovariectomy reverses the normal sex difference in dominance in male–female encounters, and also decreases both the aggressiveness and dominance of females interacting with intact female controls. To the extent that they exist, these differences may reflect indirect effects mediated by changes in the stimulus quality of the test female (Floody & Pfaff, 1977). However, more basic questions are raised by recent studies in which Payne (1974a,b) describes similar levels of aggression and dominance in intact and ovariectomized females facing male opponents. On balance, the evidence cited in Table VIII

TABLE VIII

Hormonal Effects on Aggression in Female Hamsters

Manipulation	Effect on female aggression or dominance		
	No change	Increase	Decrease
Ovariectomy or combined ovariectomy and adrenalectomy	Carter, 1972 (tested increase in aggression that normally follows mating) Floody & Pfaff, 1977 Payne, 1974a,b Vandenbergh, 1971		Kislak & Breach, 1955 Payne & Swanson, 1971, 1972 Tiefer, 1970 (but no statistics presented for this comparison)
Treatment of ovariectomized females:			
With androgen (testosterone or testosterone propionate)	Floody & Pfaff, 1977 Grelk, Papson, Cole, & Rowe, 1974 (individually housed females) Nobel & Alsum, 1975 (fight measure) Payne, 1974a,b Payne & Swanson, 1971, 1972 Tiefer, 1970 Vandenbergh, 1971	Grelk *et al.*, 1974 (group-housed females)	Nobel & Alsum, 1975 (threat & scarring measures)

With estrogen (estradiol or estradiol benzoate)	Floody & Pfaff, 1977 Grelk *et al.*, 1974 Nobel & Alsum, 1975 (threat and scarring measures) Payne & Swanson, 1971, 1972 Tiefer, 1970 Vandenbergh, 1971	Kislak & Beach, 1955 Nobel & Alsum, 1975 (fight measure)	Ciaccio, Lisk, & Reuter, 1979 Perkins, Perkins, & Bunnell, 1979
With progesterone	Floody & Pfaff, 1977 Grelk *et al.*, 1974 Kislak & Beach, 1955	Payne & Swanson, 1971, 1972	
With estrogen plus progesterone		Ciaccio *et al.*, 1979 (delayed effect)	Ciaccio *et al.*, 1979 (immediate effect) Floody & Pfaff, 1977 Kislak & Beach, 1955 Tiefer, 1970
Hypophysectomy, with or without LH, FSH, or ACTH treatment	Floody & Pfaff, 1977		
Prolactin treatment, with or without hypophysectomy	Floody & Pfaff, 1977	Buntin, Catanzaro, & Lisk, 1981 (pituitary transplant) Wise & Pryor, 1977 (depression of prolactin by ergocornine, with or without prolactin replacement)	

suggests that ovariectomy has, at best, minimal direct effects on hamster aggression. In turn, these studies suggest strongly that ovarian hormones are not necessary for the stimulation of the high levels of aggression typical of nonreceptive female hamsters.

Effects of Individual Sex Steroids

Treatment with androgens does not seem to cause reliable changes in the aggressive behaviors of ovariectomized hamsters (Table VIII). Disagreement on this point is relatively mild, though some of the differences across studies raise the possibility that conditions of housing that themselves depress aggression also set the stage for androgen-induced increases in fighting (Grelk *et al.,* 1974). Nevertheless, the limited ability of androgens to affect the aggressive behavior of female hamsters reliably is consistent with both the minimal changes that follow ovariectomy and the relatively high levels of aggression normally shown by female, as opposed to male, hamsters (Table II).

Like testosterone, estrogen alone most often has been seen to have no significant effect on aggression in ovariectomized or ovariectomized-adrenalectomized hamsters (Table VIII). Noble and Alsum (1975) described the facilitation of aggression by estrogen, but just on the basis of one of several measures. Kislak and Beach (1955) found more consistent changes in the same direction, but used an unusually high dose of hormone (0.225 mg/day of estradiol benzoate). In contrast, Ciaccio *et al.* (1979) have provided the only detailed description of inhibitory effects of estrogen on aggression in female hamsters. This study, however, provided convincing evidence for experiential changes in female aggressive behavior, yet included no control group to help distinguish between experience and changes due to estradiol treatment. This omission, together with the lack of a consistent relationship between estradiol dose and aggression, seriously weakens the case for any significant effect of estrogen alone on aggression in female hamsters.

Effects of progesterone on hamster aggression also have been the subject of some controversy (Table VIII). Here, however, the data are most consistent with the view that progesterone can stimulate aggression or dominance in at least some situations. Specifically, we have suggested that consistent effects of progesterone may occur in encounters with moderately aggressive opponents (Floody & Pfaff, 1977). Furthermore, we believe that at least some of these effects are indirect, requiring mediation by changes in aggressive behavior by the opponent in response to hormone-induced changes in a test female's stimulus quality. Finally, we have speculated that these effects of progesterone might function normally to protect females from aggression by males. While consistent with the data now available, these speculations clearly require additional evaluation. One strategy that would be especially useful in future research is that of using anosmic opponents to help distinguish between direct and indirect effects of progesterone (and other hormones) on hamster aggression.

Effects of Combined Treatment with Estradiol and Progesterone

Of the effects summarized in Table VIII, the ability of combined treatment with estrogen and progesterone to decrease aggression in female hamsters is the only phenomenon that has been tested repeatedly and remains undisputed. In all likelihood, the magnitude and clarity of this effect reflect an equally compelling biological function, that is, the precise synchronization of reduced levels of aggression with other physiological and behavioral preparations for successful reproduction.

Effects of Pituitary Hormones

Pituitary hormones apparently are not necessary for the high levels of aggression shown by nonestrous female hamsters. Neither hypophysectomy nor subsequent treatment with LH, FSH, or ACTH led to any significant change from levels of aggression comparable to those shown by intact controls (Floody & Pfaff, 1977). The same study also found no consistent effects of prolactin on the behavior of hypophysectomized females. However, more recent studies have stimulated impressive changes in aggressive behavior with manipulations that probably alter prolactin levels more selectively, preserving other aspects of the hormonal milieu that might be required for the full expression of prolactin's effects. For example, Wise and Pryor (1977) showed that treatment of lactating females with the prolactin inhibitor ergocornine depresses the elevated levels of aggression normally associated with lactation (Wise, 1974). Furthermore, this effect was antagonized by treatment with prolactin itself. More recently, Buntin, Catanzaro, and Lisk (1981) have used pituitary transplants to chronically elevate prolactin levels in ovariectomized females. Consistent with the results of Wise and Pryor, these transplants caused significant increases in female aggression in both intra- and intersexual encounters. Together, these studies suggest strongly that prolactin is responsible for the elevated aggression normally shown by lactating female hamsters.

Studies of Other Rodents

Table IX summarizes studies of the hormonal mechanisms controlling female aggression in rodents other than house mice and hamsters. Although we will focus below on cases in which hormones do affect aggression, perhaps the most impressive feature of Table IX is the large proportion of species and studies that have shown no significant effects of hormonal manipulations on female aggression. This raises the possibility that aggression in females often functions in a manner unrelated to (or simply not restricted to) reproduction, and, therefore, unregulated by gonadal hormones (Crook & Butterfield, 1968). Limitations on the estrogen- and progesterone-sensitivity of aggression in female mice and hamsters (see preceding sections) also may relate to functional constraints of this sort.

TABLE IX

Effects of Sex Hormones on Aggression in Females of Other Rodent Species

Species	Source	Treatment	Effect
Guinea pig, *Cavia porcellus*	Goldfoot, 1979	Ovariectomy plus 0.5 mg/day testosterone propionate (TP); 0.5 mg/day dihydrotestosterone propionate (DHTP); 1.0 μg/day estradiol benzoate (EB); or 0.5 mg/day DHTP plus 1.0 μg/day EB	TP ⩾ DHTP + EB > DHTP > oil = EB.
Collared lemming, *Dicrostonyx groenlandicus*	Hasler & Banks, 1976	Ovariectomy with or without treatment with 0.01–10.0 μg/day EB or daily 0.1 μg EB + single injection of 250 μg of progesterone (P)	No effect on attacks.
Brown lemming, *Lemmus trimucronatus*	Huck, Carter, & Banks, 1979	Ovariectomy, with or without 0.5 μg/day EB, 0.5 mg/day P, or EB plus single injection of P	Increase in threat-leaps after prolonged P treatment. Increase more immediate when EB-primed female treated with P.
Mongolian gerbil, *Meriones unguiculatus*	Anisko, Christenson, & Buehler, 1973	Ovariectomy, with or without 40 μg/day TP	TP depressed aggression.
	Lumia, Canino, & Drozdowski, 1976	Ovariectomy, with or without 100 μg/day TP	TP increased aggression.
Woodmouse, *Peromyscus leucopus*	Gleason, Michael, & Christian, 1979	Ovariectomy, with or without single injection of 0.1–20.0 μg EB, 100–500 μg P, or 10–200 μg TP	Aggression decreased after ovariectomy. EB had no effect; but high doses of P, and, to lesser extent, TP caused increased aggression.
	Monder, Yasukawa, & Christian, 1980	Injections of prolactin or drugs (morphine and naloxone) that affect prolactin levels	Increased prolactin levels associated with increased aggressive behavior.
Deer mouse, *P. maniculatus*	Whitsett, Gray, & Bediz, 1979	Ovariectomy	No effect.
Laboratory rat, *Rattus norvegicus*	Barr, Gibbons & Moyer, 1976	Ovariectomy, with or without 0.5 mg/day TP	No effect.
	DeBold & Miczek, 1981	Ovariectomy	Tendency for increased aggression.
	Erskine, Barfield, & Goldman, 1980	Hypophysectomy	No effect.

Several of the studies summarized in Table IX have examined effects of androgens on aggression in ovariectomized females. For example, Goldfoot (1979) has found that both testosterone propionate (TP) and the nonaromatizable androgen dihydrotestosterone propionate (DHTP) can stimulate aggression and dominance in all-female groups of guinea pigs. In addition, the effectiveness of DHTP was enhanced by concurrent estradiol benzoate (EB) treatment, though EB alone had no significant effect. These results suggest that androgens can exert direct control over aggressive behavior in female guinea pigs. However, is this an instance of "male-like" aggression, or do androgens in fact control the agonistic behaviors of normal females? To help place these results in perspective, additional research is needed to examine guinea pigs for (a) possible sex differences in aggression, (b) changes in female aggression across different reproductive states, and (c) effects of hormone depletion (ovariectomy or combined ovariectomy and adrenalectomy) on female behavior.

Studies by Anisko, Christenson, and Buehler (1973) and by Lumia, Canino, and Drozdowski (1976) have described opposite effects of TP on the aggressive behavior of female Mongolian gerbils. It is not clear how to resolve this conflict. However, Yahr and Coquelin (1980) have emphasized that the effects of castration on aggression in male gerbils depend on the time of castration relative to puberty. Whereas males castrated postpuberally showed no significant change in aggression, those castrated prepuberally actually showed increased levels of fighting. Similar contingencies could operate in females and might help explain the discrepant results described by Lumia *et al.* (postpuberal ovariectomy) and Anisko *et al.* (ovariectomy before or at about the time of puberty). Again, additional studies clearly are needed to explore this and other possible resolutions of the conflict.

Gleason, Michael, and Christian (1979) recently have reported that single injections of TP or progesterone (P) can produce significant increases in intrasexual aggression among female woodmice. The effect of P was considerably more dramatic than that of TP, suggesting that P might be instrumental in the different levels of aggression shown by intact proestrous and diestrous females (Gleason *et al.,* 1979; also see Table IV). However, Gleason *et al.* also raise the possibility that neither P nor TP affect aggression directly. Instead, they suggest that the effects of each of these steroids is mediated by changes in prolactin levels, changes that function primarily to stimulate the elevated levels of aggression presumed to accompany pregnancy and lactation in wood mice. This hypothesis is consistent with recent data suggesting that increasing prolactin levels by prolactin injections or by treatment with 2 mg/kg of morphine can increase female–female aggression in *P. leucopus* (Monder, Yasukawa, & Christian, 1980).

Dramatic effects of P on female aggression also are apparent in the brown lemming (*Lemmus trimucronatus;* Huck, Carter, & Banks, 1979). Here, treatment of estrogen-primed females with P causes a rapid and dramatic increase in the frequency of "threat-leaps" at the male partner. P alone also can increase the frequency of aggressive behavior, but only after prolonged treatment. The evidence suggests strongly that the first of these responses to P is part of a mechanism that normally functions to terminate estrus in response to stimulation received during mating.

Studies of Primates

Female rhesus monkeys tend to be involved in more male–female aggressive interactions after ovariectomy than when intact (Michael & Zumpe, 1970). However, this change is small and probably not due to direct effects of ovariectomy on aggression by females. Instead, 52% of the attacks by females represented responses to male aggression whereas 30% were associated with the refusal of male mount attempts. Together with consistent increases in the rate at which male mount attempts were resisted by less violent means, these results suggest that the changes in aggressive behavior shown by ovariectomized females were secondary to changes in male behavior, and that both, ultimately, were linked to decreases in female sexual receptivity.

The study by Michael and Zumpe suggests that females are especially prone to attack when levels of ovarian steroids have been reduced by ovariectomy. This sort of change is consistent with reports of increased wounding suffered by rhesus females near the time of menstruation (Carpenter, 1942; Rowell, 1963; Sassenrath *et al.*, 1973). Nevertheless, important questions concerning effects of ovariectomy on aggression in female primates remain unanswered. First, is it true, as claimed by Michael and Zumpe, that ovariectomy provokes a greater change in aggression in the male partner than in the test female? In fact, two of the three pairs studied by Michael and Zumpe showed similar relative levels of aggression before and after ovariectomy: Much of the apparent sex difference in the extent of change in aggression following ovariectomy may have been due to overemphasis in the statistical analysis of data from the pair with the greatest relative change in male aggression, together with underemphasis of the pair with the greatest relative change in female aggression. Second, even presuming their reliability, would the changes in male–female behavior described by Michael and Zumpe be seen in more natural situations in which nonreceptive females could better control the frequency of their interactions with males? Third, questions of spacing aside, are the effects of ovariectomy on interactions with a presumably dominant male representative of possible effects on interactions with more appropriate targets of female aggression (subordinate males and females)? These questions indicate clearly the need for additional work to clarify the effects of ovariectomy on aggression in female primates. In this regard, it is unfortunate that studies examining effects of exogenous hormones on the behavior of ovariectomized animals have not devoted more attention to the effects of ovariectomy *per se*.

Effects of Androgens

Previous sections of this chapter have emphasized the possibility of interactions between hormonal and social mechanisms controlling aggression and dominance. Such interactions can obscure hormonal effects on aggression, especially in animals with rigid social hierarchies. Accordingly, it seems unreasonable to expect treatment with any hormone to completely reverse well-established dominance relations in primates. Instead, effects of hormones should be more prominent in tests that use more subtle measures to assess aggressiveness, and in situations in which the dominance relations themselves are more fluid. As a result, significant hormonal effects may be most likely in tests using (a) animals that are too young to have established rigid dominance orders, (b) adults selected specifically for similar initial levels of dominance or aggression, or (c) animals that are housed separately except during tests, limiting the opportunity for the formation of inflexible relations.

These considerations seem relevant in interpreting previous descriptions of the effects of exogenous androgens on aggression in female primates. For example, Green, Whalen, Rutley, and Battie (1972) observed feeding orders in two mixed-sex groups of squirrel monkeys *(Saimiri sciureus)* before, during, and after treating the highest ranking female in each group with 10 mg/week of testosterone enanthate. A significant effect would have required not just that the treated females engage in intersexual aggression, but that they do so at levels sufficient to reverse apparently stable and quite polar dominance relations. Not surprisingly, no significant effect of androgen treatment was observed.

Mirsky (1955) also failed to observe consistent effects of androgen treatment on success at food competition within a group of five young female rhesus monkeys. The explanation of this outcome is unclear, though it does appear that these animals were housed together most of the time and had developed stable dominance relations that may have canceled any hormonal effect.

Significant effects of androgens on aggression in female rhesus monkeys have been found in other studies using more sensitive measures of aggression and groups that were young or in contact just during experimental tests (Cochran & Perachio, 1977; Joslyn, 1973; Trimble & Herbert, 1968). In addition, Birch and Clark (1946) found that androgen treatment of the second-ranking of three ovariectomized chimpanzees led to a significant increase in success at a food-competition task. This treatment, however, did not produce an actual reversal of dominance. Furthermore, similar treatment of the least dominant female had no detectable effect, perhaps because of greater initial differences in aggressiveness.

The studies summarized above show that androgens can increase the aggressiveness of female primates. The relevance of this to the aggressive behaviors normally performed by intact female primates, including female humans, is unclear. However, Persky (1974) has described reliable multiple correlations between measures of aggression or hostility and early follicular and late luteal testosterone levels in a rigorously screened group of female college students. This suggests that individual differences in aggression, at least during some phases of the menstrual cycle, might be due to individual differences in androgen levels. On the other hand, possible differences in aggression across phases of the menstrual cycle do not seem attributable to changing testosterone levels, since Persky found significant changes in hormone levels but was unable to detect cyclic fluctuations in mood. In addition, Rees (1953) argued that methyltestosterone was an effective treatment for severe premenstrual tension, obviously suggesting that androgens did not cause the increased irritability that frequently accompanied the syndrome.

Effects of Estrogens

The classic studies by Birch and Clark (1946, 1950) remain the most detailed attempts to tie estrogens to particular roles in the control of aggression in female primates. The basic findings were similar to those described above in connection with androgen treatment. Specifically, estrogen treatment of the second- but not third-ranked of three female chimpanzees led to a significant increase in success at a food competition task. Ordinarily, this might be interpreted as a direct stimulatory effect on female aggression. However, based in part on their impression of estrogen's effects on aggression in males, Birch and Clark instead argued that estrogen acts directly to inhibit aggression, but that the irritation arising from estrogen-induced tumescence of the female sex skin overcomes any inhibition and provokes a net increase in aggression. This interpretation was supported by results showing that delaying the peripheral response to estrogen with concurrent progesterone treatment prevented any increase in dominance by an initially submissive female. In fact, similar treatment tended to decrease the competitive success of her initially dominant partner.

The studies of Birch and Clark are admirable both for their recognition of the variety of mechanisms that can underlie hormonal effects on aggression and for their ingenuity in devising experimental means of dissecting centrally and peripherally mediated effects. On the other hand, the results and interpretations they describe are not entirely convincing. First, an explanation in terms of estrogen-induced irritability seems to demand that estrogen-treated animals behave in an irritable manner, that is, that they show some increase in overt aggression. However, the changes in dominance described by Birch and Clark (1950) apparently did not involve any actual combat. In addition, virtually all of the evidence used by Birch and Clark to support their hypothesis requires the assumptions that (a) effects of estrogen alone are similar to those of estrogen administered together with progesterone and (b) effects of combined estrogen and progesterone treatment do not change dramatically with time. Unfortunately, these assumptions now seem untenable, since much work on aggressive and sexual behaviors has shown that effects of estrogen can differ completely from those of estrogen plus progesterone (see previous sections of this chapter) and that effects of the latter can change drastically with time (e.g., Ciaccio *et al.*, 1979). As a result, the interpretations suggested by Birch and Clark should be reconsidered in future work, ideally in studies using direct central or peripheral hormone treatments to separate better the possible mechanisms of estrogen action.

Questions of mechanism aside, relatively few other studies have even claimed significant effects of estrogens on the aggressive behaviors of female primates. For example, Dixson and Herbert (1977) reported that ovariectomized talapoin monkeys typically dominate conspecific males and that this relation is not affected by treatment with 17.5 μg/week of estradiol monobenzoate. Dixson (1978) later described significant decreases in aggression by female galagos *(Galago crassicaudatus crassicaudatus)* treated with 40 μg/day of EB; however, the changes in rates of threat and attack probably were by-products of the increased sexual receptivity that also accompanied estrogen treatment. On the other hand, Michael and Zumpe (Michael, 1969; Michael & Zumpe, 1970; Zumpe & Michael, 1970) reported that estrogen treatment stimulates redirected threats and possibly some forms of more direct female–male aggression by ovariectomized rhesus monkeys. At least the second of these, a reported increase in the frequency of threats that were not associated with male mount attempts, was thought to represent a direct hormonal effect on neural mechanisms for aggression. However, both effects are questionable, though for quite different reasons. The change in rates of redirected aggression (threats directed *away* from the male partner) may be reliable, but its function and relevance to the intact animal is questioned by the failure of redirected aggression to appear at all in intact females (Zumpe & Michael, 1970). The effects on direct intersexual aggression by females are subtle and were apparent only in rather indirect analyses. Furthermore, the reported ability of estrogen to affect these behaviors is inconsistent with the results of previous studies in which no estrogenic effect was observed (Mirsky, 1955; Trimble & Herbert, 1968). Again, additional clarification is required, especially from studies pairing females with targets that are more appropriate than dominant males.

Finally, it is quite unclear how the data summarized above relate to possible menstrual cycles in the aggressive behaviors of nonhuman primates or of humans. In the latter, of course, there are no neatly controlled experimental studies to help isolate the effects of individual hormones. Among the women studied by Paige (1971), those using oral contraceptives containing just estrogen generally experienced lower levels of hostility than those cycling normally (Figure 3). It is tempting, in fact, to interpret the decline in hostility seen over the first 16 days of the cycle as a decline toward a baseline determined by endogenous estrogen levels. This speculation, however, cannot account for the failure of women on "estrogen-alone" contraceptives to show increased levels of hostility either late in the cycle, when estrogen treatment was supplemented by progesterone, or very early in the cycle, when levels of both hormones should have been low. Although factors such as estrogen dose or the timing of changes in estrogen and progesterone levels might account for these discrepancies, such an interpretation is far from compelling at present.

Effects of Progestins and of the Combination of Estrogen and Progesterone

Treatment of ovariectomized females with progesterone has no reliable effect on aggression in rhesus monkeys (Michael & Zumpe, 1970) or chimpanzees (Birch & Clark, 1950). Using a very different experimental design and human females, Persky (1974) also questioned the dependence of aggression on progestins by showing significant changes over the menstrual cycle in progesterone levels but not in levels of aggression or hostility.

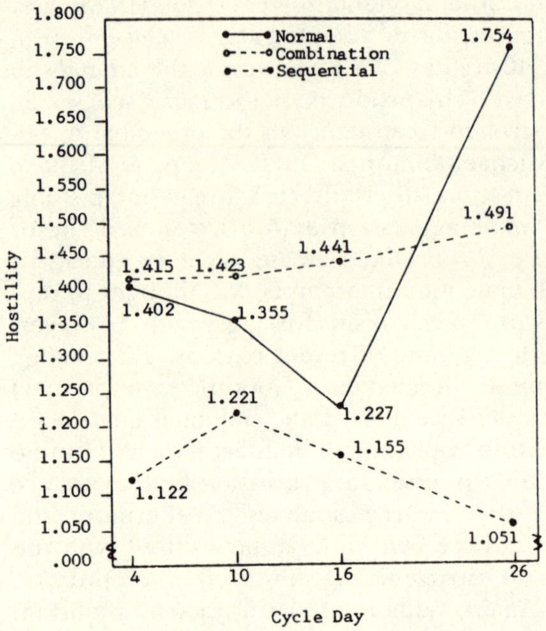

FIGURE 3. Effects of menstrual cycle phase on hostility scores derived from the analysis of speech samples using the Gottschalk technique. The Normal group includes women who were not using oral contraceptives. Women in the Combination and Sequential groups used oral contraceptives, but of different types. For members of the Combination group, each tablet included both an estrogen and progestin. For women in the Sequential group, the first 15–16 tablets in each cycle included just an estrogen; the next 5 tablets in the sequence included both an estrogen and progestin. (From "Effects of Oral Contraceptives on Affective Fluctuations Associated with the Menstrual Cycle" by K. E. Paige, *Psychosomatic Medicine*, 1971, *33*, 515–537. Copyright 1971 by Elsevier North Holland. Reprinted by permission.)

These results stand in rather stark contrast to the very strong claims that progesterone treatment effectively reduces or eliminates the irritability and increased aggressiveness that accompany the "premenstrual syndrome" in women (Dalton, 1969, 1975; Greene & Dalton, 1953; Rees, 1953). This claim, however, is clearly a topic of controversy. One source of this controversy is apparent in some of the original papers supporting the utility of progesterone treatment. For example, Greene and Dalton (1953) argued that progesterone reduced premenstrual symptoms in virtually all patients. The implication clearly was that for irritability, as for other premenstrual complaints, progesterone was the treatment of choice. However, among the histories describing major symptoms of the 78 women, only 2 specified irritability as having been experienced, treated, and alleviated by progesterone. Clearly, a more extensive presentation of data bearing specifically on the responsiveness of premenstrual irritability or aggressiveness to progestin treatment would be most helpful. In the apparent absence of such an account, Tonks (1975) recently concluded that the utility of progestin treatment is doubtful.

Combinations of estrogen and progesterone have not been tested extensively in primates. Michael and Zumpe (Michael, 1969; Michael & Zumpe, 1970; Zumpe & Michael, 1970) found that estradiol plus progesterone did not affect redirected aggression by female rhesus monkeys. The same combination, however, did tend to stimulate direct aggression by females. This effect was viewed as consistent with the increasing levels of aggression seen in pregnant females. At the same time, however, part of the effect was attributed to hormonal effects on sexual responsiveness. Specifically, Michael and Zumpe suggested that most of the increase in aggression represented attacks by females in response to male mount attempts, suggesting that the changes in aggression were largely secondary to changes in sexual receptivity.

Finally, oral contraceptives that contain fixed quantities of an estrogen and progestin sometimes have been associated with levels of hostility that are intermediate between the extremes seen at different phases of the human menstrual cycle (Paige, 1971). In such cases, it is tempting to see the increasing levels of hostility shown by normal subjects as approaching, then overshooting, baseline levels determined by a combination of estrogen and progesterone (Figure 3). However, other studies have reported relatively low or unchanged levels of aggression on the parts of women taking the same contraceptive preparation (Silbergeld, Brast, & Noble, 1971). Obviously, the precise relationship of estrogen and progesterone levels to average levels of human aggression, and to mood changes associated with the menstrual cycle, are open questions.

Conclusions

It is premature to speculate as to general rules governing the hormonal dependence of aggression in females. The conclusions suggested by studies summarized in the preceding sections are limited, though it is difficult to know the extent to which these limits reflect just the relative scarcity of studies versus the subtlety of hormonal effects that some of the studies suggest. Nevertheless, it is instructive that even these few studies have described some of the most dramatic hormonal effects on aggression to be found in behavioral endocrinology. On the one hand, females of species that repeatedly have been characterized as passive and unaggressive can be stimulated to show intense aggression by treatment with hormones of pregnancy and lactation together with exposure to tactile or olfactory stimulation from infants (e.g., Erskine *et al.*, 1978; Politch & Herrenkohl, 1980). Conversely, females that ordinarily are intensely aggressive sometimes can be transformed into remarkably passive mating partners by combined treatment with estrogen and progesterone (e.g., Floody & Pfaff, 1977).

Such examples confirm both that female mammals do behave aggressively and that their aggressive behaviors can depend on reproductive hormones. Furthermore, the hormonal mechanisms suggested by these examples differ radically from the relations emphasized in previous descriptions based on studies of males or androgenized females. Accordingly, the studies summarized here and in the preceding sections do promise to extend significantly the range of relationships between hormones and aggression in a way that could not have been accomplished just by additional studies of hormones and the aggressive behavior of males. These studies provide ample justification for the relatively modest level of attention currently devoted to aggression in females. More importantly, they indicate clearly the need for much additional work in this emerging area.

Acknowledgments

Special thanks go to Drs. Alan I. Leshner, Martha K. McClintock, and J. Mal Whitsett for helpful comments on a draft of the chapter, and to K. R. Ocker and R. A. Robenalt for typing the manuscript.

REFERENCES

Abbott, D. H., & Hearn, J. P. Physical, hormonal, and behavioural aspects of sexual development in the marmoset monkey, *Callithrix jacchus*. *Journal of Reproduction and Fertility*, 1978, *53*, 155–166.

Abbott, D. H., & Hearn, J. P. The effects of neonatal exposure to testosterone on the development of behaviour in female marmoset monkeys. *CIBA Foundation Symposium*, 1979, *62*, 299–316.

Anisko, J. J., Christenson, T., & Buehler, M. G. Effects of androgen on fighting behavior in male and female Mongolian gerbils *(Meriones unguiculatus)*. *Hormones and Behavior*, 1973, *4*, 199–208.

Archer, J. The organization of aggression and fear in vertebrates. In P. P. G. Bateson & P. H. Klopfer (Eds.), *Perspectives in ethology* (Vol. 2). New York: Plenum Press, 1976.

Ayer, M. L., & Whitsett, J. M. Aggressive behaviour of female prairie deer mice in laboratory populations. *Animal Behaviour*, 1980, *28*, 763–771.

Banks, E. M., Mankovich, N. J., & Huck, U. W. Female interspecific aggression in two species of lemmings. *Behavioral and Neural Biology*, 1979, *26*, 372–378.

Barkley, M. S., & Goldman, B. D. Testosterone-induced aggression in adult female mice. *Hormones and Behavior*, 1977, *9*, 76–84.

Barkley, M. S., & Goldman, B. D. Studies on opponent status and steroid mediation of aggression in female mice. *Behavioral Biology*, 1978, *23*, 118–123.

Barr, G. A., Gibbons, J. L., & Moyer, K. E. Male-female differences and the influence of neonatal and adult testosterone on intraspecies aggression in rats. *Journal of Comparative and Physiological Psychology*, 1976, *90*, 1169–1183.

Bernstein, I. S. Primate status hierarchies. In L. A. Rosenblum (Ed.), *Primate behavior, developments in field and laboratory research* (Vol. 1). New York: Academic Press, 1970.

Bernstein, I. S., Gordon, T. P., & Rose, R. M. Aggression and social controls in rhesus monkey *(Macaca mulatta)* groups revealed in group formation studies. *Folia Primatologica*, 1974, *21*, 81–107.

Bernstein, I. S., Rose, R. M., & Gordon, T. P. Behavioural and hormonal responses of male rhesus monkeys introduced to females in the breeding and non-breeding seasons. *Animal Behaviour*, 1977, *25*, 609–614.

Birch, H. G., & Clark, G. Hormonal modification of social behavior. II. The effects of sex-hormone administration on the social dominance status of the female-castrate chimpanzee. *Psychosomatic Medicine*, 1946, *8*, 320–331.

Birch, H. G., & Clark, G. Hormonal modification of social behavior: IV. The mechanism of estrogen-induced dominance in chimpanzees. *Journal of Comparative and Physiological Psychology*, 1950, *43*, 181–193.

Bowman, L. A., Dilley, S. R., & Keverne, E. B. Suppression of oestrogen-induced LH surges by social subordination in talapoin monkeys. *Nature*, 1978, *275*, 56–58.

Brain, P. F. Effects of isolation/grouping on endocrine function and fighting behavior in male and female golden hamsters (*Mesocricetus auratus* Waterhouse). *Behavioral Biology*, 1972, 7, 349–357.

Brain, P. F. *Hormones and aggression*. Montreal: Eden Press, 1977.

Brain, P. F., & Haug, M. Sex steroid influences on attack directed by groups of gonadectomized male and female mice towards lactating intruders. *Journal of Endocrinology*, 1979, *81*, 136P–137P (abstract).

Bronson, F. H., & Desjardins, C. Aggression in adult mice: Modification by neonatal injections of gonadal hormones. *Science*, 1968, *161*, 705–706.

Bronson, F. H., & Desjardins, C. Neonatal androgen administration and adult aggressiveness in female mice. *General and Comparative Endocrinology*, 1970, *15*, 320–325.

Bujalska, G. The role of spacing behaviour among females in the regulation of reproduction in the bank vole. *Journal of Reproduction and Fertility Supplement*, 1973, *19*, 465–474.

Buntin, J. D., Catanzaro, C., & Lisk, R. D. Facilitatory effects of pituitary transplants on intraspecific aggression in female hamsters. *Hormones and Behavior*, 1981, *15*, 214–225.

Carpenter, C. R. Sexual behavior of free ranging rhesus monkeys *(Macaca mulatta):* I. Specimens, procedures and behavioral characteristics of estrus. *Journal of Comparative Psychology*, 1942, *33*, 113–142.

Carpenter, C. R. *Naturalistic behavior of nonhuman primates*. University Park, Pa.: Pennsylvania State University Press, 1964.

Carter, C. S. Postcopulatory sexual receptivity in the female hamster: The role of the ovary and adrenal. *Hormones and Behavior,* 1972, *3,* 261–265.

Chapais, B., & Schulman, S. R. An evolutionary model of female dominance relations in primates. *Journal of Theoretical Biology,* 1980, *82,* 47–89.

Ciaccio, L. A., Lisk, R. D., & Reuter, L. A. Prelordotic behavior in the hamster: A hormonally modulated transition from aggression to sexual receptivity. *Journal of Comparative and Physiological Psychology,* 1979, *93,* 771–780.

Clarke, J. R. The aggressive behaviour of the vole. *Behaviour,* 1956, *9,* 1–23.

Cochran, C. A., & Perachio, A. A. Dihydrotestosterone propionate effects on dominance and sexual behaviors in gonadectomized male and female rhesus monkeys. *Hormones and Behavior,* 1977, *8,* 175–187.

Collias, N. E. Aggressive behavior among vertebrate animals. *Physiological Zoology,* 1944, *17,* 83–123.

Coppen, A., & Kessel, N. Menstruation and personality. *British Journal of Psychiatry,* 1963, *109,* 711–721.

Crawford, M. P. The relation between social dominance and the menstrual cycle in female chimpanzees. *Journal of Comparative Psychology,* 1940, *30,* 483–513.

Crook, J. H., & Butterfield, P. A. Effects of testosterone propionate and luteinizing hormone on agonistic and nest building behaviour of *Quelea quelea. Animal Behaviour,* 1968, *16,* 370–384.

Dalton, K. Menstruation and crime. *British Medical Journal,* 1961, *2,* 1752–1753.

Dalton, K. *The menstrual cycle.* New York: Pantheon Books, 1969.

Dalton, K. Paramenstrual baby battering. *British Medical Journal,* 1975, *2,* 279.

DeBold, J. F., & Miczek, K. A. Sexual dimorphism in the hormonal control of aggressive behavior of rats. *Pharmacology, Biochemistry and Behavior,* 1981, *14,* 89–93.

Dixson, A. F. Effects of ovariectomy and estradiol replacement therapy upon the sexual and aggressive behavior of the greater galago *(Galago crassicaudatus crassicaudatus). Hormones and Behavior,* 1978, *10,* 61–70.

Dixson, A. F., & Herbert J. Gonadal hormones and sexual behavior in groups of adult talapoin monkeys *(Miopithecus talapoin). Hormones and Behavior,* 1977, *8,* 141–154.

Downhower, J. F., & Armitage, K. B. The yellow-bellied marmot and the evolution of polygamy. *American Naturalist,* 1971, *105,* 355–370.

Drickamer, L. C. A ten-year summary of reproductive data for free-ranging *Macaca mulatta. Folia Primatologica,* 1974, *21,* 61–80.

Dunbar, R., & Dunbar, P. Social dynamics of Gelada baboons. *Contributions to Primatology,* 1975, *6,* 1–157.

Dunbar, R. I. M., & Dunbar, E. P. Dominance and reproductive success among female gelada baboons. *Nature,* 1977, *266,* 351–352.

Edwards, D. A. Mice: Fighting by neonatally androgenized females. *Science,* 1968, *161,* 1027–1028.

Edwards, D. A. Early androgen stimulation and aggressive behavior in male and female mice. *Physiology and Behavior,* 1969, *4,* 333–338.

Edwards, D. A. Post-neonatal androgenization and adult aggressive behavior in female mice. *Physiology and Behavior,* 1970, *5,* 465–467.

Edwards, D. A., & Burge, K. G. Estrogenic arousal of aggressive behavior and masculine sexual behavior in male and female mice. *Hormones and Behavior,* 1971, *2,* 239–245.

Eibl-Eibesfeldt, I. Zur ethologie des hamsters (*Cricetus cricetus* L.). *Zeitschrift fur Tierpsychologie,* 1953, *10,* 204–254.

Erpino, M. J. Androgen-induced aggression in neonatally androgenized female mice: Inhibition by progesterone. *Hormones and Behavior,* 1975, *6,* 149–157.

Erskine, M. S., Barfield, R. J., & Goldman, B. D. Intraspecific fighting during late pregnancy and lactation in rats and effects of litter removal. *Behavioral Biology,* 1978, *23,* 206–218.

Erskine, M. S., Barfield, R. J., & Goldman, B. D. Postpartum aggression in rats: I. Effects of hypophysectomy. *Journal of Comparative and Physiological Psychology,* 1980, *94,* 484–494.

Erwin, J., & Anderson, B. Agonistic behavior of pregnant female monkeys *(Macaca nemestrina):* Possible influences of fetal gonadal hormones. *Psychological Reports,* 1975, *36,* 699–702.

Ewer, R. F. The biology and behaviour of a free-living population of black rats *(Rattus rattus). Animal Behaviour Monographs,* 1971, *4,* 127–174.

Fischer, R. B., & Nadler, R. D. Status interactions of captive female lowland gorillas. *Folia Primatologica,* 1977, *28,* 122–133.

Floody, O. R., & Pfaff, D. W. Steroid hormones and aggressive behavior: Approaches to the study of hormone-sensitive brain mechanisms for behavior. In S. H. Frazier (Ed.), *Aggression (Research publications, Association for Research in Nervous and Mental Disease,* Vol. 52). Baltimore: Williams & Wilkins, 1974.

Floody, O. R., & Pfaff, D. W. Aggressive behavior in female hamsters: The hormonal basis for fluctuations in female aggressiveness correlated with estrous state. *Journal of Comparative and Physiological Psychology,* 1977, *91,* 443–464.

Fredericson, E. Aggressiveness in female mice. *Journal of Comparative and Physiological Psychology,* 1952, *45,* 254–257.

Frodi, A. Experimental and physiological responses associated with anger and aggression in women and men. *Journal of Research in Personality,* 1978, *12,* 335–349.

Frodi, A., Macaulay, J., & Thome, P. R. Are women always less aggressive than men? A review of the experimental literature. *Psychological Bulletin,* 1977, *84,* 634–660.

Gandelman, R. Mice: Postpartum aggression elicited by the presence of an intruder. *Hormones and Behavior,* 1972, *3,* 23–28.

Gleason, P. E., Michael, S. D., & Christian, J. J. Effects of gonadal steroids on agonistic behavior of female *Peromyscus leucopus. Hormones and Behavior,* 1979, *12,* 30–39.

Goldfoot, D. A. Sex-specific, behavior-specific actions of dihydrotestosterone: Activation of aggression, but not mounting in ovariectomized guinea pigs. *Hormones and Behavior,* 1979, *13,* 241–255.

Gottschalk, L. A., Kaplan, S. M., Gesser, G. C., & Winget, C. M. Variations in magnitude of emotion: A method applied to anxiety and hostility during phases of the menstrual cycle. *Psychosomatic Medicine,* 1962, *24,* 300–311.

Gray, J. A. Sex differences in emotional behaviour in mammals including man: Endocrine bases. *Acta Psychologica,* 1971, *35,* 29–46.

Gray, L. E. Jr., Whitsett, J. M., & Ziesenis, J. S. Hormonal regulation of aggression toward juveniles in female house mice. *Hormones and Behavior,* 1978, *11,* 310–322.

Green, R., Whalen, R. E., Rutley, B., & Battie, C. Dominance hierarchy in squirrel monkeys *(Saimiri sciureus). Folia Primatologica,* 1972, *18,* 185–195.

Greene, R., & Dalton, K. The premenstrual syndrome. *British Medical Journal,* 1953, *1,* 1007–1014.

Gregory, B. A. J. C. The menstrual cycle and its disorders in psychiatric patients: II. Clinical studies. *Journal of Psychosomatic Research,* 1957, *2,* 199–224.

Grelk, D. F., Papson, B. A., Cole, J. E., & Rowe, F. A. The influence of caging conditions and hormone treatments on fighting in male and female hamsters. *Hormones and Behavior,* 1974, *5,* 355–366.

Hardy, D. F. Sexual behavior in continuously cycling rats. *Behaviour,* 1972, *41,* 288–297.

Hasler, J. F., & Banks, E. M. The behavioral and somatic effects of ovariectomy and replacement therapy in female collared lemmings *(Dicrostonyx groenlandicus). Hormones and Behavior,* 1976, 7, 59–74.

Haug, M., & Brain, P. F. Effects of treatments with testosterone and oestradiol on the attack directed by groups of gonadectomized male and female mice towards lactating intruders. *Physiology and Behavior,* 1979, *23,* 397–400.

Haug, M., & Mandel, P. Sex differences in aggressive behaviour in various strains of mice. *Aggressive Behavior,* 1978, *4,* 353–364.

Hedricks, C., & Daniels, C. E. Agonistic behavior between pregnant mice and male intruders. *Behavioral and Neural Biology,* 1981, *31,* 236–241.

Huck, U. W., Carter, C. S., & Banks, E. M. Estrogen and progesterone interactions influencing sexual and social behavior in the brown lemming, *Lemmus trimucronatus. Hormones and Behavior,* 1979, *12,* 40–49.

Hyde, J. S., & Sawyer, T. F. Estrous cycle fluctuations in aggressiveness of house mice. *Hormones and Behavior,* 1977, *9,* 290–295.

Ivey, M. E., & Bardwick, J. M. Patterns of affective fluctuation in the menstrual cycle. *Psychosomatic Medicine,* 1968, *30,* 336–345.

Jolly, A. *Lemur behavior.* Chicago: University of Chicago Press, 1966.

Joslyn, W. D. Androgen-induced social dominance in infant female rhesus monkeys. *Journal of Child Psychology and Psychiatry,* 1973, *14,* 137–145.

Keverne, E. B. Sexual and aggressive behaviour in social groups of talapoin monkeys. *CIBA Foundation Symposium,* 1979, *62,* 271–286.

Kislak, J. W., & Beach, F. A. Inhibition of aggressiveness by ovarian hormones. *Endocrinology,* 1955, *56,* 684–692.

Lester, D., & Cheses, K. T. Effects of deprivation upon aggression in rats. *Psychological Reports,* 1968, *22,* 1129–1133.

Lumia, A. R., Canino, G., & Drozdowski, B. Effects of androgen on scent marking and aggressive behavior of neonatally bulbectomized female Mongolian gerbils *(Meriones unguiculatus). Hormones and Behavior,* 1976, *7,* 461–471.

Maccoby, E. E., & Jacklin, C. N. *The psychology of sex differences.* Stanford, Ca.: Stanford University Press, 1974.

Madden, J. R. Female territoriality in a Suffolk County, Long Island, population of *Glaucomys volans. Journal of Mammalogy,* 1974, *55,* 647–652.

Mallow, G. K. The relationship between aggression and cycle stage in adult female rhesus monkeys *(Macaca mulatta). Dissertation Abstracts,* 1979, *39,* 3194.

Marques, D. M., & Valenstein, E. S. Individual differences in aggressiveness of female hamsters: Response to intact and castrated males and to females. *Animal Behaviour,* 1977, *25,* 131–139.

Metzgar, L. H. Behavioral population regulation in the woodmouse, *Peromyscus leucopus. American Midland Naturalist,* 1971, *86,* 434–448.

Michael, R. P. Effects of gonadal hormones on displaced and direct aggression in pairs of rhesus monkeys of opposite sex. In S. Garattini & E. B. Sigg (Eds.), *Aggressive behaviour.* New York: Wiley, 1969.

Michael, R. P., & Zumpe, D. Aggression and gonadal hormones in captive rhesus monkeys *(Mucaca mulatta). Animal Behaviour,* 1970, *18,* 1–10.

Mirsky, A. F. The influence of sex hormones on social behavior in monkeys. *Journal of Comparative and Physiological Psychology,* 1955, *48,* 327–335.

Monder, H., Yasukawa, N., & Christian, J. J. *Opiates and aggression in female deermice.* Presented at the annual meeting of the Society for Neuroscience, Cincinnati, Ohio, November 1980. (Abstract)

Moos, R. H. The development of a menstrual distress questionnaire. *Psychosomatic Medicine,* 1968, *30,* 853–867.

Moos, R. H., Kopell, B. S., Melges, F. T., Yalom, I. D., Lunde, D. T., Clayton, R. B., & Hamburg, D. A. Fluctuations in symptoms and moods during the menstrual cycle. *Journal of Psychosomatic Research,* 1969, *13,* 37–44.

Morton, J. H., Additon, H., Addison, R. G., Hunt, L., & Sullivan, J. J. A clinical study of premenstrual tension. *American Journal of Obstetrics and Gynecology,* 1953, *65,* 1182–1191.

Moyer, K. E. Sex differences in aggression. In R. C. Friedman, R. M. Richart, & R. L. Vande Wiele (Eds.), *Sex differences in behavior.* New York: Wiley, 1974.

Moyer, K. E. *The psychobiology of aggression.* New York: Harper & Row, 1976.

Mykytowycz, R., & Dudzinski, M. L. Aggressive and protective behaviour of adult rabbits *Oryctolagus cuniculus* (L.) towards juveniles. *Behaviour,* 1972, *43,* 97–120.

Mykytowycz, R., & Hesterman, E. R. An experimental study of aggression in captive European rabbits, *Oryctolagus cuniculus* (L.). *Behaviour,* 1975, *52,* 104–118.

Noble, R. G., & Alsum, P. B. Hormone dependent sex dimorphisms in the golden hamster *(Mesocricetus auratus). Physiology and Behavior,* 1975, *14,* 567–574.

Noirot, E., Goyens, J., & Buhot, M. C. Aggressive behavior of pregnant mice toward males. *Hormones and Behavior,* 1975, *6,* 9–17.

Nowlis, V. Sexual status and degree of hunger in chimpanzee competitive interaction. *Journal of Comparative Psychology,* 1942, *34,* 185–194.

Paige, K. E. Effects of oral contraceptives on affective fluctuations associated with the menstrual cycle. *Psychosomatic Medicine,* 1971, *33,* 515–537.

Parlee, M. B. The premenstrual syndrome. *Psychological Bulletin,* 1973, *80,* 454–465.

Payne, A. P. Neonatal androgen administration and aggression in the female golden hamster during interactions with males. *Journal of Endocrinology,* 1974, *63,* 497–506. (a)

Payne, A. P. The aggressive response of the male golden hamster towards males and females of differing hormonal status. *Animal Behavior,* 1974, *22,* 829–835. (b)

Payne, A. P., & Swanson, H. H. Agonistic behaviour between pairs of hamsters of the same and opposite sex in a neutral observation area. *Behaviour,* 1970, *36,* 259–267.

Payne, A. P., & Swanson, H. H. Hormonal control of aggessive dominance in the female hamster. *Physiology and Behavior,* 1971, *6,* 366–357.

Payne, A. P., & Swanson, H. H. The effect of sex hormones on the aggressive behaviour of the female golden hamster (*Mesocricetus auratus* Waterhouse). *Animal Behaviour,* 1972, *20,* 782–787.

Pearson, O. P. Reproduction in the shrew (*Blarina brevicauda* Say). *American Journal of Anatomy,* 1944, *75,* 39–93.

Perkins, M. N., Perkins, M. S., & Bunnell, B. N. *Copulation in golden hamsters: Behavior of intact vs. ovariectomized females given hormone replacement.* Presented at the annual meeting of the Eastern Conference on Reproductive Behavior, New Orleans, June 1979. (Abstract)

Persky, H. Reproductive hormones, moods, and the menstrual cycle. In R. C. Friedman, R. M. Richart, & R. L. Vande Wiele (Eds)., *Sex differences in behavior.* New York: Wiley, 1974.

Politch, J. A., & Herrenkohl, L. R. *Maternal aggression in mice: The role of prolactin.* Presented at the annual meeting of the Eastern Conference on Reproductive Behavior, New York, June 1980. (Abstract)

Price, E. O., & Belanger, P. L. Maternal behavior of wild and domestic stocks of Norway rats. *Behavioral Biology,* 1977, *20,* 60–69.

Ralls, K. Mammals in which females are larger than males. *Quarterly Review of Biology,* 1976, *51,* 245–276.

Rees, L. The premenstrual tension syndrome and its treatment. *British Medical Journal,* 1953, *1,* 1014–1016.

Robinson, J. G. Vocal regulation of use of space by groups of titi monkeys *Callicebus moloch. Behavioral Ecology and Sociobiology,* 1979, *5,* 1–15.

Rowell, T. E. Behaviour and female reproductive cycles of rhesus macaques. *Journal of Reproduction and Fertility,* 1963, *6,* 193–203.

Rowell, T. E., Female reproductive cycles and the behavior of baboons and rhesus macaques. In S. A. Altmann (Ed.), *Social communication among primates.* Chicago: University of Chicago Press, 1967.

Rowell, T. E. Baboon menstrual cycles affected by social environment. *Journal of Reproduction and Fertility,* 1970, *21,* 133–141.

Rowell, T. E. Organization of caged groups of cercopithecus monkeys. *Animal Behaviour,* 1971, *19,* 625–645.

Rowell, T. E. Female reproduction cycles and social behavior in primates. *Advances in the Study of Behavior,* 1972, *4,* 69–105.

Ruble, D. N. Premenstrual symptoms: A reinterpretation. *Science,* 1977, *197,* 291–292.

Saayman, G. S. Aggressive behaviour in free-ranging chacma baboons *(Papio ursinus). Journal of Behavioral Science,* 1971, *1,* 77–83.

Sassenrath, E. N., Powell, T. E., & Hendrickx, A. G. Perimenstrual aggression in groups of female rhesus monkeys. *Journal of Reproduction and Fertility,* 1973, *34,* 509–511.

Schonberg, W. B., Costanzo, D. J., & Carpenter, R. S. Menstrual cycle: Phases and reaction to frustration. *Psychological Record,* 1976, *26,* 321–325.

Seward, J. P. Aggressive behavior in the rat: I. General characteristics: age and sex differences. *Journal of Comparative Psychology,* 1945, *38,* 175–197. (a)

Seward, J. P. Aggressive behavior in the rat: II. An attempt to establish a dominance hierarchy. *Journal of Comparative Psychology,* 1945, *38,* 213–224. (b)

Seyfarth, R. M. Social relationships among adult female baboons. *Animal Behaviour,* 1976, *24,* 917–938.

Silbergeld, S., Brast, N., & Noble, E. P. The menstrual cycle: A double-blind study of symptoms, mood and behavior, and biochemical variables using Enovid and placebo. *Psychosomatic Medicine,* 1971, *33,* 411–428.

Simon, N. G., & Gandelman, R. The estrogenic arousal of aggressive behavior in female mice. *Hormones and Behavior,* 1978, *10,* 118–127.

Smith, C. C. The adaptive nature of social organization in the genus of three squirrels *Tamiasciurus. Ecological Monographs,* 1968, *38,* 31–63.

Struhsaker, T. T. Social structure among vervet monkeys *(Cercopithecus aethiops)*. *Behaviour,* 1967, *29,* 83–121.

Suchowsky, G. K., Pegrassi, L., & Bonsignori, A. Steroids and aggressive behaviour in isolated male and female mice. *Psychopharmacologia,* 1971, *21,* 32–38.

Sutherland, H., & Stewart, I. A critical analysis of the premenstrual syndrome. *Lancet,* 1965, *1,* 1180–1183.

Svare, B., & Gandelman, R. Postpartum aggression in mice: Experiential and environmental factors. *Hormones and Behavior,* 1973, *4,* 323–334.

Svare, B., & Gandelman, R. Postpartum aggression in mice: Inhibitory effect of estrogen. *Physiology and Behavior,* 1975, *14,* 31–35.

Svare, B., & Gandelman, R. Suckling stimulation induces aggression in virgin female mice. *Nature,* 1976, *260,* 606–608.

Svare, B., Davis, P. G., & Gandelman, R. Fighting behavior in female mice following chronic androgen treatment during adulthood. *Physiology and Behavior,* 1974, *12,* 399–403.

Swanson, H. H. Sex differences in behaviour of the Mongolian gerbil *(Meriones unguiculatus)* in encounters between pairs of same and opposite sex. *Animal Behaviour,* 1974, *22,* 638–644.

Tenaza, R. R. Territory and monogamy among Kloss' gibbons *(Hylobates klossii)* in Siberut Island, Indonesia. *Folia Primatologica,* 1975, *24,* 60–80.

Thiessen, D. D. *The evolution and chemistry of aggression.* Springfield, Ill.: Charles C. Thomas, 1976.

Thompson, D. C. The social system of the grey squirrel. *Behaviour,* 1978, *64,* 305–328.

Tiefer, L. Gonadal hormones and mating behavior in the adult golden hamster. *Hormones and Behavior,* 1970, *1,* 189–202.

Tollman, J., & King, J. A. The effects of testosterone propionate on aggression in male and female C57BL/10 mice. *British Journal of Animal Behaviour,* 1956, *4,* 147–149.

Tonks, C. M. Premenstrual tension. *Contemporary Psychiatry,* 1975, *9,* 399–408.

Trimble, M. R., & Herbert, J. The effect of testosterone or oestradiol upon the sexual and associated behaviour of the adult female rhesus monkey. *Journal of Endocrinology,* 1968, *42,* 171–185.

Vandenbergh, J. G. The effects of gonadal hormones on the aggressive behaviour of adult golden hamsters *(Mesocricetus auratus)*. *Animal Behaviour,* 1971, *19,* 589–594.

Vom Saal, F. S., Gandelman, R., & Svare, B. Aggression in male and female mice: Evidence for changed neural sensitivity in response to neonatal but not adult androgen exposure. *Physiology and Behavior,* 1976, *17,* 53–57.

Vom Saal, F. S., Svare, B., & Gandelman, R. Time of neonatal androgen exposure influences length of testosterone treatment required to induce aggression in adult male and female mice. *Behavioral Biology,* 1976, *17,* 391–397.

Waser, P. M., & Wiley, R. H. Mechanisms and evolution of spacing in animals. In P. Marler & J. G. Vandenbergh (Eds.), *Handbook of behavioral neurobiology* (Vol. 3). New York: Plenum Press, 1980.

Whitsett, J. M., Gray, L. E. Jr., & Bediz, G. M. Gonadal hormones and aggression toward juvenile conspecifics in prairie deer mice. *Behavioral Ecology and Sociobiology,* 1979, *6,* 165–168.

Wise, D. A. Aggression in the female golden hamster: Effects of reproductive state and social isolation. *Hormones and Behavior,* 1974, *5,* 235–250.

Wise, D. A., & Pryor, T. L. Effects of ergocornine and prolactin on aggression in the postpartum golden hamster. *Hormones and Behavior,* 1977, *8,* 30–39.

Yahr, P., & Coquelin, A. Effect of pre- versus post-puberal castration on aggression between male gerbils. *Behavioral and Neural Biology,* 1980, *28,* 496–500.

Yerkes, R. M. Social dominance and sexual status in the chimpanzee. *Quarterly Review of Biology,* 1939, *14,* 115–136.

Yerkes, R. M. Social behavior of chimpanzees: Dominance between mates, in relation to sexual status. *Journal of Comparative Psychology,* 1940, *30,* 147–186.

Yerkes, R. M. Conjugal contrasts among chimpanzees. *Journal of Abnormal and Social Psychology,* 1941, *36,* 175–199.

Young, W. C., Dempsey, E. W., & Meyers, H. I. Cyclic reproductive behavior in the female guinea pig. *Journal of Comparative Psychology,* 1935, *19,* 313–335.

Zimen, E. On the regulation of pack size in wolves. *Zeitschrift für Tierpsychologie,* 1976, *40,* 300–341.

Zumpe, D., & Michael, R. P. Redirected aggression and gonadal hormones in captive rhesus monkeys *(Macaca mulatta)*. *Animal Behaviour,* 1970, *18,* 11–19.

4

Hormonal Influences on Maternal Aggression

Bruce B. Svare and Martha A. Mann

INTRODUCTION

With the exception of work conducted with female hamsters (see Floody, this volume), the study of the relationship between hormones and female aggressive behavior has never been of paramount importance. Recently, however, a number of investigators have turned to an analysis of this little-studied dimension of agonistic behavior. The results of this research tentatively suggest that female aggression during the peripartum period, like intermale aggression, is influenced by neuroendocrine status. The purpose of the present chapter is to review the available literature concerning the role of hormones in the aggressive behavior displayed by pregnant and lactating mammals. This review is limited to work conducted with hamsters, mice, and rats.

TOPOGRAPHICAL, EXPERIENTIAL, AND ENVIRONMENTAL FACTORS

While it is not the purpose of this chapter to review topographical, experiential, and environmental determinants of maternal aggression extensively, some of the more important nonbiological factors governing the behavior in laboratory animals are briefly reviewed here. (The reader is referred to Svare, 1981, for a more extensive analysis.) Although maternal aggression has been studied most extensively in lactating animals, it should be noted that some species such as hamsters are also aggressive during gestation. Pregnancy-induced aggression has seldom been studied, however, and this review will be concerned primarily with aggression exhibited during the postpartum period.

BRUCE B. SVARE AND MARTHA A. MANN • Department of Psychology, State University of New York at Albany, Albany, New York 12222. This review was written while the first author was supported by a Research Grant from the Harry Frank Guggenheim Foundation, Grant BNS80-08546 from NSF, and Grant AGO1319 from NIA.

The topography of postpartum defense in mice, rats, and hamsters appears to be similar in that in all these species, maternal aggression is characterized by immediate attacks toward the flanks and neck of the intruder coupled with almost immediate submission on the part of the encroaching conspecific (cf. Price & Belanger, 1977; Svare & Gandelman, 1973; Wise, 1974). The behavior reaches peak intensities during early lactation (around Postpartum Day 8) and declines rapidly thereafter (cf. Erskine, Barfield, & Goldman, 1978; Svare, Betteridge, Katz, & Samuels, 1981). In mice, fighting experience retards the decline in aggression that normally occurs with advancing lactation, and isolation during pregnancy has a facilitatory effect on the behavior (Green, 1978). The extent to which neuroendocrine factors might be playing a role in these experientially induced alterations in maternal defense is unknown, but their role in similar aspects of intermale aggressive behavior has been well documented (cf. Leshner, 1979; Leshner, this volume). Finally, the hormonal status of the intruder is an important determinant of the behavior in lactating mice. Parturient CD-1 female mice attack intact male mice more often than they attack virgin or lactating females. Few gonadectomized virgin animals of either sex are attacked by CD-1 lactating females as compared with intact animals, and late lactating (Postpartum Days 16–20) female intruders are attacked more often than early lactating animals (Postpartum Days 3–8); (Rosenson & Asheroff, 1975). An important question for future researchers is to determine whether hormone-based pheromones modulate maternal aggressive behavior in the same way that they influence intermale aggression.

NEUROENDOCRINE INVOLVEMENT

Hormonal Determinants during Prenatal Life

An increasing number of reports indicate that various aspects of feminine behavior may normally be influenced by the hormonal milieu during perinatal life. For example, prenatal administration of the aromatase inhibitor ATD or the antiandrogen flutamide suppresses estrogen-induced lordosis in female rats (Gladue & Clemens, 1978). Is maternal aggression in rodents modulated by the perinatal hormone environment? Several reports suggest that this may be true.

First, vom Saal and Bronson (1978) reported that female mice positioned between two males (i.e., 2M females) *in utero* were more maternally aggressive than were females located between two females (i.e., 0M female). The 2M females spent more time than did 0M animals attacking female intruders. Interestingly, the high level of aggression in 2M animals as compared with 0M animals was limited to interactions with female intruders, since the differences were not evident if male opponents were used. It is now known that during fetal life, 2M females are exposed to higher levels of testosterone in serum and amniotic fluid than are 0M females (vom Saal & Bronson, 1980). Also, elevating prenatal T by exogenous administration (1.0 μg or 2.0 μg TP/day on Days 12, 14, and 16 of fetal life) increases the number of attacks displayed by lactating Rockland-Swiss female mice (Mann & Svare, in press). The above findings therefore suggest that female aggressive behavior, like male aggression, may be modulated by perinatal testosterone.

Another report implicating perinatal hormones in maternal aggression is one by Politch and Herrenkohl (1979). They found that restraint stress during pregnancy reduced the number of maternal attacks displayed by the female offspring during their first lactation. The possible hormonal influences are manifold but it is well known that ACTH and glucocorticoids are greatly increased following stress. If elevated pituitary–adrenal hormones during prenatal life are responsible for the above reductions in maternal aggression, then one would expect that the injection of such hormones into pregnant animals should also reduce the behavior in female offspring. Reinisch, Simon, and Gandelman (1980) recently found that prenatal exposure to prednisone, a synthetic corticosteroid, reduced the proportion of animals exhibiting maternal aggression in adulthood. The mode of action of prednisone in the Reinisch *et al.* study is not completely understood, but the authors suggest that their results may have been due to the mildly androgenic properties of the hormone and/or to a depression in estrogen-binding alpha-feto-protein (AFP). Because the Reinisch *et al.* results may represent a pharmacological effect of prednisone, it is not known whether corticosteroids normally play a role in modulating maternal aggression during the perinatal period. Regardless, it is clear that more research is needed in order to examine more thoroughly the notion that perinatal hormones modulate maternal aggression.

Hormonal Determinants during Pregnancy and Lactation

Initiation of Maternal Aggression

Pregnancy. As noted earlier, considerable between- and within-species differences exist with respect to the incidence of aggression during pregnancy. Mice of some outbred mouse strains exhibit the behavior during gestation, and rats and hamsters exhibit increased fighting behavior with advancing pregnancy. The extent to which changes in aggression during gestation are related to alterations in ovarian and/or pituitary hormone secretions is not known at present. However, virgin female mice made pseudopregnant by exposure to a vasectomized male are as aggressive as pregnant females (Hedricks & Daniels, 1981; Noirot, Goyens, & Buhot, 1975). Because estrogen is low and progesterone is high dring pregnancy and pseudopregnancy, it is interesting to speculate that this hormonal condition is a prerequisite for fighting behavior during pregnancy.

Lactation

Role of Suckling Stimulation. An important source of stimulation that the mother receives during the postpartum period is suckling stimulation from her young. We conducted a series of studies in our laboratory in order to examine the possibility that such stimulation might be important for maternal aggressive behavior in the mouse.

In the first experiment, we prevented dams from receiving suckling stimulation by removing their nipples (thelectomy) prior to parturition. In this study, as well as the others described here, we instituted a fostering procedure consisting of the daily rotation of one- to three-day-old pups generated from donor mothers mated at the same time. This was done in order to ensure that all animals received a constant source of suckling stimulation from viable, replete young. Animals were tested for aggression every other day for four tests by the placement of an adult male intruder in their cage for three minutes. Our findings (Svare & Gandelman, 1976a) showed that the thelectomized animals rarely showed aggression, while control animals exhibited normally high levels of fighting. Subsequent experiments identified a critical duration of suckling stimulation for the initiation of maternal aggression: Animals receiving less than 48 hours of suckling from young rarely fought, while more than 48 hours of this stimulation ensured fighting behavior by postparturient dams. It is interesting to note that thelectomy after Postpartum Day 5 was without effect on aggression, suggesting that suckling after the early postpartum period is not essential for the maintenance of the behavior.

Coincident with our thelectomy experiments and the conclusion that suckling is important for the initiation of maternal aggression, we had repeatedly failed in other experiments to induce aggressive behavior in virgin females by chronically fostering them newborn pups. On close examination, we found that the pups were not attaching to the dam's nipples and, hence, the mothers were not receiving suckling stimulation. We reasoned that the above failure was due to an inadequate substrate, namely, that virgin females lacked the nipple growth that normally occurs during gestation. Indeed, studies in our laboratory since have demonstrated that the nipples of R-S female mice show a fourfold increase in length during gestation (Svare, Mann, & Samuels, 1980). If nipple growth and suckling stimulation are necessary for aggression, we further reasoned that virgin females should exhibit the behavior if they are first treated with a hormone regimen that induces nipple growth followed by suckling stimulation from foster young. Under the above-hypothesized conditions, virgin females were found to receive suckling stimulation and exhibit aggression identical to that of parturient animals (Svare & Gandelman, 1976b), thus further confirming the notion that nipple growth and suckling are important for the behavior. Similar findings have been reported for the rat (Gandelman & Simon, 1980), suggesting of course that hormones are at least playing an indirect role in the behavior—namely, preparing the substrate for the suckling stimulus.

Another way in which to examine the relationships among nipple growth, suckling stimulation, and the initiation of maternal aggression is to study the aggressive behavior of female mice hysterectomized at various stages of pregnancy and fostered young. We employed this tactic because previous work had shown that pregnant animals hysterectomized (i.e., pregnancy terminated) late in gestation and fostered young exhibited normal levels of aggression (Gandelman & Svare, 1974). Because the nipples develop with advancing gestation (Svare, Mann, & Samuels, 1980), the pregnancy-termination procedure allows one to vary the level of nipple growth by varying the time during gestation when hysterectomy is performed. Thus, we predicted that fighting behavior would be present in animals hysterectomized and fostered young at a time during gestation when nipple growth is adequate to permit suckling by young. Conversely, we predicted that aggression would not be evident in animals hysterectomized and fostered young at a time during gestation when nipple growth is inadequate for suckling to take place. The results showed that pregnant animals hysterectomized and fostered young after Gestation Day 11 received suckling stimulation and exhibited aggression; suckling and aggression were not evident when hysterectomy was performed prior to Gestation Day 11 (Svare, Mann, & Samuels, 1980). The findings, therefore, further confirmed our thoughts that nipple growth and suckling stimulation are necessary conditions for the establishment of aggressive behavior.

ROLE OF PITUITARY PROLACTIN. Having established that suckling stimulation is important for the initiation of maternal aggression in rodents, we recently have addressed the important question: What suckling-induced change(s) mediate the display of the behavior? The neuroendocrine reflex arc that is activated by stimulation of the teats contains a number of significant components, including, but not limited to, changes in neurotransmitter turnover rates, hypothalamic and pituitary hormone release, and EEG activity. Suckling-induced changes in ovarian and adrenal hormones are not important for the behavior, since ovariectomized and adrenalectomized animals continue to exhibit maternal aggression identical to that of sham-operated control animals (Svare & Gandelman, 1976b, unpublished observations). Thus, because the pituitary hormone prolactin (PRL) is very sensitive to changes in suckling stimulation and has been implicated in lactation and other aspects of parental care (cf. Zarrow, Gandelman, & Denenberg, 1971), we felt that this hormone would be the most likely candidate for involvement in the maternal aggressive behavior displayed by mice. As it turns out, recent data now clearly show that PRL is not involved in the maternal aggression displayed by rats and house mice. Let us examine in some detail the findings that led to this conclusion.

In our first experiment, we examined the aggressive behavior of parturient mice who received their first suckling exposure several weeks following the delivery of young. Because such "delayed" animals receive suckling stimulation but do not lactate or secrete normal quantities of PRL, they provide an ideal preparation for separating the relative importance of these factors in maternal aggressive behavior. So as to control the onset of suckling exposure and prevent animals from receiving this stimulation at the time of parturition, the homecages were modified by the placement of a wire-mesh screen slightly above the floor of the cage. Thus, during parturition, the pups dropped through the screen. Animals were fostered young for the first time either immediately or one, two, or three weeks following parturition, with aggression testing and daily fostering of young identical to that described earlier. Parturient animals in the delay groups received suckling stimulation and exhibited aggression identical to that of the animals receiving exposure to young immediately following parturition (Svare, Mann, & Samuels, 1980). Importantly, none of the animals in the delay groups lactated, whereas all the animals receiving young at parturition lactated.

These findings of course are suggestive, but do not provide conclusive evidence, that PRL is not involved in the behavior.

High levels of circulating PRL and aggression are evident during early and middle lactation, while levels of both the hormone and behavior are low during late lactation (Sinha, Selby, & Vanderlaan, 1974; Svare and Gandelman, 1973). Thus, another way in which to examine the role of circulating PRL in the maternal aggressive behaior displayed by parturient mice is to correlate levels of the hormone with the initiation, maintenance, and decline of the behavior. Maternal aggression and circulating levels of PRL were measured in separate groups of parturient mice on Postpartum Day 0 (the day of delivery), 6, 12, or 18. The results showed that there was little correspondence between circulating levels of the hormone and behavior (Broida, Michael, & Svare, 1981; see Figure 1). On Postpartum Day 0, the dams rarely fought, but circulating PRL was at its highest level (290 ng/ml). Aggression was highest on Postpartum Days 6 and 12 and significantly declined between Postpartum Days 12 and 18. Importantly, however, plasma PRL declined well in advance of a decline in aggression, since a significant reduction in levels of the hormone was observed between Postpartum Days 6 and 12 (272 versus 112 ng/ml respectively). Also, there was no

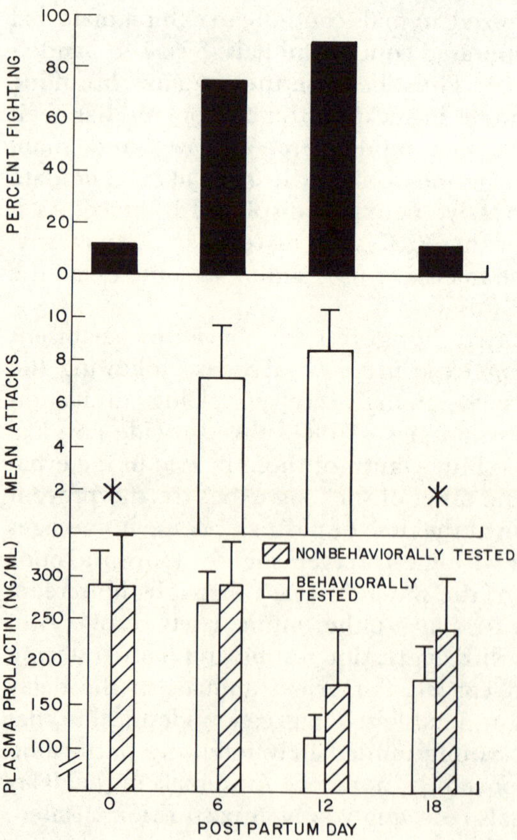

FIGURE 1. The percentage of animals exhibiting aggression (top panel), the mean number of attacks (center panel), and mean plasma prolactin (PRL) levels (bottom panel) for separate groups of parturient R-S mice tested for aggression (3-minute test) against an adult male intruder on Postpartum Day 0 (the day of delivery), 6, 12, or 18. Separate groups of animals not tested for aggression but sampled for plasma PRL also were included on each postpartum test day. Blood samples were collected by cardiac puncture without anesthesia immediately after the behavioral test or at the same time of day for the nontested controls. The asterisks (center panel) indicate that too few animals responded to report an average. (Adapted from Broida, Michael, & Svare, 1981.)

significant correlation between circulating PRL and the intensity of fighting behavior.

Once again, the above data are suggestive that PRL is not related to aggression in the house mouse. We further investigated the potential role of PRL in maternal aggression by observing the effects of specific PRL inhibitors (i.e., ergot drugs) on circulating PRL, lactation, and aggressive behavior (Mann, Michael, & Svare, 1980). In order to prevent animals from experiencing a suckling-induced rise in PRL prior to drug treatment, we again modified the homecages of timed-mated pregnant animals by placing a wire-mesh screen slightly above the floor of the cage. In this way, during parturition, the pups dropped through the screen. Following parturition, the animals received a daily injection of oil or 0.5 mg of ergocornine or bromocryptine for 10 days. Fostering of newborn young commenced on the third day of injections and continued as described earlier. Blood samples were collected 24 hours following the last aggression test. The results showed that the ergot-treated animals received suckling stimulation but did not lactate as measured by continued weight loss of the foster young. Moreover, the ergot-treated animals exhibited sevenfold and twofold reductions, respectively, in plasma PRL (152.0, 22.8, and 73.9 ng/ml for oil-, ergocornine-, and bromocryptine-treated animals, respectively). Importantly, however, aggressive behavior in the ergot-treated group was indistinguishable from that of controls, suggesting, of course, that PRL may be unnecessary for the behavior.

The best way in which to test whether PRL is involved in maternal defense is to examine the fighting behavior of hypophysectomized parturient mice. We again utilized the aforementioned wire-mesh screen for the delivery of young and the prevention of suckling stimulation. Soon after partuirition, the dams were either hypophysectomized by the parapharyngeal approach, sham-operated, or not operated. The foster litter procedure, method of aggression testing, and blood collections were identical to those of the previous experiments. The results (Mann, Broida, Michael, & Svare, 1980; Svare, Mann, Broida, & Michael, in press) showed that the groups did not differ with respect to any measure of aggression (see Table I). However, plasma PRL levels of hypophysectomized dams were extremely low, while those of control dams were very high.

The above series of experiments clearly show that PRL is not necessary for maternal aggression in the mouse. Work with rats also supports the notion that the hormone is not involved in the behavior. Hypophysectomy performed on Postpartum Day 5 in that species reduces circulating PRL but does not disrupt maternal aggression. In contrast to the above findings with rats and mice, work with hamsters and wild mice supports the hypothesis that PRL is involved in the behavior. Twice-daily injections of 0.5 mg ergocornine for three days inhibits the behavior in recently parturient female hamsters, while PRL replacement therapy reverses the trend (Wise & Pryor, 1977). Also, virgin female hamsters made hyperprolactinemic with ectopic pituitary grafts exhibit elevated levels of aggression (Buntin, Catanzaro, & Lisk, 1981). Working with female *peromyscus* mice, Gleason, Michael, and Christian (1981) reported that PRL injections elevated aggression in ovariectomized, adrenalectomized virgins. It is difficult to explain the above differences except to suggest that there may be a species

TABLE I
Maternal Aggressive Behavior and Plasma Prolactin (PRL) Levels of Parturient Rockland-Swiss Albino Female Mice That Were Hypophysectomized (HYPOX), Sham-Operated (SHAM), or Nonoperated (NOC)[a]

Group	Proportion fighting	Mean (±*SEM*) latency (in test days) to exhibit fighting[b]	Mean (± *SEM*) number of tests fighting[b]	Mean (±*SEM*) number of attacks[b]	Mean plasma prolactin (ng/ml) levels
NOC	14/24 (58%)	2.6 ± 0.3	2.3 ± 0.3	2.6 ± 0.6	159.4 ± 46.2
SHAM	19/32 (59%)	2.3 ± 0.3	2.7 ± 0.3	3.9 ± 0.5	124.5 ± 28.6
HYPOX	15/22 (68%)	1.7 ± 0.2	2.8 ± 0.3	2.9 ± 0.5	5.3 ± 1.3

[a] Surgery was performed within 8 days following parturition. To prevent any suckling-induced rise in PRL release prior to surgery, the animals delivered their young through double wire mesh partitions inserted in the floors of the cages just prior to delivery. Beginning on Postpartum Day 9, the animals fostered five 2–8 day-old replete pups on a daily basis for 9 days. Three-minute tests for aggression against an adult male intruder were conducted every other day beginning on Postpartum Day 12 for a total of four tests. (Adapted from Svare, Mann, Broida, & Michael, in press.)
[b] Only for those animals exhibiting aggression.

(genetic) difference in the degree of PRL involvement in the behavior. Clearly, additional comparative data is needed before we can make conclusions concerning the extent to which PRL mediates maternal aggression (see Table II for a summary of PRL effects on maternal aggression).

ROLE OF PEPTIDES AND NEUROTRANSMITTERS. If the initiation of maternal aggression in mice and rats is independent of pituitary control, then the involvement of other suckling-related biochemical changes must be examined. Hypothalamic hormones such as corticotropic-releasing factor (CRF), thyroid hormone-releasing factor (TRF), and luteinizing hormone-releasing factor (LRF) are obvious candidates for study. At this point, however, it is equally likely that maternal aggression is nonhormonally mediated. Suckling stimulation is known to profoundly alter the turnover rates of hypothalamic monoamines (cf. Tindall, 1978), and specific agonists and antagonists of this system may be found in the future to alter maternal defense. Because experimental evidence strongly links serotonergic function to rodent sociosexual behavior (cf. Meyerson, Palis, & Sietnieks, 1979) and because turnover tates of serotonin (5 HT) dramatically increase with suckling stimulation (cf. Tindall, 1978), we recently have been exploring the involvement of this system in the maternal aggression displayed by lactating mice. Preliminary data (Mann & Svare, unpublished observations) indicate that a daily dose of 400 mg/kg of the serotonin antagonist D-L-parachlorophenylalanine (PCPA) significantly reduces the proportion of lactating R-S animals exhibiting aggression. PCPA in the above study was administered on six consecutive days during early lactation, with behavioral testing 90 minutes following the injection on Days 4, 5, and 6 of the treatment period. It is interesting to note that the postpartum administration of testosterone propionate (TP) and estradiol benzoate (EB), steroids known to reduce serotonergic activity, also reduce fighting behavior in lactating mice (Svare, 1980; Svare & Gandelman, 1975). While much remains to be done in order to explore systematically the postulated relationships between hypothalamic monoamine activity and postpartum aggression, the initial findings at least tentatively suggest involvement of the serotonergic system.

Finally, maternal aggression appears to be different from other agonistic behaviors with respect to the action of the frequently used antidepressant lithium chloride (LiCl). Brain and Al-Maliki (1979) found that LiCl treatment had no effect on postpartum aggression in TO albino mice but depressed intermale aggression. We are unaware of any other studies examining the influence of drugs on this form of aggression. Clearly this is an area that should receive a great deal of attention in the near future.

Maintenance of Maternal Aggression

While suckling stimulation is necessary for the initiation of maternal aggression, the continued presence of this stimulus apparently is not a prerequisite for the short-term maintenance of the behavior. For the parturient mouse, the placement of the dam's litter for five hours behind a double wire-mesh partition in the homecage maintains the behavior in a manner identical to that of mothers

TABLE II
Prolactin (PRL) Involvement in Maternal Aggression

Species	Manipulation	Result	Reference
Laboratory mouse *(mus musculus)*	Postpartum ergot drug treatment	PRL reduced, aggression unaltered	Mann, Michael, & Svare, 1980
Laboratory mouse *(mus musculus)*	Postpartum hypophysectomy	PRL reduced, aggression unaltered	Svare, Mann, Broida, & Michael, in press
Laboratory rat *(rattus norvegicus)*	Postpartum hypophysectomy	PRL reduced, aggression unaltered	Erskine, Barfield, & Goldman, 1980
Woodmouse *(peromyscus leucopus)*	PRL injections in adx-ovx virgins	PRL not measured, aggression elevated	Gleason, Michael, & Christian, 1981
Golden hamster *(mesocricetus auratus)*	Postpartum ergot drug treatment	PRL not measured, aggression reduced/reversed with PRL treatment	Wise & Pryor, 1977
Golden hamster *(mesocricetus auratus)*	Ectopic pituitary grafts in ovx virgins	PRL not measured, aggression elevated	Buntin, Catanzaro, & Lisk, 1981

in direct contact with their young. Removal of young from the homecage for five hours, but not one hour, eliminates the behavior (cf. Gandelman, 1972; Svare & Gandelman, 1973). Similar findings have been reported in the rat (Erskine *et al.,* 1978). It is not known what mechanism is responsible for the decline in aggression with the removal of young and its exteroceptive maintenance in the absence of any direct physical contact with pups. We speculated in earlier reports that the effects could be related to corticosterone and prolactin release (cf. Svare, 1977). In view of the fact that these hormones are not needed for the initiation of the behavior, it would seem unlikely that they are involved in its short-term maintenance. Perhaps, once again, this stage is modulated by neurotransmitter function.

Although the short-term maintenance of maternal aggression in mice is not mediated by suckling, the long-term maintenance of the behavior *is* influenced by this stimulus. The decline in aggression that occurs with advancing lactation is retarded when newborn pups are fostered on a daily basis to lactating mice beginning early in the postpartum period (Svare, 1977). Because young pups elicit high levels of nursing activity and older young stimulate comparatively lower levels of the behavior, it is reasonable to assume that the normal decline in aggression is related to a reduction in the quality or quantity of suckling and its consequent effect on the physiology of the mother. Obviously, the above hypothesis can be addressed more directly as soon as more is known about the biological basis of the initiation of the behavior.

GENETIC FACTORS

The use of inbred mice is a tactic that only recently has been adopted for research in maternal aggression. Lactating DBA/2 and BALB female mice are known to exhibit high levels of maternal aggression, while C57BL/6 and C3H mice rarely exhibit the behavior (St. John & Corning, 1973). Recently, we have been utilizing nonaggressive C57BL/6J mice and aggressive DBA/2J animals in order to explore the nature of strain differences in maternal defense. Our findings indicate that the strain differences in the behavior are not related to maternal environmental factors, since reciprocally crossed animals exhibit DBA/2J-like levels of high aggression and cross-fostered animals perform true to their genotype (Broida & Svare, 1982). These findings apparently suggest that inherent biological differences between the strains may be responsible for differences in aggressive behavior. The nature of these differences is unknown at the present time, but it is interesting to speculate that differences in neuroendocrine and/or serotonergic function may, in part, be responsible.

The use of selectively bred mice is another strategy that may eventually help us to understand the manner in which genes and hormones operate to alter maternal aggression. For example, in wild female *mus musculus* selectively bred for spontaneous interfemale aggressive behavior, the exhibition of postpartum aggression was also differentially affected by the selection process (Hyde & Sawyer, 1979). These findings therefore demonstrate that maternal defensive behavior is a correlated characteristic of spontaneous female aggression. As yet, physiological correlates such as neuroendocrine and neurotransmitter function have not been examined.

CONCLUSIONS AND FUTURE PERSPECTIVES

The findings reviewed here indicate that neuroendocrine factors modulate the appearance of maternal aggression in rodents. Unlike other aspects of agonistic behavior, like intermale aggression, the role of hormones appears to be more indirect and less dramatic. Because the study of this behavior is not as advanced as research on other forms of aggression reported in this volume, our conclusions must be highly tentative.

We can list at least three strategies for future research that should yield important information on the possible neuroendocrine control of this behavior. First, maternal aggression is exhibited by a wide variety of vertebrate species. Comparative studies are critically needed in order to establish the generalizability and evolutionary significance of the finding already generated in rodents. Second, the utilization of inbred strains of mice, with their natural genetic variation in neuroendocrine function, provides an especially effective tool for exploring the biological basis of this behavior. These strains should be routinely adopted for future studies in this area. Finally, more basic psychopharmacological work is needed in order to explore whether and to what extent neurotransmitters modulate this behavior. One hopes that the new questions raised by this review can be answered by utilization of the above tactics.

REFERENCES

Brain, P. F., & Al-Maliki, S. Effects of lithium chloride injections on rank-related fighting, maternal aggression, and locust-killing responses in naive and experienced TO strain mice. *Pharmacology, Biochemistry, and Behavior,* 1979, *10,* 663–669.

Broida, J., & Svare, B. Postpartum aggression in C57BL/6J and DBA/6J mice: Experiential and environmental influences. *Behavioral and Neural Biology,* 1982, *35,* 76–83.

Broida, J., Michael, S. D., & Svare, B. Plasma prolactin levels are not related to the initiation, maintenance, and decline of postpartum aggression in mice. *Behavioral and Neural Biology,* 1981, *32,* 121–125.

Buntin, J. D., Catanzaro, C., & Lisk, R. D. Facilitatory effects of pituitary transplants on intraspecific aggression in female hamsters. *Hormones and Behavior,* 1981, *15,* 214–225.

Erskine, M., Barfield, R. J., & Goldman, B. D. Intraspecific fighting during late pregnancy and lactation in rats and effects of litter removal. *Behavioral Biology,* 1978, *23,* 206–218.

Gandelman, R. Mice: Postpartum aggression elicited by the presence of an intruder. *Hormones and Behavior,* 1972, *3,* 23–28.

Gandelman, R., Simon, N. G. Postpartum fighting in the rat: Nipple development and the presence of young. *Behavioral and Neural Biology,* 1980, *28,* 350–360.

Gandelman, R., & Svare, B. Pregnancy termination, lactation and aggression. *Hormones and Behavior,* 1974, *5,* 397–405.

Gladue, B. A., & Clemens, L. G. Androgenic influences on feminine sexual behavior in male and female rats: Defeminization blocked by prenatal antiandrogen treatment. *Endocrinology,* 1978, *103,* 1702–1709.

Gleason, P. E., Michael, S. D., & Christian, J. J. Prolactin-induced aggression in female *peromyscus leucopus*. *Behavioral and Neural Biology,* 1981, *33,* 243–248.

Green, J. A. Experiential determinants of postpartum aggression in mice. *Journal of Comparative and Physiological Psychology,* 1978, *92,* 1179–1187.

Hedricks, C., & Daniels, C. E. Agonistic behavior between pregnant mice and male intruders. *Behavioral and Neural Biology,* 1981, *31,* 236–241.

Hyde, J. S., & Sawyer, T. F. Correlated response in selection for aggressiveness in female mice: II. Maternal aggression. *Behavior Genetics,* 1979, *9,* 571–577.

Leshner, A. I. Kinds of hormonal effects on behavior: A new view. *Neuroscience and Biobehavioral Reviews,* 1979, *3,* 69–73.

Mann, M., & Svare, B. Prenatal testosterone exposure elevates postpartum aggression in mice. *Physiology and Behavior,* in press.

Mann, M., Broida, J., Michael, S. D., & Svare, B. *Prolactin is not necessary for maternal aggression in mice.* Paper presented at the meeting of the International Society for Developmental Psychobiology, Cincinnati, November 1980.

Mann, M., Michael, S. D., & Svare, B. Ergot drugs suppress plasma prolactin and lactation but not aggression in parturient mice. *Hormones and Behavior,* 1980, *14,* 319–328.

Meyerson, B. J., Palis, A., & Sietnieks, A. Hormone-monoamine interaction and sexual behavior. In C. Beyer (ed.), *Endocrine control of sexual behavior.* New York: Raven Press, 1979.

Noirot, E., Goyens, J., & Buhot, M. Aggressive behavior of pregnant mice toward males. *Hormones and Behavior,* 1975, *6,* 9–17.

Politch, J. A., & Herrenkohl, L. R. Prenatal stress reduces maternal aggression by mice offspring. *Physiology and Behavior,* 1979, *23,* 415–418.

Price, E. O., & Belanger, P. L. Maternal behavior of wild and domestic stocks of Norway rats. *Behavioral Biology,* 1977, *20,* 60–69.

Reinisch, J. M., Simon, N. G., & Gandelman, R. Prenatal exposure to prednisone permanently alters fighting behavior of female mice. *Pharmacology, Biochemistry, and Behavior,* 1980, *12,* 213–216.

Rosenson, L. M., & Asheroff, A. K. Maternal aggression in CD-1 mice: Influence of the hormonal condition of the intruder. *Behavioral Biology,* 1975, *15,* 219–224.

St. John, R. S., & Corning, P. A. Maternal aggression in mice. *Behavioral Biology,* 1973, *9,* 635–639.

Sinha, Y. N., Selby, F. W., & Vanderlaan, W. P. Relationship of prolactin and growth hormone to mammary function during pregnancy and lactation in the C3H/ST mouse. *Journal of Endocrinology,* 1974, *61,* 219–229.

Svare, B. Maternal aggression in mice: Influence of the young. *Biobehavioral Reviews,* 1977, *1,* 151–164.

Svare, B. Testosterone propionate inhibits maternal aggression in mice. *Physiology and Behavior,* 1980, *24,* 435–439.

Svare, B. Maternal aggression in mammals. In D. J. Gubernick & P. Klopfer (Eds.), *Parental care in mammals.* New York: Plenum Press, 1981.

Svare, B., & Gandelman, R. Postpartum aggression in mice: Experiential and environmental factors. *Hormones and Behavior,* 1973, *4,* 323–334.

Svare, B., & Gandelman, R. Postpartum aggression in mice: Inhibitory effect of estrogen. *Physiology and Behavior,* 1975, *14,* 31–36.

Svare, B., & Gandelman, R. Postpartum aggression in mice: The influence of suckling stimulation. *Hormones and Behavior,* 1976, *7,* 407–416. (a)

Svare, B., & Gandelman, R. Suckling stimulation induces aggression in virgin female mice. *Nature,* 1976, *260,* 606–608. (b)

Svare, B., Mann, M., & Samuels, O. Mice: Suckling stimulation but not lactation important for maternal aggression. *Behavioral and Neural Biology,* 1980, *29,* 453–462.

Svare, B., Betteridge, C., Katz, D., & Samuels, O. Some situational and experiential determinants of maternal aggression in mice. *Physiology and Behavior,* 1981, *26,* 253–258.

Svare, B., Mann, M., Broida, J., & Michael, S. Maternal aggression exhibited by hypophysectomized parturient mice. *Hormones and Behavior,* in press.

Tindall, J. S. Neuroendocrine control of lactation. In B. L. Larson (Ed.), *Lactation: A comprehensive treatise* (Vol. IV). New York: Academic Press, 1978.

vom Saal, F. S., & Bronson, F. H. *In utero* proximity of female mouse foetuses to males: Effect on reproductive performance during later life. *Biology of Reproduction,* 1978, *19,* 842–853.

vom Saal, F. S., & Bronson, F. H. Sexual characteristics of adult females correlate with their blood testosterone levels during prenatal development in mice. *Science,* 1980, *208,* 597–599.

Wise, D. A. Aggression in the female golden hamster: Effects of reproductive state and social isolation. *Hormones and Behavior,* 1974, *5,* 235–250.

Wise, D. A., & Pryor, T. L. Effects of ergocornine and prolactin on aggression in the postpartum golden hamster. *Hormones and Behavior,* 1977, *8,* 30–39.

Zarrow, M. X., Gandelman, R., & Denenberg, V. H. Prolactin: Is it an essential hormone for maternal behavior in the mammal. *Hormones and Behavior,* 1971, *2,* 343–354.

5

Hormones and Infanticide

Ronald Gandelman

Jonathan Swift, in his essay "A Modest Proposal for Preventing the Children of Poor People in Ireland from Being a Burden to Their Parents or Country, and for Making Them Beneficial to the Public," satirically suggested that the young of the poor be eaten. "I have been assured," he wrote, "by a very knowing American of my acquaintance in London, that a young healthy child well nursed is at a year old a most delicious, nourishing, and wholesome food, whether stewed, roasted, baked, or boiled; and I make no doubt that it will equally serve in a fricassee or a ragout."

Swift, of course, chose his subject because something sufficiently shocking and revolting was required to make people consider and perhaps attempt to alter the economic structure of Ireland and hence the plight of the poor. It is safe to assume, I think, that infanticide and child abuse in general continue to produce shock and revulsion. In fact, they are largely viewed as being the most heinous of crimes. (Unfortunately, the rather lenient sentences often meted out by judges in our society somewhat belie this statement.)

It is interesting that infanticide is displayed not only by us but also by members of other species. The use of other species, therefore, permits us to study infanticide experimentally in an attempt to elucidate the mechanism(s) responsible for its occurrence. Two approaches to the experimental study of infanticide generally have been taken. In one, an attempt is made to eliminate the behavior in animals that exhibit it spontaneously. The second approach is the converse of the first: to establish infanticide in animals that normally do not display the behavior. Most researchers have opted for the latter approach, since it often is difficult to obtain a large enough sample of animals that kill their young spontaneously. In our Rockland-Swiss mice, for example, less than 5% of adult females kill young.

The purpose of this presentation is to discuss the possibility that the en-

---

RONALD GANDELMAN • Department of Psychology, Rutgers University, New Brunswick, New Jersey 08903. The writing of this paper was supported by Grant MH-28660 from the National Institute of Health, by Grant BNS-07347 from the National Science Foundation, and by a grant from the Charles and Johanna Busch Memorial Fund.

docrine system, especially gonadal hormones, is a prime or contributory factor in the mediation of infanticidal behavior. Before proceeding, however, it would be useful to discuss briefly the function or functions of infanticide.

FUNCTIONS OF INFANTICIDE

Population Regulation

Perhaps the most prevalent idea regarding the function of infanticide is that it is one of a number of population-regulating or -controlling mechanisms that are presumed to be inherent to many species. Breder and Coates (1932), for example, reported that infanticide among the fish *Lebistes reticularis Peters* is directly proportional to the concentration of the population. Similarly, Calhoun (1962) found that the killing of young increases markedly in rats when they are crowded. It thus would appear that infanticide can serve to reduce population size during periods of high population density. The benefit of this reduction to the population is obvious.

Reproductive Strategy

It has been proposed that infanticide is a reproductive strategy. In a fascinating account of infanticide among the Hanuman langur (Hrdy, 1977), it is stated that oftentimes a male, on coming into possession of a female, will kill its young. Hrdy writes, "According to the explanatory hypothesis offered here for langurs, infant-killing is a reproductive strategy whereby the usurping male increases his own reproductive success at the expense of the former leader (presumably the father of the infant killed), the mother, and the infant" (p. 48). This strategy, therefore, should be advantageous only to species such as the langur, in which the females resume ovulating soon after the young are removed. Also, infanticide as a reproductive strategy would be advantageous to the group if one assumes that the infanticidal-usurping male is stronger and healthier than the dead infants' father. The dissemination of the usurper's genes thus would benefit the group in the long run.

Optimization of Postnatal Environment

It has been shown that infanticide can be induced by increasing the size of the litter soon after parturition. Day and Galef (1977), for instance, observed that pup-killing in hamsters increased proportionately to the number of young added to an existing litter. We (Gandelman & Simon, 1978) gave newly parturient mice either 12 or 16 one-day-old pups. The dams killed young during Days 3 through 7 of the lactation period, killing more in the 16-pup than in the 12-pup litter. Also, they appeared to "select" the young to be killed on the basis of body weight, killing significantly more young that were in the bottom than in the top half of the weight distribution for their particular litter. A summary of the data is given in Table I. The fact that the number of surviving offspring

TABLE I
The Average Number of Surviving Young for Lactating Mice Given 12 or 16 Foster Young or Allowed to Keep Their Own Young[a]

Treatment condition	Average number of surviving young
Dam + 12 foster young	9.0
Dam + 16 foster young	9.0
Dam + own young[b]	8.0

[a] After Gandelman and Simon (1978).
[b] Range of pups in litter = 4 to 15.

was similar across a number of experimental conditions suggests that the killing behavior functions to reduce litters to a degree permitting all of the surviving young to be adequately nourished. Because mice have 10 nipples, 10 or fewer pups would be optimal. Day and Galef (1977), in fact, view infanticide as "a carefully organized and regulated aspect of maternal behavior in the golden hamster, which allows the parturient female to adjust the number of young which she must nurture" (p. 1188).

FEEDING BEHAVIOR

A final idea as to the function of infanticide concerns the possibility that the young, at times, are treated as a food source. It should be mentioned that at least for rodents, infanticide routinely is accompanied by the eating of the young. It therefore is possible that under certain circumstances, such as a limited food supply, infanticide will increase in frequency. We have found (unpublished observations) that infanticide and cannibalism can be elicited readily by challenging mice with a restricted feeding regimen. Killing and cannibalism cease to be displayed when the animals again are permitted unlimited access to food.

SUMMARY

It is apparent that infanticide can serve a number of functions. However, whether any of the environmental contingencies that provide the behavior with functional significance actually prompted the evolution of infanticide is unknown. Additional analyses of infanticidal behavior in populations of animals living in natural or seminatural environments must be undertaken.

HORMONES AND INFANTICIDE

Rosenberg and his associates have provided much of the data concerning infanticide in the laboratory rat, while we have obtained a good deal of information regarding infanticide in the mouse *(Mus musculus).* As will become apparent, while there is both convergence and divergence with regard to the nature of the role of the endocrine system, the data strongly suggest that hormones are involved in the display of infanticide.

Male–Female Differences

Rosenberg, Denenberg, Zarrow, and Frank (1971) stated that in addition to other behavioral differences between males and females, such as in activity, a difference exists in the propensity to kill young. Specifically, the behavior is much more prevalent in the adult male than in the adult female. We have found a similar dimorphism in mice. In one experiment (Gandelman, 1972), 50 male and 50 virgin female mice were confronted with one-day-old young. Forty percent of the males and only 10% of the females killed them. Thus data from rats and mice show a clear sex difference in the propensity to exhibit infanticide.

A likely explanation of any sexually dimorphic behavior involves a consideration of differential hormonal milieus; relative to females, males have high levels of androgen. It is possible, then, that androgen might mediate the sexual dimorphism in killing behavior.

Induction of Infanticide in Females by Testosterone Treatment

An obvious tactic in the verification of a putative hormone-mediated dimorphism in behavior is to administer the hormone in question to a sample of animals for which levels of that hormone normally are low. The prediction, of course, would be that the behavior will appear with greater frequency.

The testosterone-administration approach was taken by Rosenberg and Sherman (1974). Some pregnant rats were injected with testosterone propionate (TP) on Days 16–20 of gestation, while others received only the sesame oil vehicle. The female offspring subsequently were injected with TP on Postnatal Days 3,5,7, and 9. All animals were ovariectomized on Day 40 and implanted subcutaneously with a pellet of testosterone. A third group of females were never exposed to androgen of exogenous origin. The results were clear: Groups of females exposed to testosterone (T) pre- and postnatally or only postnatally contained significantly more killers than did the group of animals never having been exposed to the hormone. The percentages of killers were as follows: pre- and postnatal, 0.93; postnatal, 0.80; vehicle only, 0.15. These data indicate that *postnatal* exposure to T is sufficient to augment levels of pup-killing behavior.

An approach similar to that of Rosenberg and Sherman (1974) has been taken with mice (Gandelman, 1972). Two groups of one-day-old females were formed. One group of animals was injected with 500 μg TP and the other only with the vehicle. Ovariectomy was performed on Day 21, and on Day 60 the two groups were divided into two subgroups, one of which received 10 daily injections of TP and the other only the vehicle. A test for pup-killing was given on the next day. A significant increase in the proportion of animals that killed was observed only in those groups of mice that received the hormone during *adult* life. Neonatal TP exposure alone did not increase subsequent levels of killing.

Also, neonatal exposure in combination with adult TP treatment did not yield a significantly greater proportion of killers than did exposure to the steroid during only adult life.

Data from female rats and mice show clearly that exposure to T can stimulate infanticide. To be effective, such exposure must occur during adult life, presumably simultaneous with behavioral testing. In this regard, Rosenberg *et al.* (1971) reported that the administration of T only on Postnatal Day 2 was ineffective in stimulating pup-killing in female rats. Thus it appears that in female rats and mice, pup-killing is produced by the concurrent or activational property of T. An alteration ("organization") of the developing nervous system by androgen apparently is not involved in the establishment of T-induced infanticidal behavior.

Finally it should be mentioned that T does not produce a permanent change in the behavior of the adult toward young. We have found that pup-killing ceases to be displayed following the cessation of hormone treatment (unpublished observations).

Elimination of Infanticide in Males by Castration

Another method of demonstrating the involvement of a hormone in a given behavior is to eliminate the major source of that hormone, with the expectation that the behavior eventually will cease to be exhibited. Rosenberg *et al.* (1971) castrated male rats on Day 2, 10, or 15 of life. Others were given sham operations. Each animal was tested twice in adulthood for killing. There were no differences among the three castration groups with regard to the proportion of killing, and these data were combined. The percentage of gonadectomized males that killed pups on the first test was 8.3 and on the second, 11.1. The percentage of intact controls that killed was 61.7 and 55.6 on Tests 1 and 2, respectively. It also was reported that castration in adult life did not reduce pup-killing behavior of both experienced killers and naive males. It appears, then, that castration is effective in reducing infanticide in male rats only if it is performed during prepuberal development.

A discrepancy seems to exist in the rat regarding the role of androgen in the female and male. As discussed above, exposure to androgen during early development of the female does not influence its later behavior toward young. Also, T-activated pup-killing no longer will be exhibited following termination of hormone exposure. "Termination" of hormone exposure in the adult male by gonadectomy, however, does not influence killing, yet gonadectomy during early development is effective. This discrepancy between the effects of pre- and postpuberal T exposure and genetic sex may reside, in part, in the rather obvious differences in levels of endogenous T and those resulting from exogenous application of the hormone. It would be useful to administer T to prepuberal females in a manner approximating endogenous exposure in males.

Data derived from mice (Gandelman & vom Saal, 1975) are somewhat at variance with those from rats in that castration of males during neonatal *and* adult life reduced the later incidence of pup-killing. Specifically, 0% of male

TABLE II
Number of Male Mice of Various Ages That Killed, Ignored, or Cared for One-Day-Old Young[a]

Age	Number of animals (percentage)		
	Killed	Ignored	Cared for
22 days (n = 40)	0 (00)	21 (52)	19 (47)
32 days (n = 40)	13 (32)	6 (15)	21 (52)
42 days (n = 40)	15 (37)	0 (00)	25 (62)
52 days (n = 40)	11 (27)	9 (22)	20 (50)
62 days (n = 40)	16 (40)	4 (10)	20 (50)

[a] After Gandelman (1972).

mice gonadectomized at birth displayed infanticide as opposed to 35% of controls. Gonadectomy performed in adulthood significantly reduced killing provided that the animals had had no prior pup-killing experience. Therefore, castration blocks the initial appearance of the behavior but does not eliminate infanticide once it has been displayed. This suggests that experiential factors exert a modulatory influence with regard to the infanticide-activating property of T. (More will be said about experiential factors later.)

Other data provide additional evidence supportive of a role for androgen in pup-killing behavior of the male (Gandelman, 1973b). Male mice were tested for killing at 22, 32, 42, 52, and 62 days old. The data are shown in Table II. It is apparent that killing was not displayed until Day 32, a period of rapidly rising plasma T levels (Jean-Faucher, Berger, de Turekheim, & Veyssiere, 1978).

Although it appears that pup-killing in male rats and mice is influenced by androgen, it should be noted that differential T levels cannot be used to account for individual differences in behavior directed toward young. That is, there is no difference in plasma T levels between males that do and those that do not spontaneously kill young (Svare, Bartke, & Gandelman, 1977). Androgen apparently provides animals with a bias toward killing young but an additional factor or factors is responsible for eliciting the behavior from the individual animal.

ADMINISTRATION OF TESTOSTERONE TO NONKILLING MALES

Gandelman and vom Saal (1975) injected intact and gonadectomized adult male and female mice with T. As expected, T activated infanticide in the intact and gonadectomized female. However, the hormone did not augment killing in either the intact or gonadectomized male. These data suggested that exposure to relatively high levels of androgen during early development may *inhibit* the infanticide-activating property of T when the steroid is administered during adult life. This possibility was tested in a subsequent series of experiments (Gandelman & vom Saal, 1977). Male mice were gonadectomized on Day 0, 4, 10, or 60 of life. They were tested for infanticide on Day 90, and those that exhibited

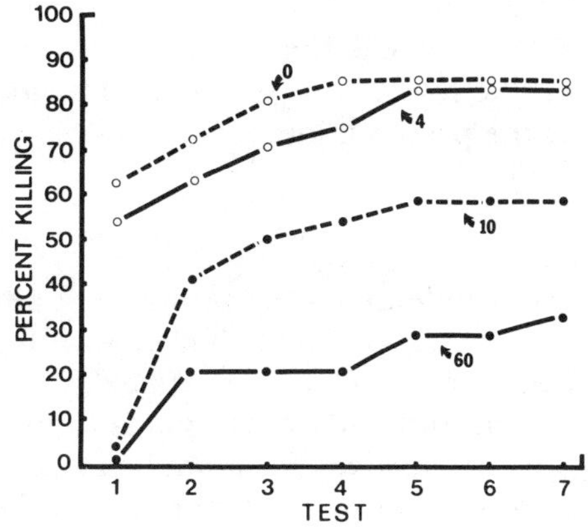

FIGURE 1. The cumulative percentage of adult male mice gonadectomized on Day 0, 4, 10, or 60 of life that killed pups in response to TP treatment later in life. Seven tests for pup-killing were given, each following five daily injections of 500 μg TP. (From "Exposure to Early Androgen Attenuates Androgen-Induced Pup-Killing in Male and Female Mice" by R. Gandelman and F. S. vom Saal, *Behavioral Biology*, 1977, *20*, 252–260. Copyright 1977 by *Behavioral Biology*. Reprinted by permission.)

the behavior were eliminated from the experiment. The remaining animals were injected daily with 500 μg TP and tested for pup-killing seven times, each test given following five injections. The data are shown in Figure 1. Testosterone treatment was highly effective in stimulating killing in males having been gonadectomized on Days 0 and 4. The hormone was moderately effective in activating infanticide from animals castrated on Day 10 and least effective for subjects gonadectomized in adulthood (Day 60). These data show that exposure to androgen of endogenous origin sometime beyond Day 4 of life reduces the effectiveness of androgen in stimulating or activating the killing of young.

The inhibitory influence of early exposure to androgen as seen in the male was supported by data derived from females. Female mice administered TP on Day 0, 4, or 8 required a significantly longer period of TP treatment in adult life to activate pup-killing than did animals injected with TP on Day 12 or 21.

The data suggest that early exposure of males and females to androgen inhibits later responsiveness to the infanticide-activating property of T. This finding is surprising for two reasons. First, it disagrees with a host of other data demonstrating that early androgen exposure facilitates later behavioral responsiveness to T, while the lack of such exposure leads to a lesser degree of responsiveness (cf. Beach, Noble, & Orndoff, 1969; Edwards, 1971; Gandelman, 1980). Second, since males always are exposed to relatively high levels of androgen during early development, why should any of them kill young? It is possible that an inverse relationship exists between prepubertal T levels and later killing behavior.

AROMATIZATION, GONADAL STEROIDS, AND INFANTICIDE

It is well known that many of the effects of T on physiology and behavior can also be obtained with estrogen (E) (cf. Doughty, Booth, McDonald, & Parrott, 1975; Edwards & Burge, 1971; Edwards & Herndon, 1970; Whalen & Nadler,

1963). Since T has been shown to be intraneuronally converted (aromatized) to E, it is possible that many of the effects ascribed to T actually are mediated by E. The role, if any, of E in infanticidal behavior will be considered in this section of the presentation.

Activation of Infanticide

Gandelman (1973a) injected adult virgin female mice that previously had displayed maternal behavior with 0.5, 5, 50, or 100 μg estradiol benzoate (EB). Injections were given for five days and pup-killing was assessed on the sixth day. The 50 and 100 μg dosages activated pup-killing whereas the other two dosages were ineffective. Estradiol benzoate, like T, produced a temporary effect, since maternal behavior rather than killing was displayed two weeks following the cessation of EB treatment. Since infanticide was not activated by progesterone treatment, the infanticide-activating property of gonadal steroids is specific to T and E.

The data described above and the fact that T is aromatized to E suggest that the steroid ultimately responsible for activating pup-killing is E. Svare (1979) tested this possibility. He reported, as expected, that both TP and EB activated pup-killing in ovariectomized virgin mice. However, it also was found that the nonaromatizable androgen dihydrotestosterone (DHT) was as potent as T in activating infanticide. Moreover, subthreshold dosages of EB and DHT (dosages that by themselves did not activate killing) synergized to activate infanticide. Svare concluded that the hormonal activation of pup-killing in mice depends on both the reduction of T to DHT and the aromatization of T to E.

Androgenic Inhibition of Infanticide

As discussed earlier, we found (Gandelman & vom Saal, 1977) that neonatal exposure to T inhibits the infanticide-activating property of T. Samuels, Jason, Mann, and Svare (1981) recently asked whether this inhibition is mediated by E derived from the aromatization process. They reported that fewer female mice having been exposed to TP or EB on the day of birth killed pups in response to TP treatment during adult life than did neonatally oil-exposed controls. Also, animals given early exposure to DHT or dihydrotestosterone propionate did not differ from controls. This suggests that E is responsible for the inhibition of T-activated pup-killing observed in animals having been exposed to relatively high levels of T during early development.

Samuels *et al.* (1981), in an attempt to substantiate the inhibitory role of E, injected newborn females with either TP alone or in combination with an anti-estrogenic compound (MER-25). Every animal, when adult, was exposed chronically to TP and tested for infanticide. The data supported an inhibitory role for E in that significantly more females administered TP in conjunction with the anti-estrogen killed young in response to adult T treatment than did animals given TP alone during neonatal life.

MODULATION OF INFANTICIDAL BEHAVIOR

This section considers factors that modulate the display of infanticidal behavior. As with a variety of behaviors that depend to some degree on a particular hormonal milieu, infanticide can be affected by a number of experiential and environmental factors. These will now be explored.

Prior Pup-Killing Experience

Rosenberg and Sherman (1975) gave 30-day-old castrated male rats injections of TP on alternate days until Day 60 and then daily to Day 90. Other animals received only the vehicle. On Day 90, half the animals of each group were allowed to kill young. (Testosterone-treated rats that failed to kill and oil-injected controls that did kill were eliminated from the experiment.) The treatment conditions then were reversed: Animals having received TP were given daily injections of oil and vice versa. The rats were tested again for killing on Day 120. The results showed that prior experience with young exerted a significant influence on the effectiveness of TP treatment. Experienced killers (animals having formerly received TP and having been permitted to kill) killed more when TP was discontinued than did inexperienced animals (TP-treated and not given killing experience). Also, experienced nonkillers (rats having formerly received oil and permitted exposure to young) killed less when switched to TP treated than did inexperienced controls (rats given oil but not exposed to young). Rosenberg and Sherman also reported that male rats having had killing experience evidenced no reduction in the incidence of the behavior following castration. This also has been reported for the mouse (Gandelman & vom Saal, 1975).

It thus appears that the behavior of the adult toward young exerts a long-lasting influence on pup-killing behavior. Animals having killed young will tend to continue to do so following discontinuation of exposure to relatively high levels of T as a result of castration or the cessation of hormone treatment. Conversely, animals not having killed young will tend not to do so following the imposition of T treatment. Experiential factors apparently can override the influence of the endocrine system on the induction of behavior. How experience exerts its rather marked effects is not known. An answer to the question would constitute a very important advance in our understanding of hormone–behavior interactions.

Ongoing Maternal Behavior

Fuller, Zarrow, Anderson, and Denenberg (1970) administered TP to pregnant rabbits. Hormone treatment on Days 17 through 29 of gestation severely disrupted maternal behavior: The animals failed to deliver in the nest box, scattered the young, did not nurse, and killed young. Killing of young was interpreted to be a behavior activated by T, while scattering and failure to nurse were assumed to result from T-induced deactivation of behavior. The fact that infanticide was produced in parturient animals by TP treatment prompted us to examine the effects of such treatment on lactating mice (Gandelman & Davis, 1973).

Primiparous mice were injected daily with TP throughout the course of the 21-day lactation period. Other animals received only the oil vehicle. Injections were continued for an additional 7 days following weaning. On the morning after the final injection, two one-day-old pups were presented to each female. The administration of TP during lactation neither resulted in infanticide nor disrupted any aspect of maternal behavior. However, when tested with one-day-olds one week after the weaning of their own young, 66% of them engaged in pup-killing. None of the oil-treated animals killed.

The inability of TP to activate killing when administered during the lactation period can be explained in two ways. First, a hormonal change (or changes) accompanying pregnancy, parturition, or lactation may inhibit the infanticide-activating property of TP. Second, TP may be incapable of disrupting ongoing maternal behavior. These possibilities were examined in a subsequent experiment by administering TP to virgin females that were housed continuously with pups that were exchanged each day for replete young. It should be noted that most adult virgin female mice are spontaneously maternal, exhibiting pup retrieval, licking the anogenital region, and assuming nursing postures over the young. Injections and pup presentation were performed concurrently for 10 days. Injections were continued for an additional 5 days, and on the morning following the last injection, each animal was presented with one-day-olds.

The results paralleled those derived from lactating mice in that pup-killing was not exhibited by animals while with young but was displayed by 50% of them when tested five days following removal of the continuously presented pups. One would conclude that the failure of TP to stimulate killing in the lactating animal is not due to physiological changes accompanying pregnancy, parturition, or lactation. Rather, TP apparently is incapable of activating pup-killing in animals while they are engaged in maternal activities. That Fuller *et al.* (1970) obtained pup-killing in the lactating rabbit probably resulted from the fact that TP was administered during pregnancy, prior to the onset of young-directed activities.

Sexual Experience

Vom Saal (1980) reported that prior sexual experience appears to inhibit pup-killing. Sexually experienced male mice were less likely to kill young than were sexually naive males. That this inhibition is caused by sexual experience rather than by cohabitation *per se* was demonstrated by sexually naive males that had cohabited with nonreceptive females displaying normal levels of infanticide. It also was reported that a small proportion of males that had previously killed pups killed young following engagement in sexual behavior as compared with sexually naive mice. This is a striking finding since it has been shown that castration of experienced pup-killers does not reduce subsequent killing (Gandelman & vom Saal, 1975; Rosenberg & Sherman, 1975). Therefore, a component of sexual behavior is a more potent inhibitor of infanticide in experienced killers than is removal of the major source of androgen.

AGE

Gandelman (1973c) presented one-day-old young to variously aged groups of female mice. The data are summarized in Table III. It is apparent that the dominant response of 22-day-olds is to ignore the young throughout the entire 15-minute session. Ignoring of the young declines until it practically disappears in 72-day-olds. Maternal responding shows an opposite profile, being displayed by few 22-day-olds and the majority of 72-day-olds. Pup-killing peaks in 32-day-olds and declines thereafter. The results of an additional experiment showed that in contrast to intact females, the proportion of ovariectomized mice that killed young did not decline following its marked rise from Day 22 to 32. This suggests that alterations in levels of ovarian hormones that accompany puberty are responsible for the decline in pup-killing behavior. The factor or factors responsible for the increase in killing from Day 22 to 32 is not known.

UTERINE POSITION

It has been reported recently that male mice having developed *in utero* contiguous to two male fetuses are less likely to kill pups than are males having resided adjacent to two female fetuses (vom Saal, 1980). Based on other data, it has been suggested that T produced by the fetal testes diffuses from the amniotic fluid surrounding the male across the amniotic membranes into the compartment of the proximate fetus (Clemens, 1974; Gandelman, vom Saal, & Reinisch, 1977). Levels of T in the amniotic fluid surrounding females contiguous to two males have been found to be elevated (vom Saal & Desjardins, 1979).

TABLE III
Percentage of Different Groups of Variously Aged Female Mice That Killed, Ignored, or Were Maternal toward One-Day-Old Young[a,b]

	Percentage		
Age	Kill	Ignore	Maternal
22 days	2	84	14
32 days	55	30	15
42 days	44	23	33
52 days	29	21	50
62 days	13	16	71
72 days	2	10	88

[a] For each age group, $n = 30$.
[b] After Gandelman (1973c).

Further examination of the possible interamniotic diffusion of hormones must be undertaken. However, if this diffusion of T does occur, vom Saal's data would agree with those showing that neonatal exposure to T reduces later pup-killing behavior (Gandelman & vom Saal, 1977; Samuels *et al.,* 1981).

CONCLUSION

Although the amount of information regarding the relationship between hormones and infanticide is small relative to that concerned with other aspects of aggression, I believe one can conclude that androgen plays a role in the expression of pup-killing behavior. However, since there is no relationship between the presence and absence of killing by an individual and that individual's plasma T levels (Svare *et al.,* 1977), other factors must determine whether killing actually will be exhibited. These extrahormonal factors probably include those described in the previous section and others as yet unspecified. It is apparent that additional work is required both to specify the additional modulatory influences and to determine whether hormones are involved in the infanticidal behavior of species other than rat and mouse.

Ultimately, I imagine, one would ask whether the endocrine system is involved in human child abuse. As one ascends the so-called phylogenetic scale and reaches the primates, one finds less reliance on hormones for the arousal and activation of behavior. Therefore, it would be unwise to suppose that the endocrine system possesses the same type and degree of control of behavior as it does in the rodent. Nevertheless, it is possible that a relationship of some type exists between the endocrine system and child abuse. In this regard it is interesting that, to my knowledge, data do not exist concerning assessment of physiological variables or an examination of retrospective health-related information from human individuals that abuse their young. Although it may seem unlikely that the behavior is in any way related to variables of an endocrine nature, one should bear in mind that the results of a number of experiments with humans tentatively suggest that exposure to certain hormones during early development affects subsequent personality development (cf. Meyer-Bahlberg & Ehrhardt, 1980).

REFERENCES

Beach, F. A., Noble, R. G., & Orndoff, R. K. Effects of perinatal androgen treatment on responses of male rats to gonadal hormones in adulthood. *Journal of Comparative and Physiological Psychology,* 1969, *68,* 490–497.

Breder, C. M., & Coates, C. W. A preliminary study of population stability and sex ratio of *Lebistes. Copeia,* 1932, *3,* 147–155.

Calhoun, J. Population density and social pathology. *Scientific American,* 1962, *206,* 139–148.

Clemens, L. G. Neurohormonal control of male sexual behavior. In W. Montagna & W. A. Sadler (Eds.), *Reproductive behavior.* New York: Plenum Press, 1974.

Day, C. S. D., & Galef, B. G. Pup cannibalism: One aspect of maternal behavior in golden hamsters. *Journal of Comparative and Physiological Psychology,* 1977, *91,* 1179–1189.

Doughty, C., Booth, J. E., McDonald, P. G., & Parrott, R. F. Effects of oestradiol-17, oestradiol benzoate, and the synthetic oestrogen RU-2858 on sexual differentiation in the neonatal female rat. *Journal of Endocrinology,* 1975, *67,* 419–424.

Edwards, D. A. Neonatal administration of androstenedione, testosterone, or testosterone propionate: Effects on ovulation, sexual receptivity, and aggressive behavior in female mice. *Physiology and Behavior,* 1971, *6,* 223–228.

Edwards, D. A., & Burge, K. G. Estrogenic arousal of aggressive behavior and masculine sexual behavior in male and female mice. *Hormones and Behavior,* 1971, *2,* 239–245.

Edwards, D. A., & Herndon, J. Neonatal estrogen stimulation and aggressive behavior in female mice. *Physiology and Behavior,* 1970, *5,* 993–995.

Fuller, G. B., Zarrow, M. X., Anderson, C. O., & Denenberg, V. H. Testosterone propionate during gestation in the rabbit: Effect on subsequent maternal behavior. *Journal of Reproduction and Fertility,* 1970, *23,* 285–290.

Gandelman, R. Induction of pup killing in female mice by androgenization. *Physiology and Behavior,* 1972, *9,* 101–102.

Gandelman, R. Maternal behavior in the mouse: Effect of estrogen and progesterone. *Physiology and Behavior,* 1973, *10,* 153–155. (a)

Gandelman, R. The development of cannibalism in male Rockland-Swiss mice and the influence of olfactory bulb removal. *Developmental Psychobiology,* 1973, *6,* 159–164. (b)

Gandelman, R. The ontogeny of maternal responsiveness in female Rockland-Swiss mice. *Hormones and Behavior,* 1973, *4,* 257–268. (c)

Gandelman, R. Gonadal hormones and the induction of intraspecific fighting in mice. *Neuroscience and Biobehavioral Reviews,* 1980, *4,* 133–140.

Gandelman, R., & Davis, P. G. Spontaneous and testosterone-induced pup killing in female Rockland-Swiss mice: The effect of lactation and the presence of young. *Developmental Psychobiology,* 1973, *6,* 251–257.

Gandelman, R., & Simon, N. G. Spontaneous pup-killing by mice in response to large litters. *Developmental Psychobiology,* 1978, *11,* 235–241.

Gandelman, R., & vom Saal, F. S. Pup-killing in mice: The effects of gonadectomy and testosterone administration. *Physiology and Behavior,* 1975, *15,* 647–651.

Gandelman, R., & vom Saal, F. S. Exposure to early androgen attenuates androgen-induced pup-killing in male and female mice. *Behavioral Biology,* 1977, *20,* 252–260.

Gandelman, R., vom Saal, F. S., & Reinisch, J. M. Contiguity to male foetuses affects morphology and behavior of female mice. *Nature,* 1977, *266,* 722–724.

Hrdy, S. B. Infanticide as a primate reproductive strategy. *American Scientist,* 1977, *65,* 38–47.

Jean-Faucher, Ch., Berger, M., de Turckheim, M., Veyssiere, G., & Jean, Cl. Developmental patterns of plasma and testicular testosterone in mice from birth to adulthood. *Acta Endocrinology,* 1978, *89,* 780–788.

Meyer-Bahlberg, H. F. L., & Ehrhardt, A. A. Neurobehavioral effects of prenatal origin: Sex hormones. In R. H. Schwarz & S. J. Yaffe (Eds.), *Drug and chemical risks to the fetus and newborn.* New York: Alan R. Liss, 1980.

Rosenberg, K. M., & Sherman, G. F. Testosterone induced pup-killing behavior in the ovariectomized female rat. *Physiology and Behavior,* 1974, *13,* 697–699.

Rosenberg, K. M., & Sherman, G. F. Influence of testosterone on pup killing in the rat is modified by prior experience. *Physiology and Behavior,* 1975, *15,* 669–672.

Rosenberg, K. M., Denenberg, V. H., Zarrow, M. X., & Frank, B. L. Effects of neonatal castration and testosterone on the rat's pup-killing behavior and activity. *Physiology and Behavior,* 1971, *7,* 363–368.

Samuels, O., Jason, G., Mann, M., & Svare, B. Pup-killing behavior in mice: Suppression by early androgen exposure. *Physiology and Behavior,* 1981, *26,* 473–477.

Svare, B. Steroidal influences on pup-killing behavior in mice. *Hormones and Behavior,* 1979, *13,* 153–164.

Svare, B., Bartke, A., & Gandelman, R. Individual differences in the maternal behavior of male mice: No evidence for a relationship to circulating testosterone levels. *Hormones and Behavior,* 1977, *8,* 372–376.

vom Saal, F. S. *The regulation of infanticide in male mice.* Paper presented at the meeting of the International Society for Developmental Psychobiology, Cincinnati, November 1980.

vom Saal, F. S., & Desjardins, C. Female sexuality: Prediction of fetal levels of testosterone in serum and amniotic fluid of mice. *Neuroscience Abstracts,* 1979, *5,* 183.

Whalen, R. E., & Nadler, R. D. Suppression of the development of female mating behavior by estrogen administered in infancy. *Science,* 1963, *140,* 273–274.

6

Hormonal Influences on Shock-Induced Fighting

Robert L. Conner, Anthony P. Constantino, and George C. Scheuch

While it is recognized that aggression is not a unitary concept, the general assumption is that data derived from studies using laboratory models of aggression among infrahuman species will ultimately contribute to our understanding of the causes of aggression and, potentially, to the prevention of aggression among humans. The extent to which laboratory findings may be generalized beyond the animal model depends, in part, on an appreciation of the adequacy of the animal model. Among several others, the shock-induced fighting paradigm is a commonly used laboratory model of aggressive behavior among infrahuman species. The present chapter will describe some of the strengths and weaknesses of this particular model.

Administration of trains of inescapable and unavoidable electric shock through the footpads of an individual rat results in marked skeletal-motor responses and various autonomic (Eichelman, Orenberg, Seagraves, & Barchas, 1976; Kaplan & Hobart, 1965), neurochemical (MaCarty & Kopin, 1978; Stolk, Conner, Levine, & Barchas, 1974; Thierry, Javoy, Glowinski, & Kety, 1968), and neuroendocrine responses (Fortier, deGroot, & Hartfield, 1959; Rossier, French, Rivier, Ling, Guillemin, & Bloom, 1977). Under specific conditions, organized patterns of behavior (e.g., sex: Barfield & Sachs, 1968; Caggiula, 1972; or consummatory behavior: Hutchinson & Emley, 1977) may also be facilitated by electric shock. When trains of footshocks are delivered to paired rats, however, a strikingly different pattern of responding is usually observed. Typically, one of the rats will initially adopt an upright posture raised up on its hind legs, then the other rat will assume a similar posture, and the animals will begin to make apparently directed slapping movements at one another with their forepaws and biting movements that appear to be directed primarily toward the head and nose region of each other. Among other labels, this stereotyped pattern of skeletal-motor

ROBERT L. CONNER, ANTHONY P. CONSTANTINO, AND GEORGE C. SCHEUCH • Department of Psychology, Bowling Green State University, Bowling Green, Ohio 43403.

responding has been referred to as "pain-elicited aggression" (Ulrich, 1966), "shock-elicited aggression" (Ulrich & Azrin, 1962), "shock-induced fighting" (e.g., Conner, Levine, Wertheim, & Cummer, 1969), or "irritable aggression" (Moyer, 1968). More recently, it has been suggested that the behavior might be described as "shock-induced defense" (Blanchard, Blanchard, & Takahashi, 1977). The issue of how its particular skeletal-motor pattern of responding in the presence of footshock might relate to other patterns of behavior labeled aggressive will be discussed later in this chapter. For the present, the descriptive label *shock-induced fighting,* or SIF, will be used.

Through the judicious choice of various parameters, the behavior of rats in the SIF situation may be brought under greater stimulus control than is usually the case with laboratory models of aggressive behavior, with the possible exception of centrally evoked (stimulus-bound) aggression. This may partly account for the wide use of the SIF paradigm in behavioral studies of brain lesion and/or drug effects. The possibility that behavioral effects may be mediated by drug- or lesion-induced hormonal changes should of course be kept in mind. As implied by the title, however, the present chapter will focus specifically on hormonal influences on SIF. Following a discussion of the SIF phenomenon (following section), the literature relating hormones to SIF will be reviewed (second section). Finally, the question of interpretation will be addressed (third section).

THE SIF PHENOMENON

The SIF phenomenon was initially reported by O'Kelly and Steckle in 1939, and subsequently by others (e.g., Daniel, 1943). The phenomenon was described in some detail during the 1960s through a series of systematic studies reported by Azrin and Ulrich and their colleagues. Three review articles (Hutchinson, 1972; Ulrich, 1966, 1967) summarize the effects of a number of variables on the behavior.

The Behavior

Despite some variation among different investigators in the definition of a fight response, reported interobserver reliabilities in each laboratory are typically quite high (Duncan & Powell, 1975; Knutson & Hynan, 1972; Powell, Holley, & Buchanan, 1976; Ulrich & Azrin, 1962). Following Ulrich and Azrin (1962), a fight response is typically scored when either animal, while standing on its hind legs in the stereotyped "boxing" posture, makes a striking motion with the fore-paws or a biting movement. Some investigators have required that both animals rear up and make physical contact by striking or biting movements before scoring a fight response (e.g., Hutzell & Knutson, 1972). Other investigators have routinely reported frequencies of both position responses (scored on each shock trial when both animals are in the stereotyped boxing posture but physical contact does not occur) and fight reponses (e.g., Conner *et al.,* 1969; Erskine, Stern, & Levine, 1975). A few investigators have reported frequencies of even more discrete behaviors, including bites and vocalizations as well as the latencies and durations of these and other specific responses (e.g., Crabtree & Moyer, 1973; Duncan & Powell, 1975; Powell *et al.,* 1976; Sbordone, Wingard, Elliott, & Jervey, 1978). Because the correlations between different responses seen in the SIF situation vary over a broad range (Crabtree & Moyer, 1973), comparisons between studies would be facilitated if future investigators reported as many discrete dependent measures as can be reliably differentiated. In most studies, the mean frequency of fighting (or mean probability of fighting, or percentage fighting) within each group constitutes the primary and often the only data reported in any detail. Consequently, in the present review it will usually be possible to discuss hormonal influences on SIF behavior only as revealed by that measure.

In most reports, pairs of similarly treated animals are observed and data from individual animals are not provided. In some cases, however, experimentally treated animals have been paired with untreated stimulus animals (e.g., Rodgers, 1979). In other cases, results of observer judgments of the dominance or submissiveness of individual animals have also been reported (e.g., Ader, 1975). Dominance–submission information is not usually reported for rat subjects in the SIF situation, probably because of the difficulty in developing unambiguous criteria for making such ratings.

In 1964, Azrin, Hutchinson, and Sallery reported that a single semirestrained squirrel monkey would grasp a ball and sink its teeth into it when a train of electric shocks was delivered through tail electrodes. A severe tail-pinch was also found to evoke such biting behavior (Azrin, Hake, & Hutchinson, 1965). Automated methods of recording bites of inanimate targets during administration of tail-shock to individual animals were subsequently described for both monkey (Hutchinson, Azrin, & Hake, 1966) and rat (Azrin, Rubin, & Hutchinson, 1968). It has been argued (Azrin *et al.,* 1968) that this single animal shock-induced biting (SIB) paradigm has the advantage of permitting even greater stimulus control of the behavior than the paired animal SIF paradigm. This line of research has been extended to human subjects with whom measurement of jaw clenching has been proposed as a useful index of human anger (Hutchinson, Pierce, Emley, Proni, & Sauer, 1977).

While the possible role of hormones in the expression of SIB has not been systematically investigated, they would appear to play little or no role. Hutzell and Knutson (1972) studied the effects of sex and rearing conditions in the individual-animal SIB and the paired-animal SIF situations. Whereas both sex and rearing conditions were potent factors in determining the frequency of fights among rats in the SIF situation, neither of these variables had any effect on the frequency of biting in the SIB situation. Since Hutzell and Knutson did not separately score biting behavior in the SIF situation, a direct comparison of biting across these two situations is not possible. Nonetheless, their results show that some variables that affect SIF do not necessarily affect SIB in the same way. Other evidence indicates that different variables, including the effects of repeated sessions (Azrin *et al.,* 1968; Hutzell & Knutson, 1972) and parameters of electric shock (Hutchinson, Azrin, & Renfrew, 1968) do produce similar effects on both SIF and SIB. But since there is no evidence implicating hormonal variables in SIB, that phenomenon will not be further discussed in the present chapter.

More recently, a single-animal procedure that is similar to the SIB paradigm but does not involve shock has been proposed as another useful animal model of biting attack behavior. Wagner, Beuving, and Hutchinson (1979) reported that male mice of the Upjohn CF-1 strain would frequently bite an inanimate target positioned in front of the mouth when they were restrained in a small tube-like apparatus. Male mice show such target-biting behavior much more then females do (Wagner, Beuving, & Hutchinson, 1980). Further, biting frequency by males decreased following castration and was reinstated by injection of testosterone. Neither ovariectomy, estrogen injections, nor testosterone injections significantly influenced the frequency of target-biting behavior in female mice (Wagner *et al.,* 1980). These findings would suggest that this paradigm might be useful for further analyses of hormonal influences on biting behavior.

Factors Influencing SIF

Before considering hormonal influences, the numerous factors shown to influence SIF will be briefly reviewed.

Species

Rats have been studied most often in investigations using the SIF paradigm, but the phenomenon of fighting in response to electric shock appears to have some species generality. It has been observed in mice (Gregor, Smith, Simons, & Parker, 1972; Tedeschi, Tedeschi, Mucha, Cook, Mattis, & Fellows, 1959), hamsters (Ulrich & Azrin, 1962), cats (Ulrich, Wolff, & Azrin, 1964), ground squirrels (Turner, Boice, & Powers, 1973), gerbils (Boice & Pickering, 1973), and squirrel monkeys (Azrin, Hutchinson, & Hake, 1963). Unlike rats, gerbils and squirrel monkeys may persist in fighting for up to one hour after shock termination (Azrin *et al.,* 1963; Dunstone, Cannon, Chickson, & Burns, 1972). During presentations of electric shock, gerbils exhibit a considerable amount of striking and biting without adopting the stereotyped upright posture (Dunstone *et al.,* 1972). This behavioral pattern has also been seen in mice (Levine, Hensen, & Conner, unpublished observation) but only rarely in rats. Also unlike rats, clear dominance–submission relationships have been reported to emerge during SIF between pairs of squirrel monkeys (Azrin, Hutchinson, & Sallery, 1964), gerbils (Boice & Pickering, 1973), and ground squirrels (Turner *et al.,* 1973). Fighting in various species can also be induced by delivering shock through the tail (mice: Puglisi-Allegra & Renzi, 1977), through the tail and one leg (turtles: Fraser & Spigel, 1971), or through subdermal electrodes (rats: Duncan & Powell, 1975). Birds have not been observed to fight in response to shock delivered either to the base of the wings (Plymouth Rock hens: Bracco & Miller, 1972) or through wires implanted around the pubis bone (White Carneaux pigeons: Rashotte, Dove, & Looney, 1974), nor do guinea pigs fight in response to footshock

(Ulrich & Azrin, 1962). Results from several studies have been interpreted as supportive of the view that "pain-elicited aggression" also occurs in human subjects (Heacock, Thurber, & Vale, 1975; Ulrich, 1966).

Parameters of the Electric Shock

The effects of varying shock intensity, duration, and rate of presentation have been examined in some detail. A train of shocks of 2.0 mA intensity, each of 0.5 sec duration, delivered at the rate of approximately 20 per minute, appears to be optimal for the continued elicitation of SIF in Sprague-Dawley rats (Azrin, Ulrich, Hutchinson, & Norman, 1964; Ulrich & Azrin, 1962). Somewhat different parameters may be optimal for different rat strains (Ulrich, 1966) or different species (Azrin *et al.,* 1963).

Effects of Shock Experience

The frequency and duration of SIF responses generally increase both within and across sessions (Ulrich & Craine, 1964). On the other hand, prior exposure of individual animals to inescapable and unavoidable footshock of high intensity ("preshock") produces a decrement in the frequency of subsequent SIF responses (Anderson, Crowell, Lupo, & Sarmina, 1977). Prior escape or avoidance training also suppresses subsequent levels of SIF (e.g., Powell, Francis, Francis, & Schneiderman, 1972). Not surprisingly, punishment of fight responses suppresses fighting (Follick & Knutson, 1978). The SIF response can allegedly be classically conditioned (Farris, Gideon, & Ulrich, 1970), but numerous conditioning trials are required and other possible interpretations of the reported conditioning evidence (e.g., sensitization) have not been unequivocally eliminated. If given the opportunity, animals will escape rather than engage in SIF behavior (Azrin, Hutchinson, & Hake, 1967). Under certain conditions, the opportunity to engage in SIF behavior can be used to reinforce other behavior (Dreyer & Church, 1970).

Effects of Social Experience

The available data describing the role of the prior social history of rats and mice on SIF behavior are inconsistent. It has been reported (Hutchinson, Ulrich, & Azrin, 1965; Hutzell & Knutson, 1972) that socially isolated rats exhibit less SIF than do communally housed subjects. Other investigators (Creer, 1974, 1975; Thor & Ghiselli, 1974) have reported just the opposite, with communally housed rats showing higher levels of SIF, and socially isolated subjects showing more stable levels of SIF after fewer test sessions (Creer & Powell, 1971). These inconsistent findings may be a function of the age at which the subjects were isolated, with Thor and Ghiselli (1974) and Creer (1974, 1975) isolating rats at 45 and 60 days of age, respectively, while Hutchinson *et al.* (1965) and Hutzell and Knutson (1972) isolated their subjects at weaning. It is possible that subjects not isolated until 45 or 60 days of age had previously formed dominance–submission relationships in the communal housing situation. Such social relationships have been shown to increase SIF in mice (Kimbrell & Chesler, 1971) and to affect certain components of the fight sequence differentially (Legrand & Fielder, 1973). On the other hand, Ader (1975) has shown that rats raised from 21–92 days of age in a noncompetitive environment showed more SIF than rats reared in a competitive environment and suggested that this result was obtained because rearing in a competitive environment promoted the emergence of a stable dominance order.

Other Factors

Other factors increasing levels of SIF in rats include food or water deprivation (Cahoon, Crosby, Dunn, Herrin, Hill, & McGinnis, 1971), rapid-eye-movement sleep deprivation (Mollenhour, Voorhees, & Davis, 1977), and environmental temperature (Berry & Jack, 1971). Factors decreasing SIF in rats include maintenance in a persistent light environment (Kane & Knutson, 1976), ablation of the vibrissae (Thor, 1976a), or anesthetization of the vibrissal pads (Ghiselli & Thor, 1974). Various pharmacological agents produce either a decrease (Jones, Barchas, & Eichelman, 1976), increase (Eichelman, Orenberg, Hackley, & Barchas, 1978), or no change in SIF behavior. Similarly, rats with brain lesions may exhibit more (Jacobs & Cohen, 1976) or less (Miczek, Brykczynski, & Grossman, 1974) SIF than control animals, depending on the lesion site. (See elsewhere in this volume for a more extensive discussion of drug and lesion effects on various models of aggressive behavior.) Finally, the size of the testing cage has a very potent effect on SIF (Thor, 1976b; Ulrich, 1966). The frequency of SIF behavior between paired animals decreases dramatically as the size of the testing chamber is increased.

HORMONAL INFLUENCES ON SIF

Ideally, the literature to be reviewed in this section would be organized around the questions of which hormones are involved in the expression of SIF, where they act, and how they produce their effects. Unfortunately, currently available data do not justify such an organization. There is still considerable controversy about the question of which hormones are involved in aggressive behavior, but it is becoming increasingly clear that different hormones may be implicated in the expression of different kinds of aggressive behavior among various species (cf. Moyer, 1968, 1976). With respect to SIF in particular, there are hardly any data specifically bearing on the questions of where in the brain or by which mechanisms hormones influence the behavior. Because hormonal influences have more often been studied using the isolation-induced fighting (IIF) paradigm with mice, specific topics will often be introduced and placed in broad context by a very brief summary of documented hormonal influences on IIF in mice.

Pituitary-Gonadal System

There are several lines of evidence bearing on the possible role that hormones of the pituitary-gonadal system play in SIF. Most of the available literature dealing with hormonal influences explicitly or implicitly deals with the role of androgens in the SIF of rats. While several reports have implicated gonadotropins in other models of aggressive behavior (e.g., stimulus-bound attack in cats: Inselman-Temkin & Flynn, 1973; dominance in starlings: Mathewson, 1961), there are no reported effects of gonadotropic hormones on SIF behavior.

Sex Differences

The fact that the males of most species generally exhibit more aggressive behavior than the females is often taken as presumptive evidence for the importance of androgens in aggressive behavior. Without additional support, this evidence would not by itself be very compelling. In fact, contradictory evidence exists. In the first place, the females of some species are sometimes more aggressive than the males (e.g., Golden hamsters; Kislak & Beach, 1955). Second, the data most often cited in support of the statement that males are more aggressive than females are obtained from mice. In fact, female mice of some strains readily exhibit aggressive behavior (Ginsburg & Allee, 1942; Gleason, Michael, & Christian, 1979; Hyde & Sawyer, 1977; Weltman, Sackler, Schwartz, & Owens, 1968), and female mice vigorously protect their young (Gandelman, 1972). Third, the behavioral testing procedure that is employed can determine the magnitude of observed sex differences in the aggressive behavior of mice (Edwards, 1969). Finally, males are generally more responsive to a given dose of exogenously administered testosterone than females. Nevertheless, if large doses of testosterone propionate are repeatedly administered, normally nonaggressive female mice will vigorously attack intruders (Svare, Davis, & Gandelman, 1974). Thus, the sex difference encompasses not only a behavioral difference but also a difference in sensitivity to exogenously administered hormones. Even though females normally produce less androgen than males, the above observations indicate that it is unlikely that differences between the sexes in aggressive behavior can be completely accounted for on the basis of differences in androgen production.

With respect to sex differences in SIF, only limited data are available because most investigators have used only male subjects. Turning first to sex differences among species other than rats, there is a report that male gerbils show higher levels of SIF than females (Boice & Pickering, 1973), but male and female deermice do not differ (Gregor *et al.*, 1972). While there are marked strain differences among mice (Southwick & Clark, 1968), male mice of many strains exhibit higher levels of IIF than females. Under the typically employed conditions in the IIF paradigm, female mice of most commonly used strains rarely fight. In the SIF paradigm, however, while there may be a sex difference, it does not appear to be as great as is often seen in the IIF paradigm. For example, in unpublished research, Levine, Hensen, and Conner observed that the males of the BALB/c and CBA strains showed much higher levels of SIF than the females, but the females did show some fighting in response to electric footshock.

In studies comparing male and female rats in SIF, every possible result has been reported. Some studies have reported no difference (Powell, Silverman, Francis, & Schneiderman, 1970; Powell, Francis, & Schneiderman, 1971; Ulrich & Azrin, 1962); others have found that males show higher levels and different patterns of SIF than females (Anderson *et al.*, 1977; Conner *et al.*, 1969). In one study, females were observed to fight more than males (Creer & Powell, 1971), and in at least one report, the sex of the rats studied was not stated at all (Roediger & Stevens, 1970). While there does not appear to be any large or consistent difference in levels of SIF between hooded and albino rats (Conner *et al.*, 1969; Milligan, Powell, & Borasio, 1973), Wistar rats have been observed to fight less than rats of others strains (Ulrich & Azrin, 1962), and in at least one study, Long-Evans males were reported to show more biting and striking behavior than Sprague-Dawley males (Duncan & Powell, 1975). Nonetheless, strain does appear to be an important factor for the demonstration of sex differences in SIF behavior. A sex difference has usually been found in hooded rats, with males showing more SIF than females (Conner *et al.*, Hutzell & Knutson, 1972; Knutson & Hynan, 1972, 1973; Milligan *et al.*, 1973; Thor, Ghiselli, & Ward, 1974). A sex difference in fighting responses was not found in a study where Sprague-Dawley × Long-Evans subjects were used, but males did show more stereotyped posture responses (with no physical contact occurring) than females (Erskine *et al.*, 1975). Further evidence suggests that differences between Sprague-Dawley males and females might be more readily seen at a moderately high shock intensity (2.0 to 2.5 mA) than at lower intensities (Conner *et al.*, 1969; Milligan *et al.*, 1973). Still other evidence indicates an interaction between sex and repeated testing sessions. In one study (Thor *et al.*, 1974), the sex difference in Long-Evans rats was very small or not apparent at all during the first testing session. But during further testing, Long-Evans males markedly increased fighting while levels in the females tended to remain low. It is concluded, therefore, that while it may not be large and it may appear only under specific conditions in certain strains, there does appear to be a modest sex difference in SIF levels of rats. As noted, while this conclusion is consistent with the hypotheses that sex hormones are involved in this difference, the mere existence of a sex difference cannot be taken as strong evidence for the involvement of pituitary-gonadal hormones in the expression of this behavior.

Developmental Changes

If hormones of the pituitary-gonadal system are involved in the expression of aggressive behavior, one might expect to find a correlation between the functional maturation of the pituitary-gonadal system and the appearance of aggressive behavior. Several observations made on developing animals suggest such a relationship. Increased levels of aggressive behavior among the males of some species (e.g., mice: Fredericson, 1950; Levy & King, 1953; and monkeys: Goy & Goldfoot, 1974) are roughly correlated with increases in circulating levels of testosterone during development (Barkley & Goldman, 1977; Resko, 1967; Selmanoff, Goldman, & Ginsburg, 1977). Further, administration of androgens to prepuberal male mice may accelerate the initial appearance of IIF behavior (Levy & King, 1953). That such treatment is without effect in female mice (Levy, 1954) indicates a differential sensitivity to exogenously administered testosterone in prepuberal male and female mice.

Several studies have assessed SIF in male rats as a function of age. Hutchinson *et al.* (1965) found that levels of SIF behavior among independent pairs of male rats increased over an age range from 24 to 93 days of age. Unlike the situation with IIF in mice, however, they found that SIF occurred less frequently in isolated rats than those raised in community conditions, although the isolated rats also exhibited substantial levels of SIF. Hutchinson *et al.* (1965) interpreted the finding that SIF occurs in socially isolated rats as support for the view that SIF is to a considerable extent an unlearned reaction, the expression of which is mediated by age-related changes in androgen activity. Powell and Creer (1969) have also observed that SIF increases as a function of age. However, because they found that age interacted with prior SIF experience in determining levels of SIF, they have suggested that while the expression of SIF is related to physiological development, it may also be influenced by learning and experience and is not therefore purely reflexive. In a subsequent study, Creer and Powell (1971) did not find age to be an important determinant of SIF among 30-, 60-, and 90-day-old rats, possibly because, according to those investigators, any effects of age were confounded by the variability produced by paired or isolation housing conditions. Sbordone *et al.* (1978) studied SIF of male rats in three age groups (40–50, 90–110, and 180–200 days old) and three strains (Sprague-Dawley, Wistar, and Long-Evans). They found that fighting frequency increased as a function of age and that animals in the two younger age groups exhibited no bites and shorter durations of fighting than the animals in the oldest age group. Thus, there is some evidence for a relationship between age and SIF among male rats, but because this relationship may be modified by experiental factors (e.g., prior shock experience), it is not altogether clear that it is necessarily linked to androgen activity.

Effects of Castration

In models of aggressive behavior other than SIF, castration experiments on males have often provided strong evidence for a relation between androgens and aggressive behavior. It may be noted, however, that male castration rarely produces an abrupt cessation of sex behavior. Castration effects on sex behavior may appear only gradually or become apparent only after careful analysis. For example, coital behavior in dogs (Beach, 1970) and cats (Rosenblatt & Aronson, 1958) may persist for prolonged periods following castration, but the frequency of intromission and the length of time that erection is maintained both decrease within six months after castration (Beach, 1970). Similarly, male mice have been observed to continue to show some fighting following castration, and the behavior may persist for a long time in some animals (Bevan, Bevan, & Williams, 1958;

Uhrich, 1938). Castration effects on aggressive behavior among rhesus monkeys have been described by Wilson and Vessey (1968). They reported a gradual drop over a period of several years in the dominance rank of six out of ten castrates. In gerbils, castration has been reported to increase intermale aggression (Anisko, Christenson, & Buehler, 1973). But typically, the levels of intermale aggressive behavior among male rodents are observed to decrease following castration during adulthood. Learning and past experience might, in part, account for the gradual rather than abrupt decline in aggressive behavior following castration.

Because the initial emergence of aggressive behavior in mice roughly coincides with the increase in circulating testosterone levels at puberty, it might be expected that aggressive behavior would never appear in prepuberally castrated mice. Data reported by Heilman, Brugmans, and Strainer (1977) however, indicate that the development of aggressive behavior in prepuberally castrated mice is only delayed in onset. After 23 weeks in isolation, prepuberally castrated male mice of the CF-1 strain showed attack behavior equivalent to that exhibited by intact control mice after only 3 weeks of isolation. At least two points can be raised on the basis of these findings. First, gonadal androgens are not a necessary condition for the appearance of IIF in mice, though the possibility has not been eliminated that in their absence adrenal androgens may be required. Second, in studies of mouse aggression, isolation is a frequently employed procedure that itself needs further study. Isolation is known to produce a variety of effects on brain biogenic amines in mice (Welch & Welch, 1971), but it is not clear how these isolation-induced physiological changes are related to isolation-induced aggression, if at all (cf. Simon, 1979).

To the extent that SIF is androgen dependent, it should decrease following castration. SIF has been studied following castration of adult, prepuberal, or neonatal male rats. When adult rats are castrated, they do not immediately show lower levels of SIF than do intact controls. Rather, castrates tend to continue to exhibit levels of SIF comparable to intact animals for three to four weeks following castration (Bernard and Paolino, 1975; Conner *et al.*, 1969; Hutchinson *et al.*, 1965). Beyond this time, however, intact rats tend to show an increase in SIF as a function of repeated testing and/or age, while castrates show either continued stable levels (Bernard & Paolino, 1975) or a gradual decline (Hutchinson *et al.*, 1965) in levels of SIF. It appears therefore, that rather than abolishing or even drastically reducing the behavior, the most prominent effect of adult castration of rats is to prevent the increase in SIF that is typically seen as a function of repeated testing and/or age in intact rats. It is important to note that changes in SIF behavior following castration of rats cannot be accounted for on the basis of any changes in sensitivity to footshock (Bernard & Paolino, 1975).

Rats castrated prepuberally and tested for SIF in adulthood show either lower levels of fighting (Conner *et al.*, 1969; Hutchinson *et al.*, 1965) or a slower rate of increase in SIF over repeated sessions (Milligan *et al.*, 1973) than do intact animals. Knutson, Hynan, and Kane (1976) have reported that prepuberally castrated rats maintained on a 12–12 light–dark regime show lower levels of SIF than intact animals, but this difference is not seen if the animals are maintained under persistent light conditions, possibly because persistent light suppresses levels of SIF. Thus, the limited data avaliable suggest that prepuberal castration more effectively decreases levels of SIF than adult castration, either because of the early age of castration or because of the long interval between castration and adult testing.

Studies employing the neonatal castration procedure have usually been conducted in the context of hypotheses concerning the organizational role of androgens in sexual differentiation. There have been many studies of the effects of neonatal castration of males and perinatal androgen treatment of females on adult patterns of reproductive behavior (cf. Goy & McEwen, 1980), and some studies describe the effects of these manipulations on nonreproductive behaviors (Beatty, 1979), including various kinds of aggression (see elsewhere in this volume). Again, there is only limited information about the possible effects of such perinatal manipulations of hormone levels on adult patterns of SIF behavior. Baenninger (1974) paired Long-Evans males castrated on the first day of life with intact males in the SIF situation when they were 177 days of age. He reported that the neonatally castrated males fought with the intact animals, exhibiting both posturing and biting behaviors, but they were generally subordinate to the intacts. Almost all of the castrated animals finally rolled on their backs and the intact opponents placed their forepaws on the stomach of the castrates. Unfortunately, it cannot be determined if this result was uniquely due to neonatal castration as opposed to castration at some later age. In the only relevant study, SIF behavior was studied during adulthood in four groups of rats: neonatally castrated males, weanling-age castrated males, sham-operated males, and untreated females (Conner *et al.*, 1969). Conner *et al.* reported that both the position and fight frequencies among the neonatally castrated males were very similar to those shown by the females. Weanling-age castrated males, on the other hand, showed higher levels of both position and fight responses than the animals in the neonatally castrated male or intact female groups but lower levels than intact males. These findings were interpreted as additional support for the view that androgen plays an organizational role during the neonatal period in the rat.

Effects of Androgen Treatment

The accumulated evidence from castrated infrahuman subjects indicates that androgen replacement by testosterone injections is effective in restoring intermale aggressive behavior to precastration levels. More recent information about the biosynthetic capacities of various androgen target tissues, including brain, indicates that in specific tissues, testosterone (or androstenedione) may serve as a substrate (or prohormone) for other biologically active steroids (e.g., dihydrotestosterone or 17β-estradiol). It has been suggested that such metabolites of testosterone, rather than testosterone itself, might mediate some of the reported effects of testosterone. Along the lines of this hypothesis (Naftalin, Ryan, & Petro, 1971), Gandelman (1980) has discussed the evidence that implicates estrogens in the expression of aggressive behavior by male mice. There is no evidence that estrogens exert any influence on behavior in the SIF situation.

The effects of administration of testosterone on SIF have been examined in intact, adult castrated, prepuberally castrated, and neonatally castrated animals. Although the administration of testosterone to intact immature male mice accelerates the onset of IIF behavior (Levy & King, 1953), suggesting that levels of circulating androgen may underlie the expression of that type of aggressive behavior, such an effect has not been seen for SIF behavior. Powell *et al.* (1971) administered daily testosterone propionate injections (1 mg) to male Sprague-Dawley rats, beginning at 17 days of age, but did not observe any effect on SIF behavior at 30 days of age, even though descent of the testes was accelerated. Conner *et al.* (1969) failed to observe any effect of testosterone propionate administered during adulthood on the SIF behavior of intact or ovariectomized female rats.

The most compelling evidence that SIF is androgen-dependent would be provided by the combined demonstration of a decrement in the behavior following castration and a reversal of that decrement by the administration of an androgen. However, since castration of adult male rats does not typically produce a dramatic decrement in the behavior, it is not surprising that there are no reported androgen effects on the SIF behavior of adult castrated rats. As we have noted, lowered levels of SIF behavior have been seen in prepuberally castrated rats, and androgen replacement effects have been examined in prepuberally castrated rats in two studies. Milligan *et al.* (1973) administered 2 mg of testosterone propionate daily for 10 days to prepuberally castrated male rats with prior SIF experience before beginning weekly tests of SIF. They found that this regimen of testosterone produced a slight increment over pre-injection baseline levels of fighting in the castrates. After the third weekly session, during which daily testosterone injections were continued, there was no difference in levels of SIF between testosterone-treated castrates and noninjected intact control animals. Over this same interval, however, oil-injected castrates showed a marked decrement from preinjection baseline levels of fighting behavior. Other evidence that androgen plays a role in SIF behavior was reported by Conner *et al.* (1969). By seven daily injections of 200 μg of testosterone propionate, they reinstated levels of SIF shown by males castrated at weaning to those shown by intact controls.

Based on the hypothesized organizational role of androgens during early development, testosterone administered neonatally to females might be expected to result in increased levels of SIF behavior during adulthood. The effects of injecting 0.5 mg testosterone propionate within 24 hours after birth on SIF behavior during adulthood was studied in male and female Sprague-Dawley rats by Powell *et al.* (1971). Neonatal testosterone injections were found to increase the general rate of fighting. Further, testosterone-treated animals showed a greater increase in SIF behavior than did oil-treated control animals across the 10 daily testing sessions. While this experiment apparently did not find an overall sex difference, it was unfortunately impossible to determine if these reported effects of neonatal testosterone injections were present in both sexes. Further, while the Powell *et al.* (1971) data do show that prior testosterone injections increase SIF behavior during adulthood, it is unclear from that study whether testosterone must be administered neonatally in order to produce the reported effects. A further prediction from the hypothesized organizational role of androgens is that androgen-dependent behaviors of male rats castrated on the first day of life would not be reinstated by exogenous androgen replacement during adulthood. As noted above, rats castrated prepuberally show SIF behavior increased to the level of intact controls in response to testosterone injections. In contrast, in that same study (Conner *et al.*, 1969), it was found that rats castrated neonatally respond to 200 μg testosterone propionate administered over seven consecutive days with only a small increase in SIF levels, which were still well below the levels shown by intact controls. Conner *et al.* interpreted their data as strong evidence that androgens exert an organizational influence on patterns of SIF behavior in rats. Taken together, then, evidence from the studies reviewed in this section indicates that androgens play both an organizational and a modest activational role in the expression of SIF behavior.

Endogenous Levels of Androgens

One possibility raised by the sex difference and the developmental effect on SIF is that endogenous levels of androgens underlie the behavior. Although the effects of castration and androgen replacement therapy on SIF were seen to be somewhat ambiguous, there is at least some evidence that these manipulations influence the behavior. Additional data are needed to clarify our understanding of these influences of pituitary-gonadal hormones on SIF behavior. Toward that end, one additional kind of data might be derived from measures of circulating levels of endogenous androgens. Except for annual variations in seasonal breeders, it was previously thought that androgen levels in males were fairly stable and acyclic. It has now been established, however, that there are variations in endogenous testosterone levels both diurnally (Howland, 1975; Kinson & Liu, 1973) and in association with sexual stimuli (Kamel, Mock, Wright, & Frankel, 1975; Rose, Gordon, & Bernstein, 1972). Rose, Holaday, and Bernstein (1971) have reported a rough positive relationship between endogenous testosterone levels and dominance status in rhesus monkeys. However, this finding has not been replicated in Japanese macaque monkeys (Eaton & Resko, 1974) or male mice (Barkley & Goldman, 1977). In another report, Rose *et al.* (1972) provided data showing that endogenous testosterone levels are subject to the experiential influences of victory or defeat. Monkeys high in the dominance hierarchy showed a decrease in baseline levels of plasma testosterone after they fell in dominance rank, and a rise in plasma testosterone when they were afforded access to sexually receptive females.

While the results of Rose and his colleagues suggest that measures of circulating testosterone might contribute to the discovery of additional hormone–behavior interactions, the following observations should dampen any hope that these interactions will be simple. First of all, administration of testosterone may not increase the sex behavior of intact male rats that normally exhibit only very low levels of sex behavior (Whalen, Beach, & Kuehn, 1961). This observation indicates that the presence of even high levels of the hormone may not be a sufficient condition to promote sex behavior in male rats. Data reviewed above indicate that androgen is neither a sufficient nor even a necessary condition for male rats to show SIF behavior. Second, measurement of circulating testosterone in rats is complicated by a pulsatile release that produces marked and irregular changes in plasma levels (Bartke, Steele, Musto, & Caldwell, 1973). Third, while social isolation of rats results in increased plasma testosterone levels (Dessì-Fulgheri, Lupo di Prisco, & Verdarelli, 1975), as noted earlier, the literature describing the effect of social isolation on SIF behavior is contradictory, with both suppression and facilitation of SIF behavior having been reported as a function of isolation. Fourth, diurnal variations in SIF (Knutson *et al.*, 1976) and plasma testosterone levels in rats (Kinson & Liu, 1973) overlap only very roughly. Finally, exposure of animals to stressful stimuli, such as the electric shock used in the SIF paradigm, produces a stress response reflected, in part, by an elevation of pituitary-adrenal hormones and a suppression of pituitary-gonadal hormones.

In closing this section, it is clear that SIF is androgen-dependent only to a modest degree. It appears to fall somewhere between Moyer's (1968) category of intermale aggression, which is androgen dependent to a large degree, and predatory aggression, which seems to be largely androgen independent (see Moyer, 1976). This moderate degree of androgen-dependency probably accounts for the lack of studies that might provide more detailed information about hormonal influences on SIF—for example, progesterone inhibition of androgen-mediated behaviors (Erpino & Chappelle, 1971). Before further experimental analyses along these lines can be pursued, it would first be very desirable to determine exactly which and to what degree components of the complex pattern of responding seen in the SIF situation are androgen dependent. At this time, the avaliable data permit only general statements about the androgen dependency of the behavior. First, organizational effects of androgens that are present early in development have been shown for the SIF behavior of male rats (Conner *et al.*, 1969). Second, there are modest activational effects of androgens on SIF behavior during adulthood, but evidence for this is not always found in the reported studies. On the basis of studies of adult and prepuberal castrated rats, androgens do not appear to be a necessary condition for the occurrence of the behavior. The possible contribution of adrenal androgens in the appearance or maintenance of SIF behavior in castrated males, or in females, has not been examined. When a decrease or a slower rate of increase over sessions is seen in the SIF behavior of castrated rats, testosterone injections effectively reinstate the behavior. The possible roles of aromatized or 5α-reduced products of testosterone in mediating this effect have not been studied. It would appear that the optimal conditions for demonstrating an effect of castration on SIF behavior include individual housing of neonatally or prepuberally castrated Long-Evans male rats that are tested weekly beginning at least six weeks after castration, using shocks of 2.0 mA intensity and 0.5 second duration presented 20 per minute. A between-groups design with age controls would provide for the assessment of the possible effects of weekly bouts in the SIF situation.

Pituitary-Adrenal System

There is considerable evidence implicating both adrenal steroid hormones and adrenocorticotropic hormone (ACTH) in kinds of aggression other than SIF (see elsewhere in this volume). Except for one report that corticosterone increased SIF among mice (Kostowski, 1967), there were no other data that pituitary-adrenal hormones influence SIF behavior until recently. In 1978, Rodgers and Semple reported that high doses (12.5 IU) of subcutaneously administered ACTH suppressed mean fight frequencies, while hydrocortisone (500 mg) or low doses (6.25 IU) of ACTH increased them. The testing procedure here differed from that of most investigators in that treated animals were paired with untreated opponent animals, and only the behavior of the treated animals were used in data analyses. A fight was scored when a treated animal made directed forward movements toward the opponent either by lunging with its whole body or by striking with the forepaws. Like the high dose of ACTH, adrenalectomy was also found to result in a lowered level of SIF. While this effect of adrenalectomy might have been due to a rise in plasma levels of ACTH consequent to adrenalectomy, this interpretation is compromised by the additional finding that hypophysectomy also suppressed SIF. In a further study, Rodgers (1979) administered 25, 50, or 100 μg of hydrocortisone to male Sprague-Dawley rats through a unilaterally implanted guide cannula into the lateral ventricle and tested for effects on SIF one hour later, again using the procedure of pairing treated with untreated opponent animals. He reported that both the low and intermediate doses of hydrocortisone increased SIF, while the high dose potently suppressed it. Erskine and Levine (1973), on the other hand, did not find any effect of placing 250 ± 25 μg of crystalline hydrocortisone directly into the median eminence of male Long-Evans rats when they paired like-treated animals in the SIF situation five to seven days after implantation. Since Erskine and Levine also demonstrated that both ACTH and adrenocortical secretions were effectively blocked by hydrocortisone implants, it would appear that the previously noted effect of hypophysectomy in suppressing SIF (Rodgers & Semple, 1978) might be due to the elimination of hormones other than those of the pituitary-adrenal system. Rodgers (1979), however, cites unpublished observations that peripherally administered hydrocortisone restores fighting to near-normal levels in hypophysectomized animals and suggests that hydrocortisone acts through an ACTH-independent mechanism on limbic systems mediating avoidance and attack. Whatever future research reveals the mechanism to be, these reported effects of manipulating pituitary-adrenal hormones on SIF cannot be accounted for on the basis of changes in thresholds for responding to electric shock (Rodgers, 1979; Rodgers and Semple, 1978).

Beach (1974) has emphasized that the possibility of behavioral influences on hormonal activity should also be recognized, and the possible importance of behavior-produced hormonal changes has been further emphasized by Leshner (1979). There is evidence that SIF produces a pattern of pituitary-adrenal activity different from that seen when animals are exposed to the same electric shock stimulation individually. While both pair-shocked and individually shocked rats show an elevation of plasma ACTH, individually shocked rats show a faster, significantly higher, and longer-lasting elevation than pair-shocked fighting rats (Conner, Vernikos-Danellis, & Levine, 1971; Weinberg, Erskine, & Levine, 1980). It is of interest to note that not only is the pattern of skeletal-motor responding very obviously different when trains of footshocks are delivered to paired rats rather than isolates, but the autonomic and neurochemical as well as the neuroendocrine effects are also very different. Williams and Eichelman (1971) found that rats shocked in pairs, "attacking after approximately 67 percent of their shocks" (p. 613), showed a significant drop in mean tail blood pressure, while rats shocked in isolation showed significant elevations. This effect of social setting on shock-induced changes in tail blood pressure is apparently mediated by brain catecholamines, since it is not seen in rats injected intracisternally with the neurotoxin 6-hydroxy-dopamine (6-OH-DA) (Williams, Eichelman, & Ng, 1972), even though intracisternal injection of 6-OH-DA itself results in an increase in SIF (Eichelman, Thoa, & Ng, 1972). Further, Weiss, Pohorecky, Salman, and Gruenthal (1976) have reported that pair-shocked rats develop fewer gastric lesions than individually shocked rats. This effect is seen even when a barrier allowing visual but not physical contact is placed between two rats exhibiting the stereotyped posture in response to tail shocks. Different patterns of brain norepinephrine metabolism were seen by Stolk *et al.* (1974) in pair-shocked "fighting" and individually shocked rats. Rats shocked in isolation showed an increase in deamination of norepinephrine in the medulla–pons region five minutes after shock, while paired rats showed a slight increase in norepinephrine synthesis. Higher levels of cAMP have also been reported in the brains of pair-shocked as compared with individually shocked rats (Eichelman *et al.*, 1976). Use of constant-current shock generators and other specific control procedures (see Conner *et al.*, 1971; Weiss *et al.*, 1976) make it unlikely that these effects are simply due to attenuation of the shock intensity when two animals are delivered shock from a single shock source. Clearly, then, the SIF situation produces a very different pattern of behavioral and physiological responses from that seen when rats are administered the same train of shock stimuli in isolation. Miller (1969) and his colleagues have reported that the behavioral and physiological responses of animals permitted to cope with electric shock by escape or avoidance behavior are also very different from those shown by yoked controls that cannot cope through escape or avoidance responses. Conner *et al.* (1971) have suggested that the act of engaging in the SIF response might also be considered a coping response. Weiss *et al.* (1976) have noted that an effective coping response provides relevant feedback consisting of stimuli that follow the response and are not paired with the stressor. While it is not clear how engaging in SIF behavior provides such relevant feedback, they have suggested, among other possibilities, that fighting may inherently provide its own relevant feedback.

As noted earlier, exposure of animals to stressful stimuli such as electric shock produces a broad spectrum of effects that are reflected in part by changes in hormone levels. Evidence reviewed above indicates that the pattern and/or magnitude of changes seen in ACTH levels and some other physiological systems in response to electric shock depends on the social context in which the shock is administered. Further research is needed to determine if shock-produced changes in pituitary-gonadal as well as other hormonal systems are also dependent on the social context in which they occur. Toward that end, the SIF paradigm would appear to be particularly useful.

THE QUESTION OF INTERPRETATION

Questions of how and/or if SIF relates to other kinds of behavior labeled aggressive have often been raised. These questions seem to arise from either of two basic concerns. First, variables that affect other kinds of aggressive behavior do not always affect SIF in the same way, if at all. Second, detailed comparisons indicate that the skeletal-motor patterns of attack behavior seen under more naturalistic conditions differ from the posturing, forepaw striking, and biting seen in the SIF situation. With respect to the first concern, the most pertinent observations in the present context are that androgens exert less influence on SIF behavior in rats than on the IIF behavior of mice, or possibly stimulus-bound attack of cats (Inselman-Temkin & Flynn, 1973), but they influence predatory behavior even less than SIF behavior. Such differences pose no problem for the view that aggression is not a unitary concept and that different external and internal variables are indeed involved in different kinds of aggression (Moyer, 1968).

With respect to the second concern, it is clear that rats exhibit a narrower range of responses in the SIF situation than typically has been described for species-specific intermale aggression in rats. This point has been made by several authors (e.g., Crabtree & Moyer, 1973; Reynierse, 1971), but the most detailed comparison has been provided by Blanchard and Blanchard (1977). Blanchard and his colleagues have conducted extensive studies of the aggressive behavior patterns observed when an intruder rat is introduced into an enclosure housing an established colony of laboratory rats, and they have presented a convincing argument that patterns of aggressive behavior shown by dominant resident and subordinate intruder rats in the colony situation are representative of the spontaneously occurring patterns of species-specific attack and defensive behaviors that have been described for rats by others (Barnett, 1975). Further, Blanchard and Blanchard have clearly pointed out that none of the three most prominent kinds of behavior shown by rats in the SIF situation are characteristic of resident dominant rats in the colony situation. Specifically, frontal posturing (boxing) and biting of the snout and facial region are characteristic of intruders (though intruders rarely bit at all), while dominant males typically launch "lateral attacks" by moving laterally toward a boxing intruder and quickly darting around to bite the back and shoulder regions of the intruder repeatedly. Like intruder rats, rats in the SIF situation predominantly show face-to-face posturing (boxing) and deliver bites primarily to the snout and facial region. In contrast to the SIF situation, attacks are not launched from the frontal boxing posture by dominant males in the colony situation, and favored target sites for bites are different in the two situations. The forepaw striking movements that are frequently observed in the SIF situation, especially when the animals are in the stereotyped boxing posture, are less frequently shown by either dominant or intruder rats in the colony situation. But based on the observation that administration of electric shock through the hind-limbs of individual rats harnessed in an upright position evokes nondirected forepaw striking movements (Blanchard *et al.,* 1977), Blanchard and Blanchard suggest that most of the forepaw striking seen in the SIF situation may only represent reflexive jerks to shock. The analyses by Blanchard and his colleagues make it clear that while the behavior of rats in the SIF situation bears some resemblance to the defensive behaviors of intruders in the colony situation, it is decidedly different from the attack behaviors of dominant residents. Rather than "shock-elicited aggression," the phenomenon might therefore be better characterized as "shock-elicited defense" (Blanchard *et al.,* 1977).

In a somewhat different approach to arriving at a descriptive label, Moyer has classified SIF behavior as one example of "irritable aggression." Moyer (1968, 1976) has argued that it is possible to classify most different kinds of aggression on the basis of the types of stimuli that elicit them and the particular topography of the response. After completing such a classification, Moyer finds it necessary to add one category (irritable aggression) for a set of aggressive behaviors that do not fit into any other category. Among other examples, Moyer discusses stimulus-bound rage, frustration-induced aggression, and pain- or shock-induced fighting under the heading of Irritable Aggression. Evidence is cited that the stereotyped boxing posture is not necessarily associated with forepaw striking or biting among rats in the SIF situation (Kimbrell, 1969; Knutson & Hynan, 1972). Further, results from a factor-analytic study of nine intercorrelated behavioral measures obtained during SIF of rats reveal that the stereotyped boxing posture, along with other noncontact measures, is associated with a "defense-threat" rather than a "fight" factor (Crabtree & Moyer, 1973). Consistent with the Blanchard and Blanchard view, Moyer therefore concludes that the stereotyped boxing posture so prominently seen in the SIF situation represents a defensive behavior rather than an aggressive response leading to damage or destruction of another organism. The "fight" factor that emerged from the Crabtree and Moyer factor-analytic study, on the other hand, included active body contact measures such as biting, wounding, and fighting (where fighting was defined as attack by one animal on the other, including contact in the form of biting or wrestling). It would thus appear that administration of electric footshock to paired rats might produce an admixture of predominantly defensive and some attack behavior.

There appears to be little doubt that the behavior of paired rats in the SIF situation is primarily defensive rather than aggressive. For a number of reasons, however, we believe that there may also be a small component of aggressive behavior in the SIF situation. As noted by Milligan *et al.* (1973), it is difficult to rationalize a sex difference in defensive behavior on an evolutionary basis, because mechanisms for defense against predators and other dangers would presumably be adaptive for both males and females. Indeed, Scott (1971) has suggested that more complex kinds of aggression most likely evolved from basic defensive reactions to painful stimuli. On this basis, we believe that the evidence of a modest sex difference in SIF suggests that the behavior has an aggressive component. It also seems reasonable that androgens would more likely exert an influence on aggressive behaviors, which are more often seen in males, than on defensive behaviors, which might be expected to occur with equal frequency in males and females. Again, the evidence that androgens exert some influence on SIF behavior further suggests an aggressive component of the behavior. Further, the admixture view permits an account of several otherwise perplexing observations. For example, Long-Evans males have been reported to exhibit higher levels of biting and striking in the SIF situation than Sprague-Dawley males (Duncan & Powell, 1975). To the extent that these particular responses are aggressive rather than defensive, this might account for the more frequently reported sex difference in studies of SIF behavior using Long-Evans rather than Sprague-Dawley rats. Presumably, the females of both strains exhibit the same levels of defensive behaviors (e.g., posturing) in the SIF situation. This behavioral difference between the strains might also make castration effects easier to see in Long-Evans than in Sprague-Dawley males in the SIF situation. Further, the prepotency of the defensive boxing posture in the animals of both strains may account for the difficulty among some investigators in making unequivocal dominant–subordinate judgments among rats in the SIF situation. Similarly, sex differences more readily appear when high shock intensities are employed in the SIF situation (Conner *et al.,* 1969; Milligan *et al.,* 1973), and high shock also evokes more striking and biting. While Blanchard *et al.* (1977) saw relatively little biting among rats in the SIF situation, others using higher shocks presented at a faster rate have reported more frequent biting as well as severe injuries in the SIF situation (cf. Duncan & Powell, 1975; Knutson & Hynan, 1972). As noted earlier, biting is also frequently shown by squirrel monkeys and gerbils in the SIF situation. Legrand and Fielder (1973) have reported frequent biting attacks among mice as well in the SIF situation, especially by mice found to be dominant on the basis of independent tests. In addition, while nondirected forepaw striking may be reflexively evoked by shocking semirestrained rats individually, it is unlikely that it is also merely a reflexive jerk in response to shock when an animal abruptly bounds across the chamber to launch an apparently well-directed forepaw striking and biting attack on the other animal in the SIF situation. The finding that some forepaw-striking behavior also occurs during encounters between rats in the colony situation (Blanchard *et al.,* 1977) would seem to compromise further the interpretation of all striking behavior in the SIF situation as reflexive jerks to shock. It would thus appear that defensive behavior (boxing posture) is readily evoked in the SIF situation under a broad range of electric-shock parameters, but the appearance of aggressive behaviors (striking and biting) requires brief, intense shocks delivered at a rapid rate.

Typically, defensive posturing initially emerges in the SIF situation, irrespective of the electric shock parameters. With optimal shock parameters (2.0 to 2.5 mA intensity, 0.5 second duration, presented at the rate of about 20 per minute), aggressive responses usually appear shortly after and continue along with defensive posturing. Activation of brain circuits underlying the expression of defensive behaviors may normally be accompanied by an activation of attack systems as well or may lower the threshold for the appearance of attack behaviors. Or alternatively, the simultaneous appearance of both defense and attack behaviors with optimal shock parameters in the SIF situation may reflect a unique response pattern seen under these unique conditions, and it may bear little relation to the behavior patterns of rats seen under more naturalistic conditions. In fact, several other related aspects of the behavior seen in the SIF situation might also be unique. In the first place, the simultaneous display of a defensive posture by two rats has not often been reported in accounts of rat encounters under more naturalistic conditions, yet this is frequently seen in the SIF situation. Second, adoption of the defensive posture clearly does not inhibit further attack by either rat in the SIF situation (though the signal function of defensive postures in rats even under more naturalistic conditions has been questioned by Blanchard and Blanchard). Finally, the defensive behavior of intruder rats in the colony situation includes a high proportion of time "on-the-back," while rats spend relatively little time in that position in the SIF situation (Blanchard *et al.,* 1977). These considerations further suggest that the behavior of rats seen in the SIF situation might be a unique admixture of defensive and aggressive responses rather than only species-typical defensive behavior. As both Crabtree and Moyer (1973) and Blanchard *et al.* (1977) have already pointed out, specification of the kinds of behaviors seen in the SIF situation does not mean that the paradigm is useless. However, we would not encourage the use of the paradigm by the investigator who is specifically interested in elucidating hormonal influences of the pituitary-gonadal system on attack behaviors. Such influences on SIF behavior can be shown, but their magnitude is generally small. We suggest that SIF behavior is androgen dependent to the same degree that administration of intense footshock to paired rats produces some aggressive responses. The behavior is only androgen dependent to a moderate degree because such shock produces predominantly defensive behaviors. Based on the observation that resident rats are typically dominant when an intruder is introduced into the home cage (Barfield, Busch, & Wallen, 1972), it would be of interest to observe the behaviors of paired rats when the electric shock is administered in the home cage of the resident animal. This situation should yield higher levels of aggressive responding (striking and biting) by the resident animal than would be seen in the usual SIF situation, when shock is administered in a neutral arena. Studies along this line might permit a more precise identification of which behavioral patterns in the SIF situation are aggressive and androgen dependent and which are defensive. However, our own view is that the greatest potential value of the paradigm might be a useful research tool in further studies of the differential autonomic, neurochemical, and neuroendocrine responses shown by individually and pair-shocked animals.

Finally, with respect to terminology, it has previously been argued that the term "aggression" usually refers to attack behavior while the term "fighting" includes both attack and defensive behaviors (Scott, 1958). Consistent with that view, we would suggest that "shock-induced fighting" remains a satisfactory label for the admixture of attack and defensive behaviors seen in the paired-animal shock situation.

REFERENCES

Ader, R. Competitive and noncompetitive rearing and shock-elicited aggression in the rat. *Animal Learning and Behavior,* 1975 *3,* 337–339.

Anderson, D. C., Crowell, C. R., Lupo, J. V., & Sarmina, T. The effects of sex and exposure to inescapable shock on shock-elicited fighting in albino rats. *Physiological Psychology,* 1977, *5,* 445–448.

Anisko, J. J., Christenson, T., & Buehler, M. G. Effects of androgen on fighting behavior in male and female mongolian Gerbils *(Mercones unguiculatus). Hormones and Behavior,* 1973, *4,* 199–208.

Azrin, N. H., Hutchinson, R. R., & Hake, D. F. Pain-induced fighting in the squirrel monkey. *Journal of the Experimental Analysis of Behavior,* 1963, *6,* 620–621.

Azrin, N. H., Hutchinson, R. R., & Sallery, R. D. Pain aggression toward inanimate objects. *Journal of the Experimental Analysis of Behavior,* 1964, *7,* 223–228.

Azrin, N. H., Ulrich, R. E., Hutchinson, R. R., & Norman, D. G. Effect of shock duration on shock-induced fighting. *Journal of the Experimental Analysis of Behavior,* 1964, *7,* 9–11.

Azrin, N. H., Hake, D. F., & Hutchinson, R. R. Elicitation of aggression by a physical blow. *Journal of the Experimental Analysis of Behavior,* 1965, *8,* 55–57.

Azrin, N. H., Hutchinson, R. R., & Hake, D. F. Attack, avoidance, and escape reactions to aversive shock. *Journal of the Experimental Analysis of Behavior,* 1967, *10,* 131–148.

Azrin, N. H., Rubin, H. B., & Hutchinson, R. R. Biting attack by rats in response to aversive shock. *Journal of the Experimental Analysis of Behavior,* 1968, *11,* 633–639.

Baenninger, R. Effects of day 1 castration on aggressive behaviors of rats. *Bulletin of the Psychonomic Society,* 1974, *3,* 189–190.

Barfield, R. J., & Sachs, B. D. Sexual behavior: Stimulation by painful electric shock to skin of male rats. *Science,* 1968, *161,* 392–395.

Barfield, R. J., Busch, D. E., & Wallen, K. Gonadal influence on agonistic behavior in the male domestic rat. *Hormones and Behavior,* 1972, *3,* 247–249.

Barkley, M. S., & Goldman, B. D. A quantitative study of serum testosterone, sex accessory organ growth, and the development of intermale aggression in the mouse. *Hormones and Behavior,* 1977, *8,* 208–218.

Barnett, S. A. *The rat: A study in behavior.* Chicago: University of Chicago Press, 1975.

Bartke, A., Steele, R. E., Musto, N., & Caldwell, B. V. Fluctuations in plasma testosterone levels in adult male rats and mice. *Endocrinology,* 1973, *92,* 1223–1228.

Beach, F. A. Coital behavior in dogs: VI. Long-term effects of castration upon mating in the male. *Journal of Comparative and Physiological Psychology Monograph,* 1970, *70*(3, Pt. 2), 1–32.

Beach, F. A. Behavioral endocrinology and the study of reproduction. *Biology of Reproduction,* 1974, *10,* 2–18.

Beatty, W. W. Gonadal hormones and sex differences in nonreproductive behaviors in rodents: Organizational and activational influences. *Hormones and Behavior,* 1979, *12,* 112–163.

Bernard, B. K., & Paolino, R. M. Temporal effects of castration on emotionality and shock-induced aggression in adult male rats. *Physiology and Behavior,* 1975, *14,* 201–206.

Berry, R. M., & Jack, C. E. The effect of temperature upon shock-elicited aggression in rats. *Psychonomic Science,* 1971, *23,* 341–343.

Bevan, J. M., Bevan, W., & Williams, B. F. Spontaneous aggressiveness in young castrate C_3H male mice treated with three dose levels of testosterone. *Physiological Zoology,* 1958, *31,* 284–288.

Blanchard, R. J., & Blanchard, D. C. Aggressive behavior in the rat. *Behavioral Biology,* 1977, *21,* 197–224.

Blanchard, R. J., Blanchard, D. C., & Takahashi, L. K. Reflexive fighting in the albino rat: Aggressive or defensive behavior? *Aggressive Behavior,* 1977, *3,* 145–155.

Boice, R., & Pickering, A. Shock-elicited behaviors in paired gerbils. *Bulletin of the Psychonomic Society,* 1973, *2,* 221–224.

Bracco, H. F., & Miller, L. L. Effects of varying shock intensities on pain-elicited aggression in the domestic fowl. *Perceptual and Motor Skills,* 1972, *34,* 719–723.

Caggiula, A. R. Shock-elicited copulation and aggression in male rats. *Journal of Comparative and Physiological Psychology,* 1972, *80,* 393–397.

Cahoon, D. D., Crosby, R. M., Dunn, S., Herrin, M. S., Hill, C. C., & McGinnis, M. The effect of food deprivation on shock elicited aggression in rats. *Psychonomic Science,* 1971, *22,* 43–44.

Conner, R. L., Levine, S., Wertheim, G. A., & Cummer, J. F. Hormonal determinants of aggressive behavior. *Annals of the New York Academy of Sciences,* 1969, *159,* 760–776.

Conner, R. L., Vernikos-Danellis, J., & Levine, S. Stress, fighting and neuroendocrine function. *Nature,* 1971, *234,* 564–566.

Crabtree, J. M., & Moyer, K. E. Sex differences in fighting and defense induced in rats by shock and d-amphetamine during morphine abstinence. *Physiology and Behavior,* 1973, *11,* 337–343.

Creer, T. L. Communal housing and shock-induced aggression. *Bulletin of the Psychonomic Society,* 1974, *4,* 51–53.

Creer, T. L. Effects of previous housing conditions on shock-induced aggression. *Journal of the Experimental Analysis of Behavior,* 1975, *23,* 451–456.

Creer, T. L., & Powell, D. A. Effects of age and housing conditions on shock-induced aggression. *Psychonomic Science,* 1971, *22,* 259–261.

Daniel, W. J. An experimental note on the O'Kelley-Steckle reaction. *Journal of Comparative Psychology,* 1943, *35,* 267–268.

Dessì-Fulgheri, F., Lupo di Prisco, C. & Verdarelli, P. Influence of long-term isolation on the production and metabolism of gonadal sex steroids in male and female rats. *Physiology and Behavior,* 1975, *14,* 495–499.

Dreyer, P. I., & Church, R. M. Reinforcement of shock-induced fighting. *Psychonomic Science,* 1970, *18,* 147–148.

Duncan, S. D., & Powell, D. A. Agonistic postures elicited by subdermal shock in Sprague-Dawley and Long-Evans rats. *Psychological Reports,* 1975, *36,* 327–337.

Dunstone, J. J., Cannon, J. T., Chickson, J. T., & Burns, W. K. Persistence and vigor of shock-induced aggression in gerbils *(Meriones unguiculatus). Psychonomic Science,* 1972, *28,* 272–274.

Eaton, G. G., & Resko, J. A. Plasma testosterone and male dominance in a Japanese macaque *(Macaca fuscata)* troop compared with repeated measures of testosterone in laboratory males. *Hormones and Behavior,* 1974, *5,* 251–259.

Edwards, D. A. Early androgen stimulation and aggressive behavior in male and female mice. *Physiology and Behavior,* 1969, *4,* 333–338.

Eichelman, B. S., Thoa, N. B., & Ng, K. Y. Facilitated aggression in the rat following 6-hydroxydopamine administration. *Physiology and Behavior,* 1972, *8,* 1–3.

Eichelman, B., Orenberg, E., Seagraves, E., & Barchas, J. Influence of social setting on the induction of brain cyclic AMP in response to electric shock in the rat. *Nature,* 1976, *263,* 433–434.

Eichelman, B., Orenberg, E., Hackley, P., & Barchas, J. Methyxanthine-facilitated shock-induced aggression in the rat. *Psychopharmacology,* 1978, *56,* 305–308.

Erpino, M. J., & Chappelle, T. C. Interactions between androgens and progesterone in mediation of aggression in the mouse. *Hormones and Behavior,* 1971, *2,* 265–272.

Erskine, M. S., & Levine, S. Suppression of pituitary-adrenal activity and shock-induced fighting in rats. *Physiology and Behavior,* 1973, *11,* 787–790.

Erskine, M. S., Stern, J. M., & Levine, S. Effects of prepubertal handling on shock-induced fighting and ACTH in male and female rats. *Physiology and Behavior,* 1975, *14,* 413–420.

Farris, H. E., Gideon, B. E., & Ulrich, R. E. Classical conditioning of aggression: A developmental study. *The Psychological Record,* 1970, *20,* 63–67.

Follick, M. J., & Knutson, J. F. Punishment of irritable aggression. *Aggressive Behavior,* 1978, *4,* 1–17.

Fortier, C., deGroot, J., & Hartfield, J. E. Plasma free corticosteroid response to faradic stimulation in the rat. *Acta Endocrinologica,* 1959, *30,* 219–221.

Fraser, D., & Spigel, I. M. Shock-induced threat and biting by the turtle. *Journal of the Experimental Analysis of Behavior,* 1971, *16,* 349–353.

Fredericson, E. The effects of food deprivation upon competitive and spontaneous combat in C57 black mice. *Journal of Psychology,* 1950, *29,* 89–100.

Gandelman, R. Mice: Postpartum aggression elicited by the presence of an intruder. *Hormones and Behavior,* 1972, *3,* 23–28.

Gandelman, R. Gonadal hormones and the induction of intraspecific fighting in mice. *Neuroscience and Biobehavioral Reviews,* 1980, *4,* 133–140.

Ghiselli, W. B., & Thor, D. H. The rodent model of irritable aggression: A method for analysis of individual roles in paired fighting. *Bulletin of the Psychonomic Society,* 1974, *4,* 17–19.

Ginsburg, B., & Allee, W. C. Some effects of conditioning on social dominance and subordination in inbred strains of mice. *Physiological Zoology,* 1942, *15,* 485–506.

Gleason, P. E., Michael, S. D., & Christian, J. J. Effects of gonadal steroids on agonistic behavior of female *Peromyscus leucopus. Hormones and Behavior,* 1979, *12,* 30–39.

Goy, R. W., & Goldfoot, D. A. Experiential and hormonal factors influencing development of sexual behavior in the male rhesus monkey. In F. O. Schmitt & F. G. Worden (Eds.), *The neurosciences: Third study program.* Cambridge, Mass.: MIT Press, 1974.

Goy, R. W., & McEwen, B. S. *Sexual differentiation of the brain.* Cambridge, Mass.: MIT Press, 1980.

Gregor, G. L., Smith, R. F., Simons, L. S., & Parker, H. B. Behavioral consequences of crowding in the deermouse (Peromyscus maniculatus). *Journal of Comparative and Physiological Psychology,* 1972, *79,* 488–493.

Heacock, D., Thurber, S., & Vale, D. Shock elicited aggression by human subjects. *The Journal of Social Psychology,* 1975, *95,* 55–59.

Heilman, R. D., Brugmans, M., & Strainer, S. M. Delayed development of aggressive behavior in castrate isolated male mice. *Hormones and Behavior,* 1977, *9,* 107–111.

Howland, B. E. Effects of short periods of fasting and time of day on serum levels of gonadotropins, testosterone and glucose in male rats. *Hormone and Metabolic Research,* 1975, *7,* 40–43.

Hutchinson, R. R. The environmental causes of aggression. In J. Cole & D. Jensen (Eds.), *Nebraska Symposium on Motivation* (Vol. 20). Lincoln, Neb.: University of Nebraska Press, 1972.

Hutchinson, R. R., & Emley, G. S. Electric shock produced drinking in the squirrel monkey. *Journal of the Experimental Analysis of Behavior,* 1977, *28,* 1–12.

Hutchinson, R. R., Ulrich, R. E., & Azrin, N. H. Effects of age and related factors on the pain aggression reaction. *Journal of Comparative and Physiological Psychology,* 1965, *59,* 365–369.

Hutchinson, R. R., Azrin, N. H., & Hake, D. F. An automatic method for the study of aggression in squirrel monkeys. *Journal of the Experimental Analysis of Behavior,* 1966, *9,* 233–237.

Hutchinson, R. R., Azrin, N. H., & Renfrew, J. W. Effects of shock intensity and duration on the frequency of biting attack by squirrel monkeys. *Journal of the Experimental Analysis of Behavior,* 1968, *11,* 83–88.

Hutchinson, R. R., Pierce, G. E., Emley, G. S., Proni, T. J., & Sauer, R. A. The laboratory measurement of human anger. *Biobehavioral Reviews,* 1977, *1,* 241–259.

Hutzell, R. R., & Knutson, J. F. A comparison of shock-elicited fighting and shock-elicited biting in rats. *Physiology and Behavior,* 1972, *8,* 477–480.

Hyde, J. S., & Sawyer, T. F. Estrous cycle fluctuations in aggressiveness of house mice. *Hormones and Behavior,* 1977, *9,* 290–295.

Inselman-Temkin, B. R., & Flynn, J. P. Sex-dependent effects of gonadal and gonadotropic hormones on centrally-elicited attack in cats. *Brain Research,* 1973, *60,* 393–410.

Jacobs, B. L., & Cohen, A. Differential behavioral effects of lesions of the median or dorsal raphe nuclei in rats: Open field and pain-elicited aggression. *Journal of Comparative and Physiological Psychology,* 1976, *90,* 102–108.

Jones, A. B., Barchas, J. D., & Eichelman, B. Taming effects of *p*-chlorophenylalanine on the aggressive behavior of septal rats. *Pharmacology, Biochemistry and Behavior,* 1976, *4,* 397–400.

Kamel, F., Mock, E. J., Wright, W. N., & Frankel, A. I. Alterations in plasma concentrations of testosterone, LH, and prolactin associated with mating in the male rat. *Hormones and Behavior,* 1975, *6,* 277–288.

Kane, N. L., & Knutson, J. F. Influence of colony lighting conditions on home-cage spontaneous aggression. *Journal of Comparative and Physiological Psychology,* 1976, *90,* 889–897.

Kaplan, S., & Hobart, J. L. New technique of recording skin resistance. *The American Journal of Medical Electronics,* 1965, *4,* 117–120.

Kimbrell, G. McA. Relationship of the upright agonistic posture in the foot shock situation to dominance-submission in male C57BL/6 mice. *Psychonomic Science,* 1969, *16,* 167–168.

Kimbrell, G. McA., & Chesler, D. Role of dominance status in shock-stimulated agonistic behavior. *Psychological Reports,* 1971, *29,* 768–770.

Kinson, G. A., & Liu, C. Diurnal variation in plasma testosterone of the male laboratory rat. *Hormone and Metabolic Research,* 1973, *5,* 233–234.

Kislak, J. W., & Beach, F. A. Inhibition of aggressiveness by ovarian hormones. *Endocrinology*, 1955, *56*, 684–692.

Knutson, J. F., & Hynan, M. T. Influence of upright posture on shock-elicited aggression in rats. *Journal of Comparative and Physiological Psychology*, 1972, *81*, 297–306.

Knutson, J. F., & Hynan, M. T. Predatory aggression and irritable aggression: Shock-induced fighting in mouse-killing rats. *Physiology and Behavior*, 1973, *11*, 113–115.

Knutson, J. F., Hynan, M. T., & Kane, N. L. Influence of home-cage lighting conditions on shock-induced fighting. *Journal of Comparative and Physiological Psychology*, 1976, *90*, 877–888.

Kostowski, W. The actions of hydrocortisone, deoxycorticosterone and G-strophanthin on aggressive behavior in mice. *Dissertationes Pharmaceuticae et Pharmacologicae*, 1967, *19*, 619–623.

Legrand, R., & Fielder, R. Role of dominance-submission relationships in shock-induced fighting of mice. *Journal of Comparative and Physiological Psychology*, 1973, *82*, 501–506.

Leshner, A. I. Kinds of hormonal effects on behavior: A new view. *Neuroscience and Biobehavioral Reviews*, 1979, *3*, 69–73.

Levy, J. The effects of testosterone propionate on fighting behavior in C57/10 young female mice. *Proceedings of the West Virginia Academy of Science*, 1954, *26*, 14.

Levy, J. V., & King, J. A. The effects of testosterone propionate on fighting behavior in young male C57/10 mice. *Anatomical Record*, 1953, *117*, 562–563.

Mathewson, S. F. Gonadotropic control of aggressive behavior in starlings. *Science*, 1961, *134*, 1522–1523.

McCarty, R., & Kopin, I. J. Changes in plamsa catecholamines and behavior of rats during the anticipation of footshock. *Hormones and Behavior*, 1978, *11*, 248–257.

Miczek, K. A., Brykczynski, T., & Grossman, S. P. Differential effects of lesions in the amygdala, periamygdaloid cortex, and stria terminalis on aggressive behaviors in rats. *Journal of Comparative and Physiological Psychology*, 1974, *87*, 760–761.

Miller, N. E. Psychosomatic effects of specific types of training. *Annals of the New York Academy of Sciences*, 1969, *159*, 1025–1040.

Milligan, W. L., Powell, D. A., & Borasio, G. Sex variables and shock-elicited aggression. *Journal of Comparative and Physiological Psychology*, 1973, *83*, 441–450.

Mollenhour, M. N., Voorhees, J. W., & Davis, S. F. Sleepy and hostile: The effects of REM sleep deprivation on shock-elicited aggression. *Animal Learning and Behavior*, 1977, *5*, 148–152.

Moyer, K. E. Kinds of aggression and their physiological basis. *Communications in Behavioral Biology*, 1968, *2*, 65–87.

Moyer, K. E. *The psychobiology of aggression*. New York: Harper & Row, 1976.

Naftalin, F., Ryan, K. J., & Petro, Z. Aromatization of androstenedione by the anterior hypothalamus of adult male and female rats. *Endocrinology*, 1971, *90*, 295–297.

O'Kelly, L. I., & Steckle, L. C. A note on long enduring emotional responses in the rat. *Journal of Psychology*, 1939, *8*, 125–131.

Powell, D. A., & Creer, T. L. Interaction of developmental and environmental variables in shock-elicited aggression. *Journal of Comparative and Physiological Psychology*, 1969, *69*, 219–225.

Powell, D. A., Silverman, T., Francis, J., & Schneiderman, N. E. The effects of sex and prior experience with fighting on shock-induced aggression. *Communications in Behavioral Biology*, 1970, *5*, 51–56.

Powell, D. A., Francis, J., & Schneiderman, N. The effects of castration, neonatal injections of testosterone and previous experience with fighting on shock-elicited aggression. *Communications in Behavioral Biology*, 1971, *5*, 371–377.

Powell, D. A., Francis, M. J., Francis, J., & Schneiderman, N. Shock-induced aggression as a function of prior experience with avoidance, fighting, or unavoidable shock. *Journal of the Experimental Analysis of Behavior*, 1972, *18*, 323–332.

Powell, D. A., Holley, J. R., & Buchanan, S. L. Agonistic postures in the rat: Reliability of human observations. *Bulletin of the Psychonomic Society*, 1976, *7*, 409–412.

Puglisi-Allegra, S., & Renzi, P. A technique for the measurement of aggressive behavior in mice. *Behavior Research Methods and Instrumentation*, 1977, *9*, 503–504.

Rashotte, M. E., Dove, L. D., & Looney, T. A. Absence of shock-elicited aggression in pigeon. *Journal of the Experimental Analysis of Behavior*, 1974, *21*, 267–275.

Resko, J. A. Plasma androgen levels of the rhesus monkey: Effects of age and season. *Endocrinology,* 1967, *81,* 1203–1212.

Reynierse, J. H. Submissive postures during shock-elicited aggression. *Animal Behavior,* 1971, *19,* 102–107.

Rodgers, R. J. Alterations in shock-induced fighting and locomotor activity following intracerebroventricular injection of hydrocortisone in the rat. *Aggressive Behavior,* 1979, *5,* 31–40.

Rodgers, R. J., & Semple, J. M. Pituitary-adrenocortical axis and shock-induced fighting in rats. *Physiology and Behavior,* 1978, *20,* 533–537.

Roediger, H. L., III, & Stevens, M. C. The effects of delayed presentation of the object of aggression on pain-induced fighting. *Psychonomic Science,* 1970, *21,* 55–56.

Rose, R. M., Holaday, J. M., & Bernstein, I. S. Plasma testosterone, dominance rank, and aggressive behavior in male rhesus monkeys. *Nature,* 1971, *231,* 366–368.

Rose, R. M., Gordon, T. P., & Bernstein, I. S. Plasma testosterone levels in the male rhesus: Influences of sexual and social stimuli. *Science,* 1972, *178,* 643–645.

Rosenblatt, J., & Aronson, L. R. The influence of experience on the behavioral effects of androgen in prepubertally castrated male cats. *Journal of Animal Behaviour,* 1958, *6,* 171–182.

Rossier, J., French, E. D., Rivier, C., Ling, N., Guillemin, R., & Bloom, F. E. Foot-shock induced stress increases β-endorphin levels in blood but not brain. *Nature,* 1977, *270,* 618–620.

Sbordone, R. J., Wingard, J. A., Elliott, M. L., & Jervey, J. Mescaline produces pathological aggression in rats regardless of age or strain. *Pharmacology, Biochemistry and Behavior,* 1978, *8,* 543–546.

Scott, J. P. *Aggression.* Chicago: University of Chicago Press, 1958.

Scott, J. P. Theoretical issues concerning the origin and causes of fighting. In B. E. Eleftheriou & J. P. Scott (Eds.), *The physiology of aggression and defeat.* New York: Plenum Press, 1971.

Selmanoff, M. E., Goldman, B. D., & Ginsburg, B. E. Developmental changes in serum luteinizing hormone, follicle stimulating hormone, and androgen levels in males of two inbred mouse strains. *Endocrinology,* 1977, *100,* 122–127.

Simon, N. G. The genetics of intermale aggressive behavior in mice: Recent research and alternative strategies. *Neuroscience and Biobehavioral Reviews,* 1979, *3,* 97–106.

Southwick, C. H., & Clark, L. H. Interstrain differences in aggressive behavior and exploratory activity of inbred mice. *Communications in Behavioral Biology,* 1968, *1,* 49–59.

Stolk, J. M., Conner, R. L., Levine, S., & Barchas, J. D. Brain norepinephrine metabolism and shock-induced fighting behavior in rats: Differential effects of shock and fighting on the neurochemical response to a common footshock stimulus. *The Journal of Pharmacology and Experimental Therapeutics,* 1974, *190,* 193–209.

Svare, B., Davis, P. G., & Gandelman, R. Induction of fighting behavior in female mice following chronic androgen treatment during adulthood. *Physiolgoy and Behavior,* 1974, *12,* 339–403.

Tedeschi, R. E., Tedeschi, D. H., Mucha, A., Cook, L., Mattis, P. A., & Fellows, E. J. Effects of various centrally acting drugs on fighting behavior of mice. *Journal of Pharmacology and Experimental Therapeutics,* 159, *125,* 28–34.

Thierry, A. M., Javoy, F., Glowinski, J., & Kety, S. S. Effects of stress on the metabolism of norepinephrine, dopamine and serotonin in the central nervous system of the rat: I. Modifications of norepinephrine turnover. *Journal of Pharmacology and Experimental Therapeutics,* 1968, *163,* 163–171.

Thor, D. H. Enhanced social docility in male hooded rats by dermal cautery of the vibrissal pads. *Aggressive Behavior,* 1976, *2,* 39–53. (a)

Thor, D. H. Intraspecific aggression in rats and dimensions of enclosure. *Psychological Reports,* 1976, *38,* 1253–1254. (b)

Thor, D. H., & Ghiselli, W. B. Visual and social determinants of shock-elicited aggressive responding in rats. *Animal Learning and Behavior,* 1974, *2,* 74–76.

Thor, D. H., Ghiselli, W. B., & Ward, T. B. Infantile handling and sex differences in shock-elicited aggressive responding of hooded rats. *Developmental Psychobiology,* 1974, *7,* 273–279.

Turner, J. L., Boice, R., & Powers, P. C. Behavioral components of shock-induced aggression in ground squirrels *(Citellus tridecemlineatus). Animal Learning and Behavior,* 1973, *1,* 254–262.

Uhrich, J. The social hierarchy in albino mice. *Journal of Comparative Psychology,* 1938, *25,* 373–413.

Ulrich, R. Pain as a cause of aggression. *American Zoologist,* 1966, *6,* 643–662.

Ulrich, R. E. Pain-aggression. In G. A. Kimble (Ed.), *Foundations of conditioning and learning.* New York: Appleton-Century-Crofts, 1967.

Ulrich, R. E., & Azrin, N. H. Reflexive fighting response to aversive stimulation. *Journal of the Experimental Analysis of Behavior,* 1962, *5,* 511–520.

Ulrich, R. E., & Craine, W. H. Behavior: Persistence of shock-induced aggression. *Science,* 1964, *143,* 971–972.

Ulrich, R. E., Wolff, P. C., & Azrin, N. H. Shock as a elicitor of intra- and interspecies fighting behavior. *Animal Behaviour,* 1964, *12,* 14–15.

Wagner, G. C., Beuving, L. J., & Hutchinson, R. R. Androgen dependency of aggressive target-biting and paired fighting in male mice. *Physiology and Behavior,* 1979, *22,* 43–46.

Wagner, G. C., Beuving, L. J., & Hutchinson, R. R. The effects of gonadal hormone manipulations on aggressive target-biting in mice. *Aggressive Behavior,* 1980, *6,* 1–7.

Weinberg, J., Erskine, M., & Levine, S. Shock-induced fighting attenuates the effects of prior shock experience in rats. *Physiology and Behavior,* 1980, *25,* 9–16.

Weiss, J. M., Pohorecky, L. A., Salman, S., & Gruenthal, M. Attenutation of gastric lesions by psychological aspects of aggression in rats. *Journal of Comparative and Physiological Psychology,* 1976, *90,* 252–259.

Welch, A. S., & Welch, B. L. Isolation, reactivity and aggression: Evidence for an involvement of brain catecholamines and serotonin. In B. E. Elefthetiou & J. P. Scott (Eds.), *The physiology of aggression and defeat.* New York: Plenum Press, 1971.

Weltman, A. S., Sackler, A. M., Schwartz, R., & Owens, H. Effects of isolation stress on female albino mice. *Laboratory Animal Care,* 1968, *18,* 426–435.

Whalen, R. E., Beach, F. A., & Kuehn, R. E. Effects of exogenous androgen on sexually responsive and unresponsive male rats. *Endocrinology,* 1961, *68,* 373–380.

Williams, R. B., & Eichelman, B. Social setting: Influence on the physiological response to electric shock in the rat. *Science,* 1971, *174,* 613–614.

Williams, R. B., Eichelman, B., & Ng, L. K. Y. Depletion of brain amines reverses blood pressure response to footshock in rat. *Nature,* 1972, *240,* 276–277.

Wilson, A. P., & Vessey, S. H. Behavior of free-ranging castrated rhesus monkeys. *Folia Primatologica,* 1968, *9,* 1–14.

7

Hormonal Influences on Territorial Marking Behavior

Pauline Yahr

SCENT-MARKING IN THE CONTEXT OF COMMUNICATION AND AGGRESSION

COMMUNICATION AND THE COST–BENEFIT RATIO OF FIGHTING

The benefits of fighting accrue strictly to the winner. Drawbacks, on the other hand, are more evenly divided. While the combatants are fighting, their mates may wander off or be herded away by other rivals (Clutton-Brock, Albon, Gibson, & Guinness, 1979). After the fight, both opponents are likely to be exhausted and both may suffer wounds that can impair sensory abilities or movement. Thus, even the winner may be less able to forage, to mate, to avoid predation, and to fend off competitors after an aggressive encounter than before. These risks are what make communication such an important part of aggression.

Communication is beneficial in the context of aggression when sending signals increases the probability that the signaler will win or when receiving signals enables the recipient to decrease risks. The first step in decreasing risks is the careful selection of opponents. No animal should run the risks of a fight unless the opponent presents a threat to its reproductive success. It is therefore helpful if the animal can distinquish categories of conspecifics that pose threats from those that do not.

The categories of conspecifics that are threatening vary from one species or population to another depending on ecology and social structure (Emlen & Oring, 1977; Trivers, 1972). In monogamous species, for example, it is often essential that both parents invest considerable time and energy in caring for their young. In these species, a male's reproductive success can be threatened by adults of either sex. Only the males, of course, present the problem of cuckoldry. They therefore pose the greatest threat to a monogamous male when his

---

PAULINE YAHR • Department of Psychobiology, School of Biological Sciences, University of California, Irvine, California 92717. This review was prepared while the author was supported by research grant MH-26481 from the National Institute of Mental Health.

mate is receptive. But afterwards, feeding and protecting the young become the top priorities. Both males and females can undermine this effort by taking limited food resources, occupying available nest sites, or cannibalizing young. Juveniles and subadults usually do not pose comparable threats. Thus, adult males of monogamous species should direct their aggressiveness primarily toward other adults, but should aggress roughly equally against both sexes except when females are fertile.

In polygynous species, the situation is quite different. Here, males often contribute nothing to parental care. Instead, their reproductive success is determined almost completely by the number of females they fertilize. Accordingly, threats to a male's reproductive success come primarily from other reproductively competent males. It is no surprise then that in polygynous species, adult males direct nearly all of their aggression toward each other (Clutton-Brock *et al.*, 1979; LeBoeuf, 1974). Clearly, animals can restrict their aggressiveness this way only if they are sensitive to social signals that distinguish the critical subsets of the population from the others.

Restricting aggressiveness to certain opponents within a category is also beneficial. This involves sensitivity to cues that provide information about the opponent's strength, stamina, aggressive experience, and state of arousal and motivation (Clutton-Brock & Albon, 1979). If possible, an animal should initiate or escalate a fight only when it is likely to win.

The second step in decreasing the risks of aggression is to ritualize the fight, that is, to avoid the use of potentially lethal weapons and, essentially, to transform the aggression into an exchange of information (Geist, 1971; Tinbergen, 1951; Wilson, 1975). Yet whether the aggression is ritualized or not, an animal benefits from any signals it produces that bias the outcome of the encounter in its favor. The very best strategy is for the animal to produce signals that enable it to avoid fighting altogether and to defeat or to intimidate its opponent by communication alone. It is in the context of altering the probability and/or outcome of fights that the analysis of scent-marking is most directly related to the study of aggression.

The Suitability of Odors as Social Cues and the Evolution of Scent-Marking

As animals undergo internal physiological changes, such as the hormonal changes that accompany reproductive cycles or social stress, the chemical composition of their urine, feces, sweat, and other bodily secretions changes accordingly (Epple, 1974; Leon, 1974; Michael, 1975; Müller-Schwarze, 1974). Physiological differences due to species, sex, genotype, or diet are similarly reflected in chemicals released to the outside of an animal's body (Beauchamp, 1976; Epple, 1974; Kessler, Harmatz, & Gerling, 1975; Müller-Schwarze, 1974; Skeen & Thiessen, 1977). These excretions and secretions therefore contain considerable latent information about the organism's internal state and hence about the behavior patterns it is likely to show.

The well-developed olfactory systems of mammals enable them to detect many of these chemical differences and to use the available information to good advantage during social interactions. If the signal producer benefits, in turn, from the efficient use of its signals by conspecifics, the stage is set for selection to act back on the signal producer to improve the accuracy of the chemical message and/or to enhance its dissemination with specialized scent-marking behaviors. Scent-marking would therefore reflect an advanced stage in the evolution of olfactory communication. It would also be reasonable to assume that the chemicals deposited during scent-marking contain odors that are actually used for intraspecific communication to the mutual benefit of both the signal sender and receiver.

Types of Odors and Specialized Behaviors That Mammals Use to Scent-Mark Their Environment

Mammals use several types of body chemicals and many distinctive behavior patterns to scent-mark their environments. Perhaps the most familiar examples of scent-marking are the urinary-marking habits of household pets. Everyone has seen male dogs use the specialized leg-lift posture to direct urine onto prominent vertical objects. Cat owners may also recognize the posture cats assume when spraying urine onto upholstery or drapes. In each case, the amount of urine released is small relative to the volume of the bladder, allowing the animal to distribute its signal over several signposts. Many other species also practice urinary scent-marking, including wild felids and canids (Henry, 1980; Kleiman, 1966; Mech & Peters, 1977; Schaller, 1967), ungulates (Grau, 1976), primates (Epple, 1974), and rodents (Bronson, 1976; Brown, 1975; Price, 1975).

Other species scent mark with specialized skin glands that have been modified during evolution for this purpose (Quay, 1977). These glands may be on any part of the body. Their secretions can be deposited passively (i.e., while the animal is involved in other activities, such as sand-bathing or walking through heavy brush), but are usually rubbed onto objects or the ground with special motor patterns (Mykytowycz, 1970; Ralls, 1971; Thiessen & Rice, 1976). Gerbils and woodrats, for example, press their bellies to the ground to deposit secretions from a ventral gland (Fleming & Tambosso, 1980; Howe, 1977; Thiessen, Blum, & Lindzey, 1970; Thiessen, Wallace, & Yahr, 1973). Hamsters and water voles scent-mark with flank glands. Hamsters rub the glands directly on vertical objects (Johnston, 1975a). Water voles rake the soles of their hind feet over the gland and then scent-mark by drumming their feet on the ground (Jannett & Jannett, 1974). Some wild canids rub the roof of their lairs with a gland located just above the base of their tail (Ewer, 1968). Some lagomorphs, marsupials, and ungulates wipe facial or cephalic glands on grasses, bushes, tree boughs, and each other (Grau, 1976; Mykytowycz, 1968; Schaller, 1967; Schultze-Westrum, 1969). Marmosets scent-mark branches with glands near the genitalia (Epple, 1970, 1980). Lemurs wipe objects with scent glands on the arm (Jolly, 1966).

Feces are also well suited for scent-marking. Rabbits, vicuñas, and wolves are examples of species that do this. For rabbits, the feces simply provide a vehicle for depositing secretions from glands surrounding the anus (Mykytowycz,

1968). The rabbits coat some, but not all, of their feces with anal gland secretions and selectively deposit the marked fecal pellets in mounds at crossroads of trails traversing their warren. Uncoated pellets are not piled up this way. Vicuñas display a ritualized pattern of sniffing and kneading before defecating on communal dung piles (Franklin, 1980). Wolves defecate in prominent places such as at the tops of snow banks (Peters & Mech, 1975). Such dung piles and scat marks on snow provide visual as well as olfactory cues.

Problems in Identifying Particular Scent-Marking Behaviors with Aggression or Territorial Defense

Scent-marking behaviors can serve many functions. Odors deposited around an animal's living area may attract mates, repel intruders, enable young to regain the nest, or simply reassure the resident. Even a single scent could serve all these purposes if it contained a complex of odors or if the same odor elicited different responses from different categories of conspecifics. This is one reason why it is often difficult to determine the function of a particular scent-marking behavior. Nonetheless, clues can be gathered by collecting data on the conditions under which scent-marking occurs and by assessing the responses of conspecifics to marked areas.

If a marking response is shown most often by lactating females near the nest and if the young approach the scent, then we can be confident that the marking behavior facilitates mother–young reunions. The same scent may, of course, repel conspecific adults since it would be a good indicator that the marked area is occupied by an adult prone to vigorous defense. The mother and her young would benefit from this repellent effect if intruders attempt cannibalism, consume resources, or draw predators' attention to the nest. In this context, we could say that the marking behavior is related to territoriality or aggression. From either perspective, the critical thing is that the distributed odors denote the presence of a lactating female. The appropriate response to this information varies from one individual to another.

The same caveats apply when we consider scent-marking by adult males. If the marking behavior is shown primarily by sexually mature males during the breeding season and if the scent attracts receptive females, then it appears that the marking behavior subserves mating. Yet the same odors may influence aggressive interactions between males. Both functions of the scent-marks derive from the fact that the marks reliably inform conspecifics that the resident is a reproductively competent male likely to mate or attack depending on the sex of the intruder it encounters. Müller-Schwarze (1977) recommends use of the term "informer pheromone" in recognition of the fact that odors used in mammalian communication often provide information about the sender to the recipient without automatically eliciting physiological or behavioral responses, as pheromones do in invertebrates.

Perhaps the thorniest problem is to evaluate the role of scent-marks in territorial defense. Scent-marking would certainly be advantageous for territorial residents if the marks did form a true barrier to intrusions. This would enable the residents to avoid the risks of fighting in the service of territorial defense. Yet to benefit the residents, the scent-marks need not repel intruders. It may suffice that they limit the intruding animal's behavior. A resident male, for example, may tolerate intruders, even adult males, provided they move about cautiously and do not scent-mark, approach receptive females, consume limited resources, or dawdle unduly. In essence, observing that intruders enter marked areas does not mean that the scent-marks were ignored, nor does it imply that the marks do not function in the context of territorial defense, for they may decrease the probability of a fight or bias the outcome in the resident's favor.

Even in cases where residents invariably attack intruders, some intrusions would be expected because of selection pressures operating on the individuals that lack territories. From their perspective, the potential benefits of acquiring a breeding site may justify the risks of probing a marked territory to test the vigor of the resident. After all, intruders benefit if they avoid residents to whom they would lose but challenge residents they might be able to defeat.

Surely there are limits to how precisely scent-marks can predict the outcome of a fight. However, an intruder's experiences within marked areas can increase the information content of the marks through associative learning. A sound defeat by a resident should enhance the repellent power of the resident's scent-marks for that intruder. The intruder's reaction to scent-marks of other residents may also change through stimulus generalization. Similarly, encounters with other residents could improve discrimination. It is becoming increasingly clear that mammals often rely on experience for the normal development of responses to intraspecific olfactory signals (Cheal, 1975; Leon, 1980; Nyby & Whitney, 1980; Nyby, Thiessen & Wallace, 1970; Skeen & Thiessen, 1977; Yahr, 1977a). Mammals may actually be predisposed to associate odors with the consequences of social interactions, as they are to associate taste with gastrointestinal distress (Garcia, Hankins, & Rusiniak, 1974).

CONTROL OF SCENT-MARKING BY GONADAL STEROID HORMONES

Advantages to Hormonal Control of Marking Behavior

Earlier, I suggested that scent-marking evolved because it improved the dissemination of odors already used by conspecifics to monitor the animal's internal state and behavioral disposition. Secretions and excretions containing volatile compounds controlled by gonadal steroids are preadapted for such monitoring functions in regard to reproduction and aggression. Steroid hormone secretion directly reflects the functional state of the gonads and maintains the accessory sex organs. It also modulates courtship, copulatory behavior, and aggression and varies as a result of the animal's social interactions (see review by Leshner, 1978). Therefore odors controlled by these same hormones are useful cues for gauging the animal's reproductive state, predicting its social behavior, and assessing its social history. The accuracy of these signals and, therefore, their value for communication are maintained and even enhanced when the scent-marking behavior is controlled by the same hormone.

Sex Differences in Scent-Marking

If scent-marking behaviors are controlled by gonadal steroids, sex differences are expected. Observations in both the field and the laboratory suggest that sexual dimorphism is indeed one of the most common and conspicuous characteristics of mammalian scent-marking (see reviews by Johnson, 1973; Mykytowycz, 1970; Ralls, 1971; Thiessen & Rice, 1976). In species that employ specialized skin glands for scent-marking, the glands are often present in only one sex or are distinctly larger and/or more active in one sex than the other. Typically, though not always, the gland is more prominent in the male. Often the frequency of scent-marking also shows large sex differences, again usually favoring the males. Male gerbils, woodrats, rabbits, hares, and antelope all have more prominent scent glands and scent mark more often than conspecific females (Fleming & Tambosso, 1980; Gilbert, 1974; Howe, 1977; Moy, 1970; Mykytowycz, 1965; Ralls, 1971; Schaller, 1967; Thiessen, Blum, & Lindzey, 1970).

Sex differences are common in urinary and fecal marking as well. Male dogs urinate more often than females and are much more likely to raise their legs to direct urine onto targets (Sprague & Anisko, 1973). Urine-spraying is more often a problem with tomcats than with females (Hart, 1974a). Similarly, male mice deposit urine in many small spots whereas females leave fewer and larger pools (Desjardins, Maruniak, & Bronson, 1973; Maruniak, Owen, Bronson, & Desjardins, 1975). Male rabbits have larger anal glands than females and are primarily responsible for establishing the dunghills within a warren (Mykytowycz, 1968). Vicuñas of both sexes defecate and urinate exclusively on communal dung piles, but only the male incorporates this behavior into ritualized displays for territorial defense (Franklin, 1980).

Sex differences in scent-marking are not universal, though, and do not always favor males when they occur. Several laboratories report that female hamsters flank-mark as often as male hamsters, although their flank glands are smaller (Johnston, 1975a, 1977; Tiefer, 1970; Vandenbergh, 1971). In contrast, another report indicates that male hamsters mark considerably more than females do (Payne & Swanson, 1970). Female marmosets scent-mark as much or more often than males do (Epple, 1970, 1978).

Pubertal and Seasonal Changes in Scent-Marking

Field and laboratory observations also suggest that many marking behaviors develop at puberty and, in some species, show seasonal fluctuations that correspond closely to changes in gonadal activity (see discussions by Mykytowycz, 1970; Steiner, 1974; Thiessen & Rice, 1976). Most of the developmental data, though, are not detailed. A notable exception is the analysis of changes in scent-marking behavior of male gerbils approaching sexual maturity (Lee & Estep, 1971; Lindzey, Thiessen, & Tucker, 1968). The marking frequencies of gonadally intact male gerbils increase gradually between 75 and 100 days of age. This period of behavioral development corresponds to or lags slightly behind the time when blood titers of testosterone are increasing and when secretory activity of the ventral scent gland begins (Glenn & Gray, 1965; Rice, 1975). The increase in testosterone secretion causes the onset of scent-marking in male gerbils, as indicated by the fact that prepubertal castration prevents development of marking behavior and by the fact that injections of testosterone propionate (TP) induce the behavior prematurely.

In rabbits, sex differences in the size of the anal and submandibular (chin) glands also develop at puberty under the influence of the testes and ovaries (Mykytowycz, 1965, 1968; Wales & Ebling, 1971). If male rabbits are castrated before puberty, their chin and anal glands remain small and inactive. If they receive testosterone, their glands enlarge and secrete more sebum. Testosterone also stimulates the scent glands of female rabbits, but ovarian steroids play an inhibitory role. Females ovariectomized before puberty develop larger chin and anal glands than controls. At puberty, the sex difference in marking behavior also begins. Like development of the scent glands, it is apparently under hormonal control. Prepuberal castration decreases chin-gland marking by 90% in males. Prepuberally castrated females do not chin-mark at all.

In male tree shrews, a scent gland in the throat and upper chest area develops at puberty. Scent-marking with the gland begins then, too (von Holst & Buergel-Goodwin, 1975). In ground squirrels, scent-marking, which involves a variety of glands, is shown by juveniles and adults alike, but adult males scent-mark more often than females, and only adults make "marking trips" (Steiner, 1974).

Pronghorn antelope are an example of a species in which scent-marking and scent-gland activity show pronounced seasonal variations. In this species, males but not females have a gland beneath each ear that becomes enlarged and highly secretory in the spring. Changes in the scent-gland size parallel the annual cycle of spermatogenesis and antler growth (Moy, 1970; O'Gara, Moy, & Bear, 1971). Before and during the rut, the older bucks establish territories and breeding herds and scent-mark frequently with their subauricular glands (Gilbert, 1974).

Chin-gland marking in rabbits also occurs more often in the breeding season than at other times of year (Mykytowycz, 1965). The same is true for the various marking behaviors of ground squirrels. In this case, marking behavior can be enhanced out of season by exposing animals to artificially long days in the laboratory (Steiner, 1974).

Androgen Control of Scent-Marking in Adult Males

The data on sexual dimorphism and the development of scent-marking are consistent with the notion that gonadal hormones stimulate these behaviors. Proving this requires quantifying changes in scent-marking following direct ma-

TABLE I

Examples of the Effects of Testosterone on the Frequency of Scent-Marking by Adult Males

Species	Odor source	Intact	Castrated	Testosterone	Reference
Mouse	Urine	400/hr	30/hr	—	Maruniak, Desjardins, & Bronson, 1977
Rat, wild	Urine	31–35/24 hr	10–11/24 hr	36/24 hr	Price, 1975
Rat, domestic	Urine	18–20/24 hr	3–5/24 hr	38/24 hr	Price, 1975
Dog	Urine	7–8/min	5–7/5 min	—	Hart, 1974b
Hamster	Flank gland	13/8 min	4/8 min	15/8 min	Johnston, 1981
Mongolian gerbil	Ventral gland	31–42/10 min	1/10 min	38/10 min	Turner, 1975
Israeli gerbil	Ventral gland	5/5 min	0.5/5 min	3/5 min	Thiessen, Wallace, & Yahr, 1973
Woodrat	Ventral gland	0.4–3.1/6 min	0.4–1.6/6 min	4.6/6 min	Fleming & Tambosso, 1980
Rabbit	Chin gland	14/10 min	<1/10 min	—	Mykytowycz, 1965
Musk shrew	Flank gland	19–25/15 min	2/15 min	23–43/15 min	Balakrishnan & Alexander, 1976
Musk shrew	Perineal gland	10–11/15 min	1/15 min	12–17/15 min	Balakrishnan & Alexander, 1976
Musk shrew	Throat gland	6–8/15 min	0–2/15 min	2–6/15 min	Balakrishnan & Alexander, 1976
Tree shrew	Chin gland	4–25/10 min	0–2/10 min	2–13/10 min	von Holst & Buergel-Goodwin, 1975
Marmoset	Genital gland	8–9/10 min	9/10 min	—	Epple, 1978

TABLE II
Effects of Anosmia on the Frequency of Scent-Marking by Adult Males

Species	Control	Anosmic	Reference
Mouse	318–408/hr	103–123/hr	Maruniak, Darney, & Bronson, 1975
Gerbil	12–22/15 min	0–1/15 min	Baran & Glickman, 1970
Gerbil	8–9/10 min	1–2/10 min	Lumia, Westervelt, & Rieder, 1974
Gerbil	25–31/10 min	3/10 min	Wallen & Glickman, 1974
Dog	7/5 min	5/5 min	Hart & Haugen, 1971

nipulation of hormone levels. The most basic technique of endocrinology, endocrine-gland removal combined with hormone replacement, is appropriate here and has been used to study the role of testicular androgens in the control of scent-marking in males of several species. Some of this research has gone on to address questions of broader interest in the area of behavioral endocrinology, such as comparing the potencies of different steroid hormones, identifying the locus of hormone action within the central nervous system, and analyzing the role of hormones in the development of sex differences. Tables I, II, and III provide examples of the data discussed in this section.

Urine-Marking in House Mice

Male house mice scent-mark by repeatedly dabbing or dragging their penes on the substrate while releasing small amounts of urine. The behavior is measured by allowing a mouse to mark in a box lined with filter paper. Marking frequency is then determined by counting the blue fluorescent spots and trails that appear when the filter paper is viewed under ultraviolet (UV) light. When placed in a novel apparatus, a gonadally intact male mouse may mark 300–400 times in an hour (Bronson, 1976; Desjardins *et al.*, 1973).

Castration reduces marking frequency by 85%–90% (Bronson & Desjardins, 1974; Maruniak, Desjardins, & Bronson, 1977), to a level at or below that characteristic of adult females (Maruniak, Owen, Bronson, & Desjardins, 1975). Subcutaneous implants of testosterone that maintain plasma concentrations of

TABLE III
Frequency of Scent-Marking by Adult Males as a Function of Social Status

Species	Dominant	Subordinate	Reference
Mouse	400/hr	20/hr	Maruniak, Desjardins, & Bronson, 1977
Mouse	1,000/12 hr	2–4/12 hr	Desjardins, Maruniak, & Bronson, 1973
Woodrat	3/6 min	0.05/6 min	Fleming & Tambosso, 1980
Hamster	23/10 min	1–2/10 min	Johnston, 1975a
Gerbil	15–25/5 min	2–10/5 min	Yahr, 1977a
Gerbil	21–47/30 min	2–14/30 min	Thiessen, Lindzey, & Nyby, 1970
Rabbit	6–8/hr	0–1/hr	Mykytowycz, 1965
Marmoset	5–55/hr	1–3/hr	Epple, 1970

this hormone at or above those found in intact males prevent this behavioral change. Thus the decrease in urine-marking after castration is caused by the loss of androgen.

Other data show that hormonal stimulation is necessary but not sufficient for high levels of marking to occur. The marking rates of gonadally intact male mice drop off quickly unless the test situation is novel (Maruniak, Owen, Bronson, & Desjardins, 1974). The olfactory system seems particularly important for detecting either biological or physical novelty since anosmia (produced by perfusing the nostrils with zinc sulfate) inhibits the marking response to either a novel environment or an estrous female (Maruniak, Darney, & Bronson, 1975).

Social rank is also critical. Social subordination suppresses scent-marking as effectively as castration or more so (Desjardins *et al.*, 1973; Maruniak *et al.*, 1977). Although the stress of defeat can reduce testosterone secretion in mice, subordination does not inhibit scent-marking through this mechanism. Castrated male mice bearing testosterone implants show the same precipitous drop in scent-marking when defeated as do gonadally intact males, despite the fact that only the males with testes can show changes in testosterone secretion (Maruniak *et al.*, 1977).

Urine-Marking in Rats

Wild and domesticated Norway rats also use urine to scent-mark. Two different procedures have been developed to measure this response. One procedure involves placing a wire-mesh block covered with a paper towel in the male rat's home cage and later determining the area of toweling covered with urine. The male marks the block by rubbing his hindquarters over the top of it, by using a leg-lift standing posture to urinate onto it, or by urinating while walking alongside the block with one foot on top of it (Brown, 1975). The second procedure for assessing scent-marking consists of suspending a threaded metal rod a few centimeters above the floor of the male's home cage and later scoring, under UV light, the number of points along the length of the rod that are covered with urine. In this case, the male deposits the urine by arching its back slightly (thereby lowering its penis) while crossing over the rod at right angles (Price, 1975).

Using either technique, scent-marking scores decrease after castration but increase again if the castrates are given injections of TP or subcutaneous implants of testosterone (Brown, 1977, 1978; Price, 1975; Scouten, Burrell, Palmer, & Cegavske, 1980). As is the case for mice, no data are available on dose–response relationships for maintaining or reinstating the behavior. It is worth noting, though, that the hormone doses tested in rats (0.5–1.5 mg TP/day or capsules 60 mm long) exceed those needed to reinstate copulatory behavior in males of the same strain by at least an order of magnitude (Damassa, Smith, Tennett, & Davidson, 1977; Davidson & Bloch, 1969), although the 60-mm capsules produce plasma concentrations of testosterone only slightly above the range of gonadally intact males (Damassa *et al.*, 1977).

Injections of estradiol benzoate (EB) also reinstate urine-marking in castrated male rats (Brown, 1978). Progesterone neither enhances this effect nor promotes scent-marking on its own.

Urine-Spraying in Cats

Urine-spraying by male cats is difficult to analyze quantitatively because it occurs relatively infrequently. It appears, though, that this behavior is controlled by gonadal androgens. This conclusion is based on data from a telephone survey of cat owners whose pets were neutered in an attempt to control objectionable behaviors, including roaming, fighting, and urine-spraying (Hart & Barrett, 1973). According to the owners, castration eliminated or decreased spraying in most cases. Most (78%) reported rapid changes. A few (9%) said the behavior declined gradually over months. Others observed no change. Needless to say, hormonal reinstatement of spraying behavior was not attempted in this study.

Urine-Marking in Dogs

An exception to the rule that castration decreases urinary scent-marking comes from research on domestic dogs. In this species, castration in adulthood alters neither the frequency nor the posture of male urination behavior (Beach, 1974; Hart, 1974b). Males castrated as juveniles, just prior to puberty, also develop the normal masculine urination style. Nonetheless, treatment of castrates with gonadal steroids increases marking frequency two- to fivefold. This is true whether the castrates receive TP or EB plus progesterone.

Dogs are also unusual in that anosmia has minimal effects on marking. Although male dogs typically smell objects before marking them and prefer urinating where dogs have marked before (Sprague & Anisko, 1973), anosmia (tracheostomy plus occlusion of the nares) increases the latency and decreases the frequency of urine-marking only slightly (Hart & Haugen, 1971).

Scent-Gland Marking in Golden Hamsters

Male hamsters scent-mark with pigmented sebaceous glands on their flanks. This behavior is usually studied in the laboratory by placing a male hamster into a cage soiled by another hamster or by pairing the male with an adult of either sex. One mark is scored every time the hamster arches its back and presses one of the glands against a cage wall while moving forward.

The marking behavior induced by exposure to another male's odors is androgen dependent, as is the integrity of the scent glands themselves. Castration decreases the size and pigmentation of the scent gland (Hamilton & Montagna, 1950; Whitsett, 1975) to female levels and reduces marking frequency to 25% of precastration values (Johnston, 1981). Implanting castrated males with silastic capsules of TP that generate plasma testosterone levels comparable to those of intact males (Turek, 1977) completely restores the behavioral response (Johnston, 1981).

Social stimuli are at least as important as gonadal hormones, though, in the control of hamster flank-marking. Like male mice, male hamsters show a large and rapid drop in scent-marking after social subordination (Johnston, 1975a). Stress-induced decreases in testosterone secretion may contribute to this, but it is unlikely that they can fully account for the decline since marking ceases more quickly and more completely after defeat than after castration (Johnston, 1975a, 1981). Similarly, intact males do not flank mark when odors of estrous females are present, even though these odors increase testosterone secretion in males (Johnston, 1975a; Macrides, Bartke, Fernandez, & D'Angelo, 1974). In the context of aggression, male hamsters show similar levels of flank-marking whether castrated or gonadally intact (Tiefer, 1970; Whitsett, 1975). Thus, testicular androgens are neither sufficient nor necessary for flank-marking to occur. It would be interesting to know whether castrates with aggressive experience would continue to scent-mark at intact male levels when not directly exposed to an opponent. As in the case of surgical castration, suppressing testosterone secretion by exposing male hamsters to short photoperiods does not affect flank-marking (in aggressively experienced males), but the scent glands regress (Garrett & Campbell, 1980).

Scent-Gland Marking in Gerbils

Gerbils scent-mark by rubbing a ventral gland on objects as they move forward over them. This behavior is usually measured by observing the number of times a gerbil marks marbles or small pegs attached to the floor of a simple test arena. The frequency of scent marking does not habituate over tests as it does in mice, nor is it noticeably affected by conspecific odors in the test arena unless those odors have been associated with defeat (Nyby *et al.*, 1970; Thiessen, Lindzey, & Nyby, 1970; Yahr, 1977a; Yahr, Jackson, Newman, Stephens, & Clancy, 1980).

More is known about hormonal control of scent-marking in gerbils than in any other species. In males, the frequency of scent-marking and the integrity of the scent gland are both maintained by gonadal androgens. Castration causes a decline in scent-marking and scent-gland size that can be prevented or reversed by injecting TP or by implanting silastic capsules of testosterone under the skin (Anisko, Christenson, & Buehler, 1973; Lumia, Westervelt, & Rieder, 1975; Thiessen, Friend, & Lindzey, 1968; Turner, 1975; Yahr, Newman, & Stephens, 1979). The capacity of testosterone for reinstating scent-marking in castrates is a function of the dose (Blum & Thiessen, 1971; Thiessen *et al.*, 1968; Yahr *et al.*, 1979). Very large doses of testosterone can even bring marking frequencies of castrates above the levels of intact controls. As in the case of sexual behavior, more testosterone is needed to restore marking behavior once it has ceased than is needed to prevent the decline (Yahr *et al.*, 1979).

As noted earlier, the normal development of scent-marking at puberty is stimulated by increased secretion of testosterone from the testes (Lindzey *et al.*, 1968). Among adult males, marking frequency correlates positively, but not always significantly, with the amount of testosterone in the blood. The correlation between plasma testosterone concentration and scent-marking frequency varies from 0.23 to 0.72 for males 90 days of age or older. Nonetheless, males with high titers of testosterone outscore males with low titers of testosterone when only the test closest to blood-sampling is considered (Rice, 1975). Marking frequency correlates positively and significantly ($p = 0.54$) with seminal vesicle weight (Blum & Thiessen, 1971). The size of this accessory sex organ probably provides a better index of the average concentration of testosterone in the blood than do the moment-to-moment fluctuations. Yet even prolonged decreases in testosterone secretion that are large enough to decrease seminal vesicle weight do not necessarily affect marking behavior. Male gerbils that must satisfy their water requirements from food metabolism have smaller testes, seminal vesicles, and ventral scent glands than males that have water freely available, but their marking scores are not significantly reduced (Yahr & Kessler, 1975).

Individual differences in scent-marking are also related to differences in sexual activity among the males. Males that mount receptive females during standardized tests scent-mark more often than males that do not mount (Yahr *et al.*, 1980). Since sexual behavior is androgen dependent in male gerbils (Yahr *et al.*, 1979), these data suggest that individual differences in scent-marking may be affected by differences in circulating levels of testosterone and by differences in sensitivity of the neural substrate to this hormone. Support for the notion that both factors are involved comes from the effects of TP on males that do not scent-mark while gonadally intact. These males have smaller scent glands than males that do mark (Yahr, 1976a), which suggests that their endogenous androgen levels are low. This is consistent with the fact that TP increases scent-gland size in these males but not in spontaneously marking males (Thiessen *et al.*, 1968; Yahr, 1976a). In contrast, TP does not significantly improve their marking scores. Thus, the neural sensitivity of these low-marking males to testosterone is also low.

A variety of steroids besides testosterone have been tested for their ability to stimulate scent-marking in male gerbils. Among the effective ones is estradiol. Whether injected or implanted in silastic capsules, estradiol is as potent as testosterone for maintaining and reinstating scent-marking in castrated males (Yahr & Stephens, unpublished observations; Yahr & Thiessen, 1972). In contrast, EB is less potent than TP (Nyby & Thiessen, 1971). In gonadally intact males, EB decreases marking scores (Thiessen, Lindzey, & Nyby, 1970). This is apparently due to the fact that EB suppresses endogenous androgen secretion through negative feedback and is unable to compensate fully for the androgen loss. Yet even when EB and TP are injected together, EB partially counteracts TP's behavioral effects (Nyby & Thiessen, 1971). The explanation offered is that estradiol competes with testosterone for receptor sites but is less able to activate the behavior once bound. This explanation does not fit well, though, with the fact that estradiol and testosterone are equally effective in their nonesterified forms, even when ceiling effects are not involved.

Two other estrogens, estriol and estrone, stimulate scent-marking in castrated males, although at the doses tested, estrone was active only when applied directly to the brain (Yahr & Thiessen, 1972). Neither progesterone nor 17α-hydroxyprogesterone reinstates marking behavior, nor does progesterone alter

EB's effects on the marking behavior of gonadally intact males (Thiessen, Lindzey, & Nyby, 1970).

Three other androgens that activate scent-marking in male gerbils are 19-nortestosterone, 6α-fluorotestosterone, and 5α-dihydrotestosterone (DHT). The effectiveness of these hormones provides useful clues about how androgens act on the brain to modify marking behavior. In particular, they are helpful in evaluating the role of hormone metabolism. In the analysis of other androgen-dependent behaviors, such as copulatory behavior or aggression (see, for example, Brain and Bowden's chapter in this volume), considerable effort has been devoted to testing the hypothesis that testosterone affects the brain and behavior through its metabolism to estradiol and/or DHT. The fact that 19-nortestosterone is as potent as testosterone for maintaining and reinstating scent-marking (Yahr & Stephens, unpublished observations; Yahr & Thiessen, 1972) suggests that testosterone's aromatization to estradiol cannot account for testosterone's effects. In standard assay systems, 19-nortestosterone is aromatized only 20% as efficiently as testosterone (Ryan, 1960). However, the data should be interpreted cautiously. It may very well be that hormone-accumulating cells in the gerbil's brain aromatize 19-nortestosterone more effectively than the placental microsomes used in the standard assay.

The same precautions are needed in evaluating the effects of 6α-fluorotestosterone. This hormone is not aromatized in human placenta (Gual, Morato, Hayano, Gut, & Dorfman, 1962), but it apparently is aromatized in rat brain (Nordeen & Yahr, 1981). Therefore, its ability to stimulate scent-marking in castrated male gerbils (Yahr, 1976b) may support rather than contradict the idea that metabolism to estrogen is involved. Nonetheless, the fact that estradiol is no more potent than testosterone for activating scent-marking (indeed, it may be less so) make it unlikely that estradiol is testosterone's active form.

The same consideration makes it unlikely that testosterone activates scent-marking through its metabolism to DHT. We only recently determined that DHT can promote scent-marking in males. Earlier research suggested that DHT was without effect because it did not facilitate marking when implanted into the brain (Thiessen, Yahr, & Owen, 1973). Yet when given systemically in silastic capsules, DHT maintains and reinstates scent-marking in castrated males. Its ability to do this suggests again that aromatization is not essential for androgenic stimulation of marking behavior, since DHT is considered to be nonaromatizable (Ryan, 1960). The fact that DHT is no more potent than testosterone also argues against the possibility that DHT is the behaviorally active form of testosterone. The possibility remains that two metabolites of testosterone, estradiol and DHT, synergize to produce its behavioral effects.

Other androgens that have been tested but that lack behavioral effects are androstenedione, androstenediol, dehydroepiandrosterone, epitestosterone, and 1α-methyltestosterone (Yahr, 1976b; Yahr & Thiessen, 1972). All of these steroids except dehydroepiandrosterone stimulate scent-gland growth.

In gerbils, as in mice, gonadal androgens are necessary but not sufficient to insure that frequent scent-marking occurs. Social stimuli play an equally important role. For example, housing male gerbils in groups of four or more decreases marking scores (Hull, Langan, & Rosselli, 1973; Thiessen, Lindzey, Blum, & Wallace, 1970; Yahr *et al.*, 1980). Decreasing testosterone secretion probably contributes to this since grouping lowers testis and seminal vesicle weight and diminishes scent-gland size (Hull *et al.*, 1973; Thiessen, Lindzey, Blum, & Wallace, 1970). It is doubtful, though, that changes in testosterone secretion can fully explain the dramatic variations in scent-marking that accompany social victory and defeat (Nyby *et al.*, 1970; Thiessen, Lindzey, Blum, & Wallace, 1970; Thiessen, Owen, & Lindzey, 1971; Yahr, 1977a). Like mice, gerbils show large drops in marking frequency when socially subordinated, but unless the dominant and subordinate males are living with each other, this socially induced decrease is limited to situations that contain cues associated with defeat. Such behavioral changes associated with setting may well reflect hormonal change, but testosterone is probably not the critical hormone. The same comments apply in regard to the rapid increase in scent marking that winners display in a newly acquired territory (Thiessen *et al.*, 1971).

Gerbils also resemble mice in that impairing the sense of smell disrupts marking behavior. Males made anosmic by olfactory bulbectomy stop scent-marking completely, and even massive amounts of TP can only partially override this effect (Baran & Glickman, 1970; Lumia *et al.*, 1975; Thiessen, Lindzey, & Nyby, 1970). The decreases in marking after destruction of the olfactory epithelium (by zinc sulfate perfusion) are nearly as severe (Wallen & Glickman, 1974). These deficits may be temporary though. No decrease in scent-marking was noted after olfactory bulbectomy or peripheral deafferentation of the olfactory bulbs when testing was delayed until two months after surgery (Hull, Hamilton, Engwall, & Rosselli, 1974).

Scent marking in a species of gerbil, *Meriones tristrami,* native to Israel, is also controlled by gonadal androgens (Thiessen, Wallace, & Yahr, 1973). Marking frequency of *M. tristrami* males declines after castration but increases again if the males are fed methyltestosterone.

Scent-Gland Marking in Woodrats

Desert woodrats, like gerbils, scent-mark by rubbing objects with a ventral gland. The gland and the behavior are apparently both controlled by gonadal androgens in males (Clarke, 1975; Fleming and Tambosso, 1980); however, as in other species, the extent to which hormones regulate scent-marking is constrained by social factors. Isolated males lacking sexual and aggressive experience obtain low marking scores, similar to those of females, although there is no reason to suspect that their testosterone secretion is impaired. In contrast, isolated males with both sexual and aggressive experience scent-mark considerably more often than females or inexperienced males. The suggestion that testicular androgens influence scent-marking in male woodrats is based on the observation that silastic implants of testosterone stimulate scent-marking in castrated males, provided the males are isolated or socially dominant. As in mice, subordinate male woodrats do not scent-mark even though testosterone levels are artificially maintained. The stimulatory effects of testosterone are unusual, though, in that they last for a month or more after the hormone capsule is removed. Estradiol capsules do not facilitate scent-marking in woodrat males.

Scent-Gland Marking in Musk Shrews

Musk shrews scent-mark with urine and with the secretions of several skin glands (Balakrishnan & Alexander, 1976). Both sexes rub vertical objects with glands on their flanks. Males (and females?) also rub throat and perineal glands on the ground and deposit trails of urine by urinating while moving forward and dragging their hindquarters along the ground. In male musk shrews, all of the marking behaviors that involve scent glands are androgen dependent, as are the glands themselves (Dryden & Conway, 1967). After castration, the frequencies of flank-gland marking, throat-rubbing, and perineal rubbing gradually decline, but they can be restored to precastration levels or above by feeding the shrews methyltestosterone linguits or powdered ethinylestradiol (Balakrishnan & Alexander, 1976). The frequency of urination is apparently unaffected by either castration or hormone treatment.

Scent-Gland Marking in Tree Shrews

The tree shrews native to Thailand scent-mark tree branches and other objects by extending their heads and using a back-and-forth motion to apply secretions from a jugulosternal gland. Because of the posture involved, this behavior has been dubbed "chinning." Male tree shrews chin-mark often, especially in novel environments, although individual differences in chinning frequency are large. Chinning is controlled by gonadal androgens in males, as is the scent gland itself (von Holst & Buergel-Goodwin, 1975). After castration, chinning frequency declines to 10%–25% of precastration values. The behavior can be restored in a dose-dependent fashion by injecting castrated males with TP or testosterone depots.

The degree to which males continue to scent-mark after castration and their maximal response to hormone therapy are related to the amount of chinning they displayed while gonadally intact. The fact that castrated males given large doses of TP scent-mark about as often as they did before castration suggests that the amount of testosterone normally circulating in the blood does not limit marking behavior. This suggestion is supported by the observation that supplemental injections of TP do not appreciably elevate chinning scores of gonadally intact males.

The observation that some males continue to scent-mark after castration suggests that either some degree of chinning is independent of hormonal control or that the residual levels of chinning are maintained by other hormones, for instance, androgens secreted by the adrenal cortex. The latter possibility is supported by the fact that injections of ACTH (adrenocorticotropic hormone) cause a brief increase in castrates' chinning scores.

Both EB and progesterone also stimulate scent-marking in castrated males, provided odors of other males are absent. In contrast, the chinning behavior of gonadally intact males does not change in response to EB or progesterone unless odors of other males are present, in which case chinning frequency declines.

Scent-Gland Marking in Marmosets

Saddle back tamarins scent-mark with large skin glands in the circumgenital–suprapubic area. These glands involve the skin of the external genitalia and extend forward along the midline as darkly pigmented, hairless cushions. Marking behavior involves rubbing these glands on perches or other objects, on each other, or on the ground. The tamarins often urinate at the same time.

Gonadal androgens play a relatively small role in the control of scent-marking in male tamarins. Castration in adulthood does not affect how often males scent-mark either objects or their female partners (Epple, 1978). Prepuberal castration reduces marking, but even its effects are slight. Prepuberally castrated male tamarins living in pairs with adult females scent-mark less often than paired, gonadally intact males when an unfamiliar conspecific intrudes into their living areas (Epple, 1981). However, when intruders are absent, scent-marking of novel objects is not significantly reduced. Prepuberal castration also impairs scent-gland development, but castration in adulthood does not (Epple, 1978, 1981).

Estrogen and Progesterone Control of Scent-Marking in Adult Females

Considerably less information is available on hormonal control of scent-marking behaviors of females. This is not surprising given that, in many species, females scent-mark much less often than males. The data that are available for most species are easily summarized.

Female house mice scent-mark with urine as males do, but they mark less often and their behavior is not gonadally controlled. Marking frequency does not change with the estrous cycle or after ovariectomy (Maruniak, Owen, Bronson, & Desjardins, 1975). In contrast, ovariectomy virtually eliminates the low levels of urine-marking in female rats and the low levels of ventral-gland marking in female woodrats (Brown, 1977, 1978; Fleming & Tambosso, 1980). Testosterone implants (woodrats) or TP injections (rats) stimulate scent-marking in both these species, but ovarian steroids (estradiol implants in woodrats; EB, progesterone, or EB plus progesterone injections in rats) do not. In female woodrats, as in males, the stimulatory effects of testosterone capsules persist for weeks after they are removed. In ovariectomized female dogs, injections of TP or EB plus progesterone increase urination frequency but do not affect urination posture.

Flank-gland marking by female hamsters does not vary from one stage of the estrous cycle to another (Johnston, 1977). However, female hamsters also scent-mark with vaginal secretions, which they deposit using a behavior similar to a perineal drag. This behavior does change across the female's reproductive cycle in a manner suggesting that it serves as an advance advertisement of sexual receptivity (Johnston, 1977, 1979). Females show the highest levels of vaginal marking the day before estrus. Marking levels are low at estrus *per se* and remain low, if the female becomes pregnant, until the end of lactation.

In female lemurs, genital marking peaks at estrus (Jolly, 1966), suggesting that it serves sexual rather than aggressive functions. In saddle back tamarins, females scent-mark more than males, but ovarian steroids apparently are not involved. Preliminary data indicate that ovariectomy does not alter the frequency of marking behavior (Epple, unpublished observations discussed in Epple, 1978).

More detailed information is available on hormonal control of scent-marking behavior in female gerbils. In this species, as in many others, scent-marking frequencies of most females are low. However, a few females scent-mark as much as males, and nearly all females show pronounced increases in scent-marking during pregnancy and lactation. These high levels of scent-marking, when they occur, are controlled by ovarian hormones. Ovariectomy reduces marking frequency in postpartum females and in cycling females selected for high marking scores (Owen, Wallace, & Thiessen, 1974; Wallace, Owen, & Thiessen, 1973; Yahr & Thiessen, 1975). Marking behavior can be restored after ovariectomy in spontaneously high-marking females by injections of EB (Yahr & Thiessen, 1975). Progesterone does not enhance the effectiveness of EB in these females, nor is it ever effective alone (Owen & Thiessen, 1973, 1974).

The low levels of scent-marking characteristic of most cycling female gerbils are not under steroid hormone control. They do not change as a result of ovariectomy, adrenalectomy, or a combination of the two (Thiessen *et al.*, 1971; Turner, 1975; Whitsett & Thiessen, 1972). These low-marking females can be stimulated to scent-mark by being supplied with exogenous hormones, but the effectiveness of these treatments is variable. TP is the most consistently effective steroid, but it does not necessarily make females mark as much as males (Lumia, Canino, & Drozdowski, 1976; Owen & Thiessen, 1973; Thiessen & Lindzey, 1970; Turner, 1975; Yahr, 1976a). EB alone does not promote marking, nor does it enhance TP's effects (Owen & Thiessen, 1973, 1974; Thiessen & Lindzey, 1970; Yahr, 1976a; Yahr & Thiessen, 1972). EB is also ineffective when injected simultaneously with progesterone (Owen & Thiessen, 1973); however, sequential injections of EB followed by progesterone sometimes elicit large increases in scent-marking (Owen & Thiessen, 1974). The effects do not vary predictably, though, with the dose of either steroid, and sometimes scent-marking does not increase at all (Yahr, 1976a; Yahr & Thiessen, 1975). The lactation-induced increase in scent-marking is also blunted in these spontaneously low-marking females (Yahr, 1976a).

Other steroids that have been studied for their effects on scent-marking in (randomly selected) ovariectomized female gerbils are pregnenolone, 17α-hydroxyprogesterone, estrone, and androstenedione (Owen & Thiessen, 1973). None of these hormones increases marking scores. Androstenedione is also ineffective in ovariectomized high-marking females when implanted directly into the brain (Owen *et al.*, 1974); however, it does elevate marking scores when given systemically to low-marking females that are gonadally intact. Why androstenedione acts only in the presence of the ovary is unknown. This difference is not observed with TP (Turner, 1975).

Somatotropin, a growth-promoting hormone from the pituitary, does not affect scent-marking frequency in intact females (Thiessen & Lindzey, 1970).

In female gerbils, as in males, anosmia (neonatal olfactory bulbectomy) virtually eliminates marking behavior (Lumia *et al.*, 1976). In this case, TP injections did not overcome the deficit, but the dose used was considerably below the dose that partially restored scent-marking in males (Thiessen, Lindzey, & Nyby, 1970).

Neuroendocrine Control of Scent-Marking in Adults

Role of the Hypothalamus–Preoptic Area

Attempts to identify the neural substrates of scent-marking in males have been limited to four species—gerbils, rats, cats, and dogs—but the results reveal surprising consistency. In each case, the medial preoptic area (MPOA)–anterior hypothalamus (AH) or their projections to the medial forebrain bundle are essential for marking to occur and/or mediate androgen control of the response.

This portion of the brain attracted attention because it affects other sexually dimorphic behavior patterns, most notably copulatory behaviors, that are sensitive to gonadal androgens in males. For example, lesions of the MPOA–AH eliminate or severely disrupt sexual behavior in adult males despite adequate hormonal stimulation (Ginton & Merari, 1977; Hart, 1974c; Hart, Haugen, & Peterson, 1973; Slimp, Hart, & Goy, 1978; Twiggs, Popolow, & Gerall, 1978; Wheeler & Crews, 1978). Implanting testosterone directly into this area, but not others, reinstates sexual activity in castrated males (Barfield, 1969, 1971; Morgantaler & Crews, 1978; Smith, Damassa, & Davidson, 1977; Wada & Gorbman, 1977). Cells in this area of the brain selectively accumulate steroid hormones from the blood (Morrell, Kelley, & Pfaff, 1975; Stumpf & Sar, 1978). This region also displays a number of anatomical sex differences in rodents (Gorski, Gordon, Shryne, & Southam, 1978; Greenough, Carter, Steerman, & DeVoogd, 1977; Raisman & Field, 1973).

In gerbils, the MPOA–AH is the neural locus for androgen control of scent-gland marking in males. Castrated male gerbils resume scent-marking when testosterone, TP, estradiol, or other behaviorally active steroids are implanted unilaterally into this area (Thiessen & Yahr, 1970; Thiessen, Yahr, & Owen, 1973; Yahr & Thiessen, 1972). The effectiveness of TP implants in the MPOA–AH are a function of implant size, but even large implants are ineffective at other brain sites.

The portions of the MPOA–AH that control scent-marking in male gerbils can be distinguished from those controlling sexual behavior (Yahr, Commins, Jackson, & Newman, 1982). Lesions confined to the middle (anterior–posterior) of the MPOA severely disrupt copulatory behavior in male gerbils but produce only temporary decrements in scent-marking that can be overridden by testosterone. The critical regions for the control of marking behavior seem to be anterior and/or posterior to this site, for example, near the diagonal band of Broca or AH (Yahr *et al.*, 1982). The anterior sites, at least, are distinct from those that seem to control secretion of LH (luteinizing hormone) in males (Turner, 1973).

In female gerbils, the preoptic area, AH, and septum have been implicated in the control of marking behavior (Owen *et al.*, 1974). Unilateral implants of TP or EB into the MPOA–AH reinstate scent-marking in ovariectomized females that originally had high marking scores. Progesterone does not potentiate the effect of either steroid, and TP and EB are no more effective together than alone. When EB is implanted bilaterally, marking levels are highest when the hormone is in the AH. The lateral preoptic area, near the medial forebrain bundle, is the next most effective site, followed by the medial septum.

In rats, the MPOA–AH may mediate androgen control of urinary scent-marking behavior (Scouten *et al.*, 1980). In this species, castration reduces scent-marking scores of males to roughly half of control values, whereas sexual behavior ceases entirely. Severing the outputs of the MPOA–AH that travel laterally toward the medial forebrain bundle mimics these castration effects. These data suggest that the role of the MPOA–AH is limited to mediating the androgenic effects. On the other hand, it may simply be that the cells in the MPOA–AH that mediate the residual (androgen-insensitive) component of scent marking have axons that project in a different plane. It would be interesting to know whether lesions of the MPOA–AH itself would produce larger deficits in scent-marking.

In dogs it is clearly the case that the MPOA–AH exerts a greater influence on the control of scent-marking than can be accounted for in terms of androgenic effects. In dogs, scent-marking is not under hormonal control, in the sense that castration of juvenile or adult males does not affect marking behavior. Nonetheless, lesions of the MPOA–AH in adult males greatly reduce marking scores (Hart, 1974c). Lesioned males continue to use the leg-lift urination style, but the frequency of urination declines. The extent of the deficit in marking behavior corresponds closely to the extent of disruption in sexual behavior produced by the same lesions. Interestingly, when the lesions are made earlier in development, scent-marking and sexual behavior are differentially affected (Hart & Ladewig, 1979). In adult dogs that received MPOA–AH lesions before puberty, copulatory behavior is severely impaired but urine-marking behavior is spared.

Lesions of the MPOA–AH also reduce urine-spraying in male cats (Hart & Voith, 1978). Because of the difficulty of studying urine-spraying in the lab, these data were collected by pet owners who accepted neurosurgery as an alternative to euthanasia for stopping urine-spraying of male cats in which the problem behavior persisted despite castration and treatment with long-acting progestin (B. L. Hart, personal communication, 1978). The owners reported that urine-spraying stopped almost completely after the lesions were made. The only exceptions occurred when the cats were taken to strange surroundings (e.g., a relative's home). In cats, as in other species, such lesions also disrupt male sexual behavior (Hart *et al.*, 1973). Interestingly, seemingly identical lesions do not reduce urine-spraying in female cats that show the behavior (Hart & Voith, 1978).

Mechanism of Androgen Action on the Brain

Research on how androgens act within the brain to control marking behavior has been undertaken only in gerbils. The data are consistent with the notion that androgens activate scent marking by modulating gene expression (RNA and protein synthesis) within MPOA–AH cells. This conclusion is based primarily on the ability of antibiotics to block the behavioral effects of testosterone or TP when implanted into the brain (Thiessen, Yahr, & Owen, 1973). Actinomycin D, for example, is an antibiotic that blocks transcription of RNA from DNA. When mixed with testosterone and implanted unilaterally into the MPOA–AH of castrated male gerbils, actinomycin D completely blocks the induction of scent-marking that occurs when the hormone is implanted alone. Puromycin, an antibiotic that prevents translation of RNA into protein, produces identical inhibitory effects.

Other research, again limited to gerbils, has sought clues as to which neurotransmitters are involved. The data here suggest that variation in production or turnover of acetylcholine, rather than serotonin or catecholamines, may mediate hormonal control of marking behavior (Yahr, 1972, 1977b). When applied to preoptic area cells, pilocarpine nitrate, a drug that mimics the effects of acetylcholine on postsynaptic receptors, decreases the ability of testosterone to induce scent-marking in castrated males. Atropine sulfate, a competitive inhibitor of acetylcholine, potentiates the hormone's behavioral effects. Thus testosterone may stimulate scent-marking in male gerbils by decreasing the amount of acetylcholine produced or released or by otherwise interfering with its postsynaptic effects.

Sexual Differentiation of Scent-Marking

Sex differences in scent-marking behaviors could result simply from sex differences in hormone secretion among adults. Alternatively, males and females may also differ in their capacities to respond to hormonal stimulation. In mammals, sex differences in behavior often involve sex differences in sensitivity to gonadal steroids that develop as a result of differential exposure to testicular androgens during a critical period of sexual differentiation before or shortly after birth. The effects of early exposure to gonadal androgens on the development of scent-marking have been studied in gerbils and dogs.

Although many studies have examined hormonal stimulation of scent-marking in gerbils, only three have directly compared the responses of males and females (Clancy, 1978; Holman, 1977; Turner, 1975). Each of these studies indicates that when the dose of testosterone is held constant, ovariectomized females obtain lower marking scores than castrated males. The greater sensitivity of males to testosterone reflects their exposure to gonadal androgens around the time of birth. Sensitivity to testosterone within each sex can be raised or lowered by manipulating androgen levels neonatally.

For example, male gerbils castrated on the day they are born, or one day later, scent-mark no more when given TP or testosterone in adulthood than normal females do (Clancy, 1978; Holman, 1977; Turner, 1975). If castration is delayed until three, four, or five days after birth, the males scent-mark more than females but less than normal males when given TP as adults (Turner, 1975). Conversely, giving female gerbils TP neonatally increases their sensitivity to TP as adults. The magnitude of this effect also depends on how old the developing female is when exposure to androgen occurs. Females scent-mark as often as males when given TP in adulthood if they received an injection of TP four days after birth or earlier (Holman, 1977; Turner, 1975). Delaying the neonatal TP injection until seven or ten days after birth lessens the adult response (Turner, 1975).

Precisely when the critical period terminates is unknown. Turner (1975) reported that female gerbils given TP (100 μg) five days after birth scent-mark as often as normal males when given TP injections as adults. In contrast, Clancy (1978) found that females given TP (100 μg) five days after birth display no more scent-marking in adulthood than normal females when implanted subcutaneously with capsules of testosterone. The differences between the two studies may reflect differences in the method of hormone administration to the adults (injections versus implants). They could also be due to differences in exposure to ovarian steroids before and during hormone treatment in adulthood. Clancy (1978) ovariectomized his female subjects. Turner (1975) did not. Thus the females in Turner's study were exposed either to continuously high levels of estrogen (if neonatal exposure to TP caused anovulatory sterility) or to cyclic changes in estrogen and progesterone.

The role of the MPOA–AH in sexual differentiation of scent-marking has also been studied in gerbils. Considering the importance of this brain region in the hormonal control of scent-marking in adults, and considering that the anatomy of the MPOA–AH develops differently in males and females of other species as a result of exposure to androgens neonatally (Gorski *et al.*, 1978; Greenough *et al.*, 1977; Raisman & Field, 1973), it is reasonable to expect that anatomical changes in the MPOA–AH of gerbils may accompany sexual differentiation of marking behavior.

Sex differences do exist in the sizes of cell nuclei in the MPOA–AH of gerbils (Clancy, 1978). On the average, cell nuclei in the MPOA–AH of male gerbils are smaller than those in females. The relationship of this sex difference to scent-marking is not clear, though, since cell nuclear size did not change in males castrated on the day of birth. Cell nuclear size did not change in females either when they were given TP five days after birth; however, in this study neonatal TP treatment did not affect the females' marking behavior. Interestingly, this anatomical sex difference is not confined to the MPOA–AH. Female gerbils also have larger cell nuclei than males in the neocortex. This is not true in the basal ganglia, anterior amygdala, or medial septum (Clancy, 1978).

The possibility that masculinization of scent-marking in gerbils involves androgen action in the MPOA–AH is supported by data on the effects of TP implants directly into the brains of neonates. Female gerbils given unilateral implants of TP into the hypothalamus or preoptic area on the day of birth scent-mark more often in adulthood in response to systemic TP injections than do females that received cholesterol implants neonatally (Holman, 1977).

In dogs, exposure to androgens during early development affects the postures that males and females use to urinate but does not affect urination frequency (Beach, 1974). Female dogs exposed to TP during their second trimester of gestation and again after birth use the leg-lift urinary posture much more often than normal females but much less often than neonatally castrated males. Limiting androgen exposure to the neonatal period produces a smaller but noticeable effect. In contrast, limiting androgen exposure to the second trimester produces no masculinization at all. Castrating male dogs neonatally only slightly decreases their use of the leg-lift stance. However, when these males do use another posture, it usually is not the squat posture employed by most females. Instead, the neonatally castrated males revert to the stand-lean posture typical of juvenile males. This posture is rarely used by females even if they have been exposed to TP during development. In dogs, these behavioral effects of early androgenization appear in the complete absence of gonadal hormones in adulthood and do not intensify when hormones are supplied.

Indirect Hormonal Control of Scent-Marking

Direct stimulation of the brain is only one way hormones could act to control marking behavior. They could also regulate scent marking by altering feedback from the scent-marking act itself. For example, hormones could alter the tactile sensitivity of a skin gland. Or, detection of hormone-dependent odors deposited during scent-marking may be necessary to maintain the behavior (positive feedback) or to stop it (negative feedback).

In both gerbils and hamsters, these possibilities have been tested directly by assessing scent-marking frequencies of males after their scent glands were surgically removed. In gerbils, excision of the ventral scent gland affects neither the time-course nor the frequency of scent marking, if the wound is given plenty of time to heal (Baran & Glickman, 1970; Blum & Thiessen, 1970; Goist, Twiggs, Schwartz, & Christenson, 1977). Thus hormonal control of marking behavior is not mediated via the scent gland in male gerbils. The same conclusion seems to hold for hamsters as well. Male hamsters whose flank glands are removed obtain scent marking scores within the normal range for gonadally intact males (Johnston, 1975b). Some increase or decrease cannot be ruled out, though, because the marking frequencies of the male hamsters prior to surgery was unknown.

Another important way in which hormones can influence scent-marking is by altering odors or other social signals that elicit the behavior or by altering sensitivity to those cues. Males may scent-mark more, for example, in the presence of scent-marks laid down by another male. The effects of olfactory and other stimuli that fall into this category of hormonal influence are too many and varied to be covered here. Suffice it to say that many researchers have systematically manipulated stimuli, many of which are hormone dependent, to which animals are exposed during scent-marking tests. (For the interested reader, a few examples from the literature already cited include Brown, 1977, 1978; Epple,

1978, 1980, 1981; Fleming & Tambosso, 1980; Hull *et al.,* 1973; Johnston, 1975a,b, 1977, 1979; Maruniak *et al.,* 1974; Maruniak, Owen, Bronson, & Desjardins, 1975; Maruniak, Darney, & Bronson, 1975; von Holst & Buergel-Goodwin, 1975.)

CONCLUSIONS

Communication plays an important role in the control of aggression. By providing conspecifics with information about each other, social signals can decrease the frequency of fighting. The risks involved in fighting are high, even for the winner, so avoiding or reducing these risks clearly benefits both potential combatants.

In mammals, scent-marking is a form of communication that can function in this context. As the research summarized in this review shows, scent-marking behaviors of mammals are often under hormonal control. Testicular androgens, in particular, are among the most important factors for stimulating scent-marking behaviors in males. This influence of testosterone coordinates scent-marking seasonally and developmentally with reproductive activity and aggression. Androgen control of scent-marking suggests, however, a different role for testosterone in the control of aggression than is usually described.

Most of the literature on androgens and aggression emphasize stimulation of aggression by testosterone. Yet through its effects on scent-marking, testosterone channels social interactions away from attacks and thus acts as an aggression inhibitor. Granted, testosterone increases the likelihood that the animal will attack an opponent that ignores the social signals or intrudes despite them. Nonetheless, the same hormone promotes behaviors that bridle this potential aggressiveness and provide social restraint. Other forms of communication, such as visual or auditory displays, that serve similar functions are also controlled by gonadal androgens as suggested by sexual dimorphism, seasonality, pubertal development, and/or direct sensitivity to testosterone (Adkins-Regan, 1981; Arnold, 1981; Crews & Greenberg, 1981; Dizinno & Whitney, 1977; Floody, 1981; Green, 1981; Harding, 1981; Silver, O'Connell, & Saad, 1979; Wada & Gorbman, 1977). This suggests that the combined inhibition and stimulation of aggression by testosterone may be a widespread adaptation. Our understanding of this dual hormonal influence on aggression depends on analyses of communication.

REFERENCES

Adkins-Regan, E. Hormone specificity, androgen metabolism and social behavior. *American Zoologist,* 1981, *21,* 257–271.

Anisko, J. J., Christenson, T., & Buehler, M. G. Effects of androgen on fighting behavior in male and female Mongolian gerbils *(Meriones unguiculatus). Hormones and Behavior,* 1973, *4,* 199–208.

Arnold, A. P. Logical levels of steroid hormone action in the control of vertebrate behavior. *American Zoologist,* 1981, *21,* 233–242.

Balakrishnan, M., & Alexander, K. M. Hormonal control of scent marking in the Indian musk shrew, *Suncus murinus viridescens* (Blyth). *Hormones and Behavior,* 1976, *7,* 431–439.

Baran, D., & Glickman, S. E. "Territorial marking" in the Mongolian gerbil: A study of sensory control and function. *Journal of Comparative and Physiological Psychology,* 1970, *71,* 237–245.

Barfield, R. J. Activation of copulatory behavior by androgen implanted into the preoptic area of the male fowl. *Hormones and Behavior,* 1969, *1,* 37–52.

Barfield, R. J. Activation of sexual and aggressive behavior by androgen implantation into the male ring dove brain. *Endocrinology,* 1971, *89,* 1470–1476.

Beach, F. A. Effects of gonadal hormones on urinary behavior in dogs. *Physiology and Behavior,* 1974, *12,* 1005–1013.

Beauchamp, G. K. Diet influences attractiveness of urine. *Nature,* 1976, *263,* 587–588.

Blum, S. L., & Thiessen, D. D. Effect of ventral gland excision on scent marking in the male Mongolian gerbil. *Journal of Comparative and Physiological Psychology,* 1970, *73,* 461–464.

Blum, S. L., & Thiessen, D. D. The effect of different amounts of androgen on scent marking in the male Mongolian gerbil. *Hormones and Behavior,* 1971, *2,* 279–285.

Bronson, F. H. Urine marking in mice: Causes and effects. In R. L. Doty (Ed.), *Mammalian olfaction, reproductive processes, and behavior.* New York: Academic Press, 1976.

Bronson, F. H., & Desjardins, C. Relationships between scent marking by male mice and the pheromone-induced secretion of the gonadotropic and ovarian hormones that accompany puberty in female mice. In W. Montagna & W. A. Sadler (Eds.), *Reproductive behavior: Advances in behavioral biology* (Vol. 2). New York: Plenum Press, 1974.

Brown, R. E. Object-directed urine-marking by male rats *(Rattus norvegicus). Behavioral Biology,* 1975, *15,* 251–254.

Brown, R. E. Odor preference and urine-marking scales in male and female rats: Effects of gonadectomy and sexual experience on responses to conspecific odors. *Journal of Comparative and Physiological Psychology,* 1977, *91,* 1190–1206.

Brown, R. E. Hormonal control of odor preferences and urine-marking in male and female rats. *Physiology and Behavior,* 1978, *20,* 21–24.

Cheal, M. Social olfaction: A review of the ontogeny of olfactory influences on vertebrate behavior. *Behavioral Biology,* 1975, *15,* 1–25.

Clancy, A. N. *Hormonal differentiation of ventral gland marking in the Mongolian gerbil (Meriones unguiculatus).* Unpublished doctoral dissertation, University of Texas at Austin, 1978.

Clarke, J. W. Androgen control of the ventral scent gland in *Neotoma floridana. Journal of Endocrinology,* 1975, *64,* 393–394.

Clutton-Brock, T. H., & Albon, S. D. The roaring of red deer and the evolution of honest advertisement. *Behaviour,* 1979, *27,* 750–760.

Clutton-Brock, T. H., Albon, S. D., Gibson, R. M., & Guinness, F. E. The logical stag: Adaptive aspects of fighting in red deer (*Cervus elaphus* L.). *Animal Behaviour,* 1979, *27,* 211–225.

Crews, D., & Greenberg, N. Function and causation of social signals in lizards. *American Zoologist,* 1981, *21,* 273–294.

Damassa, D. A., Smith, E. R., Tennett, B., & Davidson, J. M. The relationship between circulating testosterone levels and male sexual behavior in rats. *Hormones and Behavior,* 1977, *8,* 275–286.

Davidson, J. M., & Bloch, G. J. Neuroendocrine aspects of male reproduction. *Biology of Reproduction,* 1969, *1,* 67–92.

Desjardins, C., Maruniak, J. A., & Bronson, F. H. Social rank in house mice: Differentiation by ultraviolet visualization of urinary marking patterns. *Science,* 1973, *182,* 939–941.

Dizinno, G., & Whitney, G. Androgen influence on male mouse ultrasounds during courtship. *Hormones and Behavior,* 1977, *8,* 188–192.

Dryden, G. L., & Conway, C. H. The origin and hormonal control of scent production in *Suncus murinus. Journal of Mammalogy,* 1967, *48,* 420–428.

Emlen, S. T., & Oring, L. W. Ecology, sexual selection, and the evolution of mating systems. *Science,* 1977, *197,* 215–223.

Epple, G. Quantitative studies on scent marking in the marmoset *(Callithrix jacchus). Folia Primatologica,* 1970, *13,* 48–62.

Epple, G. Primate pheromes. In M. C. Birch (Ed.), *Pheromones.* Amsterdam: North-Holland Publishing Company, 1974.

Epple, G. Lack of effects of castration on scent marking, displays, and aggression in a South American primate *(Saguinus fuscicollis). Hormones and Behavior,* 1978, *11,* 139–150.

Epple, G. Relationships between aggression, scent marking and gonadal state in a primate, the tamarin *Saguinus fuscicollis.* In D. Müller-Schwarze & R. M. Silverstein (Eds.), *Chemical signals: Vertebrates and aquatic invertebrates.* New York: Plenum Press, 1980.

Epple, G. Effects of prepubertal castration on the development of the scent glands, scent marking, and aggression in the saddle back tamarin (*Saguinus fuscicollis,* Callithricidae, Primates). *Hormones and Behavior,* 1981, *15,* 54–67.

Ewer, R. F. *Ethology of mammals.* New York: Plenum Press, 1968.

Fleming, A. S., & Tambosso, L. Hormonal and sensory control of scent-marking in the desert woodrat *(Neotoma lepida lepida). Journal of Comparative and Physiological Psychology,* 1980, *94,* 564–578.

Floody, O. R. The hormonal control of ultrasonic communication in rodents. *American Zoologist,* 1981, *21,* 129–142.

Franklin, W. L. Territorial marking behavior by the South American vicuña. In D. Müller-Schwarze & R. M. Silverstein (Eds.), *Chemical signals: Vertebrates and aquatic invertebrates.* New York: Plenum Press, 1980.

Garcia, J., Hankins, W. G., & Rusiniak, K. W. Behavioral regulation of the milieu interne in man and rat. *Science,* 1974, *185,* 824–831.

Garrett, J. W., & Campbell, C. S. Changes in social behavior of the male golden hamster accompanying photoperiodic changes in reproduction. *Hormones and Behavior,* 1980, *14,* 303–318.

Geist, V. *Mountain sheep.* Chicago: University of Chicago Press, 1971.

Gilbert, B. K. Scent marking and territoriality in pronghorn *(Antilocapra americana)* in Yellowstone National Park. *Mammalia,* 1974, *37,* 25–33.

Ginton, A., & Merari, A. Long range effects of MPOA lesions on mating behavior in the male rat. *Brain Research,* 1977, *120,* 158–163.

Glenn, E. M., & Gray, J. Effect of various hormones on the growth and histology of the gerbil *(Meriones unguiculatus)* abdominal sebaceous gland pad. *Endocrinology,* 1965, *76,* 1115–1123.

Goist, K. C., Jr., Twiggs, D. G., Schwartz, B. D., & Christenson, T. E. Role of cutaneous feedback in ventral rubbing in the male Mongolian gerbil. *Physiology and Behavior,* 1977, *19,* 581–583.

Gorski, R. A., Gordon, J. H., Shryne, J. E., & Southam, A. M. Evidence for a morphological sex difference within the medial preoptic area of the rat brain. *Brain Research,* 1978, *148,* 333–346.

Grau, G. A. Olfaction and reproduction in ungulates. In R. L. Doty (Ed.), *Mammalian olfaction, reproductive processes, and behavior.* New York: Academic Press, 1976.

Green, S. M. Sex differences and age gradation in vocalizations of Japanese and lion-tailed monkeys (*Macaca fuscata* and *Macaca silenus*). *American Zoologist,* 1981, *21,* 165–183.

Greenough, W. T., Carter, C. S., Steerman, C., & DeVoogd, T. J. Sex differences in dendritic patterns in hamster preoptic area. *Brain Research,* 1977, *126,* 63–72.

Gual, C., Morato, T., Hayano, M., Gut, M., & Dorfman, R. I. Biosynthesis of estrogens. *Endocrinology,* 1962, *71,* 920–925.

Hamilton, J. B., & Montagna, W. The sebaceous glands of the hamster: I. Morphological effects of androgens on integumentary structures. *American Journal of Anatomy,* 1950, *86,* 199–233.

Harding, C. F. Social modulation of circulating hormone levels in the male. *American Zoologist,* 1981, *21,* 223–231.

Hart, B. L. Gonadal androgen and sociosexual behavior of male mammals: A comparative analysis. *Psychological Bulletin,* 1974, *81,* 383–400. (a)

Hart, B. L. Environmental and hormonal control of urine marking behavior in the male dog. *Behavioral Biology,* 1974, *11,* 167–176. (b)

Hart, B. L. Medial preoptic-anterior hypothalamic area and sociosexual behavior of male dogs: A comparative neuropsychological analysis. *Journal of Comparative and Physiological Psychology,* 1974, *86,* 328–349. (c)

Hart, B. L., & Barrett, R. E. Effects of castration on fighting, roaming, and urine spraying in adult male cats. *Journal of the American Veterinary Medical Association,* 1973, *173,* 290–292.

Hart, B. L., & Haugen, C. M. Scent marking and sexual behavior maintained in anosmic male dogs. *Communications in Behavioral Biology,* 1971, *6,* 131–135.

Hart, B. L., & Ladewig, J. Effects of medial preoptic-anterior hypothalamic lesions on development of sociosexual behavior in dogs. *Journal of Comparative and Physiological Psychology,* 1979, *93,* 566–573.

Hart, B. L., & Voith, V. L. Changes in urine spraying, feeding and sleep behavior of cats following medial preoptic-anterior hypothalamic lesions. *Brain Research,* 1978, *145,* 406–409.

Hart, B. L., Haugen, C. M., & Peterson, D. M. Effects of medial preoptic-anterior hypothalamic lesions on mating behavior of male cats. *Brain Research,* 1973, *54,* 177–191.

Henry, J. D. The urine marking behavior and movement patterns of red foxes *(Vulpes vulpes)* during a breeding and post-breeding season. In D. Müller-Schwarze & R. M. Silverstein (Eds.), *Chemical signals: Vertebrates and aquatic invertebrates.* New York: Plenum Press, 1980.

Holman, S. D. *Developmental influences of hormones on sexual behaviour in the Mongolian gerbil.* Unpublished doctoral thesis, Cambridge University, 1977.

Howe, R. J. Scent marking behavior in three species of woodrats in captivity. *Journal of Mammalogy,* 1977, *58,* 685–688.

Hull, E. M., Langan, C. J., & Rosselli, L. Population density and social, territorial, and physiological measures in gerbils. *Journal of Comparative and Physiological Psychology,* 1973, *84,* 414–422.

Hull, E. M., Hamilton, K. L., Engwall, D. B., & Rosselli, L. Effects of olfactory bulbectomy and peripheral deafferentiation on reactions to crowding in gerbils *(Meriones unguiculatus). Journal of Comparative and Physiological Psychology,* 1974, *86,* 247–254.

Jannett, F. J., Jr., & Jannett, J. Z. Drum-marking by *Arvicola richardsoni* and its taxonomic significance. *American Midland Naturalist,* 1974, *92,* 230–234.

Johnson, R. P. Scent marking in mammals. *Animal Behaviour,* 1973, *21,* 521–535.

Johnston, R. E. Scent marking by male golden hamsters *(Mesocricetus auratus):* I. Effects of odors and social encounters. *Zeitschrift für Tierpsychologie,* 1975, *37,* 75–98. (a)

Johnston, R. E. Scent marking by male golden hamsters *(Mesocricetus auratus):* II. The role of the flank gland scent in the causation of marking. *Zeitschrift für Tierpsychologie,* 1975, *37,* 138–144. (b)

Johnston, R. E. The causation of two scent-marking behaviour patterns in female hamsters *(Mesocricetus auratus). Animal Behaviour,* 1977, *25,* 317–327.

Johnston, R. E. Olfactory preferences, scent marking, and "proceptivity" in female hamsters. *Hormones and Behavior,* 1979, *13,* 21–39.

Johnston, R. E. Testosterone dependence of scent marking by male hamsters *(Mesocricetus auratus). Behavioral and Neural Biology,* 1981, *31,* 96–99.

Jolly, A. *Lemur behavior: A Madagascar field study.* Chicago: University of Chicago Press, 1966.

Kessler, S., Harmatz, P., & Gerling, S. A. The genetics of pheromonally mediated aggression in mice: I. Strain differences in the capacity of male urinary odors to elicit aggression. *Behavior Genetics,* 1975, *5,* 233–238.

Kleiman, D. Scent marking in Canidae. *Symposium of the Zoological Society of London,* 1966, *18,* 167–177.

LeBoeuf, B. J. Male-male competition and reproductive success in elephant seals. *American Zoologist,* 1974, *14,* 163–176.

Lee, C. T., & Estep, D. The developmental aspect of marking and nesting behaviors in Mongolian gerbils *(Meriones unguiculatus). Psychonomic Science,* 1971, *22,* 312–313.

Leon, M. Maternal pheromone. *Physiology and Behavior,* 1974, *13,* 441–453.

Leon, M. Development of olfactory attraction by young Norway rats. In D. Müller-Schwarze & R. M. Silverstein (Eds.), *Chemical signals: Vertebrates and aquatic invertebrates.* New York: Plenum Press, 1980.

Leshner, A. I. *An introduction to behavioral endocrinology.* New York: Oxford University Press, 1978.

Lindzey, G., Thiessen, D. D., & Tucker, A. Development of hormonal control of territorial marking in the male Mongolian gerbil *(Meriones unguiculatus). Developmental Psychobiology,* 1968, *1,* 97–99.

Lumia, A. R., Westervelt, M. O., & Rieder, C. A. Effects of olfactory bulb ablation and androgen on marking and agonistic behavior in male Mongolian gerbils *(Meriones unguiculatus). Journal of Comparative and Physiological Psychology,* 1975, *89,* 1091–1099.

Lumia, A. R., Canino, G., & Drozdowski, B. Effects of androgen on scent marking and aggressive behavior of neonatally bulbectomized female Mongolian gerbils *(Meriones unguiculatus). Hormones and Behavior,* 1976, *7,* 461–471.

Macrides, F., Bartke, A., Fernandez, F., & D'Angelo, W. Effects of exposure to vaginal odor and receptive females on plasma testosterone in the male hamster. *Neuroendocrinology,* 1974, *15,* 355–364.

Maruniak, J. A., Owen, K., Bronson, F. H., & Desjardins, C. Urinary marking in male house mice: Responses to novel environmental and social stimuli. *Physiology and Behavior,* 1974, *12,* 1035–1039.

Maruniak, J. A., Owen, K., Bronson, F. H., & Desjardins, C. Urinary marking in female house mice: Effects of ovarian steroids, sex experience, and type of stimulus. *Behavioral Biology,* 1975, *13,* 211–217.

Maruniak, J. A., Darney, K. J., Jr., & Bronson, F. H. Olfactory perception of the nonsocial environment by male house mice. *Behavioral Biology,* 1975, *14,* 237–240.

Maruniak, J. A., Desjardins, C., & Bronson, F. H. Dominant-subordinate relationships in castrated male mice bearing testosterone implants. *American Journal of Physiology,* 1977, *233,* E495–E499.

Mech, L. D., & Peters, R. P. The study of chemical communication in free-ranging mammals. In D. Müller-Schwarze & M. M. Mozell (Eds.), *Chemical signals in vertebrates.* New York: Plenum Press, 1977.

Michael, R. P. Hormonal steroids and sexual communication in primates. *Journal of Steroid Biochemistry,* 1975, *6,* 161–170.

Morgantaler, A., & Crews, D. Role of the anterior hypothalamus-preoptic area in the regulation of reproductive behavior in the lizard, *Anolis carolinensis:* Implantation studies. *Hormones and Behavior,* 1978, *11,* 61–73.

Morrell, J. I., Kelley, D. B., & Pfaff, D. W. Sex steroid binding in the brains of vertebrates. In K. M. Knigge & D. E. Scott (Eds.), *Brain-endocrine interaction* (Vol. II): *The ventricular system.* Basel: Karger, 1975.

Moy, R. F. Histology of the subauricular and rump glands of the pronghorn (*Antilocapra americana* Ord). *American Journal of Anatomy,* 1970, *129,* 65–88.

Müller-Schwarze, D. Olfactory recognition of species, groups, individuals and physiological states among mammals. In M. C. Birch (Ed.), *Pheromones.* Amsterdam: North-Holland Publishing Company, 1974.

Müller-Schwarze, D. Complex mammalian behavior and pheromone bioassay in the field. In D. Müller-Schwarze & M. M. Mozell (Eds.), *Chemical signals in vertebrates.* New York: Plenum Press, 1977.

Mykytowycz, R. Further observations on the territorial function and histology of the submandibular cutaneous (chin) glands in the rabbit, *Oryctolagus cuniculus* (L). *Animal Behaviour,* 1965, *13,* 400–411.

Mykytowycz, R. Territorial marking by rabbits. *Scientific American,* 1968, *218*(5), 116–126.

Mykytowycz, R. The role of skin glands in mammalian communication. In J. W. Johnston, Jr., D. G. Moulton, & A. Turk (Eds.), *Communication by chemical signals: Advances in chemoreception* (Vol. 1). New York: Appleton Century Crofts, 1970.

Nordeen, E. J., & Yahr, P. Activation and differentiation of sexual behavior and translocation of hypothalamic estrogen receptors in rats by 6α-fluorotestosterone. *Hormones and Behavior,* 1981, *15,* 123–140.

Nyby, J., & Thiessen, D. D. Singular and interactive effects of testosterone and estrogen on territorial marking in castrated male Mongolian gerbils *(Meriones unguiculatus). Hormones and Behavior,* 1971, *2,* 279–285.

Nyby, J., & Whitney, G. Experience affects behavioral responses to sex odors. In D. Müller-Schwarze & R. M. Silverstein (Eds.), *Chemical signals: Vertebrates and aquatic invertebrates.* New York: Plenum Press, 1980.

Nyby, J., Thiessen, D. D., & Wallace, P. Social inhibition of territorial marking in the Mongolian gerbil *(Meriones unguiculatus). Psychonomic Science,* 1970, *21,* 310–312.

O'Gara, B. W., Moy, R. F., & Bear, G. D. The annual testicular cycle and horn casting in the pronghorn *(Antilocapra americana). Journal of Mammalogy,* 1971, *52,* 537–544.

Owen, K., & Thiessen, D. D. Regulation of scent marking in the female Mongolian gerbil *Meriones unguiculatus*. *Physiology and Behavior*, 1973, *11*, 441–445.

Owen, K., & Thiessen, D. D. Estrogen and progesterone interaction in the regulation of scent marking in the female Mongolian gerbil *(Meriones unguiculatus)*. *Physiology and Behavior*, 1974, *12*, 351–355.

Owen, K., Wallace, P., & Thiessen, D. Effects of intracerebral implants of steroid hormones on scent marking in the ovariectomized female gerbil *(Meriones unguiculatus)*. *Physiology and Behavior*, 1974, *12*, 755–760.

Payne, A. P., & Swanson, H. H. Agonistic behaviour between pairs of hamsters of the same or opposite sex in a neutral observation area. *Behaviour*, 1970, *36*, 259–269.

Peters, R., & Mech, L. D. Scent-marking in wolves. *American Scientist*, 1975, *63*, 628–637.

Price, E. O. Hormonal control of urine-marking in wild and domestic Norway rats. *Hormones and Behavior*, 1975, *6*, 393–397.

Quay, W. B. Structure and function of skin glands. In D. Müller-Schwarze & M. M. Mozell (Eds.), *Chemical signals in vertebrates*. New York: Plenum Press, 1977.

Raisman, G., & Field, P. M. Sexual dimorphism in the neuropil of the preoptic area of the rat and its dependence on neonatal androgen. *Brain Research*, 1973, *54*, 1–29.

Ralls, K. Mammalian scent marking. *Science*, 1971, *171*, 443–449.

Rice, M. G. *Relationships among sex behavior, scent marking, aggression, and plasma testosterone in the male Mongolian gerbil (Meriones unguiculatus)*. Unpublished doctoral dissertation, University of Texas at Austin, 1975.

Ryan, K. J. Estrogen formation by the human placenta: Studies on the mechanism of steroid aromatization by mammalian tissue. *Acta Endocrinologica*, 1960, *35*, (Suppl. 51) 697–698.

Schaller, G. B. *The deer and the tiger*. Chicago: University of Chicago Press, 1967.

Schultze-Westrum, T. G. Social communication by chemical signals in flying phalangers. In C. Pfaffman (Ed.), *Olfaction and taste: Proceedings of the third international symposium*. New York: Rockefeller University Press, 1969.

Scouten, C. W., Burrell, L., Palmer, T., & Cegavske, C. F. Lateral projections of the medial preoptic area are necessary for androgenic influence on urine marking and copulation in rats. *Physiology and Behavior*, 1980, *25*, 237–243.

Silver, R., O'Connell, M., & Saad, R. The effect of androgens on the behavior of birds. In C. Beyer (Ed.), *Endocrine control of sexual behavior*. New York: Raven Press, 1979.

Skeen, J. T., & Thiessen, D. D. Scent of gerbil cuisine. *Physiology and Behavior*, 1977, *19*, 11–14.

Slimp, J. C., Hart, B. L., & Goy, R. W. Heterosexual and sociosexual behavior of adult male rhesus monkeys with medial preoptic anterior hypothalamic lesions. *Brain Research*, 1978, *142*, 105–122.

Smith, E. R., Damassa, D. A., & Davidson, J. M. Plasma testosterone and sexual behavior following intracerebral implantation of testosterone propionate in the castrated male rat. *Hormones and Behavior*, 1977, *8*, 77–87.

Sprague, R. H., & Anisko, J. J. Elimination patterns in the laboratory beagle. *Behaviour*, 1973, *47*, 257–267.

Steiner, A. L. Body-rubbing, marking and other scent-related behavior in some ground squirrels *(Sciuridae)*, a descriptive study. *Canadian Journal of Zoology*, 1974, *52*, 889–906.

Stumpf, W. E., & Sar, M. Anatomical distribution of estrogen, androgen, progestin, corticosteroid and thyroid hormone target sites in the brain of mammals: Phylogeny and ontogeny. *American Zoologist*, 1978, *18*, 435–445.

Thiessen, D. D., & Lindzey, G. Territorial marking in the female Mongolian gerbil: Short-term reactions to hormones. *Hormones and Behavior*, 1970, *1*, 157–160.

Thiessen, D., & Rice, M. Mammalian scent gland marking and social behavior. *Psychological Bulletin*, 1976, *83*, 505–539.

Thiessen, D. D., & Yahr, P. Central control of territorial marking in the Mongolian gerbil. *Physiology and Behavior*, 1970, *5*, 275–278.

Thiessen, D. D., Friend, H. C., & Lindzey, G. Androgen control of territorial marking in the Mongolian gerbil. *Science*, 1968, *160*, 26–30.

Thiessen, D. D., Blum, S. L., & Lindzey, G. A scent marking response associated with the ventral sebaceous gland of the Mongolian gerbil *(Meriones unguiculatus)*. *Animal Behaviour*, 1970, *18*, 26–30.

Thiessen, D. D., Lindzey, G., Blum, S. L., & Wallace, P. Social interactions and scent marking in the Mongolian gerbil *(Meriones unguiculatus). Animal Behaviour,* 1970, *19,* 505–513.

Thiessen, D. D., Lindzey, G., & Nyby, J. The effects of olfactory deprivation and hormones on territorial marking in the male Mongolian gerbil *(Meriones unguiculatus). Hormones and Behavior,* 1970, *1,* 315–326.

Thiessen, D. D., Owen, K., & Lindzey, G. Mechanisms of territorial marking in the male and female Mongolian gerbil *(Meriones unguiculatus). Journal of Comparative and Physiological Psychology,* 1971, *77,* 38–47.

Thiessen, D. D., Wallace, P., & Yahr, P. Comparative studies of glandular scent marking in *Meriones tristrami* an Israeli gerbil. *Hormones and Behavior,* 1973, *4,* 143–147.

Thiessen, D. D., Yahr, P. I., & Owen, K. Regulatory mechanisms of territorial marking in the Mongolian gerbil. *Journal of Comparative Physiological Psychology,* 1973, *82,* 382–393.

Tiefer, L. Gonadal hormones and mating behavior in the adult golden hamster. *Hormones and Behavior,* 1970, *1,* 189–202.

Tinbergen, N. *The study of instinct.* New York: Oxford University Press, 1971.

Trivers, R. L. Parental investment and sexual selection. In B. G. Campbell (Ed.), *Sexual selection and the descent of man, 1871–1971.* Chicago: Aldine, 1972.

Turek, F. W. The interaction of photoperiod and testosterone in regulating serum gonadotropin levels in castrated male hamsters. *Endocrinology,* 1977, *101,* 1210–1215.

Turner, J. W., Jr. Effect of hypothalamic lesions and androgens on plasma LH levels in castrated male gerbils. *Endocrinology,* 1973, *92,* 227–234.

Turner, J. W., Jr. Influence of neonatal androgen on the display of territorial marking behavior in the gerbil. *Physiology and Behavior,* 1975, *15,* 265–270.

Twiggs, D. G., Popolow, H. B., & Gerall, A. A. Medial preoptic lesions and male sexual behavior: Age and environmental interactions. *Science,* 1978, *200,* 1414–1415.

Vandenbergh, J. G. The effects of gonadal hormones on the aggressive behavior of adult golden hamsters *(Mesocricetus auratus). Animal Behaviour,* 1971, *19,* 589–594.

von Holst, D., & Buergel-Goodwin, U. The influence of sex hormones on chinning by male *Tupaia belangeri. Journal of Comparative Physiology,* 1975, *103,* 123–151.

Wada, M., & Gorbman, A. Relation of mode of administration of testosterone to evocation of male sexual behavior in frogs. *Hormones and Behavior,* 1977, *8,* 310–319.

Wales, N. A. M., & Ebling, F. J. The control of apocrine glands of the rabbit by steroid hormones. *Journal of Endocrinology,* 1971, *51,* 763–771.

Wallace, P. K., Owen, K., & Thiessen, D. D. The control and function of maternal scent marking in the Mongolian gerbil *(Meriones unguiculatus). Physiology and Behavior,* 1973, *10,* 463–466.

Wallen, K., & Glickman, S. E. Effect of peripheral anosmia on ventral rubbing in the gerbil. *Behavioral Biology,* 1974, *11,* 569–572.

Wheeler, J. M., & Crews, D. Role of the anterior hypothalamus-preoptic area in the regulation of male reproductive behavior in the lizard, *Anolis carolinensis:* Lesion studies. *Hormones and Behavior,* 1978, *11,* 42–60.

Whitsett, J. M. The development of aggressive and marking behavior in intact and castrated male hamsters. *Hormones and Behavior,* 1975, *6,* 47–57.

Whitsett, J. M., & Thiessen, D. D. Sex differences in the control of scent marking in the Mongolian gerbil *(Meriones unguiculatus). Journal of Comparative and Physiological Psychology,* 1972, *78,* 381–385.

Wilson, E. O. *Sociobiology: The new synthesis.* Cambridge, Mass.: Belknap Press, 1975.

Yahr, P. I. *Molecular regulation of territorial marking in the Mongolian gerbil (Meriones unguiculatus).* Unpublished doctoral dissertation, University of Texas at Austin, 1972.

Yahr, P. Effects of hormones and lactation on gerbils that seldom scent mark spontaneously. *Physiology and Behavior,* 1976, *16,* 395–399. (a)

Yahr, P. The role of aromatization in androgen stimulation of gerbil scent marking. *Hormones and Behavior,* 1976, *7,* 259–265. (b)

Yahr, P. Social subordination and scent-marking in male Mongolian gerbils *(Meriones unguiculatus). Animal Behaviour,* 1977, *25,* 292–297. (a)

Yahr, P. Central control of scent marking. In D. Müller-Schwarze & M. M. Mozell (Eds.), *Chemical signals in vertebrates.* New York: Plenum Press, 1977. (b)

Yahr, P., Commins, D., Jackson, J. C., & Newman, A. Independent control of sexual and scent marking behaviors of male gerbils by cells in or near the medial preoptic area. *Hormones and Behavior,* 1982, *16,* 304–322.

Yahr, P., & Kessler, S. Suppression of reproduction in water-deprived Mongolian gerbils *(Meriones unguiculatus). Biology of Reproduction,* 1975, *12,* 249–254.

Yahr, P., & Thiessen, D. D. Steroid regulation of territorial scent marking in the Mongolian gerbil *(Meriones unguiculatus). Hormones and Behavior,* 1972, *3,* 359–368.

Yahr, P., & Thiessen, D. D. Estrogen control of scent marking in female Mongolian gerbils *(Meriones unguiculatus). Behavioral Biology,* 1975, *13,* 95–101.

Yahr, P., Newman, A., & Stephens, D. R. Sexual behavior and scent marking in male gerbils: Comparison of changes after castration and testosterone replacement. *Hormones and Behavior,* 1979, *13,* 175–184.

Yahr, P., Jackson, J. C., Newman, A., Stephens, D. R., & Clancy, A. N. Pardigm for comparing sexual behavior and scent marking in male gerbils. *Physiology and Behavior,* 1980, *24,* 263–266.

II

MECHANISMS OF HORMONE ACTION AND AGGRESSIVE BEHAVIOR

8

Genetics, Hormones, and Aggression

Stephen C. Maxson, Paul Shrenker, and Lynne C. Vigue

INTRODUCTION

It has been proposed (Brain, 1979a,b; Moyer, 1968, 1976) that there are several types of aggression that differ in response patterns, stimulus conditions, and physiological mechanisms. Genetic (Lagerspetz & Lagerspetz, 1975; Maxson, 1981; Scott, 1966; Scott & Fredericson, 1951; Simon, 1979) and hormonal (see reviews in this volume) influences on most if not all of these types of aggression have been studied in one or more vertebrate species. However, the intersection of genetic and hormonal factors has been explored primarily for androgens and intermale aggression in mice *(Mus musculus).* This review will consider the evidence that some gene effects on the expression and/or development of intermale aggression are mediated, at least in part, by androgens.

GENOTYPE, HORMONES, AND THE ACTIVATION OF AGGRESSION

For several decades, it has been known that testosterone influences the occurrence and severity of fighting between adult male mice (Gandelman, 1980). In 1947, Beeman showed that castration abolishes intermale aggression in mice and that such aggression is restored by testosterone replacement therapy. Similar findings have been reported by Tollman and King (1956) and by Ulrich (1938). Also, in 1969, Edwards observed that there is a positive relationship between dose of testosterone-replacement therapy and the percentage of fighting by castrates, and it has been shown that relatively nonaggressive mice will become very aggressive after injections of testosterone (Banarjee, 1971). These and similar results suggest that some strain differences in intermale aggression of mice may be due to gene effects on the adult level of serum testosterone. The possible role of adult serum testosterone in the intermale aggression of four pairs of inbred strains of mice is reviewed in this section.

STEPHEN C. MAXSON, PAUL SHRENKER, AND LYNNE C. VIGUE • Department of Biobehavioral Sciences, University of Connecticut, Storrs, Connecticut 06268.

Selmanoff, Maxson, and Ginsburg (1976) reported that the propensity to fight and the severity of fighting was greater in DBA/2 than in C57BL/10 males. This strain difference in intermale aggression has also been observed by Calhoun (1956) and by van Abeelen (1966). Such strain differences may be due to cytoplasmic, maternal, or genic factors (Maxson, 1981; Wahlsten, 1979). It was also shown by Selmanoff *et al.* (1976) that the reciprocal F_1 hybrids of these two strains were as pacific as males of the C57BL/10 strain. This finding provides evidence that this strain difference is indeed genic and that X- and Y-chromosomal genes are not involved. In this study, intermale aggression was measured in a homogeneous set test (Fuller & Hahn, 1976) over three days, beginning at 50 days of age.

On the basis of the following evidence, Bartke and his colleagues proposed that C57BL/10 males are androgen-deficient relative to DBA/2 males, and that this androgen deficiency may account, in part, for this strain's pacific behavior. At 8 and 11 weeks of age, C57BL/10J males have a higher level of esterified cholesterol in the testes, as well as smaller testes, than DBA/2J males; and at 7 weeks of age, C57BL/10J males have smaller submandibular glands and kidneys than DBA/2J males (Bartke & Shire, 1972). Also, at 100 days of age, DBA/2J males have larger submandibular glands and testes than C57BL/10J males (West, Evans, & Hamilton, 1980). However, between 64 and 100 days of age, these strains do not differ in seminal vesicle weights (Bartke & Shire, 1972; Selmanoff, Abreu, Goldman, and Ginsburg, 1977; West *et al.*, 1980). Shire and Bartke (1972) also showed that C57BL/10J males have spermatogenesis with a lower yield than DBA/2J males, and that DBA/2J males have twice as many germinal cells in the testes as do C57BL/10J males. This difference in the number of germinal cells could account for the strain difference in spermatogenesis and in testis weight. Some behavioral studies are also consistent with the hypothesis that C57BL/10J males are androgen-deficient relative to males of the DBA/2J and some other strains. C57BL/10J males do not show the pheromonally mediated Bruce effect of pregnancy-blocking by a strange male (Chapman & Whitten, 1969) and are slower to recover "sex drive" after mating (Shire & Bartke, 1972). Bartke and Shire (1972) speculated that these strain differences in biochemistry, morphology, and behavior may be due to gene effects on either circulating levels of testosterone, circulating levels of other androgens, binding capacity of target tissues for androgenic steroids, or sensitivity of target tissues to androgenic steroids.

Several studies have shown that DBA and C57BL strains differ in adult levels of serum testosterone. Bartke (1974) reported that at 10 weeks of age, the levels of circulating testosterone as measured by radioimmunoassay (Bartke, Steele, Musto, & Caldwell, 1973) was 4.57 ng/ml in DBA/2J and 1.32 ng/ml in C57BL/10J. Similar strain differences have been obtained for DBA/2J and C57BL/6J males. The C57BL/10J and C57BL/6J strains are closely related and may differ in alleles at no more than 100 loci (Bailey, 1978). Eleftheriou and Lucas (1974) reported that at 8 weeks of age, the level of serum testosterone is higher in DBA/2J (3.66 ng/ml) than in C57BL/6J (1.06 ng/ml) males. Also, at this age, C57BL/6J males have smaller testes than DBA/2J males, and males of the two strains have the same seminal vesicle weights. Similar findings were obtained for testis and seminal vesicle weights at 100 days of age for DBA/2J and C57BL/6J males (West *et al.,* 1980). C57BL/6Bg males are also as pacific as C57BL/10Bg males, and reciprocal F_1 males of DBA/2Bg and C57BL/6Bg strains are as pacific as C57BL/6Bg males (Selmanoff *et al.,* 1976). Batty (1978) also measured serum testosterone levels by radioimmunoassay at 9 to 10 weeks of age in DBA/2J, C57BL/6J, and their reciprocal F_1 hybrid males. The level of circulating testosterone was higher in DBA/2J males (5.5 ng/ml) than in C57BL/6J males (3.2 ng/ml). Also, the level of circulating testosterone in the reciprocal F_1 hybrid males of these two strains was not significantly different and was slightly lower than that in the C57BL/6J males. However, Selmanoff, Goldman, and Ginsburg (1977) reported that at 8 to 9 weeks of age, there were no significant differences between DBA/1Bg or DBA/2Bg and C57BL/10Bg or C57BL/6Bg males in levels of circulating testosterone, as measured by radioimmunoassay. Selmanoff and Ginsburg (1981) attribute this finding to the marked individual variation in levels of serum testosterone, which may be due to its pulsatile release (Bartke *et al.,* 1973). They argue that this variation masked significant mean differences in adult levels of circulating testosterone between DBA/1Bg or DBA/2Bg and C57BL/10Bg or C57BL/6Bg males.

On the basis of the above findings, there appears to be a correlation between adult level of serum testosterone and intermale aggression for DBA/2, C57BL/6, and C57BL/10 mice and for their F_1 hybrids. DBA/2 males are more aggressive than C57BL/10, C57BL/6, and F_1 hybrids of DBA/2 and either C57BL/10 or C57BL/6 males (Selmanoff *et al.,* 1976), and DBA/2 males have a higher level of adult serum testosterone than C57BL/10, C57BL/6, and F_1 hybrids of DBA/2 and C57BL/6 (Bartke, 1974; Batty, 1978). However, the biochemical and morphological differences between DBA/2 and C57BL/10 strains may not be related to the differences in intermale aggression or adult level of serum testosterone. In reciprocal hybrids of DBA/2J and C57BL/10J strains, the F_1 males resembled the DBA/2J males in percentage of esterified cholesterol in the testes, in weight of testes and kidneys, and in number of germinal cells in the testes (Shire & Bartke, 1972); whereas the F_1 males resembled C57BL/10Bg males in their aggressive behavior (Selmanoff *et al.,* 1976). Also, the F_1 males of DBA/2J and C57BL/6J parents resembled C57BL/6J males in adult level of serum testosterone (Batty, 1978); the C57BL/10J and C57BL/6J are closely related strains.

The strain correlation for serum testosterone level and intermale aggression in DBA and C57BL stains may be due to causally pleiotropic effects of the same gene(s) on both phenotypes. Selmanoff, Abreu, Goldman, and Ginsburg (1977) tested this hypothesis, in part, by a study of the effects of castration and testosterone replacement on intermale aggression of DBA/2Bg and C57BL/10Bg mice. Between 60 and 70 days of age, the mice were orchidectomized and then implanted with 0.75 or 6.0 mg of testosterone in silastic tubing. The 0.75-mg dose maintained seminal vesicle at 14% (DBA/2) or 46% (C57BL/10), and coagulating glands at 29% (DBA/2) or 61% (C57BL/10) of intact control weights, and the 6.0 mg dose maintained seminal vesicle at 120% (DBA/2) or 170% (C57BL/10) and coagulating glands at 130% (DBA/2) or 120% (C57BL/10) of intact control weights. By these measures, the 0.75-mg implant provided a hypophysiological and the 6.0-mg implant a hyperphysiological maintenance of the androgen target-organ weights. Twenty days after castration and implant, a separate group of mice was given a homogeneous set test for intermale aggression. In the DBA/2 strain, the group with the hypophysiological dose of testosterone had a lower frequency and severity of aggression than intact controls, whereas the group with the hyperphysiological dose had the same frequency and severity of aggression as intact controls. A different pattern was observed for the C57BL/10 strain. In this strain, the hypophysiological and hyperphysiological doses substantially increased the frequency but not the severity of aggression in comparison with intact controls. These findings suggest that in the DBA/2, the incidence and severity of fighting depends on serum testosterone levels, and that this strain difference in the incidence but not the severity of fighting may be due to gene effects on the adult levels of serum testosterone.

Since some of the difference in intermale aggression of DBA and C57BL mice may be due to gene effects on adult serum testosterone levels, it is relevant to determine the possible site(s) of gene action in the hypothalamic-pituitary-gonadal axis. For this purpose, Bartke, Roberson, and Dalterio (1977) measured the levels of gonadotropins in the plasma and the responsiveness of the testes *in vitro* to gonadotropic stimulation in two-month-old males of the DBA/2J and C57BL/10J strains. They reported that the concentration of luteinizing hormone (LH) in the peripheral plasma was lower in C57BL/10J than in DBA/2J males, and that the testes of males of both strains had equivalent responses to either human chorionic gonadotropin or pregnant mare serum gonadotropin. Similar experiments were conducted by Selmanoff and Ginsburg (1981). They also reported that at 60 or 70 days of age, DBA/1Bg males had a markedly higher titer of circulating LH than C57BL/10Bg males or reciprocal F_1 hybrid males, and that at 60 days of age, each genotype had an equivalent *in vivo* output of serum testosterone after treatment with either human chorionic gonadotropin or pregnant mare serum gonadotropin. In addition, following castration, there was a greater increase in serum LH levels in DBA/1Bg than in C57BL/10Bg males. Both Bartke *et al.* (1977) and Selmanoff and Ginsburg (1981) concluded on the basis of these data that the low level of plasma testosterone in C57BL/10 males is not due to impaired steroidogenesis of the Leydig cells of the testes in response to LH stimulation, but rather to inadequate elaboration or release of LH by the pituitary gland. Thus, genes with effects on the difference between DBA and C57BL males in adult serum level of testosterone, and consequently in intermale aggression, do not act in the testes but may act in the pituitary gland or hypothalamus.

The above genetic and endocrine research on intermale aggression of DBA and C57BL strains used the method of homogeneous set tests. It has been suggested (Fuller & Hahn, 1976; Simon, 1979) that strain differences in a homogeneous set test may be due not only to gene effects on the individual's propensity for fighting, but also on the occurrence of stimuli for eliciting fighting from conspecifics. In fact, there is evidence that some of the strain differences for DBA/2 and C57BL mice in the homogeneous set test may be due to gene effects on urinary pheromones that elicit the fighting response of males. Kessler, Harmatz, and Gerling (1975) studied strain differences in the capacity of male urinary odors to evoke aggressive behavior. The strains used were DBA/2J, CBA/J, and C57Br/cdJ. The C57Br/cdJ is closely related to the C57BL/10 and C57BL/6 strains, and may differ from them in alleles at no more than 100 loci (Bailey, 1978). Castrated DBA/2J males were daubed with either water or urine collected from one of these strains. Intermale aggression was tested by pairing an intact DBA/2J male in its home cage with one of the four stimulus animals. The latency to attack was shorter in response to castrates scented with DBA/2 urine than to castrates scented with either water, CBA urine, or C57Br urine. Also, the mean number of attacks was significantly greater in response to castrates scented with DBA/2 or CBA urine than with water or C57Br urine, and similarly, the accumulated attack time was significantly greater against castrates scented with DBA/2 or CBA urine than with water or C57Br urine. However, intact DBA/2 males significantly attacked any stimulus animal regardless of the solution daubed on it. Thus, it appears that the level of fighting in DBA/2 males in a homogeneous set test may be due not only to gene effects on the fighting propensity *per se*, but also on the elicitation of fighting by the strain's potent urinary pheromones. The pheromones for the elicitation of aggression are known to be testosterone-dependent (Mugford, 1974; Mugford & Nowell, 1972; Thiessen, 1976). Consequently, in the DBA/2 strain, if some of the gene effects on intermale aggression in a homogeneous set test are mediated by effects on the adult level of serum testosterone, these may act, in part, on the elaboration of potent pheromones for the arousal of fighting behavior.

The possibility that gene effects on intermale aggression are mediated, at least in part, by effects on adult levels of serum testosterone have also been explored for the CBA strain of mouse. The CBA strain was derived in 1920 from crosses of the DBA and BALB/c strains (Staats, 1976), and thus the DBA and CBA strains should share some of their alleles. Both strains are very aggressive (Selmanoff *et al.*, 1976; Stewart, Manning, & Batty, 1980), and both strains produce potent urinary pheromones for the elicitation of intermale aggression (Kessler *et al.*, 1975). Since CBA males are very aggressive in a standard stimulus test against repeatedly defeated BALB/c males, the intermale aggression of CBA mice in a homogeneous set test may be due not only to this strain's response to its own potent pheromones for evoking aggression, but also to its fighting propensity *per se*. Batty (1978) has reported that at 70 days of age, CBA males have a higher level of serum testosterone (5.3 ng/ml), as measured by radioimmunoassay, than C57BL/6 males (3.2 ng/ml). Similarly, Stewart *et al.* (1980) have shown that at about 80 days of age, the level of serum testosterone as measured by radioimmunoassay is higher in a double backcross population with predominantly CBA autosomes (3.8 ng/ml) than in one with predominantly C57BL/6 autosomes (2.5 ng/ml). Thus, fighting propensity *per se* and the elicitation of fighting by urinary pheromones in CBA males are associated with an elevated level of serum testosterone. In contrast to DBA strains, CBA males have smaller testes than C57BL/6 males (Stewart *et al.*, 1980). The strain difference is detected before puberty and continues into maturity; it is not due to an effect on the number of germinal cells or interstitial tissue, but rather to an effect on the length, but not diameter, of the seminiferous tubules (Hayward & Shire, 1974). Also, CBA males have normal sperm morphology and spermatogenesis and are fully fertile.

Intermale aggression in another strain also appears to be associated with the level of serum testosterone as measured by radioimmunoassay. BALB/cJ mice are more aggressive than A/J mice in a dangler test using a CBA male as the standard stimulus (Kessler, Elliott, Orenberg, & Barchas, 1977), and BALB/cJ mice are more aggressive than C57BL/6J mice in a homogeneous set test (Eleftheriou, Bailey, & Denenberg, 1974; Hahn & Haber, 1982). In addition, BALB/c mice, like CBA mice, have a higher level of serum testosterone than C57BL/6 mice (5.8 ng/ml versus 3.2 ng/ml; Batty, 1978). Since the CBA strain is derived from a cross of DBA and BALB/c mice (Staats, 1976), the similar association of intermale aggression and adult level of serum testosterone in CBA and BALB/c mice may involve the same allelic variants. However, it should be noted that BALB/c mice are not always more aggressive than C57BL mice (Lagerspetz & Lagerspetz, 1975). The reversal in strain ranking of BALB/c and C57BL mice for aggression may be due to effects of the conditions and methods of testing (Simon, 1979).

Endocrinological aspects of intermale aggression have also been examined for the TA (aggressive) and TNA (nonaggressive) strains. The TA and TNA strains were established by selective breeding for and against the severity of fighting (Lagerspetz, 1961); the base population for the selective breeding program was a stock of Swiss albino mice. The difference between the two selected strains is not mediated by postnatal maternal factors (Lagerspetz & Wuorinen, 1965) and may be due to genetic factors. Although the seminal vesicle weights are the same in TA and TNA males, the testes are 26% to 32% larger in TA than in TNA males (Lagerspetz, Tirri, & Lagerspetz, 1968). The difference in the size of the testes does not involve the interstitial tissue, and this observation suggests that androgen production may be the same in both strains. Lagerspetz and Lagerspetz (1975) also found that injection of adult females with testosterone propionate (TP) induced fighting in TA but not TNA mice. Also, combined neonatal and adult androgen treatment was ineffective in inducing aggression in TNA females. These findings are compatible with the hypothesis that the difference in fighting between TA and TNA males is not due to gene effects on the adult level of serum testosterone. It is thus clear that not all strain differences in intermale aggression are associated with strain differences in adult serum testosterone levels and that some, but not all, gene effects on intermale aggression may be mediated by effects on adult levels of serum testosterone.

Strain differences in male sexual behavior and in the levels of serum testosterone have also been investigated for DBA/2, CBA, BALB/c, C57BL/6, TA, and TNA strains. For DBA/2, CBA, BALB/c, and C57BL/6 strains, there is an inverse correlation between the level of serum testosterone and average percentage of tests with mounting, intromission, or ejaculation, and a direct correlation between the level of serum testosterone and latency to mounting or intromission (Batty, 1978). McGill (1978) reported similar findings for latency to ejaculation. Thus, for these strains, high levels of aggressiveness are associated with high levels of serum testosterone, and low levels of "sex drive" are associated with high levels of serum testosterone. Lagerspetz and Lagerspetz (1975) report that the strain difference in intermale aggression between TA and TNA strains does not appear to involve gene effects on the levels of serum testosterone and is not associated with a difference in male sexual behaviors as measured by mean number of mounts of estrous females of the same strain. Taken together, these findings suggest the following hypothesis: If gene effects on intermale aggression are mediated by effects on the level of serum testosterone, these same genes will have an effect on male sexual behavior, and the intermale aggression and male sexual behavior will be inversely correlated for these genotypes. However, if gene effects on intermale aggression do not involve effects on the levels of serum testosterone, there is no *a priori* reason to expect these gene(s) to also affect male sexual behaviors. It has been reported that the strain difference in intermale aggression of AKR/J and C3H/HeJ mice is positively correlated with a strain difference in male sexual behaviors (Mosig, 1974). The above hypothesis predicts that AKR/J and C3H/HeJ strains will not differ in adult levels of serum testosterone.

GENOTYPE, HORMONES, AND THE DEVELOPMENT OF AGGRESSION

Androgenic hormones are involved not only in the activation of the aggression response of adult males, but also in the development of intermale aggression (Gandelman, 1980). Neonatal castration abolishes the development of aggression in male mice (Edwards, 1969), and neonatal exposure to exogenous testosterone induces pugnacity in castrated male and ovariectomized or intact female mice (Edwards, 1968, 1969, 1970). There have been two major interpretations of these findings. On the one hand, it has been suggested that early exposure to testosterone "organizes" or causes the differentiation of neural systems involved in intermale aggression (Edwards, 1969). On the other hand, it has been proposed that early exposure to testosterone lowers the threshold of the neural systems involved in intermale aggression to activation by testosterone in the adult (Gandelman, 1980). In this context, three aspects of the intersection of genetics and hormones will be reviewed for the development of intermale aggression in mice. The use of the testicular feminization mutation (Tfm) to study the role of neonatal androgens in the development of intermale aggression will be discussed. Also, the evidence will be considered supporting the hypotheses that the effects of some Y-chromosomal gene(s) on intermale aggression are mediated by the pubertal rise in the level of serum testosterone, and that some strain differences in intermale aggression are due to interactions in development of the strain's genotype with androgenic steroids.

The mutation for testicular feminization (Tfm) has been detected in mice, rats, cattle, and man, and it affects the development of androgen-dependent secondary sexual characteristics (Attardi, 1976; Ohno, 1971, 1979). Testicular feminized individuals are of the XY karyotype and possess undescended, testosterone-producing testes. Also, they have neither Wolffian nor Mullerian duct derivatives and externally resemble females of their species. The effect of this mutation on the development of secondary sexual characteristics is due to its rendering target tissues insensitive to androgen, and this androgen-insensitivity of target tissues is, in turn, due to a reduction in the affinity of the 8S cytosol androgen receptor. In mice, the target tissues of Tfm individuals contain about 20% to 30% as much androgen binding capacity as normal individuals, and androgen insensitivity of target tissues such as kidney and submaxillary gland is complete. However, in rats, the androgen-insensitivity of some target tissues in Tfm individuals is incomplete, and some of these target tissues have near-normal binding capacities of the 8S cytosol receptor (Attardi, 1976). Ohno (1979) has argued that the testicular feminization mutations occur in the structural gene for the 8S cytosol receptor protein, whereas Attardi (1976) has suggested that the Tfm mutations occur in a gene regulating the activity of the structural gene for the 8S cytosol receptor protein. In either case, the testicular feminization gene in mice has been shown to be X-linked, and it is expected to be X-linked in all mammalian species. On the basis of findings on X-linked genes in two or more mammalian species of one or more mammalian orders, Ohno (1979) has hypothesized that if a gene is found to be X-linked in one mammalian species, it will be X-linked in all mammalian species. This X-linked Tfm gene has been used in one study with mice and two studies with rats to assess the role of neonatal androgens in masculinization of the brain.

Male, female, and Tfm mice were used in the study of Ohno, Geller, and Lai (1974). Some of the mice in this study were given a single treatment with 10 mg of testosterone, estradiol 17β, and 5α dihydrotestosterone at 7 to 25 days before testing. In this study, there were two tests of aggression and two tests of sexual behavior. One test was for aggressiveness toward an intruder male mouse. In it, a normal, adult male mouse was introduced into the home cage of a mouse to be tested. The cage had not been cleaned for 24 hours prior to testing. The intruder was attacked by normal males, but not by Tfm mice either untreated with hormones or treated with any of the three hormones. Thus, intermale aggression was not elicited from Tfm mice. Another test was a measure of the provocativeness of an intruder mouse. This was the reverse of the previously described test. In it, the resident mouse was a normal, adult male. The resident male attacked normal male intruders, but not Tfm mouse intruders either untreated with hormones or treated with any of the three hormones. Thus, intermale aggression is not elicited by intruder Tfm mice. Also in this test situation, normal males mounted Tfm mice about as often as they mounted diestrous females and castrated males; however, the Tfm mice displayed no female sexual behaviors. The third test measured sexual behavior toward a normal adult estrous female, and the fourth test measured sexual behavior toward a normal adult diestrous female. Both tests were conducted in the female's home cage. In these tests, the Tfm mice either untreated with hormones or treated with one of the three hormones showed no male sexual behaviors. Thus, Tfm mice do not display either male or female sexual behavior patterns. On the basis of these findings, Ohno *et al.* (1974) proposed that neonatal hormonal imprinting is a prerequisite not only for masculinization but also for feminization of the central nervous system of mice, and that Tfm blocks masculinization of the mouse CNS by interfering with the induction by testosterone of the aromatizing enzyme. It is an assumption of this hypothesis that in the neonate, high intracellular concentrations of estradiol cause masculinization of the CNS of mice and low circulating levels of estradiol cause feminization of the mouse CNS. Also, on the basis of findings with sex-reversed (Srx/+) males and with neural mosaicism for Tfm, these authors concluded that a small pool of neurons is involved in the neonatal hormonal imprinting for a masculine or feminine CNS in mice, and that this pool of neurons is not the same for intermale aggression and for male sexual behavior.

Different conclusions on the role of neonatal hormones in the development of intermale aggression have come from recent studies of experimental manipulation of castration and hormone replacement (Gandelman, 1980). These studies suggest that neonatal hormones are not required for the development of intermale aggression. Similarly, in contrast to the findings with mice (Ohno *et al.*, 1974), two recent studies with male, female, and Tfm rats have also concluded that binding neonatal androgens to the androgenic receptor is not required for the development of male sexual behaviors. Beach and Buehler (1977) measured the percentage of mounts, mounts and thrusts, intromission patterns, and ejaculation patterns in normal males and Tfm rats in response to an estrous female. Tfm rats had lower levels of all behaviors than normal males, but had a substantial frequency (15% to 38%) of mounts only or mounts plus thrusts and had little or no (3% to 4%) intromission pattern or ejaculation pattern. Similarly, Shapiro, Levine, and Adler (1980) measured female and male sexual patterns in intact male, female, and Tfm rats and in castrated male, female, and Tfm rats either untreated with hormones or after 7 days of injection with testosterone (10 mg/day) or a single injection of estradiol (0.5 mg). Intact male and female rats showed the homotypical sexual behaviors and intact Tfm rats showed neither male nor female sexual behaviors. Female sexual behaviors were measured as percentage of animals responding with lordosis, and the lordosis quotient and male sexual behaviors were measured by the number of animals responding with and the mean number of intromission patterns and ejaculation patterns. Hormonally treated Tfm rats showed the same female and male sexual behaviors as similarly treated normal males. On the basis of these findings, the authors concluded that in the rat, the brain can be masculinized for sexual behavior in the absence of the androgenic receptor and that the occurrence of a partially defeminized and masculinized brain in Tfm rats may be due to estrogens from the testes or to testosterone subsequently aromatized in the brain. It is an assumption of this hypothesis that induction of the aromatizing enzyme does not depend on normal affinity of the 8S cytosol receptor for testosterone. The discrepancy between the findings of Ohno *et al.* (1974) for mice and Shapiro *et al.* (1980) for rats may be due to differences in hormone treatment or to differences, as suggested by McGill (1978), in procedures for inducing and testing sexual behavior. It would be of interest to test Tfm rats for intermale aggression and to replicate the study of

Ohno *et al.* (1974) with procedures similar to those used in the study of Shapiro *et al.* (1980).

The previous section reviewed the intersection of a X-chromosomal gene and androgens in the development of intermale aggression; this one will consider the intersection of the Y-chromosome and androgens in the development of intermale aggression. The aggressive behavior of DBA/1Bg, C57BL/10Bg, and their reciprocal hybrids were measured in three studies using a homogeneous set test (Maxson, Ginsburg, & Trattner, 1979; Selmanoff, Jumonville, Maxson, & Ginsburg, 1975; Selmanoff *et al.,* 1976; see Table I). In each study, there was a significant difference between reciprocal F_1 hybrids, with B10D1F_1 males being more aggressive than D1B10F_1 males. Since the more aggressive F_1 receives its Y-chromosome from the highly aggressive DBA/1 parent and its ovarian cytoplasm and X-chromosome from the less aggressive C57BL/10 parent, as well as being exposed to the uterine and postnatal maternal environment of the C57BL/10 parent, it was hypothesized that Y-chromsomal gene(s) of the DBA/1Bg strain make an incremental contribution of approximately 10 to 14 units (see Maxson, 1981; Selmanoff & Ginsburg, 1981) to the intermale aggression of DBA/1Bg and B10D1F_1 hybrid mice. In order to test this hypothesis critically, a congenic stock of C57BL/10 mice carrying the DBA/1Bg Y-chromosome was developed (Maxson *et al.,* 1979). The stock is designated as C57BL/10.DBA/1-Y or BL10.D1-Y. The C57BL/10 strain and the BL10.D1-Y congenic stock have the same intermale aggression score. This finding does not support the hypothesis that, irrespective of genetic background, the Y-chromosome of the DBA/1Bg strain makes an incremental contribution to intermale aggression scores. However, it is conceivable that the influence of the DBA/1 Y-chromosome on intermale aggression depends on an interaction between it and DBA/1Bg autosomal gene(s). Although DBA/1, B10D1F_1, and C57BL/10.DBA/1-Y mice are identical in their Y-chromosomes, they differ in proportion of DBA/1 autosomes. This proportion is 100% for DBA/1, 50% for B10D1F_1, and essentially 0% for BL10.D1-Y mice. The interaction hypothesis was tested by comparing the mean intermale aggression score of D1/B10.D1-Y F_1 and B10D1F_1 mice, and both of these with that of D1B10F_1 mice. The D1/B10.D1-Y F_1 and B10D1F_1 are identical in autosomes and Y-chromosome, but differ in X-chromosome, maternal environments, and ovarian cytoplasm; whereas the D1/B10.D1-Y F_1 and D1B10F_1 are identical in autosomes, X-chromosomes, maternal environments, and ovarian cytoplasm, but differ in Y-chromosome. The D1/B10.D1-Y F_1 and B10D1F_1 had the same high intermale aggression score, and both of these had a higher intermale aggression score than the D1B10F_1 mice. These findings are consistent with the hypothesis that an interaction of Y-chromosomal and autosomal genes is involved in the inheritance and development of intermale aggression in *these* mice.

It is possible that the Y-chromosome of the DBA/1, in interacting with this strain's autosomes, influences the development of intermale aggression by an effect on the pubertal (but not adult) levels of serum testosterone. Selmanoff, Goldman, and Ginsburg (1977) measured serum testosterone levels of DBA/1Bg and C57BL/10Bg strains from birth to 70 days of age. Hormone levels were determined by radioimmunoassay at 3-day intervals from 1 to 30 days postbirth,

Table I
Reciprocal Effects on Intermale Aggression

	DBA/1Bg	D1/B10.D1-Y F_1	B10D1 F_1	D1B10 F_1	B10.D1-Y	C57BL/10Bg
Study 1[a]	34.3 ± 5.1 (10)[b]	—	34.0 ± 3.0 (10)	21.5 ± 5.5 (10)	—	5.0 ± 2.1 (10)
Study 2[a]	35.4 ± 4.6 (20)	—	19.8 ± 3.0 (43)	10.7 ± 2.2 (42)	—	7.3 ± 2.3 (22)
Study 3[a]	39.7 ± 5.3 (15)	25.8 ± 4.2 (23)	21.9 ± 6.3 (22)	10.9 ± 3.6 (20)	4.9 ± 1.6 (22)	8.5 ± 2.4 (21)

[a] Study 1 was carried out at the University of Chicago between 1967 and 1968 (Selmanoff, Jumonville, Maxson, & Ginsburg, 1975); Study 2 was carried out at the University of Connecticut between 1972 and 1973 (Selmanoff et al., 1975; Selmanoff, Maxson, & Ginsburg, 1976); and Study 3 was carried out at the University of Connecticut between 1977 and 1978 (Maxson, Ginsburg, & Trattner, 1979).

[b] $\overline{X}$ ± SEM (N) of "any aggression" score. Underlined means are statistically equivalent by Neuman-Keuls analysis of ordered means (Winer, 1962).

and at 35, 40, 50, 60, and 70 days of age. There were no differences in androgen levels between DBA/1Bg and C57BL/10Bg at 1 to 30 days of age at 50 to 70 days of age. However, the serum testosterone levels were lower in C57BL/10Bg than in DBA/1Bg at 35 and 40 days of age, and the pubertal increase in serum testosterone levels was markedly steeper for DBA/1Bg than for C57BL/10Bg males. In order to test the hypothesis that the Y-chromosome had an influence on the pubertal rise and levels of serum testosterone, the levels of circulating testosterone were assayed in $D1B10F_1$ and $B10D1F_1$ reciprocal hybrids at 30, 35, 40, 50, 60, and 70 days of age (Selmanoff, Goldman, Maxson, & Ginsburg, 1977; see Figure 1). The $B10D1F_1$ had higher levels of serum testosterone on day 35 than did C57BL/10Bg males, whereas the $D1B10F_1$ had the same level of serum testosterone on day 35 as C57BL/10Bg males. Similarly, comparisons of the regression coefficients for serum testosterone levels spanning 30 to 50 days of age indicate that the slope of the pubertal surge in serum testosterone levels in $B10D1F_1$ (but not $D1B10F_1$) is steeper than that of C57BL/10Bg males. These findings are consistent with the hypothesis that the DBA/1Bg Y-chromosome influences not only intermale aggression but also the pubertal rise in serum testosterone levels. However, at days 35 and 40, there were no significant differences in serum testosterone levels between either DBA/1 and the two reciprocal F_1 hybrids of $D1B10F_1$ and $B10D1F_1$. Similarly, there was no significant difference in slope of the pubertal rise in serum testosterone for DBA/1Bg, $B10D1F_1$, and $D1B10F_1$ males. That the additional reciprocal differences were not obtained may be due to the large variability in serum testosterone levels. This variability has been attributed to the pulsatile release of testosterone (Bartke *et al.*, 1973). The hypothesis of a possible association for Y-chromosomal effects on intermale aggression and pubertal rise in serum testosterone is consistent with the findings in several studies of an effect of the pubertal rise in serum testosterone levels on the development of intermale aggression (Barkley & Goldman, 1977a,b,c; McKinney & Desjardins, 1972; Owen, Peters, & Bronson, 1973).

The above hypothesis does not imply that the Y-chromosome is involved in every strain difference for intermale aggression, or that all Y-chromosomes

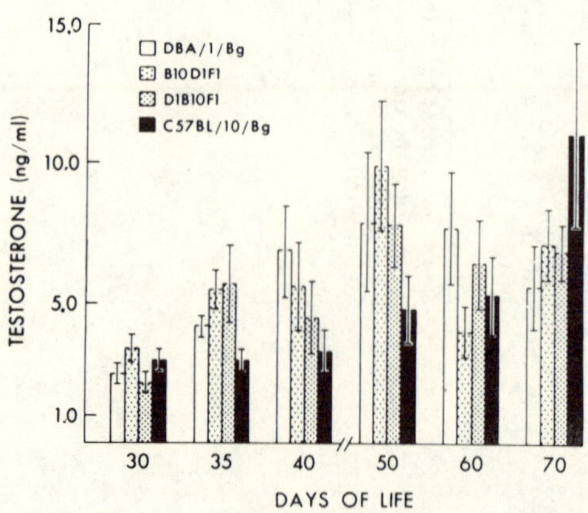

FIGURE 1. Developmental serum testosterone in males of the parental strains and reciprocal hybrids. (From "Correlated Effects of the Y-Chromosome of Mice on Developmental Changes in Testosterone Levels and Intermale Aggression" by M. K. Selmanoff, B. D. Goldman, S. C. Maxson, and B. E. Ginsburg, *Life Sciences*, 1977, *20*, 359–366. Copyright 1977 by Pergamon Press, Ltd. Reprinted by permission.)

influence intermale aggression, or that all Y-chromosomes influence intermale aggression by an effect on the pubertal rise in serum testosterone. For example, Kessler *et al.* (1975) did not obtain a reciprocal difference for intermale aggression of F_1 hybrids of BALB/cJ and A/J strains in a dangler test, and Selmanoff *et al.* (1976) did not find a reciprocal difference in intermale aggression of DBA/2Bg and C57BL/10Bg in a homogeneous set test. Thus, it appears that in these strains and hybrids, the Y-chromosome has no influence on intermale aggression. Lagerspetz and Lagerspetz (1975) have also argued that the difference between TA and TNA strains in intermale aggression does not involve Y-chromosomal gene(s). However, it has also been shown that the Y-chromosomes of the PHH and PHL strains (Weir, 1976; Weir & Hogle, 1973), of the CBA/FaCam and C57BL/6J strains (Stewart *et al.,* 1980), and of the Rb/1Bg and C57BL/10Bg strains (Maxson, Platt, Shrenker, & Trattner, 1982) differ in their effect on intermale aggression of mice. Unfortunately, there is no information on the effect of these Y-chromosomes on the pubertal rise in serum testosterone levels. However, Stewart *et al.* (1980) have shown that the CBA/FaCam Y-chromosome has no effect on adult levels of serum testosterone. This is similar to the finding of Selmanoff, Goldman, and Ginsburg (1977) for the effect of the DBA/1Bg Y-chromosome on adult levels of serum testosterone. Although the Y-chromsomes of the DBA/1, PHH, and CBA strains have the same effect on intermale aggression, they have dissimilar effects on other phenotypes. For example, the PHH Y-chromosome also has an effect on male sexual behavior (Weir, 1976) and no effect on testis weight (Herrick & Wolfe, 1977); whereas the CBA Y-chromosome has no effect on male sexual behavior and a decremental effect on testis weight (Stewart *et al.,* 1980). Maxson (1981) has reviewed the effects of the DBA/1, PHH, and CBA Y-chromosomes on traits other than intermale aggression.

The effect of testosterone on the expression of genes influencing the de velopment of intermale aggression is another possible intersection of genotype and hormones. Here, strain differences in intermale aggression would be due to genotype–hormone interactions. The genotype of one strain would be responsive to testosterone effects, and the genotype of the other strain would be nonresponsive to testosterone effects for the development of this behavior. This possible intersection of genotype and hormones has been explored by comparing the order of inbred strains for male rank-related aggression and for rank-related aggression of females treated as neonates and/or as adults with testosterone. Males and females of the same strain share the same alleles at each locus except for those on the Y-chromosome. Consequently, if the strain difference in intermale aggression were due to genotype–hormone interaction, there should be the same strain rank order of aggression for females treated with testosterone as for untreated males. Two studies have made use of this approach.

Vale, Ray, and Vale (1972) compared the rank-related aggression of females injected neonatally with testosterone propionate (TP) and of males injected neonatally with oil from the BALB/cJ, C57BL/6J, and A/J strains. At 70 to 90 days of age, the mice were tested in a homogeneous set test with animals paired for same genotype, sex, and treatment. In this test, there were treatment but no treatment-by-genotype effects on latency to attack and mean number of attacks.

The neonatal androgen treatment decreased the latency to attack and increased the mean number of attacks for females of all three strains. However, on both measures, males treated with oil were more aggressive than females treated with TP. Next, the mice were given a standard stimulus test using A/J males in the dangler paradigm. In this test, there were strain-by-treatment interactions. Female BALB/cJ mice treated neonatally with TP had a lower latency to attack and a higher mean number of attacks than oil controls of that strain or than treated females of the C57BL/6J and A/J strains for both the first- and second-day tests. Similarly, oil-treated males of the BALB/cJ strain had lower latency to attack and higher mean number of attacks on this test than oil-treated males of the C57BL/6J and A/J strains. Thus, the strain rank order for aggression in this test is the same for oil-treated males and testosterone-treated females. Vale *et al.* (1972) suggested on the basis of these findings that some of the strain differences in intermale aggression of BALB/c versus either C57BL/6 or A/J may be due to genotype–hormone interactions and that neonatal androgens activate the genes influencing aggression in male and female BALB/c mice, but not C57BL/6 or A/J mice. Similary, Vale, Ray, and Vale (1972, 1973a) reported that treatment of females with neonatal androgens or estrogens decreased the latency to attack and increased the mean number of attacks against a male attempting to mount for BALB/c but not for C57BL/6 or A/J mice.

For TA and TNA strains, Lagerspetz and Lagerspetz (1975) compared the rank-related aggression of females treated as neonates and/or as adults with testosterone with that of males similarly treated with oil. At six months of age, the TA females (but not TNA females) treated with either testosterone regime had more biting attacks and tail rattles against mice of either sex than did control females. Thus, there was a parallel strain difference for TA and TNA males and for testosterone-treated TA and TNA females. As with the study of Vale *et al.* (1972), results from this study are consistent with the hypothesis that some of the difference in aggression between TA and TNA mice is due to genotype and hormone interaction, and that neonatal or adult androgens may activate the genes influencing aggression in males and females of the TA but not the TNA strain.

Both Vale, Ray, and Vale (1973a,b) and Lagerspetz and Lagerspetz (1975) examined the effect of these hormone treatments not only on aggression, but also on male-like sexual behaviors of females, and compared the strain rank order of untreated males and treated females for both aggression and male sexual behavior. In the study of Vale *et al.* (1973b), females of the BALB/cJ, C57BL/6J, and A/J strains were injected neonatally with either oil, TP, or estradiol benzoate and tested for male-like sexual behaviors with an untreated estrous C57BL/6J female at 75 to 90 days of age. Except for number of mounts and latency to mount, the rank order of the three strains on male sexual behaviors was the same for neonatally androgen-treated females and for neonatally oil-treated males. Testosterone-treated BALB/c females had fewer mounts than A/J females, whereas oil-treated male BALB/c mice had more mounts than similarly treated A/J males. Also, the latency to mount was shorter in treated BALB/c females than in treated A/J females, whereas oil-treated male BALB/c mice have a longer latency to mount than similarly treated A/J males. However, treated C57BL/6 females and oil-treated C57BL/6 males are more active in mating than either albino strain. These findings are consistent with the hypothesis that some but not all of the strain differences in male sexual behavior are due to genotype–hormone interactions, and that neonatal androgens may activate genes influencing the development of male sexual behaviors. However, these studies do not provide evidence on whether the genotype–hormone interactions for intermale aggression and male sexual behavior in these strains involve the same or different genetic loci. Conversely, Lagerspetz and Lagerspetz (1975) did not find that there was a genotype–hormone interaction for male-like sexual behavior in TA and TNA females treated as neonates and/or as adults with testosterone. At six months of age, both TA and TNA females treated with testosterone mounted receptive (estrous) but not nonreceptive (nonestrous) females and rarely mounted males. Thus, in TA and TNA female mice, the effect of testosterone treatment on male sexual behavior is not genotype dependent, whereas the effect on aggression is genotype dependent. On this basis, Lagerspetz and Lagerspetz (1975) speculate that the genes involved in the difference between TA and TNA mice for intermale aggression are not involved in male sexual behavior as measured by frequency of mounting a receptive female. However, TA females treated with testosterone are more persistent in attempts at mounting than are similarly treated TNA females.

PROSPECTUS AND SUMMARY

The research described in this review has provided evidence for the following intersections of genotypic and hormonal factors in the development and expression of intermale aggression: (1) interactions of genotype and androgens in the neonate and adult; (2) Y-chromosomal effects on pubertal rise in serum testosterone levels; and (3) autosomal effects on adult levels of serum testosterone. Since most of the studies on androgens and aggression have used inbred strains and their F_1 hybrids, it is possible that some of the reported hormone and behavior correlations are due to accidental associations, and/or to linkage of different genes with effects on each phenotype rather than due to pleiotropism of the same genes with effects on each phenotype. This issue should now be studied using segregated populations of F_2 and backcrosses, diallel crosses, heterogeneous stocks, recombinant inbred strains, and congenic or coisogenic strains (Fuller & Thompson, 1978). These techniques may also be used to isolate, identify, and map individual genes with effects on both hormonal and behavioral phenotypes. Such studies would also provide a more substantial data base for understanding the effect of genotype on the development of intermale aggression and for using genotype as a tool to study the role of androgens in the development of intermale aggression. These same genetic methods may be used to analyze the role of other hormones in intermale aggression and of all hormones in other types of aggression. Gene effects on other hormones (Hamburg, 1967; Shire, 1968; Vale, Vale, & Harley, 1971) and hormonal effects on other types of aggression (see reviews in this volume) are known.

REFERENCES

Attardi, B. Genetic analysis of steroid hormone action. *Trends in Biochemical Sciences,* 1976, *1,* 241–244.

Bailey, D. W. Sources of subline divergence and their relative importance for sublines of six major inbred strains of mice. In H. C. Morse, III (Ed.), *Origins of inbred mice.* New York: Academic Press, 1978.

Banarjee, U. The influence of some hormones and drugs on isolation induced aggression in male mice. *Communications in Behavioral Biology,* 1971, *6,* 163–170.

Barkley, M. S., & Goldman, B. D. A quantitative study of serum testosterone, sex accessory organ growth, and the development of intermale aggression in the mouse. *Hormones and Behavior,* 1977, *8,* 208–218. (a)

Barkley, M. S., & Goldman, B. D. The effects of castration and silastic implants of testosterone on intermale aggression in the mouse. *Hormones and Behavior,* 1977, *9,* 32–48. (b)

Barkley, M. S., & Goldman, B. D. Testosterone-induced aggression in adult female mice. *Hormones and Behavior,* 1977, *9,* 76–84. (c)

Bartke, A. Increased sensitivity of seminal vesicles to testosterone in a mouse strain with low plasma testosterone levels. *Journal of Endocrinology,* 1974, *60,* 145–148.

Bartke, A., & Shir, J. G. M. Differences between mouse strains in testicular cholesterol levels and androgen target organs. *Journal of Endocrinology,* 1972, *55,* 173–184.

Bartke, A., Steele, R. E., Musto, N., & Caldwell, B. V. Fluctuations in plasma testosterone levels in adult male rats and mice. *Endocrinology,* 1973, *92,* 1223–1228.

Bartke, A., Roberson, C., & Dalterio, S. Concentration of gonadotrophins in the plasma and testicular responsiveness to gonadotrophic stimulation in androgen-deficient C57BL/10J mice. *Journal of Endocrinology,* 1977, *75,* 441–442.

Batty, J. Plasma levels of testosterone and male sexual behaviour in strains of the house mouse *(Mus musculus). Animal Behaviour,* 1978, *26,* 339–348.

Beach, F. A., & Buehler, M. G. Male rats with inherited insensitivity to androgen show reduced sexual behavior. *Endocrinology,* 1977, *100,* 197–200.

Beeman, E. A. The effect of male hormone on aggressive behavior in mice. *Physiological Zoology,* 1947, *20,* 373–405.

Brain, P. F. *Hormones and behavior* (Vol. 2). St. Albans, England: Eden Medical Research, 1979. (a)

Brain, P. F. *Hormones, drugs and behavior* (Vol. 3). St. Albans, England: Eden Medical Research, 1979. (b)

Calhoun, J. B. A comparative study of the social behavior of two inbred strains. *Ecological Monograph,* 1956, *26,* 81–103.

Chapman, W. V., & Whitten, W. K. The occurrence and inheritance of pregnancy blocking in inbred mice. *Genetics,* 1969, *61,* S9.

Edwards, D. A. Mice: Fighting by neonatally androgenized females. *Science,* 1968, *161,* 1027–1028.

Edwards, D. A. Early androgen stimulation and aggressive behavior in male and female mice. *Physiology and Behavior,* 1969, *4,* 333–338.

Edwards, D. A. Post-neonatal androgenization and adult aggressive behavior in female mice. *Physiology and Behavior,* 1970, *5,* 465–467.

Eleftheriou, B. E., & Lucas, L. A. Age-related changes in testes, seminal vesicles and plasma testosterone levels in male mice. *Gerontologia,* 1974, *20,* 231–238.

Eleftheriou, B. E., Bailey, D. W., & Denenberg, V. H. Genetic analysis of fighting behavior in mice. *Physiology and Behavior,* 1974, *13,* 773–777.

Fuller, J. L., & Hahn, M. E. Issues in the genetics of social behavior. *Behavior Genetics,* 1976, *6,* 391–406.

Fuller, J. L., & Thompson, W. R. *Foundations of behavior genetics.* St. Louis: C. V. Mosby, 1978.
Gandelman, R. Gonadal hormones and the induction of intraspecific fighting in mice. *Neuroscience and Biobehavioral Reviews,* 1980, *4,* 133–140.
Hahn, M. E., & Haber, S. B. The inheritance of agonistic behavior in male mice: A diallel analysis. *Aggressive Behavior,* 1982, *8,* 19–38.
Hamburg, D. A. Genetics of adrenocortical hormone metabolism in relation to psychological stress. In J. Hirsch (Ed.), *Behavior-genetic analysis.* New York: McGraw-Hill, 1967.
Hayward, P., & Shire, J. G. M. Y chromosome effects on adult testis size. *Nature,* 1974, *250,* 499–500.
Herrick, C. S., & Wolfe, H. G. Effect of Y-chromosome on testis size in the mouse *(Mus musculus). Genetics,* 1977, *86,* S27.
Kessler, S., Harmatz, P., & Gerling, S. A. The genetics of pheromonally mediated aggression in mice: I. Strain differences in the capacity of male urinary odors to elicit aggression. *Behavior Genetics,* 1975, *5,* 233–238.
Kessler, S., Elliott, G. R., Orenberg, E. K., & Barchas, J. D. A genetic analysis of aggressive behavior in two strains of mice. *Behavior Genetics,* 1977, *7,* 313–321.
Lagerspetz, K. Genetic and social causes of aggressive behavior in mice. *Scandinavian Journal of Psychology,* 1961, *2,* 167–173.
Lagerspetz, K. M. J., & Lagerspetz, K. Y. H. The expression of the genes in aggressiveness in mice: The effect of androgen on aggression and sexual behavior in females. *Aggressive Behavior,* 1975, *1,* 291–296.
Lagerspetz, K. M. J., & Wuorinen, K. W. A cross-fostering experiment with mice selectively bred for aggressiveness and non-aggressiveness. *Reports of the Psychological Institute, University of Turku,* 1965, *17,* 1–6.
Lagerspetz, K. Y. H., Tirri, R., & Lagerspetz, K. M. J. Neurochemical and endocrinological studies of mice selectively bred for aggressiveness. *Scandinavian Journal of Psychology,* 1968, *9,* 157–160.
Maxson, S. C. The genetics of aggression in vertebrates. In P. F. Brain & D. Benton (Eds.), *The biology of aggression.* Alphen aan den Rijn, Netherlands: Sijthoff en Noordhoff, 1981.
Maxson, S. C., Ginsburg, B. E., & Trattner, A. Interaction of Y-chromosomal and autosomal gene(s) in the development of intermale aggression in mice. *Behavior Genetics,* 1979, *9,* 219–226.
Maxson, S. C., Platt, T., Shrenker, P., & Trattner, A. The influence of the Y-chromosome of Rb/1Bg mice on agonistic behaviors. *Aggressive Behavior,* 1982, *8,* 285–292.
McGill, T. E. Genetic factors influencing the action of hormones on sexual behavior. In J. B. Hutchinson (Ed.), *Biological determinants of sexual behavior.* New York: Wiley, 1978.
McKinney, T. D., & Desjardins, C. Postnatal development of testis, fighting behavior, and fertility in the house mouse. *Biology and Reproduction,* 1972, *9,* 279–294.
Mosig, D. W. *Sexual and aggressive behaviors in two inbred strains of mice.* Unpublished doctoral dissertation, University of Florida, 1974.
Moyer, K. E. Kinds of aggression and their physiological basis. *Communications in Behavioral Biology,* 1968, *2A,* 65–87.
Moyer, K. E. *The psychobiology of aggression.* New York: Harper & Row, 1976.
Mugford, R. A. Androgenic stimulation of aggression eliciting cues in adult opponent mice castrated at birth, weaning or maturity. *Hormones and Behavior,* 1974, *5,* 93–109.
Mugford, R. A., & Nowell, N. W. The dose response to testosterone propionate of preputial glands, pheromones, and aggression in mice. *Hormones and Behavior,* 1972, *3,* 39–46.
Ohno, S. An argument for the simplicity of mammalian regulatory systems: Single gene determination of male and female phenotypes. *Nature,* 1971, *234,* 134–138.
Ohno, S. *Major sex determining genes.* New York: Springer-Verlag, 1979.
Ohno, S., Geller, L. N., & Lai, D. J. Tfm mutation and masculinization versus feminization of mouse central nervous system. *Cell,* 1974, *3,* 235–242.
Owen, K., Peters, P. J., & Bronson, F. H. Differential responsiveness to replacement therapy of pre- and postpubertally castrated mice with respect to intermale aggression. *Hormones and Behavior,* 1973, *4,* 301–306.
Scott, J. P. Agonistic behavior of mice and rats: A review. *American Zoologist,* 1966, *6,* 683–701.
Scott, J. P., & Fredericson, E. The causes of fighting in mice and rats. *Physiological Zoology,* 1951, *24,* 273–309.

Selmanoff, M. K., & Ginsburg, B. E. Genetic variability in aggression and endocrine function in inbred strains of mice. In P. F. Brain & D. Benton (Eds.), *Multidisciplinary approaches to aggression research.* Amsterdam: Elsevier/North-Holland Biomedical Press, 1981.

Selmanoff, M. K., Jumonville, J. E., Maxson, S. C., & Ginsburg, B. E. Evidence for a Y chromosomal contribution to an aggressive phenotype in inbred mice. *Nature,* 1975, *253,* 529–530.

Selmanoff, M. K., Maxson, S. C., & Ginsburg, B. E. Chromosomal determinants of intermale aggressive behavior in inbred mice. *Behavior Genetics,* 1976, *6,* 53–69.

Selmanoff, M. K., Abreu, E., Goldman, B. D., & Ginsburg, B. E. Manipulation of aggressive behavior in adult DBA/2/Bg and C57BL/10/Bg male mice implanted with testosterone in silastic tubing. *Hormones and Behavior,* 1977, *8,* 377–390.

Selmanoff, M. K., Goldman, B. D., & Ginsburg, B. E. Developmental changes in serum luteinizing hormones, follicle stimulating hormone and androgen levels in males of two inbred mouse strains. *Endocrinology,* 1977, *100,* 122–127.

Selmanoff, M. K., Goldman, B. D., Maxson, S. C., & Ginsburg, B. E. Correlated effects of the Y-chromosome of mice on developmental changes in testosterone levels and intermale aggression. *Life Sciences,* 1977, *20,* 359–366.

Shapiro, B. H., Levine, D. C., & Adler, N. T. The testicular feminized rats: A naturally occurring model of androgen independent brain masculinization. *Science,* 1980, *209,* 418–420.

Shire, J. G. M. Genes, hormones and behavioural variation. In J. M. Thoday & A. S. Parkes (Eds.), *Genetic and environmental influences on behavior.* New York: Plenum Press, 1968.

Shire, J. G. M., & Bartke, A. Strain differences in testicular weight and spermatogenesis with special reference to C57BL/10J and DBA/2J mice. *Endocrinology,* 1972, *55,* 163–171.

Simon, N. G. The genetics of intermale aggressive behavior in mice: Recent research and alternative strategies. *Neuroscience and Biobehavioral Reviews,* 1979, *3,* 97–106.

Staats, J. Standardized nomenclature for inbred strains of mice: Sixth listing. *Cancer Research,* 1976, *36,* 4333–4377.

Stewart, A. D., Manning, A., & Batty, J. Effects of Y-chromosome variants on the male behaviour of the mouse *Mus musculus. Genetic Research, Cambridge,* 1980, *35,* 261–268.

Thiessen, D. D. *The evolution and chemistry of aggression.* Springfield, Ill.: Charles C Thomas, 1976.

Tollman, T., & King, J. A. The effect of testosterone propionate on aggression in male and female C57BL/10 mice. *British Journal of Animal Behaviour,* 1956, *4,* 147–149.

Ulrich, J. The social heredity in albino mice. *Journal of Comparative Psychology,* 1938, *25,* 373–413.

Vale, J. R., Vale, C. A., & Harley, J. P. Interaction of genotype and population number with regard to aggressive behavior, social grooming, and adrenal and gonadal weight in male mice. *Communications in Behavioral Biology,* 1971, *6,* 209–221.

Vale, J. R., Ray, D., & Vale, C. A. The interaction of genotype and exogenous neonatal androgen: Agonistic behavior in female mice. *Behavioral Biology,* 1972, *7,* 321–334.

Vale, J. R., Ray, D., & Vale, C. A. Interaction of genotype and exogenous neonatal estrogen: Aggression in female mice. *Physiology and Behavior,* 1973, *10,* 181–183. (a)

Vale, J. R., Ray, D., & Vale, C. A. The interaction of genotype and exogenous neonatal androgen and estrogen: Sex behavior in female mice. *Developmental Psychobiology,* 1973, *6,* 319–327. (b)

van Abeelan, J. H. F. Effects of genotype on mouse behaviour. *Animal Behaviour,* 1966, *14,* 218–225.

Wahlsten, D. A critique of the concepts of heritability and heredity in behavioral genetics. In J. R. Royce & L. P. Mos (Eds.), *Theoretical advances in behavior genetics.* Germantown, Md.: Sijthoff & Noordhoff, 1979.

Weir, J. A. Allosomal and autosomal control of sex ratio in PHH and PHL mice. *Genetics,* 1976, *84,* 755–764.

Weir, J. A., & Hogle, G. A. Influence of the Y chromosome on sex ratio and mating behavior in PHH and PHL mice. *Genetics,* 1973, *74,* S294.

West, W. T., Evans, M. I., & Hamilton, J. B. Strain differences in target organ weight changes among mice treated with androgens. *Growth,* 1980, *44,* 36–45.

Winer, B. J. *Statistical principles in experimental design.* New York: Plenum Press, 1962.

9

Models of Early Hormonal Effects on Intrasex Aggression in Mice

Frederick S. vom Saal

INTRODUCTION

Aggression is a term that has been applied to many different types of behaviors, from hostile language and gestures to physical violence (Buss, 1961). Aggression, as used here, will refer to a category of behaviors associated with physical attack by one house mouse (male or female) on another mouse of the same sex. This type of aggression is therefore most appropriately referred to as "intrasex" aggression. But conspecific aggression in mammals is often referred to as "intermale" aggression, since fighting is generally considered to be a trait characteristic of males and uncharacteristic of females. An exception to this general statement has been the recognition that many, but not all, female mammals exhibit aggression during the postpartum period in defense of their young (Svare, 1981). It is important to recognize that other types of conspecific aggression (e.g., fear-induced aggression, sex-related aggression, infanticide) are commonly observed in many mammals. These other forms of conspecific aggression are influenced by variables different from those that influence either intermale or interfemale aggression (Moyer, 1974). For example, when a nonreceptive female mouse is placed into the home cage of a sexually experienced adult male, the male will attempt to mount the female and the mounting attempts will be rejected. As a result, the male often becomes highly aroused and attacks the female (vom Saal & Bronson, 1978). Male mice normally do not attempt to mount females that are not in estrus, but a male that is isolated, except for occasional opportunities to mate with a receptive female, will become highly aroused if any female is placed in its cage. Aggression by a male toward a nonreceptive female may thus reflect the response of a highly aroused male to being frustrated in its attempts to mate. It is therefore necessary to specify the particular type of aggression one is referring to when discussing aggression in mammals.

FREDERICK S. VOM SAAL • Division of Biological Sciences and Department of Psychology, University of Missouri, Columbia, Missouri 65211.

Scott and Fredericson (1951) have termed behaviors that occur when two animals come in conflict "agonistic behaviors." The two most obvious agonistic behaviors are attack and escape. Under the appropriate conditions, aggression between mice can be very easily identified and quantified. Intrasex aggression in house mice, *Mus musculus,* can be divided into a number of distinct behaviors. When placed together, two male or two female mice will sniff and groom each other, with the grooming by one mouse often becoming more and more intense. At the same time, an increase in general arousal may occur, as evidenced by an increase in general activity. Tail rattling (vibrating of the tail) quite commonly precedes an attack, which is typically aimed at the flanks of the opponent.

MASCULINIZATION AND DEFEMINIZATION

Sexual differentiation commences in mammals during the early fetal period (Jost, 1972). If the gonads have differentiated into testes as a result of the presence of a Y chromosome (Wachtel, 1977), androgens are secreted by the interstitial cells. Exposure of males to suprathreshold levels of androgens during the time that specific target tissues are differentiating profoundly influences the development of these tissues. It has now been demonstrated that within the preoptic area of the hypothalamus of rats, prenatal and neonatal androgen exposure alters both the timing and number of cell divisions in an area referred to as the "sexual dimorphic nucleus" (Gorski, Harlan, Jacobson, Shryne, & Southam, 1980; Jacobson & Gorski, 1980; Jacobson, Shryne, Shapiro, & Gorski, 1980). The pattern of synaptic connections within the preoptic area is also different in male and female rats as a result of perinatal androgen exposure (Raisman & Field, 1973). Thus, androgen exposure during the period of sexual differentiation influences brain morphology in rats. It seems reasonable to speculate that these morphological differences could lead to differences in brain function and thus in behavior.

Two intensively studied behaviors that are modulated by prenatal and neonatal androgen titers in mice are copulatory behavior and intermale aggression. Since androgens influence both brain development and differentiation of the external and internal reproductive organs, one problem that immediately became apparent to researchers was that of separating central from peripheral effects of androgen exposure during early life on adult sexual performance (Beach, 1945; Luttge, 1979). In contrast, intermale aggression provides a model behavior that is sexually dimorphic, is influenced by different variables in males and females, and is not dependent on specific anatomical structures, such as a penis or vagina, to be exhibited normally. Aggression therefore has been utilized as a model behavior to assess the role of androgens in the ontogeny of masculine behavior in species such as the house mouse in which adult males of most strains are found to be highly aggressive when placed together (Simon, 1979). While useful for studying the process of masculinization, the study of the ontogeny of intermale aggression cannot be utilized to assess the role of gonadal steroids in the defeminization process, that is, the loss of the potential to exhibit female behaviors. Defeminization has been proposed to be an independent process that does not merely reflect the degree to which an animal is masculinized (Beach, 1975; Whalen, 1974).

The following section will briefly review experiments supporting some different models concerning the role of androgen exposure during early life on aggression between male mice. Research that I have conducted concerning the influence of intrauterine positioning of male and female fetuses on fetal hormone titers and adult behavior of both male and female mice will then be discussed in relation to models concerning the role of the early hormonal environment on both intermale and interfemale aggression.

THE ONTOGENY OF INTERMALE AGGRESSION IN MICE

Androgen secretion in male mouse fetuses has been reported to commence on about Day 13 of gestation (Block, Lew, & Klein, 1971; Rugh, 1968). Differentiation of the Wolffian ducts and external genitalia into the male pattern commences during prenatal life and continues into the early neonatal period. At birth, androgen levels in male mice are low and do not begin to increase again until approximately one month of age. At this time, plasma androgen levels begin increasing and in many strains reach maximal values between 60 and 70 days of age and then decline somewhat (McKinney & Desjardins, 1973; Selmanoff, Goldman, & Ginsburg, 1977; vom Saal & Bronson, 1980a). The testes, seminal vesicles, prostate, and preputial glands go through a period of development at puberty (the onset of the capacity to reproduce), which occurs between Days 40 and 50 after birth. It has been reported that the onset of intermale aggression in intact male mice is correlated with the increase in androgen output by the testes at puberty. The onset of intermale aggression has been reported to occur around Day 35 in intact male mice, somewhat before the successful insemination of females. However, the aggression exhibited by juvenile males is less intense (as measured by the latency to attack, number of attacks and duration of fighting) than that observed in older males (Barkley & Goldman, 1977a; Fredericson, 1950; McKinney & Desjardins, 1973; McKinney & Ingle, 1974).

The onset of aggressive behavior in male mice is correlated with an increase in the percentage of circulating androgen that is present as testosterone rather than a 5α-reduced androgen (e.g., dihydrotestosterone). By adulthood, testosterone accounts for about 75% of the androgens in the plasma (McKinney & Desjardins, 1973). One of the earliest studies relating testosterone levels to aggression was reported by Beeman (1947). She found that castrating male mice at ages ranging from 23 to 83 days resulted in a complete absence of fighting (as well as of other behaviors associated with fighting, e.g., rough grooming and tail rattling) in male mice if a minimum of 25 days elapsed between castration and the onset of behavior testing against a male opponent. When implanted with pellets containing testosterone propionate (TP), a longer-acting synthetic androgen than testosterone, the previously nonaggressive castrated males exhibited aggression after four days of exposure to TP. Removal of the TP implant resulted in the animals again exhibiting no aggression when tested four days following removal of the implant. Subsequent experiments have confirmed the importance of circulating androgen for the induction of aggression in male mice that have not had previous experience fighting. But it has also been found that after male mice have become dominant or subordinate, regardless of the presence or absence of testosterone, dominant males will continue to be aggressive and subordinate males will exhibit submissive behaviors (Maruniak, Desjardins, & Bronson, 1977; Scott & Fredericson, 1951; vom Saal, unpublished observation). Thus, circulating titers of testosterone do not correlate with aggressiveness in mice that have had previous fighting experience (see Leshner, 1975, for a model proposing such a correlation). Learning also plays a major role in determining aggressiveness and dominance status in rhesus monkeys (Joslyn, 1973; Koyama, 1967; Trimble & Herbert, 1968). In addition, prior experience fighting (and thus dominance status) influences other androgen-dependent behaviors such as infanticide and urine-marking (Maruniak, Desjardins, & Bronson, 1977; vom Saal, in press d; vom Saal & Howard, 1982).

Circulating testosterone induces a male to exhibit specific behaviors in response to appropriate environmental cues. In adult male mice, the exhibition of aggression is thus also influenced by the physical environment (e.g., is it the animal's home territory? Barfield, Busch, & Wallen, 1972; Charpentier, 1969), the social environment (e.g., isolation versus group housing; Valzelli, 1969), and the perception of pheromonal cues (Dixon & Mackintosh, 1975; Mugford & Nowell, 1970, 1971).

The Organization/Activation Model of Intermale Aggression

Numerous experiments have provided support for the concept that exposure to androgen during the neonatal period is necessary to organize the neural substrate mediating aggression so that aggression can be elicited by appropriate environmental cues when the organism is activated by exposure to androgen again in adulthood. The organization/activation model proposes that the stimuli for aggression would be ineffective in eliciting aggressive behavior without both organization during the neonatal critical period of neural differentiation and adult activation of the previously organized neural substrate by androgen. According to this model, therefore, male- and female-typical behavior patterns are determined by the hormonal environment during development but are not exhibited until activated by the appropriate hormones later in life (cf. Beatty, 1979).

The organization/activation model was borrowed from developmental biology. Of particular importance was the finding of Pfeiffer (1936) that female rats exposed to androgen through transplanted testicular tissue during the neonatal period were acyclic (i.e., did not exhibit estrous cycles) in adulthood, since implanted ovaries did not show any evidence of corpora lutea formation indicative of ovulation. Similar results were obtained with male rats castrated in adulthood and then implanted with ovaries. Castration of a male rat at birth, however, resulted in ovaries implanted in adulthood showing evidence of a cyclic appearance of follicles and corpora lutea indicative of the female pattern of cyclic gonadotropin release (cf. Harris, 1964; Harris & Jacobson, 1952). Exposure to androgen during a critical period shortly after birth thus results in the absence of normal estrous cycles in rats (Gorski, 1979) and mice (Barraclough & Leathem, 1954). This phenomenon, referred to as the "androgen sterility syndrome," demonstrated that the course of development of both neural and peripheral target tissues was altered by androgen during critical (restricted) periods in early life.

Behavioral scientists were attracted by the possibility that androgen-mediated differentiation of neural tissues governing sexually dimorphic behaviors such as aggression might occur during critical periods in development similarly to the differentiation of neuroendocrine mechanisms and peripheral androgen-sensitive tissues (e.g., the genitals and accessory sex organs, liver, kidneys, etc.; De Moor, Verhoeven, & Heyns, 1973). The organization/activation model was thus applied to the development of the neural substrates mediating sexually dimorphic behaviors. The application of this model to behavior was first proposed by Young, Goy, and Phoenix (1964), based on experiments designed to determine the effect of neonatal androgen exposure on adult sex behavior in rats. Harris (1964) also proposed that in rats, differentiation of the genitalia and accessory sex organs occurs as a result of exposure to androgen during the prenatal period, but that differentiation of the neural substrate mediating sex behavior occurs as a result of exposure to androgen during the neonatal period. This model was subsequently applied to the ontogeny of intermale aggressive behavior in mice.

Numerous experiments were conducted, utilizing mice, that were reported to demonstrate that neonatal androgen exposure was a necessary prerequisite for aggressive behavior to be activated by adult exposure to testosterone. One problem with all these experiments that became apparent later on was that behavior testing was not conducted for a sufficiently long period of time. The critical period for organizing the neural substrate mediating aggression into the "male" pattern was hypothesized to commence at birth and terminate sometime during the second or third week after birth. However, the brain was hypothesized to become progressively less sensitive to the organizing action of androgen subsequent to birth.

In the study on the influence of androgen exposure during the neonatal period on adult aggressiveness in mice, females injected with androgen after birth have been utilized as model organisms since the hormonal milieu during development rather than genotype is presumed to be the mediator of sex differences in behavior. Specifically, Edwards (1969) reported that, when the females were again exposed to increasing doses of TP in adulthood, fewer female mice injected with 500 μg TP on Day 10 after birth exhibited aggression than female mice injected with TP on the day of birth. Bronson and Desjardins (1970) found that, when injected with 200 μg TP in adulthood, fewer female mice injected with 400 μg TP/gm when 24 days old exhibited aggression (and that fighting animals exhibited fewer attacks) than females injected with the same dose when they were younger than 12 days old. The important aspect of these data is that 24-day-old females usually weigh around 13 grams, which means that they would have been receiving over 5 mg of TP. This is a very large dose for a mouse and was considered strong evidence that the brain was no longer capable of being organized by Day 24 in mice. Also, female mice injected with 500 μg TP when 21 days old did not exhibit aggression (Levy, 1954), whereas male mice injected with 500 μg TP when 18 days old were induced to exhibit aggression (Levy & King, 1953). These data indicated that by Day 18, mice are sufficiently mature to exhibit aggressive behavior if plasma testosterone titers increase above normal levels for this age, and if they have previously been exposed to androgen during early life.

Conversely, the issue of the effect of early androgen exposure on adult aggressive behavior has been addressed by castrating males at different times in life. As mentioned previously, for aggression to be exhibited between male mice without previous fighting experience, androgen must be present in the circulation during the time of behavior testing. Therefore, in studies involving castration of naive male mice, replacement of androgen in adulthood is necessary to induce aggression. Edwards (1969) reported that, when administered increasing doses of TP in adulthood, more male mice gonadectomized on Day 10 of life exhibited aggression than males gonadectomized on the day of birth. Peters, Bronson, and Whitsett (1972) castrated male mice on Day 0, 2, 6, 12, or 40 of life and then administered all animals varying doses of TP in adulthood. A greater percentage of animals from the groups castrated on Day 6 or later showed evidence of wounding when grouped four per cage for 3 days than did the animals castrated before Day 6 after birth. Observation of the animals for the first two hours following grouping revealed a clear difference between the Day 2 and 4 castrates, which exhibited virtually no aggression, and the Day 6 castrates, which did exhibit aggression.

The conclusion drawn from the above studies utilizing both male and female mice was that organization of the neural tissue mediating aggression occurred during the first 2 to 6 days following birth. During this period a marked decrease in the capacity of the neural substrate to be organized into the male pattern appeared to occur, with the termination of the critical period for organization supposedly occurring sometime prior to Day 24 of life.

The Sensitivity Model of Intermale Aggression

The validity of the hypothesis that androgen exposure during a critical period in early life is a necessary prerequisite for aggressive behavior in mice to be induced by testosterone in adulthood was challenged by Edwards (1970). He reported that injections of 100 μg TP to female mice from Days 30 to 49 after birth resulted in more of the TP-treated females exhibiting aggression than non-TP-treated females when all mice were exposed to increasing doses of TP and tested for aggression at 95 days of age. This finding provided clear evidence against the concept of androgen-induced neural organization during a critical neonatal period being a necessary prerequisite for the induction of adult aggressive behavior in response to testosterone. However, this finding did not rule out the possibility that neural organization at some prior period was a necessary prerequisite for activation of aggressive behavior later in life.

The results of a number of experiments have demonstrated that prior organization of the neural substrate mediating aggression in mice is not necessary for the activation of aggressive behavior by testosterone exposure in adulthood. For example, Svare, Davis, and Gandelman (1974) reported that female mice exposed to 500 μg TP for the first time in adulthood would exhibit aggression against an intruder male if the exposure period to TP continued for a longer period of time than had previously been attempted. Brain and Evans (1975) reported similar findings utilizing silastic implants of testosterone. Vom Saal, Gandelman, and Svare (1976) sought to determine whether exposing female mice to 500 μg TP for the first time in adulthood, until they exhibited aggression, would reduce the duration of TP treatment needed to reinduce aggression after a 60-day interval without exposure to TP. The results suggested that no organization or permanent change in the sensitivity of the neural substrate mediating aggression to testosterone had occurred during the first adult TP-exposure period, since no reduction in the duration of TP exposure required to reinduce aggressive behavior was observed. But in this experiment each animal was administered TP during the initial test phase only until it exhibited aggression, with the result that the sensitization period of TP treatment differed in length for each animal. In contrast, Barkley and Goldman (1977b) reported that following the induction of aggression in adult female mice (not previously androgenized) by implanting a capsule containing 10 mg of testosterone (a high dose), a capsule containing 0.3 mg testosterone was effective in maintaining aggression. Without prior exposure to the 10-mg capsule of testosterone, however, the 0.3-mg capsule was ineffective in inducing aggression in female mice. These data indicate that sensitization of the neural substrate mediating aggression to testosterone can occur in adult mice. It seems unlikely that the mouse brain suddenly loses the capacity to be sensitized by exposure to androgen around Day 60 after birth, as the results of vom Saal *et al.* (1976) suggested. Instead, the findings of Edwards (1970), Svare *et al.* (1974), and Barkley and Goldman (1977b) indicate that, with respect to the influence of androgen on intermale aggression in mice, there is no defined critical period during which sensitization of the brain by androgen can occur and after which it cannot occur. In addition, sensitization to androgen during early life is not a necessary prerequisite for female mice to be induced to exhibit intermale-like aggression when treated with androgen for the first time in adulthood.

The demonstration that organization of neural tissue prior to adulthood is not a necessary prerequisite to the induction of aggression toward a male or androgenized female opponent in female mice treated with TP for the first time in adulthood leads to the conclusion that the organization/activation model is not appropriate for describing the mechanism of action of androgen on the neural substrate mediating intermale aggression in mice. As an alternative, a model based on the concept of a continuous decrease in the sensitivity to androgen of specific neural areas from birth through adulthood seems more appropriate. According to this sensitivity model, exposure to androgen at any time in life results in an increase in the sensitivity of the neural substrate mediating aggression to subsequent exposure to androgen, with the sensitivity of the neural substrate to androgen being an inverse function of the time following birth that exposure to androgen first occurred. Without prior sensitization by exposure to androgen, the neural substrate is relatively insensitive to the aggression-inducing action of testosterone in adulthood, with the result that an extended period of exposure to a high dose of testosterone is required to induce aggression. This concept is not novel, since Beach (1945), writing with regard to sex behavior in rodents, had stated:

> The sex hormones, in so far as they enter the behavioral picture, are best regarded, not as stimuli or as organizing agents, but as chemical sensitizers which alter the stimulability of critical mechanisms within the central nervous system. (p. 400)

The model based on diminishing sensitivity, as opposed to that based on the concept of a critical organizational period, led to the prediction that exposure to androgen during prenatal life, as well as during postnatal life, would alter the sensitivity of the neural substrate mediating aggression to androgen in adulthood. Since it had been reported that males castrated at birth do not exhibit aggression in response to androgen administration in adulthood (cf. Peters *et al.*, 1972), it was assumed that prenatal exposure to androgen, which commences on Day 13 of gestation in male mice, was without effect on subsequent responsivity to androgen in adulthood. As described above, this conclusion was based on experiments which did not allow the animals to be exposed to androgen for a sufficient period of time prior to aggression testing. To test whether prenatal androgen exposure influenced adult aggression, 1.5 mg TP was injected daily into pregnant mice from Day 12 through Day 16 of pregnancy to masculinize the female fetuses and possibly supermasculinize the male fetuses. Other pregnant females were administered oil or remained undisturbed. All males and females from these litters were then gonadectomized at birth and in adulthood implanted with silastic capsules containing 5 mg testosterone and tested for aggression against an olfactory-bulbectomized male opponent. The duration of exposure to testosterone required to induce aggression against a male opponent was significantly longer in the females not exposed to TP prenatally than in males or the females that had been exposed to androgen during prenatal life and were thus more sensitive to androgen in adulthood. As part of this experiment, offspring of both sexes from TP-treated or untreated mothers were gonadectomized at birth and injected with 100 μg of TP on the next day. The TP injection on the day following birth had the effect of rendering all of these mice more sensitive to the aggression-inducing action of testosterone in adulthood than mice experiencing only prenatal androgen exposure. It appears, therefore, that while prenatal exposure to androgen can alter subsequent responsivity to androgen, the period during which the brain is most sensitive in this regard is around the time of birth (vom Saal, 1979).

PRENATAL ANDROGEN EXPOSURE: DIRECT (ORGANIZATIONAL) AND LATENT (SENSITIZING) EFFECTS ON BEHAVIOR

One aspect of the sensitivity model which may be incorrect is that exposure to gonadal steroids during early life in mice has only latent effects on behavior. It now appears that during prenatal life in mice, gonadal steroids have both direct (organizational) effects on behavior that are observed in the absence of gonadal secretions in adulthood and latent (sensitizing) effects on behavior that are apparent only after "activation" of the sensitized neural areas by specific hormones during later life. In both dogs and monkeys, some sexually dimorphic behaviors have been found to be organized by androgen during prenatal life and not to require activation by androgen to be exhibited, while other behaviors are observed only after activation by androgen. For example, Beach (1975) has reported that in beagles, the leg-lift urination posture is organized into the male pattern by exposure to androgen during early life and does not require adult exposure to testosterone to be exhibited. Mounting, on the other hand, is exhibited only by beagles that have testes or are being treated with testosterone, and animals that have been sensitized by exposure to testosterone during early life are more likely to exhibit mounting than are animals not exposed to testosterone during early life. In infant rhesus monkeys, rough-and-tumble play, threat displays, and infantile sex behavior (rubbing and thrusting) are organized by androgen into the male pattern during prenatal life and do not require functional testes or treatment with testosterone to be exhibited by infant males. In contrast, in monkeys as well as dogs, rats, and mice, adult male copulatory behavior is influenced by both prenatal and adult androgen exposure, but activation by androgen is required for male sex behavior to be exhibited (Goy, 1968; Harlow, 1965). It has been proposed that in primates, androgen exposure during early life has both direct and latent effects, while in rodents, only latent, organizational effects are observed (Eaton, Goy, & Phoenix, 1973).

THE INTRAUTERINE POSITION PHENOMENON AND INTERMALE AGGRESSION

The positioning of fetuses during intrauterine development is an important source of variation in terms of the hormone titers that both male and female fetuses are exposed to in mice and rats and possibly in other polytocous (multiple-birth) species. I have termed this the intrauterine-position phenomenon. To obtain animals from known intrauterine positions, female mice are time-mated, and their offspring are delivered by cesarean section and raised by foster mothers (see vom Saal, 1981, for methods). The purpose of describing experiments in which both male and female mice from known intrauterine positions were compared for their behavior is that aggressiveness among males and among females has been found to be influenced by their prior intrauterine position. This phenomenon has thus proven to be a unique, naturally occurring experimental model system for examining the relationship between fetal hormone titers and adult behavior without the necessity of pharmacological intervention. In this regard, it is also important to remain aware that studies involving comparisons of animals from different intrauterine positions are correlational and do not allow for firm cause-and-effect conclusions to be drawn.

On Day 17 of gestation, differentiation of the sex accessory ducts, external genitals, and hypothalamus (Rugh, 1968) is occurring in fetal mice, and male mouse fetuses have three times as much circulating testosterone as do female fetuses at this time (vom Saal & Bronson, 1980a). In addition, on Day 17 of gestation, female mouse fetuses were found to have about 80% higher amniotic fluid titers of estradiol than male fetuses, and males located between two female fetuses (0M males) had about 50% higher amniotic fluid titers of estradiol than did males that developed between two male fetuses (2M males; vom Saal, Grant, McMullen, & Laves, in press). 0M female fetuses also tended to have higher amniotic fluid titers of estradiol than 2M female fetuses (see Figure 1), although the difference was not statistically significant (vom Saal & Bronson, 1980a). 0M and 2M male mouse fetuses did not differ significantly in their blood or amniotic fluid titers of testosterone, although 2M male fetuses did tend to have somewhat

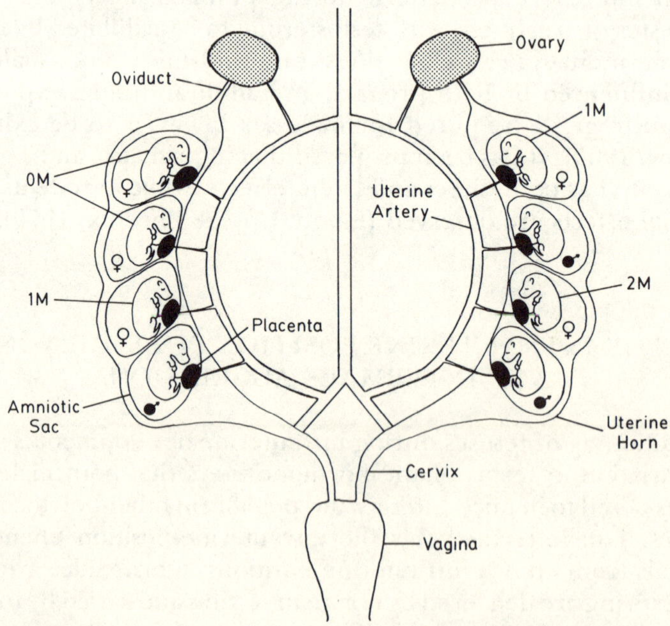

FIGURE 1. Schematic diagram of a mouse uterus showing the three potential intrauterine positions that female fetuses can occupy relative to male fetuses: 2M female, between 2 males; 1M female, next to 1 male; OM female, not next to a male. To determine the intrauterine positioning of fetuses, mice are time-mated and the fetuses are delivered by cesarean section just before normal parturition. The young are then placed with foster mothers.

higher blood titers of testosterone than 0M male fetuses (vom Saal *et al.,* in press). During the fetal period of sexual differentiation in humans, males also have significantly higher blood titers of testosterone than do females, and female fetuses have significantly higher blood titers of estradiol than do males (Belisle & Tulchinsky, 1980). Thus, the same pattern of sex differences in fetal steroid titers occurs in humans and mice.

The mechanism of transport of hormones between contiguous fetuses remains unknown. Based on behavioral data, Meisel and Ward (1981) concluded that in rats, steroids are transported between fetuses via the uterine blood vessels. Meisel and Ward claimed that blood flowed from the cervix toward the ovary in both the loop artery and the loop vein feeding each uterine horn (see Figure 1). This belief led to a comparison of animals that were labeled as upstream or downstream in this vascular scheme. Blood flow in the uterine arteries and veins in both the rat and the mouse is actually bidirectional (McLaren & Michie, 1960; vom Saal, unpublished observation). Meisel and Ward based their analysis system on a misinterpretation of an article (Del Campo & Ginther, 1972) that, in fact, did not address the issue of the direction of uterine blood flow as Meisel and Ward stated; their conclusion concerning the mechanism underlying the passage of hormones between fetuses is therefore invalid. It is considered likely that androgen and estrogen can diffuse across the amniotic and chorionic membranes surrounding each fetus. The amniotic fluid circulates through the gut during fetal life, and amniotic fluid titers of estradiol and testosterone are thought to be in equilibrium with the titers of these steroids in the fetal blood stream (Belisle & Tulchinsky, 1980). An important related observation is that during the last four to five days of gestation, mouse fetuses are tightly packed in each uterine horn, thus providing a large area of contact between the chorionic membranes of contiguous fetuses. In cattle, freemartins (sterile offspring) are found when male and female fetuses develop together *in utero,* a phenomenon thought to be due to vascular anastomoses between contiguous fetuses. Vascular anastomoses between contiguous fetuses have not been found in mice, however (Marcum, 1974; McLaren & Michie, 1959, 1960; Ohno, 1969; Ohno, Christian, Wachtel, & Koo, 1976; vom Saal, 1981).

The results of comparisons of males from different intrauterine positions suggest that exposure to different titers of estradiol results in variation in numerous characteristics relating to reproduction. In many experiments, only 0M and 2M males have been compared, since 1M males have always been found to be intermediate between 0M and 2M males in phenotype. In a comparison of their sexual performance, gonadally intact (i.e., not treated with hormones) 0M male mice exhibited significantly more mounts and intromissions than 2M males when placed for 30 minutes with a sexually receptive female. In rats, 0M males were also more sexually active than 2M males: 0M males exhibited more ejaculations to satiety (30 minutes without a mount) than did 2M males. There thus appears to be a decrement in sexual performance in 2M males relative to 0M males in both rats and mice (vom Saal *et al.,* in press).

Of primary relevance to the current topic is the finding that intrauterine position influences different types of aggression in adult male mice, namely intermale aggression and aggression toward infants (infanticide). Intermale aggression was examined in 0M and 2M male mice that had been castrated within one hour of cesarean delivery. The purpose of castrating the males at birth was (1) to eliminate any potential postnatal differences between 0M and 2M males in the titers of gonadal hormones to which they would have been exposed and (2) to compare 0M and 2M males for their aggressiveness while being treated with a known quantity of testosterone administered via silastic capsules. These 0M and 2M males were implanted with a silastic capsule containing 5 mg testosterone and were tested every other day for 10 minutes for aggressiveness (biting and chasing) toward a 1M male opponent that had been rendered anosmic via olfactory bulbectomy. The 2M males required a significantly shorter period of exposure to testosterone to exhibit aggression than did the 0M males. In addition, when autopsied after 35 days of exposure to testosterone, the 2M males had larger seminal vesicles than the 0M males (vom Saal *et al.,* in press). These findings indicate that adult 2M males are more sensitive than 0M males to the activational effects of testosterone in both the neural substrate mediating intermale aggression and in the seminal vesicles.

The duration of exposure to testosterone required to induce aggression provides an index of aggressiveness in mice. Thus, 2M male mice appear to be more aggressive toward other males than do 0M male mice. But a correlate of an increased tendency to be aggressive in 2M males appears to be a decrease in sexual performance relative to 0M males. Taken together, these findings suggest that among male mice, the males that are the most aggressive toward other males should have the lowest sexual performance.

Another form of aggression, attack directed toward an infant by an adult (infanticide), has also been found to be influenced by intrauterine position in male mice. An adult mouse usually only attacks another adult of the same sex. In contrast, male CF-1 mice in my colony that do commit infanticide are equally likely to commit infanticide when placed with either newborn male or newborn female mice (vom Saal & Howard, 1982). Gonadally intact 0M, 1M, and 2M male mice were compared for their behavior toward two newborn mice that were placed into each male's home cage for 30 minutes. The proportion of 0M, 1M, and 2M males that committed infanticide (killed the young), behaved parentally (retrieved the young to a nest and hovered over them), or ignored the young differed significantly: most 0M males committed infanticide, most 2M males behaved parentally, and 1M males were equally likely to behave parentally or commit infanticide (vom Saal, in press a).

Infanticide is similar to intermale aggression in that unless the testes are present or testosterone is administered following gonadectomy, naive male mice do not commit infanticide, regardless of prior intrauterine position. When 0M and 2M male mice were castrated at cesarean delivery and treated with testosterone via silastic capsules in adulthood as described above, more 0M than 2M males began committing infanticide, while more 2M than 0M males behaved parentally toward newborn mice (vom Saal, in press a). Thus, to observe differences between 0M and 2M male mice in behaviors that could influence reproductive fitness (infanticide, intermale aggression, and sexual performance), either the testes have to be present or exogenously administered testosterone has to be present in the circulation. All of these findings are thus consistent with the sensitivity model described previously in that differences in behavior among adult males due to prior intrauterine position appear to reflect a latent difference in sensitivity to testosterone.

The finding that 0M males are the most likely to commit infanticide yet the least aggressive toward other adult males might at first seem surprising. But previous experiments have revealed that hormonal manipulations at the time of birth that lead to an increase in intermale aggressiveness lead to a decrease in infanticide: Elevating testosterone levels at birth enhances intermale aggressiveness and decreases infanticide in mice (Gandelman & vom Saal, 1977; Samuels, Jason, Mann, & Svare, 1981; vom Saal, 1979). It is thus not surprising to find that there is a negative correlation in male mice between the tendency to be aggressive toward other adult males and the tendency to commit infanticide, although the adaptive significance of such a relationship is not readily apparent.

THE AROMATIZATION HYPOTHESIS

Estradiol Facilitates the Ontogeny of Male Sexual Behavior

The aromatization hypothesis refers to the proposition that testosterone serves as a prohormone that is converted (aromatized) to estradiol within target cells which contain aromatase enzymes. The estradiol that is produced by intracellular aromatization of testosterone is thought to interact with cytoplasmic estrogen receptors and then somehow to be translocated into the nucleus where organizational or sensitizing effects of the hormone–receptor complex are exerted on the cell. The presumption has been that testosterone is required as a prohormone due to the fact that circulating estrogens are bound and thus inactivated by blood-binding proteins. Alpha-fetoprotein (AFP) is a protein that has been found in the blood of fetal and neonatal mice and rats, and AFP does bind estradiol with high capacity (MacClusky & Naftolin, 1981; McEwen, 1980).

Because of the hypothesis that circulating estrogens were inhibited from entering cells as a result of being bound by plasma proteins, circulating estrogens were not thought to influence the course of sexual differentiation in any mammal. Thus, while it has been known for some time that there are sex differences in estradiol titers in human fetuses (Reyes, Boroditsky, Winter, & Faiman, 1974), this difference has not previously been thought to influence sexual differentiation in humans. One problem with ignoring the finding that there are sex differences in circulating estrogens during fetal life in humans in that with the exception of rodents, plasma proteins that bind estrogens have not been identified in other mammalian species (MacClusky & Naftolin, 1981). Also, AFP has actually been found inside brain cells in rats (Benno & Williams, 1978; McEwen, Plapinger, Chaptal, Gerlach, & Wallach, 1975), raising the possibility that even in rats and mice, AFP that is bound to estrogens may actually be actively transported into brain cells (perhaps by endocytotic mechanisms into specific neural areas). I have proposed, therefore, that sex differences in fetal estradiol titers (female fetuses have higher titers than males) interact with sex differences in gonadal secretions of androgens (male fetuses have higher titers than females) in regulating sexual differentiation. In addition, variation in phenotype among male and among female mice due to intrauterine position may be due to differences in both testosterone and estradiol titers (vom Saal, in press c). Thus, the intrauterine position phenomenon is proposed to be mediated by both estrogen and androgen, rather than just androgen as previously believed (vom Saal & Bronson, 1980a).

Considerable research has been conducted to test the aromatization hypothesis. The effects of estrogen agonists and antagonists as well as inhibitors of aromatase enzyme activity (which block the conversion of testosterone to estradiol) have been examined. The dependent measure in virtually all of these experiments has been some aspect of sexual behavior: the rate or frequency of mounting, intromitting, or ejaculating or the capacity to exhibit proceptive or receptive behaviors typical of female rodents in estrus. The evidence is convincing that testosterone acts on the neural substrate mediating male sexual behaviors after conversion to estrogen within brain cells (MacClusky & Naftolin, 1981). For example, Södersten (1979) has reported that exposure to estradiol during perinatal life appears to be necessary for the normal development of ejaculatory behavior in male rats. Thus, the findings that 0M male fetuses have higher titers of estradiol than 2M male fetuses and that adult 0M male rats exhibit more ejaculations to satiety than 2M males are consistent with the hypothesis that exposure during early life to elevated titers of estradiol enhances the development of male sexual behavior. Presumably, more estradiol passes from the circulation into brain cells in 0M male fetuses than in 2M male fetuses, thus increasing the intracellular pool of estradiol in 0M males. While 0M and 2M male fetuses do not differ significantly in their blood testosterone titers, 0M males thus appear as if they had been exposed to high titers of androgens relative to 2M males during fetal life, at least in terms of adult sexual behavior.

Estradiol Inhibits the Ontogeny of Intermale Aggression

Adult 2M male mice were more sensitive to testosterone in terms of the induction of intermale aggression and seminal vesicle growth, a finding that suggests that 2M males were exposed to higher titers of testosterone than were 0M males during fetal life (exactly the opposite of the previous statement concerning sexual behavior). Since 0M and 2M male mice do not differ in their blood or amniotic fluid titers of testosterone, but 0M males have elevated amniotic fluid titers of estradiol, I have proposed that exposure to elevated concentrations of estradiol interferes with the effects of testosterone in both the neural tissues mediating intermale aggression and in the seminal vesicles during fetal life in males (vom Saal, in press c). Thus, 0M males have high titers of estradiol relative to 2M males, and 0M males appear as if they had been exposed to an antiandrogen during fetal life. For example, administration of antiandrogens such as cyproterone acetate during fetal life reduces aggressiveness in adult male mice (vom Saal, 1976).

The above hypothesis might appear to be totally at odds with current ideas concerning the masculinizing effects of estrogens during early life (i.e., the aromatization hypothesis), but this is not in fact the case. As mentioned above, virtually all previous research concerning the effects of estrogens on sexual differentiation has involved examining sexual behavior rather than aggression. In the studies which have reported that an injection of estradiol benzoate into neonatal mice had the same effect on adult aggressiveness as testosterone (a finding that, if valid, would provide support for the hypothesis that testosterone is aromatized to estradiol in the neural areas mediating intermale aggression), very high doses of estradiol benzoate have been utilized. For example, Edwards and Herndon (1970) reported that an injection of 50 μg estradiol benzoate into newborn mice enhanced sensitivity to the aggression-inducing action of testosterone in adulthood. They concluded that during early life in mice, estradiol acted to organize the neural substrate mediating aggression, thus extending the aromatization hypothesis to include the ontogeny of aggressive behavior as well as sexual behavior. The conclusion drawn by Edwards and Herndon (1970) is obviously exactly opposite to what I have just proposed to be the effect of exposure to elevated titers of estrogens during early life on intermale aggression in male mice. However, sexual behavior can be induced in an adult, 25-gram female mouse with an injection of 5 μg estradiol benzoate (a 0.2 mg/Kg dose). A newborn mouse weighs 1.5 grams, thus making a 50 μg injection equal to a dose of 33.3 mg/Kg (more than 150 times the dose required to activate sexual behavior in adult mice). Exposing animals at any time in life to massive doses of a steroid may produce effects that do not reflect the actual physiological role of the steroid. This is particularly true during perinatal life, since mice are thought to pass through a period of heightened sensitivity to the permanent effects of steroids on behavior around the time of birth (vom Saal, 1979). Conclusions drawn from studies in which massive doses of a steroid are utilized are thus highly suspect, and it is possible that had doses of estradiol benzoate within a physiological range been utilized by Edwards and Herndon (1970), exactly the opposite results might have been obtained.

MODEL TESTING AND EXPERIMENTER BIAS

The bias inherent in model-testing often leads researchers to overlook or reject information that is not consistent with the current models. Such has been the case with the study of aggression in mice. Models concerning the hormonal control of aggression have strongly influenced both the types of questions that have been asked and the paradigms of the experiments designed to test empirically the validity of the models being tested. For example, it was hypothesized that male mice castrated at birth and female mice not treated with TP at birth would not be aggressive. Without this expectation, these animals might have been tested for a longer period of time, and the hypothesis would have been rejected (cf. Edwards, 1969). There is little doubt that any observation involves some subjective component, and the experimenter bias that is inherent in observational research can be markedly influenced by current social theories. There is an interesting parallel, for instance, between male and female sex roles during the 1950s and early 1960s and the concept inherent in the organization/activation model that without the presence of testes or testicular secretions, females would be passive. Harlow, Joslyn, Senko, and Dopp (1966) exemplified this view in their description of the receptive posture assumed by female monkeys as indicating that the female role is one of a passive sexual partner submitting to the more dominant, active, sex-initiating male. The receptive posture in monkeys and many other primates is, at least superficially, similar to the posture assumed by a subordinate animal that is threatened by a more dominant animal. However, the behavior is also observed to be displayed as a means of recalling an infant that has strayed too far, by dominant males as a reassuring posture, and by animals that are soliciting grooming. Other views on this subject were expressed, of course. Inherent in Beach's (1945) argument—that a model based on sensitization rather than organization of neural areas during early life was appropriate for describing hormonal effects on behavior development—was the notion that there existed in adult males and females considerable plasticity in sex-related behaviors. Birch and Clark (1946) also argued that the belief that females were passive, based on the impression of an observer that the receptive posture resembled the submissive posture, was a blatant "projection of the Victorian idea of human sexuality into the relam of biological theory" (p. 330). They observed, for instance, that when a sexually experienced female chimpanzee presented to an inexperienced male and the male failed to mount her immediately, he was viciously attacked. It hardly seems reasonable to consider the receptive posture (which in this case is called soliciting behavior) as indicating passivity in light of this observation.

There is now little argument that females play an active, initiating role in copulation in many mammals that have been studied (Beach, 1976). It has also been proposed that in rodents, females may play an important role in the regulation of population size (vom Saal, 1981, in press b). The current view among psychologists and biologists that in many species females play an important role in initiating copulation, as well as the belated recognition by population biologists that in many species females as well as males influence social structure and population dynamics, may thus actually be a reflection of, rather than the driving force behind, the changing attitude toward the role of women in Western societies over the last 10 to 15 years. In particular, the view of women as passive, submitting sex partners has changed radically.

With this thought in mind, a repeated observation led me to question both the notion that intact female mice are not aggressive and the relevance of the sensitivity model to aggression between female mice. I had observed than when adult female Rockland-Swiss or CF-1 albino mice were grouped together for the first time, at least one of the females would invariably attack and/or mount the other females. Initially, I was surprised by the consistency with which aggression was observed between adult female mice after group-housing, as well as by the intensity of the aggression observed. Typically, it is stated that nonlactating female rodents are not aggressive (Scott & Fredericson, 1951). A survey of the literature revealed anecdotal reports that female mice had been observed fighting. But, because of the general belief that female mice were nonaggressive, these observations were usually discounted (cf. Anderson & Hill, 1965; vom Saal, in press b; White, Mayo, & Edwards, 1969). In some studies involving reproduction of mice in seminatural environments, females have been observed exhibiting aggression and taking part in territorial defense (Lidicker, 1976; Lloyd & Christian, 1969; Reimer & Petras, 1967). In general, though, male–male aggressive interactions have continued to be the focus in studies utilizing rodents.

THE INTRAUTERINE POSITION PHENOMENON AND INTERFEMALE AGGRESSION

I hypothesized that the female mice that were exhibiting aggression toward other females might have been rendered aggressive due to exposure to elevated concentrations of testosterone during early life. This hypothesis was prompted by the report of Clemens (1974) that female rats that developed *in utero* between male fetuses had significantly longer anogenital spaces at birth than did female rats that did not develop contiguous to a male. In both rats and mice, anogenital distance at birth provides a sensitive bioassay for prenatal exposure to androgen, and the length of the space in males is almost twice that of females. In mice, anogenital distance of females at birth was also found to increase as a function of the number of male fetuses contiguous to a female during prenatal life: Females located between two male fetuses (2M females) showed greater distances than females located next to one male fetus (1M females), which in turn showed greater distances than females not located next to a male fetus (0M females; see Figure 1; vom Saal, 1976; vom Saal & Bronson, 1978). Subsequently, it was verified that 2M female fetuses have significantly higher titers of testosterone in both their amniotic fluid and blood than do 0M females during prenatal but not postnatal life (vom Saal & Bronson, 1980a).

The hypothesis based on the sensitivity model that adult 2M females should be more sensitive to testosterone as a result of exposure to elevated titers of testosterone during prenatal life was examined by Clemens, Gladue, and Coniglio (1978). They found that 2M female rats exhibited higher mount frequencies than did 0M females when injected daily with 250 μg TP and tested with sexually receptive females at weekly intervals. This finding provided support for the sensitivity model described above, but the experiment was based on the assumption that adult females would have to be administered exogenous testosterone to activate androgen-sensitive neural areas and to observe the otherwise latent differences between 0M and 2M females. On the other hand, the observation that some intact female mice were highly aggressive while others appeared to be nonaggressive raised the intriguing possibility that prior intrauterine position, and thus differential exposure to testosterone (and estrogen) during prenatal life, might influence the tendency to exhibit interfemale aggression in mice. To answer this question, adult, intact 0M and 2M female mice were matched for age and weight and housed in a cage divided into two living areas by a wooden partition. When both females were in diestrus, the partition was raised and the females were observed for 30 minutes. Of 28 pairs of females, biting attacks were observed in 14 pairs, and in 12 of these pairs the 2M female exhibited the attacks and established clear dominance over the 0M female. In 6 more pairs, tail rattling, chasing, and rough (aggressive) grooming were observed, and in 5 of these pairs the 2M female exhibited these behaviors. Thus, in 17/20 pairs (85%) in which some aspect of aggression was observed, the 2M female exhibited aggression toward the 0M female (Figure 2).

The intensity of aggression observed in intact female mice varies as a function of the estrous cycle. Hyde and Sawyer (1977) reported observing the highest levels of aggression (which, unfortunately, was not clearly defined) between females that were in proestrus and metestrus, which correlates with the time of the cycle during which progesterone levels are highest (Michael, 1976), suggesting a possible causal relationship. But 0M and 2M females were compared for aggression when in early diestrus—that is, when both estrogen and progesterone levels are lowest. Although as yet untested, it is proposed that the difference in aggressiveness of 0M and 2M females does not reflect a differential sensitivity to the activational effects of gonadal and/or adrenal steroids.

Lactating 0M and 2M female mice were also compared for the tendency to attack an ovariectomized female intruder into their home territory on Day 7 following delivery of a litter. The intensity of the aggression exhibited toward

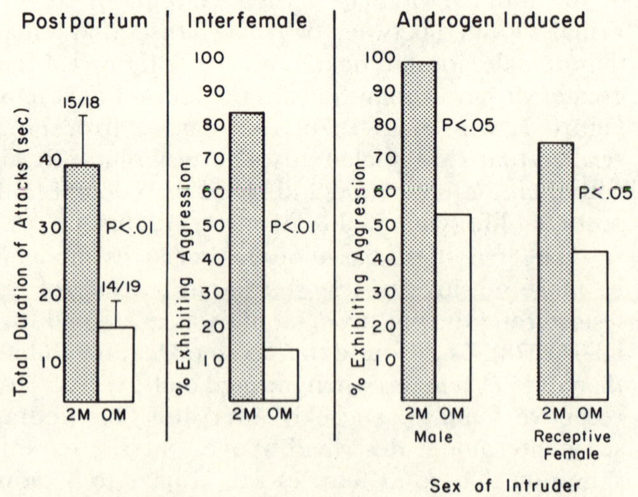

FIGURE 2. Comparison of aggressiveness in 0M and 2M female mice. Postpartum aggression—the mean duration of time (± SEM) that 14 0M and 15 2M females that did fight spent biting and chasing an intruder female during a 10-minute period when tested on postpartum Day 7 while nursing their young. Interfemale aggression—the percent of 0M and 2M females that attacked and established dominance when 0M and 2M females were paired for 30 minutes while in diestrus (based on 20 pairs in which aggressive behaviors were observed). Androgen-induced aggression—the cumulative percent of 0M and 2M females that exhibited biting and chasing during 28 days of exposure to 5 mg testosterone contained in a silastic capsule. 0M and 2M females were tested for 10 minutes every other day with a male intruder (15 2M and 9 0M; vom Saal, 1976), or at weekly intervals with a sexually receptive female intruder (25 females per group; vom Saal & Bronson, 1978). Different groups of 0M and 2M females were utilized in these experiments.

a female intruder by the 2M females was significantly greater than that exhibited by the 0M females (Figure 2). Thus, both nonlactating and lactating 2M female mice are highly aggressive toward other females, while 0M females appear nonaggressive (vom Saal & Bronson, 1978).

Another interesting finding was obtained with 0M and 2M female mice were ovariectomized and implanted with silastic capsules containing testosterone. In this study, the 0M and 2M females were paired once per week for four weeks with a sexually receptive female in an attempt to replicate, in mice, the finding of Clemens *et al.* (1978) in rats. The proportion of 0M and 2M females that mounted was not significantly different, although somewhat more 2M females (91%) mounted after 28 days of testosterone exposure than did 0M females (64%; $p = 0.08$). But significantly more of the 2M females (74%) than the 0M females (44%) exhibited biting attacks toward the sexually receptive female (vom Saal & Bronson, 1978). In a previous experiment, ovariectomized 2M female mice had been found to require a shorter period of testosterone exposure to exhibit aggression toward a bulbectomized male than did 0M females (vom Saal, 1976). When compared in terms of the percentage of 0M and 2M females that exhibited aggression after 28 days of testosterone treatment when paired with either a sexually receptive female (vom Saal & Bronson, 1978) or a bulbectomized male (vom Saal, 1976), the data from these two studies are surprisingly similar (Figure 2). Since intact male mice do not usually exhibit aggression toward sexually receptive females, these findings raise questions concerning the validity of experiments that have utilized female mice as model animals for the study of hormonal influences on intermale aggression. It is possible that the stimuli that elicit aggression in intact male mice are different than the stimuli that elicit aggression in intact or testosterone-treated female mice. Barkley and Goldman (1978) also found that more adult female mice that were injected with TP or that received testosterone through silastic implants exhibited aggression toward a female opponent than did similarly treated males.

When viewed *in toto,* all of the data from comparisons of the aggressiveness of 0M and 2M females suggest a fundamental biasing (organization) of the potential to exhibit aggression toward other females as a result of exposure to high concentrations of testosterone during prenatal life. For example, it appears that postpartum aggression in mice may be induced by direct neural stimulation as a result of suckling by young, and that ovarian and pituitary hormones do not induce postpartum aggressiveness (Svare, 1981). Thus, differences in the intensity of postpartum aggression based on prior intrauterine position cannot be viewed within the sensitivity model in which differences in adult behavior are presumed to reflect differences in sensitivity to the activational effects of hormones on the neural substrate mediating aggression. Instead of invoking the concept of differential sensitivity to hormones to account for differences in aggressiveness between 0M and 2M females, a more parsimonious approach is to propose that while the characteristics of the aggression exhibited by a female toward another female may vary depending on the physiological state of the attacking female, elevated titers of androgen during prenatal life exert organizational effects on specific neural areas, so that adult 2M females are more aggressive than 0M females whenever the situation is such that aggression is observed. Intermale aggression, on the other hand, involves exposure to androgen during early life modulating the sensitivity of the neural substrate mediating aggression to the aggression-inducing effects of androgen in adulthood. Again, no such simple explanation can account for all of the results of comparisons of 0M and 2M female mice presented in Figure 2. Finally, it seems incorrect to view the action of testosterone as exerting a masculinizing effect on a 2M female's brain during prenatal life, since aggression toward females is not a typical masculine characteristic. Prenatal exposure to testosterone does not behaviorally masculinize 2M females. The 2M females are more aggressive than the 0M females, but unless they are lactating or are treated with exogenous testosterone in adulthood, this aggressiveness is directed *only* at other females and not at males (vom Saal & Bronson, 1978). Findings based on comparisons of 0M and 2M females therefore appear most relevant to the study of the ontogeny of variability in "normal" female characteristics rather than as a model system for studying the process of masculinization.

It is interesting to speculate about the implications of the intrauterine position phenomenon to the reproductive ecology of mice. The findings from laboratory experiments conducted with 0M and 2M female mice have led to the prediction that the female mice that are dispersed from high-density populations in the wild are most likely 0M females. Since the great majority of dispersing animals are presumed to die (Lidicker, 1975), being highly aggressive should confer a substantial reproductive advantage of 2M females at high population densities (vom Saal, 1981). A critical related finding is that the aggressive 2M females and nonaggressive 0M females do not differ in their basic capacity to produce and raise healthy young (von Saal & Bronson, 1978). But 0M and 2M females do differ in the timing of puberty (vom Saal & Bronson, 1978), the length of estrous cycles during adolescence and in adulthood (vom Saal & Bronson, 1980b; vom Saal, Pryor, & Bronson, 1981), the age at which they cease producing live offspring (vom Saal, Moyer, & Rines, 1982), their attractiveness to and their capacity to sexually arouse males (vom Saal & Bronson, 1978), and their rates of urine-marking behavior. Urine marking serves to delineate territorial boundaries in mice (Harrington, 1976) and is correlated with dominance status (Maruniak *et al.*, 1977), and 2M female mice mark a novel environment at significantly higher rates than do 0M females (vom Saal & Bronson, 1978). Thus, 2M females appear to be more aggressive and possibly also more territorial than 0M females. While only 0M and 2M female mice have been compared in most studies, in all experiments in which 1M females have also been examined, 1M females have been intermediate in their characteristics—anogenital distance at birth, the timing of puberty, and aggressiveness when treated with testosterone—between 0M and 2M females (vom Saal, 1976, unpublished observations).

Intrauterine position thus influences the morphology, physiology, and behavior of female mice. Since positioning of fetuses by sex in a uterine horn is a random process (vom Saal, 1981), differences in aggressiveness due to intrauterine position cannot be related to differences in genotype. In contrast, individual differences in aggressiveness and sexual performance have been proposed to be due to differences in genotype rather than differences in fetal hormone titers (cf. Chitty, 1967; Krebs, 1978). When taken together with the observation of Lloyd and Christian (1969) that only a few aggressive, dominant females in high-density populations of mice successfully produced surviving offspring (the subordinate females either failed to mate, failed to reach term if inseminated, or were unable to protect their young if they did carry the litter to term), my findings suggest that at high population density, the females that would be most likely to establish and defend a nest area and reproduce successfully would be the 2M females.

CONCLUSION

Based on the assumption that male mice are aggressive while female mice are passive, most researchers examining the effects of hormone exposure during early life on adult aggressiveness have focused on intermale aggression. It was assumed that male mice castrated at birth and female mice not treated with TP at birth would not be aggressive. This led numerous experiments to be conducted with too short a period of testosterone treatment and aggression testing. These experiments thus provided support for the organization/activation model of intermale aggression in mice. Subsequent experiments utilizing longer periods of treatment with testosterone have revealed that prior organization of the neural substrate mediating intermale aggression is not a necessary prerequisite for adult mice to be induced to exhibit intermale aggression when treated with testosterone. But exposure to testosterone during early life renders a mouse more sensitive to the aggression-inducing action of testosterone in adulthood, with the period of development during which testosterone has the greatest sensitizing effect being around the time of birth. These findings provide support for a sensitivity model of intermale aggression—that is, exposure to androgen during early life exerts a latent (sensitizing) effect on the neural substrate mediating intermale aggression. Testosterone must be present in the circulation to induce (activate) a mouse to respond to the presence of a male opponent by exhibiting aggression.

The intrauterine position phenomenon—the capacity of fetuses to have their development modified by exposure to steroids secreted by contiguous fetuses—has proven to be a valuable model system for examining the relationship between endogenous differences in gonadal steroid concentrations during fetal development and behavior during later life. Relevant to the belief that females are generally nonaggressive is the finding that female mice that develop *in utero* between two male fetuses have elevated blood titers of testosterone during prenatal life and are highly aggressive toward other females but not males in adulthood. These females appear to have been organized into an aggressive phenotype by exposure to elevated titers of testosterone during prenatal life, but they cannot be considered to have been masculinized, at least in terms of aggressiveness. Male mice attack other males but not females, while the prenatally androgenized female mice are more aggressive only toward other females, not males. It is hypothesized that during early life in mice, androgens exert latent (sensitizing) effects on intermale aggression and direct (organizational) effects on interfemale aggression. Thus, intermale and interfemale aggression are under different hormonal and stimulus control in mice.

REFERENCES

Anderson, P. K., & Hill, J. L. *Mus musculus:* Experimental induction of territory-formation. *Science,* 1965, *148,* 1753–1755.

Barfield, R. J., Busch, D. E., & Wallen, L. Gonadal influence on agonistic behavior in the male domestic rat. *Hormones and Behavior,* 1972, *3,* 247–259.

Barkley, M. S., & Goldman, B. D. A quantitative study of serum testosterone, sex accessory organ growth, and the development of intermale aggression in the mouse. *Hormones and Behavior,* 1977, *8,* 208–218. (a)

Barkley, M. S., & Goldman, B. D. Testosterone-induced aggression in adult female mice. *Hormones and Behavior,* 1977, *9,* 76–84. (b)

Barkley, M. S., & Goldman, B. D. Studies on opponent status and steroid mediation of aggression in female mice. *Behavioral Biology,* 1978, *23,* 118–123.

Barraclough, C. A., & Leathem, J. H. Infertility induced in mice by a single injection of testosterone propionate. *Proceedings of the Society of Experimental Biology and Medicine,* 1954, *85,* 673–674.

Beach, F. A. Bisexual mating behavior in the male rat: Effects of castration and hormone administration. *Physiological Zoology,* 1945, *18,* 390–402.

Beach, F. A. Hormonal modification of sexual dimorphic behavior. *Psychoneuroendocrinology,* 1975, *1,* 3–23.

Beach, F. A. Sexual attractivity, proceptivity, and receptivity in female mammals. *Hormones and Behavior,* 1976, *7,* 105–138.

Beatty, W. Gonadal hormones and sex differences in nonreproductive behaviors in rodents: Organizational and activational influences. *Hormones and Behavior,* 1979, *12,* 112–163.

Beeman, E. The effect of male hormone on aggression behavior in mice. *Physiological Zoology,* 1947, *20,* 373–405.

Belisle, S., & Tulchinsky, D. Amniotic fluid hormones. In D. Tulchinsky & K. Ryan (Eds.), *Maternal-fetal endocrinology.* Philadelphia: Saunders, 1980.

Benno, R., & Williams, T. Evidence for intracellular localization of alpha-fetoprotein in the developing rat brain. *Brain Research,* 1978, *142,* 182–186.

Birch, H. G., & Clark, G. Hormonal modification of social behavior: II. The effects of sex-hormone administration on the social dominance status of the female-castrate chimpanzee. *Psychosomatic Medicine,* 1946, *8,* 320–331.

Block, E., Lew, M., & Klein, M. Studies on the inhibition of fetal androgen formation: Testosterone synthesis by fetal and newborn mouse testes *in vitro. Endocrinology,* 1971, *88,* 41–46.

Brain, P. F., & Evans, C. M. Attempts to influence fighting and threat behaviors in adult isolated female CFW mice in standard opponent aggression tests using injected and subcutaneously implanted androgens. *Physiology and Behavior,* 1975, *14,* 551–556.

Bronson, F. H., & Desjardins, C. Neonatal androgen administration and adult aggressiveness in female mice. *General and Comparative Endocrinology,* 1970, *15,* 320–325.

Buss, A. H. *The psychology of aggression.* New York: Wiley, 1961.

Charpentier, J. Analysis and measurement of aggressive behaviour in mice. In S. Garattini & E. B. Sigg (Eds.), *Aggressive behavior.* New York: Wiley, 1969.

Chitty, D. The natural selection of self-regulatory behavior in animal populations. *Proceedings of the Ecological Society of Australia,* 1967, *2,* 51–78.

Clemens, L. G. Neurohormonal control of male sexual behavior. In W. Montagua & W. Sadler (Eds.), *Reproductive behavior.* New York: Plenum Press, 1974.

Clemens, L., Gladue, B. & Coniglio, L. Prenatal endogenous androgenic influences on masculine sexual behavior and genital morphology in male and female rats. *Hormones and Behavior,* 1978, *10,* 40–53.

Del Campo, C., & Ginther, O. Vascular anatomy of the uterus and ovaries and the unilateral luteolytic effect of the uterus: guinea pigs, rats, hamsters, and rabbits. *American Journal of Veterinary Research,* 1972, *33,* 2561–2570.

De Moor, P., Verhoeven, G., & Heyns, W. Permanent effects of foetal and neonatal testosterone secretion on steroid metabolism and binding. *Differentiation,* 1973, *1,* 241–253.

Dixon, A., & Mackintosh, J. The relationship between the physiological condition of female mice and the effects of their urine on the social behavior of adult males. *Animal Behaviour,* 1975, *23,* 513–520.

Eaton, G. G., Goy, R. W., & Phoenix, C. H. Effect of testosterone treatment in adulthood on sexual behavior of female pseudohermaphrodite Rhesus monkeys. *Nature, New Biology,* 1973, *242,* 119–120.

Edwards, D. Early androgen stimulation and aggressive behavior in male and female mice. *Physiology and Behavior,* 1969, *4,* 333–338.

Edwards, D. A. Post-neonatal androgenization and adult aggressive behavior in female mice. *Physiology and Behavior,* 1970, *5,* 465–467.

Edwards, D. A., & Herndon, J. Neonatal estrogen stimulation and aggressive behavior in female mice. *Physiology and Behavior,* 1970, *5,* 993–995.

Fredericson, E. The effects of food deprivation and competitive and spontaneous combat in C57 black mice. *Journal of Psychology,* 1950, *29,* 89–100.

Gandelman, R., & vom Saal, F. S. Exposure to early androgen attenuates androgen-induced pup-killing in male and female mice. *Behavioral Biology,* 1977, *20,* 252–260.

Gorski, R. The neuroendocrinology of reproduction: An overview. *Biology of Reproduction,* 1979, *20,* 111–127.

Gorski, R., Harlan, R., Jacobson, C., Shryne, J., & Southam, A. Evidence for the existence of a sexually dimorphic nucleus in the preoptic area of the rat. *Journal of Comparative Neurology,* 1980, *193,* 529–539.

Goy, R. W. Organizing effects of androgen on the behavior of rhesus monkeys. In R. P. Michael (Ed.), *Endocrinology and human behaviour.* New York: Oxford University Press, 1968.

Harlow, H. Sexual behavior in the rhesus monkey. In F. Beach (Ed.), *Sex and behavior.* New York: Wiley, 1965.

Harlow, H. F., Joslyn, W. P., Senke, M. G., & Dopp, A. Behavioral aspects of reproduction in primates. *Journal of Animal Science,* 1966, *25,* 49–67.

Harrington, J. E. Recognition of territorial boundaries by olfactory cues in mice (*Mus musculus* L.). *Zeitschrift für Tierpsychologie,* 1976, *41,* 295–306.

Harris, G. W. Sex hormones, brain development and brain function. *Endocrinology,* 1964, *45,* 627–651.

Harris, G. W., & Jacobson, D. Functional grafts of the anterior pituitary gland. *Proceedings of the Royal Society (Series B),* 1952, *139,* 263–276.

Hyde, J. S., & Sawyer, T. F. Estrous cycle fluctuations in aggressiveness of house mice. *Hormones and Behavior,* 1977, *9,* 290–295.

Jacobson, C., & Gorski, R. *Neurogenesis of the sexually dimorphic nucleus of the preoptic area.* Paper presented at the 10th meeting of the Society for Neuroscience, Cincinnati, 1980.

Jacobson, C., Shryne, J., Shapiro, F., & Gorski, R. Ontogeny of the sexually dimorphic nucleus of the preoptic area. *Journal of Comparative Neurology,* 1980, *193,* 541–548.

Joslyn, W. D. Androgen-induced social dominance in infant female Rhesus monkeys. *Journal of Child Psychology and Psychiatry,* 1973, *14,* 137–145.

Jost, A. A new look at the mechanisms controlling sex differentiation in mammals. *Johns Hopkins Medical Journal,* 1972, *130,* 38–53.

Koyama, N. On dominance, rank and kinship of the wild Japanese monkey troop in Arashiyama. *Primates,* 1967, *8,* 189–216.

Krebs, C. A review of the Chitty Hypothesis of population regulation. *Canadian Journal of Zoology,* 1978, *56,* 2463–2480.

Leshner, A. A model of hormones and agonistic behavior. *Physiology and Behavior,* 1975, *15,* 225–235.

Levy, J. V. The effects of testosterone propionate on fighting behavior in C57 BL/10 young female mice. *Proceedings of the West Virginia Academy of Science,* 1954, *26,* 14.

Levy, J. V., & King, J. A. The effects of testosterone propionate on fighting behavior in young male C57 BL/10 mice. *Anatomical Record,* 1953, *117,* 562–563.

Lidicker, W. Z. The role of dispersal in the demography of small mammals. In F. Golley, K. Petrusewicz, & L. Ryszkowski (Eds.), *Small mammals: Their productivity and population dynamics.* Cambridge: Cambridge University Press, 1975.

Lidicker, W. Z. Social behaviour and density regulation in house mice living in large enclosures. *Journal of Animal Ecology,* 1976, *45,* 677–697.

Lloyd, J. A., & Christian, J. Reproductive activity of individual females in three experimental freely growing populations of house mice *(Mus musculus). Journal of Mammalogy,* 1969, *50,* 49–59.

Luttge, W. Endocrine control of mammalian male sexual behavior: An analysis of the potential role of testosterone metabolites. In C. Beyer (Ed.), *Endocrine control of sexual behavior.* New York: Raven Press, 1979.

MacClusky, N., & Naftolin, F. Sexual differentiation of the central nervous system. *Science,* 1981, *211,* 1294–1303.

Marcum, J. The freemartin syndrome. *Animal Breeding Abstracts,* 1974, *42,* 227–242.

Maruniak, J., Desjardins, C., & Bronson, F. Dominant-subordinate relationships in castrated male mice bearing testosterone implants. *American Journal of Physiology,* 1977, *233,* 495–499.

McEwen, B. Gonadal steroids and brain development. *Biology of Reproduction,* 1980, *22,* 43–48.

McEwen, B., Plapinger, L., Chaptal, C., Gerlach, J., & Wallach, G. Role of fetoneontal estrogen binding proteins in the associations of estrogen with neonatal brain cell nuclear receptors. *Brain Research,* 1975, *96,* 400–406.

McKinney, T. D., & Desjardins, C. Postnatal development of the testis, fighting behavior, and fertility in house mice. *Biology of Reproduction,* 1973, *9,* 279–294.

McKinney, T. D., & Ingle, M. K. Intermale aggression in mice: Modification by postneonatal castration. *Proceedings of the Society of Experimental Biology and Medicine,* 1974, *145,* 340–342.

McLaren, A., & Michie, D. Experimental studies on placental fusion in mice. *Journal of Experimental Zoology,* 1959, *141,* 47–73.

McLaren, A., & Michie, D. Control of prenatal growth in mammals. *Nature,* 1960, *187,* 363–365.

Meisel, R., & Ward, I. Fetal female rats are masculinized by male littermates located caudally in the uterus. *Science,* 1981, *213,* 239–242.

Michael, S. D. Plasma prolactin and progesterone during the estrous cycle in the mouse. *Proceedings of the Society of Experimental Biology and Medicine,* 1976, *153,* 254–257.

Moyer, K. E. Sex differences in aggression. In R. Friedman, R. Richart, & R. Vande Wiele (Eds.), *Sex differences in behavior.* New York: Wiley, 1974.

Mugford, R. A., & Nowell, N. W. The aggression of male mice against androgenized females. *Psychonomic Science,* 1970, *20,* 191–192.

Mugford, R. A., & Nowell, N. W. The relationship between endocrine status of female opponents and aggressive behaviour of male mice. *Animal Behaviour,* 1971, *19,* 153–155.

Ohno, S. The problem of the bovine freemartin. *Journal of Reproduction and Fertility,* 1969, *Suppl. 7,* 53–61.

Ohno, S., Christian, L., Wachtel, S., & Koo, G. Hormone-like role of H-Y antigen in bovine freemartin gonad. *Nature,* 1976, *261,* 597–599.

Peters, P. J., Bronson, F. H., & Whitsett, J. M. Neonatal castration and intermale aggression in mice. *Physiology and Behavior,* 1972, *8,* 265–268.

Pfeiffer, C. A. Sexual differences of the hypophyses and their determination by the gonads. *American Journal of Anatomy,* 1936, *58,* 195–225.

Phoenix, C. Prenatal testosterone in the nonhuman primate and its consequences for behavior. In R. Friedman, R. Richart, & R. Vande Wiele (Eds.), *Sex differences in behavior.* New York: Wiley, 1974.

Raisman, G., & Field, P. Sexual dimorphism in the neuropil of the preoptic area of the rat and its dependence on neonatal androgen. *Brain Research,* 1973, *54,* 1–29.

Reimer, J. D., & Petras, M. L. Breeding structure of the house mouse, *Mus musculus,* in a population cage. *Journal of Mammalogy,* 1967, *48,* 88–99.

Reyes, F., Boroditsky, R., Winter, J., & Faiman, C. Studies on human sexual development: II. Fetal and maternal serum gonadotropin and sex steroid concentrations. *Journal of Clinical Endocrinology and Metabolism,* 1974, *38,* 612–617.

Rugh, R. *The mouse: Its reproduction and development.* Minneapolis: Burgess Publishing, 1968.

Samuels, O., Jason, G., Mann, M., & Svare, B. Pup-killing behavior in mice: Suppression by early androgen exposure. *Physiology and behavior,* 1981, *26,* 473–477.

Scott, J. P., & Fredericson, E. The causes of fighting in mice and rats. *Physiological Zoology,* 1951, *24,* 273–309.

Selmanoff, M. K., Goldman, B. D., & Ginsburg, B. E. Developmental changes in serum luteinizing hormone, follicle stimulating hormone and androgen levels in two inbred mouse strains. *Endocrinology,* 1977, *100,* 122–127.

Simon, N. G. The genetics of intermale aggressive behavior in mice: Recent research and alternative strategies. *Neuroscience and Biobehavioral Reviews,* 1979, *3,* 97–106. (a)

Södersten, P. Role of estrogen in the display and development of sexual behavior in male rats. In C. Beyer (Ed.), *Endocrine control of sexual behavior.* New York: Raven Press, 1979.

Svare, B. B. Maternal aggression in mammals. In D. Gubernik & P. Klopfer (Eds.), *Parental care in mammals.* New York: Plenum Press, 1981.

Svare, B. B., Davis, P. G., & Gandelman, R. Fighting behavior in female mice following chronic androgen treatment during adulthood. *Physiology and Behavior,* 1974, *12,* 399–403.

Trimble, M. R., & Herbert, J. The effect of testosterone or oestradiol upon the sexual and associated behavior of the adult female rhesus monkey. *Journal of Endocrinology,* 1968, *42,* 171–185.

Valzelli, L. Aggressive behavior induced by isolation. In S. Garattini & E. B. Sigg (Eds.), *Aggressive behavior.* New York: Wiley, 1969.

vom Saal, F. S. *Prenatal exposure to androgen influences intraspecific aggression in male and female mice.* Unpublished doctoral thesis, Rutgers University, New Brunswick, New Jersey, 1976.

vom Saal, F. S. Prenatal exposure to androgen influences morphology and aggressive behavior of male and female mice. *Hormones and Behavior,* 1979, *12,* 1–11.

vom Saal, F. S. Variation in phenotype due to random intrauterine positioning of male and female fetuses in rodents. *Journal of Reproduction and Fertility,* 1981, *62,* 633–650.

vom Saal, F. S. Variation in infanticide and parental behavior in male mice due to prior intrauterine proximity to female fetuses: Elimination by prenatal stress. *Physiology and Behavior,* in press. (a)

vom Saal, F. Intrauterine position influences aggression and population dynamics in mice. In R. Blanchard, C. Blanchard, & K. Flannelley (Eds.), *Biological perspectives on aggression.* New York, Liss, in press. (b)

vom Saal, F. The interaction of circulating estrogens and androgens in regulating mammalian sexual differentiation. In J. Balthazart, E. Prove, & Gilles (Eds.), *Hormones and behavior in higher vertebrates.* Springer Verlag, in press. (c)

vom Saal, F. Proximate and ultimate causes of infanticide in male house mice. In G. Hausfater, S. Blaffer-Hrdy, & C. Vogel (Eds.), *Infanticide in animals and man.* Springer Verlag, in press. (d)

vom Saal, F. S., & Bronson, F. H. *In utero* proximity of female mouse fetuses to males: Effect on reproductive performance during later life. *Biology of Reproduction,* 1978, *19,* 842–853.

vom Saal, F. S., & Bronson, F. H. Sexual characteristics of adult female mice are correlated with their blood testosterone levels during prenatal development. *Science,* 1980, *208,* 597–599. (a)

vom Saal, F. S., & Bronson, F. H. Variation in length of the estrous cycle in mice due to former intrauterine proximity to male fetuses. *Biology of Reproduction,* 1980, *22,* 777–780. (b)

vom Saal, F., & Howard, L. The regulation of infanticide and parental behavior: Implications for reproductive success in male mice. *Science,* 1982, *215,* 1270–1272.

vom Saal, F. S., Gandelman, R., & Svare, B. B. Aggression in male and female mice: Evidence for changed neural sensitivity in response to neonatal but not adult androgen exposure. *Physiology and Behavior,* 1976, *17,* 53–57.

vom Saal, F. S., Pryor, S., & Bronson, F. H. Effects of prior intrauterine position and housing on oestrous cycle length in adolescent mice. *Journal of Reproduction and Fertility,* 1981, *62,* 33–37.

vom Saal, F., Moyer, C., & Rines, J. *Prenatal effects on aging in female mice.* Paper presented at the meeting of the Society for the Study of Reproduction, East Lansing, Michigan, 1982.

vom Saal, F. S., Grant, W., McMullen, C., & Laves, K. High fetal estrogen titers correlate with enhanced adult sexual performance and decreased aggression in male mice. *Science,* in press.

Wachtel, S. H-Y antigen and the genetics of sex determination. *Science,* 1977, *198,* 797–799.

Whalen, R. E. Sexual differentiation: Models, methods and mechanisms. In R. Friedman, R. Richart, & R. Vande Wiele (Eds.), *Sex differences in behavior.* New York: Wiley, 1974.

White, M., Mayo, S., & Edwards, D. A. Fighting in female mice as a function of the size of the opponent. *Psychonomic Science,* 1969, *6,* 14–15.

Young, W. C., Goy, R. W., & Phoenix, C. H. Hormones and sexual behavior. *Science,* 1964, *143,* 212–218.

10

Hormone–Brain Interactions and Their Influence on Agonistic Behavior

David B. Adams

INTRODUCTION

Although most specialists would agree that hormones affect aggressive behaviors by acting on the brain mechanisms of those behaviors, beyond that there would be little agreement on details. The best that can be done at this stage of our knowledge is to build hypothetical models of the brain and put them to the test of experiments. Having recently published a hypothetical model of the neural circuitry of offense, defense, and submission (Adams, 1979a), I am in a position to speculate on how specific hormones may act on that neural circuitry.

Rather than using the term "aggressive behavior" and dealing with the behaviors usually included in it, I will use the term "agonistic behavior" and deal with a related but somewhat different group of behaviors. The reason for this is that "aggressive behaviors" are a group of behaviors that in some cases have no functional relationships, while "agonistic behaviors" refer to behaviors that are all functionally related. Brain (1979) has distinguished five types of "aggression" and noted that they are very different types of behavior: (1) self-defense ("fear"-mediated attack), (2) maternal aggression, (3) predatory aggression, (4) pup-killing, and (5) social aggression (offensive attack). Some authors consider that predatory behavior should not be considered aggression at all. Agonistic behaviors, on the other hand, are all related; they include those behaviors related to physical conflict between adult members of the same species (Scott, 1966). Among the types of behavior considered agonistic by Scott are fighting, escape, threat, freezing, defensive postures, dominance, and subordination. To this list should be added the behaviors of patrol and marking, which are critical for territory maintenance (Adams, 1980).

The present chapter should be considered in relation to a general picture

DAVID B. ADAMS • Department of Psychology, Wesleyan University, Middletown, Connecticut 06457.

of agonistic behavior in muroid rodents that is contained in a series of reviews. In addition to this chapter and the paper noted above on brain mechanisms, there is a paper concerning the effects of learning and imprinting on agonistic behaviors in muroid rodents (Adams, 1979b), and a paper that reviews field studies and laboratory experiments on agonistic behavior in muroid rodents (Adams, 1980). I have concentrated for the most part on muroid rodents because they are a single superfamily (Wood, 1965) that has only recently diverged in terms of geologic time (Wood, 1959), and they include many of the best-studied laboratory animals, such as the rat, mouse, gerbil, and hamster.

Through consideration of brain mechanisms, I have suggested that agonistic behavior is mediated by four motivational systems that may be distinguished in terms of their neural substrates but that are closely interrelated in their behavioral functions. The four systems are offense, defense, submission, and patrol/marking. The rationales for the distinctions are outlined in the following sections. Maternal aggression, it should be noted, may be considered as a special case involving the simultaneous activation of both offense and defense systems.

OFFENSE VERSUS DEFENSE

The offense motivational system may be distinguished from the defense motivational system on both neural and behavioral grounds in muroid rodents. Lesions of the medial and lateral hypothalamus affect them differentially; medial hypothalamic lesions enhance defense and do not affect offense, while lateral hypothalamic lesions abolish offense and do not affect defense (Adams, 1971). Behavioral studies have shown that offense includes a "bite-and-kick" attack directed at the back or flank of the opponent after assumption of a stereotyped full aggressive posture, while defense is characterized by a "lunge-and-bite" attack directed at the face or other protruding part of the opponent and launched from a position some distance from the opponent with the forepaws extended to push the attacker back from the target as soon as the bite is delivered. The offensive bite-and-kick attack is used by a resident animal, while the defensive lunge-and-bite attack is used by an intruder (Adams, 1976; Blanchard, Fukunaga, Blanchard, & Kelley, 1975; Lehman & Adams, 1977). As a result, intruders are bitten on the back and flank, while residents are bitten on the face (as has been observed in *Microtus agrestis:* Clarke, 1956; *Dicrostonyx groenlandicus:* Bowen & Brooks, 1978; *Microtus ochrogaster:* Rose & Gaines, 1976; and *Rattus norvegicus:* Blanchard *et al.,* 1975; Calhoun, 1962, p. 181; Crabtree & Moyer, 1973; Thor, 1976). There are many other differences between offense and defense, such as different displays associated with the two (Grant & Mackintosh, 1963; Johst, 1967).

Maternal aggression appears to be the simultaneous activation of both offense and defense. No studies have yet been done to determine the neural substrate of maternal aggression, but the presence of both bite-and-kick attack and lunge-and-bite attack in the enhanced aggressiveness of the lactating female suggests that both offense and defense are involved (Adams, 1980).

DEFENSE VERSUS SUBMISSION

Whereas the distinction between offense and defense is now agreed on by many researchers of the brain mechanisms of aggression, the distinction between submission and defense is more controversial (cf. commentaries in Adams, 1979a, especially those of Blanchard and Blanchard). As evidence for the distinction, I have pointed to the fact that muroid rodents are much more likely to engage in a lunge-and-bite attack after lesions of the septum and ventromedial hypothalamus. Usually, the lunge-and-bite attack is used during intraspecific fighting by wild animals and not by laboratory animals; in fact, the behavior of animals with septal and ventromedial hypothalamic lesions is similar to the normal defense behavior of wild animals. Not only the lunge-and-bite attack but also various other motor patterns of defense are more vigorous in laboratory animals after these particular lesions and are more similar to the defense of wild animals.

To explain the effects of lesions and the differences between wild and laboratory animals, I have postulated that there are two related but alternative motivational mechanisms, submission and defense, that are under the control of a *consociate modulator* (Adams, 1979a,b). The motor patterns activated by the two motivational mechanisms are similar in most respects, although defense includes the lunge-and-bite attack, while submission includes ultrasound and specific submissive postures. Since both defense and submission can be abolished by lesions of the midbrain central gray and adjoining tegmentum (Edwards & Adams, 1974), I have suggested that this is the location of the two motivational mechanisms. The consociate modulator is responsible for shifting the animal's "fear-induced" behavior from defense to submission when the opponent is a familiar animal—a consociate. The consociate modulator may be located in the ventromedial hypothalamus, from which it inhibits defense and facilitates submissions.

PATROL/MARKING

There are a number of behaviors involving locomotion and scent-marking that are functionally related to physical conflict between adult members of the same species and that should be included within the definition of agonistic behavior in muroid rodents. These behaviors have been variously described as "patrol" (Crowcroft, 1955; Mackintosh, 1970), "passage-guarding" (Anderson & Hill, 1965; Calhoun, 1962), and "sentry" (Crowcroft, 1955). From these behavioral descriptions it would appear that dominant, territorial animals appear to seek out confrontations with intruders into their territory. Along with scent-marking behaviors, that are also functionally related, I have included these behaviors in a motivational system called *patrol/marking* (Adams, 1980).

NEURAL CIRCUITRY OF AGONISTIC BEHAVIOR

The neural circuitry of agonistic behaviors may be considered as the substrate on which circulating hormones influence these behaviors. From this it follows that the more we know of this circuitry, the better we will understand how hormones affect agonistic behavior.

The analysis of the neural circuitry of agonistic behavior is still hypothetical and rather general. The analysis of the circuitry at the level of recording from individual neurons has been attempted in only a few studies (Adams, 1968; Mink, Adams, & Sinnamon, in press; Pond, Sinnamon, & Adams, 1977; Sandner, Schmitt, & Karli, 1979). Our knowledge, instead, is still limited to data from lesion and electrical-stimulation techniques that tell us which gross brain areas are involved, such as the amygdala, septum, midbrain central gray, etc. At this level the picture is complicated, because it appears that the various motivational systems of agonistic behavior involve neural substrates with substantial overlap in these regions. Offense, defense, and submission all involve forebrain pathways that process motivating stimuli in the septum, basal forebrain, and amygdala; convey these influences to the hypothalamus; and then relay them caudally to motivational mechanisms in the midbrain (Adams, 1979a). Patrol/marking may also involve similar overlapping pathways.

MOTIVATING STIMULI AND MOTIVATIONAL MECHANISMS

The neural circuitry of agonistic behavior may be analyzed in terms of different components of their motivational system. I have distinguished four such components (Adams, 1979a; Adams, 1980). Central to a motivational system is a set of homogenous neurons (still only hypothetical since the neuronal basis of behavior is not well known) that I have called a *motivational mechanism.* Activity of these neurons determines the motivational state of the animal, for example, whether it is offensive, defensive or submissive at that particular moment. The actual motor patterns that the animal performs are primed by the firing of inputs from the motivational mechanism to the neural substrate of the motor pattern, which I have called the *motor-patterning mechanism.* Whether or not a motor pattern is produced, however depends on simultaneously active inputs from releasing and directing stimuli as well.

Sensory influences on motivational systems may be distinguished in terms of the level on which they act on their neural circuitry. If sensory influences act on the motivational mechanism, then they may be considered *motivating stimuli.* These stimuli affect all of the motor patterns of the motivational system equally. Thus, for example, the motivating stimulus of pain for the defense motivational system activates not only a lunge-and-bite attack but also the defensive sideways posture, defensive upright posture, flight locomotion, freezing, etc. If, on the other hand, sensory influences act on individual motor patterns and not on related motor patterns, then they may be considered *releasing and directing stimuli.* Such stimuli release and direct individual motor patterns. Thus, for example, vibrissal inputs function as releasing and directing stimuli of the defensive upright posture (Kanki & Adams, 1978), but they do not appear to be motivating stimuli that can activate other motor patterns of defense such as squealing, freezing, etc.

HOW CIRCULATING HORMONES MAY AFFECT AGONISTIC BEHAVIOR

A schematic view of the neural circuitry of agonistic behavior in muroid rodents is shown in Figure 1. In the schema one can distinguish four motivational systems: patrol/marking, offense, submission, and defense. Each motivational system is divided into four components: sensory analyzers and synthesizers for motivating stimuli, motivational mechanism, motor patterning mechanisms, and sensory analyzers and synthesizers for releasing and directing stimuli. It may be seen that there may be considerable overlap and interrelationship among the various components and motivational systems. Most types of motivating and releasing and directing stimuli may affect more than one motivational mechanism, and most motor-patterning mechanisms may be activated by more than one motivational mechanism.

Although the schema may give the impression that the agonistic behavior of muroid rodents is "hard-wired" and inborn, it has been pointed out elsewhere that there are many ways in which it is modified and developed as a function of experience (Adams, 1979b, 1980). In fact, as will be documented below, hormonal effects are among the important ways that the systems are modified as a function of experience.

The schema enables one to make specific hypotheses about hormonal effects. First, which motivational system or systems does it affect? Does it affect patrol/marking, offense, defense, or submission? And if it affects the latter two systems, does it act directly on the consociate modulator that switches the animal from defense to submission? Second, which component of a motivational system does it affect? Does it affect the motivational mechanism itself, the motor-patterning mechanism, or a particular sensory analyzer or synthesizer?

Ultimately, the answers to these questions will require appropriate single-neuron recording techniques, but in the meantime, tentative answers may be obtained from careful behavioral analysis.

To distinguish which motivational systems are influenced by a specific hormone, it is possible to rely on certain motor patterns that are characteristic of a particular system. Certain ambivalent motor patterns such as approach locomotion and sideways or upright postures are not specific enough for this method. Others, however, are not ambivalent and reflect the activation of one particular motivational mechanism. Thus, most of the scent-marking patterns are specific

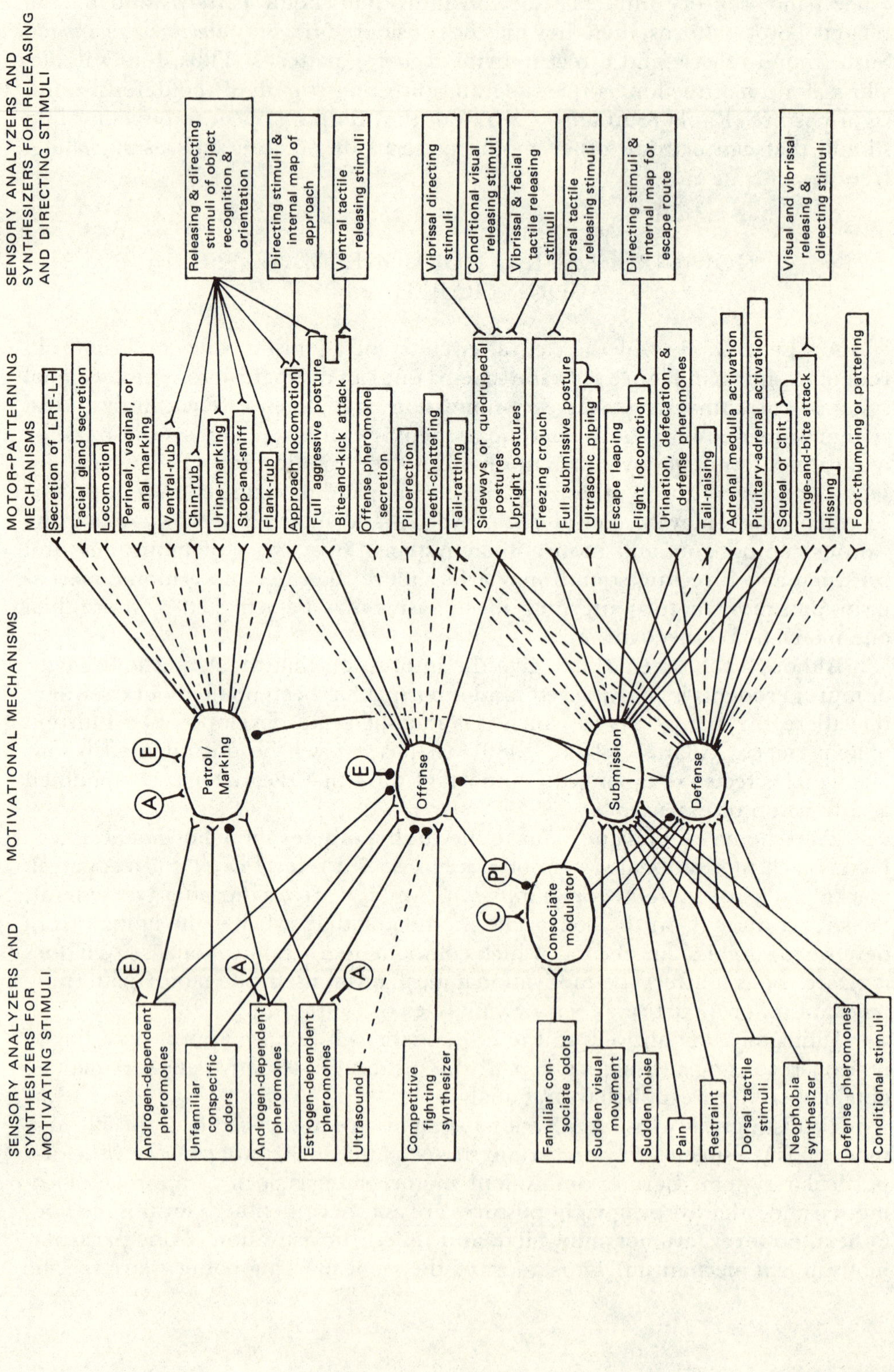

SENSORY ANALYZERS AND SYNTHESIZERS FOR MOTIVATING STIMULI
MOTIVATIONAL MECHANISMS
MOTOR-PATTERNING MECHANISMS
SENSORY ANALYZERS AND SYNTHESIZERS FOR RELEASING AND DIRECTING STIMULI
Androgen-dependent pheromones
Unfamiliar conspecific odors
Androgen-dependent pheromones
Estrogen-dependent pheromones
Ultrasound
Competitive fighting synthesizer
Familiar consociate odors
Sudden visual movement
Sudden noise
Pain
Restraint
Dorsal tactile stimuli
Neophobia synthesizer
Defense pheromones
Conditional stimuli
E
A
A
A
E
E
PL
C
Patrol/ Marking
Offense
Submission
Defense
Consociate modulator
Secretion of LRF-LH
Facial gland secretion
Locomotion
Perineal, vaginal, or anal marking
Ventral-rub
Chin-rub
Urine-marking
Stop-and-sniff
Flank-rub
Approach locomotion
Full aggressive posture
Bite-and-kick attack
Offense pheromone secretion
Piloerection
Teeth-chattering
Tail-rattling
Sideways or quadrupedal postures
Upright postures
Freezing crouch
Full submissive posture
Ultrasonic piping
Escape leaping
Flight locomotion
Urination, defecation, & defense pheromones
Tail-raising
Adrenal medulla activation
Pituitary-adrenal activation
Squeal or chit
Lunge-and-bite attack
Hissing
Foot-thumping or pattering
Releasing & directing stimuli of object recognition & orientation
Directing stimuli & internal map of approach
Ventral tactile releasing stimuli
Vibrissal directing stimuli
Conditional visual releasing stimuli
Vibrissal & facial tactile releasing stimuli
Dorsal tactile releasing stimuli
Directing stimuli & internal map for escape route
Visual and vibrissal releasing & directing stimuli

to patrol/marking, the bite-and-kick attack is specific to offense, the lunge-and-bite attack is specific to defense, and the full submissive posture may be specific to submission.

To distinguish which component of a motivational system is influenced by a specific hormone, it is necessary to test for a number of different motor patterns under a number of conditions involving different motivating stimuli. If the hormone acts on an analyzer or synthesizer of motivating stimuli, it will affect only the behaviors elicited by those motivating stimuli and not others. If it acts on the motivational mechanism, the hormone will affect the frequencies of all motor patterns in response to all types of conditions involving all of the various motivating stimuli. If it acts on a sensory analyzer for releasing and directing stimuli, or on a motor-patterning mechanism, it will affect only the particular motor pattern in question and not the other motor patterns activated by the same motivational system.

ANDROGENS

Androgens facilitate both the offense and patrol/marking motivational systems in muroid rodents. This is indicated by the higher levels of the bite-and-kick attack (offense) and scent-marking motor patterns (patrol/marking) in dominant intact males than in subordinates, castrates, or females, and by the restoration of higher levels in castrates by administration of androgens. In the case of offense, the effect of androgens is well known and has often been reviewed (Bronson & Desjardins, 1971; Edwards & Rowe, 1975; Leshner, 1975; Whalen, 1974). In the case of patrol/marking it has not been as extensively reviewed but has been shown experimentally in rats (Price, 1975), gerbils (Thiessen, Friend, & Lindzey, 1968), and mice (Bronson, 1976).

There is little evidence that androgens affect the defense and submission motivational systems. In the laboratory rat there are higher rates of upright posture and boxing during shock-elicited fighting in males, an effect that is dependent on neonatal androgens (Conner & Levine, 1969) and reduced by castration (Hutchinson, Ulrich, & Azrin, 1965). However, we have found no difference in the rate of upright posture and boxing between males and females

FIGURE 1. Effects of hormones on the hypothetical neural circuitry that organizes and activates the agonistic behaviors of muroid rodents. Sites of action of the following hormones are shown: androgens (A), estrogens (E), corticosteroids (C), and hormones of pregnancy and lactation (PL). Four motivational systems are shown: patrol/marking, offense, submission, and defense. Four components are shown for each motivational system: the motivational mechanism, motor-patterning mechanisms, and two types of sensory analyzers and synthesizers—one for motivating stimuli and one for releasing and directing stimuli. Rationale for this organization may be found in the text. Facilitative effects are illustrated as small forks at the tips of projections from one structure to another. Inhibitory effects are illustrated as filled circles at the tips of projections. Projections that are present in all or most species of muroid rodents are illustrated as solid lines, while those present in only some species are shown as dotted lines. (Modified from Adams, 1980).

in response to bite-and-kick attack by a partner during competitive fighting (Adams, unpublished observations). Also, the defensive responses to handling by an experimenter do not differ between the sexes (Korn & Moyer, 1968), nor do the submissive postures during defeat (Nock & Leshner, 1976).

The effect of androgens on offense is apparently exerted on sensory analyzers for motivating stimuli rather than on the motivational mechanism. This is indicated by the fact that offense due to certain motivating stimuli, such as those associated with competitive fighting, is not dependent on androgens in the mouse (Fredericson & Birnbaum, 1954) or rat (Zook & Adams, 1975). Also, there is offense by females as well as males in wild rats (Telle, 1966), wild mice (Ebert & Hyde, 1976), and other muroid rodent species such as the gerbil (Swanson, 1974) and hamster (Vandenbergh, 1971). It has been suggested that within the laboratory, the mouse and rat are special cases in which other motivating stimuli are not effective in eliciting offense and only the sensory analyzer dependent on androgens is functional (Adams, 1980).

The location of the neural substrate for the action of androgen on the offense motivational system has been determined. Studies by Bermond (1978) and by Bean and Conner (1978) have shown that injection of androgen into the preoptic hypothalamus facilitates the intermale offense of the laboratory rat after it has been reduced by prior castration. An earlier study by Owen, Peters, and Bronson (1974) had found that septal implants of androgen were effective, but they did not test implants in the preoptic area, which is adjacent and which might have been affected by spread of the hormones from the septal implant. Neurons of the preoptic area show a differential uptake of androgen (Stumpf & Sar, 1976), and their firing patterns in response to olfactory stimulation are altered by androgen (Pfaff & Pfaffman, 1969).

The intermale fighting of laboratory rats and mice, it may be hypothesized, depends on an androgen-organized and androgen-activated sensory analyzer, located at least partly in the preoptic hypothalamus, that is tuned to androgen-dependent pheromones and that activates the offense motivational mechanism. Part of the basis for this hypothesis derives from the preceding paragraphs. Other relevant evidence comes from the fact that the fighting depends on olfaction (Adams, 1976) and on the presence of androgen in the opponent (Mugford & Nowell, 1970; Thor & Flannelly, 1976). Also, androgens must be present not only in the adult (i.e., an activating role) but also in the perinatal period (an organizing role), as has been shown in the mouse (Edwards, 1969) and rat (Barr, Gibbons, & Moyer, 1976).

There is another aspect of the intermale fighting of muroid rodents that may be androgen-dependent: an inhibition of attack on adult conspecific females. In experiments using urine from a donor animal, the inhibition depends on estrogen-dependent pheromones of the donor animal (Haag, Jerhoff, & Kirkpatrick, 1974). One may suppose that the effect is due to the action of these pheromones on an androgen-activated sensory analyzer that is tuned to estrogen-dependent pheromones and that inhibits the offense motivational mechanism. Its neural substrate has not been investigated, but it may be related to the other mechanism described above.

The effects of androgens on patrol/marking are apparently exerted at two points, both on the motivational mechanism itself and on sensory analyzers for motivating stimuli of patrol/marking.

The facilitation of the motivational mechanism of patrol/marking by androgens is indicated by experiments in which effects on motivating stimuli are ruled out. Thus, androgens are apparently responsible for the "spontaneous" patrolling of males that occurs even in the absence of relevant motivating stimuli (Shillito, 1963). Also, androgen administration can reinstate the ventral-rub marking of gerbils after its motivating stimuli have been rendered ineffective by olfactory deafferentation (Thiessen, Lindzey, & Nyby, 1970).

There are also effects of androgens on the sensory analyzers of motivating stimuli for patrol/marking. There may be two such analyzers, similar to or identical with the analyzers that process motivating stimuli of offense: (1) a sensory analyzer tuned to odors of androgen-dependent pheromones, activated by androgens, and inhibiting the patrol/marking motivational mechanism; and (2) a sensory analyzer tuned to odors of estrogen-dependent pheromones, actived by androgens and facilitating the patrol/marking motivational mechanism. They determine the fact that intact males explore and scent-mark in reaction to conspecific females but not in reaction to males. The data for various muroid rodent species has recently been reviewed in detail (Adams, 1980).

The effects of androgens on patrol/marking, like their effects on offense, apparently take place in the preoptic hypothalamus. Therefore, this brain area may contain either the motivational mechanism of patrol/marking, its sensory analyzers, or both. Lesions of this area abolish the approach locomotion that is normally enhanced by androgen in males (Singer, 1968), and implantation of androgen in this area reinstates the ventral-rub scent-marking of male gerbils after it has been reduced by castration (Yahr, 1977). Neurons of this area respond differentially to urine odors that may normally function as motivating stimuli for patrol/marking, and they may be sensitized by the administration of androgen (Pfaff & Pfaffman, 1969). Also, they have been shown to have a high uptake of radioactive androgen, as noted above (Stumpf & Sar, 1976). We have obtained data showing that medial preoptic neurons are differentially active during approach by a male rat toward an estrous female (Mink *et al.*, in press).

ESTROGENS

The effect of estrogens on offense is apparently exerted, at least in part, directly on the offense motivational mechanism, since offense is reduced during estrus against all types of opponents. Apparently the effect is mediated by estrogen and facilitated by progestins (Floody & Pfaff, 1977). One set of contradictory data have been obtained in a preliminary study in our laboratory that suggested that competitive fighting by females is not reduced during estrus. It is possible, however, that the motivating-stimulus inputs for competitive fighting are facilitated by estrogen, since females usually show more competitive fighting than males, and that during estrus there are two opposing influences of estrogens that more or less cancel each other out.

More work remains to be done on the effects of estrogens upon the neural substrate of offense in female muroid rodents. The major data come from only one species, the hamster. Although there are data on the mouse and rat, the reduction of offense during estrus in the one study on the mouse was not highly significant for chasing and attack (Hyde & Sawyer, 1977), and preliminary reports on reduction of offense during estrus on the rat have not yet been fully published. Furthermore, there may be an additional inhibitory effect depending on male gonadal hormones in the opponent, which could explain why females of many wild rodent species are less likely to attack strange males than they are to attack strange females.

The effects of estrogens on patrol/marking, like those of androgens, are apparently exerted at two points, both on the motivational mechanism itself and on sensory analyzers for motivating stimuli.

The facilitation of the motivational mechanism of patrol/marking by estrogens is indicated by experiments in which possible effects on motivating stimuli are apparently ruled out. When female muroid rodents of various species are in estrus, they show higher levels of locomotion, apparently due to patrol/marking activation (Richards, 1966; Stinson, 1952; Vick & Banks, 1969; Wang, 1923). This locomotion increase may be obtained from females that are isolated in cages that exclude the normal motivating stimuli of patrol/marking.

Estrogens also may activate a sensory analyzer that is tuned to androgen-dependent pheromones and that facilitates patrol/marking in the female when she encounters another male. This may be the same analyzer that is responsible for her decreased offense against such a male. In the rat, adult females approach, urine-mark, and sniff more in the presence of male odors than in the presence of the odors of castrates or females, and the effect is abolished by ovariectomy (Brown, 1977). In the mouse, females are attracted to the odors of adult male preputial glands and the effect is enhanced by estrogen injections (Caroom & Bronson, 1971). Also, they locomote more in response to odors from intact males than to odors from castrates (Ropartz, 1970), and they urine-mark more in the presence of male odors than of castrate odors (Maruniak, Owen, Bronson, & Desjardins, 1975). Although the latter authors found no differences as a function of the estrous cycle or ovariectomy, their data have recently been contradicted and an estrous effect has been described that would be consistent with estrogen activation (Wolff & Powell, 1979).

Estrogens, like androgens, apparently exert at least some of their effects on patrol/marking by acting on neurons of the preoptic region. Lesions of the preoptic region reduce approach locomotion that is produced by systemic estrogen injections (King, 1979). Implantation of estrogen into the preoptic region facilitates locomotor activity of ovariectomized rats (Colvin & Sawyer, 1969; Wade & Zucker, 1970). Neurons in this region have also been shown to have a high uptake of radioactive estrogens (Pfaff, Lewis, Diakow, & Keiner, 1973). Whether the region contains the motivational mechanism of patrol/marking or its sensory analyzers or both is a question that cannot yet be answered.

HORMONES OF PREGNANCY AND LACTATION

The lactating female is especially aggressive in muroid rodents. This is true not only in laboratory species, but also in many wild species, including *Rattus rattus* (Ewer, 1971), *Cricetomys gambianus* (Ewer, 1967), *Micromys* (Frank, 1957), *Lemmus lemmus* (Arvola, Ilmén, & Koponen, 1962), *Microtus agrestis* (Clarke, 1956), *Neotoma fuscipes* (Linsdale & Tevis, 1951), *Dicrostonyx groenlandicus* (Brooks & Banks, 1973), and *Apodemus sylvaticus* (Gurnell, 1977).

The pregnant or pseudopregnant female may also be especially aggressive. Noirot, Goyens, and Buhot (1975) have described high rates of "impulsive attacking" against male intruders by pregnant and pseudopregnant mice. Erskine, Barfield, and Goldman (1980b) have noted increased aggressiveness by rats during what appeared to be pseudopregnancy. And in our laboratory we have seen increased rates of competitive fighting (offense) by females during pseudopregnancy (Adams, unpublished observations).

The enhanced aggressiveness of pregnant and lactating females may be due to activation of both the defense and offense motivational systems. In laboratory species, various reports have described both lunge-and-bite and bite-and-kick attacks by lactating females. In lactating *Mus musculus,* the bite-and-kick attack has been described as "grasping and rolling about while biting" (Rosenson & Asheroff, 1975, p. 221) and the lunge-and-bite attack as "the lactating female would immediately lunge and bite" (Gandelman, 1972, p. 24). In lactating *Peromyscus leucopus,* the bite-and-kick attack has been described as an "attack-fight sequence" while the lunge-and-bite was defined as a "lunge" attack (Rowley & Christian, 1976). In *Rattus norvegicus,* we have noted in our laboratory that the lactating female may alternate between lunge-and-bite attacks (defense) and bite-and-kick attacks with offensive sideways posture (offense) against an unfamiliar intruder.

The aggressiveness of pregnant or pseudopregnant females may also involve both enhanced defense and offense. As noted above, we have seen enhanced offense during pseudopregnancy in the rat. The observations of Erskine *et al.* (1980b) on pseudopregnant rats are consistent with an increase in offense, since they described offensive sideways posture and bites that were directed at the dorsal neck area of the intruder. On the other hand, the "impulsive attacking" described by Noirot *et al.* (1975) may have been a defensive lunge-and-bite attack.

Two hormones, progestins and prolactin, have been investigated with regard to their possible role in the aggressiveness of pregnancy and lactation. Progestin levels are high during pregnancy and low during lactation, while prolactin levels increase periodically during pregnancy and reach their highest levels during lactation.

Despite some reports that prolactin is critical for aggressiveness during lactation, other recent work has cast doubt on this notion. In favor of a role for prolactin are data that administration of ergocornine, which inhibits the release of prolactin, also suppresses maternal aggression in the hamster (Wise & Pryor, 1977) and mouse (Politch & Herrenkohl, 1980). Also it has been reported that prolactin injection can increase maternal aggression in mice (Hansult & Kessler, 1977), and that ectopic pituitary grafts that raise prolactin levels can increase aggressiveness in hamsters (Buntin, Catanzaro, & Lisk, 1979). On the other hand, the effectiveness of ergocornine as a suppressor of maternal aggression in mice has been contradicted by Mann, Michael, and Svare (1980), and it has been found that maternal aggression survives hypophysectomy, which removes the major source of circulating prolactin (Erskine, Barfield, & Goldman, 1980a).

Progestins may play a role in the enhanced aggressiveness of pregnant and pseudopregnant females. In certain species of muroid rodents, the aggresssiveness of the female may be reduced by ovariectomy and restored by progestin, but not estrogen, administration. In a study on *Lemmus trimucronatus* (Huck, Carter, & Banks, 1979), the attacks were clearly defensive lunge-and-bite ("female assumes a posture with head elevated, mouth open, and teeth prominently displayed. In this posture female vocalizes and lunges towards the male." p. 42). In similar studies involving *Peromyscus leucopus* (Gleason, Michael, & Christian, 1979) and *Mesocricetus auratus* (Payne & Swanson, 1971), it was not clear from the published accounts whether the attacks affected by progestins were bite-and-kick or lunge-and-bite attacks. In the former species, there was a simultaneous inhibitory effect of the progestin on submissive behavior (Gleason *et al.*, 1979).

Progestin has another effect: It synergizes with estrogen in the estrous inhibition of offense by muroid rodent females against male opponents. Thus, injection of progestin leads to a greater effect than injection of estrogen alone, described earlier (Floody & Pfaff, 1977). Although progestin administration has been reported to inhibit offense in males (Soares, Kalberer, & Erpino, 1978), this may not have any behavioral significance since progestins do not normally circulate in significant levels in the male (Bartke, Smith, Michael, Peron, & Dalterio, 1977).

In view of a recent report that the posterior pituitary hormone secreted during lactation, oxytocin, may be involved in stimulating other behaviors associated with lactation (Pedersen, 1980), one would want to determine if it may also play a role in the enhanced aggressiveness of the lactating female.

It may be hypothesized that the hormones of pregnancy and lactation act on offense and defense/submission by influencing the offense motivational mechanism and the consociate modulator, as illustrated in Figure 1. The effect on the offense motivational mechanism would be excitatory and that on the consociate modulator would be inhibitory, thereby releasing defense from its suppressive modulation.

The evidence for the sites of action of progestin in the brain are consistent with its hypothesized roles in offense and defense/submission. Progestins are taken up in two locations in the hypothalamus, the preoptic region and the ventromedial region (Stumpf & Sar, 1976). The effects on offense might be expected to take place in the preoptic hypothalamus, as noted above in the section on estrogens. An inhibitory effect of progestin on the consociate modulator might be expected to occur in the ventromedial hypothalamus, which may be the site of the neurons of the consociate modulator (Adams, 1979a).

ACTH AND CORTICOSTEROIDS

Activation of the pituitary-adrenal axis and their hormones ACTH and corticosteroids have been implicated in the increased submissiveness and decreased offense shown by male muroid rodents after repeated defeat. Leshner and colleagues performed a series of experiments from which they concluded that the effects on offense are due to ACTH and those on submission are due to corticosteroids. Leshner, Walker, Johnson, Kelling, Kreisler, and Svare (1973) reported that supranormal levels of ACTH reduce the frequency of intermale fighting (i.e., offense) of mice despite manipulations that hold constant other hormones, including androgens and corticosteroids. The effect of ACTH on offense is complex, however, being facilitative when acute and inhibitory when chronic (Brain & Evans, 1977; Matte, 1979). Corticosteroids have been reported to increase the passive avoidance by a mouse of a previously victorious opponent (Moyer & Leshner, 1976) and to increase the rate of submissive upright posture in response to attack (Nock & Leshner, 1976). The effect is not due to ACTH, since there is no effect if ACTH levels are altered and corticosteroid levels are kept constant (Moyer & Leshner, 1976). The effects are also not due to altered levels of androgens (Leshner & Moyer, 1975).

The site of action of corticosteroids and ACTH on the neural substrates of submission and offense have not been directly investigated. Corticosteroids are taken up by neurons in the hippocampus, septum, amygdala, and pyriform and cingulate cortex (Stumpf & Sar, 1976). These are the structures that have been implicated by lesion and brain stimulation studies in the control of submissive behavior (Figure 2 in Adams, 1979a). ACTH, on the other hand, does not seem to cross the blood–brain barrier at normal concentrations in the circulation (Krieger & Liotta, 1979), so the mechanism of its effects on the neural substrate of offense is problematic.

It appears likely that the pituitary-adrenal hormones do not affect the neural substrates of defense, except insofar as the behavior of a subordinate animal is shifted from defense to submission in confrontation with a conspecific, as noted above. In a situation that does not involve a conspecific opponent, the novel open-field test situation, the suppression of pituitary-adrenal activity by hypothalamic implants of hydrocortisone does not affect freezing behavior in rats (Stern, Erskine, & Levine, 1973).

In summary, it is possible to account for most of the data regarding the affects of normal levels of ACTH and corticosteroids on agonistic behavior by supposing that corticosteroids facilitate the consociate modulator and/or neural structures that project to it.

EPINEPHRINE AND NOREPINEPHRINE

Since epinephrine and norepinephrine are secreted by the adrenal medulla in response to various motivating stimuli of defense—including handling and transfer to a strange cage, painful shock, and anticipation of painful shock (McCarty & Kopin, 1978) and fighting, defeat, and stimuli associated with defeat (Hucklebridge & Nowell, 1974)—it might be expected that these hormones would facilitate defense or submissive behaviors. As far as I know, however, there are no studies investigating this question in social interactions. Instead, there are indirect data from studies on avoidance or escape from shock, indicating that these behaviors may be enhanced by the hormones of the adrenal medulla. Removal of the adrenal medulla retards active avoidance (Conner & Levine, 1969) and injection of epinephrine may enhance active avoidance (Conner & Levine, 1969), escape (Kamano, 1968), or passive avoidance (Leventhal & Killackey, 1968).

In cold-blooded vertebrates, the hormones of the adrenal medulla may have a critical facilitative effect on offense and defense mediated by a rise in body temperature. It is possible that the secretion of hormones from the adrenal medulla into the peripheral circulation may contribute to the rapid and dramatic rise in body temperature that makes a cold-blooded animal ready for effective flight or attack (Engbretson & Livezey, 1972).

HORMONES AND REPRODUCTIVE STATES

Considered in a broad ecological context, the various hormonal effects on the neural substrates of agonistic behavior may be understood as part of a more general role of hormones that establish and maintain the reproductive state of the individual. One may suppose that each individual has only a limited number of possible reproductive states and that at any one time in its life it can be in only one—in other words, that these are not graded conditions, but all-or-none states. In terms of systems theory they are "complexes of stable, rhythmic limit cycles" (Iberall & McCulloch, 1969, p. 290). Circulating hormones play a critical role in the stability, rhythmicity, and limitation of the reproductive states; each state is established and maintained by a particular hormone that is secreted as part of a positive feedback chain of endocrine interactions.

The "courtship reproductive state" is established and maintained by high levels of the gonadal hormones, androgens in the male and estrogens in the female. These hormones facilitate the patrol/marking motivational system (both motivational mechanism and sensory analyzers) and the motivational systems of male and female sexual behavior. In males, androgens facilitate offense against other males but inhibit offense against females, and androgens facilitate patrol/marking in response to females but inhibit it in response to males. In females, estrogens have the opposite effects; they inhibit offense against males and facilitate patrol/marking in response to males. Thus, males patrol and exclude other males from their territories and they seek out and mate with females, while females patrol during estrus and seek out and mate with males. Courtship is not interrupted or suppressed by offense, because males are inhibited from attacking estrous females and estrous females are inhibited from attacking males. Although these effects sound complex when considered in a behavioral context, they may be mediated by only four sites of hormonal action in the brain, as shown in Figure 1. Of course, there are other effects of gonadal hormones besides those shown; for example, they facilitate the sexual motivational systems and stimulate production of pheromones from scent glands, which are then spread by the scent-marking behaviors activated by the patrol/marking motivational system. The courtship reproductive state may be maintained by a positive feedback chain that may be visualized in Figure 1: Activation of patrol/marking stimulates the release of gonadotropins as a motor pattern, gonadotropins stimulate the production of gonadal hormones, and gonadal hormones, in turn, facilitate both sensory analyzers and the motivational mechanism of patrol/marking.

The "pregnancy reproductive state" is established and maintained by high levels of progestins in the female. As a result, there is an increased level of defense, offense, and nest-building, by which the female prepares and defends the site for the nest that will house the young following paturition. This state might also be called a "luteal reproductive state" since the progestins are secreted by the corpora lutea, which are induced by copulation either directly or indirectly, and if the copulation is not successful, the animal undergoes a pseudopregnancy with many of the same behavioral characteristics as pregnancy. In most species of muroid rodents, the corpora lutea are directly stimulated by copulation; in the microtines, however, they are caused only indirectly by copulation, which causes ovulation after which the corpora lutea develop spontaneously (Dewsbury, 1977).

The "lactation reproductive state" is characterized by high levels of prolactin and other hormones in the female. Possibly as a result of these hormones, there is an increased level of offense and defense by which the female protects her young. Also, as a result of the prolactin, the female produces milk for the young, who in turn stimulate the parental motivational system. As a result, the motor patterns of the parental motivational system are activated, including nest-building, nursing, pup-retrieval, and grooming of the young. There are apparently several positive feedback chains involved in the establishment and maintenance of this reproductive state, but they have not yet been fully understood.

There are other states that may be called the "reproductive postponement states." One such state, triggered by repeated defeat (submission) or predator pressure (defense) is established and maintained by the pituitary-adrenal hormones ACTH and corticosterone. These hormones facilitate submission, inhibit defense, and suppress the gonadal hormone system. As a result, the animal does not engage in patrol/marking (i.e., does not maintain a territory) or engage in sexual behavior or offense. Also, and perhaps most crucial, because of the reduced levels of gonadal hormones, males in this state do not produce the androgen-dependent pheromones that increase their "attackability" (Leshner, 1975). Males in this state are not attacked. Females in this state do not reproduce; thus they do not run the greater risk and waste their metabolic energy on pregnancy and lactation under conditions that are not propitious for successful reproduction. Since pituitary-adrenal activation not only activates the consociate modulator but also is a motor pattern of submission (Figure 1), there is a potential here for a positive feedback chain that may establish and maintain the reproductive postponement state.

Another type of "reproductive postponement state" occurs during winter for muroid rodents that live in temperate climates. Like the state mediated by corticosteroids, it includes suppression of gonadal hormone production, and therefore it suppresses the offense and patrol/marking motivational systems. Its hormonal basis is not yet well understood (Ellis & Turek, 1980; Morin, Fitzgerald, Rusak, & Zucker, 1977), although it has been determined that the adrenal cortex is not essential (Bittman & Goldman, 1979; Ellis & Turek, 1980).

The fitness value of reproductive postponement states, from the standpoint of evolutionary genetics, presumably derives from the enhanced ability of an animal to reproduce at a later period of its life following the end of the postponement state. Documentation of this under natural conditions is difficult, because it requires the long-term observation of large numbers of individually recognizable animals. One such study has been done on the arctic ground squirrel (Carl, 1971). Over a period of two seasons, approximately half of the adult males obtained territories and the other half remained as "floaters," subordinate males who had apparently entered a reproductive postponement state. Of the 22 floater males, 5 were able to obtain territories later on the disappearance of the former territory holder.

There may also be a reproductive state associated with emigration, although if there is such a state its hormonal basis is not yet known. The behavior of emigrating animals, including locomotor activity and defense, is different from that of animals in the other reproductive states described above (Fairbairn, 1978; Myllymäki, Aho, Lind, & Tast, 1962). By analogy to birds, which have a migrational reproductive state associated with specific hormone changes (Meier, 1973), one might look for hormonal differences in emigrating muroid rodents that might underlie the difference between their behavior and that of the animals who remain behind.

The process by which an animal switches from one reproductive state to another may be initiated by behavioral events and mediated by motivational systems such as patrol/marking and submission. The courtship reproductive state may be initiated by social stimuli and mediated by the following process: Unfamiliar conspecific odors, particularly those of the opposite sex, activate the patrol/marking motivational mechanism, which activates, as one of its motor patterns, the release of gonadotropin, which initiates, in turn, the courtship reproductive state. The reproductive postponement state may also be initiated by another type of social stimuli and mediated by the following process: Severe or repeated defeat by a conspecific activates the submission motivational mechanism, which activates, as one of its motor patterns, the release of ACTH and corticosteroids, which in turn initiate the reproductive postponement state. Once a reproductive state is initiated, it maintains itself by positive feedback chains of hormone secretion until behavioral or other types of environmental stimuli (e.g., seasonal changes) initiate a different state.

CONCLUSION

Most of the effects of hormones on agonistic behavior in muroid rodents can be accounted for in terms of their actions on the six hypothetical neural mechanisms illustrated in Figure 1. They are the motivational mechanisms of patrol/marking and offense, the consociate modulator, and three sets of sensory analyzers tuned to pheromones that depend for their secretion on gonadal hormones. Although these mechanisms are still hypothetical, there are some data that suggest what may be their anatomical substrates. The uptake of corticosteroids and progestins in the region of the ventromedial hypothalamus is consistent with the suggestion that neurons of that nucleus are involved in the hypothetical consociate modulator. And the uptake of the gonadal hormones in the preoptic region of the hypothalamus is consistent with the suggestion that neurons in that region are involved in the motivational mechanism of patrol/marking and sensory analyzers for motivating stimuli of patrol/marking and offense.

The picture of hormonal action presented in Figure 1 is still very hypothetical. It depends on a neural organization of agonistic behavior consisting of four motivational systems and distinctions between components of motivational systems that have been hypothesized only recently and that have not yet received the hard tests of experimentation (Adams, 1979a). It also depends on data concerning the effects of hormones on agonistic behavior, which are still incomplete. For example, data on the effects of progestin on agonistic behavior of females has been obtained only in the past few years and is only preliminary. Similarly, the data on prolactin are still highly controversial and the data on the effects of estrogen on female offense are sketchy. No doubt, new experiments will force a revision of our knowledge of how hormones affect behavior.

Even if the picture presented here is correct in its basic outlines, it will no doubt turn out to be an oversimplification. We can expect that hormones act at more than six points in the motivational systems of agonistic behavior. For example, there are effects of hormones on peripheral receptive fields, mediated perhaps by the sympathetic nervous system (Bereiter, Stanford, & Barker, 1980). And we can expect that hormones besides those illustrated have effects on agonistic behavior. For example, there are reports that vasopressin (Leshner & Roche, 1977) and alpha-melanocyte stimulating hormone (Nowell, Thody, & Woodley, 1980) may have effects on submission and patrol/marking, respectively.

The picture is also oversimplified because the neural mechanisms of agonistic behavior operate within a larger framework of behaviors that are influenced by and play a role in establishing and maintaining the reproductive states of the individual. Not only agonistic behaviors but also sexual and parental behaviors are involved in these reproductive states and need to be considered because they may interact with agonistic behaviors.

Like all such theoretical formulations, the present one is intended to pose hypotheses and stimulate the design of critical experiments. Even if the picture presented here turns out to be incorrect, it will be useful if it leads to the generation of critical data that can be used to draw a new picture of the mechanisms of agonistic behavior that corresponds even more closely to its anatomical basis.

REFERENCES

Adams, D. B. The activity of single cells in the midbrain and of the cat during affective defense behavior. *Archives Italiennes de Biologie,* 1968, *106,* 243–269.

Adams, D. B. Defence and territorial behaviour dissociated by hypothalamic lesions in the rat. *Nature,* 1971, *232,* 573–574.

Adams, D. B. The relation of scent-marking, olfactory investigation, and specific postures in the isolation-induced fighting of rats. *Behaviour,* 1976, *56,* 286–297.

Adams, D. B. Brain mechanisms for offense, defense, and submission. *The Behavioral and Brain Sciences,* 1979, *2,* 201–241. (a)

Adams, D. B. Inborn and acquired aspects of offense and defense motivational systems in muroid rodents: Role of memory. In T. Oniani (Ed.), *Neurophysiological bases of memory.* Tbilisi, USSR: Metsniereba, 1979. (b)

Adams, D. B. Motivational systems of agonistic behaviour in muroid rodents: A comparative review and neural model. *Aggressive Behavior,* 1980, *6,* 295–346.

Anderson, P. K., & Hill, J. L. *Mus musculus:* Experimental induction of territory formation. *Science,* 1965, *148,* 1753–1755.

Arvola, A., Ilmén, M., & Koponen, T. On the aggressive behaviour of the Norwegian lemming *(Lemmus lemmus)* with special reference to the sounds produced. *Annales Zoologici Societatis Zoologicae Botanicae Fennicae "Vanamo"* 1962, *17,* 80–101.

Barr, G. A., Gibbons, J. L., & Moyer, K. E. Male-female differences and the influence of neonatal and adult testosterone on intraspecies aggression in rats. *Journal of Comparative and Physiological Psychology,* 1976, *90:* 1169–1183.

Bartke, A., Smith, M. S., Michael, S. D., Peron, F. G., & Dalterio, S. Effects of experimentally-induced chronic hyperprolectinemia on testosterone and gonadotrophin levels in male rats and mice. *Endocrinology,* 1977, *100,* 182–186.

Bean, N. J., & Conner, R. Central hormonal replacement and home-cage dominance in castrated rats. *Hormones and Behavior,* 1978, *11,* 100–109.

Bereiter, D. A., Stanford, L. R., & Barker, D. J. Hormone-induced enlargement of receptive fields in trigeminal mechanoreceptive neurons: II. Possible mechanisms. *Brain Research,* 1980, *184,* 411–423.

Bermond, B. *Neurohormonal regulation of aggressive and sexual behavior in the rat.* Unpublished doctoral dissertation, University of Amsterdam, 1978.

Bittman, E. L., & Goldman, B. D. Serum levels of gonadotrophins in hamsters exposed to short photoperiods: Effects of adrenalectomy and ovariectomy. *Journal of Endocrinology,* 1979, *83,* 113–118.

Blanchard, R. J., Fukunaga, K., Blanchard, D. C., & Kelley, M. M. Conspecific aggression in the laboratory rat. *Journal of Comparative and Physiological Psychology,* 1975, *39,* 1204–1209.

Bowen, D. W., & Brooks, R. J. Social organization of confined male collared lemmings (*Dicrostonyx groenlandicus* Traill). *Animal Behaviour,* 1978, *26,* 1126–1135.

Brain, P. F. Dividing up aggression and considerations in studying the physiological substrates of these phenomena. *The Behavioral and Brain Sciences,* 1979, *2,* 216.

Brain, P. F., & Evans, A. E. Acute influences of some ACTH-related peptides on fighting and adrenocortical activity in male laboratory mice. *Pharmacology Biochemistry & Behavior,* 1977, *7,* 425–433.

Bronson, F. H. Urine marking in mice: Causes and effects. In R. L. Doty (Ed.), *Mammalian olfaction, reproductive processes, and behavior.* New York: Academic Press, 1976.

Bronson, F. H., & Desjardins, C. Steroid hormones and aggressive behavior in mammals. In B. E. Eleftheriou & J. P. Scott (Eds.), *The physiology of aggression and defeat.* New York: Plenum Press, 1971.

Brooks, R. J., & Banks, E. M. Behavioural biology of collared lemming *(Dicrostonyx groenlandicus):* An analysis of acoustic communication. *Animal Behaviour Monographs,* 1973, *6,* 1–83.

Brown, R. E. Odor preference and urine-marking scales in male and female rats: Effects of gonadectomy and sexual experience on response to conspecific odors. *Journal of Comparative and Physiological Psychology,* 1977, *91,* 1190–1206.

Buntin, J. D., Catanzaro, C., & Lisk, R. D. *Elevation of prolactin levels by pituitary transplantation: Effects on intraspecific aggression in female gold hamsters (Mesocricetus auratus).* Paper presented at the meeting of the Animal Behavior Society, New Orleans, June 1979.

Calhoun, J. B. *The ecology and sociology of the Norway rat* (U. S. Department of Health, Education, and Welfare P. H. S. Monograph No. 1008). Washington, D. C.: U. S. Government Printing Office, 1962.

Carl, E. A. Population control in arctic ground squirrels. *Ecology,* 1971, *52,* 395–413.

Caroom, D., & Bronson, F. H. Response of female mice to preputial attractant: Effects of sexual experience and ovarian hormones. *Physiology and Behavior,* 1971, *7,* 659–662.

Clarke, J. R. The aggressive behaviour of the vole. *Behaviour,* 1956, *9,* 1–23.

Colvin, G. B. & Sawyer, C. H. Induction of running activity by intracerebral implants of estrogen in ovariectomized rats. *Neuroendocrinology,* 1969, *4,* 309–320.

Conner, R. L., & Levine, S. The effects of adrenal hormones on the acquisition of signaled avoidance behavior. *Hormones and Behavior,* 1969, *1,* 73–83.

Crabtree, J. M., & Moyer, K. E. Sex differences in fighting and defense induced in rats by shock and d-amphetamine during morphine abstinence. *Physiology and Behavior,* 1973, *11,* 337–343.

Crowcroft, P. Territoriality in wild house mice, *Mus musculus* L. *Journal of Mammalogy,* 1955, *36,* 299–301.

Dewsbury, D. A. Estrous cycles of nine species of muroid rodents. *Journal of Mammalogy,* 1977, *58,* 89–92.

Ebert, P. D., & Hyde, J. S. Selection for agonistic behavior in wild female *Mus musculus. Behavior Genetics,* 1976, *6,* 291–304.

Edwards, D. A. Early androgen stimulation and aggressive behavior in male and female mice. *Physiology and Behavior,* 1969, *4,* 333–338.

Edwards, D. A., & Rowe, F. A. Neural and endocrine control of aggressive behavior. In B. E. Eleftheriou & R. S. Sprott (Eds.), *Hormonal correlates of behavior.* New York: Plenum Press, 1975.

Edwards, M. A., & Adams, D. B. Role of midbrain central gray in pain-induced defensive boxing of rats. *Physiology and Behavior,* 1974, *13,* 113–121.

Ellis, G. B., & Turek, F. W. Photoperiodic regulation of serum luteinizing hormone, and follicle-stimulating hormone in castrated and castrated-adrenalectomized male hamsters. *Endocrinology,* 1980, *106,* 1338–1344.

Engbretson, G. A., & Livezey, R. L. The effects of aggressive display on body temperature in the fence lizard *Sceleporus occidentalis occidentalis* Baird and Girard. *Physiological Zoology,* 1972, *45,* 247–254.

Erskine, M. S., Barfield, R. J., & Goldman, B. D. Postpartum aggression in rats: I. Effects of hypophysectomy. *Journal of Comparative and Physiological Psychology,* 1980, *94,* 484–494. (a)

Erskine, M. S., Barfield, R. J., & Goldman, B. D. Postpartum aggression in rats: II. Dependence on maternal sensitivity to young and effects of experience with pregnancy and parturition. *Journal of Comparative and Physiological Psychology*, 1980, *94*, 495–505. (b)

Ewer R. F. The behaviour of the African giant rat (*Cricetomys gambianus* Waterhouse). *Zeitschrift für Tierpsychologie*, 1967, *24*, 6–79.

Ewer, R. F. The biology and behaviour of a free-living population of black rats *(Rattus rattus)*. *Animal Behaviour Monographs*, 1971, *4*, 127–174.

Fairbairn, D. J. Behaviour of dispersing deer mice *(Peromyscus maniculatus)*. *Behavioral Ecology and Sociobiology*, 1978, *3*, 265–282.

Floody, O. R., & Pfaff, D. W. Aggressive behavior in female hamsters: The hormonal basis for fluctuations in femal aggressiveness correlated with estrous state. *Journal of Comparative and Physiological Psychology*, 1977, *91*, 443–464.

Frank, F. The causality of microtine cycles in Germany. *The Journal of Wildlife Management*, 1957, *21*, 113–121.

Fredericson, E., & Birnbaum, E. A. Competitive fighting between mice with different hereditary backgrounds. *The Journal of Genetic Psychology*, 1954, *85*, 271–280.

Gandelman, R. Mice: Postpartum aggression elicited by the presence of an intruder. *Hormones and Behavior*, 1972, *3*, 23–28.

Gleason, P. E., Michael, S. D., & Christian, J. J. Effects of gonadal steroids on agonistic behavior of female *Peromyscus leucopus*. *Hormones and Behavior*, 1979, *12*, 30–39.

Grant, E. C., & Mackintosh, J. H. A comparison of the social postures of some common laboratory rodents. *Behaviour*, 1963, *21*, 246–259.

Gurnell, J. Neutral cage behavioural interaction in wild Wood mice, *Apodemus sylvaticus* (Linne, 1758). *Saugetierkundliche Mitteilung*, 1977, *25*, 57–66.

Haag, C., Jerhoff, B., & Kirkpatrick, J. F. Ovarian hormones and their role in aggression inhibition among male mice. *Physiology and Behavior*, 1974, *13*, 175–177.

Hansult, C. D., & Kessler, S. Role of hormones in maternal aggression and infant care. *Behavior Genetics*, 1977, 7, 65.

Huck, U. W., Carter, C. S., & Banks, E. M. Estrogen and progesterone interactions influencing sexual and social behavior in the brown lemming, *Lemmus trimucronatus*. *Hormones and Behavior*, 1979, *12*, 40–49.

Hucklebridge, F. H., & Nowell, N. W. Plasma catecholamine response to physical and psychological aspects of fighting in mice. *Physiology and Behavior*, 1974, *13*, 35–40.

Hutchinson, R. R., Ulrich, R. E., & Azrin, N. H. Effects of age and related factors on the pain-aggression reaction. *Journal of Comparative and Physiological Psychology*, 1965, *59*, 365–369.

Hyde, J. S., & Sawyer, T. F. Estrous cycle fluctuations in aggressiveness of house mice. *Hormones and Behavior*, 1977, *9*, 290–295.

Iberall, A. S., & McCulloch, W. S. The organizing principle of complex living systems. *Journal of Basic Engineering*, 1969, *91*, 290–294.

Johst, V. V. Vergleichende Untersuchung des agonistischen Verhaltens einiger Arten von *Clethrionomys*. *Zeitschrift für Tierpsychologie*, 1967, *24*, 558–579.

Kamano, D. K. Enhancement of learned fear with epinephrine. *Psychonomic Science*, 1968, *12*, 331.

Kanki, J. P., & Adams, D. B. Ventrobasal thalamus necessary for visually-released defensive boxing of rat. *Physiology and Behavior*, 1978, *21*, 7–12.

King, J. M. Effects of lesions of the amygdala, preoptic area, and hypothalamus on estradiol-induced activity in the female rat. *Journal of Comparative and Physiological Psychology*, 1979, *93*, 360–367.

Korn, J. H., & Moyer, K. E. Behavioral effects of isolation in the rat: The role of sex and time of isolation. *The Journal of Genetic Psychology*, 1968, *113*, 263–273.

Krieger, D. T., & Liotta, A. S. Pituitary hormones in brain: Where, how, and why? *Science*, 1979, *205*, 366–372.

Lehman, M. N., & Adams, D. B. A statistical and motivational analysis of the social behaviors of the male laboratory rat. *Behaviour*, 1977, *61*, 238–275.

Leshner, A. I. A model of hormones and agonistic behavior. *Physiology and Behavior*, 1975, *15*, 225–235.

Leshner, A. I., & Moyer, J. A. Androgens and agonistic behavior in mice: Relevance to aggression and irrelevance to avoidance-of-attack. *Physiology and Behavior,* 1975, *15,* 695–699.

Leshner, A. I., & Roche, K. E. Comparison of the effects of ACTH and lysine vasopressin on avoidance-of-attack in mice. *Physiology and Behavior,* 1977, *18,* 879–883.

Leshner, A. I., Walker, W. A., Johnson, A. E., Kelling, J. S., Kreisler, S. J., & Svare, B. B. Pituitary adrenocortical activity and intermale aggression in isolated mice. *Physiology and Behavior,* 1973, *11,* 705–711.

Leventhal, G. S., & Killackey, H. Adrenaline stimulation and preference for familiar stimuli. *Journal of Comparative and Physiological Psychology,* 1968, *65,* 152–155.

Linsdale, J. M., & Tevis, L. P. *The dusky footed wood rat (Neotoma fuscipes).* Los Angeles: University of California Press, 1951.

Mackintosh, J. H. Territory formation by laboratory mice. *Animal Behaviour,* 1970, *18,* 177–183.

Mann, M., Michael, S., & Svare, B. Ergot drugs suppress plasma prolactin and lactation but not aggression in parturient mice. *Hormones and Behavior,* 1980, *14,* 319–328.

Maruniak, J. A., Owen, K., Bronson, F. H., & Desjardins, C. Urinary marking in female house mice: Effects of ovarian steroids, sex experience, and type of stimulus. *Behavioral Biology,* 1975, *13,* 211–217.

Matte, A. C. Biphasic and dissociated effects of ACTH on motor activity, aggression and "emotionality" in mice. *Psychoneuroendocrinology,* 1979, *4,* 21–27.

McCarty, R., & Kopin, I. J. Changes in plasma catecholamines and behavior of rats during the anticipation of footshock. *Hormones and Behavior,* 1978, *11,* 248–257.

Meier, A. H. Daily hormone rhythms and seasonal conditions in a sparrow. *American Scientist,* 1973, *61,* 184–191.

Mink, J. W., Adams, D. B., & Sinnamon, H. M. Activity of neurons in the rat during basal forebrain motivated behaviors. *Behavioral Brain Research,* in press.

Morin, L. P., Fitzgerald, K. M., Rusak, B., & Zucker, I. Circadian organization and neural mediation of hamster reproductive rhythms. *Psychoneuroendocrinology,* 1977, *2,* 73–98.

Moyer, J. A., & Leshner, A. I. Pituitary-adrenal effects on avoidance-of-attack in mice: Separation of the effects of ACTH and corticosterone. *Physiology and Behavior,* 1976, *17,* 297–301.

Mugford, R. A., & Nowell, N. W. Pheromones and their effect on aggression in mice. *Nature,* 1970, *226,* 967–968.

Myllymäki, A., Aho, J., Lind, E. A., & Tast, J. Behaviour and daily activity of the Norwegian lemming, *Lemmus lemmus* during autumn migration. *Annales Zoologici Societatis Zoologicae Botanicae Fennicae "Vanamo,"* 1962, *24,* 1–31.

Nock, B. L., & Leshner, A. I. Hormonal mediation of effects of defeat on agonistic responding in mice. *Physiology and Behavior,* 1976, *17,* 111–119.

Noirot, E., Goyens, J., & Buhot, M. Aggressive behavior of pregnant mice toward males. *Hormones and Behavior,* 1975, *6,* 9–17.

Nowell, N. W., Thody, A. J., & Woodley, R. Alpha-melanocyte stimulating hormone and aggressive behavior in male mouse. *Physiology and Behavior,* 1980, *24,* 5–9.

Owen, K., Peters, P. J., & Bronson, F. H. Effects of intracranial implants of testosterone proprionate on intermale aggression in the castrated male mouse. *Hormones and Behavior,* 1974, *5,* 83–92.

Payne, A. P., & Swanson, H. H. Hormonal control of aggressive dominance in the female hamster. *Physiology and Behavior,* 1971, *6,* 355–357.

Pedersen, C. A. *Induction of maternal behavior by posterior pituitary hormones.* Paper presented at the Eastern Conference on Reproductive Behavior, New York, June 1980.

Pfaff, D. W., & Pfaffman, C. Olfactory and hormonal influences on the basal forebrain of the male rat. *Brain Research,* 1969, *15,* 137–156.

Pfaff, D., Lewis, C., Diakow, C., & Keiner, M. Neurophysiological analysis of mating behavior responses as hormone-sensitive reflexes. *Progress in Physiological Psychology,* 1973, *5,* 253–297.

Politch, J. A., & Herrenkohl, L. R. *Maternal aggression in mice: The role of prolactin.* Paper presented at the Eastern Conference on Reproductive Behavior, 1980.

Pond, F. J., Sinnamon, H. M., & Adams, D. B. Single unit recording in the midbrain of rats during shock-elicited fighting behavior. *Brain Research,* 1977, *120,* 469–484.

Price, E. O. Hormonal control of urine-marking in wild and domestic Norway rats. *Hormones and Behavior,* 1975, *6,* 393–397.

Richards, M. M. Activity measured by running wheels and observation during the oestrus cycle, pregnancy, and pseudopregnancy in the golden hamster. *Animal Behaviour,* 1966, *14,* 450–458.

Ropartz, P. Effet de l'odeur de l'urine de male sur des groupes de souris femelles. *Comptes Rendus de l'Academie des Sciences, Paris,* 1970, *271:* 1571–1574.

Rose, R. K., & Gaines, M. S. Levels of aggression in fluctuating populations of the prairie vole, *Microtus ochrogaster,* in eastern Kansas. *Journal of Mammalogy,* 1976, *57,* 43–57.

Rosenson, L. M., & Asheroff, A. K. Maternal aggression in CD-1 mice: Influence of the hormonal condition of the intruder. *Behavioral Biology,* 1975, *15,* 219–224.

Rowley, M. H., & Christian, J. J. Intraspecific aggression of *Peromyscus leucopus. Behavioral Biology,* 1976, *17,* 249–253.

Sandner, G., Schmitt, P., & Karli, P. Central gray and medial hypothalamic stimulation: Correlation between escape behavior and unit activity. *Brain Research,* 1979, *170,* 459–474.

Scott, J. P. Agonistic behavior of mice and rats: A review. *American Zoologist,* 1966, *6,* 683–701.

Shillito, E. E. Exploratory behaviour in the short-tailed vole *Microtus agrestis. Behaviour,* 1963, *21,* 145–154.

Singer, J. J. Hypothalamic control of male and female sexual behavior in female rats. *Journal of Comparative and Physiological Psychology,* 1968, *66,* 738–742.

Soares, M. J., Kalberer, W. D., & Erpino, M. J. Progesterone: Effects on investigatory preferences, aggression, and olfaction in orchidectomized, testosterone-treated mice. *Behavioral Biology,* 1978, *23,* 260–266.

Stern, J. M., Erskine, M. S., & Levine, S. Dissociation of open-field behavior and pituitary-adrenal function. *Hormones and Behavior,* 1973, *4,* 149–162.

Stinson, R. H. *Effects on some environmental factors on the behavior of Peromyscus.* Unpublished doctoral dissertation, University of Toronto, 1952.

Stumpf, W. E., & Sar, M. Steroid hormone target sites in the brain: The differential distribution of estrogen, progestin, androgen and gluco-corticosteroid. *Journal of Steroid Biochemistry,* 1976, *7,* 1163–1170.

Swanson, H. H. Sex differences in behaviour of the Mongolian gerbil *(Meriones unguiculatus)* in encounters between pairs of same or opposite sex. *Animal Behaviour,* 1974, *22,* 638–644.

Telle, H. J. Contribution to the knowledge of behavioural patterns in two species of rats, *Rattus norvegicus* and *Rattus rattus. Zeitschrift für Angewandte Zoologie,* 1966, *53,* 129–196. (Translated by National Research Council of Canada. Technical Translation 1608.)

Thiessen, D. D., Friend, H. C., & Lindzey, G. Androgen control of territorial marking in the Mongolian gerbil. *Science,* 1968, *160,* 432–434.

Thiessen, D. D., Lindzey, G., & Nyby, J. The effects of olfactory deprivation and hormones on territorial marking in the male Mongolian gerbil *(Meriones unguiculatus). Hormones and Behavior,* 1970, *1,* 315–326.

Thor, D. H. Enhanced social docility in male hooded rats by dermal cautery of the vibrissal pads. *Aggressive Behavior,* 1976, *2,* 39–53.

Thor, D. H., & Flannelly, K. J. Intruder gonadectomy and elicitation of territorial aggression in the rat. *Physiology and Behavior,* 1976, *17,* 725–727.

Vandenbergh, J. G. The effects of gonadal hormones on the aggressive behaviour of adult golden hamsters *(Mesocricetus auratus). Animal Behaviour,* 1971, *19,* 589–594.

Vick, L. H. & Banks, E. M. The estrous cycle and related behavior in the Mongolian gerbil *(Meriones unguiculatus). Communications in Behavioral Biology,* 1969, *3,* 117–124.

Wade, G. N., & Zucker, I. Modulation of food intake and locomotor activity in female rats by diencephalic hormone implants. *Journal of Comparative and Physiological Psychology,* 1970, *72,* 328–336.

Wang, G. H. The relation between "spontaneous" activity and estrous cycle in the white rat. *Comparative Psychology Monographs,* 1923, *2,* 1–27.

Whalen, R. E. Hormone-behavior analysis. In R. E. Whalen (Ed.), *The neuropsychology of aggression.* New York: Plenum Press, 1974.

Wise, D. A., & Pryor, T. L. Effects of ergocornine and prolactin on aggression in postpartum golden hamster. *Hormones and Behavior,* 1977, *8,* 30–39.

Wolff, P. R., & Powell, A. J. Urination patterns and estrous cycling in mice. *Behavioral and Neural Biology,* 1979, *27,* 379–383.

Wood, A. E. Eocene radiation and phylogeny of the rodents. *Evolution,* 1959, *13,* 354–361.

Wood, A. E. Grades and clades among rodents. *Evolution,* 1965, *19,* 115–130.

Yahr, P. Central control of scent marking. In D. Müller-Schwarze & M. M. Mozell (Eds.), *Chemical signals in vertebrates.* New York: Plenum Press, 1977.

Zook, J. M., & Adams, D. B. Competititve fighting in the rat. *Journal of Comparative and Physiological Psychology,* 1975, *88,* 418–423.

11

Molecular Mechanisms of Steroid Hormone Actions in the Brain

William G. Luttge

As evidenced by the diversity and comprehensiveness of the other chapters in this book, the hypothesis that endogenous and exogenous steroid hormones can have direct effects on neurons in specific regions of the brain that may in turn precipitate and/or regulate the expression of aggressive behaviors is now widely accepted, even though we still know surprisingly little about the molecular mechanisms of these hormone actions. This paucity of information is striking in view of the long history of experimental behavioral neuroendocrinology dating back for literally thousands of years, to when castration was first used to curb sexual and aggressive behaviors in man and other animals (see Luttge, 1971). Over a century ago, Berthold (1849/1944) was the first to report that castration and subsequent testis transplantation could reversibly inhibit sexual, aggressive, and crowing behaviors in roosters, thus setting the pattern for future studies on the endocrine bases of these and other behaviors in many species. The present chapter reviews recent literature on the molecular mechanisms of estrogen, progestin, androgen, and glucocorticoid hormone actions in the mammalian brain. Each of these classes of steroids has been shown to influence the display of agonistic behaviors in mammals. No attempt will be made to relate the biochemical and behavioral actions of these steroids, because this material is reviewed in other chapters in this book. This chapter also excludes detailed discussions of the involvement of steroid hormones in the regulation of nonagonistic behaviors or the secretion of adenohypophysial hormones, since these subjects have been thoroughly reviewed in other recent books (e.g., Adler & Pfaff, 1983; Beyer, 1979; Motta, 1980; Nemeroff & Dunn, 1983). It is hoped that this brief summary will stimulate further research into the neurochemical mechanisms of steroid actions in aggression.

WILLIAM G. LUTTGE • Department of Neuroscience and Center for Neurobiological Sciences, University of Florida College of Medicine, Gainesville, Florida 32610.

OVERVIEW OF STEROID HORMONE MECHANISMS OF ACTION IN THE BRAIN

Figure 1 presents a diagrammatic summary of steroid actions in the brain. Since extensive work with gonadal and adrenal hormones in peripheral tissues strongly supports a "genomic" mechanism of action, it is not surprising that these responses have also received considerable attention in studies with brain tissues. The genomic mechanism requires that steroids enter the cell in order to initiate the required nuclear biochemical events. This uptake occurs in "target" (i.e., high-affinity, receptor-containing) and nontarget cells, appears to be mediated by diffusion (and possibly carrier-mediated transport—see Allera, Rao, & Breuer, 1980), and is dependent on the relative concentration of the free steroids in the CSF and vascular sources as well as on their solubility in the lipid-rich plasma membranes. Steroids bound to albumin readily dissociate, thereby facilitating their transport across the blood–brain barrier and neuronal plasma membranes, while steroids bound to certain transport proteins (e.g., corticosteroid- and tes-

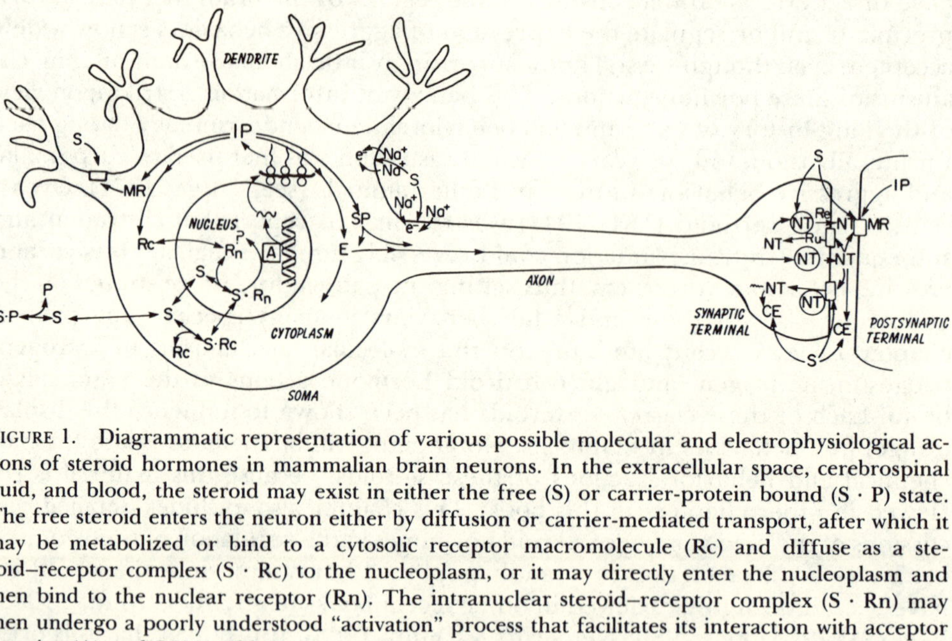

FIGURE 1. Diagrammatic representation of various possible molecular and electrophysiological actions of steroid hormones in mammalian brain neurons. In the extracellular space, cerebrospinal fluid, and blood, the steroid may exist in either the free (S) or carrier-protein bound (S · P) state. The free steroid enters the neuron either by diffusion or carrier-mediated transport, after which it may be metabolized or bind to a cytosolic receptor macromolecule (Rc) and diffuse as a steroid–receptor complex (S · Rc) to the nucleoplasm, or it may directly enter the nucleoplasm and then bind to the nuclear receptor (Rn). The intranuclear steroid–receptor complex (S · Rn) may then undergo a poorly understood "activation" process that facilitates its interaction with acceptor proteins (A) and DNA. This S · Rn · A interaction in the chromatin may in turn precipitate the production of specific mRNAs that code for the production of specific "induced proteins" (IP) and enzymes (E). Examples of IPs include the 39K- to 45K-dalton E_2IP, structural proteins (SP) (e.g., tubulin), membrane receptor proteins (MR) (e.g., MRs for DA, β-adrenergic agents, catechol estrogens, and glucocorticoids), cytoplasmic receptor proteins for steroids (Rc) (e.g., E_2 and P_4 receptors), and various anabolic and catabolic enzymes (CE) (e.g., ChAcT). Release (Re) and reuptake (Ru) of neurotransmitters (NT) may also be modulated by the direct actions of steroids. Last, steroids may interact directly with the neuronal soma or axonal plasma membrane to modulate the transmembrane diffusion or distribution of ions (e.g., Na^+) and thereby produce an electrophysiological response (e^-) or alter the conduction velocity of a response.

tosterone-binding globulins) can be effectively excluded from the brain (e.g., Pardridge & Meitus, 1979). Once in the neuronal cytoplasm, the hormones may undergo a variety of metabolic transformations (see Figures 2 and 3), which may increase or decrease their genomic effectiveness and thus provide additional regulatory possibilities for the target cells. The steroid may then form noncovalent high-affinity attachments with specific "receptor" sites on soluble protein-rich cytoplasmic macromolecules. The differential binding affinities and capacities of these receptors for various steroids provides the basis of the target cell- and steroid-specificity of the genomic mechanism. The transiently established hormone–receptor complex can diffuse through the nuclear pores into the nucleoplasm and undergo an as yet poorly understood "activation" process (which may also occur in the cytoplasm), thus facilitating subsequent interactions with nuclear "acceptors" that regulate the production of various species of RNA, including those coding for proteins required for the elicitation and/or maintenance of hormone-dependent behaviors. The steroid, receptor, and acceptor molecules may then be catabolized, conjugated, or in some other way "processed" in the nucleoplasm, or they may be returned to the cytoplasm (or nucleoplasm) for reutilization. Receptor reutilization (or even their initial utilization following *de novo* synthesis) may, however, require "capacitation" in either the cyto- or nucleoplasm prior to its involvement in hormone binding.

Even though the genomic mechanisms of steroid actions are clearly the most widely studied, there are viable alteratives with potential importance in the brain, including the possibility that steroids may interact with neuronal membrane receptors in much the same manner as peptidergic hormones and neurotransmitters. Compared with cytosolic and nuclear receptors, there is little direct evidence for intracerebral membrane-associated receptors, although there is a wealth of data on steroid-mediated alterations in postsynaptic firing and presynaptic neurotransmitter release rates. Pre- and postsynaptic modulations in the efficacy of chemical neurotransmission can also result from steroid-induced changes in the activity of neurotransmitter synthetic and catabolic enzymes. Last, the considerable energy requirements for these genomic and nongenomic mechanisms of hormone action must be supplied by intracerebral carbohydrate metabolism, which in turn also appears to be directly and/or indirectly modified by the actions of steroid hormones. Each of these aspects of the intracerebral mechanisms of steroid action will be discussed more fully below. This material has also been extensively reviewed in other recent publications (e.g., Luttge, 1979, 1983; Whalen, Yahr, & Luttge, 1983).

NEUROANATOMICAL DISTRIBUTION OF STEROID-RECEPTORS

The recent development of high-resolution autoradiographic techniques for diffusible substances, together with the increased availability of high-specific-activity tritiated steroid, has facilitated the proliferation of detailed investigations of the neuroanatomical distribution of estrogen-, progestin-, androgen-, and glucocorticoid-retaining cells in a wide variety of vertebrates (see reviews by Kim,

TABLE I
Steroid Hormone Names and Abbreviations

IUPAC names	Abbreviations	Trivial names
5α-Androstan-3,17-dione	5αAA	5α-Androstanedione
5α-Androstan-3α,17β-diol	3α-Adiol	3α-Androstanediol
5α-Androstan-3β,17β-diol	3β-Adiol	3β-Androstanediol
5α-Androstan-3α-ol-17-one	A	Androsterone
5α-Androstan-17β-ol-3-one	DHT	Dihydrotestosterone
5β-Androstan-3,17-dione	5β-AA	5β-Androstanedione
4-Androsten-17β-ol-3-one	T	Testosterone
4-Androsten-17α-ol-3-one[a]	Epi-T	Epitestosterone
4-Androsten-17β,19-diol-3-one	19-OH-T	19-Hydroxytestosterone
4-Androsten-17β-ol-3,19-dione	19-OXO-T	19-Oxotestosterone
4-Androsten-2β,17β-diol-3,19-dione	2-OH-19-OXO-T	2-Hydroxy-19-testosterone
4-Androsten-3,17-dione	AE	Androstenedione
4-Androsten-19-ol-3,17-dione	19-OH-AE	19-Hydroxyandrostenedione
4-Androsten-3,17,19-trione	19-OXO-AE	19-Oxoandrostenedione
4-Androsten-2β-ol-3,17,19-trione	2-OH-19-OXO-AE	2-Hydroxy-19-oxoandrostenedione
5-Androsten-3β-ol-17-one	DHEA	Dehydroepiandrosterone
5-Androsten-3β,17β-diol	AEDiol	5-Androstendiol
1,3,5(10)-Estratrien-3-ol-17-one	E_1	Estrone
1,3,5(10)-Estratrien-3,17α-diol[a]	17α-E_2	17α-Estradiol
1,3,5(10)-Estratrien-3,17β-diol	E_2	Estradiol
1,3,5(10)-Estratrien-3,16α-17β-triol	E_3	Estriol

1,3,5(10)-Estratrien-2,3-diol-17-one	2-OH-E_1	2-Catechol estrone
1,3,5(10)-Estratrien-3,4-diol-17-one	4-OH-E_1	4-Catechol estrone
1,3,5(10)-Estratrien-2,3,17β-triol	2-OH-E_2	2-Catechol estradiol
1,3,5(10)-Estratrien-3,4,17β-triol	4-OH-E_2	4-Catechol estradiol
4-Estren-17β-ol-3-one[a]	19-Nor-T	19-Nortestosterone
5α-Pregnan-3,20-dione	5α-DHP	5α-Dihydroprogesterone
5α-Pregnan-3α-ol-20-one	3α-OH-5α-DHP	3α-Hydroxydihydroprogesterone
5α-Pregnan-3α,20α-diol	3α,20α-OH-5α-DHP	3α,20α-Dihydroxydihydroprogesterone
5α-Pregnan-20α-ol-3-one	20α-OH-5α-DHP	20α-Hydroxydihydroprogesterone
5β-Pregnan-3,20-dione	5β-DHP	5β-Dihydroprogesterone
4-Pregnen-3,20-dione	P_4	Progesterone
4-Pregnen-11β,17α,21-triol-3,20-dione	F	Cortisol
4-Pregnen-11β,21-diol-3,20-dione	B	Corticosterone
4-Pregnen-11β,21-diol-3,18,20-trione	Aldo	Aldosterone
4-Pregnen-17α-ol-3,20-dione	17α-OH-P_4	17α-Hydroxyprogesterone
4-Pregnen-17α,21-diol-3,20-dione	DOF	Cortexolone
4-Pregnen-17α,21-diol-3,11,20-trione	E	Cortisone
4-Pregnen-20α-ol-3-one	20α-OH-P_4	20α-Hydroxyprogesterone
4-Pregnen-21-ol-3,20-dione	DOC	Desoxycorticosterone
5-Pregnen-3β-ol-20-one	Pg	Pregnenolone
5-Pregnen-3β,17α-diol-20-one	17α-OH-Pg	17α-Hydroxypregnenolone
1,4-Pregnadien-9-fluoro-11β,16α,17α,21-tetrol-3,20-dione,16,17-acetonide[a]	TA	Triamcinolone acetonide
1,4-Pregnadien-9-fluoro-16α-methyl-11β,17α,21-triol-3,20-dione[a]	Dex	Dexamethasone
19-nor-4,9-Pregnadien-17,21-dimethyl-3,20-dione[a]	R5020	Promegestone

[a] Structure not shown in Figure 2 or Figure 3.

Stumpf, Sar, & Martinez-Vargas, 1978; Morrell & Pfaff, 1978; Stumpf & Sar, 1978; Warembourg, 1977; and the book by Stumpf & Grant, 1975). Thus, following the systemic administration of [^{3}H]estradiol (E_2), [^{3}H]testosterone (T), or [^{3}H]dihydrotestosterone (DHT) (see Figure 3 for structural formulas and Table I for IUPAC names and abbreviations), silver grains in the autoradiographs concentrate over neuronal nuclei primarily in certain hypothalamic (e.g., AH, Arc, PeVN, PM, and VM) (see Table II for neuroanatomical abbreviations), preoptic (e.g., MPOA, POA-PeVN, and POA-SCN), septal (e.g., DB, LS, and NIST), and amygdaloid (e.g., CAmg and MAmg) nuclei or in the homologous structures in nonmammalian species (see Table III for rat autoradiographic patterns). The recent discovery that progestin-receptor concentrations are greatly magnified following estrogen-priming (see below), coupled with the synthesis of an extremely potent progestin ligand, R5020 (e.g., Raynaud, 1977), has dramatically increased our knowledge of the distribution of progestin-receptors in the brain (see Table III for rat examples). Studies with the naturally occurring progestins, progesterone (P_4), and dihydroprogesterone (5α-DHP) (see Table I and Figure 2) have been plagued by the comparative instability and short half-life of the steroid-receptor complexes. Thus, in recent work with [^{3}H]R5020 (and [^{3}H]P_4), estrogen-sensitive, limited capacity, progestin-labeled neuronal nuclei were found to be localized in many of the same hypothalamic, preoptic, and septal, but not amygdaloid, nuclei as described earlier for estrogens and androgens (see Table III).

The autoradiographic evidence suggests that glucocorticoid-receptor distribution in the brain is radically different from that described for estrogens, androgens, and progestins (see Table III). Following systemic injections of [^{3}H]corticosterone (B) (see Table I and Figure 2) in adrenalectomized rats, silver

TABLE II
Abbreviations for Neuroanatomical Structures

Abbreviation	Structure
AH	Anterior hypothalamus
Amg	Amygdala
AP	Area postrema
APS	Anterior perforated substance
Arc	Arcuate nucleus (hypothalamus)
BS	Brain stem
CA	Cornu ammonis (hippocampus)
CAmg	Cortical amygdaloid nucleus
Cb	Cerebellum
CDB	Caudal diagonal band nucleus
CG	Central gray (NE-containing areas)
CMAmg	Corticomedial amygdaloid nucleus
CN	Caudate nucleus
CP	Cerebral peduncle
Cx	Cerebral cortex
DB	Diagnonal band nucleus
DG	Dentate gyrus (hippocampus)

TABLE II (Continued)

Abbreviation	Structure
DHpc	Dorsal hippocampus
DM	Dorsomedial nucleus (hypothalamus)
DVDB	Dorsoventral diagonal band nucleus
FB	Forebrain
Hb	Habenula
Hpc	Hippocampus
HTh	Hypothalamus
IP	Interpeduncular nucleus
LDAR-A8	Lateral dopamine region A-8
LH	Lateral hypothalamus
LNER-A7	Lateral norepinephrine region A-7
LPOA	Lateral preoptic area
LS	Lateral septal nucleus
LVM	Lateral ventromedial nucleus (hypothalamus)
MB	Midbrain
MBH	Medial basal hypothalamus
MDAR-A10	Medial dopamine region A-10
MDH	Medial dorsal hypothalamus
ME	Median eminence
Med	Medulla
Mes	Mesencephalon
MFB	Medial forebrain bundle
MH	Middle hypothalamus
MPOA	Medial preoptic area
MVM	Medial ventromedial nucleus (hypothalamus)
NA	Nucleus accumbens
NISM	Nucleus interstitialis stria medullaris
NIST	Nucleus interstitialis stria terminalis
OB	Olfactory bulb
PAG	Periaqueductal gray
PaVN	Paraventricular nucleus (hypothalamus)
PeVN	Periventricular nucleus (hypothalamus)
PFCx	Prefrontal cortex
PH	Posterior hypothalamus
PM	Premammillary nucleus (hypothalamus)
PMAmg	Posterior medial amygdaloid nucleus
POA	Preoptic area
Raphe	Raphe nuclei
RDB	Rostral diagonal band nucleus
SCN	Suprachiasmatic nucleus
Sep	Septal area
SM	Supramammillary nucleus
SN	Substantia nigra
SON	Supraoptic nucleus
Str	Striatum
TD	Telencephalon–diencephalon
Th	Thalamus
VM	Ventromedial nucleus (hypothalamus)
VMB	Ventral midbrain
VS	Ventral subiculum
VTA	Ventral tegmental area

TABLE III

Representative Autoradiographic Patterns of Steroid Hormone Localization in Rat Brain[a]

	Hypothalamus							Preoptic area			Septal Area			Amygdala		Hippocampus		
Steroid	PeVN	VM	Arc	PM	AH	LH	PH	MPOA	SCN	PeVN	LS	NIST	DB	CAmg	MAmg	CA	DG	References
$[^3H]E_2$	++	++	++	++	++	+	+	++	++	++	+	++	+	++	++	–	–	Pfaff & Keiner, 1973
$[^3H]$DHT	++	+	+	++[b]	+[c]	+[c]	+[c]	++[b]	+	+	+[b]	++	+	+	++	+	–	Sar & Stumpf, 1977
$[\alpha\text{-}^3H]$T[e]	++	++	++	++	++	+	+	++	++	++	+	++	+	+	++	–	–	Sheridan, 1979
$[\beta\text{-}^3H]$T[d]	–	–	–	+[e]	+[f]	–	–	–	–	–	–	++	–	–	+	–	–	Sheridan, 1979
$[^3H]$R5020	++	++	++	++	+	–	–	+	++	++	–	++	+	–	–	–	–	Warembourg, 1978
$[^3H]$B	–	–	–	–	–	–	–	–	–	–	+	–	–	+	–	++	+	Warembourg, 1975a

[a] See Tables I and II for steroid and neuroanatomical abbreviation definitions.
[b] Labeling density reduced following pretreatment with E_2.
[c] Only dispersed labeled cells found in this nucleus.
[d] The dramatically different autoradiographic patterns derived with $[\alpha\text{-}^3H]$T and $[\beta\text{-}^3H]$T occur because the $2\beta\text{-}^3H$ is lost during the aromatization of T to E_2. Thus, the autoradiographic pattern with $[\beta\text{-}^3H]$ should be more representative of true androgen binding (i.e., presumably similar to that of $[^3H]$DHT) than that of $[\alpha\text{-}^3H]$T, which is confounded with aromatization-produced $[^3H]E_2$. Note, however, that $[^3H]$DHT and $[\beta\text{-}^3H]$T labeling patterns are very different in several nuclei (e.g., PeVN, VM, Arc, SCN, DB, CAmg, and MAmg), while in other nuclei the reduction in the $[^3H]$DHT labeling with E_2 pretreatment (e.g., PM, MPOA, and LS) suggests that some of the binding may be due to association with E_2 receptors.
[e] Labeled cells in this nucleus found in only three of the eight rats injected with $[\beta\text{-}^3H]$T.
[f] Labeled cells found in only the dorsolateral portion of this nucleus.

grains in the autoradiographs are not found to concentrate over neuronal nuclei in the hypothalamus and preoptic area; instead, they concentrate in the hippocampus (primarily over the pyramidal cells in the CA_1 and CA_2 fields of Ammon's horn and the granule neurons of the dentate gyrus) and to a lesser extent in the septal area (e.g., LS), amygdala (e.g., CAmg), and in certain areas of the cortex (e.g., prefrontal). The autoradiographic distribution pattern of the synthetic glucocorticoid [^{3}H]dexamethasone (Dex) differs from that of [^{3}H]B, since considerable concentrations of silver grains are found in the autoradiographs over medial basal hypothalamic (MBH) nuclei, while only sparcely labeled cells are found in the hippocampus (e.g., Rees, Stumpf, & Sar, 1975; Rhees, Grosser, & Stevens, 1975; Warembourg, 1975b). The recent finding that [^{3}H]aldosterone (Aldo) is concentrated by nuclei in many of the same regions as [^{3}H]B (e.g., Birmingham, Stumpf, & Sar, 1979; Ermish & Ruhle, 1978; Moguilewsky & Raynaud, 1980) has further complicated the interpretation of the distribution of glucocorticoid-receptor-containing neurons in the brain. Estrogen-, androgen-, and glucocorticoid-concentrating cells have also been found in lower brain-stem regions and in the spinal cord, with androgen- and glucocorticoid-labeled cells predominating in certain somatomotor cranial nerves and spinal-cord regions, while estrogen-labeled cells predominate in certain sensory cranial nerves and spinal-cord regions (e.g., Heritage, Stumpf, Sar, & Grant, 1980; Stumpf & Sar, 1979).

PHYSICOCHEMICAL PROPERTIES AND MOLECULAR/ PHYSIOLOGICAL REGULATION OF INTRACEREBRAL STEROID RECEPTORS AND ACCEPTORS

Receptor Physico- and Biochemical Properties

The first reports on cytoplasmic steroid-receptors in the brain appeared over 12 years ago (e.g., Eisenfeld, 1970; Kahwanago, Heinrichs, & Herrmann, 1970). These early *in vitro* studies utilized the 100,000*g* supernatant (i.e., cytosol) fraction from brain homogenates incubated with [^{3}H]E_2 at 0–4°C for several hours. Macromolecular-associated radioactivity was found using gel chromatography and in the 7S range, using no- or low-salt sucrose gradient ultracentrifugation separations. Treatment of the cytosol with agents that block sulfhydryl groups or with proteolytic enzymes, but not with RNase or DNase, dramatically reduced its binding capacity, suggesting that estrogen-receptors were proteins. The regional distribution of these receptors paralleled the patterns revealed with autoradiography. In competition studies, these receptors appeared to be specific for estrogens since androgens, progestins, and glucocorticoids failed to reduce the maximum binding capacity (ca. 20 fmol/mg protein) or affinity (ca. 0.1 nM) (e.g., Kato, 1975). In more recent work conducted at higher salt concentrations, it now appears that the estrogen-receptor exists *in situ* as a 4S monomer species (ca. 80K daltons) that normally must combine with a second macromolecule (or a complex of macromolecules) to form a 5S complex (ca. 130K daltons) prior to its interaction with nuclear acceptors (e.g., Fox & Johnston, 1974; Murayama, Fukai, & Yamamoto, 1980; Weichman & Notides, 1979). This 4S to 5S transformation (i.e., activation) is facilitated by DNA (e.g., Fox, 1977; Yamamoto, 1974), occurs in cytosol as well as nucleoplasm (e.g., Jungblut, Kallweit, Sierralta, Truitt, & Wagner, 1978; Linkie & Siiteri, 1978), proceeds at different rates in different tissues (e.g., Fox, 1977; Linkie, 1977), and is both time and temperature dependent (e.g., Fox, 1977; Gorski & Gannon, 1976). The rate of estrogen dissociation from the activated receptor complex, and from the nucleus, closely correlates with the hormones' biological activity (e.g., Jungblut, Hughes, Gaues, Kallweit, Marschler, Parl, Sierralta, Szendro, & Wagner, 1979; Weichman & Notides, 1980).

The intracerebral progestin-receptors appear to be quite similar to those for estrogen. In 1977, Seiki, Haruki, Imanishi, and Enomoto published the first clear evidence for a macromolecular binding species for $[^3H]P_4$ in MBH cytosol from estrogen-primed ovariectomized rats. Scatchard plots yielded a K_d of 6 nM and a B_{max} of 20 fmol/mg protein, while sucrose gradient separation yielded sedimentation coefficients of 7S in low-salt and 4S in higher salt concentration buffers (e.g., Seiki, Haruki, Imanishi, & Enomoto, 1979). Other workers have found lower K_d values (e.g., 1.9 nM using spheroidal hydroxylapatite chromatography) and have shown that these progestin receptors are susceptible to proteolysis, are specific for progestins, and have a half-life on the order of minutes (37–160 min) rather than hours to days as observed with the estrogen- and androgen-receptor complexes (e.g., Lee, Davies, & Ryan, 1979; Thrower & Lim, 1980). More extensive studies conducted with $[^3H]$R5020 have confirmed and expanded on the results obtained with $[^3H]P_4$ (e.g., Kato & Onouchi, 1979; MacLusky & McEwen, 1980; Moguilewsky & Raynaud, 1979a; however, see Thrower & Lim, 1980). Salt extraction of target cell nuclei, labeled by an exchange reaction with $[^3H]$R5020 and unlabeled P_4, releases nuclear progestin-receptor complexes with binding affinities and capacities similar to the cytosol receptors but with sedimentation coefficients in the 5–6S range, suggesting that they may be homologous or heterologous aggregates of the 4S cytosolic species (e.g., Blaustein & Feder, 1980; Kato & Onouchi, 1979). While these results clearly suggest a 4S to 5S activation transformation similar to that described earlier for estrogens, much work remains to be done before this analogy can be accepted.

In vitro studies with hypothalamic and preoptic area cytosol preparations from rats, mice, and other mammals reveal that $[^3H]$T and $[^3H]$DHT associate with high-affinity (K_d = 0.3–8.0 nM), limited capacity (B_{max} = 2–14 fmol/mg protein), negatively charged (pI = 5.8), protease and sulfhydryl reagent-sensitive macromolecules sedimenting as 5–8S aggregates in low-salt sucrose gradients and as 4S (monomer?) species in gradients with higher salt concentrations (e.g., Clark & Nowell, 1979; Gustafsson, Pousette, & Svensson, 1976; Kato, 1976). These receptors appear to be specific for androgens and their distribution parallels the patterns observed in autoradiographic studies. In contrast to the results obtained with the estrogen and progestin intracerebral steroid-receptors, but in agreement with results obtained with peripheral tissue androgen-receptors (e.g., Wilson & French, 1979), the nuclear-extracted androgen-receptors appear to sediment in sucrose gradients with the same rate (i.e., 3–4S) as that of the cytosol receptor (e.g., Lieberburg, MacLusky, & McEwen, 1977), clearly suggesting that the activation of these receptors does not involve homologous or heterologous aggregation with other macromolecules.

In vitro studies with cytosol obtained from either Hpc or whole brain homogenates have revealed high-affinity macromolecular receptors for natural and synthetic glucocorticoids that are destroyed by proteolytic enzymes and sulfhydryl reagents, but not by DNase, RNase, lipase, neuraminidase, and hyaluronidase (e.g., Grosser, Stevens, & Reed, 1973; Tsuboi, Kawashima, Tomioka, Nakata, Sakamoto, & Fujita, 1979). The receptor (stabilized with 20 mM Na_2MoO_4, 1 mM DTT, and 10% glycerol) sediments at 9.2S in 150 mM KCl sucrose gradients, has a Stokes radius of 7.7 nm (Sephacryl S-300), and has a calculated molecular weight of 300K daltons ([^{3}H]triamcinolone acetonide; Luttge & Gray, unpublished observations) and, when precautions are taken to eliminate proteolysis and contamination with transcortin, has an isoelectric value of 6.1 ([^{3}H]Dex and [^{3}H]B) (e.g., MacLusky, Turner, & McEwen, 1977; Wrange, 1979). Recent work in our laboratory has shown that the binding affinity (K_d) for [^{3}H]B is slightly less than for [^{3}H]Dex in whole brain cytosol from ovariectomized/adrenalectomized female mice (7.0 ± 0.6 versus 2.3 ± 0.3 nM), whereas the binding capacity (B_{max}) for the natural glucocorticoid is nearly twice that for the synthetic steroid in the same samples (620 ± 40 versus 340 ± 30 fmol/mg protein). The Hill coefficients for [^{3}H]B and [^{3}H]Dex were unity, suggesting the absence of cooperativity during steroid–receptor association (Gray & Luttge, unpublished observations). We also found that [^{3}H]Dex was competitively displaced (K_{di}) by B (7.5 ± 1.9 nM), DOC (11 ± 1.9 nM), F (16 ± 3.8 nM), and P_4 (14 ± 1.6 nM); to a lesser extent by Aldo (27 ± 3.1 nM) and cortexolone (40 ± 3.4 nM); and very little by T (540 ± 81 nM). The high affinities of P_4 and DOC suggest that they may exert significant physiological effects by interacting directly with the glucocorticoid receptors either as antagonists (P_4) or perhaps as agonists (DOC). The observations that P_4 facilitates the rate of [^{3}H]Dex-receptor dissociation (Gray & Luttge, unpublished observations) and inhibits B-stimulated synthesis of an apparently specific 54K-dalton protein (Etgen, Martin, Gilbert, & Lynch, 1980; see below) further supports a direct interaction between P_4 and the brain glucocorticoid receptor.

Raising the temperature of brain cytosol to 25°C for short periods of time results in the activation of glucocorticoid-receptor complexes, as assessed by increased binding to DNA-cellulose columns (e.g., de Kloet & McEwen, 1976). This activation also causes a dramatic change in the size and surface characteristics of the complex. For example, we have recently shown that activation at 22°C for as little as 3–6 min (and subsequent stabilization with 20 mM Na_2MoO_4) reduces the sedimentation coefficient to 3.8S in 150 mM KCl sucrose gradients, the Stokes radius to 5.8 nm, and the calculated molecular weight to 92K daltons (Luttge & Gray, unpublished observations). These conformational changes are further reflected in a dramatic increase in the surface hydrophobicity (GF/C filter and PEG/Dextran phase partition assays) and density of positive charges (DEAE anion-exchange filter and column assays) (Luttge & Gray, unpublished observations; Sakaue & Thompson, 1977). The increased hydrophobicity may facilitate the transport of the activated complex across the nuclear membrane whereas the increased number of positive charges on the surface may facilitate the binding of the complex with DNA and acceptor molecules. For further information on glucocorticoid receptors in the brain see Rees and Gray (1983).

Receptor Utilization and Replenishment

The possibility of hormone-induced fluctuations in steroid-receptor density has stimulated many studies. For example, early work indicated that the total number of available (i.e., unbound) estrogen-binding sites in the hypothalamus increased during periods with low circulating estrogen titers (e.g., diestrus, metestrus, and ovariectomy), while they declined during periods of high estrogen titers (e.g., proestrus, estrus, and exogenous estrogen replacement) (e.g., Ginsberg, MacLusky, Morris, & Thomas, 1975; Luttge, 1972). Recent work with exchange assays, however, indicates that total binding capacity for estrogens (i.e., unbound and bound receptor density) often fluctuates in parallel with estrogen titers (e.g., White, Thrower, & Lim, 1978). In these same studies it was shown that cytosolic estrogen-receptor density in the hypothalamus, but not in the uterus, still decreases during proestrus, presumably as a result of the redistribution of these receptors to the nucleoplasm. It thus appears that the uterus has a replenishment/synthesis system capable of homeostatically maintaining cytosolic estrogen-receptor density, whereas in the hypothalamus, cytosol receptor depletion exceeds the rate of replenishment (at least during the estrous cycle).

In ovariectomized rats, the magnitude of cytosolic estrogen-receptor depletion correlates with the dose of exogenous estrogen, typically reaches a peak within one hour, and is not affected by pretreatment with the protein synthesis inhibitor cycloheximide (e.g., Cidlowski & Muldoon, 1974, 1976). Cytosolic receptor replenishment, however, may require 15 hours or more, and it is susceptible to cycloheximide inhibition. As the dose of estrogen increases, the relative proportion of receptor replenishment *not affected* by the cycloheximide also increases. These and other results suggest that the bulk of cytosolic estrogen-receptor replenishment in the hypothalamus and uterus that results from interactions with *endogenous* estrogens is due to *de novo* synthesis of new receptors (or enzymes required for the capacitation of apo-receptors), rather than to the recycling of old receptors from the nucleoplasm back into the cytoplasm.

This conclusion is also consistent with the intranuclear degradation or "processing" of estrogen receptors recently demonstrated in brain (White & Lim, 1980) and uterus (Jungblut *et al.*, 1979), and it may serve as a stimulant for the *de novo* synthesis of new cytoplasmic estrogen receptors. Although estrogen stimulation is not absolutely required for this synthesis (i.e., estrogen-receptor synthesis continues in ovariectomized-hypophysectomized rats), the rate of synthesis is clearly facilitated by estrogen-induced receptor processing (e.g., Little, Szendro, Hughes, & Jungblut, 1975). The observation that these newly formed receptors are detectable first in the microsomal fraction (e.g., Little *et al.*, 1975;

Watson & Muldoon, 1977) has contributed to the development of the following proposed pathway for *in vivo de novo* estrogen-receptor synthesis: 3.5S "basic" microsomal receptor → 3.5S "acidic" microsomal receptor → 4S "acidic" cytosolic receptor → 5S "acidic" cytosolic and nuclear receptor (e.g., Jungblut *et al.,* 1978; Little *et al.,* 1975).

As stated earlier, the intracerebral progestin-receptor appears to be an estrogen-dependent protein. Hypothalamic and preoptic (but not cerebral cortical, midbrain, and amygdaloid) progestin receptor concentrations decrease after gonadectomy or removal of exogenous estrogen sources in gonadectomized animals (receptor half-life approximates one day); increase in a dose- and time-dependent manner after estrogen stimulation (24–48 hours required for maximal effect) in female and, to a lesser extent, male rats and other animals; and fail to respond to estrogen after treatment with anisomycin (e.g., Blaustein & Feder, 1979a,b; MacLusky & McEwen, 1980; Moguilewsky & Raynaud, 1979a,b; Parsons, MacLusky, Krey, Pfaff & McEwen, 1980; Rainbow, Davis, & McEwen, 1980; Roy, Schmit, McEwen, & Wade, 1979). These findings suggest that progestin-receptor levels are increased by estrogen-receptor-acceptor-dependent *de novo* protein synthesis. A positive interaction also exists between the dose of estrogen used to stimulate the appearance of the cytoplasmic receptor and the dose of progestin used to stimulate the redistribution of that receptor to the nucleus (e.g., Blaustein & Feder, 1980; Blaustein & Wade, 1978). Coincident with progestin-induced nuclear progestin-receptor accumulation, there is a dose-dependent, long-duration (at least 24 hours) reduction in cytosolic progestin-receptor concentrations (Blaustein & Feder, 1979b, 1980; Moguilewsky & Raynaud, 1979a). This failure to regulate progestin-receptor titers homeostatically is similar to that described above for estrogen-receptors, and it further suggests that receptor "processing" may also be required for progestin-receptor genomic actions. When progestins are given concurrently with estrogen, there is a dose-dependent negative interaction between the two steroids resulting in a progestin-induced reduction in estrogen-stimulated progestin-receptor appearance in the cytosol (Blaustein & Feder, 1979b; Moguilewsky & Raynaud, 1979b). Since this inhibition does not appear to involve a direct interaction between the progestin- and the estrogen-receptors, the mechanism of this effect is unknown.

In vivo studies with male rats indicate that [^{3}H]DHT-binding nuclear macromolecules in the hypothalamus are saturated in the gonadally intact animal and become apparent only following castration (often accompanied with adrenalectomy) (Lieberburg *et al.,* 1977). [^{3}H]B-binding capacity by hippocampal cytosolic and nuclear receptors also increases after removal of the endogenous hormone source (i.e., adrenalectomy)—rapidly during the first 2 hours (paralleling the decline in endogenous glucocorticoid titers and presumably reflecting the dissociation of preexisting steroid-receptor complexes), remaining at the new level for the next 12 hours, followed by a slow increase for the next 6 hours (e.g., McEwen, Wallach, & Magnus, 1974). Diurnal and stress-induced increases in plasma and brain levels of endogenous glucocorticoids decrease [^{3}H]B *in vitro* cytosol binding in most brain regions (e.g., Butte, Kakihana, & Noble, 1976; Carroll, Heath, & Jarrett, 1975; McEwen *et al.,* 1974; Stevens, Reed, Erickson,

& Grosser, 1973). The observation that [^{3}H]B binding in the hippocampus after the *in vivo* administration of the steroid displays a diurnal rhythm even in adrenalectomized rats (Angelucci, Valeri, Palmery, Paracchioli, & Catalani, 1980), suggests that there is a steroid-independent aspect to glucocorticoid-receptor regulation.

In contrast to estrogen- and progestin-receptor regulation, there is no net depletion of total hippocampal *in vitro* [^{3}H]B cytosolic binding capacity (bound and unbound) following *in vivo* administration of high doses of labeled and unlabeled glucocorticoid, even though there is a clear dose-dependent increase in the nuclear accumulation of glucocorticoid-receptor complexes (Turner & McEwen, 1980). The *in vivo* measured [^{3}H]B hippocampal nuclear binding capacity was further shown to be about 40% of the *in vitro* measured cytosolic binding capacity. These and other results support the findings in non-neural tissues suggesting that there is a large reserve of apo-receptors in many glucocorticoid target tissues. These apo-receptors appear to require "capacitation" (possibly a phosphorylation-induced modification) prior to their involvement in steroid binding. Cytoplasmic and/or nuclear "decapacitation" or processing (possibly a phosphatase-induced modification) has also been demonstrated in neural and non-neural glucocorticoid target tissues (e.g., de Kloet, Wallach, & McEwen, 1975; Luttge & Densmore, unpublished observations; Sando, Hammond, Stratford, & Pratt, 1979; Sando, LaForest, & Pratt, 1979; see also review by Rees & Gray, 1983).

Unoccupied Nuclear Receptors

The concept that in order for a steroid-receptor to bind with its nuclear acceptors, it must be bound, and thereafter remain bound, to its appropriate ligand has come under increasing attack in recent years. Isolation of target cell nuclei from ovariectomized and ovariectomized-adrenalectomized animals, followed by cytosol-free incubation with [^{3}H]E_2, has revealed the presence of unoccupied 4S and 5S estrogen-receptors within the nucleus (e.g., Jungblut *et al.*, 1979; Levy, Mortel, Eychenne, Robel, & Baulieu, 1980; White & Lim, 1980). Unoccupied nuclear progestin-receptors have also been recently demonstrated in the guinea pig uterus (Saffran & Loeser, 1979). Failure to detect these unoccupied receptors in earlier studies may be the result of methodological problems (see Martin & Sheridan, 1980). It now appears possible that in some target cell systems, most probably those with high nucleoplasm-to-cytoplasm ratios (e.g., many neurons), a majority of the unbound receptors in a cell may reside *in situ* in the nucleoplasm rather than in the cytoplasm. This distribution may not be due solely to diffusion (although the nuclear pores do not present a barrier to steroid-receptors), since the hydrophilic receptors should be preferentially retained with the nucleoplasm, where unoccupied water space is reported to be more abundant than in the cytoplasm (Horowitz & Moore, 1974). The biological significance of unoccupied nuclear receptors is still unclear; however, it should be noted that unbound hypothalamic estrogen-receptors can bind to DNA- and oligo(dT)cellulose (Fox, 1977), suggesting that they may be involved in genomic responses similar to those documented for the activated bound receptor (see Lippman, Bolan, Monaco, Pinkus, & Engel, 1976; Zava & McGuire, 1977). It is conceivable that steroid-binding and steroid-receptor activation function primarily to increase the avidity, or nucleotrophy, of the receptor for its nuclear acceptors. It has even been proposed that the unoccupied rather than the occupied nuclear estrogen-receptor may be the active species, since the concentration of unoccupied nuclear receptors in the adult female rat hypothalamus is reduced if these females are rendered sterile and behaviorally less responsive to E_2 by treating them with TP on the fourth day after birth (White & Lim, 1980).

Nuclear Acceptors

Work on a variety of steroid target tissue systems has suggested that steroid-receptor complexes must interact with both DNA and chromatin to selectively increase or decrease the synthesis of specific mRNAs coding for specific proteins (e.g., Fox, 1977; Kumar, Beach, & Dickerman, 1980; Perry & Lopez, 1978; Whalen & Olsen, 1978; see below). Following the initial observation that not all estrogen-receptor complexes could be extracted with hypertonic KCl from uterine cell nuclei, several investigators suggested that the "salt-resistant" fraction represents a specific class of nuclear receptors tightly bound to a specific class of acceptors that are essential for the genomic response to estrogen (e.g., Clark & Peck, 1976). Salt-extractable and salt-resistant nuclear binding in hypothalamic, septal, and other limbic brain structures has also been reported for androgens (e.g., Kato, 1975; Lieberburg & McEwen, 1977; Lieberburg *et al.*, 1977; Monbon, Loras, Reboud, & Bertrand, 1974), estrogens derived from the intracerebral metabolic aromatization of androgens (e.g., Krey, Kamel, & McEwen, 1980; Lieberburg & McEwen, 1975; see below), and progestins (e.g., Kato & Onouchi, 1979; Seiki *et al.*, 1979). Although there is evidence supporting the distinction between salt-resistant and salt-extractable steroid-receptor–acceptor nuclear complexes, several recent studies have argued that this dichotomy has been over emphasized (e.g., Traish, Muller, & Wotiz, 1977). For example, since virtually all of the nuclear estrogen-receptor complexes found in hypothalamic samples can be extracted by increasing the concentration of the KCl extraction buffer to 0.4–0.6*M* (e.g., Roy & McEwen, 1979), salt-extractability appears to be a relative and not an absolute classification. The observation that steroid nuclear receptor (and acceptor?) dissociation rates vary inversely with the salt concentration required for receptor extraction (e.g., Roy *et al.*, 1979) probably reflects the differential ability of agonists and antagonists to induce conformational changes in the receptor that alter its affinity for the nuclear acceptors.

Affinity chromatography studies have suggested that nuclear acceptors for estrogen-, progestin-, androgen-, and glucocorticoid-receptors in non-neural tissues are basic (pI $\simeq$ 8.6–9.3), nonhistone proteins (ca. 70–85K daltons), with high-affinity (e.g., K_d = 0.2 nM for E_2), saturable and, to varying extents, tissue- and hormone-specific binding for steroid-receptor complexes (e.g., Mainwaring, Symes, & Higgins, 1976; Puca, Nola, Hibner, Cicala, & Sica, 1975). For example, uterine nonhistone basic proteins coupled to Sepharose were found to retain nearly four times more [^{3}H]E_2-labeled uterine receptor complexes than did similarly prepared acceptor proteins extracted from rat liver and prostate. Uterine receptor–acceptor-Sepharose binding was enhanced if the receptors were preexposed to E_2, but not with similar exposure to P_4, T, or F. Addition of native or denatured DNA reduced total binding capacity but not affinity, suggesting that there is an intrinsic affinity of the acceptors for DNA. If DNA-Sepharose is used, it is possible to create estrogen-receptor–acceptor-DNA-Sepharose complexes by sequentially binding the various components. Since receptor–acceptor-Sepharose binding saturation can be achieved with relatively low protein concentrations, it is unlikely that the *in vitro* saturation of the acceptors is an artifact produced by highly concentrated cytosol preparations (e.g., Chamness, Jennings, & McGuire, 1974) known to contain low- and high-molecular-weight inhibitors of receptor activation and receptor–acceptor binding (e.g., Asai, Yu, & Leung, 1979; Sato, Noma, Nishizawa, Nakao, Matsumoto, & Yamamura, 1980; Shen, Thrower, & Lim, 1979). The marked physicochemical similarity of non-neural acceptors for estrogen- (uterine), progestin- (uterine), androgen- (prostate), and glucocorticoid- (liver) receptors (e.g., Mainwaring *et al.,* 1976) suggests that the intracerebral acceptors for these steroid receptor complexes will also be similar.

INTRACEREBRAL STEROID METABOLISM

Estrogens, progestins, androgens, and glucocorticoids can all be metabolized to varying extents within the brain (and peripheral target tissues) to form steroids that may have quantitatively and/or qualitatively different biological activities from their parent hormones. Some of these metabolic pathways have been the subject of extensive examination in studies of the hormonal regulation of agonistic behavior (e.g., the aromatization and 5α-reduction of T to E_2 and DHT, respectively; see chapter by Brain and Bowden). The followoing sections briefly review the major routes of steroid metabolism in the mammalian brain (see Figures 2 and 3 and Table I, and the more detailed discussions presented in Luttge, 1979, 1983; Whalen *et al.,* 1983).

Estrogens

Estradiol (E_2) has long been considered the major estrogen secreted by the ovary as well as the principle estrogen active in neural and non-neural tissues. Recent studies have shown, however, that estrone (E_1), estriol (E_3), and the catechol estrogens 2-OH-E_2, 2-OH-E_1, 4-OH-E_2, and 4-OH-E_1 can also be found in peripheral plasma and/or in intracerebral extracts and that these metabolites may participate in mediating certain behavioral, neuroendocrine, and somatic estrogenic actions.

In early *in vivo* studies, [^{3}H]E_1 was detected in certain brain regions of male and female rats given [^{3}H]E_2, while [^{3}H]E_2 was detected in the brains of rats given [^{3}H]E_1 (e.g., Luttge & Whalen, 1972). Metabolically produced E_1 was

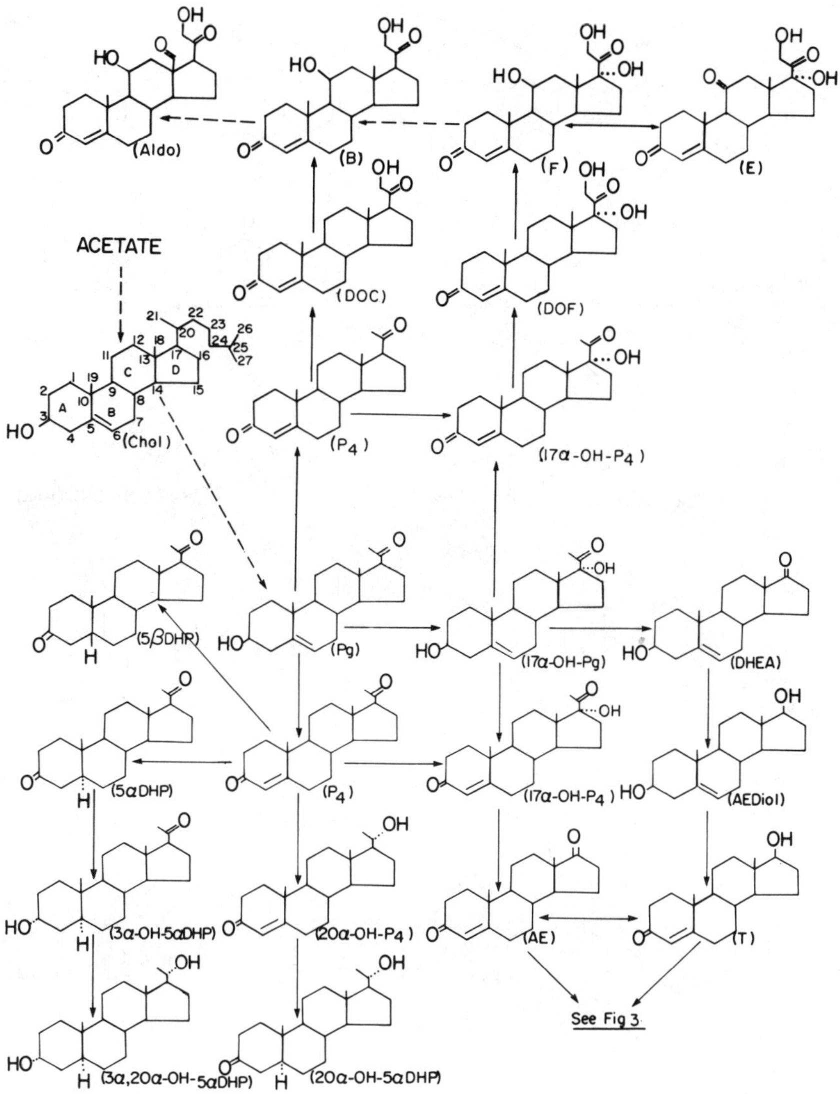

FIGURE 2. Structural formulas and probable metabolic pathways for androgen, progestin, and glucocorticoid synthesis and metabolism in mammalian brain and/or peripheral organs such as the testis, ovary, and adrenal gland. Dashed arrows indicate multistep conversions and/or pathways that are established in peripheral tissues but not yet observed in brain tissue. Abbreviations are defined in Table I.

comparatively enriched in PH and IP samples, while metabolically produced E_2 was most prevalent in AH and POA samples (see Table II for neuroanatomical name abbreciations). Very little E_3 was found in these studies, but in later *in vitro* studies, $[^3H]E_3$ was conclusively identified in rat brain incubations with $[^3H]E_2$ and, to a much lesser extent, $[^3H]E_1$ (Reddy, 1979a). A similar preferential *in vivo* conversion of $[^3H]E_1$ to $[^3H]E_2$ by hypothalamic, and to a lesser extent

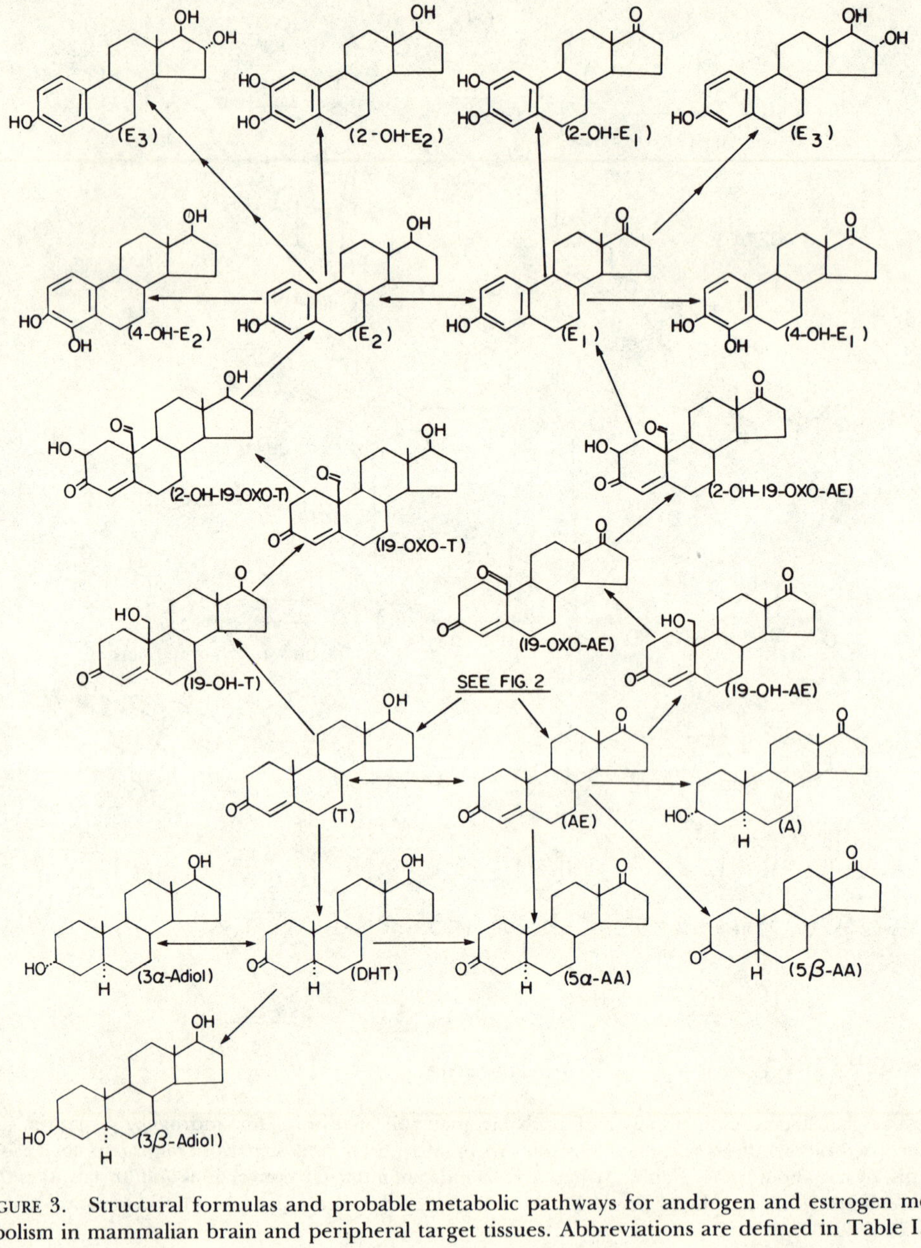

FIGURE 3. Structural formulas and probable metabolic pathways for androgen and estrogen metabolism in mammalian brain and peripheral target tissues. Abbreviations are defined in Table I.

septal, amygdaloid and cerebral cortical samples has recently been demonstrated in adult male and female guinea pigs (Landau & Feder, 1980). The soluble 17β-oxidoreductase enzyme has been isolated from rat, rabbit, and other mammalian brains (e.g., Kazama & Longcope, 1974; Reddy, 1979b).

After systemic injections of $[^3H]E_2$, isolated nuclei from rat brain tissue pools containing HTh, POA, Sep, Amg, and OB regions retain mostly E_2 (86%), although significant amounts of E_1 (4%) and E_3 (2%) can also be detected (Zigmond & McEwen, 1970). Nuclei from tissue pools containing Hpc and Cx regions retain comparatively less E_2 (75%), more E_3 (8%), and a similar amount of E_1 (4%) to that detected in the pooled limbic brain regions. E_3 retention by HTh nuclei after injections of $[^3H]E_3$ in ovariectomized guinea pigs has also been demonstrated (Landau & Feder, 1977). Although the $[^3H]E_3$ in this study was not converted to either $[^3H]E_2$ or $[^3H]E_1$, the free estrogens accounted for less than 1% of the total radioactivity, suggesting that most of it was probably associated with water-soluble conjugated estrogens.

Intracerebral estrogen metabolism is fairly consistent across mammalian species, with E_2 accounting for 76%–90% of the HTh radioactivity after systemic injections of $[^3H]E_2$ in mice, hamsters, guinea pigs, and rats (e.g., Feder, Siegel, & Wade, 1974; Luttge, 1972; Luttge, Gray, & Hughes, 1976). E_1 accounts for the majority of the remaining radioactivity. Lastly, since $[^3H]E_2$-benzoate (E_2B) appears to be metabolized to $[^3H]E_2$ before entering the brain (Eaton, Goy, & Resko, 1975), the intracerebral metabolism of this behaviorally useful synthetic estrogen is essentially the same as that for E_2, with the exception that the time course is delayed due to the slow rate of release from peripheral depot storage sites.

Although once thought to be of importance mainly in the catabolism of estrogens in the liver, the recent demonstrations that E_2 and E_1 can readily be metabolized in the brain to their 2- and 4-hydroxy- or catechol-derivatives (see Figure 3) has stimulated increased interest in these novel estrogens (for reviews see Ball & Knuppen, 1980; Luttge, 1983; Paul, Hoffman, & Axelrod, 1980). Catechol estrogen formation is usually greater in the hypothalamus than in the cerebral cortex, is equivalent in male and female brains, is equally prevalent at the 2 and 4 positions in brain, increases dramatically in brain (but not liver) during proestrus in rats, is mediated by a microsomal cytochrome P-450 mixed function oxidase enzyme, and is more effective with E_2 than E_1 as the enzyme substrate. The intracerebrally formed catechol estrogens are rapidly turned over, forming highly reactive superoxide intermediates that covalently bind to brain microsomal proteins and compete with $[^3H]E_2$ for high-affinity binding to brain cytosol estrogen-receptors and with norepinephrine and other biogenic amines for high-affinity interactions with brain tyrosine hydroxylse, catecholamine-O-methyl-transferase and other enzymes. The limited evidence on the actions of catechol estrogens suggests that 4-OH-E_2 is the most potent, often being as effective as E_2 in studies on female sexual behavior and LH secretion (e.g., Franks, Ball, Naftolin, & Ruf, 1980; Jellnick, Davis, Krey, Luine, Roy, & McEwen, 1979; Luttge & Jasper, 1977). The potential role of catechol estrogens in aggressive behaviors has not been established.

PROGESTINS

Progesterone (P_4) is generally regarded as the major circulating progestin in mammals, but 20α-OH-P_4 and, to a lesser extent, 5α-pregnan-3α,20α-diol and 5α-pregnan-20α-ol-3-one also contribute, in a major way, to the progestins available to the brain (for review see Whalen *et al.*, 1983). All of these progestins should cross the blood–brain barrier (Pardridge & Mietus, 1980) and thus be susceptible to intracerebral metabolic transformation (for review see Luttge, 1983). Early studies examining intracerebral progestin metabolites indicated that 1–4 hours after systemic injections of [^{3}H]P_4, only 20%–30% of the organic solvent-extractable radioactivity from rat and mouse brains was isopolar with authentic P_4, while in guinea pigs, this figure often approached 70%. The principal metabolite was shown to be 5α-DHP, the 5α-reductase product of P_4. Injections of [^{3}H]5α-DHP did not result in significant intracerebral accumulation of P_4; rather, most of the radioactivity remained associated with 5α-DHP, suggesting that the 5α-reduction of P_4 is essentially irreversible.

Although *in vitro* studies with hypothalamic and mesencephalic tissue incubated with radiolabeled P_4 have revealed a wide range of metabolites, 5α-DHP is still the most abundant product. The further reduction of 5α-DHP to 5α-pregnan-3α-ol-20-one may account for 2%–10% of the extracted radioactivity in rat brains, while other metabolites are less well established (for review see Luttge, 1983). The 3β-hydroxy-steroid oxidoreductase, 5-ene-isomerase catalyzed conversion of pregnenalone to P_4 in various brain regions (Amg > Sep > Hpc > HTh > Cx) has also been very recently demonstrated (Weidenfeld, Siegel, & Chowers, 1980). Subcellular studies indicate that there are at least two 3α-hydroxysteroid oxidoreductase enzymes, a cytosolic NADPH-linked form and a plasma membrane NADH-linked form (Krause & Karavolas, 1980). Kinetic studies on microsomal 5α-reductase have revealed that P_4 (Km = 0.48 μM) and 20α-OH-P_4 (Km = 0.86 μM) are both more effective substrates than T (Km = 16 μM) (Cheng & Karavolas, 1975), a difference that may prove to be important in studies of androgen actions in aggression where the 5α-reduction of T to DHT is of major concern (see Chapter 1 by Brain). The intracerebral conversion of P_4 to 20α-OH-P_4 and its 5α-reductase product, 5α-pregnan-20α-ol-3-one, has potential importance in female sexual behavior (e.g., Kubli-Garfias & Whalen, 1977), but little is known of their actions in agonistic behavior.

Androgens

Testosterone (T) is the predominant circulating androgen in most adult male mammals, but androstenedione (AE), dehydroepiandrosterone (DHEA), dihydrotestosterone (DHT), 3α- and 3β-androstanediol (3α- and 3β-Adiol), and epitestosterone (epi-T) (see Table I and Figure 3) can also be found in blood samples taken from a large number of mammalian species (for review see Whalen *et al.*, 1983). Peripheral target and nontarget tissue metabolism, coupled with the release of these metabolites into blood, and their further metabolism by blood-borne enymes further increase the total number of androgen metabolites available for uptake and further metabolism by the brain (for reviews see Luttge, 1979, 1983; Whalen *et al.*, 1983). The magnitude of this non-neural androgen metabolism clearly complicates the interpretation of behavioral studies employing the systemic administration of androgens.

The interconversion of T and AE by cytosolic 17β-oxidoreductase has been demonstrated *in vivo* and *in vitro* in many brain regions taken from a wide range of mammalian species. The 5α-reduction of T to DHT has also been demonstrated *in vivo* and *in vitro* in brain from numerous mammalian species, but unlike AE, DHT appears to be concentrated in the hypothalamus to a much greater extent than in other brain regions. The microsomal/nuclear localized 5α-reductions of both T and P_4 are essentially irreversible and they may in fact be mediated by the same enzyme, with P_4 acting as the preferred substrate in direct competition studies. Of the numerous other androgenic intracerebral metabolites, including 3α- and 3β-Adiol, 5α- and 5β-androstanedione (5α- and 5β-AA), and androsterone (A) (see Table I and Figure 3), the interconversion of DHT and 3α-Adiol by 3α-oxidoreductase, a soluble enzyme found throughout the brain, has clearly been the most actively studied (for reviews on these and other intracerebral androgenic metabolites see Luttge, 1979, 1983; Whalen *et al.,* 1983).

The aromatization of C-19 androgens to C-18 estrogens is probably the most extensively studied intracerebral androgen metabolic pathway. The conversion of T and AE to E_2 and E_1, respectively, has been demonstrated *in vivo* and *in vitro* in many amammalian species, but there are exceptions, including the guinea pig (for review see Callard, Petro, & Ryan, 1978). The highest activities in male rat brains are found in the MPOA, PeVN-POA, and MAmg, with much lower levels found in the LPOA, MBH, and LH (Kobayashi & Reed, 1977; Selmanoff, Brodkin, Weiner, & Siiteri, 1977). Nonlimbic brain regions usually lack aromatase activity in the adult. These findings agree with *in vivo* findings of comparatively high estrogen levels in purified nuclei from HTh, POA, Amg, and Sep tissue pools from male and female rats injected systemically with [^{3}H]T (e.g., Lieberburg & McEwen, 1975, 1977; Rezek, 1977).

The human placental microsomal system has provided the bulk of the evidence on the actual metabolic sequences involved in the aromatization of androgens (for review see Luttge, 1983; Whalen *et al.,* 1983). A single cytochrome P-450 requiring enzyme catalyzes a series of three hydroxylations (see Figure 3); the final product then undergoes a nonenzymatic collapse within the A-ring to form an estrogen. Due to the lack of direct evidence for a similar pathway in brain, the tacit acceptance of this scheme in behavioral and neurochemical studies must be viewed with caution. For example, even though the division of androgens into aromatizable (i.e., androgens with an unsaturated A-ring such as T, AE, DHEA, 19-OH-T, and 19-OH-AE) and nonaromatizable (i.e., androgens with a saturated A-ring such as DHT, 5α- and 5β-AA, 3α- and 3β-Adiol, and A) classes may actually be valid for the placenta, definitive support for the generalizability of this dichotomy to brain is largely nonexistent. The apparent irreversibility of the intracerebral 5α-reduction of T to DHT does indirectly support the assumption that at least this androgen (and presumably its 3α-, 3β-, and 17β-oxidoreductase products) is probably not readily aromatized by brain enzymes, but it may still be converted to an estrogen *in situ* through the action of gut bacterial flora enzymes (e.g., Hill, Goddard, & Williams, 1971).

Glucocorticoids

The principal circulating glucocorticoid in some mammals is corticosterone (B) (e.g., rat, mouse, and rabbit), while in other mammals, cortisol (F) predominates (e.g., hamster, sheep, and primate). This dichotomy may not, however, extend to the brain. For example, in recent work with baboons it was discovered that even though F is clearly the major circulating glucocorticoid (i.e., B/F ratio ≤ 0.1), B appears to be selectively concentrated (i.e., B/F ratio ≥ 0.5) by nuclear receptors in several brain areas (e.g., POA-Sep, HTh, Amg, and Hpc) (Turner, Smith, & Carroll, 1979). The locus of this apparent metabolic conversion of F to B is unknown. *In vivo* and *in vitro* studies have also shown that the brain can readily interconvert F and cortisone (E) (e.g., Grosser & Axelrod, 1968), while similar studies with B indicate that little metabolism occurs prior to the binding of this glucocorticoid with brain cytosolic and nuclear receptors (e.g., McEwen, Magnus, & Wallach, 1972). The recent demonstration of a specific mineralocorticoid receptor in rat brain (Moguilewsky & Raynaud, 1980) clearly suggests that the potential intracerebral conversion of B to aldosterone (Aldo) should be closely examined.

NEUROCHEMICAL ACTIONS OF STEROID HORMONES

The genomic model of steroid hormone action (see Figure 1) clearly predicts *de novo* synthesis of specific mRNA's and proteins in response to steroid-receptor–acceptor activation of the transcriptional activity in the nucleus. In peripheral target tissues there are several well-established examples of steroid-induced genomic responses (e.g., Chan, Means, & O'Malley, 1978), while in brain, nucleic acid and protein synthetic responses to steroids have only recently been detected (see below). Indirect support for these responses comes from studies demonstrating the inhibition of behavioral, neuroendocrine, and developmental actions of steroids by systemic or intracerebral administration of nucleic acid or protein-synthesis-blocking drugs (for review see Luttge, 1983). The recent demonstration that steroid hormones can either stimulate or inhibit synaptogenesis during development and axonal sprouting after neural damage (e.g., Brawer & Naftolin, 1979; Loy & Milner, 1980; Matsumoto & Arai, 1979, 1980; Scheff, Thorne, Sasvary, Bernardo, & Cotman, 1979; see also Luttge, 1983) also supports the genomic model of steroid action in the brain. The following sections briefly review some of the direct tests of that hypothesis (further details and references found in Luttge, 1983; Rees & Gray, 1983).

Nucleic Acids

Since intracerebral neurogenesis in mammals is restricted to the early perinatal period (e.g., Altman & Bayer, 1978), the lack of effect of steroids on DNA synthesis in the adult brain is predictable. Correlative support for steroid-induced increases in neuronal RNA synthesis in the adult brain has been obtained from *in vivo* and *in vitro* studies purporting to demonstrate an increase in either total RNA content or [^{3}H]uridine labeling of RNA during the proestrus and/or estrus stages of the rat estrous cycle (e.g., Foreman, Wickersham, & Anthony,

1977). While these changes do coincide temporally with increased E_2 and P_4 blood titers, they also synchronize with other hormones, including glucocorticoids and gonadotropins. The latter point is especially relevant in view of the recent demonstrations of increased [^{3}H]uridine labeling in Hpc in respone to exogenous B and in HTh in response to LH-RH (e.g., Biro, 1978; Dokas, 1979). Furthermore, not all workers have been able to find these estrous-cycle-related effects on RNA content and synthesis (e.g., Luck, 1975). More direct support for steroid effects on RNA synthesis comes from recent *in vitro* studies in which were found close temporal, regional, and subcellular correlations between a transient estrogen-induced increase in HTh DNA-dependent RNA polymerase II activity (i.e., that associated with mRNA synthesis) and estrogen nuclear receptor occupancy (e.g., Greenstein, 1980; Peck, Miller, & Kelner, 1979). Similar, albeit less definitive, findings have also been reported for glucocorticoids (e.g., Stith & Bottoms, 1972). In spite of these encouraging results, much more work clearly remains to be done on the nucleic acid aspects of the genomic model of steroid actions in the brain.

Proteins and Peptides

There have been numerous *in vivo* and *in vitro* studies examining the effects of gonadectomy, adrenalectomy, hypophysectomy, estrous-cycle and diurnal fluctuations, stress, and exogenous steroid hormone replacement on radiolabeled amino acid incorporation into total protein in whole or subregions of brain (for reviews see Luttge, 1983; Rees & Gray, 1983). Although these demonstrations of steroid-induced modulation of total brain protein synthesis are important, they provide no evidence for the induction of specific proteins that may be related to the behavioral and/or neuroendocrine actions of gonadal and adrenal hormones. Studies with rat uterus, using a double-labeling amino acid incorporation technique and sodium dodecyl sulfate (SDS) polyacrylamide gel electrophoresis, led to the discovery of a soluble estrogen-induced protein (E_2IP) with an approximate molecular weight of 39–45K daltons and a pI of 4.5. The synthesis of E_2IP was maximal during proestrus in intact adult female rats and at one hour after exogenous E_2 in immature or ovariectomized adult female rats. Uterine E_2IP synthesis induction was blocked by estrogen antibodies, antiestrogens, and protein and/or RNA synthesis inhibitors (e.g., Manak, Wertz, Slabaugh, Denari, Wang, & Gorski, 1980). Remarkably, the use of similar techniques has revealed the existence of a hypothalamic E_2IP, which appears to be identical to uterine E_2IP in terms of kinetics of production, mobility on polyacrylamide gels, pI, peptide maps, and immunological cross-reactivity (e.g., Beinfeld & Packman, 1976; Reiss & Kaye, 1979; Vertes, Vertes, & Kovacs, 1978; Walker, Negreanu, Gozes, & Kay, 1979). E_2IP synthesis has not been observed in nontarget regions of the brain (as defined by the presence of high-affinity receptors in adulthood). While the function of E_2IP is still a mystery, the intertissue similarity strongly suggests that it plays a fundamental regulatory role in estrogen-receptor–acceptor interactions. This role could include the synthesis of the 4S estrogen-receptor, the second component of the 5S estrogen-receptor complex, the nuclear nonhistone acceptor, and even the cytosol factors that inhibit estrogen-receptor activation/aggregation and receptor-acceptor binding (see Luttge, 1983).

There is considerable biochemical, physiological, and behavioral evidence supporting the existence of a second intracerebral/uterine E_2IP, namely the high-affinity progestin receptor. As described previously, hypothalamic, preoptic, and uterine progestin-receptor concentrations decrease following ovariectomy and increase in an estrogen-specific, dose-dependent, anisomycin-sensitive fashion after hormone-replacement therapy. The time course of synthesis (24–48 hours versus 1 hour), the isoelectric points (5.8 versus 4.5), and other factors make it unlikely that the progestin-receptor and E_2IP are the same protein. The adrenal and gonadal steroid-induced activation and/or synthesis of brain enzymes involved in neurotransmitter and carbohydrate metabolism will be discussed later.

Recent *in vitro* studies with rat Hpc slices has revealed the existence of a glucocorticoid-induced protein with an approximate molecular weight of 54K daltons (Etgen, Lee, & Lynch, 1979; Etgen *et al.*, 1980). A 12% increase in $[^3H]$leucine incorporation into this protein(s) was obtained in slices incubated for 30–120 minutes with $10^{-9}M$ B. This response was inhibited when either cycloheximide or P_4 was added to the incubation media, suggesting that it is mediated by a glucocorticoid-receptor–acceptor activation of *de novo* protein synthesis. The function of this glucocorticoid-induced intracerebral protein is unknown. Adrenalectomy has also recently been found to decrease the concentration of the peptide angiotensinogen in several brain regions (e.g., POA, AH, PAG, and AP) (Wallis & Printz, 1980). Replacement therapy with B reverses these effects and even increases angiotensinogen concentrations in the MBH and Sep.

Polyamines and GABA

The diamine putrescine (PUT) and the polyamines spermidine (SPD) and spermine (SPM) are ubiquitously distributed in mammalian tissues, while the neurotransmitter gamma-aminobutyric acid (GABA) is confined almost exclusively to the nervous system (for reviews see Luttge, 1983; Roberts, Chase, & Tower, 1976; Russell, 1980). Even though our knowledge of polyamine actions is incomplete, there is strong support for a regulatory role in nucleic acid and protein synthesis and membrane function. In their non-neural target tissues, trophic hormones characteristically stimulate a rapid increase in ornithine decarboxylase (ODC, a rate-limiting enzyme in PUT biosynthesis) activity, polyamine concentrations, and, in some cases, tissue hypertrophy and/or hyperplasia. Polyamines have also been implicated in various aspects of DNA and protein synthesis in the adult and perinatal brain (e.g., Fleischer-Lambropoulos, Sarkander, & Brade, 1975; Goertz, 1979; Jasper, 1980; Stipek, Crkovska, Trojan, & Prokes, 1978). GABA is present in the hypothalamus in comparatively high concentrations; iontophoretic application of GABA can inhibit ongoing neural electrical activity in several hypothalamic nuclei (e.g., PaVN, SON, VM, and LH), while iontophoretic application of the GABA-receptor-blocker picrotoxin can abolish the recurrent inhibition seen after Arc stimulation (e.g., Curtis & Johnston, 1974; Fahn, 1976; Yagi & Sawaki, 1975). Although glutamic acid decarboxylase (GAD, a rate-limiting enzyme in GABA biosynthesis) activity is unevenly distributed within the hypothalamus, the rates of GABA synthesis in this brain region generally exceed those of other areas (with several prominent exceptions, including the substantia nigra) (e.g., Tappaz, Brownstein, & Kopin, 1977; Wallis & Luttge, 1980). GABAergic neurons appear to have a facilitatory influence on the release of LH and prolactin and in the induction of sexual receptivity in female rats (e.g., McGinnis, Gordon, & Gorski, 1980b; Mioduszewski, Grandison, & Meites, 1976; Wallis, 1976), while their potential actions in other steroid-dependent behaviors remain largely unexplored.

Studies on steroid-induced changes in the biosynthesis and turnover of the polyamines and GABA are confounded by the fact that many of these compounds are interconvertible and by the fact that they display a variety of end-product inhibition- and facilitation-modulations of the synthetic enzymes (see Luttge, 1983). In spite of these limitations, several recent studies have provided provocative findings. For example, a single injection of 20 μg E_2B in ovariectomized rats was found to increase the concentrations of PUT in the MBH and POA-AH at 12 hours postinjection, whereas there were no effects on PUT in Cx or on SPD and SPM in any of the three brain regions (Gray, Jasper, Luttge, Shukla, & Rennert, 1980). The elevated PUT concentrations in these two estrogen-concentrating brain regions returned to baseline values within the next 12 hours. A similar, although more dramatic, transient increase in pituitary PUT concentrations was also observed in this study. It is as yet unknown whether these effects are the result of an estrogen-induced synthesis or activation of ODC and/or an increase in the conversion of SPD to PUT. In a similar study with glucocorticoids, exogenous B failed to influence telencephalic polyamine levels; however, adrenalectomy did prevent an ACTH-induced increase in telencephalic PUT concentration (Tintner, Dunn, Iuvone, Shukla, & Rennert, 1980). Further work on steroid-induced modifications in polyamine synthesis are currently underway in our laboratory.

The data and references on steroid hormone effects on intracerebral GABA concentrations, synthesis, catabolism, uptake, and reuptake are summarized in Table IV (further details are reviewed in Luttge, 1983). It is clear that there are no simple generalizations to describe the diversity of these effects. In some cases the physiological significance of the results is questionable due to the use of extremely high doses of steroid (e.g., Sadasivudu, Rao, & Murphy, 1977), whereas in others the use of low systemic or intracerebral doses strongly suggests that these steroids may have direct effects *in situ* on GABAergic neurons. It is important to note that these effects can be at several different stages in the life-cycle of the transmitter (i.e., synthesis, release, reuptake, catabolism, receptors, etc.) and that they may elicit very different effects (both direct and indirect—note inhibition of GAD activity in the SN following E_2 implantation in the Amg or CN; McGinnis, Gordon, & Gorski, 1980a) in different regions of the brain.

TABLE IV

Representative Effects of Steroid Hormones on Brain GABA Neurons[a]

Treatment	Effect(s)[b]	References
Content		
♂ → ♂̸	[GABA] ↓ in Amg & Sep; ↑ in MB	Early & Leonard, 1978
♂̸ + TD	[GABA] ↑ in Amg & Sep; — in MB	Early & Leonard, 1978
♂̸ + E_2D	[GABA] ↑ in Amg, Sep, & MB	Early & Leonard, 1978
♂̸ + DHTD	[GABA] — in Amg & Sep; ↓ in MB	Early & Leonard, 1978
♀ → ♀̸	[GABA] ↑ in FCx	Saad, 1970
♂ → Adx ♂	[GABA] ↓ in Cx	Pandalfo & Macaione, 1964
♂ or ♀ + F (5 days at 10 mg/day)	[GABA] ↓ in BS, Cb, & Cx (mouse)[c]	Sadasivudu, Rao, & Murphy, 1977
Synthesis		
♀̸ + E_2B	GAD ↓ in AH, Arc, CMAmg, NISM, NIST, SM, SN, VM, & VTA; — in Cx, DHpc, LS, MPOA, OB, & VS	Gordon, Nance, Wallis, & Gorski, 1977, 1979; Wallis & Luttge, 1980
♀̸ + P_4	GAD ↓ in AH, Arc, DHpc, MPOA, NISM, NIST, OB, VM, VS, & VTA; — CMAmg, Cx, LS, SM, & SN	Wallis & Luttge, 1980
♀̸ + E_2B + P_4	GAD ↓ in LS; — AH, Arc, CMAmg, Cx, DHpc, MPOA, NISM, NIST, OB, SM, SN, VM, VS, & VTA	Wallis & Luttge, 1980
♀̸ + E_2 in Amg or CN	GAD ↓ in SN; — in VTA	McGinnis, Gordon, & Gorski, 1980a
♀̸ + E_2 in NA	GAD — in SN; ↓ in VTA	McGinnis *et al.*, 1980a
♀̸ + E_2 in POA or VM	GAD — in SN & VTA	McGinnis *et al.*, 1980a

♂ → Adx ♂	GAD — in Amg, Arc, Hpc, PaVN, & SON	Acs, Palkovits, & Stark, 1980
Adx ♂ + Dex	GAD ↑ in SON; ↓ in Amg; — in Arc, Hpc, & PaVN	Acs *et al.*, 1980
Adx ♂ + B	GAD — in Hpc or whole brain	Edwards & Rousseau, 1980; Meyer, Luine, Khylcheveskaya, & McEwen, 1979
♂ + B	GAD — in AH, Cb, Cx, Hpc, PH, Sep, Str, & Th (mouse)	Dunn, Gildersleeve, & Gray, 1978
Catabolism		
♂ → Adx ♂	GABA-T ↓ in Cx; — in whole brain	Edwards & Rousseau, 1980; Pandalfo & Macaione 1964
Adx ♂ + B	GABA-T — in Hpc & whole brain	Edwards & Rousseau 1980; Meyer *et al.*, 1979
Uptake and Reuptake		
♀̶ + E_2B	[GABA] ↑ in OB; ↓ in Arc, Amg, & Hpc; — in AH, Cx, SM, & Sep minces	Wallis, 1976
♀̶ + E_2B	[GABA] ↓ in Hpc; — in Cx synaptosomes	Wallis, 1976
♂ → Adx ♂	[GABA] ↑ in Hpc; — in Cb & FCx synaptosomes	Miller, Chaptal, McEwen, & Peck, 1978
Adx ♂ + B	[GABA] ↓ in Hpc synaptosomes	Miller *et al.*, 1978
Media + B or E_2	[GABA] — in Hpc synaptosomes	Miller *et al.*, 1978; Wallis, 1976

[a] Abbreviations: ♂, intact male; ♂̶, castrated ♂; Adx ♂, adrenalectomized ♂; ♀, intact female; ♀̶, ovariectomized ♀; ↑, increase; ↓, decrease; —, no change. All other abbreviations are defined in the text and/or in Tables I and II.

[b] Unless otherwise indicated all data refer to experiments with rats.

[c] Controversial results—see text for explanation.

Catecholamines

The catecholamine neurotransmitters dopamine (DA), norepinephrine (NE), and, to a lesser extent, epinephrine have been widely implicated in the mediation of certain aspects of steroid hormone actions in neural development and aging, adenohypophyseal hormone secretion regulation, and agonistic, sexual, parental, and other behaviors (for reviews see Adler & Pfaff, 1983; Beyer, 1979; Nemeroff & Dunn, 1983; and other chapters in the present volume). The following discussion briefly reviews the actions of steroid hormones on catecholamine content, turnover, synthesis catabolism, release, reuptake, and receptor binding in mammalian brain (for summaries and references see Table V; for further details see Luttge, 1983). Because each steroid may have facilitatory and/or inhibitory effects at any step of catecholamine neurotransmission, because the steroids may interact with each other in a facilitatory and/or inhibitory fashion and, finally, because there is extensive regional specificity to the effects, summarization of these various relationships is difficult and the extrapolation of these effects to behavior is speculative at best.

For example, following castration there are transient increases in NE concentrations in the AH and ME accompanied by transient decreases in DA concentrations in the AH without similar effects in the ME. These changes in catecholamine content are paralleled by a castration-induced increase in DA and no effect on NE turnover (a pharmacological index of neuronal activity in which an electrophysiologically active system is believed to display a higher turnover rate, as measured by a greater degree of catecholamine depletion following synthesis inhibition, than a less active system) in the ME and an increase in tyrosine hydroxylase (TH, rate-limiting enzyme in the biosynthesis of catecholamines, responsible for conversion of tyrosine to DOPA) activity in the ME without a similar effect in several other hypothalamic nuclei. Replacement therapy with T (and in some cases DHT and AE) following castration restores or prevents castration-induced catecholamine content changes, while it reduces TH activity in the ME as well as turnover of NE in the MBH and of DA in the POA. Dopamine-β-hydroxylase (DBH, enzyme responsible for conversion of DA to NE) activity in these same brain regions was unaffected by castration and exogenous androgen treatment. These effects suggest that endogenous androgens have differential negative feedback effects on DA- and NE-containing neurons in different regions of the hypothalamus and limbic system. While some of these effects may be mediated by the direct actions of androgens on hypothalamic nerve terminals (e.g., suppressing NE turnover in the MBH following implantation of T into the MBH), other effects may be mediated by (genomic) actions on the cell bodies both within and outside the hypothalamus (e.g., Heritage & Grant, 1979).

In the AH of ovariectomized female rats, NE concentrations are decreased and turnover increased by exogenous E_2 replacement therapy. NE turnover in the CG, LS, and NIST and DA turnover in the DB are decreased with the same treatment. The addition of P_4 to the E_2-primed ovariectomized rat further decreases NE concentrations in the AH and increases NE turnover in the structures

TABLE V
Representative Effects of Steroid Hormones on Brain Catecholamine Neurons[a]

Treatment	Effect(s)[b]	References
Content		
♂ → ♂̸ (5 days)	[DA] — in AH, FCx, MH, & PH [NE] — in AH, FCx, MH, & PH	Donoso, Stefano, Biscardi, & Cukier, 1967
♂ → ♂̸ (10–20 days)	[DA] ↓ in AH; — in FCx, ME, MH, & PH [NE] ↑ in AH; — in FCx, ME, MH, & PH	Donoso *et al.*, 1967; Iuvone, Morasco, & Dunn, 1977
♂ → ♂̸ (13 days)	[NE] ↓ in OB (rat); — in OB (hamster)	Cornwell-Jones & Marasco, 1980
♂ → ♂̸ (4–8 hr)	[DA] — in ME[c] [NE] ↑ in ME	Chiocchio, Negro-Vilar, & Tramezami, 1976
♂ → ♂̸ (12 hr)	[DA] — in ME[c] [NE] — in ME	Chiocchio *et al.*, 1976
♂ → ♂̸ + TP (8–12 hr)	[DA] — in ME [NE] — in ME	Chiocchio *et al.*, 1976
♂ + TP, DHT, or AE	[DA] ↓ in Amg & HTh; — in Mes & Str [NE] — in Amg, HTh, Mes, & Str	Vermes, Varszegi, Toth, & Telegdy, 1979
Met ♀ → Di ♀	[DA] ↑ in LS; ↓ in DB; — in AH, CG, CMAmg, CN, ME, MPOA, NA, NIST, OB, PaVN, PeVN, SON, & VM [NE] ↓ in CG; — in AH, CMAmg, CN, DB, LS, ME, MPOA, NA, NIST, PaVN, PeVN, SON, & VM	Crowley, O'Donohue, & Jacobowitz, 1978
Di ♀ → Pro ♀	[DA] ↓ in LS; — in AH, CG, CMAmg, CN, DB, ME, MPOA, NA, NIST, OB, PaVN, PeVN, SON, & VM [NE] ↓ in PaVN; — AH, CG, CMAmg, CN, DB, LS, ME, MPOA, NA, NIST, PeVN, SON, & VM	Crowley, O'Donohue, & Jacobowitz, 1978
Pro ♀ → Est ♀	[DA] ↑ in ME; ↓ in CN; — in AH, CG, CMAmg, DB, LS, MPOA, NA, NIST, OB, PaVN, PeVN, SON, & VM [NE] — in AH, CG, CMAmg, CN, DB, LS, ME, MPOA, NA, NIST, PaVN, PeVN, SON, & VM	Crowley, O'Donohue, & Jacobowitz, 1978
Est ♀ → Met ♀	[DA] ↑ in DB; — in AH, CG, CMAmg, CN, LS, ME, MPOA, NA, NIST, OB, PaVN, PeVN, SON & VM [NE] ↑ in CG, LS, MPOA & PaVN; — in AH, CMAmg, CN, DB, ME, NA, NIST, PeVN, SON, & VM	Crowley, O'Donohue, & Jacobowitz, 1978
Met ♀ → Pro ♀[d]	[DA] ↑ in CN; ↓ in MPOA [NE] ↓ in MPOA	Crowley, O'Donohue, & Jacobowitz, 1978
Est ♀ → Di ♀[d]	[DA] ↑ in CN	Crowley, O'Donohue, & Jacobowitz, 1978
♀̸ + E_2B	[NE] ↓ in AH	Nagle & Rosner, 1980

(Continued)

TABLE V (*Continued*)

Treatment	Effect(s)[b]	References
E_2B ♀̶ + P_4	[NE] ↓ in AH	Nagle & Rosner, 1980
♂ → Adx ♂ (2 days)	[DA] — in BS & HTh	Shen & Ganong, 1976
	[NE] — in BS & HTh	
♂ → Adx ♂ (15 days)	[DA] ↓ in BS & Str; — in HTh	Rastogi & Singhal, 1978a
	[NE] — in BS; ↓ in Str & HTh	
Adx ♂ + B (4 or 24 hr)	[DA] — in BS & HTh	Shen and Ganong, 1976
	[NE] — in BS & HTh	
Adx ♂ + B (7 days)	[DA] ↑ in BS & Str; — in HTh	Rastogi & Singhal, 1978a
	[NE] — in BS; ↑ in Str & HTh	
♂ + B (10–180 min)	[DA] ↑ in ME; — in LC, SON, VM, & VTA	Fekete, Stark, Herman, Palkovits, & Kanyicska, 1978
	[NE] ↑ in ME & SON; — in LC, VM, & VTA	
Turnover[e]		
♂ → ♂̸ (10 days)	[DA] ↓ in ME[e]	Kizer, Humm, Nicholson, Greeley, & Youngblood, 1978
	[NE] — in ME	
♂̸ + T	[DA] ↑ in POA; — in MBH	Simpkins, Kalra, & Kalra, 1980
	[NE] ↑ in MBH; — in POA	
♂̸ + T in MBH	[DA] — in MBH & POA	Simpkins *et al.*, 1980
	[NE] ↑ in MBH; — in POA	
Di ♀ → Pro ♀	[NE] ↓ in MPOA	Honma & Wuttke, 1980
♀̶ + E_2B	[DA] ↑ in DB; — in CN, LS, ME, NA, NIST, OB & PeVN	Crowley, O'Donohue, Wachslicht, & Jacobowitz, 1978
	[NE] ↑ in CG, LS & NIST; ↓ in AH; — in Arc, DB, LVM, MAmg, ME, MPOA, PaVN, PeVN, & SON	
E_2B ♀̶ + P_4	[DA] ↓ in DB & NIST; — in CN, LS, ME, NA, OB, & PeVN	Crowley, O'Donohue, Wachslicht, & Jacobowitz, 1978
	[NE] ↑ in AH, LVM, PaVN & PeVN; ↓ in Arc, CG, LS, NIST, & SON; — in DB, MAmg, ME, & MPOA	
♂ → Adx ♂ (2 days)	[NE] ↓ in HTh[f]	Shen & Ganong, 1976
Synthesis		
♂ → ♂̸ (7 days)	TH ↑ in ME; — in Arc, DM, MFB, PaVN, PeVN, & POA	Kizer, Palkovits, Zivin, Brownstein, Saavedra, & Kopin, 1974
	DBH — in Arc, DM, ME, MFB, PaVN, PeVN & POA	
♂ → ♂̸ + TP (7 days)	TH — in Arc, DM, ME, MFB, PaVN, PeVN, & POA	Kizer *et al.*, 1974
	DBH — in Arc, DM, ME, MFB, PaVN, PeVN & POA	
Di ♀ → Pro ♀	[^{3}H]Tyr → [^{3}H]NE ↑ in HTh (*in vivo*)	Donoso & Moyano, 1970
♀ → ♀̶ (3 months)	TH — in AMBH, MPOA, PMBH, & NA	Krieger & Wuttke, 1980
	DBH ↑ in MPOA & PMBH; — in AMBH & NA	

♀ + E_2[Low]	TH ↑ in HTh	Tobias, Carr, & Vooft, 1979
♀ + E_2[High]	TH ↓ in MBH	Luine, McEwen, & Black, 1977
E_2B ♀ + P_4	TH ↓ in HTh	Beattie, Rodgers, & Soyka, 1972
♀ + E_2B + P_4	[^{3}H]Tyr → [^{3}H]NE ↓ in AH; — in PH *(in vivo)*	Bapna, Neff, & Costa, 1971
♂ → Adx ♂ (1 or 2 days)	DBH ↓ in BS & HTh; — in Hpc	Shen & Ganong, 1976
♂ → Adx ♂ (5 days)	DBH ↓ in BS; — in HTh	Shen & Ganong, 1976
♂ → Adx ♂ (7 days)	TH ↓ in Arc, DM, ME; — in MFB, PaVN, PeVN, & POA DBH — in Arc, DM, ME, MFB, PaVN, PeVN, & POA	Kizer *et al.*, 1974
♂ → Adx ♂ (7 or 15 days)	TH ↑ in Str	Rastogi & Singhal, 1978a
♂ → Adx ♂ (2 days)	[^{3}H]DA → [^{3}H]NE ↑ in BS & HTh *(in vivo)*	Shen & Ganong, 1976
Adx ♂ + B (4 or 24 hr)	DBH — in BS & HTh	Shen & Ganong, 1976
Adx ♂ + B[High] (4 hr)	DBH ↑ in BS & HTh; — in Hpc	Shen & Ganong, 1976
Adx ♂ + Dex (7 days)	TH ↑ in ME; — in Arc, DM, MFB, PaVN, PeVN, & POA DBH — in Arc, DM, ME, MFB, PaVN, PeVN, & POA	Kizer *et al.*, 1974
♂ + B	[^{3}H] Tyr → [^{3}H]DA & [^{3}H]NE ↑ in HTh & whole brain; — in BS, Cx, Hpc, Str, & Sep (*in vivo* mouse)	Dunn, Gildersleeve, & Gray, 1978; Iuvone *et al.*, 1977
♂ + Dex	PNMT ↑ in HTh & Med	Moore & Phillipson, 1975
Media + P_4,17α-OH-P_4, or 20α-OH-P_4	TH ↓ in HTh	Beattie & Soyka, 1973
Media + 2-OH-E_2, or 2-OH-E_1	TH ↓ in Amg, Hpc, HTh, POA, & Str	Foreman & Porter, 1980; Lloyd & Ebersole, 1980
Media + E_2 or E_1	TH — in Amg, Hpc, HTh, POA, & Str	Foreman & Porter, 1980; Lloyd & Ebersole, 1980; Luine *et al.*, 1977
Catabolism[g]		
♂ + TP	MAO ↑ in MPOA & HTh + POA + Amg	Luine, Khylcheveskaya, & McEwen, 1975b; Luine, MacLusky, & McEwen, 1979
Di ♀ → Est ♀	MAO ↑ in Amg, FCx, & HTh	Zolovick, Pearse, Boehlke, & Eleftheriou, 1966
♀ → ♀ (18 days)	MAO ↑ in HTh	Kato & Minaguchi, 1964
♀ + E_2 or E_2B	MAO ↓ in Amg & MBH	Luine *et al.*, 1975b, 1979
♂ → Adx ♂ (15 days)	MAO ↑ in Cx COMT — in Cx	Rastogi & Singhal, 1978a
Adx ♂ + B (7 days)	MAO ↓ in Cx COMT — in Cx	Rastogi & Singhal, 1978a
♂ + B	MAO ↓ in Cb, HTh, MBr, & Pons + Med; — in OB	Veals, Korduba, & Symchowicz, 1977
Media + 2-OH-E_2	COMT ↓ in whole brain & neuroblastoma culture	Breuer & Koster, 1974; Lloyd, Weisz, & Breakfield, 1978

(Continued)

TABLE V (*Continued*)

Treatment	Effect(s)[b]	References
Release		
E_2B ♀ + P_4	[NE] ↑ in jugular vein	Nagle & Rosner, 1980
♀ + E_2	[DA] ↓ in tuberohypophysial vessels	Cramer, Parker, & Porter, 1979a
E_2 ♀ + P_4	[DA] ↑ in tuberohypophysial vessels	Cramer, Parker, & Porter, 1979b
Media + E_2	[DA] & [NE] ↑ in efflux from HTh cultures	Paul, Axelrod, Saavedra, & Skolnick, 1979
Media + E_1, E_3, or B	[DA] & [NE] — in efflux from HTh cultures	Paul *et al.*, 1979
Reuptake		
♀ + E_2	[DA] — in HTh minces	Enderly & Wilson, 1974
	[NE] ↑ in HTh minces	
♀ + E_2 + P_4	[DA] ↓ in HTh minces	Enderly & Wilson, 1974
	[NE] ↑ in HTh minces	
♀ + E_2	[DA] ↑ in AH; — MBH & PH synaptosomes	Cardinali & Gomez, 1977
	[NE] ↑ in MBH & PH; — in AH synaptosomes	
♂ + F	[NE] ↑ in Cx slices	Maas & Mednieks,1971
Receptors		
♂ + E_2V	[DA-receptors] ↑ in CN	Hruska & Silbergeld, 1979
♀ + E_2	[DA-receptors] ↑ in FCx, Str, & NA + OB	Paolo, Labrie, Dupont, Barden, & Langelier, 1979
♀ + E_2 or E_2B	[α-receptors] — in HTh & MBH	Vacas & Cardinali, 1980; Wilkinson, Herdon, Pearce, & Wilson, 1979
	[β-receptors] ↑ in HTh & MBH	
Media + E_2 or E_1	[DA and NE] ↓ in brain P_2 membranes	Inaba & Kamata, 1979
Media + T or DHT	[DA and NE] ↑ in brain P_2 membranes	Inaba & Kamata, 1979
Media + B, F, or P_4	[DA and NE] — in brain P_2 membranes	Inaba & Kamata, 1979

[a] Abbreviations: Met, metestrous; Di, diestrous; Pro, proestrous; Est, estrous; TH, tyrosine hydroxylase; DBH, dopamine-β-hydroxylase; PNMT, phenylethanolamine-N-methyltransferase; MAO, monoamine oxidase; COMT, catechol-O-methyltransferase. All other abbreviations are defined in the text and/or in Tables I, II, or IV.

[b] Unless otherwise indicated all data refer to experiments with rats.

[c] Comparison made with sham ♉, which is itself decreased from ♂ controls.

[d] Note that intermediate day's values were not different from either group in this special comparison. Null effect values and signficant differences between intermediate day and either of the other days have been omitted.

[e] Turnover rates are expressed as the rate of loss of the catecholamine after TH inhibition with αMPT (except as noted below). Thus, an increase in the rate of loss (i.e., [DA] or [NE] ↓) is interpreted as an increase in the turnover rate.

[f] Turnover rate determined from the disappearance rate of [^{3}H]NE and its O-methylated metabolites following the intraventricular injection of [^{3}H]NE.

[g] Controversial result (MAO types 1 and 2 not differentiated)—see text for further details. See Table VI for additional data on MAO.

showing NE turnover decreases after E_2 priming alone and decreases NE turnover in structures previously displaying NE turnover increase or no effects with E_2 alone. P_4 treatment also reverses the E_2-induced decrease in DA turnover in the DB and NIST. The effects of ovarian hormones on TH activity are further complicated by dose-related differential responses, such that with low doses of E_2, TH activity increases in the HTh, whereas with higher doses it decreases in the MBH as it does in the HTh with the combination of E_2 and P_4. NE release and reuptake by the hypothalamus are both increased with ovarian hormone replacement therapy, and while DA release is also increased from hypothalamic organ cultures with E_2 treatment, it is decreased when measured in the hypophysial portal vessels in *in vivo* studies. DA release into the portal vessels increases following P_4 injections in the E_2-primed ovariectomized rat, and while the reuptake of this catecholamine decreases in hypothalamic minces treated with these steroids, it is increased in AH synaptosomes. The increased density of DA- and β-receptors in many regions following estrogen treatment combined with the direct competition of E_2 and E_1 for NE and DA binding to P_2 membrane fragments adds still further levels of complexity.

Adrenalectomy results in a slow-onset decrease in DA content in the BS and Str, and in NE content in the Str and HTh. There was no effect on DA content in the HTh or on NE in the BS. These effects are paralleled by a decrease in TH and DBH activity in many brain regions, although there are increases in TH activity in the Str, in the rate of DA conversion to NE in the BS and HTh, and in the turnover of NE in the HTh. Adrenalectomy also results in an increase in monoamine oxidase (MAO, type A or B not specified, thus effects could be related to indolamine catabolism), but not catechol-O-methyl-transferase (COMT) activity in the Cx (other brain regions not tested). In many cases glucocorticoid replacement therapy will reverse the effects of adrenalectomy on catecholamine neurons, often leading to an increase in catecholamine biosynthetic activity and a decrease in its catabolism. The glucocorticoids appear to have little effect on the density of catecholamine receptors.

In attempting to interpret these sometimes conflicting results, it should be noted that some of the effects may only represent pharmacological anomalies. While it may seem more physiological to consider only the catecholamine system during natural endocrine fluctuations such as the estrous cycle and diurnal rhythms, it must be remembered that these endogenous hormone changes are not monohormonal but multihormonal, thus adding to their own interpretational complexities. Even though it may be difficult to sort out the multiple possibilities, it is likely that at least some of changes in the catecholamine system result from direct actions of steroid hormones on the catecholamine-containing neurons themselves, or on neurons that synaptically interact with catecholamine-containing neurons. This conclusion is supported by histo- and biochemical data illustrating a close correspondence between steroid-concentrating neurons and the catecholamine nerve terminals and cell bodies (e.g., Heritage & Grant, 1979; Heritage, Grant, & Stumpf, 1977; Towle, Sze, & Lauder, 1979). This conclusion is challenged, however, by the observations that adenohypophyseal hormones may also influence catecholamine activity, presumably reflecting their behavioral and short feedback loop regulatory mechanisms (e.g., Anton-Tay, Pelham, & Wurtman, 1969; Dunn, Gildersleeve, & Gray, 1978).

It is likely that both genomic and nongenomic responses are involved in mediating steroid actions on the catecholamine system. For example, the estrogen-induced increase in DA- and β-receptor density could well reflect a genomic mechanism, whereas the competitive interaction between P_4 and 2-OH-E_2 on TH, as well as the competitive interaction between 2-OH-E_2 and NE and DA on COMT, are clear examples of nongenomic mechanisms of hormone actions.

Indolamines

The principal mammalian cerebral indolamines, serotonin (5HT) and melatonin, have both been implicated in the regulation of steroid hormone-influenced behaviors and adenohypophyseal hormone secretion (for reviews see Adler & Pfaff, 1983; Beyer, 1979; Nemeroff & Dunn, 1983; Reiter, 1980; and other chapters in the present book). The following discussion briefly reviews the actions of steroid hormones on brain serotonin content, turnover, synthesis, catabolism, uptake, and receptors (for summaries and references see Table VI; for further details see Luttge, 1983; Rees & Gray, 1983). As in the case of the previously discussed steroid effects on GABAergic and catecholaminergic neurons, steroid hormones influence virtually every step in serotonergic neurotransmission, have different effects in different regions of the brain, and have negative and positive interactive effects with combinations of steroids.

Castration of adult rats can result in an increase in 5HT concentrations in certain hypothalamic and nonhypothalamic brain regions, while replacement therapy with TP can result in a dose-dependent decrease in 5HT content. These results contrast with the observation of either no effect or an increase in tryptophan hydroxylase (TPH, rate-limiting enzyme in serotonin biosynthesis) activity in certain brain regions following castration and a reversal of these effects with exogenous TP treatment. These latter findings (Engel, Ahlenius, Almgren, Carlsson, Larsson, & Sodersten, 1979), however, have been questioned by other investigators (Alderson, Starr, & Baum, 1979). Exogenous TP treatment has also been shown to increase MAO type A (i.e., indolamine substrate preference) activity in the MPOA (but not in the Amg and HTh), while pharmacological concentrations of T can be shown to inhibit 5HT reuptake by whole brain synaptosomes.

The effects of ovarian hormones on brain serotonergic neurons are complicated by the often opposite or nullifying interactive effects of E_2 and P_4 replacement therapy. In certain brain regions, estrogen appears to decrease 5HT content and turnover, decrease MAO activity, and increase 5HT uptake by synaptosomal preparations. In several cases the addition of P_4 eliminates these effects. Estrous-cycle-related effects also suggest a direct role of endogenous steroids in the regulation of brain serotonergic neurons.

The effects of adrenalectomy and exogenous glucocorticoid treatment on brain serotonergic neurotransmission are clearly the most controversial and difficult to replicate. Numerous essentially opposite findings are summarized in

TABLE VI
Representative Effects of Steroid Hormones on Brain Serotonin Neurons[a]

Treatment	Effect(s)[b]	References
Content		
♂ → ♂̸ (6 wk)	[5HT] ↑ in brain (minus HTh); — in HTh	Bernard & Paolino, 1974
♂ → ♂̸ (1 wk)	[5HT] ↑ in MFB & VM; — in AH, Arc, DM, MPOA, & SCN	van der Kar, Levine, & Van Orden, 1978
♂̸ + TP (5–10 mg/2 days)	[5HT] ↓ in AH, Arc, MFB, SCN, & VM; — in DM & MPOA	van der Kar *et al.*, 1978
♂̸ + TP (1 mg/day)	[5HT] — in AH, Amg, Arc, CN, DM, Hpc, MFB, MPOA, SCN, & VM	van der Kar *et al.*, 1978
Met ♀ (AM) → Pro ♀ (AM)	[5HT] ↑ in LPOA; — in AH, CAmg, Cx, Hpc, LDAR-A8, LH, LNER-A7, MBH, MDAR-A10, MDH, MPOA, Raphe, Sep, SN, & Str	Kueng, Wirz-Justice, Menzi, & Chappuis-Arndt, 1976
Pro ♀ (AM) → Pro ♀ (PM)	[5HT] ↓ in LDAR-A8, LNER-A7, LPOA, MBH & MDH; — in AH, CAmg, Cx, Hpc, LH, MDAR-A10, MPOA, Raphe, Sep, SN, & Str	Kueng *et al.*, 1976
Pro ♀ (PM) → Est ♀ (AM)	[5HT] — in AH, CAmg, Cx, Hpc, LDAR-A8, LH, LNER-A7, LPOA, MBH, MDAR-A10, MDH, MPOA, Raphe, Sep, SN, & Str	Kueng *et al.*, 1976
Est ♀ (AM) → Est ♀ (PM)	[5HT] ↑ in AH, CAmg, LDAR-A8, LNER-A7, LPOA, MBH, MDAR-A10, & Sep; — in Cx, Hpc, LH, MDH, MPOA, Raphe, SN, & Str	Kueng *et al.*, 1976
Est ♀ (PM) → Met ♀ (AM)	[5HT] ↓ in AH, LNER-A7 & MDAR-A10; — in CAmg, Cx, Hpc, LDAR-A8, LH, LPOA, MBH, MDH, MPOA, Raphe, Sep, SN, & Str	Kueng *et al.*, 1976
♀̸ + E_2B + P_4	[Try] ↑ in AH & PH	Bapna, Neff, & Costa, 1971
♀̸ + E_2B	[5HT] ↓ in AH[c]; — in DB, DM, ME, MPOA, NIST, PaVN, SCN, SON, & VTA	Crowley, O'Donohue, Muth, & Jacobowitz, 1979; Munaro, 1978
♀̸ + P_4	[5HT] ↑ in DB & VTA; — in AH, DM, ME, MPOA, NIST, PaVN, SCN, & SON	Crowley *et al.*, 1979
E_2B ♀̸ + P_4	[5HT] ↑ in ME; — in AH, DB, DM, MPOA, NIST, PaVN, SCN, SON, & VTA	Crowley *et al.*, 1979; Munaro, 1978

(Continued)

TABLE VI (*Continued*)

Treatment	Effect(s)[b]	References
♂ → Adx ♂ (7 days)	[Try] — in BS (SD rat) [5HT] ↓ in BS & Str (SD rat); — in HTh (SD rat); ↓ in HTh (RA rat) [5HIAA] ↑ in BS (SD rat); — in HTh & Str (SD rat)	Rastogi & Singhal, 1978b; Vermes, Telegdy, & Lissak, 1973
♂ → Adx ♂ (10 days)	[5HT] — in BS & TD (SD rat)	Bapna *et al.*, 1971
♂ → Adx ♂ (15 days) (SD rat)	[Try] ↓ in BS [5HT] ↓ in BS & Str; — in HTh [5HIAA] ↑ in BS & Str; — in HTh	Rastogi & Singhal, 1978b
Adx ♂ + B (7 days at 1 mg/kg/day) (SD rat)	[Try] ↑ in BS [5HT] ↑ in BS & Str; — in HTh [5HIAA] ↓ in BS & Str; — in HTh	Rastogi & Singhal, 1978b
Adx ♂ + B (1 mg/kg) (RA rat)	[5HT] ↑ in HTh (30–60 min); — in HTh (7 days)	Vermes *et al.*, 1973
♂ + B (1 mg/kg) (RA and CFY rats)	[5HT] ↑ in HTh & Mes (30 min); — in HTh & Mes (1, 2, or 24 hr)	Kovacs, Telegdy, & Lissak, 1977; Vermes *et al.*, 1973
♂ + B (30 min) (CFY rat)	[5HT] ↑ in HTh & Mes (1 mg/kg); — in HTh & Mes (5 mg/kg); ↓ in HTh & Mes (10 mg/kg)	Kovacs *et al.*, 1977
♂ + DOC (1 or 25 mg/kg) (RA and CFY rats)	[5HT] — in HTh & Mes (30–60 min)	Kovacs *et al.*, 1977; Vermes *et al.*, 1973
♂ + FA (5 mg/kg) (6 hr) (SD rat)	[5HT] ↓ in whole brain (26, 46, & 66 days of age) — in whole brain (107 & 225 days of age)	Green & Curzon, 1975
♂ + FA (20 mg/kg) (90 min) (mouse)	[5HT] ↓ in whole brain (2 months of age)	Neckers & Sze, 1975
♂ + BA in MBH (SD rat)	[5HT] ↑ in MBH	Ulrich, Yuwiler, & Geller, 1975
Turnover		
♀ + E_2B	[5HIAA] ↑ in AH[c,d]	Munaro, 1978
♀ + E_2B + P_4	[5HIAA] — AH[d]	Munaro, 1978
♀ + P_4	[5HT] ↑ in HTh, Raphe, & Sep[e]	Ladisich, 1974

♂ + FA	[5HT] ↑ in whole brain (mouse)[e]	Neckers & Sze, 1975
Synthesis		
♂ → ♂̸ (9 days) ± TP	TPH — in Amg, Arc, Hpc, ME, MFB, MPOA, PeVN, Raphe, SCN, SON, & VM	Kizer, Palkovits, Kopin, Saavedra, & Brownstein, 1976
♂ → ♂̸ (50 days)	TPH ↑ in diencephalon & limbic forebrain[f]	Engel, Ahlenius, Almgren, Carlsson, Larsson, & Sodersten, 1979
♂̸ (50 days) + TP	TPH ↓ in diencephalon & limbic forebrain[f]	Engel *el al.*, 1979
♀̸ + E_2B + P_4	[^{3}H or ^{14}C]Try → [^{3}H – ^{14}C]5HT — in brain *(in vivo)*	Bapna *et al.*, 1971; Hyyppa, Cardinali, Baumgarten, & Wurtman, 1973
♂ → Adx ♂ (9 days) ± Dex	TPH — in Amg, Arc, Hpc, ME, MFB, MPOA, PeVN, Raphe, SCN, Sep, SON, & VM	Kizer *et al.*, 1976
♂ → Adx ♂ (1–8 wk) ± B	TPH — in whole brain (mouse)	Sze, Neckers, & Towle, 1976
♂ → Adx ♂	TPH — in BS (7 days); ↓ in BS (15 days)	Rastogi & Singhal, 1978b
Adx ♂ + B (7 days)	TPH ↑ in BS	Rastogi & Singhal, 1978b
♂ → Adx ♂ (1–2 wk)	TPH ↓ in MB	Azmitia & McEwen, 1969
Adx ♂ + B (4 hr–5 days)	TPH ↑ in MB	Azmitia & McEwen, 1969
♂ + B (1–4 hr)	TPH ↑ in FB & MB	Azmitia & McEwen, 1974, 1976
♂ + FA (1 hr)	TPH — in whole brain (mouse) 5HTP-DC — in whole brain (mouse)	Neckers & Sze, 1975
♂ → Adx ♂ (10 days)	[^{3}H]Try → [^{3}H]5HT ↓ in BS; — in TD *(in vivo)*	Azmitia, Algeri, & Costa, 1970
♂ + B (10 min)	[^{14}C]Try → [^{14}C]5HT ↑ in whole brain *(in vivo)*	Millard, Costa & Gall, 1972
Media + B (10^{-4}*M*)	TPH — in MB & whole brain	Azmita & McEwen, 1974; Sze *et al.*, 1976
Catabolism[g]		
♂̸ + TP	MAO ↑ in MPOA; — in Amg & HTh	Luine, Khylcheveskaya, & McEwen, 1975b
♂̸ + E_2B	MAO — in Amg, HTh, & MPOA	Luine *et al.*, 1975b
Pro ♀ (AM) → Pro ♀ (PM)	MAO ↓ in AH, MBH, & Raphe; — in CMAmg, Cx, Hpc, LDAR-A8, LNER-A7, LPOA, MDAR-A10, MDH, MPOA, Sep, SN, & Str	Kueng *et al.*, 1976
Est ♀ (AM) → Est ♀ (PM)	MAO ↑ in AH, MBH, & Raphe; — in CMAmg, Cx, Hpc, LDAR-A8, LNER-A7, LPOA, MDAR-A10, MDH, MPOA, Sep, SN, & Str	Kueng *et al.*, 1976

(Continued)

TABLE VI (*Continued*)

Treatment	Effect(s)[b]	References
♀ + TP	MAO — in CMAmg, Cx, MBH, & MPOA	Luine *et al.*, 1975b
♀ + E_2B	MAO ↓ in CMAmg & MBH; — in Cx & MPOA	Luine & McEwen, 1977a; Luine *et al.*, 1975b
Uptake and Reuptake		
♀ + E_2	[5HT] ↑ in AH[c] & PH; — in MBH synaptosomes	Cardinali & Gomez, 1977
♀ + $E_2 \pm P_4$	[5HT] — in HTh minces	Enderly & Wilson, 1974
Media + T ($10^{-4}M$)	[5HT] ↓ in whole brain synaptosomes[f]	Nixon, Janowsky, & Davis, 1974
Media + E_2 or P_4	[5HT] — in whole brain synaptosomes	Nixon *et al.*, 1974
Media + FA ($10^{-7}M$)	[Try] ↑ in whole brain synaptosomes	Neckers & Sze, 1975
Media + B ($10^{-5}M$)	[Try] ↑ in whole brain synaptosomes	Neckers & Sze, 1975
Media + DHT or 17α-OH-P_4	[Try] — in whole brain synaptosomes	Neckers & Sze, 1975
Media + Dex or F	[Try] — in Cx slices (Adx & non-Adx)	Kiely, 1980
Receptors		
Di ♀ → Pro ♀	[5HT] ↑ in membrane fragments from basal forebrain; — in membrane fragments from CN, Cx, & Hpc	Biegon, Bercovitz, & Samuel, 1980

[a] Abbreviations: TPH, tryptophan hydroxylase; 5HTP-DC, 5-hydroxytryptophan decarboxylase; Try, tryptophan, CFY, RA, and SD, albino rat strains. All other abbreviations are defined in the text and/or in Tables I, II, IV, or V.

[b] Unless otherwise indicated all data refer to experiments with rats.

[c] Morning (AM) samples only—no differences in afternoon (PM) samples.

[d] Turnover rate is expressed as the rate of decrease in 5HIAA concentration after MAO inhibition with pargyline. Thus, a reduction in the rate of loss (i.e., [5HIAA] ↑) is interpreted as a decrease in the rate of 5HT turnover.

[e] Turnover rate is expressed as the rate of increase in 5HT concentration after MAO inhibition with pargyline. Thus, an increase in the rate of 5HT accumulation (i.e., [5HT] ↑) is interpreted (e.g., Ladisich, 1974) as an increase in the rate of 5HT turnover. Other workers (e.g., Neckers & Sze, 1975), however, have used this same assay as an index of 5HT synthetic activity.

[f] Controversial results—see text for explanation.

[g] See Table V for additional steroid effects on intracerebral MAO activity.

Table VI. Some of these apparent discrepancies may be due to differences in age, species, strain, hormone treatment paradigm (i.e., dose, route, and/or duration of administration), and dissection criteria, while others remain to be resolved.

ACETYLCHOLINE

Acetylcholine (ACh) is a predominantly excitatory neurotransmitter implicated in the mediation of certain steroid hormone effects on adenohypophyseal hormone secretion and various behaviors (for reviews see Frohman & Berelowitz, 1983; Ganong, 1974; Meyerson & Eliasson, 1977; and other chapters in the present book). Several recent studies have indicated that androgens, estrogens, and, to an apparently much lesser extent, glucocorticoids may directly or indirectly influence the synthesis and catabolism of this important neurotransmitter (for summaries and references see Table VII; for further details see Luttge, 1983; Rees & Gray, 1983).

Castration of adult male rats results 15 days later in increased ACh concentrations in the MPOA, RDB, and VTA and choline acetyltransferase (ChAcT, rate-limiting enzyme in ACh biosynthesis) activity in the MPOA and PMAmg, while it has no effect on these parameters in many other brain regions. Replacement therapy with TP started two days after castration partially prevented the increases in ACh and ChAcT and even led to a reduction of ChAcT activity in the RDB. When hormone treatment was delayed until one week after castration, however, ChAcT activity in the MPOA was further increased by TP but not E_2B treatment. The apparent brain region and steroid/sex specificity of these androgenic effects is interesting since some of the affected brain regions are rich in androgen- and estrogen-receptors (e.g., MPOA), while others are not (e.g., VTA), and while still other brain regions rich in both steroid-receptors fail to display a cholinergic response to hormone treatment (e.g., MBH). These somewhat surprising exceptions are further illustrated by experiments with ovariectomized, ovariectomized/adrenalectomized, and hypophysectomized adult female rats in which three days of treatment with E_2B but not TP increased ChAcT activity in the estrogen-concentrating MPOA and CMAmg, but not in the MBH, which of course also markedly concentrates estradiol. At higher doses of E_2B, ChAcT activity was also increased in the estrogen-receptor-poor Hpc. This latter effect may have been due to the actions of estrogen in the septal region (an estrogen-concentrating nuclear group with strong cholinergic projections to the Hpc) rather than to the direct actions of this hormone in the Hpc.

Neither E_2B nor P_4 treatment alone altered ACh concentrations in the brains of ovariectomized female rats, but when P_4 was given in combination with E_2B, ACh concentrations were decreased in the VTA and PeVN. Since anti-estrogen (MER-25) administration (not shown in Table VII) can block the effects of E_2B on ChAcT in female rats (Luine & McEwen, 1977b; Luine, Khylcheveskaya, & McEwen, 1975b), it is probable that they are mediated through a steroid-receptor–acceptor interaction. The possibility that the mechanism is genomic was further strengthened by the observations that there is no direct cell-free effect

TABLE VII
Representative Effects of Steroid Hormones on Brain Acetylcholine Neurons[a]

Treatment	Effect(s)[b]	References
Content		
♂ → ☿ (15 days)	[ACh] ↑ in MPOA, RDB & VTA; — in AH, Arc, CAmg, CDB, CG, CN, DM, DVDB, Hb, Hpc, IP, LS, ME, NA, OB, PaVN, PeVN, PMAmg, SON, SN, & VM	Muth, Crowley, & Jacobowitz, 1980
♂ → ☿ + TP (12 days at 100 μg/day)	[ACh] — in AH, Arc, CAmg, CDB, CG, CN, DM, DVDB, Hb, Hpc, IP, LS, ME, MPOA, NA, OB, PaVN, PeVN, PMAmg, RDB, SON, SN, VM, & Vta	Muth *et al.*, 1980
☿ + E_2B or P_4	[ACh] — in AH, Arc, CAmg, CDB, CG, CN, DM, DVDB, Hpc, IP, LS, ME, MPOA, NA, OB, PaVN, PeVN, PMAmg, RDB, SON, SN, VM, & VTA	Muth *et al.*, 1980
♀ + E_2B + P_4	[ACh] ↓ in VTA; — in AH, Arc, CAmg, CDB, CG, CN, DM, DVDB, Hb, Hpc, IP, LS, ME, MPOA, NA, OB, PaVN, PeVN, PMAmg, RDB, SON, SN, & VTA	Muth *et al.*, 1980
E_2B ♀ + P_4	[ACh] ↓ in PeVN; — in AH, Arc, CAmg, CDB, CG, CN, DM, DVDB, Hb, Hpc, IP, LS, ME, MPOA, NA, OB, PaVN, PMAmg, RDB, SON, SN, VM, & VTA	Muth *et al.*, 1980
Synthesis		
♂ → ☿ (15 days)	ChAcT ↑ in MPOA & PMAmg; — in AH, Arc, CAmg, CDB, CG, CN, DM, DVDB, Hb, Hpc, IP, LS, ME, NA, OB, PaVN, PeVN, RDB, SON, Sn, VM, & VTA	Muth *et al.*, 1980
♂ → ☿ + TP (12 days at 100 μg/day)	ChAcT ↓ in RDB; — in AH, Arc, CAmg, CDB, CG, CN, DM, DVDB, Hb, Hpc, IP, LS, ME, MPOA, NA, OB, PaVN, PeVN, PMAmg, SON, SN, VM, & VTA	Muth *et al.*, 1980
☿ + TP (7 days at 28 μg/day)	ChAcT ↑ in MPOA; — in Amg & HTh	Luine, Khylcheveskaya, & McEwen, 1975b
☿ + E_2B (7 days at 30 μg/day)	ChAcT — in Amg, HTh, & MPOA	Luine *et al.*, 1975b
♀ + E_2B (5 μg) (54 hr)	ChAcT — in AH, Arc, CAmg, CDB, CG, CN, DM, DVDB, Hb, Hpc, IP, LS, ME, MPOA, NA, OB, PaVN, PeVN, PMAmg, RDB, SON, SN, VM & VTA	Muth *et al.*, 1980

♀̸ + E_2B (1–3 days at 5–100 μg/day)	ChAcT ↑ in CMAmg, Hpc[c] & MPOA; — in MBH	Luine *et al.*, 1975b; Luine, Park, Joh, Reis, & McEwen, 1980
Hypox ♀ + E_2B (3 days at 5 μg/day)	ChAcT ↑ in CMAmg, Hpc & MPOA; — in MBH	Luine *et al.*, 1975b
♀̸ or Hypox ♀ + TP	ChAcT — in CMAmg, Hpc, MBH, & MPOA	Luine *et al.*, 1975b
♀̸ + E_2 (100 μg) (3 hr)	ChAcT — in CMAmg, Hpc, MBH, & MPOA	Luine *et al.*, 1980
♀̸ + P_4 (1.5 mg) (6 hr)	ChAcT ↓ in CDB; — in AH, Arc, CMAmg, CG, CN, DM, DVDB, Hb, Hpc, IP, LS, ME, MPOA, NA, OB, PaVN, PeVN, PMAmg, RDB, SON, SN, VM, & VTA	Muth *et al.*, 1980
♀̸ + E_2B + P_4	ChAcT ↑ in SON; ↓ in CDB & PeVN; — in AH, Arc, CAmg, CG, CN, DM, DVDB, Hb, Hpc, IP, LS, ME, MPOA, NA, OB, PaVN, PMAmg, RDB, SN, VM, & VTA	Muth *et al.*, 1980
E_2B ♀̸ + P_4	ChAcT ↓ in CDB; — in AH, Arc, CAmg, CG, CN, DM, DVDB, Hb, Hpc, IP, LS, ME, MPOA, NA, OB, PaVN, PeVN, PMAmg, RDB, SON, SN, VM, & VTA	Muth *et al.*, 1980
Adx ♂ + B	ChAcT — in Hpc	Meyer, Luine, Khylcheveskaya & McEwen, 1979
Media + E_2	ChAcT — MPOA	Luine *et al.*, 1975b
Catabolism		
♀ → ♀̸ (21 days)	AChE ↓ in CB & Cx (at 9 & 29, but not 65 weeks of age)	Moudgil & Kanungo, 1973
♀̸ + E_2 (10 μg/100g) (4 hr)	AChE ↑ in Cb & Cx (at 9 & 29, but not 65 weeks of age)	Moudgil & Kanungo, 1973
♂ → Adx ♂	AChE ? in Hpc[d]	Meyer *et al.*, 1979
Adx ♂ + B	AChE — in Hpc	Meyer *et al.*, 1979
Uptake and Reuptake		
♂ and ♀ + DOCA, FA, or TDA	[Ch] ↑ in CP; — in APS, Hpc, & PFCx synaptosomes (cat)[e]	Ricker, Sastre, Baker, Roth, & Riker, 1979
Media + DOCA, FA, or TDA	[Ch] — in APS, CP, Hpc, & PFCx synaptosomes (cat)	Ricker *et al.*, 1979

[a] Abbreviations: Hypox, hypophysectomized; Ch, choline; ACh, acetylcholine; ChAcT, choline acetyltransferase; AchE, acetylcholine esterase. All other abbreviations are defined in the text and/or in Tables I, II, and IV.

[b] Unless otherwise indicated all data refer to experiments with rats.

[c] Increase in ChAcT significant only with the 30 μg/day × 3 days treatment.

[d] Actual effect is unclear since paper only indicated that Adx group was significantly different from sham Adx group.

[e] Sex not specified.

of E_2 on ChAcT activity and no increase in ChAcT activity within three hours after a single injection of E_2, whereas there is a clear increase in immunoprecipitable ChAcT (indicating an increase in the number of ChAcT molecules) following one–three days of E_2B treatment (Luine, Park, Joh, Reis, & McEwen, 1980). The failure of some workers to achieve similar results with different estrogen treatment paradigms (e.g., 5 μg E_2, rats killed for ChAcT assay 54 hours later) suggests that the effects are not robust and that much more work is needed before the hormone responsiveness of ChAcT can be fully described.

The termination of the synaptic action of ACh is primarily mediated through degradation by acetylcholinesterase (AChE). A single injection of E_2 can result four hours later in a significant increase in AChE activity in the cerebral and cerebellar cortices of immature and adult rats. Since these effects can be blocked by actinomycin-D, administered one hour before E_2 (not shown in Table VII; Moudgil & Kanugo, 1973), the observed increase in AChE activity may represent yet another example of E_2-stimulated *de novo* protein synthesis. However, since there are few E_2-concentrating cells in these two brain regions, further work is needed before this genomic possibility can be considered seriously.

Choline (Ch) uptake and reuptake by the presynaptic terminal is an important requirement for continued ACh biosynthesis. This aspect of cholinergic neurotransmission was found to be increased in the CP of adult cats with the *in vivo* but not *in vitro* administration of several synthetic glucocorticoids. These effects were not seen in several other brain regions, suggesting that although the glucocorticoid actions in the CP may not be direct, they may be regionally specific. Adrenalectomy has been reported to influence AChE activity in the Hpc, but the direction of this effect was not specified and it was apparently not reversed with glucocorticoid replacement therapy. Exogenous glucocorticoids also failed to influence ChAcT activity in the Hpc even though this region is exceptionally rich in cytosolic and nuclear glucocorticoid-receptors. Thus, the presence or absence of steroid-receptors is no guarantee of steroid responsivity.

Histamine

Although the neurotransmitter role of histamine (HA) in the mammalian brain is still controversial, the administration of this amine, or of drugs that block or mimic its actions at membrane receptors (i.e., H_1 and H_2 sites), influences adenohypophyseal hormone secretion and intracerebral adenylate cyclase activity (for reviews see Donoso, 1978; Schwartz, Pollard, & Quach, 1980). It is thus undoubtedly important that steroid hormones may influence the content, synthesis, catabolism, and actions of HA in the mammalian brain (for references and a summary of apparent steroid effects see Table VIII).

For example, castration of adult male rats leads to an increase in HTh HA concentrations, while adrenalectomy has no effect on whole mouse brain HA concentrations. The activity of histadine decarboxylase (HAD, rate-limiting enzyme in HA biosynthesis) in rabbit HTh is increased after ovariectomy (and after thyroidectomy, the effect being even greater with both operations), while it is decreased in mouse whole brain following adrenalectomy. In immature male

TABLE VIII
Representative Effects of Steroid Hormones on Brain Histamine Neurons[a]

Treatment	Effect(s)[b]	References
Content		
♂ → ♂̸ (41 days)	[HA] ↑ in HTh	Orr & Quay, 1975
♂ → Adx ♂ (5 days)	[HA] — in whole brain (mouse)	Verdiere, Rose, & Schwartz, 1977
Synthesis and Catabolism		
♀ → ♀̸ (20 days)	HAD ↑ in HTh (rabbit)	Bjorklund, Hakamson, Nobin, & Sjoberg, 1972
♂ → Adx ♂ (5 days)	HAD ↓ in whole brain (mouse) $[^3H]$His → $[^3H]$HA ↓ in whole brain (*in vivo* mouse)	Verdiere *et al.*, 1977
♂[c] + T (1 mg) (30–360 min)	$[^{14}C]$His → $[^{14}C]$HA & $[^{14}C]$m-HA — in HTh (*in vivo*)	Mosebach & Peter, 1971
♂[c] + T (5 days at 100 μg/day)	$[^{14}C]$His → $[^{14}C]$HA & $[^{14}C]$m-HA — in Cb, FCx, & HTh	Mosebach & Peter, 1971
♂[c] + 19norT (1 mg) (30–360 min)	$[^{14}C]$His → $[^{14}C]$HA — in HTh; → $[^{14}C]$m-HA ↑ in HTh	Mosebach & Peter, 1971
Receptors and Action		
♂ → ♂̸ (31 days)	[cAMP] — in HTh with HA [10^{-6} or $10^{-5}M$]	Portaleone, Genazzani, Pagnini, Crispino, & DiCarlo, 1980
♀ → ♀̸ (31 days)	[cAMP] ↑ in HTh with HA [10^{-6} or $10^{-5}M$]	Portaleone *et al.*, 1980
♂̸ + E_2 (10 days at 10 μg/kg/day)	[cAMP] ↓ in HTh with HA [$10^{-5}M$]	Portaleone *et al.*, 1980
♀̸ + mT (10 days at 100 μg/kg/day)	[cAMP] — in HTh with HA [$10^{-5}M$]	Portaleone *et al.*, (1980)
Media + E_2 or 2-OH-E_2 [10^{-6} or $10^{-5}M$]	[cAMP] ↓ in HTh with HA [$10^{-5}M$]	Portaleone *et al.*, 1980
Media + T or P_4 [10^{-6} or $10^{-5}M$]	[cAMP] — in HTh with HA [$10^{-5}M$]	Portaleone *et al.*, 1980

[a] His, histidine; HA, histamine; m-HA, methyl histamine, cAMP, cyclic adenosine monophosphate; HAD, histidine decarboxylase. All other abbreviations are defined in the text and/or in Tables I, II, or IV.
[b] Unless otherwise indicated all data refer to experiments with rats.
[c] Immature animals.

rats, exogenous T does not appear to influence the HTh synthesis of HA or its metabolism to methyl-HA (m-HA), whereas the acute administration of 19-nor-T does increase HTh production of m-HA. Last, it has been very recently shown that HA stimulation of adenylate cyclase in HTh membrane fragments from gonadectomized adult male and female rats is inhibited by the *in vivo* and *in vitro* administration of E_2 but not methyl-T (mT) or P_4. 2-OH-E_2 was also effective *in vitro*. E_2 and 2-OH-E_2 *in vitro* inhibition of HA-stimulated adenylate cyclase clearly suggests a nongenomic competitive interaction between the estrogen and HA for the HTh E_2 binding sites (Portaleone, Pagnini, Crispino, & Genazzani, 1978). This hypothesis is supported by the recent observation that E_2 and 2-OH-E_2 (without HA) actually increase HTh membrane adenylate cyclase activity (Portaleone, Genzazzini, Pagnini, Crispino, & DiCarlo, 1980). The relationship of these provocative preliminary findings to the behavioral and neuroendocrine actions of steroid hormones clearly deserves further investigation (e.g., Mazurkiewicz-Kwilecki & Prell, 1980).

Cyclic Adenosine Monophosphate

Intracerebral cyclic adenosine monophosphate (cAMP) is known often to play a central role in mediating protein and polypeptide hormone and neurotransmitter actions and synthesis (e.g., Debus & Kehr, 1979; Sundberg, Fawcett, & McCann, 1976; Williams, 1979). Several recent studies have provided direct and indirect biochemical and behavioral data suggesting that the intracerebral actions of steroids may also be mediated, at least in part, through a modulation of cAMP synthesis. For example, systemic E_2 increases and pretreatment with anti-estrogens decreases HTh cAMP concentration (Gunaga, Kawano, & Menon, 1974). Since this effect can be blocked by *in vivo* pretreatment with either α- or β-blockers, it is thought to be mediated by catecholamines. *In vitro* studies have confirmed and extended these findings, establishing that they may be confined to the estrogen-concentrating regions of the HTh, are probably mediated through an estrogen-receptor–acceptor interaction, are inhibited by DA-receptor blockers (e.g., pimozide and haloperidol) as well as α- and β-blockers, require more time to increase HTh cAMP than direct exposure to catecholamines (e.g., 50 versus 5 minutes), and facilitate NE and DA efflux and increase cAMP concentrations with a similar time course (e.g., Gunaga & Menon, 1973; Paul, Axelrod, Saavedra, & Skolnick, 1979; Weissman & Johnson, 1976; Weissman & Skolnick, 1975; Weissman, Daly, & Skolnick, 1975). These *in vitro* findings clearly support the notion that estrogen-induced increases in HTh cAMP are probably mediated by increases in DA and/or NE synaptic activity (e.g., Williams, 1979).

The correlation between estrogen-induced HTh catecholamine efflux and increased cAMP concentrations has been extended by demonstrating that E_1 and E_3 unlike E_2, do not exhibit physiologically relevant potency in either catecholamine efflux or cAMP accumulation (Paul & Skolnick, 1977; Paul *et al.*, 1979). While 2-OH-E_2 also has no effect on HTh cAMP concentrations, it inhibits E_2- and DES-stimulated HTh cAMP accumulations when given before or concurrently, but not 40 minutes after these estrogens. Although [^{3}H]2-OH-E_2 has very recently been shown to bind with high affinity to membrane receptors in HTh, Str, Cx, and other brain regions that also display high-affinity binding for DA (Schaeffer & Hseuh, 1979; Schaeffer, Stevens, & Smith, 1980), the DA receptors are probably not involved in 2-OH-E_2-induced anti-estrogenic activity: DA but not E_2 competes with 2-OH-E_2 for membrane-associated binding, yet 2-OH-E_2 inhibits E_2-induced cAMP accumulation only when given prior to the expected release of catecholamines (Paul & Skolnick, 1977; Paul *et al.*, 1979). Furthermore, since 2-OH-E_2 can effectively compete with E_2 for HTh cytosolic receptor binding (Davies, Naftolin, Ryan, Fishman, & Siu, 1975), the inhibition of this potential genomic mechanism of E_2 action may be responsible for the 2-OH-E_2-mediated loss of E_2-stimulated HTh adenylate cyclase.

The possibility that estrogens may also directly influence HTh cAMP synthesis was supported by the recent demonstration that E_2 and (unlike the results discussed above) 2-OH-E_2, but not mT or P_4, incubated with HTh membrane fragments increases adenylate cyclase activity (Portaleone *et al.*, 1980). Even though the molecular mechanisms of this effect are still unclear, they cannot involve genomic interactions since there were no intact nuclei or cells in the preparation. Other examples of estrogen involvement in brain cAMP systems include the previously discussed estrogen inhibition of HA-induced increases in HTh cAMP (Portaleone *et al.*, 1978, 1980), and the estrogen-dependent maintenance of adenylate cyclase-sensitive DA- and β-receptors in the hypothalamus and other brain regions (e.g., Kumakura, Hoffman, Cocchi, Trabucchi, Speno, & Muller, 1979; for an opposing view see Paola, Labrie, Dupont, Barden, & Langelier, 1979).

A recent study with male rats suggests that glucocorticoids may also be able to alter brain adenylate cyclase sensitivity to catecholamines (Mobley & Sulser, 1980). Adrenalectomy was found to increase and exogenous replacement therapy with B to decrease the maximal *in vitro* stimulation of FCx adenylate cyclase with NE. Since adrenalectomy failed to alter the basal levels of cAMP or the binding of $[^3H]$DHA to β-receptors, the mechanism of these effects is unknown.

Carbohydrate Metabolism

The fact that glucose is the primary, if not sole, energy source for the adult mammalian brain (under normal conditions) is consistent with the observation that natural and experimental diabetes can dramatically inhibit the receptor-binding and behavioral and neuroendocrine actions of steroids (e.g., Denari & Rosner, 1972; Florez-Lozano, Menendez-Patterson, & Marin, 1978; Gentry, Wade, & Blaustein, 1977; Schiaffini, Marin, & Foglia, 1970; Siegel & Wade, 1979). Thus, as might be predicted, direct and indirect actions of gonadal (but not as yet adrenal) steroids have also been reported to influence the tricarboxylic and glycolytic, but not pentose phosphate, cycle enzyme activities in the brain (for references and a summary of steroid effects see Table IX).

In the male rat, castration decreases oxygen consumption using either glucose, citrate, or succinate as the *in vitro* carbon source, in slices of AH and PH, while it increases oxidative activity in the Amg and Cx and has no effect in the

TABLE IX
Representative Effects of Steroid Hormones on Brain Carbohydrate Metabolism[a]

Treatment	Enzymatic analyses or metabolism substrates	Effect(s)[b]	References
♂ → ⚦	Glucose	↑ in Amg & Cx; ↓ in AH & PH; — in Hpc & MH	Moguilevsky, Kalbermann, Libertun, & Gomez, 1971; Schiaffini & Martini, 1972
	[U-^{14}C]Glucose	↓ in HTh; — in Cx	Scacchi, Moguilevsky, & Szwarcfarb, 1971
	Citrate or succinate	↓ in AH & PH; — in MH	Moguilevsky, Kalbermann, Libertun, & Gomez, 1971
	Glutamate or pyruvate	— in AH, MH, & PH	Moguilevsky, Kalbermann, Libertun, & Gomez, 1971
	CytOx	↓ in AH & PH; — in MH	Moguilevsky, Kalbermann, Libertun, & Gomez, 1971
	SDH	— in AH, MH, & PH	Moguilevsky, Kalbermann, Libertun, & Gomez, 1971
⚦ + T or TP	Glucose	↑ in AH & PH; ↓ in Amg & Cx; — in Hpc & MH	Moguilevsky, Kalbermann, Libertun, & Gomez, 1971; Schiaffini & Martini, 1972
	[U^{14}C]Glucose	↑ in HTh; — in Cx	Scacchi *et al.*, 1971
	Citrate	↑ in AH & PH; — in MH	Moguilevsky, Kalbermann, Libertun, & Gomez, 1971
	Glutamate or pyruvate	— in AH, MH, & PH	Moguilevsky, Kalbermann, Libertun, & Gomez, 1971
	CytOx	↑ in AH & PH; — in MH	Moguilevsky, Kalbermann, Libertun, & Gomez, 1971
	SDH	— in AH, MH, & PH	Moguilevsky, Kalbermann, Libertun, & Gomez, 1971
	MDH	↑ in Amg; — in MBH	Luine, Khylcheveskaya, & McEwen, 1975a
	ICDH	↑ in MBH; — in Amg	Luine *et al.*, 1975a
	G6PDH	↑ in MBH	Luine *et al.*, 1975a

☿ + E_2B	MDH	↑ in MBH; — in Amg	Luine *et al.*, 1975a
	ICDH	↑ in Amg; — in MBH	Luine *et al.*, 1975a
	G6PDH	— in MBH	Luine *et al.*, 1975a
♂ → Hypox ♂	Glucose	↑ in AH & PH; ↓ in Amg; — in Cx, Hpc, & MH	Moguilevsky, Libertun, & Foglia, 1970b; Schiaffini & Martini, 1972
☿ → Hypox ☿	Glucose	↓ in Amg; — in AH, Cx, Hpc, MH, & PH	Moguilevsky *et al.*, 1970b; Schiaffini & Martini, 1972
Hypox ☿ + TP	Glucose	↓ in AH; — in MH & PH	Moguilevsky, Libertun, & Foglia, 1970a
☿ + T *(in vitro)*	Glucose	↓ in Cx; — in AH, Amg, Hpc, MH, & PH	Moguilevsky *et al.*, 1970a; Schiaffini & Martini, 1972
♂ → Pancrex ♂	Glucose	↓ in AH; — in Cx, MH, & PH	Schiaffini, Moguilevsky, Libertun, & Foglia, 1968
♂ → Adx ♂	[^{3}H]Deoxyglucose	— in AH, AP, BS, Hpc, MBH, NA, PAG, POA, Sep, & VMB	Wallis & Printz, 1980
Adx ♂ + B	MDH, LDH, ICDH, SDH, G6PDH, 6PGDH, HK, & PK	— in Hpc	Meyer, Luine, Khylcheveskaya, & McEwen, 1979
Di_1 ♀ → Di_2 ♀	PK	↓ in Arc & PaVN; — in AH, LPOA, LVM, MPOA, MVM, PH, SCN, & SON	Packman, Bragdon, & Boshans, 1977
	HK	— in AH, Arc, LPOA, LVM, MPOA, MVM, PaVN, PH, SCN, & SON	Packman *et al.*, 1977
	PFK	— in AH, Arc, LPOA, LVM, MPOA, MVM, PaVN, PH, SCN, & SON	Packman *et al.*, 1977
Di_2 ♀ → Di_3 ♀	PK	— in AH, Arc, LPOA, LVM, MPOA, MVM, PaVN, PH, SCN, & SON	Packman *et al.*, 1977
	HK	↓ in Arc, LVM, MVM, & PaVN; — in AH, LPOA, MPOA, PH, SCN, & SON	Packman *et al.*, 1977
	PFK	↓ in PH; — in AH, Arc, LPOA, LVM, MPOA, MVM, PaVN, SCN, & SON	Packman *et al.*, 1977

(Continued)

TABLE IX (*Continued*)

Treatment	Enzymatic analyses or metabolism substrates	Effect(s)[b]	References
Di_3 ♀ → Pro ♀	PK	— in AH, Arc, LPOA, LVM, MPOA, MVM, PaVN, PH, SCN, & SON	Packman *et al.*, 1977
	HK	↓ in AH & SCN; — in Arc, LPOA, LVM, MPOA, MVM, PaVN, PH, & SON	Packman *et al.*, 1977
	PFK	↑ in PH; — in AH, Arc, LPOA, LVM, MPOA, MVM, PaVN, SCN, & SON	Packman *et al.*, 1977
Pro ♀ → Est ♀	PK	↑ in Arc, LPOA, LVM, MVM, PaVN, & PH; — in AH, MPOA, SCN, & SON	Packman *et al.*, 1977
	HK	↑ in AH, MVM, & SCN; — in Arc, LPOA, LVM, MPOA, PaVN, PH, & SON	Packman *et al.*, 1977
	PFK	— in AH, Arc, LPOA, LVM, MPOA, MVM, PaVN, PH, SCN & SON	Packman *et al.*, 1977
Est ♀ → Di_1 ♀	PK	↓ in Arc, LPOA, LVM, MVM, PaVN, & PH; — in AH, MPOA, SCN, & SON	Packman *et al.*, 1977
	HK	— in AH, Arc, LPOA, LVM, MPOA, MVM, PaVN, PH, SCN, & SON	Packman *et al.*, 1977
	PFK	— in AH, Arc, LPOA, LVM, MPOA, MVM, PaVN, PH, SCN, & SON	Packman *et al.*, 1977
Est ♀ → Di_2 ♀[c]	PK	↓ in Arc & SCN	Packman *et al.*, 1977
	HK	↑ in MPOA & PH	Packman *et al.*, 1977
Est ♀ → Di_3 ♀[c]	HK	↓ in Arc, LPOA, LVM, MVM, & PaVN	Packman *et al.*, 1977
Di_1 ♀ → Pro ♀[c]	HK	↑ in SON	Packman *et al.*, 1977
Di ♀ → Est ♀	Glucose	↑ in AH, Amg, HTh, & PH; ↓ in Hpc; — in MH	Menendez-Patterson, Florez-Lozano, & Marin, 1979; Schiaffini & Marin, 1971
Di ♀ → Est ♀	[U-^{14}C]Glucose	↑ in HTh; — in Cx	Scacchi, Moguilevsky, & Schiaffini, 1973

Di ♀ → Est ♀	Succinate	↑ in AH & PH; — in MH	Moguilevsky, Schiaffini, Szwarcfarb, & Libertun, 1971
	CytOx	— in AH, MH, & PH	Moguilevsky, Schiaffini, Szwarcfarb, & Libertun, 1971
	SDH	↑ in AH & PH; — in MH	Moguilevsky, Schiaffini, Szwarcfarb, & Libertun, 1971
Di ♀ → ♀	Glucose	↑ in Amg & Cx; ↓ in HTh; — in Hpc (hamster and rat)	Menendez-Patterson *et al.*, 1979; Schiaffini & Marin, 1971
	[U-^{14}C]Glucose	↑ in HTh; — in Cx	Scacchi *et al.*, 1973
♀ + E_2B	MDH & ICDH	↑ in Amg, HTh & MBH; — in Cx & MPOA	Luine, Khylcheveskaya, & McEwen, 1974
	G6PDH	↑ in MBH; — in Amg, Cx, HTh, & MPOA	Luine *et al.*, 1974
	6PGDH, LDH, & HK	— in Amg, Cx, HTh, MBH, & MPOA	Luine *et al.*, 1974
♀ + TP	MDH, ICDH, & G6PDH	— in Amg & MBH	Luine *et al.*, 1975a
Hypox ♀ + E_2B	MDH & ICDH	↑ in Amg & MBH	Luine *et al.*, 1975a
	G6PDH	↑ in MBH	Luine *et al.*, 1975a
Adx ♀ + E_2B	MDH & ICDH	↑ in HTh; — in Hpc	Luine *et al.*, 1974
♀ + E_2 *(in vitro)*	MDH & ICDH	— in HTh	Luine *et al.*, 1974
Di ♀ → Pancrex ♀	Glucose	↑ in Amg; ↓ in HTh; — in Hpc	Schiaffini, Marin, & Foglia 1970

[a] Abbreviations: CytOx, cytochrome oxidase; SDH, succinic dehydrogenase; MDH, malic dehydrogenase; ICDH, isocitric dehydrogenase; G6PDH, glucose-6-phosphate dehydrogenase; 6PGDH, 6-phosphogluconate dehydrogenase; LDH, lactic dehydrogenase; HK, hexokinase; PK, pyruvate kinase; PFK, phosphofructokinase; pancrex, pancreatectomy. All other abbreviations are defined in the text and/or in Tables I, II, and IV.

[b] Unless otherwise indicated all data refer to experiments with rats.

[c] Note that intermediate day's values were not different from either group in this special comparison. Null effect values and significant differences between intermediate day and either of the other days have been omitted.

MH (which includes the ME, Arc, VM, and DM) and Hpc. Similar relationships were obtained by measuring $^{14}CO_2$ production from [U-^{14}C]glucose or by direct cytochrome oxidase (CytOx) assays, but there were no significant castration effects on glutamate and pyruvate oxidative metabolism or on succinic dehydrogenase (SDH) activity. *In vivo* but not *in vitro* (except in Cx) replacement therapy with T (or TP) reverses these effects and further increases isocitric (ICDH) and glucose-6-phosphate (G6PDH) dehydrogenase activities in the MBH and malic dehydrogenase (MDH) in the CMAmg. *In vivo* replacement with E_2B produces a different pattern of effects, including increases in the activity of ICDH in the Amg and of MDH in the MBH, and no effect on G6PDH in the MBH.

The lack of *in vitro* effects with T has prompted the suggestion that the effects of castration on oxidative metabolism in the HTh and Amg are actually due to the short feedback actions of gonadotropins, which increase or decrease their secretion rates following castration and T and E_2 replacement therapy, respectively (for reviews see Luttge, 1983). In support of this hypothesis, hypophysectomy of gonadally intact but not castrated male rats increases oxidative activity in the AH and PH, decreases it in the Amg, and has no effect in the MH, Cx, and Hpc. In hypophysectomized/castrated male rats, the *in vivo* administration of LH decreases oxygen consumption in the AH but not in the PH or MH, while FSH decreases it in the PH but not in the AH or MH. The *in vitro* exposure to gonadotropin increases oxidative activity in the Amg but not in the Cx or Hpc. It should be noted, however, that administration of TP to hypophysectomized/castrated male rats can result in a reduction of oxygen consumption in the AH (but apparently not in the PH or MH) suggesting that, at least in the AH, androgens can influence carbohydrate metabolism independently of their negative feedback actions on gonadotropins.

In the female rat, ovariectomy reduces oxygen consumption in the HTh, increases it in the Amg and Cx, and has no consistent effect in the Hpc. Although these results are similar to those described earlier for males, they are not accompanied by parallel changes in $^{14}CO_2$ production of CytOx activity, but they are accompanied by changes in SDH activity, which in the male was insensitive to gonadal hormone actions. Estrogen replacement therapy in gonadectomized females also increased the activity of *both* MDH and ICDH in *both* the HTh and Amg and of G6PDH in the HTh—again, a pattern distinctly different from that observed in males. G6PDH activity in the Amg was not affected by either E_2B or TP in ovariectomized females. MDH and ICDH activities in the MBH and CMAmg were similarly unaffected by TP treatment in ovariectomized females. In gonadally intact females there are numerous estrous-cycle-related alterations in tricarboxylic and glycolytic enzyme activities (see Table IX), but, as stated earlier, these changes may or may not be directly caused by estrous-cycle-related changes in circulating steroid hormones.

It is possible that the short feedback actions of gonadotropins in females may confound the interpretation of *in vivo* studies with steroids in a similar fashion to that discussed above for males. For example, although E_2 did not alter the activity of either MDH or ICDH in HTh fragments incubated directly with the steroid, *in vivo* replacement therapy with E_2B does increase the activity of both enzymes in MBH and CMAmg tissue fragments from hypophysectomized/ovariectomized females, suggesting that estrogens (like androgens) can influence intracerebral carbohydrate metabolism independently of their actions on gonadotropin secretion. However, since estrogen-induced increases in the activity of these enzymes is greater in ovariectomized females that still have their pituitary and adrenal glands, it is likely that pituitary (gonadotropin?) and adrenal (glucocorticoid?) hormones may still have permissive effects on estrogen-potentiated energy production.

The completion of steroid-induced increases in nucleic acid and protein synthesis presumably requires the ribose sugar intermediates and NADPH-reducing equivalents normally supplied by the pentose phosphate cycle. Thus, it is curious that 6-phosphogluconate dehydrogenase (6PGDH) activity does not appear to be altered in the HTh by steroid hormones. It is possible, however, that the steroid-induced increased production of MDH- and ICDH-derived reducing equivalents may at least partially compensate for this apparent nonresponsiveness of the pentose phosphate cycle enzymes. The relative contributions of genomic (e.g., *de novo* synthesis of new enzyme) and nongenomic (e.g., allosteric activation of preexisting enzyme) mechanisms in mediating these and other steroid actions in intracerebral carbohydrate metabolism are still to be determined.

FINAL COMMENTS

The preceding sections briefly review the current literature on the molecular mechanisms of steroid hormone actions in the mammalian brain. Many of these effects appear to be mediated by steroid-receptor–acceptor genomic responses (see Figure 1; e.g., E_2 stimulation of P_4 receptor and ChAcT synthesis), while others are clearly not the result of hormone-induced mRNA and protein synthesis (e.g., 2-OH-E_2 competitive inhibition of COMT actions on NE and DA). In addition to their actions on nucleic acids, proteins, polyamines, neurotransmitters, cAMP, and carbohydrate-energy metabolism, however, estrogens, progestins, androgens, and glucocorticoids can also be shown to influence directly the electrophysiological activity of neurons (see Figure 1). This capability is evidence by changes in the electroencephalogram (EEG), slow-wave evoked potentials and single or multiple unit activity (SUA and MUA, respectively) after systemic or intracerebral *in vivo* or *in vitro* injections or iontophoretic application of steroids (for reviews see Luttge, 1983; Pfaff & Modianos, 1980; Rees & Gray, 1983). Some of these responses occur in brain regions with cytosolic and nuclear receptors for the steroids (e.g., estrogen actions on SUA in the HTh and POA—Kelly, Moss, & Dudley, 1978—and glucocorticoid actions on MUA in the Hpc—Michal, 1974), whereas others occur in classic nontarget regions (e.g., estrogen and androgen actions in Hpc slices—Teyler, Vardaris, Lewis, & Rawitch, 1980). Some of these effects are probably mediated by steroid actions at somal and/or presynaptic membrane receptors (e.g., Schaeffer *et al.*, 1980; Towle & Sze, 1978) although others appear to be acting directly on axons (e.g., Kendrick & Drewett,

1980). The molecular mechanisms of these electrophysiological steroid-induced responses are still poorly understood, but it is clear that they may have direct relevance to our understanding of the hormonal regulation of aggressive, sexual, and other behaviors.

In the course of investigating the physiological and anatomical substrates for behaviors, investigators frequently employ electrical stimulation, lesion and tract-cutting procedures, and drug administrations without apparently considering the possibility that these manipulations may also be altering the ability of steroid hormones to exert their actions. The interpretation of the findings from these studies may therefore be confounded by the multiple interactive effects of the experimental procedures. Even the often-used technique of isolation housing for facilitating the expression of intermale aggression may have these confounding problems, since this procedure has been shown to alter the intracerebral aromatization of androgens in rats (e.g., Lupo Di Prisco, Lucarini, & Dessi-Fulgheri, 1978). Intracerebral electrical stimulation (e.g., Stith, Person, & Dana, 1976) and lesion techniques (e.g., Carrillo, 1980; Nyakas, de Kloet, & Bohas, 1979) can be shown to have dramatic effects on steroid-receptor density—even in brain areas not directly affected by the procedure. Centrally acting drugs (e.g., neurotransmitter synthesis and receptor-blockers) and exogenously administered neurotransmitters (e.g., HA and epinephrine) have also been shown to have the potential for markedly altering steroid metabolism and receptor-binding in the brain (e.g., Kaneyuki, Kohsaka, & Shohmori, 1979; Nock, Blaustein, & Feder, 1980; Rosner, Denari, Castro-Vazquez, Nagle, Neuspiller, Bedes, Pedroza, Martin, & Gomez, 1972; Shani, Givant, Sulman, Eylath, & Eckstein, 1971). These observations clearly indicate that it is imperative for all future investigations of the regulation of aggressive behavior to consider the possibility that the molecular mechanisms of steroid hormone actions in the brain may be directly or indirectly affected by the experimental procedures and may therefore be of potential significance in the interpretation of the experimental findings.

The goal of the present review was to summarize briefly the current literature on the molecular mechanisms of steroid hormone actions in the mammalian brain. It is hoped that this introduction will alert researchers to the numerous complexities and potential experimental approaches to the investigation of the physiological regulation of aggressive behavior.

REFERENCES

Acs, A., Palkovits, M., & Stark, E. Changes of glutamic acid decarboxylase activity after dexamethasone in selected areas of the rat brain. *Neuroscience Letters*, 1980, *19*, 97–101.

Adler, N. T., & Pfaff, D. W. (Eds.). *Neurobiology of reproduction.* Book in preparation, 1983.

Alderson, L. M., Starr, M. S., & Baum, M. J. Effects of castration on dopamine metabolism in rat striatum and limbic forebrain. *Society for Neuroscience Abstracts*, 1979, *5*, 1471.

Allera, A., Rao, G. S., & Breuer, H. Specific interaction of corticosteroids with components of the cell membrane which are involved in the translocation of the hormone into the intravesicular space of purified rat liver plasma membrane vesicles. *Journal of Steroid Biochemistry*, 1980, *12*, 259–266.

Altman, J., & Bayer, S. A. Development of the diencephalon in the rat: I. Autoradiographic study of the time of origin and settling patterns of neurons of the hypothalamus. *Journal of Comparative Neurology*, 1978, *182*, 945–972.

Angelucci, L., Valeri, P., Palmery, M., Paracchioli, F. R., & Catalani, A. Brain glucocorticoid receptor: correlation of *in vivo* uptake of corticosterone with behavioral, endocrine and neurophamacological events. *Advances in Biochemical Psychopharmacology*, 1980, *21*, 391–406.

Anton-Tay, F., Pelham, R. W., & Wurtman, R. J. Increased turnover of ^{3}H-norepinephrine in rat brain following castration or treatment with ovine follicle-stimulating hormone. *Endocrinology*, 1969, *84*, 1489–1492.

Asai, M., Yu, W., & Leung, B. S. Cytoplasmic competitors of estrogen receptor at chromatin binding site. *Endocrine Society Abstracts*, 1979, 781.

Azmitia, E. C., & McEwen, B. S. Corticosterone regulation of tryptophan hydroxylase in midbrain of the rat. *Science*, 1969, *166*, 1274–1276.

Azmitia, E. C., & McEwen, B. S. Adrenalcortical influence on rat brain tryptophan hydroxylase activity. *Brain Research*, 1974, *78*, 291–302.

Azmitia, E. C., & McEwen, B. S. Early response of rat brain tryptophan hydroxylase activity to cycloheximide, puromycin and corticosterone. *Journal of Neurochemistry*, 1976, *27*, 773–778.

Azmitia, E. C., Algeri, S., & Costa, E. *In vivo* conversion of ^{3}H-L-tryptophan into ^{3}H-serotonin in brain areas of adrenalectomized rats. *Science*, 1970, *169*, 201–203.

Ball, P., & Knuppen, R. Catecholoestrogens (2- and 4-hydroxyoestrogens)—Chemistry, biogenesis, metabolism, occurrence and physiological significance. *Acta Endocrinologica*, 1980, *93*, Supplementum 232.

Bapna, J., Neff, N. H., & Costa, E. A method for studying norepinephrine and serotonin metabolism in small regions of rat brain: Effect of ovariectomy on amine metabolism in anterior and posterior hypothalamus. *Endocrinology*, 1971, *89*, 1345–1349.

Beattie, C. W., & Soyka, L. F. Influence of progestational steroids in hypothalamic tyrosine hydroxylase activity *in vitro*. *Endocrinology*, 1973, *93*, 1453–1455.

Beattie, C. W., Rodgers, C. H., & Soyka, L. F. Influence of ovariectomy and ovarian steroids on hypothalamic tyrosine hydroxylase activity in the rat. *Endocrinology*, 1972, *91*, 276–279.

Beinfeld, M. C., & Packman, P. M. Estrogen induction of specific soluble proteins in the hypothalamus of the immature rat. *Biochemical and Biophysical Research Communications*, 1976, *73*, 646–652.

Bernard, B., & Paolino, R. Time dependent changes in brain biogenic amine dynamics following castration in male rats. *Journal of Neurochemistry*, 1974, *22*, 951–956.

Berthold, A. A. Transplantation der Hoden. *Archiv für Anatomi, Physiologie und wissenschaftliche Medicin*, 1849, *16*, 42–46. (Trans. in *Bulletin of the History of Medicine*, 1944, *16*, 399–401.)

Beyer, C. *Endocrine control of sexual behavior*. New York: Raven Press, 1979.

Biegon, A., Bercovitz, H., & Samuel, D. Serotonin receptor concentration during the estrous cycle of the rat. *Brain Research*, 1980, *187*, 221–225.

Birmingham, M. K., Stumpf, W. E., & Sar, M. Nuclear localization of aldosterone in rat brain cells assessed by autoradiography. *Experientia*, 1979, *35*, 1240–1241.

Biro, J. The effect of luteinizing hormone releasing hormone (LH-RH) and oestrogen on RNA synthesis in anterior pituitary and different brain regions of rats. *Endokrinologie*, 1978, *72*, 285–290.

Bjorklund, A., Hakamson, R., Nobin, A., & Sjoberg, N. O. Increase in rabbit hypothalamic histidine decarboxylase activity after oophorectomy and thyroidectomy. *Experientia*, 1972, *28*, 1232–1233.

Blaustein, J. D., & Feder, H. H. Cytoplasmic progestin receptors in guinea pig brain: Characteristics and relationship to the induction of sexual behavior. *Brain Research*, 1979, *169*, 481–497. (a)

Blaustein, J. D., & Feder, H. H. Cytoplasmic progestin receptors in female guinea pig brain and their relationship to refractoriness in expression of female sexual behavior. *Brain Research*, 1979, *177*, 489–498. (b)

Blaustein, J. D., & Feder, H. H. Nuclear progestin receptors in guinea pig brain measured by an *in vitro* exchange assay after hormonal treatments that affect lordosis. *Endocrinology*, 1980, *106*, 1061–1069.

Blaustein, J. D., & Wade, G. N. Progestin binding by brain and pituitary cell nuclei and female rat sexual behavior. *Brain Research,* 1978, *140,* 360–367.

Brawer, J. R., & Naftolin, F. The effects of oestrogen on hypothalamic tissue. *Sex Hormones and Behavior,* 1979, *62,* 19–40.

Breuer, H., & Koster, G. Interaction between oestrogens and neurotransmitters at the hypophysial-hypothalamic level. *Journal of Steroid Biochemistry,* 1974, *5,* 961–968.

Butte, J. C., Kakihana, R., & Noble, E. P. Circadian rhythm of corticosterone levels in rat brain. *Journal of Endocrinology,* 1976, *68,* 235–239.

Callard, G. V., Petro, Z., & Ryan, K. Conversion of androgen to estrogen and other steroids in the vertebrate brain. *American Zoologist,* 1978, *18,* 511–523.

Cardinali, D. P., & Gomez, E. Changes in hypothalamic noradrenaline, dopamine and serotonin uptake after estradiol administration to rats. *Journal of Endocrinology,* 1977, *73,* 181–182.

Carrillo, A. J. Estrogen receptors in the medial basal hypothalamus of the rat following complete hypothalamic deafferentation. *Brain Research,* 1980, *816,* 157–164.

Carroll, B. J., Heath, B., & Jarrett, D. B. Corticosteroids in brain tissue. *Endocrinology,* 1975, *97,* 290–300.

Chamness, G. C., Jennings, A. W., & McGuire, W. L. Estrogen receptor binding to isolated nuclei. A nonsaturable process. *Biochemistry,* 1974, *13,* 327–331.

Chan, L., Means, A. R., & O'Malley, B. W. Steroid hormone regulation of specific gene expression. *Vitamins and Hormones,* 1978, *36,* 259–295.

Cheng, Y.-J., & Karavolas, H. J. Subcellular distribution and properties of progesterone (Δ4-steroid)5α-reductase in rat medial basal hypothalamus. *Journal of Biological Chemistry,* 1975, *250,* 7997–8003.

Chiocchio, S. R., Negro-Vilar, A., & Tramezami, J. H. Acute changes in norepinephrine content in the median eminence induced by orchidectomy or testosterone replacement. *Endocrinology,* 1976, *99,* 629–635.

Chytil, F., & Toft, D. O. Corticoid binding component in rat brain. *Journal of Neurochemistry,* 1972, *19,* 2877–2880.

Cidlowski, J. A., & Muldoon, T. G. Estrogenic regulation of cytoplasmic receptor populations in estrogen-responsive tissue of the rat. *Endocrinology,* 1974, *95,* 1621–1629.

Cidlowski, J. A., & Muldoon, T. G. Sex-related differences in the regulation of cytoplasmic estrogen receptor levels in responsive tissues of the rat. *Endocrinology,* 1976, *98,* 833–841.

Clark, C. R., & Nowell, N. W. Binding properties of testosterone receptors in the hypothalamic-preoptic area of the adult male mouse brain. *Steroids,* 1979, *33,* 407–426.

Clark, J. H., & Peck, E. J., Jr. Nuclear retention of receptor oestrogen complex and nuclear acceptor sites. *Nature,* 1976, *260,* 635–637.

Cornwell-Jones, C. A., & Marasco, E. M. Castration decreases olfactory bulb norepinephrine in male rats but not hamsters. *Brain Research,* 1980, *183,* 377–382.

Cramer, O. M., Parker, C. R., & Porter, J. C. Estrogen inhibition of dopamine release into hypophysial portal blood. *Endocrinology,* 1979, *104,* 419–422. (a)

Cramer, O. M., Parker, C. R., & Porter, J. C. Stimulation of dopamine release into hypophysial portal blood by administration of progesterone. *Endocrinology,* 1979, *105,* 929–933. (b)

Crowley, W. R., O'Donohue, T. L., & Jacobowitz, D. M. Changes in catecholamine content in discrete brain nuclei during the estrous cycle of the rat. *Brain Research,* 1978, *147,* 315–326.

Crowley, W. R., O'Donohue, T. L., Wachslicht, H., & Jacobowitz, D. M. Effects of estrogen and progesterone on plasma gonadotropins and on catecholamine levels and turnover in discrete brain regions of ovariectomized rats. *Brain Research,* 1978, *154,* 345–357.

Crowley, W. R., O'Donohue, T. L., Muth, E. A., & Jacobowitz, D. M. Effects of ovarian hormones on levels of luteinizing hormone in plasma and on serotonin concentrations in discrete brain nuclei. *Brain Research Bulletin,* 1979, *4,* 571–574.

Curtis, D. R., & Johnston, G. A. R. Amino acid transmitters in the mammalian central nervous system. *Ergebnisse der Physiologie,* 1974, *69,* 97–188.

Davies, I. J., Naftolin, F., Ryan, K. J., Fishman, J., & Siu, J. The affinity of catechol estrogens for estrogen receptors in the pituitary and anterior hypothalamus of the rat. *Endocrinology,* 1975, *97,* 554–557.

Debus, G., & Kehr, W. Catecholamine and 5-hydroxytryptamine synthesis and metabolism following intraventricular injection of dibutyryl cyclic AMP. *Journal of Neural Transmission,* 1979, *45,* 195–206.

Denari, J. H., & Rosner, J. M. Sexual steroid uptake in the alloxanized diabetic rat. *Steroids and Lipids Research,* 1972, *3,* 151–155.

Dokas, L. A. Corticosterone and RNA metabolism in the rat hippocampus. *Society for Neuroscience Abstracts,* 1979, *5,* 1498.

Donoso, A. O. Induction of prolactin and luteinizing hormone release by histamine in male and female rats and the influence of brain transmitter antagonists. *Journal of Endocrinology,* 1978, *76,* 193–202.

Donoso, A. O., & Moyano, M. B. G. Adrenergic activity in hypothalamus and ovulation. *Proceedings of the Society of Experimental Biology and Medicine,* 1970, *135,* 633–635.

Donoso, A. O., Stefano, F. J. E., Biscardi, A. M., & Cukier, J. Effects of castration on hypothalamic catecholamines. *American Journal of Physiology,* 1967, *212,* 737–739.

Dunn, A. J., Gildersleeve, N. B., & Gray, H. E. Mouse brain tyrosine hydroxylase and glutamic acid decarboxylase following treatment with adrenocorticotrophic hormone, vasopressin or corticosterone. *Journal of Neurochemistry,* 1978, *31,* 977–982.

Early, C. J., & Leonard, B. E. GABA and gonadal hormones. *Brain Research,* 1978, *155,* 27–34.

Eaton, G. G., Goy, R. W., & Resko, J. A. Brain uptake and metabolism of estradiol benzoate and estrous behavior in ovariectomized guinea pigs. *Hormones and Behavior,* 1975, *6,* 81–97.

Edwards, E. M., & Rousseau, G. G. Effects of adrenalectomy and corticosteroids on three enzymes involved in amino acid metabolism in the brain of adult and newborn rats. *Journal of Steroid Biochemistry,* 1980, *13,* 567–569.

Eisenfeld, A. J. ^{3}H-Estradiol: In vitro binding to macromolecules from rat hypothalamus, anterior pituitary and uterus. *Endocrinology,* 1970, *86,* 1313–1318.

Enderly, C. A., & Wilson, C. A. The effect of ovarian steroids on the accumulation of ^{3}H-labelled monoamines by hypothalamic tissue *in vitro. Brain Research,* 1974, *73,* 321–331.

Engel, J., Ahlenius, S., Almgren, O., Carlsson, A., Larsson, K., & Sodersten, P. Effects of gonadectomy and hormone replacement on brain monoamine synthesis in male rats. *Pharmacology, Biochemistry and Behavior,* 1979, *10,* 149–154.

Ermish, A., & Ruhle, H. -J. Autoradiographic demonstration of aldosterone-concentrating neuron populations in rat brain. *Brain Research,* 1978, *147,* 154–158.

Etgen, A. M., Lee, K. S., & Lynch, G. Glucocorticoid modulation of specific protein metabolism in hippocampal slices maintained in vitro. *Brain Research,* 1979, *165,* 37–45.

Etgen, A. M., Martin, M., Gilbert, R., & Lynch, G. Characterization of corticosterone-induced protein synthesis in hippocampal slices. *Journal of Neurochemistry,* 1980, *35,* 598–602.

Fahn, S. Regional distribution studies of GABA and other putative neurotransmitters and their enzymes. In E. Roberts, T. N. Chase, & D. B. Tower (Eds.), *GABA in nervous system function.* New York: Raven Press, 1976.

Feder, H. H., Siegel, H., & Wade, G. N. Uptake of [6,7,-^{3}H] estradiol-17β in ovariectomized rats, guinea pigs, and hamsters: Correlation with species differences in behavioral responsiveness to estradiol. *Brain Research,* 1974, *71,* 93–103.

Fekete, M. I., Stark, E., Herman, J. P., Palkovits, M., & Kanyicska, B. Catecholamine concentration of various brain nuclei of the rat as affected by ACTH and corticosterone. *Neuroscience Letters,* 1978, *10,* 153–158.

Fleischer-Lambropoulos, H., Sarkander, H.-I., & Brade, W. P. Effects of polyamines on amino acid incorporation into protein by cerebral and cerebellar as well as "neural" and "glial" nuclei of rat brain. *Biochemical and Biophysical Research Communications,* 1975, *63,* 792–800.

Florez-Lozano, J. A., Menendez-Patterson, A., & Marin, B. Sexual behavior of the pancreatectomized (95%) male hamster (Mesocricetus auratus). *Physiology & Behavior,* 1978, *20,* 465–468.

Foreman, M. M., & Porter, J. C. Effects of catechol estrogens and catecholamines on hypothalamic and corpus striatal tyrosine hydroxylase activity. *Journal of Neurochemistry,* 1980, *34,* 1175–1183.

Foreman, M. M., Wickersham, E. W., & Anthony, A. Cytophotometric analysis of hypothalamic RNA fluctuations during the rat estrous cycle. *Brain Research,* 1977, *119,* 471–475.

Fox, T. O. Conversion of the hypothalamic estradiol receptor to the 'nuclear' form. *Brain Research,* 1977, *120,* 580–583.

Fox, T. O., & Johnston, C. Estradiol receptors from mouse brain and uterus: Binding to DNA. *Brain Research,* 1974, *77,* 330–336.

Franks, S., Ball, P., Naftolin, F., & Ruf, K. B. Effect of catechol oestrogens on induced ovulation in the immature rat. *Journal of Endocrinology,* 1980, *86,* 263–268.

Frohman, L. A., & Berelowitz, M. The physiological and pharmacological control of anterior pituitary hormone secretion. In C. B. Nemeroff & A. J. Dunn (Eds.), *Peptides, hormones, and behavior: Molecular and behavioral neuroendocrinology.* New York: Spectrum Publications, 1983.

Ganong, W. F. The role of catecholamines and acetylcholine in the regulation of endocrine function. *Life Sciences,* 1974, *15,* 1401–1414.

Gentry, R. T., Wade, G. N., & Blaustein, J. D. Binding of [^{3}H]estradiol by brain cell nuclei and female rat sexual behavior: Inhibition by experimental diabetes. *Brain Research,* 1977, *135,* 135–146.

Ginsburg, M., MacLusky, N. J., Morris, J. D., & Thomas, P. J. Physiological variation in abundance of oestrogen specific high-affinity binding sites in hypothalamus, pituitary and uterus of the rat. *Journal of Endocrinology,* 1975, *64,* 443–449.

Goertz, B. Effect of polamines on cell-free protein synthesizing systems from rat cerebral cortex, cerebellum and liver. *Brain Research,* 1979, *173,* 125–135.

Gordon, J. H., Nance, D. M., Wallis, C. J., & Gorski, R. A. Effects of estrogen on dopamine turnover, glutamic acid decarboxylase activity and lordosis behavior in septal lesioned female rats. *Brain Research Bulletin,* 1977, *2,* 341–346.

Gordon, J. H., Nance, D. M., Wallis, C. J., & Gorski, R. A. Effect of septal lesions and chronic estrogen treatment on dopamine, GABA and lordosis behavior in male rats. *Brain Research Bulletin,* 1979, *4,* 85–89.

Gorski, J., & Gannon, F. Current models of steroid hormone action: A critique. *Annual Review of Physiology,* 1976, *38,* 425–450.

Gray, H. E., Jasper, T. W., Luttge, W. G., Shukla, J. B., & Rennert, O. M. Estrogen increases hypothalamic and pituitary polyamine levels in ovariectomized rats. *Journal of Neurochemistry,* 1980, *34,* 753–755.

Green, A. R., & Curzon, G. Effects of hydrocortisone and immobilization on tryptophan metabolism in brain and liver of rats of different ages. *Biochemical Pharmacology,* 1975, *24,* 713–716.

Greenstein, B. D. An assay for RNA polymerase activity in rat brain nuclei—Effects of injection of oestradiol benzoate. *Journal of Endocrinology,* 1980, *85,* 341–347.

Grosser, B. I., & Axelrod, L. R. Conversion of cortisol to cortisol acetate, cortisone acetate and cortisone by the developing primate brain. *Steroids,* 1968, *11,* 827–836.

Grosser, B. I., Stevens, W., & Reed, D. J. Properties of corticosterone-binding macromolecules from rat brain cytosol. *Brain Research,* 1973, *57,* 387–395.

Gunaga, K. P., & Menon, K. M. J. Effect of catecholamines and ovarian hormones on cyclic AMP accumulation in rat hypothalamus. *Biochemical and Biophysical Research Communications,* 1973, *54,* 440–448.

Gunaga, K. P., Kawano, A., & Menon, K. M. J. *In vivo* effect of estradiol benzoate on the accumulation of adenosine 3′,5′-cyclic monophosphate in the rat hypothalamus. *Neuroendocrinology,* 1974, *16,* 273–281.

Gustafsson, J. -A., Pousette, A., & Svensson, E. Sex-specific occurrence of androgen receptors in rat brain. *Journal of Biological Chemistry,* 1976, *251,* 4047–4054.

Heritage, A. S., & Grant, L. D. ^{3}H-Dihydrotestosterone in catecholamine neurons of rat brain stem. *Anatomical Record,* 1979, *193,* 564.

Heritage, A. S., Grant, L. D., & Stumpf, W. E. ^{3}H-Estradiol in catecholamine neurons of rat brain stem: Combined localization by autoradiography and formaldehyde-induced fluorescence. *Journal of Comparative Neurology,* 1977, *176,* 607–630.

Heritage, A. S., Stumpf, W. E., Sar, M., & Grant, L. D. Brainstem catecholamine neurons are target sites for sex steroid hormones. *Science,* 1980, *207,* 1377–1379.

Hill, M. J., Goddard, P., & Williams, R. E. O. Gut bacteria and aetiology of cancer of the breast. *Lancet,* 1971, *2,* 472–473.

Honma, K., & Wuttke, W. Norepinephrine and dopamine turnover rates in the medial preoptic area and medial-basal hypothalamus of the rat brain after various endocrinological manipulations. *Endocrinology,* 1980, *106,* 1848–1853.

Horowitz, S. B., & Moore, L. C. The nuclear permeability, intracellular distribution and diffusion of insulin in the amphibian oocyte. *Journal of Cellular Biology,* 1974, *60,* 405–415.

Hruska, R. E., & Silbergeld, E. K. Increased dopamine receptor sensitivity after estrogen treatment. *Society for Neuroscience Abstracts,* 1979, *5,* 236.

Hyyppa, M. T., Cardinali, D. P., Baumgarten, H. G., & Wurtman, R. J. Rapid accumulation of H^3-serotonin in brains of rats receiving intraperitoneal H^3-tryptophan: Effects of 5,6-dihydroxytryptamine or female sex hormones. *Journal of Neural Transmission,* 1973, *34,* 111–124.

Inaba, M., & Kamata, K. Effect of estradiol-17β and other steroids on noradrenaline and dopamine binding to synaptic membrane fragments of rat brain. *Journal of Steroid Biochemistry,* 1979, *11,* 1491–1497.

Iuvone, P. M., Morasco, J., & Dunn, A. J. Effects of corticosterone on the synthesis of [^{3}H]catecholamines in the brains of CD-1 mice. *Brain Research,* 1977, *120,* 571–576.

Jasper, T. W. *Polyamines in the normally developing male and female mouse brain.* Unpublished doctoral dissertation, University of Florida, 1980.

Jellink, P. H., Davis, P. G., Krey, L. C., Luine, V. N., Roy, E. J., & McEwen, B. S. Central and peripheral action of estradiol and catechol estrogens administered by continuous infusion. *Endocrine Society Abstracts,* 1979, 838.

Jungblut, P. W., Kallweit, E., Sierralta, W., Truitt, A. J., & Wagner, R. K. The occurrence of steroid-free, "activated" estrogen receptor in target cell nuclei. *Hoppe-Seyler's Zietschrift für Physiologische Chemie,* 1978, *359,* 1259–1268.

Jungblut, P. W., Hughes, A., Gaues, J., Kallweit, E., Marschler, I., Parl, F., Sierralta, W., Szendro, P. I., & Wagner, R. K. Mechanisms involved in the regulation of steroid receptor levels. *Journal of Steroid Biochemistry,* 1979, *11,* 273–278.

Kahwanago, I., Heinrichs, W. L. R., & Herrmann, W. L. Estradiol receptors in hypothalamus and anterior pituitary gland: Inhibition of estradiol binding by SH-group blocking agents and clomiphene citrate. *Endocrinology,* 1970, *86,* 1319–1326.

Kaneyuki, T., Kohsaka, M., & Shohmori, T. Sex hormone metabolism in the brain—Influence of central acting drugs on 5-alpha-reduction in rat diencephalon. *Endocrinologia Japonica,* 1979, *26,* 345–352.

Kar, van de, L., Levine, J., & Van Orden, L. S. Serotonin in hypothalamic nuclei: Increased content after castration of male rats. *Neuroendocrinology,* 1978, *27,* 186–192.

Kato, J. The role of hypothalamic and hypophyseal 5α-dihydrotestosterone, estradiol and progesterone receptors in the mechanism of feedback action. *Journal of Steroid Biochemistry,* 1975, *6,* 979–987.

Kato, J. Cytosol and nuclear receptors for 5α-dihydrotestosterone and testosterone in the hypothalamus and hypophysis, and testosterone receptors isolated from neonatal female rat hypothalamus. *Journal of Steroid Biochemistry,* 1976, *7,* 1179–1187.

Kato, J., & Minaguchi, H. Cholinergic and adrenergic mechanisms in the female rat hypothalamus with special reference to reproductive functions. *Gunma Symposium on Endocrinology,* 1964, *1,* 269–281.

Kato, J., & Onouchi, T. Nuclear progesterone receptors and characterization of cytosol receptors in the rat hypothalamus and anterior hypophysis. *Journal of Steroid Biochemistry,* 1979, *11,* 845–854.

Kazama, N., & Longcope, C. *In vivo* studies of the metabolism of estrone and estradiol-17β by the brain. *Steroids,* 1974, *23,* 469–481.

Kelly, M. J., Moss, R. L., & Dudley, C. A. The effects of ovariectomy on the responsiveness of preoptic-septal neurons to microelectrophoresed estrogen. *Neuroendocrinology,* 1978, *25,* 204–211.

Kendrick, K. M., & Drewett, R. F. Testosterone-sensitive neurons respond to oestradiol, but not to dihydrotestosterone. *Nature,* 1980, *286,* 67–68.

Kim, Y. S., Stumpf, W., Sar, M., & Martinez-Vargas, M. C. Estrogen and androgen target cells in the brain of fishes, reptiles and birds: Phylogeny and ontogeny. *American Zoologist,* 1978, *18,* 425–434.

Kiely, M. E. Effect of hypophysectomy, adrenalectomy and glucocorticoids on tryptophan accumulation by rat cerebral cortex slices. *Research Communications in Psychology, Psychiatry and Behavior,* 1980, *5,* 49–60.

Kizer, J. S., Palkovits, M., Zivin, J., Brownstein, M., Saavedra, J. M., & Kopin, J. J. The effect of endocrinological manipulations of tyrosine hydroxylase and dopamine-β-hydroxylase activities in individual hypothalamic nuclei of the adult male rat. *Endocrinology,* 1974, *95,* 799–812.

Kizer, J. S., Palkovits, M., Kopin, I. J., Saavedra, J. M., & Brownstein, M. J. Lack of effect of various endocrine manipulations on tryptophan hydroxylase activity of individual nuclei of the hypothalamus, limbic system and midbrain of the rat. *Endocrinology,* 1976, *98,* 743–747.

Kizer, J. S., Humm, J., Nicholson, G., Greeley, G., & Youngblood, W. The effect of castration, thyroidectomy and haloperidol upon the turnover rates of dopamine and norepinephrine and the kinetic properties of tyrosine hydroxylase in discrete hypothalamic nuclei of the male rat. *Brain Research,* 1978, *146,* 95–107.

de Kloet, E. R., & McEwen, B. S. Differences between cytosol receptor complexes with corticosterone and dexamethasone in hippocampal tissue from rat brain. *Biochemica et Biophysica Acta,* 1976, *421,* 124–132.

de Kloet, E. R., Wallach, G., & McEwen, B. S. Differences in corticosterone and dexamethasone binding to rat brain and pituitary. *Endocrinology,* 1975, *96,* 598–609.

Kobayashi, R. M., & Reed, K. C. Conversion of androgens to estrogens (aromatization) in discrete regions of the rat brain: Sexual differences and effects of castration. *Society for Neuroscience Abstracts,* 1977, *3,* 1115.

Kovacs, G. L., Telegdy, G., & Lissak, K. Dose-dependent action of corticosteroids on brain serotonin content and passive avoidance behavior. *Hormones and Behavior,* 1977, *8,* 155–165.

Krause, J. E., & Karavolas, H. J. Subcellular location of hypothalamic progesterone metabolizing enzymes and evidence for distinct NADH- and NADPH-linked 3α-hydroxysteroid oxidoreductase activity. *Journal of Steroid Biochemistry,* 1980, *13,* 271–280.

Krey, L. C., Kamel, F., & McEwen, B. S. Parameters of neuroendocrine aromatization and estrogen receptor occupation in the male rat. *Brain Research,* 1980, *193,* 277–283.

Krieger, A., & Wuttke, W. Effects of ovariectomy and hyperprolactinemia on tyrosine hydroxylase and dopamine-β-hydroxylase activity in various limbic and hypothalamic structures. *Brain Research,* 1980, *193,* 173–180.

Kubli-Garfias, C., & Whalen, R. E. Induction of lordosis behavior in female rats by intravenous administration of progestins. *Hormones and Behavior,* 1977, *9,* 380–386.

Kueng, W., Wirz-Justice, A., Menzi, B., & Chappuis-Arndt, E. Regional brain variation of tryptophan monoamines, monoamine oxidase activity, plasma free tryptophan and total tryptophan during the estrous cycle of the rat. *Neuroendocrinology,* 1976, *21,* 289–296.

Kumakura, K., Hoffman, M., Cocchi, D., Trabucchi, M., Speno, P. F., & Muller, E. E. Long term effect of ovariectomy on dopamine stimulated adenylate cyclase in rat striatum and nucleus accumbens. *Psychopharmacologia,* 1979, *61,* 13–16.

Kumar, S. A., Beach, T. A., & Dickerman, H. W. Specificity of oligodeoxynucleotide binding of mouse uterine cytosol estradiol receptors. *Proceedings of the National Academy of Science (U.S.A.),* 1980, *77,* 3341–3345.

Ladisich, W. Effects of progesterone on regional 5-hydroxytryptamine metabolism in rat brain. *Neuropharmacology,* 1974, *13,* 877–883.

Landau, I. T., & Feder, H. H. Whole cell and nuclear uptake of [^{3}H]estriol in neural and peripheral tissues of the ovariectomized guinea pig. *Brain Research,* 1977, *121,* 190–195.

Landau, I. T., & Feder, H. H. Uptake and metabolism of ^{3}H-estrone in neural and peripheral tissue of gonadectomized adult and neonatal guinea pigs. *Psychoneuroendocrinology,* 1980, *5,* 25–32.

Lee, H., Davies, I. J., & Ryan, K. J. Progesterone receptor in the hypothalamic cytosol of female rats. *Endocrinology,* 1979, *104,* 791–800.

Levy, C., Mortel, R., Eychenne, B., Robel, P., & Baulieu, E. E. Unoccupied nuclear oestradiol-receptor sites in normal human endometrium. *Biochemical Journal,* 1980, *185,* 733–738.

Lieberburg, I., & McEwen, B. S. Estradiol-17β a metabolite of testosterone recovered in cell nuclei from limbic areas of adult male rat brains. *Brain Research,* 1975, *91,* 171–174.

Lieberburg, I., & McEwen, B. S. Brain cell nuclear retention of testosterone metabolites, 5α-dihydrotestosterone and estradiol-17β in adult rats. *Endocrinology,* 1977, *100,* 588–597.

Lieberburg, I., MacLusky, N. J., & McEwen, B. S. 5α-Dihydrotestosterone (DHT) receptors in rat brain and pituitary cell nuclei. *Endocrinology,* 1977, *100,* 598–607.

Linkie, D. M. Estrogen receptors in different target tissues: Similarities of form—dissimilarities of transformation. *Endocrinology,* 1977, *101,* 1862–1870.

Linkie, D. M., & Siiteri, P. K. A re-examination of the interaction of estradiol with target cell receptors. *Journal of Steroid Biochemistry,* 1978, *9,* 1071–1078.

Lippman, M., Bolan, G., Monaco, M., Pinkus, L., & Engel, L. Model systems for the study of estrogen action in tissue culture. *Journal of Steroid Biochemistry,* 1976, *7,* 1045–1051.

Little, M., Szendro, P., Hughes, A., & Jungblut, P. W. Biosynthesis and transformation of microsomal and cytosol estradiol receptors. *Journal of Steroid Biochemistry,* 1975, *6,* 493–500.

Lloyd, T., & Ebersole, B. J. Feedback inhibition of tyrosine hydroxylase from five regions of rat brain by 2-hydroxyestradiol and dihydroxyphenylalanine. *Journal of Neurochemistry,* 1980, *34,* 726–731.

Lloyd, T., Weisz, J., & Breakfield, X. O. The catechol estrogen, 2-hydroxyestradiol, inhibits catechol-O-methyltransferase activity in neuroblastoma cells. *Journal of Neurochemistry,* 1978, *31,* 245–250.

Loy, R., & Milner, T. A. Sexual dimorphism in extent of axonal sprouting in rat hippocampus. *Science,* 1980, *208,* 1282–1284.

Luck, D. N. Comparison of the effects of oestrogen on macromolecular synthesis in the uterus and brain of the immature mouse. *Journal of Reproduction and Fertility,* 1975, *43,* 359–362.

Luine, V. N., & McEwen, B. S. Effect of oestradiol on turnover of type A monoamine oxidase in brain. *Journal of Neurochemistry,* 1977, *28,* 1221–1227. (a)

Luine, V. N., & McEwen, B. S. Effects of an estrogen antagonist on enzyme activities and (^{3}H)estradiol nuclear binding in uterus, pituitary and brain. *Endocrinology,* 1977, *100,* 903–910. (b)

Luine, V. N., Khylcheveskaya, R. I., & McEwen, B. S. Oestrogen effects on brain and pituitary enzyme activities. *Journal of Neurochemistry,* 1974, *23,* 925–934.

Luine, V. N., Khylcheveskaya, R. I., & McEwen, B. S. Effect of gonadal hormones on enzyme activities in brain and pituitary of male and female rats. *Brain Research,* 1975, *86,* 283–292. (a)

Luine, V. N., Khylcheveskaya, R. I., & McEwen, B. S. Effect of gonadal steroids on activities of monoamine oxidase and choline acetylase in rat brain. *Brain Research,* 1975, *86,* 293–306. (b)

Luine, V. N., McEwen, B. S., & Black, I. B. Effect of 17β-estradiol on hypothalamic tyrosine hydroxylase activity. *Brain Research,* 1977, *120,* 188–192.

Luine, V. N., MacLusky, N. J., & McEwen, B. S. Testosterone effects on enzymes in central and peripheral target sites in Tfm mutant mice. *Society for Neuroscience Abstracts,* 1979, *5,* 1529.

Luine, V. N., Park, D., Joh, T., Reis, D., & McEwen, B. S. Immunochemical demonstration of increased choline acetyltransferase concentration in rat preoptic area after estradiol administration. *Brain Research,* 1980, *191,* 273–277.

Lupo Di Prisco, C., Lucarini, N., & Dessi-Fulgheri, F. Testosterone aromatization in rat brain is modulated by social environment. *Physiology & Behavior,* 1978, *20,* 345–348.

Luttge, W. G. The role of gonadal hormones in the sexual behavior of the rhesus monkey and human: A literature survey. *Archives of Sexual Behavior,* 1971, *1,* 61–88.

Luttge, W. G. The estrous cycle of the rat: Effects on the accumulation of estrogenic metabolites in brain and peripheral tissues. *Brain Research,* 1972, *38,* 315–325.

Luttge, W. G. Endocrine control of mammalian male sexual behavior: An analysis of the potential role of testosterone metabolites. In C. Beyer (Ed.), *Endocrine control of sexual behavior.* New York: Raven Press, 1979.

Luttge, W. G. Cerebral effects of gonadal steroid hormones. In C. B. Nemeroff & A. J. Dunn (Eds.), *Peptides, Hormones and Behavior: Molecular and behavioral neuroendocrinology.* New York: Spectrum Publications, 1983.

Luttge, W. G., & Jasper, T. W. Studies on the possible role of 2-OH-estradiol in the control of sexual behavior in female rats. *Life Sciences,* 1977, *20,* 419–426.

Luttge, W. G., & Whalen, R. E. The accumulation, retention and interaction of oestradiol and oestrone in central neural and peripheral tissues of gonadectomized female rats. *Journal of Endocrinology,* 1972, *52,* 379–395.

Luttge, W. G., Gray, H. E., & Hughes, J. R. Regional and subcellular ^{3}H-estradiol localization in selected brain regions and pituitary of female mice: Effects of unlabeled estradiol and various anti-hormones. *Brain Research,* 1976, *104,* 273–281.

Maas, J. W., & Mednieks, M. Hydrocortisone-mediated increase of norepinephrine uptake by brain slices. *Science,* 1971, *171,* 178–179.

MacLusky, N. J., & McEwen, B. S. Progestin receptors in rat brain: Distribution and properties of cytoplasmic progestin-binding sites. *Endocrinology,* 1980, *106,* 192–202.

MacLusky, N. J., Turner, B. B., & McEwen, B. S. Corticosteroid binding in rat brain and pituitary cytosols: Resolution of multiple binding components by polyacrylamide gel based isoelectric focusing. *Brain Research,* 1977, *130,* 564–571.

Mainwaring, W. I. P., Symes, E. K., & Higgins, S. J. Nuclear components responsible for the retention of steroid-receptor complexes, especially from the stand-point of the specificity of hormonal responses. *Biochemical Journal,* 1976, *156,* 129–141.

Manak, R., Wertz, N., Slabaugh, M., Denari, H., Wang, J.-T., & Gorski, J. Purification and characterization of the estrogen-induced protein (IP) of the rat uterus. *Molecular and Cellular Endocrinology,* 1980, *17,* 119–132.

Martin, P. M., & Sheridan, P. J. Intracellular distribution of estrogen receptors—A function of preparation. *Experientia,* 1980, *36,* 620–622.

Matsumoto, A., & Arai, Y. Synaptogenic effect of estrogen on the hypothalamic arcuate nucleus of the adult female rat. *Cell and Tissue Research,* 1979, *198,* 427–433.

Matsumoto, A., & Arai, Y. Sexual dimorphism in wiring pattern in the hypothalamic arcuate nucleus and its modification by neonatal hormone environment. *Brain Research,* 1980, *190,* 238–242.

Mazurkiewicz-Kwilecki, I. M., & Prell, G. D. Brain histamine: Plasma corticosterone, spontaneous locomotor activity and temperature. *Pharmacology, Biochemistry and Behavior,* 1980, *12,* 549–553.

McEwen, B. S., Magnus, C., & Wallach, G. Soluble corticosterone-binding macromolecules extracted from rat brain. *Endocrinology,* 1972, *90,* 217–226.

McEwen, B. S., Wallach, G., & Magnus, C. Corticosterone binding to hippocampus: Immediate and delayed influences of the absence of adrenal secretion. *Brain Research,* 1974, *70,* 321–334.

McGinnis, M. Y., Gordon, J. H., & Gorski, R. A. Time course and localization of the effects of estrogen on glutamic acid decarboxylase activity. *Journal of Neurochemistry,* 1980, *34,* 785–792. (a)

McGinnis, M. Y., Gordon, J. H., & Gorski, R. A. Influence of γ-aminobutyric acid on lordosis behavior and dopamine activity in estrogen primed spayed female rats. *Brain Research,* 1980, *184,* 179–191. (b)

Menendez-Patterson, A., Florez-Lozano, J. F., & Marin, B. Effects of ovariectomy on the oxidative metabolism of the central nervous system and adrenal glands in female hamster (Mesocricetus auratus). *Experientia,* 1979, *35,* 349–350.

Meyer, J. S., Luine, V. N., Khylcheveskaya, R. I., & McEwen, B. S. Glucocorticoids and hippocampal enzyme activity. *Brain Research,* 1979, *166,* 172–175.

Meyerson, B. J., & Eliasson, M. Pharmacological and hormonal control of reproductive behavior. In L. L. Iversen, S. D. Iversen, & S. H. Snyder (Eds.), (Vol. 8): *Handbook of Psychopharmacology Drugs, transmitters, and behavior.* New York: Plenum Press, 1977.

Michal, E. K. Dexamethasone inhibits multi-unit activity in the rat hippocampus. *Brain Research,* 1974, *65,* 180–183.

Millard, S. A., Costa, E., & Gall, E. M. On the control of brain serotonin turnover by end product inhibition. *Brain Research,* 1972, *40,* 545–551.

Miller, A. L., Chaptal, C., McEwen, B. S., & Peck, E. J. Modulation of high affinity GABA uptake into hippocampal synaptosomes by glucocorticoids. *Psychoneuroendocrinology,* 1978, *3,* 155–164.

Mioduszewski, R., Grandison, L., & Meites, J. Stimulation of prolactin release in rats by GABA. *Proceedings of the Society of Experimental Biology and Medicine,* 1976, *151,* 44–46.

Mobley, P. L., & Sulser, F. Adrenal corticoids regulate sensitivity of noradrenaline receptor coupled adenylate cyclase in brain. *Nature,* 1980, *286,* 608–609.

Moguilevsky, J. A. Oxidative activity of different hypothalamic areas during sexual cycle in rats. *Acta Physiologica Latino Americana,* 1965, *15,* 423–424.

Moguilevsky, J. A., Libertun, C., & Foglia, V. G. Metabolic sensitivity of different hypothalamic areas to luteinizing hormone (LH), follicle stimulating hormone (FSH), and testosterone. *Neuroendocrinology,* 1970, *6,* 153–159. (a)

Moguilevsky, J. A., Libertun, C., & Foglia, V. G. Oxidative metabolism of the hypothalamus in hypophysectomized-castrated rats. *Experientia,* 1970, *26,* 421–422. (b)

Moguilevsky, J. A., Kalbermann, L. E., Libertun, C., & Gomez, C. J. Effects of ovariectomy on the amino acid incorporation into proteins of anterior pituitary and hypothalamus of rats. *Proceedings of the Society of Experimental Biology and Medicine,* 1971, *136,* 1115–1118.

Moguilevsky, J. A., Schiaffini, O., Szwarcfarb, B., & Libertun, C. Metabolic evidences of the short feed back mechanism controlling LH and FSH secretion. *Acta Endocrinologica Panamericana,* 1971, *2,* 177–186.

Moguilewsky, M., & Raynaud, J.-P. Estrogen-sensitive progestin-binding sites in the female rat brain and pituitary. *Brain Research,* 1979, *164,* 165–175. (a)

Moguilewsky, M., & Raynaud, J.-P. The relevance of hypothalamic and hypophyseal progestin receptor regulation in the induction and inhibition of sexual behavior in the female rat. *Endocrinology,* 1979, *104,* 516–522. (b)

Moguilewsky, M., & Raynaud, J.-P. Evidence for a specific mineralocorticoid receptor in rat pituitary and brain. *Journal of Steroid Biochemistry,* 1980, *12,* 309–314.

Monbon, M., Loras, B., Reboud, J. P., & Bertrand, J. Binding and metabolism of testosterone in the rat brain during sexual maturation: I. Macromolecular binding of androgens. *Journal of Steroid Biochemistry,* 1974, *5,* 417–424.

Moore, K. E., & Phillipson, O. T. Effects of dexamethasone on phenylethanolamine N-methyltransferase and adrenaline in the brains and superior cervical ganglia of adult and neonatal rats. *Journal of Neurochemistry,* 1975, *25,* 289–294.

Morrell, J. I., & Pfaff, D. W. A neuroendocrine approach to brain function: Localization of sex steroid concentrating cells in vertebrate brains. *American Zoologist,* 1978, *18,* 447–460.

Mosebach, K.-O., & Peter, H.-G. Early influences of testosterone and 19-nortestosterone on the metabolism of L-histidine-^{14}C in the brain of immature male rats. *Acta Endocrinologica,* 1971, Supplement *155,* 28.

Motta, M. *The endocrine functions of the brain.* New York: Raven Press, 1980.

Moudgil, V. K., & Kanungo, M. S. Effect of age of the rat on induction of acetylcholinesterase of the brain by 17β-estradiol. *Biochemica et Biophysica Acta,* 1973, *329,* 211–220.

Munaro, N. I. The effect of ovarian steroids on hypothalamic 5-hydroxy-tryptamine neuronal activity. *Neuroendocrinology,* 1978, *26,* 270–276.

Murayama, A., Fukai, F., & Yamamoto, T. Disorganization and in vitro assembly of the constituents of the cytoplasmic estrogen receptor system of cow uterus. *Journal of Biochemistry,* 1980, *88,* 1457–1466.

Muth, E. A., Crowley, W. R., & Jacobowitz, D. M. Effect of gonadal hormones on luteinizing hormone in plasma and on choline acetyltransferase activity and acetylcholine levels in discrete nuclei of rat brain. *Neuroendocrinology,* 1980, *30,* 329–336.

Nagle, C. A., & Rosner, J. M. Rat brain norepinephrine release during progesterone-induced LH secretion. *Neuroendocrinology,* 1980, *30,* 33–37.

Neckers, L., & Sze, P. Y. Regulation of 5-hydroxytryptamine metabolism in mouse brain by adrenal glucocorticoids. *Brain Research,* 1976, *93,* 123–132.

Nemeroff, C. B., & Dunn, A. J. (Eds.). *Peptides, hormones and behavior: Molecular and behavioral neuroendocrinology.* New York: Spectrum Publications, 1983.

Nixon, R. L., Janowsky, D. S., & Davis, J. M. Effects of progesterone, β-estradiol, and testosterone on the uptake and metabolism of ^{3}H-norepinephrine, ^{3}H-dopamine and ^{3}H-serotonin in rat brain synaptosomes. *Research Communications in Chemical Pathology and Pharmacology,* 1974, *7,* 233–236.

Nock, B., Blaustein, J. D., & Feder, H. H. Changes in noradrenergic transmission alter the concentration of cytoplasmic progestin receptors in hypothalamus. *Society for Neuroscience Abstracts,* 1980, *6,* 31.

Nyakas, C., de Kloet, E. R., & Bohas, B. Hippocampal function and putative corticosterone receptors: Effects of septal lesions. *Neuroendocrinology,* 1979, *29,* 301–312.

Orr, E., & Quay, W. The effects of castration on histamine levels and 24-hour rhythm in the male rat hypothalamus. *Endocrinology,* 1975, *97,* 481–484.

Packman, P. M., Bragdon, M. J., & Boshans, R. L. Quantitative histochemical studies of the hypothalamus: Control point enzymes during the estrous cycle. *Neuroendocrinology,* 1977, *23,* 76–87.

Pandolfo, L., & Macaione, S. Effect of adrenalectomy on activity of GABA transaminase and glutamic acid decarboxylase from rat brain cortex. *Italian Journal of Biochemistry,* 1964, *13,* 247–252.

Paolo, T. D., Labrie, F., Dupont, A., Barden, N., & Langelier, P. Effects of estrogen treatment on normal and sensitized rat striatal dopamine (DA) receptors and DA-sensitive adenylyl cyclase. *Society for Neuroscience Abstracts,* 1979, *5,* 1495.

Pardridge, W. M., & Mietus, L. J. Transport of steroid hormones through the rat blood-brain barrier. Primary role of albumin-bound hormone. *Journal of Clinical Investigation,* 1979, *64,* 145–154.

Pardridge, W. M., & Mietus, L. J. Effects of progesterone-binding globulin versus a progesterone antiserum on steroid hormone transport through the blood-brain barrier. *Endocrinology,* 1980, *106,* 1137–1141.

Parsons, B., MacLusky, N. J., Krey, L., Pfaff, D. W., & McEwen, B. S. The temporal relationship between estrogen-inducible progestin receptors in the female rat brain and the time course of estrogen activation of mating behavior, *Endocrinology,* 1980, *107,* 774–779.

Paul, S. M., & Skolnick, P. Catechol oestrogens inhibit oestrogen elicited accumulation of hypothalamic cyclic AMP suggesting role as endogenous anti-oestrogens. *Nature,* 1977, *266,* 559–561.

Paul, S. M., Axelrod, J., Saavedra, J. M., & Skolnick, P. Estrogen-induced efflux of endogenous catecholamines from the hypothalamus *in vitro. Brain Research,* 1979, *178,* 499–505.

Paul, S. M., Hoffman, A. R., & Axelrod, J. Catechol estrogens: Synthesis and metabolism in brain and other tissues. In L. Martini & W. F. Ganong (Eds.), *Frontiers in neuroendocrinology* (Vol. 6). New York: Raven Press, 1980.

Peck, E. J., Jr., Miller, A. L., & Kelner, K. L. Estrogen receptors and the activation of RNA polymerase by estrogens in the central nervous system. In T. H. Hamilton, J. H. Clark, & W. A. Sandler (Eds.), *Ontogeny of receptors and reproductive hormone action.* New York: Raven Press, 1979.

Perry, B. N., & Lopez, T. A. The binding of ^{3}H-labelled oestradiol- and progesterone-receptor complexes to hypothalamic chromatin of male and female sheep. *Biochemical Journal,* 1978, *176,* 873–883.

Pfaff, D. W., & Keiner, M. Atlas of estradiol-concentrating cells in the central nervous system of the female rat. *Journal of Comparative Neurology,* 1973, *151,* 121–158.

Pfaff, D. W., & Modianos, D. Neural mechanisms of female reproductive behavior. In N. T. Adler & D. Pfaff (Eds.), *Neurobiology of reproduction.* Book in preparation, 1983.

Portaleone, P., Pagnini, G., Crispino, A., & Genazzani, E. Histamine-sensitive adenylate cyclase in hypothalamus of rat brain: H_1 and H_2 receptors. *Journal of Neurochemistry,* 1978, *31,* 1371–1374.

Portaleone, P., Genazzani, E., Pagnini, G., Crispino, A., & DiCarlo, F. Interaction of estradiol and 2-hydroxy-estradiol with histamine receptors at hypothalamic level. *Brain Research,* 1980, *187,* 216–220.

Puca, G. A., Nola, E., Hibner, U., Cicala, G., & Sica, V. Interaction of the estradiol receptor from calf uterus with its nuclear acceptor sites. *Journal of Biological Chemistry,* 1975, *250,* 6452–6459.

Rainbow, T. C., Davis, P. G., & McEwen, B. S. Anisomycin inhibits the activation of sexual behavior by estradiol and progesterone. *Brain Research,* 1980, *194,* 548–555.

Rastogi, R. B., & Singhal, R. L. Evidence for the role of adrenocortical hormones in the regulation of noradrenaline and dopamine metabolism in certain brain areas. *British Journal of Pharmacology,* 1978, *62,* 131–136. (a)

Rastogi, R. B., & Singhal, R. L. Adrenocorticoids control 5-hydroxytryptamine metabolism in rat brain. *Journal of Neural Transmission,* 1978, *42,* 63–71. (b)

Raynaud, J. P. R5020, a tag for the progestin receptor. In W. L. McGuire, J. P. Raynaud, & E. E. Baulieu (Eds.), *Progesterone receptors in normal and neoplastic tissues.* New York: Raven Press, 1977.

Reddy, V. V. R. Estriol synthesis in rat brain and pituitary. *Brain Research,* 1979, *175,* 165–168. (a)

Reddy, V. V. R. Estrogen metabolism in neural tissues of rabbits: 17β-hydroxysteroid oxidoreductase activity. *Steroids,* 1979, *34,* 207–215. (b)

Rees, H. D., & Gray, H. E. Glucocorticoids and mineralocorticoids: actions on brain and behavior. In C. B. Nemeroff & A. J. Dunn (Eds.), *Peptides, hormones and behavior: Molecular and behavioral neuroendocrinology*. New York: Spectrum Publications, 1983.

Rees, H. D., Stumpf, W. E., & Sar, M. Autoradiographic studies with ^{3}H-dexamethasone in the rat brain and pituitary. In W. E. Stumpf & L. D. Grant (Eds.), *Anatomical neuroendocrinology*. Basel: Karger, 1975.

Reiss, N., & Kaye, A. M. Separation of a protein from rat brain resembling the uterine "estrogen-induced protein" in electrophoretic and immunologic properties. *Israel Journal of Medicine*, 1979, *15*, 545.

Reiter, R. J. The pineal and its hormones in the control of reproduction in mammals. *Endocrine Reviews*, 1980, *1*, 109–131.

Rezek, D. L. Nuclear localization of testosterone, dihydrotestosterone and estradiol-17β in basal rat brain. *Psychoneuroendocrinology*, 1977, *2*, 173–178.

Rhees, R. W., Grosser, B. I., & Stevens, W. The autoradiographic localization of [^{3}H]dexamethasone in the brain and pituitary of the rat. *Brain Research*, 1975, *100*, 151–156.

Ricker, D. K., Sastre, A., Baker, T., Roth, R. H., & Riker, W. F. Regional high-affinity [^{3}H]choline accumulation in cat forebrain: Selective increase in the caudate-putamen after corticosteroid pretreatment. *Molecular Pharmacology*, 1979, *16*, 886–899.

Roberts, E., Chase, T. N., & Tower, D. B. *GABA in nervous system function*. New York: Raven Press, 1976.

Rosner, J. M., Denari, J. H., Castro-Vazquez, A., Nagle, C. A., Neuspiller, M. R., Bedes, G. D. D. P., Pedroza, E., Martin, J. L., & Gomez, E. Oestrogen uptake by the central nervous system. *Gynecological Investigation*, 1972, *3*, 30–42.

Roy, E. J., & McEwen, B. S. Oestrogen receptors in cell nuclei of the hypothalamus–preoptic area–amygdala following an injection of oestradiol or the antioestrogen CI-628. *Journal of Endocrinology*, 1979, *83*, 285–293.

Roy, E. J., Schmit, E., McEwen, B. S., & Wade, G. N. Anti-estrogens in the central nervous system. In M. K. Agarwal (Ed.), *Antihormones*. Amsterdam: Elsevier-North Holland Biomedical Press, 1979.

Russell, D. H. Ornithine decarboxylase as a biological and pharmacological tool. *Pharmacology*, 1980, *20*, 117–129.

Saad, S. F. The effect of ovariectomy on the γ-aminobutyric acid content in the cerebral hemispheres of young rats. *Journal of Pharmacy and Pharmacology*, 1970, *22*, 307.

Sadasivudu, B., Rao, T. I., & Murphy, C. R. K. Metabolic effects of hydrocortisone in mouse brain. *Neurochemical Research*, 1977, *2*, 521–532.

Saffran, J., & Loeser, B. K. Nuclear binding of guinea pig uterine progesterone receptor in cell-free preparations. *Journal of Steroid Biochemistry*, 1979, *10*, 43–51.

Sakaue, Y., & Thompson, E. B. Characterization of two forms of glucocorticoid hormone-receptor complex separated by DEAE-cellulose column chromatography. *Biochemical and Biophysical Research Communications*, 1977, *77*, 533–541.

Sando, J. J., Hammond, N. D., Stratford, C. A., & Pratt, W. B. Activation of thymocyte glucocorticoid receptors to the steroid binding form. *Journal of Biological Chemistry*, 1979, *254*,4779–4789.

Sando, J. J., LaForest, A. C., & Pratt, W. B. ATP-Dependent activation of L cell glucocorticoid receptors to the steroid binding form. *Journal of Biological Chemistry*, 1979, *254*, 4772–4778.

Sar, M., & Stumpf, W. E. Androgen concentration in motor neurons of cranial nerves and spinal cord. *Science*, 1977, *197*, 77–79.

Sato, B., Noma, K., Nishizawa, Y., Nakao, K., Matsumoto, K., & Yamamura, Y. Mechanism of activation of steroid receptors: Involvement of low molecular weight inhibitor in activation of androgen, glucocorticoid, and estrogen receptor systems. *Endocrinology*, 1980, *106*, 1142–1148.

Scacchi, P., Moguilevsky, J. A., & Szwarcfarb, B. Effect of castration on $C^{14}O_2$ production from glucose-U-C^{14} in the hypothalamus and cerebral cortex. *Proceedings of the Society of Experimental Biology and Medicine*, 1971, *136*, 1068–1071.

Scacchi, P., Moguilevsky, J. A., & Schiaffini, O. Glucose oxidation in the hypothalamus during the sexual cycle in rats: Influence of castration. *Neuroendocrinology*, 1973, *11*, 321–327.

Schaeffer, J. M., & Hseuh, A. J. W. 2-Hydroxyestradiol interaction with dopamine receptor binding in rat anterior pituitary. *Journal of Biological Chemistry,* 1979, *254,* 5606–5608.

Schaeffer, J. M., Stevens, S. R., & Smith, R. G. 2-Hydroxyestradiol (2-OH-E_2) binding to rat anterior pituitary membranes. *Endocrine Society Abstracts,* 1980, *106,* 317.

Scheff, S. W., Thorne, D. R., Sasvary, G., Bernardo, L. S., & Cotman, C. W. Chronic glucocorticoid administration alters axonal sprouting in the rat hippocampal formation. *Society for Neuroscience Abstracts,* 1979, *5,* 1555.

Schiaffini, O., & Marin, B. Effect of ovariectomy on the oxidative activity of the hypothalamus and of the limbic system of the rat. *Neuroendocrinology,* 1971, *7,* 302–307.

Schiaffini, O., & Martini, L. The amygdala and the control of gonadotrophin secretion. *Acta Endocrinologica,* 1972, *70,* 209–219.

Schiaffini, O., Moguilevsky, J. A., Libertun, C., & Foglia, V. G. Oxidative and glycolytic metabolism of different hypothalamic areas in diabetic rat. *Acta Physiologica Latino Americana,* 1968, *18,* 257–262.

Schiaffini, O., Marin, B., & Foglia, V. G. Metabolic alterations in hypothalamus and limbic structures in female diabetic rats. *Experientia,* 1970, *26,* 610–611.

Schwartz, J. -C., Pollard, H., & Quach, T. T. Histamine as a neurotransmitter in mammalian brain: Neurochemical evidence. *Journal of Neurochemistry,* 1980, *35,* 26–33.

Seiki, K., Haruki, Y., Imanishi, Y., & Enomoto, T. Further evidence of the presence of progesterone-binding proteins in female rat hypothalamus. *Endocrinologia Japonica,* 1977, *24,* 233–238.

Seiki, K., Haruki, Y., Imanishi, Y., & Enomoto, T. Progestin binding *in vitro* by brain cell nuclei of ovariectomized oestrogen-primed rats. *Journal of Endocrinology,* 1979, *82,* 347–360.

Selmanoff, M. K., Brodkin, L. D., Weiner, R. J., & Siiteri, P. Aromatization and 5α-reduction of androgens in discrete hypothalamic and limbic regions of the male and female rat. *Endocrinology,* 1977, *101,* 841–848.

Shani, J., Givant, Y., Sulman, F., Eylath, U., & Eckstein, B. Competition of phenothazines with oestradiol receptors in the brain. *Neuroendocrinology,* 1971, *8,* 307–316.

Shen, G., Thrower, S., & Lim, L. Uterine oestrogen-receptor binding to oligo(dT)-cellulose: An inhibitor from hypothalamic cytosol. *Biochemical Journal,* 1979, *182,* 241–243.

Shen, J. T., & Ganong, W. F. Effect of variations in adrenocortical function on dopamine β-hydroxylase and norepinephrine in the brain of the rat. *Journal of Pharmacology and Experimental Therapeutics,* 1976, *199,* 639–648.

Sheridan, P. J. The nucleus interstitialis striae terminalis and the nucleus amygdaloideus medialis: Prime targets for androgen in the rat forebrain. *Endocrinology,* 1979, *104,* 130–136.

Siegel, L. I., & Wade, G. N. Insulin withdrawal impairs sexual receptivity and retention of brain cell nuclear estrogen receptors in diabetic rats. *Neuroendocrinology,* 1979, *29,* 200–206.

Simpkins, J. W., Kalra, P. S., & Kalra, S. P. Effects of testosterone on catecholamine turnover and LHRH contents in the basal hypothalamus and preoptic area. *Neuroendocrinology,* 1980, *30,* 94–100.

Stevens, W., Reed, D. J., Erickson, S., & Grosser, B. I. The binding of corticosterone to brain proteins: Diurnal variation. *Endocrinology,* 1973, *93,* 1152–1156.

Stipek, S., Crkovska, J., Trojan, S., & Prokes, J. The effects of polyamines on RNA synthesis in cell nuclei isolated from the rat brain. *Physiologica Bohemoslovaca,* 1978, *27,* 280–281.

Stith, R. D., & Bottoms, G. D. Intracellular binding of [^{3}H]cortisol and its effect on RNA polymerase activity in hypothalamus of the pig. *Brain Research,* 1972, *41,* 423–434.

Stith, R. D., Person, R. J., & Dana, R. C. Effects of hippocampal and amygdalar stimulation on the uptake and binding of ^{3}H-hydrocortisone in the hypothalamus of the cat. *Journal of Neuroscience Research,* 1976, *2,* 317–322.

Stumpf, W. E., & Grant, L. D. *Anatomical neuroendocrinology.* Basel: Karger, 1975.

Stumpf, W. E., & Sar, M. Anatomical distribution of estrogen, androgen, progestin, corticosteroid and thyroid hormone target sites in the brain of mammals: Phylogeny and ontogeny. *American Zoologist,* 1978, *18,* 435–445.

Stumpf, W. E., & Sar, M. Steroid action in CNS and anterior pituitary: I. Steroid hormone target cells in the extrahypothalamic brain stem and cervical spinal cord: Neuroendocrine significance. *Journal of Steroid Biochemistry,* 1979, *11,* 801–807.

Sundberg, D. K., Fawcett, C. P., & McCann, S. M. The involvement of cyclic-3′,5′AMP in the release of hormones from the anterior pituitary in vitro. *Proceedings of the Society of Experimental Biology and Medicine,* 1976, *151,* 149–154.

Sze, P. Y., Neckers, L., & Towle, A. C. Glucocorticoids as a regulatory factor for brain tryptophan hydroxylase. *Journal of Neurochemistry,* 1976, *26,* 169–173.

Tappaz, M. L., Brownstein, M. J., & Kopin, I. J. Glutamate decarboxylase (GAD) and γ-aminobutyric acid (GABA) in discrete nuclei of hypothalamus and substantia nigra. *Brain Research,* 1977, *125,* 109–121.

Teyler, T. J., Vardaris, R. M., Lewis, D., & Rawitch, A. B. Gonadal steroids: Effects on excitability of hippocampal pyramidal cells. *Science,* 1980, *209,* 1017–1019.

Thrower, S., & Lim, L. Characterization of rat hypothalamic progestin binding by spheroidal hydroxylapatite chromatography. *Biochemical Journal,* 1980, *186,* 295–300.

Tintner, R., Dunn, A. J., Iuvone, P. M., Shukla, J. B., & Rennert, O. M. Corticotrophin increases cerebral polyamine content. *Journal of Neurochemistry,* 1979, *33,* 1067–1073.

Tobias, H., Carr, L., & Vooft, J. Effects of estradiol on catecholamine synthesizing enzymes, luteinizing hormone (LH) and prolactin in the ovariectomized rat. *Society for Neuroscience Abstracts,* 1979, *5,* 1569.

Towle, A. C., & Sze, P. Y. Binding of corticosterone to synaptic plasma membrane from rat brain. *Society for Neuroscience Abstracts,* 1978, *4,* 356.

Towle, A. C., Sze, P. Y., & Lauder, J. M. Cytosol glucocorticoid receptors in monoaminergic cell groups. *Transactions of the American Society of Neurochemistry,* 1979, *10,* 199.

Traish, A. M., Muller, R. E., & Wotiz, H. H. Binding of estrogen receptor to uterine nuclei: Salt-extractable versus salt-resistant receptor estrogen complexes. *Journal of Biological Chemistry,* 1977, *252,* 6823–6830.

Tsuboi, S., Kawashima, R., Tomioka, O., Nakata, M., Sakamoto, N., & Fujita, T. Glucocorticoid binding proteins of human brain cytosol. *Brain Research,* 1979, *179,* 181–185.

Turner, B. B., & McEwen, B. S. Hippocampal cytosol binding capacity of corticosterone: No depletion with nuclear loading. *Brain Research,* 1980, *189,* 169–182.

Turner, B. B., Smith, E. M., & Carroll, B. J. Baboon corticosterone: Substantial brain binding of a "minor" adrenal glucocorticoid. *Society for Neuroscience Abstracts,* 1979, *5,* 462.

Ulrich, R., Yuwiler, A., & Geller, E. Effects of hydrocortisone on biogenic amine levels in the hypothalamus. *Neuroendocrinology,* 1975, *19,* 259–268.

Vacas, M. I., & Cardinali, D. P. Effect of estradiol on α- and β-adrenoceptor density in medial basal hypothalamus, cerebral cortex and pineal gland of ovariectomized rats. *Neuroscience Letters,* 1980, *17,* 73–77.

Veals, J. W., Korduba, C. A., & Symchowicz, S. Effect of dexamethasone on monoamine oxidase inhibition by iproniazid in rat brain. *European Journal of Pharmacology,* 1977, *41,* 291–299.

Verdiere, M., Rose, C., & Schwartz, J. C. Turnover of central histamine in a stressful situation. *Brain Research,* 1977, *129,* 107–119.

Vermes, I., Telegdy, G., & Lissak, K. Correlation between hypothalamic serotonin content and adrenal function during acute stress. Effect of adrenal corticosteroids on hypothalamic serotonin content. *Acta Physiologica Academiae Scientiarum Hungaricae,* 1973, *43,* 33–42.

Vermes, I., Varszegi, M., Toth, E. K., & Telegdy, G. Action of androgenic steroids on brain neurotransmitters in rats. *Neuroendocrinology,* 1979, *28,* 386–393.

Vertes, A., Vertes, M., & Kovacs, S. Hypothalamic effect of oestradiol. *Acta Physiologica Academiae Scientiarum Hungaricae,* 1978, *51,* 218–219.

Walker, M. D., Negreanu, V., Gozes, I., & Kaye, A. M. Identification of the 'estrogen-induced protein' in uterus and brain of untreated immature rats. *FEBS Letters,* 1979, *98,* 187–191.

Wallis, C. J. *Neuroendocrine influences on gamma-aminobutyric acid metabolism in rodent brain tissue.* Unpublished doctoral dissertation, University of Florida, 1976.

Wallis, C. J., & Luttge, W. G. Influence of estrogen and progesterone on glutamic acid decarboxylase activity in discrete regions of rat brain. *Journal of Neurochemistry,* 1980, *34,* 609–613.

Wallis, C. J., & Printz, M. P. Adrenal regulation of regional brain angiotensinogen content. *Endocrinology,* 1980, *106,* 337–342.

Warembourg, M. Radioautographic study of the rat brain after injection of [1,2-^{3}H]corticosterone. *Brain Research,* 1975, *89,* 61–70. (a)

Warembourg, M. Radioautographic study of the rat brain and pituitary after injection of ^{3}H-dexamethasone. *Cell and Tissue Research,* 1975, *161,* 183–191. (b)

Warembourg, M. Fixation des steroides au niveau du systeme nerveux central et de l'hypophyse chez differents mammiferes. *Annales d'Endocrinologie (Paris),* 1977, *38,* 41–54.

Warembourg, M. Radioautographic study of the rat brain, uterus and vagina after [^{3}H]R-5020. *Molecular and Cellular Endocrinology,* 1978, *12,* 67–79.

Watson, G. H., & Muldoon, T. G. Microsomal estrogen receptors in rat uterus and anterior pituitary. *Federation Proceedings,* 1977, *36,* 912.

Weichman, B. M., & Notides, A. C. Analysis of estrogen receptor activation by its [^{3}H]estradiol dissociation kinetics. *Biochemistry,* 1979, *18,* 220–225.

Weichman, B. M., & Notides, A. C. Estrogen receptor activation and the dissociation kinetics of estradiol, estriol and estrone. *Endocrinology,* 1980, *106,* 434–439.

Weidenfeld, J., Siegel, R. A., & Chowers, I. *In vitro* conversion of pregnenolone to progesterone by discrete brain areas of the male rat. *Journal of Steroid Biochemistry,* 1980, *13,* 961–963.

Weissman, B. A., & Johnson, D. F. Possible role of dopamine in diethylstilbestrol-elicited accumulation of cyclic AMP in incubated male rat hypothalamus. *Neuroendocrinology,* 1976, *21,* 1–9.

Weissman, B. A., & Skolnick, P. Stimulation of adenosine 3′,5′-monophosphate formation in incubated rat hypothalamus by estrogenic compounds: Relationship to biologic potency and blockade by anti-estrogens. *Neuroendocrinology,* 1975, *18,* 27–34.

Weissman, B. A., Daly, J. W., & Skolnick, P. Deithylstilbestrol-elicited accumulation of cyclic AMP in incubated rat hypothalamus. *Endocrinology,* 1975, *97,* 1559–1566.

Whalen, R. E., & Olsen, K. L. Chromatin binding of estradiol in the hypothalamus and cortex of male and female rats. *Brain Research,* 1978, *152,* 121–131.

Whalen, R. E., Yahr, P., & Luttge, W. G. Hormones, metabolism and sexual behavior. In N. T. Adler & D. Pfaff (Eds.), *Neurobiology of reproduction.* Book in preparation, 1983.

White, J. O., & Lim, L. Unoccupied nuclear oestrogen receptors in the female rat hypothalamus. Increases on oestrogen administration. *Biochemical Journal,* 1980, *190,* 833–837.

White, J. O., Thrower, S., & Lim, L. Intracellular relationships of the oestrogen receptor in the rat uterus and hypothalamus during the oestrous cycle. *Biochemical Journal,* 1978, *172,* 37–47.

Wilkinson, M., Herdon, H., Pearce, M., & Wilson, C. Radioligand binding studies on hypothalamic noradrenergic receptors during the estrous cycle or after steroid injection in ovariectomized rats. *Brain Research,* 1979, *168,* 652–655.

Williams, M. Protein phosphorylation in the mammalian nervous system. *Trends in Biochemical Science,* 1979, *4,* 25–28.

Wilson, E. M., & French, F. S. Effects of proteases and protease inhibitors on the 4.5S and 8S androgen receptor. *Journal of Biological Chemistry,* 1979, *254,* 6310–6319.

Wrange, O. A comparison of the glucocorticoid receptor in cytosol from rat liver and hippocampus. *Biochimica et Biophysica Acta,* 1979, *582,* 346–357.

Yagi, K., & Sawaki, Y. Recurrent inhibition and facilitation. Demonstration in the tuberinfundibular system and effects of strychnine and picrotoxin. *Brain Research,* 1975, *84,* 155–159.

Yamamoto, K. R. Characterization of the 4S and 5S forms of the estradiol receptor protein and their interaction with deoxyribonucleic acid. *Journal of Biological Chemistry,* 1974, *249,* 7068–7075.

Zava, D. T., & McGuire, W. L. Estrogen receptor: Unoccupied sites in nuclei of breast tumor cell line. *Journal of Biological Chemistry,* 1977, *252,* 3703–3708.

Zigmond, R. E., & McEwen, B. S. Selective retention of oestradiol by cell nuclei in specific brain regions of the ovariectomized rat. *Journal of Neurochemistry,* 1970, *17,* 889–899.

Zolovick, A. J., Pearse, R., Boehlke, K. W., & Eleftheriou, B. E. Monoamine oxidase activity in various parts of the rat brain during the estrous cycle. *Science,* 1966, *154,* 649.

12

Hormone–Drug Interactions and Their Influence on Aggressive Behavior

Klaus A. Miczek and Joseph F. DeBold

INTRODUCTION

The study of the chemical events that are necessary for aggressive behavior to be initiated and performed is pursued by neurochemists, endocrinologists, and pharmacologists. Endocrinological and pharmacological studies of aggression share many features of research strategy, and yet the interactive effects of drugs and hormones on aggressive behavior are rarely investigated. It is common in behavioral pharmacology and endocrinology to employ receptor agonists and antagonists to remove or block the endogenous source of a transmitter or hormone, subsequently replacing it; researchers in both fields examine how the behavioral effects of hormone or drug manipulations are dependent on the dose of the agent and on the time course of action. A further common objective is to localize the anatomical sites of action within the brain that are critical for the behavioral effects of hormones and drugs.

It has become apparent only within the last decade that certain known neurotransmitters and steroid hormones may be present and active in the same anatomical sites and even within the same neurons. The final site of action for hormones and drugs may indeed converge on one neural process. At present, such possibilities for interaction can be entertained at least for gonadal hormones and the monoamines.

In this chapter we will try to show how some drugs can alter endocrine as well as transmitter activity and how the consequent changes in aggressive behavior may have been produced by interaction of certain neurotransmitters and hormones. Most research focuses on either hormonal or pharmacological manipulations of aggressive behavior but not on their potential interactions. We

---

KLAUS A. MICZEK AND JOSEPH F. DEBOLD • Department of Psychology, Tufts University, Medford, Massachusetts 02155. Preparation of this contribution was supported in part by United States Public Health Research Grants DA-02632 and AA-05122.

will attempt to merge two lines of investigation that have remained separate so far. There are several remarkable efforts to elucidate drug–hormone interactions affecting sexual behavior, and we will use these studies as models for future investigations in aggression research. However, this chapter is not intended as a complete review of the pharmacology of sexual behavior.

Drugs as Probes of Neurotransmitter Activity

Recent histochemical and neurophysiological studies have identified neurons that use dopamine, norepinephrine, and serotonin as transmitter substances. There is also increasing evidence for the discrete localization of storage sites and receptors, and of enzymes for synthesis and degradation for acetylcholine and several neuropeptides. It is the selective action on specific neurochemical systems that makes drugs powerful tools for the study of the chemical basis of aggression (Cooper, Bloom, & Roth, 1978). Drugs can alter impulse flow in the nervous system and produce their behavioral effects in a number of ways. Drugs may modify neurotransmission by blocking axonal transport or axonal membrane permeability, increasing or decreasing the availability of the precursor, disrupting the synthesis or storage of the neurotransmitter, altering the release process, blocking the uptake mechanisms of the precursor or the released transmitter, or influencing metabolic enzymes. In addition to specific and selective effects on a defined aspect of neurotransmission, every drug exerts on processes in the CNS and peripheral organs multiple effects that may not be the focus of the investigation. It is important to recognize that most drugs that alter behavior also may be stressors and, consequently, may activate adrenal secretion.

Limitations

This chapter will not discuss two particular classes of pharmacological agents. Steroid antagonists—the antiestrogens and antiandrogens—are used mainly to assess the mediation of steroid hormone action by steroid receptors. Since they interfere primarily with the action of steroids on target cells, these agents have remained tools of the endocrinologist rather than the neuropharmacologist. Closely related to the steroid antagonists are synthetic steroids used to inhibit specific steroid metabolic pathways. Again, these agents are most useful for identifying intracellular elements of steroid action rather than the interactions between hormones and drugs. The steroid antagonists and synthetic steroids that can interfere with steroid metabolism will be considered in the chapter on cellular and molecular mechanisms of steroid hormone control of behavior. This chapter also will not deal with the effects of environmental toxins such as lead, cadmium, and DDT on aggressive behavior or the endocrine system. The effects of drugs from both of these classes can be very interesting, but they are outside the scope of this chapter.

Forms of Aggression

The term "aggression" has eluded a specific useful definition that is universally accepted. Instead, a variety of behavior patterns have been labeled "aggressive," and it has become clear that aggression is not a unitary phenomenon. Up to now, it is unclear how to categorize different aggressive behavior patterns, whether according to the situations in which they occur, the functions they might serve, or the topography with which they are displayed. Most researchers have pursued the study of aggression in the laboratory using animal species and experimental protocols that reliably engender a wide range of aggressive or defensive behavior. Studies on the hormonal basis of aggression have focused primarily on conflict between members of the same species (i.e., *intraspecies aggression*) and, in particular, on the attack behavior of adult males toward other males (i.e., *male–male aggression*). From a functional viewpoint, male–male aggression may be territorial or rival fighting. Hormonal factors in the control of defensive and submissive reactions have begun to be investigated only recently (e.g., Roche & Leshner, 1979). Also, the endocrine basis of aggression by females has received little experimental attention (i.e., *female–female aggression*).

Table I summarizes several frequently used laboratory paradigms for the study of aggression. The table lists the paradigms by their usual name and indicates the animal species most often used. The essential features of the stimulus situation and experimental manipulations are specified in the next column. The wide variety of behavior patterns generated by the different paradigms is described under "Behavioral Topography." Little is known about the biological function of most laboratory paradigms of aggression, and entries in this column are highly speculative. Most investigators choose a certain paradigm on pragmatic grounds, on the basis of convenience and ease, and out of tradition, with little functional analysis. Historically, the most often-used paradigm in research with hormonal and pharmacological manipulations is the isolation-induced aggression paradigm. At present, there are contrasting views on the biological function of aggression by isolated animals—territorial defense on the one hand and a model of a behavioral pathology on the other (Brain, 1975; Valzelli, 1973).

Models of Drug–Hormone Interactions

The neuropharmacologist often uses drugs as tools to study the influence of specific neurotransmitters on behavior. This strategy can be extended to a mechanistic approach to the study of the hormonal control of behavior.

It is clear from many of the chapters in this volume that steroids can modulate the expression of aggressive behavior. Our understanding of how steroid hormones act as neuroregulators in this system, however, is rather limited. There is general agreement that steroids exert their actions on their target cells by binding with an intracellular receptor. This steroid–receptor complex then stimulates changes in nuclear mRNA synthesis, thus altering protein synthesis (see chapter by Luttge). But how alterations in protein synthesis are translated into

TABLE I
Major Paradigms for the Study of Aggression in the Laboratory

Paradigm	Stimulus situation	Behavioral topography	Biological function
Isolation-induced aggression, mostly in mice	Prolonged isolated housing before confrontation with another isolate or group-housed animal in a test cage or home-cage of the isolate	Complete agonistic behavior pattern; isolates attack, threaten, and pursue opponent	Territorial defense or compulsive, abnormal pathological behavior
Pain-elicited or shock-induced aggression, mostly in rats	Pairs of animals are exposed to pulses of electric shock delivered through grid floor	Defensive reactions, including upright postures, bites toward face of opponent, and audible vocalizations	Some similarity to reaction toward predator or toward large opponent
Frustration- or extinction-induced aggression, mostly in pigeons	Conditioning history: operant behavior under control of a reinforcement schedule; omission or prolonged intermittency of reinforcement	Attack bites or pecks, threat displays towards suitable object or conspecific	Competition for resources (?)
Predatory aggression, mostly in rats and cats	Appearance of prey, food deprivation	Stalking, seizing, killing, and consuming (?)	Food source; "killer instinct"
Brain-lesion-induced aggression, in rats and cats	Destruction of neural tissue and subsequent social or environmental challenges	Defensive reactions; biting	Brain disease
Brain-stimulation-induced aggression, in cats, rats, and monkeys	Electrical excitation of neural tissue, mostly in diencephalon and mesencephalon, but also in other, limbic or cerebellar areas	Defensive reactions accompanied by autonomic arousal; predatory attack and kill	Predation
Resident–intruder aggression, in most animals	Confrontation with an unfamiliar adult member of the species	Full species-specific repertoire of agonistic behavior, including attack, threat, defense, submission, and flight	Territorial or group defense (?)
Maternal aggression, mostly in rodents	Lactating female, in the presence of litter, confronting an intruder male	Species-specific repertoire of attack and threat behavior	Defense of young

changes in neural activity or changes in neuronal responsiveness is far from certain. So far, steroids have been shown to be capable of altering the levels of many neurotransmitters, some enzymes controlling transmitter synthesis and degradation, that is, the binding of some transmitter agonists and transmitter release (for a review see McEwen, Davis, Parsons, & Pfaff, 1979). It is not known at present how many of or to what extent these effects are crucial to the ability of steroid hormones to affect aggression. However, questions of this kind can be approached pharmacologically. To some extent, this approach has been pursued in the study of the hormonal control of sexual behavior, but it has rarely been employed for the study of aggressive behavior (Meyerson, Palis, & Sietniks, 1979).

Hormones and drugs may interact at multiple levels in the CNS. Three such possibilities are shown schematically in their simplest form in Figure 1. A similar set of models has been suggested by Meyerson *et al.* (1979). A given hormone and drug could affect behavior by acting on the same neurons (Figure 1A). It has recently been demonstrated that some cells in the brainstem contain both estrogen-receptors and catecholamines (Heritage, Stumpf, Sar, & Grant, 1980). Demonstrations of neurons containing both neuropeptide hormones and biogenic amines are increasing (Hökfelt, Johansson, Ljungdahl, Lundberg, & Schultzberg, 1980). More commonly, there is significant overlap in the distribution of hormone target cells and known neurotransmitter systems, although not necessarily within the same cells. A given drug may act on a neuron or neurons in sequence with a hormone-sensitive cell, as is shown schematically in Figure 1B. For example, it has recently been demonstrated that altering noradrenergic activity can affect progesterone-receptors in the hypothalamus of

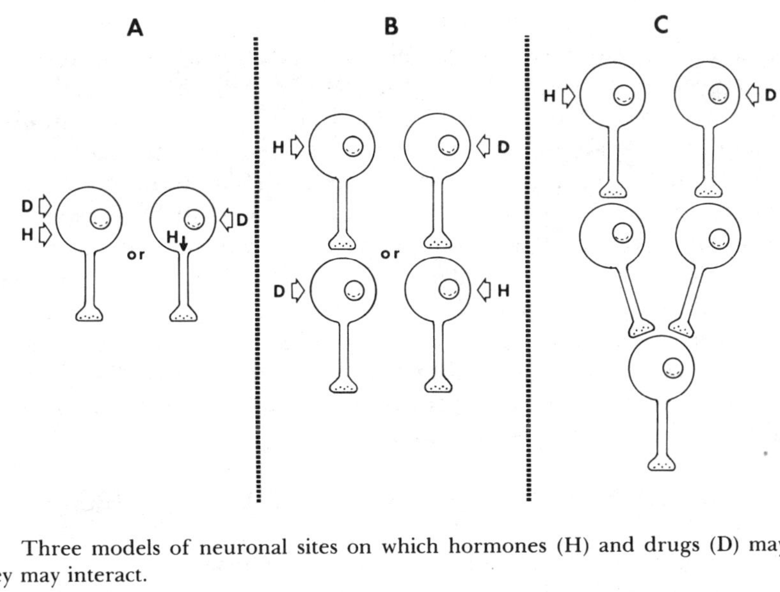

FIGURE 1. Three models of neuronal sites on which hormones (H) and drugs (D) may act and where they may interact.

female guinea pigs (Nock, Blaustein, & Feder, 1981). Conversely, estrogen treatment has been shown to alter caudate neuron sensitivity to iontophoretically applied dopamine (Arnauld, Dufy, Pestre, & Vincent, 1981).

Another possible pattern of interaction is through convergence of parallel hormone-sensitive neurons and drug-sensitive cells onto some third set of neurons (Figure 1C). These three possibilities are important for the interpretation of the results of studies investigating the interactions of drugs and hormones. They also should be kept in mind when designing and interpreting pharmacological studies of hormone-dependent behaviors. The hormone-sensitive cells in Figure 1 can also represent hormone-secreting neurons. Because many hormones originate in or are controlled by neurons, neuroactive drugs can change the hormonal state of the organism and can thus alter behavior. For example, aminergic inputs can alter hypothalamic secretion of corticotropic releasing factor (CRF), which will then affect pituitary secretion of adrenocorticotropic hormone (ACTH) (for a review see Kizer & Youngblood, 1978). In addition to having behavioral effects of its own, ACTH stimulates the release of steroid hormones from the adrenal cortex. These include the corticosteroids, androgens, estrogens, and progestins. Individual steroids within all of these classes have been reported to influence aggressive behavior and could conceivably be the basis of some drug effects.

Finally, steroid hormones determine the course of sexual differentiation that results in sexual dimorphisms in anatomy, physiology, and behavior. This forms the basis of another possible level of hormone–drug interactions: sex differences in response to pharmacological agents. Sex differences in pharmacological response may reflect neural dimorphisms due to sexual differentiation. Sex differences in the distribution of synapses, dendritic orientation, and neuron distribution have been reported for the preoptic area, and, at least to some extent, these differences are determined during sexual differentiation (Goy & McEwen, 1980). The preoptic area is not only an important site in the hormonal control of aggressive behavior, but it may also be a site of interaction between hormones and drugs. Thus, sex differences in pharmacological response could reflect male–female differences in hormone–drug–neuron interactions, and the appropriate model within Figure 1 may differ for males and females for a given system. There are sex differences in the levels and distribution of dopamine and norepinephrine within the brain that are a function of sexual differentiation (Crowley, O'Donohue, & Jacobowitz, 1978). In addition, sexual differentiation can also alter brain and peripheral metabolism of at least some drugs and hormones. This can result in sex differences in behavioral response due to differences in clearance or perhaps utilization. Virtually none of these differences have been used or exploited in the study of possible brain mechanisms for the hormonal control of aggressive behavior.

DRUGS THAT AFFECT NEUROTRANSMITTERS

Current thinking links the effects of drugs on aggressive behavior to their action on neurotransmitter activity. After surveying the major findings on drug–neurotransmitter–aggression relationships, we will consider how drugs affect hormones and how these endocrine changes could participate in the mechanisms that mediate aggressive behavior. By and large, there is no evidence relating drug action on neurotransmitter *to that* on hormonal mechanisms accounting for changes in aggressive behavior. One may extrapolate from studies on sexual behavior that the drug–hormone interactions are also relevant to the changes in aggressive behavior, but future experimentation will have to provide the evidence for this important relationship.

Initially, we will consider the drugs that act primarily on catecholamines (CA), dopamine (DA), and norepinephrine (NE); next, drugs acting on serotonin (5-hydroxytryptamine, 5-HT); and subsequently, drugs acting on acetylcholine (ACh) and on endogenous opioids.

Monoamines in General

Monoamine-depleting agents such as reserpine and tetrabenazine reduce the incidence of attack behavior in isolated mice (DaVanzo, Daugherty, Ruckart, & Kang, 1966; Sofia, 1969; Valzelli, Giacalone, & Garattini, 1967; Yen, Stanger, & Millman, 1959), and defensive reactions to electric footshock in pairs of rats and mice (Brunaud & Siou, 1959; Hintgen & Hamm, 1969; Sofia, 1969; Tedeschi, Mucha, Cook, Mattis, & Fellows, 1959). Additional measurements of motor activity indicate that monoamine-depleting drugs alter aggressive and defensive behavior only at neurotoxic doses.

The obvious problem with monoamine-depleting drugs is their pervasive action. They affect a number of neurotransmitter systems throughout the CNS. Reserpine was originally used as a pharmacological tool in studying the mechanisms of endocrine control over female sexual behavior. Reserpine could be substituted for progesterone in the stimulation of sexual receptivity in ovariectomized estrogen-primed female rats (Meyerson, 1964). This suggested initially an inhibitory role of the monoamines in the neural control of female sexual behavior that might be disinhibited by the action of ovarian steroids. However, it was discovered that reserpine, as well as many other drugs, causes pituitary–adrenal activation, releasing ACTH and a variety of steroids including progesterone from the adrenal cortex (Paris, Resko, & Goy, 1971). Adrenal activation appears to be an important aspect of reserpine action on sexual behavior, because the drug has little effect on sexual receptivity in adrenalectomized rats (Paris *et al.,* 1971). It is possible that monoamine-depleting drugs alter aggressive behavior due to their action on ACTH, and adrenal steroids are known to influence aggressive behavior (see chapter by Leshner).

Catecholamines

Inhibition of CA synthesis by *alpha*-methyl-*para*-tyrosine (AMPT) decreases all types of aggressive behavior that require pursuit and initiation of aggressive acts and displays. By contrast, AMPT has inconsistent or no effects on defensive and flight reactions. Specifically, AMPT decreases attack behavior in mice, reduces aggressive and other social behavior in macaque monkeys, and suppresses the affective aggressive display in cats (Karczmar & Scudder, 1969; Redmond, Maas, Kling, Graham, & Dkirmenjian, 1971; Reis, 1974; Welch & Welch, 1969). Similar aggression-reducing effects are seen after inhibition of dopamine-*beta*-hydroxylase (DBH) with disulfiram and FLA 63.

Inhibiting tyrosine hydroxylase with AMPT is neurochemically more specific than depleting monoamines in general, but AMPT still affects more than one neurotransmitter. Levels of both norepinephrine and dopamine are reduced unless precursors or other drugs are also used. Since these two neurotransmitters interact in at least some behavioral systems (Antelman & Caggiula, 1977), the mechanisms of action of AMPT on behavior may be difficult to interpret. For example, inhibition of CA synthesis by AMPT has been reported to increase sexual receptivity in estrogen-primed female rats (Everitt, Fuxe, Hökfelt, & Jonsson, 1975; Södersten & Ahlenius, 1972). However, some investigators find this behavioral action of AMPT, like reserpine, to be dependent on adrenal activation (Eriksson & Södersten, 1973), while others find AMPT effects to be independent of the adrenal (Everitt *et al.*, 1975). In male rats, AMPT exerts some inhibitory effects on sexual behavior (Malmnäs, 1973; Södersten, Larsson, Ahlenius, & Engel, 1976).

As with all drugs that ultimately interfere with dopamine, the possibility of increased prolactin-secretion must be recognized with AMPT. At least in female mice, prolactin can stimulate aggressive behavior (Gleason, Michael, & Christian, 1979), and this could form a part of some drug effects.

In addition to synthesis inhibition, the functions of CA in aggression have been studied by the cytotoxic agent 6-hydroxydopamine (6-OHDA). This drug can be given so that it may preferentially deplete either NE or DA. The evidence on CA depletion by 6-OHDA and consequent changes in aggression is complex.

Early reports suggested that severe CA depletion may enhance behaviors related to aggression. Otherwise placid rats and cats have been judged to be more reactive and irritable after 6-OHDA treatment (e.g., Beleslin, Samardzic, & Stefanovic-Denic, 1981; Nakamura & Thoenen, 1972), and defensive postures and bites in reaction to electric footshock are more frequent in pairs of 6-OHDA-treated rats (e.g., Eichelman, Thoa, & Ng, 1972). In nonaggressive mice, 6-OHDA facilitates attack behavior toward other mice; the magnitude of this facilitation was correlated with the depletion of NE and DA in brain (see Figure 2; Pöschlova, Masek, & Krsiak, 1976) and appears to be specifically related to DA depletion (Krsiak, Sulcova, Tomasikova, Dlohozkova, Kosar, & Masek, 1981).

By contrast, the behavior of aggressive mice remained unaffected by severe DA depletion (Krsiak *et al.*, 1981); similarly, 6-OHDA failed to alter attacks on prey and on conspecifics, elicited by hypothalamic electrical stimulation in cats (Dubinsky, Karpowicz, & Goldberg, 1973; Johansson, Paakkonen, Ruusunen, Sandstrom, & Huttunen, 1974). The social and aggressive behavior of rats housed in large colonies remained unaffected by 6-OHDA (Ellison, 1976), and the increased irritability after 6-OHDA treatment was evident only in rats housed in isolation (Ellison & Bryan, 1976). A decrease in social behavior was seen in macaque monkeys under field and laboratory conditions (Kraemer, Breese, Prange,

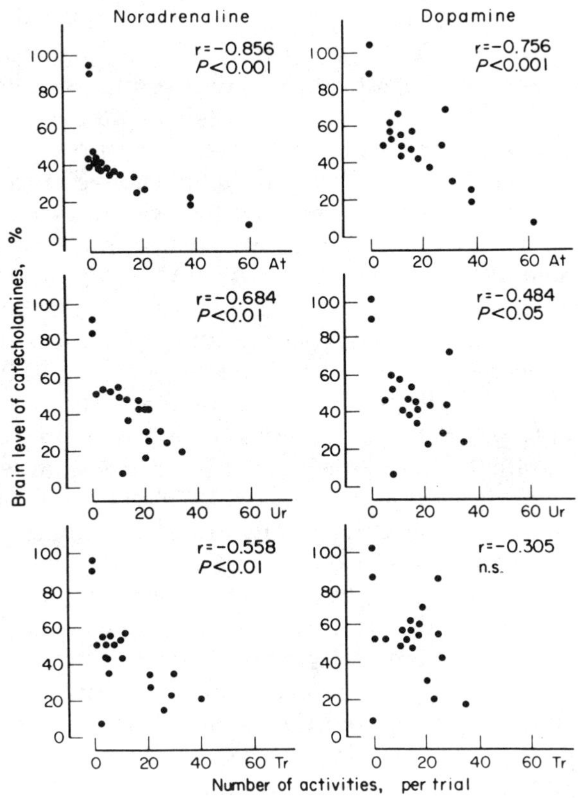

FIGURE 2. Correlation between stimulation of aggressive activities and depletion of brain catecholamines produced by 50 μg 6-OHDA in initially nonaggressive singly housed mice. Ordinate: individual level of NA and DA expressed as a percentage of mean level of respective catecholamines in the vehicle-treated animals. Abscissa: individual increase in the number of attacks (At), aggressive unrests (Ur), and of tail-rattles (Tr) in comparison with the pretreatment interaction. Spearman rank correlation coefficients are given; n.s., not significant. (From "Facilitated Intermale Aggression in the Mouse after 6-Hydroxydopamine Administration" by N. Pöschlova, K. Masek, and M. Krsiak, *Neuropharmacology*, 1976, *15*, 403–407. Copyright 1976 by Pergamon Press, Ltd. Reprinted by permission.)

Moran, Lewis, Kemnitz, Bushnell, Howard, & McKinney, 1981; Redmond, Hinrichs, Maas, & Kling, 1973).

Whether CA depletion leads to increased irritability, reactivity, defensive reactions, and attack behavior; to no change in these behavior patterns; or to an actual decrease seems to be determined by several factors. A major determinant of 6-OHDA effects is the functional state of the CA systems at the time of the cytotoxic intervention. The CA systems, traditionally linked to "arousal" and "stress," are greatly modulated by the living conditions, the history of aggressive interactions, the social context—isolated versus group housing. Of course, species differences and the degree and selectivity of CA depletion may account for some of the apparently discrepant 6-OHDA effects on aggression.

The effect of CA depletion by 6-OHDA on hormonally induced male rat sexual behavior is primarily inhibitory (Caggiula, Shaw, Antelman, & Edwards, 1976), while in females it enhances the receptivity response to estrogen and progesterone, and inhibits the proceptivity responses. These effects are seen even during exogenous hormone treatment and are unaffected by concomittant inhibition of pituitary–adrenal secretion (Caggiula, Herndon, Scanlon, Greenstone, Bradshaw, & Sharp, 1979). Caggiula and co-workers have interpreted their results as suggesting that 6-OHDA and the resulting DA depletion in particular (Clark, 1980) disrupt active behaviors while enhancing immobility responses such as lordosis. This hypothesis could be tested by a separate analysis of 6-OHDA effects on active versus immobile components of hormonally mediated aggressive and defensive behaviors.

Blockade of CA functioning by antagonizing NE- or DA-receptors has consistently decreased aggressive behavior in various animal species. *Alpha*- and *beta*-adrenergic-receptor blockers such as phenoxybenzamine and propranolol are generally less effective than the DA-receptor antagonists in reducing aggressive behavior. Haloperidol and pimozide, neuroleptics with specific DA-receptor-blocking properties, decrease attack and threat behavior and, to a lesser degree, defensive and flight reactions in mice (e.g., Hodge & Butcher, 1975; Miczek, 1981; see Figure 3). Defensive reactions in pairs of rats, evoked by electric footshock, remain unaffected by intrahypothalamic spiroperidol (Bell & Brown, 1976b); the same type of behavior is only slightly decreased by chronic treatment with haloperidol (Eichelman, 1979). The anti-aggressive effects of DA receptor blockers are not behaviorally specific and appear to be part of the general activity-reducing effects of these drugs.

Several recent experiments have reexamined the effects of *beta*-adrenergic blockers on different types of aggressive and defensive behaviors. Initial indications of potent antiaggressive effects of propranolol in isolated aggressive mice (Valzelli *et al.*, 1967) were confirmed (Miczek, 1981; Weinstock & Weiss, 1980), although the marked effects on autonomic and motor functions may, in part, be responsible for the effects on aggression (see Figure 3). Acute and chronic administrations of *beta*-receptor blockers produce opposite effects on defensive reactions in rats; propranolol, given acutely, suppresses electric footshock-induced defensive posturing in rats (Sheard, 1979; Vassout & Delini-Stula, 1977), but after 14 days of twice-daily administration, the same type of behavior is enhanced (Eichelman, 1979). Chronic blockade of *beta*-receptors may lead to changes akin to dopaminergic supersensitivity that is seen after chronic haloperidol and that is also associated with enhanced aggressive behavior in mice (Gianutsos, Drawbaugh, Hynes, & Lal, 1974).

The behavioral changes following chronic treatment with monoamine oxidase inhibitors (MAOIs) have been attributed to reduced sensitivity of *beta*-receptors. Acute administration of MAOI has only slight attack-inhibiting effects in isolated mice, unless large motor-inhibiting doses were used (e.g., Karczmar & Scudder, 1969; Valzelli *et al.*, 1967); however, repeated administration of MAOIs suppresses aggressive behavior of isolated mice at a time when the responsivity of *beta*-receptors is seen (Delini-Stula & Vassout, 1981). By contrast, chronic administration of MAOIs may increase defensive reactions in rats exposed to electric footshock (Eichelman & Barchas, 1975).

The respective roles of DA and NE in the mechanisms controlling aggressive behavior have been further investigated by administering direct and indirect agonists, in the form of precursors, releasing agents, inhibitors of metabolic enzymes, reuptake blockers, or specific receptor agonists. Higher rates of attack behavior were seen in mice maintained on a diet supplemented with the amino acid precursors L-tyrosine or L-phenylalanine (Thurmond, Lasley, Kramarcy, & Brown, 1979); L-dopa may increase attack behavior at selected low doses in mice and rats confronting an intruder (e.g., Miczek, 1977; Poshivalov, 1981); but more frequently higher L-dopa doses are used and a decrease in attacks and threats is seen (e.g., Hodge & Butcher, 1975; Miczek & O'Donnell, 1978). Defensive reactions in pairs of rats exposed to electric footshock can be slightly increased by L-dopa (e.g., Butcher & Dietrich, 1973). However, when combined with an MAOI, L-dopa can evoke rage-like behaviors with defensive and flight components in placid laboratory mice, rats, and cats (e.g., Reis, Moorhead, & Merlino, 1970; Rolinski, 1973; Zetler & Otten, 1969).

Amphetamine, a CA-releasing agent, exerts a biphasic effect on attack and threat behavior in rats (Miczek, 1974, 1979) and, under some conditions, also in mice (e.g., Welch & Welch, 1969; Winslow & Miczek, in press), low doses enhancing and higher doses decreasing the behavior. Similarly, low doses of *d*-amphetamine facilitate predatory attack elicited by hypothalamic stimulation in cats (Marini, Walters, & Sheard, 1979). Most often, however, amphetamine decreases attack behavior, even at low doses, in mice, cats, and monkeys (e.g., Crowley, Stynes, Hydinger, & Kaufman, 1974; Hoffmeister & Wuttke, 1969; Krsiak, 1979; Miczek & O'Donnell, 1978; Miczek, Wooley, Schlisserman, & Yoshimura, 1981). Amphetamine may increase defensive postures and bites in monkeys and rats, and facilitate flight in mice (e.g., Emley & Hutchinson, 1972; Krsiak *et al.,* 1981; Mukherjee & Pradhan, 1976), although this effect is not seen in some situations (Miczek, 1981; Sheard, 1981).

Apomorphine, a direct DA-receptor agonist, markedly increases the frequency of attacks and other components of aggressive behavior in mice that are timid and normally do not fight (Krsiak *et al.,* 1981), yet apomorphine and other DA agonists reduce attack and threat behavior in aggressive mice (Baggio & Ferrari, 1980; Hodge & Butcher, 1975; Krsiak *et al.,* 1981). The effects of apomorphine on defensive reactions have been studied in various preparations. In a certain proportion of placid laboratory rats, apomorphine induces defensive biting and upright postures (Dlabac, 1973; McKenzie, 1971; Senault, 1970, 1976); apomorphine readily triggers defensive reactions when rats are exposed to electric footshock or are rendered supersensitive to DA, as in morphine withdrawal (see below), after prolonged treatment with haloperidol, or after treatment with the CA neurotoxin 6-OHDA or the monoamine depletor reserpine (e.g., Anand, Gupta, & Bhargave, 1977; Gianutsos & Lal, 1978; Senault, 1970; Thoa, Eichelman, & Ng, 1972).

Bromocriptine and ergocornine can act as direct DA-receptor agonists, in addition to their *alpha*-adrenergic receptor-blocking and antiserotonergic properties (Kruk & Pycock, 1979). Both drugs have been found to reduce plasma prolactin in mice and in hamsters, but the two species differ with regard to the behavioral effects of this treatment. In hamsters, ergocornine suppression of prolactin reduces maternal aggression (Wise & Pryor, 1977), whereas female mice continued to show intact maternal aggression after prolactin inhibition with ergocornine or bromocriptine (Mann, Michael, & Svare, 1980).

Activation of *alpha*-adrenergic receptors by clonidine was found to induce

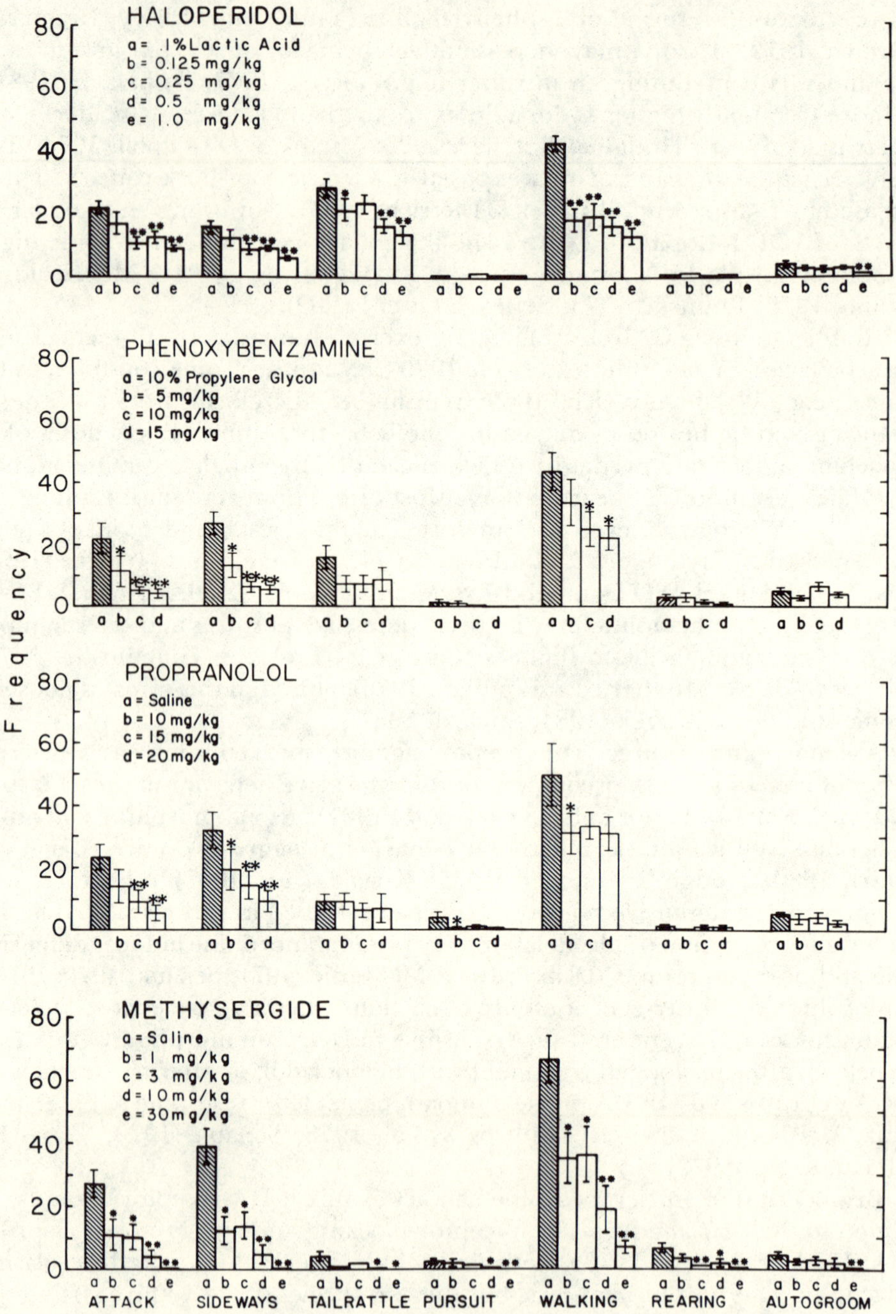

FIGURE 3. Left: The effects of haloperidol, phenoxybenzamine, propranolol, and methysergide on various components of attack behavior and nonagonistic behavior of resident mice.

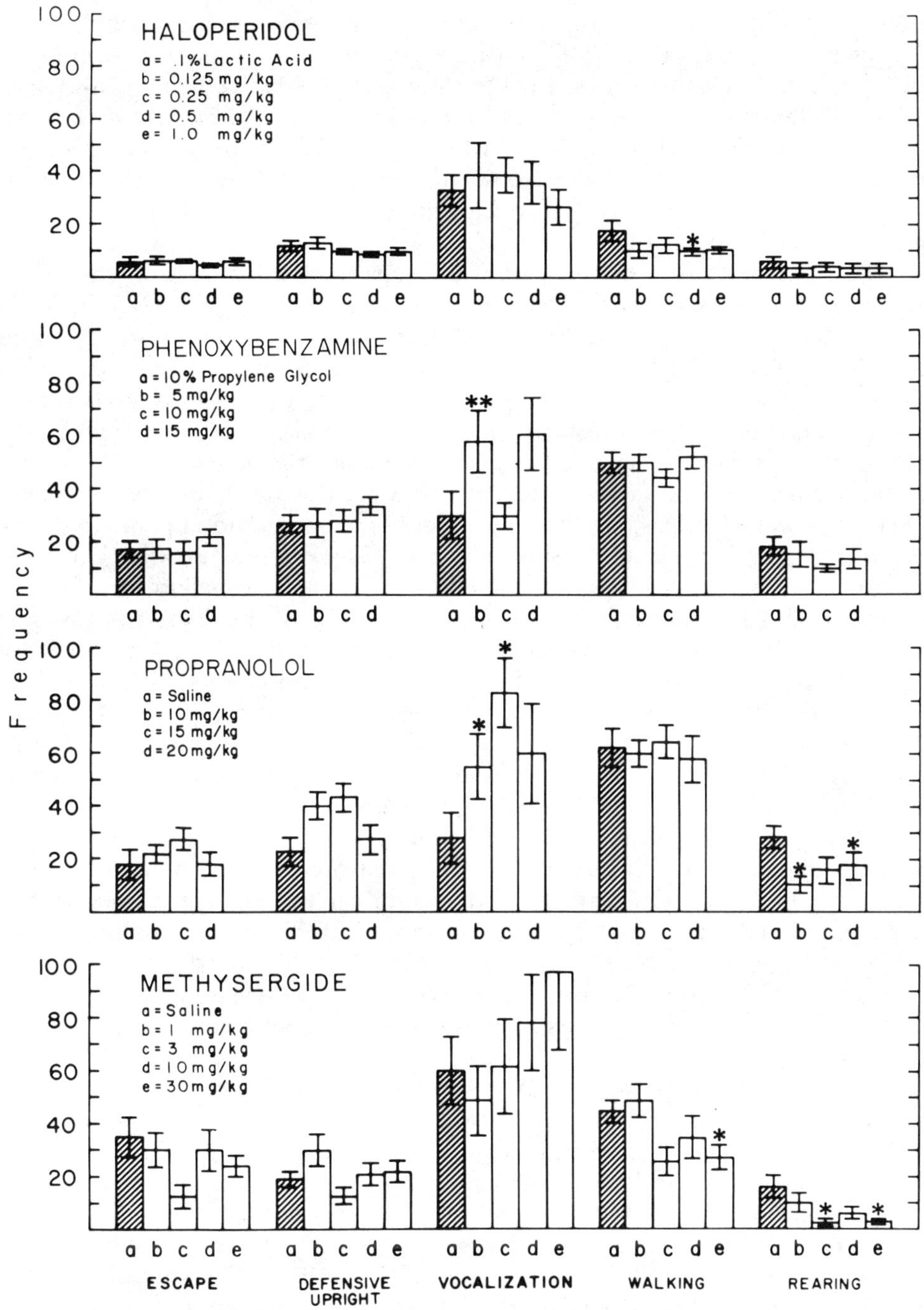

Right: Drug effects on escape and defensive reactions and nonagonistic behavior of intruder mice that were the drug recipients.

biting attacks in group-housed nonaggressive mice (Morpurgo, 1968; Ozawa, Miyauchi, & Sugawara, 1975). Clonidine-induced biting in mice can be enhanced by chronic treatment with an MAOI (Maj, Moglinicka, & Kordecka-Magiera, 1980). Defensive behavior in pairs of rats in reaction to electric footshock is decreased by low doses of clonidine, which primarily act on *alpha*-autoreceptors, whereas higher doses that activate postsynaptic receptors were ineffective; effects in the opposite direction were seen with the *alpha*-receptor blockers piperoxane and yohimbine (Sheard, 1981).

In summary, offensive behavior patterns, including pursuit, threat, and attack, require intact brain CA functioning. Pharmacological interference with the synthesis, storage, release, and receptor activation of DA and NE renders attack behavior unlikely. The results from studies with the CA neurotoxin 6-OHDA are more difficult to interpret; most of the evidence suggests that the social experience prior to destruction of CA neurons determines the resultant change in aggressive behavior. Drugs that enhance the activity of CA neuronal systems can increase attack and threat behavior; this facilitation is seen only at very low doses of releasing agents or receptor agonists and appears to depend on the predrug level of aggressive behavior. More often, aggressive behavior is disrupted after pharmacological activation of brain CA functioning.

Defensive reactions are inconsistently affected by impairing CA functions. Synthesis inhibition and DA-receptor blockade, for example, leave defensive reactions in rats unaffected, but the cytotoxic agent 6-OHDA can substantially enhance this behavior. The most promising recent findings pertain to the role of *alpha*- and *beta*-adrenergic receptors, *alpha*-autoreceptor activation selectively blocking defensive reactions in rats and *beta*-receptor unresponsiveness leading to suppressed attack behavior and heightened defensive reactions.

Specific manipulations of catecholamines also affect hormonally mediated sexual behaviors. There is strong evidence that male sexual behavior involves dopaminergic circuitry. Treatment with L-dopa can increase copulatory behavior in castrated male rats maintained on low levels of testosterone (Malmnäs, 1973, 1976). While there is little effect of noradrenergic drugs, dopamine agonists such as apomorphine stimulate male sexual behavior (Malmnäs, 1973, 1977), and, conversely, DA antagonists such as spiroperidol and clozapine inhibit male copulation (Baum & Starr, 1980).

Female sexual behavior, on the other hand, appears to be more affected by noradrenergic drugs. However, in females there are very clear species differences in the direction of the influence of *alpha*-receptors. In guinea pigs, clonidine increases the duration of sexual receptivity and phenoxybenzamine inhibits sexual behavior (Crowley, Feder & Morin, 1976). In rats, this relationship is reversed: Clonidine inhibits and phenoxybenzamine slightly facilitates sexual receptivity (Davis & Kohl, 1977; Foreman & Moss, 1978). This species difference may be related to differences in response to copulatory stimuli and in control of the duration of receptivity (Irving, Goy, Haning, & Davis, 1981).

Dopamine has generally been thought to have an inhibitory influence on female sexual behavior. Dopamine antagonists activate receptivity in estrogen-primed female rats, while DA agonists reduce receptivity in estrogen- and progesterone-treated rats (Ahlenius, Engel, Eriksson, Modigh, & Södersten, 1975; Everitt *et al.*, 1975; Michanek & Meyerson, 1975). Interestingly, dopamine agonists at low doses can actually increase receptivity, but this appears to be due to stimulation of presynaptic receptors and can be blocked with sulpiride (Everitt & Fuxe, 1977). As with the noradrenergic drugs, species differences in response to dopaminergic manipulations exist. Dopamine antagonists have little effect on the sexual behavior of female guinea pigs or hamsters (Carter, Bahr, & Ramirez, 1978; Davis & English, 1977).

It is an intriguing possibility to view male copulatory and attack behavior patterns as sharing common androgenic and dopaminergic control mechanisms. According to this view, one would predict that the evidence on the interaction of DA and androgens in male copulatory behavior may be extrapolated to mechanisms mediating attack and threat behaviors.

Serotonin

The long-postulated role for brain 5-HT in the inhibition of behavior (e.g., Brodie & Shore, 1957) may account for the changes in a variety of species-specific behavior patterns. Many studies have focused on the induction of killing behavior in laboratory rats after pharmacological blockade of serotonin synthesis, after electrolytic or cytotoxic destruction of 5-HT-containing neurons, and after exposure to a tryptophan-low diet (e.g., Barr, Gibbons, & Bridger, 1976; Kulkarni & Plotnikoff, 1978; Vergnes & Kempf, 1981).

In spite of the many and varied attempts, a direct functional relationship between brain serotonin and killing behavior remains to be established; this conclusion is particularly supported by studies that report various sequelae of serotonin depletion. One of the critical factors determining whether or not 5-HT depletion will facilitate mouse-killing in "nonkiller" rats appears to be the familiarity of the rats with the potential prey before pharmacological treatments; if the nonkiller rat was previously exposed to the prey, subsequent 5-HT depletion does not induce killing (Marks, O'Brien, & Paxinos, 1977; Vergnes & Kempf, 1981). Also, it has been difficult to relate the degree of pharmacological insult to 5-HT functioning and its time course with a graded facilitation of killing behavior. As a matter of fact, 5-HT-receptor-blocking agents or the long-term tryptophan hydroxylase-inhibitor *para*-chloroamphetamine completely fail to induce killing behavior but actually block it (e.g., Miczek, Altman, Appel, & Boggan, 1975). Also, in animals other than laboratory rats—such as cats and grasshopper mice, which are natural predators—inhibition of tryptophan hydroxylase by *para*-chlorophenylalanine is either ineffective or blocks killing behavior (Dubinsky *et al.*, 1973; McCarty, Whitesides, & Tomosky, 1976). In spite of limitations in the studies linking predatory killing, often referred to as *interspecies aggression,* to low brain 5-HT functioning, the weight of the evidence favors a significant role for this transmitter in an organism's general reactivity to sudden changes in environmental events, including interactions with a potential prey.

The experimental results with drugs affecting brain 5-HT and intraspecies aggression are considerably more complex than those on killing behavior. In general, inhibition of 5-HT synthesis with *para*-chlorophenylalanine (PCPA) or fenfluramine blocks attack behavior in mice confronting an opponent (e.g., Karczmar & Scudder, 1969; Le Douarec & Neveu, 1970; Weischer & Opitz, 1972; Welch & Welch, 1969). However, an attack-enhancing effect was reported after the administration of a high dose of *para*-chloro-*N*-methylamphetamine (20 mg/kg) in isolated mice (Hodge & Butcher, 1974); yet, *para*-chloroamphetamine and also the 5-HT uptake inhibitor zimelidine reduce aggressive behavior in isolated mice (Ogren, Holm, Renyi, & Ross, 1980). Similarly, depletion of brain serotonin with the neurotoxic agent 5,6-dihydroxytryptamine reduces aggressive and some social behavior of isolated mice toward group-housed mice (Pöschlova, Masek, & Krsiak, 1977). In rats, a decrease in "dominance" behaviors was seen after injection of 5,7-dihydroxytryptamine into the amygdaloid complex (File, James, & Macleod, 1981). Several 5-HT-receptor antagonists also decrease attack behavior by mice (Malick & Barnett, 1976; Miczek, 1981; see Figure 3), even when attackers have habituated to their opponents and show low levels of aggression (Winslow & Miczek, in press). Recently, 5-HT has been modulated by controlling dietary tryptophan, and subsequent changes in aggressive behavior have been examined. Isolated mice fight less when maintained on tryptophan-free diet (Kantak, Hegstrand, Whitman, & Eichelman, 1980), whereas tryptophan supplements to the diet were without effect on mouse aggression (Thurmond, Lasley, Conkin, & Brown, 1977).

No changes in aggressive behavior were detected in groups of macaque monkeys (*Macaca arctoides* and *Macaca fascicularis*) after prolonged treatment with the 5-HT synthesis-inhibitor PCPA (Boelkins, 1973; Redmond, Maas, Kling, Graham, & Dkirmenjian, 1971). In vervet monkeys *(Cercopithecus aethiops)*, PCPA and 5-hydroxytryptophan treatments did not affect aggressive behavior consistently but modulated other social interactions such as grooming and approaching (Raleigh, Brammer, Yuwiler, Flannery, McGuire, & Geller, 1980).

In summary, the initiation and execution of attack and threat behavior in mice and rats can be readily interfered with by treatments that lead to a blockade or depletion of brain 5-HT. Aggressive behavior in monkeys appears to remain intact after pharmacological insult of the 5-HT neurons.

Defensive and fear reactions, like killing behavior, can be enhanced by pharmacological treatments that deplete brain 5-HT. Defensive reactions in rats that are exposed to electric footshock become more frequent after 5-HT synthesis-inhibition with PCPA or *para*-chloroamphetamine (Ellison & Bresler, 1974; Sheard & Davis, 1976), although under some parameters of electric shock delivery this effect is not seen (e.g., Conner, Stolk, Barchas, Dement, & Levine, 1970). In cats, PCPA facilitates strong affective defensive reactions in the form of back arching and hissing (MacDonnell, Fessock, & Brown, 1971). Maintenance on a tryptophan-free diet that depletes brain 5-HT increases shock-induced defensive biting in rats (Kantak, Hegstrand, & Eichelman, 1980). Intracerebral injection of the neurotoxins 5,6- or 5,7-dihydroxytryptamine causes rats to be more submissive and mice more often to assume timid postures (File *et al.*, 1981; Pöschlova, Masek, & Krsiak, 1977). Infusions of 5,7-dihydroxytryptamine into the brains of rats that were exposed to electric footshock produced more frequent defensive postures and bites (Hole, Johnson, & Berge, 1977; Kantak, Hegstrand, & Eichelman, 1981). The increased frequency of defensive postures in rats exposed to electric footshock after administration of low doses of lysergic acid diethylamide (LSD) has been attributed to this drug's inhibitory action on presynaptic 5-HT receptors (Sheard, 1981). However, similar experiments with N,N-dimethyltryptamine and 5-methoxy-N,N-dimethyltryptamine did not parallel the LSD results (Walters, Sheard, & Davis, 1978). Moreover, increased submissive gestures in macaque monkeys after 5-methoxy-N,N-dimethyltryptamine were readily blocked by DA- but not 5-HT-receptor antagonists, indicating possible DA participation in the behavioral effects of this 5-HT agonist (Schlemmer & Davis, 1981).

Lithium may alter serotonin synthesis, release, uptake, and binding to receptors (e.g., Knapp & Mandell, 1973; Treiser, Cascio, O'Donohue, Thoa, Jacobowitz, & Kellar, 1981); these changes may be the basis for lithium's antiaggressive effects in humans (e.g., Marini & Sheard, 1977). Lithium suppresses attack behavior in male mice, but leaves intact attacks by lactating females toward male intruders *(maternal aggression)* and killing behavior by mice and rats (e.g., Brain & Al-Maliki, 1979).

In summary, intact 5-HT functioning is required for attack and threat behavior to occur; by contrast, blockade or depletion of 5-HT may facilitate defensive, submissive, and flight reactions, possibly due to "disinhibited" CA activity.

The case for serotonergic involvement in male sexual behavior is clearer than that for aggressive behavior. Reducing serotonin with PCPA, para-chloroamphetamine, or 5,7-dihydroxytryptamine consistently increases copulatory behavior in castrated male rats (Södersten, Berge, & Hole, 1978). Increasing serotonin activity through MAOIs, 5-HTP, fluoxetine, or 5-methoxy-N,N-dimethyltryptamine inhibits male sexual behavior (Baum & Starr, 1980; Malmnäs, 1973; Södersten *et al.,* 1976). However, there were earlier failures to demonstrate serotonin's role in male sexual behavior (Whalen & Luttge, 1970). The excitatory effects of serotonergic depletion are best seen in castrated males or in castrated animals treated with threshold amounts of testosterone as opposed to intact males (Malmnäs & Meyerson, 1971; Södersten *et al.,* 1976). In fact, serotonin levels and synthesis increase after castration (Engel, Ahlenius, Almgren, Carlsson, Larsson, & Södersten, 1979; van de Kar, Levine, & van Orden, 1978). To observe increases in a behavior that an animal normally performs, it is important to have a test animal that is behaving submaximally. It would be interesting to study the effect of serotonergic depletion or antagonism on the aggressive behavior of castrated male mice. Perhaps in castrated mice or in castrated mice treated with low levels of testosterone, the potentiating effects of serotonergic depletion on male aggression might be revealed. Since the hormonal control of these two behaviors is so similar, it might be expected that the neurotransmitter circuits involved in their control would be more similar than has thus far been demonstrated.

Serotonergic activity appears to inhibit hormonally induced sexual receptivity in female rats (Meyerson, 1964). PCPA treatment has been reported to facilitate receptivity in estrogen-primed females (Ahlenius, Engel, Eriksson, Modigh, & Södersten, 1972; Everitt *et al.,* 1975; Meyerson & Lewander, 1970; Zemlan, Ward, Crowley, & Margules, 1973), but not all laboratories can replicate this in adrenalectomized animals (Eriksson & Södersten, 1973; Gorzalka & Whalen, 1975; Hansult, Uphouse, Schlessinger, & Wilson, 1972). There is also a negative report on the effect of *para*-chloroamphetamine and 5,7-dihydroxytryptamine on receptivity (Södersten *et al.,* 1978). There is more agreement on the facilitation of lordosis by methysergide in estrogen-primed female rats (Davis & Kohl, 1978; Zemlan *et al.,* 1973). However, in guinea pigs, no effect of PCPA or methysergide has been found (Davis & English, 1977), and only weak positive effects have been described in female hamsters (Carter *et al.,* 1978). The evidence for an important serotonergic component to lordosis is less convincing than that for male copulatory behavior (Södersten *et al.,* 1978).

Acetylcholine

Historically, drugs mimicking the actions of acetylcholine (ACh) have long been found to evoke some forms of aggressive and defensive behavior (e.g., Allikmets, 1974; Karczmar, Richardson, & Kindel, 1978). Carbachol, a potent cholinomimetic drug, applied intracerebrally to various sites in the limbic forebrain, diencephalon, and mesencephalon of cats induces rage-like behavior; these defensive reactions are often referred to as *affective* aggression and resemble those evoked by electrical brain stimulation (e.g., Hull, Buchwald, & Ling, 1967; Karczmar *et al.,* 1978). Affective defensive/aggressive reactions can also be induced by drugs that inhibit acetylcholinesterase, such as physostigmine or neostigmine, administered intracerebrally, and these reactions can be blocked by antimuscarinic drugs such as atropine or scopolamine (e.g., Allikmets, 1974; Beleslin & Samardzic, 1979; Nagy & Decsi, 1974). Major limitations of the evidence on cholinergically mediated aggressive/defensive behavior are the exclusive reliance on studies in cats using single drug doses.

Physostigmine, a centrally active acetylcholinesterase-inhibitor, was found to enhance aggressive-defensive postures in pairs of rats exposed to pulses of electric footshock; similar facilitative effects were seen after systemic and intra-amygdaloid or intrahypothalamic administration of physostigmine (Bell & Brown, 1976a; Rodgers & Brown, 1973, 1976). By contrast, the antimuscarinic drugs scopolamine and atropine, given systemically or intracerebrally, block the shock-elicited defensive postures in pairs of rats (Bell & Brown, 1976a; Powell, Milligan & Walters, 1973; Rodgers & Brown, 1976). In a different species and different situations, physostigmine increases aggressive behaviors in pairs of water-deprived squirrel monkeys competing for a preferred fluid (Thomas, Caccamise, & Clark, 1978); scopolamine, by contrast, decreases aggressive responses by squirrel monkeys in a food-competition situation (Plotnik, Mollenauer, Gore, & Popov, 1975).

Atropine and scopolamine also appear to be among the most potent and selective agents in reducing intraspecies aggression in a variety of species and situations (e.g., Karczmar *et al.*, 1978; Miczek & Krsiak, 1979, 1981). Recently, for example, Krsiak and Tomasikova (1980) showed a highly effective and behaviorally specific reduction in several aggressive activities and a concurrent increase in sociable behaviors with low doses of scopolamine in isolated aggressive mice. Scopolamine was much less potent—that is, required higher doses—in reducing agonistic defensive and escape reactions. An earlier detailed ethological analysis revealed a dose-dependent decrease in aggression, defense, approach, and play after administration of scopolamine but not methylscopolamine to pairs of rats (Van der Poel & Remmelts, 1971).

Some time ago, systematic studies of nicotine effects on aggression began with Silverman's (1971) observation that a single low dose, referred to as a "smoking" dose, specifically reduces aggressive acts and postures without altering other types of behavior. Subsequently, nicotine has consistently been found to decrease defensive postures in pairs of rats presented with electric footshocks (Driscoll & Baettig, 1981; Rodgers, 1979; Waldbillig, 1980), defensive biting in reaction to electric shock in squirrel monkeys (Hutchinson & Emley, 1973), and arecoline-induced predatory-like attack in cats (Berntson, Beattie, & Walker, 1976).

Given the pervasive and widely distributed nature of ACh in the peripheral and central nervous system, it is difficult to identify the mechanism for the behavioral effects of cholinergic drugs. Moreover, most reports rely on a single set of drug and experimental parameters and are open to multiple interpretations. However, it is possible to discern some consistencies in the available evidence. Cholinergic agonists evoke defensive, rage-like reactions in cats, may increase shock-elicited aggressive/defensive behavior in rats, and may increase competitive behaviors in squirrel monkeys; antimuscarinic drugs block these aggressive/defensive behaviors. Attack, threat, and other offensive behaviors are effectively reduced by antimuscarinic agents, surprisingly, no evidence is available for an enhancement of the attack or offensive behavior pattern by cholinergic agonists. Finally, nicotine emerges as a potent antiaggressive agent in a variety of species and situations.

The same problems of interpretation apply to studies of the relationship of cholinergic systems and male sexual behavior. For the most part, both cholinergic agonists and antagonists inhibit male copulation (Bignami, 1966; Singer, 1968). However, under some conditions low doses of nicotine may stimulate male sexual behavior (Soulairac & Soulairac, 1975). In receptive female rats, muscarinic agonists reduce receptivity, but in estrogen-primed females muscarinic agonists activate receptivity (Lindstrom, 1975). Both of these effects appear to be secondary to cholinergic stimulation, since the first is blocked by reserpine or PCPA and the second by adrenalectomy or hypophysectomy (Lindstrom, 1975).

The interaction among estrogen, acetylcholine, and female sexual behavior may be more readily interpreted when the drugs are given intracerebrally. Carbachol infused into the preoptic area or the ventromedial hypothalamus facilitates receptivity in estrogen-primed female rats (Dohanich & Clemens, 1981). Receptivity could be decreased by intracerebral infusion with atropine or hemicholinium-3 (Clemens & Dohanich, 1980). Moreover, estrogen increases hypothalamic choline acetyltransferase and muscarinic receptors (Luine, Park, Tong, Reis, & McEwen, 1980; Rainbow, Degroff, Luine, & McEwen, 1980).

It is not known to what extent the effects of estrogen on aggressive behavior might specifically involve cholinergic mechanisms. The marked peripheral effects of cholinergic drugs might mask the central actions. Future studies on hormone–ACh interactions and their role in mechanisms mediating aggression should focus on central sites.

Opioids

Interest in opioids and aggression has accelerated in recent years due to the discovery of endogenous opioid peptides and their receptors (e.g., Hughes, Smith, Kosterlitz, Fathergill, Morgan, & Morris, 1975). It has long been known that morphine and other narcotic analgesic drugs, given acutely, suppress different kinds of aggressive and defensive behaviors as engendered by several experimental procedures in various species. For example, morphine decreases fighting in isolted mice; defensive behavior in reaction to electric footshock in pairs of rats; biting by squirrel monkeys after electric shock to their tail; the sham rage response in hypothalamically lesioned cats; defensive biting, jumping, and hyperreactivity of rats with lesions in the septal forebrain; and territorial aggression in cichlid fish (see for reviews Gianutsos & Lal, 1978; Miczek & Barry, 1976). The morphine doses that are necessary for the antiaggressive effect are generally sufficient also to produce pain inhibition, yet they are not debilitating. The margin in dose range separating antiaggressive and sedative effects is quite small, suggesting limited behavioral specificity.

Opioid antagonists have only recently begun to be used as independent tools in the study of aggression, since it was previously believed that they were behaviorally inactive in organisms that were not dependent on morphine. Naloxone has been found to enhance defensive postures and bites in pairs of rats and mice exposed to electric shock (Fanselow, Sigmundi, & Bolles, 1980; Puglisi-Allegra & Oliverio, 1981). On the other hand, we have recently observed a decrease in the frequency of attacks, threats, and pursuits in mice that confronted an intruder and were treated with naloxone or naltrexone (Kantak & Miczek, unpublished observations).

The interaction of endogenous opioid systems and sexual behavior seemed almost a foregone conclusion after the identification of their existence. However, elucidation of the forms of this interaction is far from complete. There is good agreement that opioid stimulation with *beta*-endorphin, d-ala^2-met-enkephalinamide, or morphine markedly inhibits copulatory behavior in male rats (Gessa, Paglietti, & Pellegrini Quaranotti, 1979; McIntosh, Vallano, & Barfield, 1980; Meyerson & Terenius, 1977). This effect parallels the earlier findings with morphine in monkeys (e.g., Crowley *et al.*, 1974).

The effect of naloxone on sexual behavior appears to depend on the predrug level of behavior of the experimental animal. In sexually inactive male rats, naloxone stimulates the full complement of copulatory behaviors (Gessa *et al.*,

1979; McIntosh *et al.,* 1980). In normal male rats, naloxone produces rather small quantitative changes that are more difficult to detect. In some experiments, there are changes in the postejaculatory interval, while in others, mount latency, intromission latency, intromission frequency or ejaculatory latency are reduced (Gessa *et al.,* 1979; McIntosh *et al.,* 1980; Myers & Baum, 1979, 1980; Sachs, Valcourt, & Flagg, 1981). Similar behavioral effects have been seen in hamsters (Noble, 1980).

Naloxone also causes increases in LHRH and LH secretion in both males and females, while morphine and met-enkephalin reduce LH secretion (Blank, Panerai, & Friesen, 1979; Bruni, Van Vugt, Marshall, & Meites, 1977; Cicero, Meyer, Gabriel, Bell, & Wilcox, 1980). This action on LHRH has been suggested as a possible basis for some of the behavioral effects of naloxone in intact males (Myers & Baum, 1980). Others have suggested that naloxone alters male sexual behavior by affecting hypothalmic catecholamine activity (McIntosh *et al.,* 1980). However, naltrexone actually inhibits sexual behavior in male Talapoin monkeys even though it does elevate LH and testosterone levels, as in rats (Meller, Keverne, & Herbert, 1980).

The suppressive effects of acute opioid stimulation on aggressive and copulatory behavior parallel each other. The pattern of results after blockade of opioid receptors appears similar to that following 5-HT depletion: Naloxone facilitates sexual behavior in inactive males and has less marked effects in fully active males. It is possible that low sexual and aggressive behavior in castrated males maintained on subthreshold doses of testosterone could be enhanced by opioid blockade, suggesting an interaction between androgens and endogenous opioids. This relationship might also include an interaction between opiate and serotonergic neurons (McConnell, Baum, & Badger, 1981), although there is no clear evidence for this at present.

A major role of endogenous opioids appears to be played in the sequelae of submission or defeat. In a series of experiments, Miczek, Thompson, and Shuster (1982) found that mice subjected to repeated attacks and showing a behavioral pattern of defeat fail to react to pain; this analgesia is completely antagonized by naloxone and naltrexone but not by quaternary naltrexone, indicating a mechanism in the CNS. Mice that are tolerant to morphine and are challenged with a defeat experience react normally to pain; conversely, mice that adapt to daily defeats fail to show analgesia when challenged with morphine (Figure 4). Single defeat experiences substantially altered *beta*-endorphin activity in the brain (Thompson, Miczek, & Shuster, 1981). Exposure to a biologically relevant stressor such as defeat appears readily to activate endorphin mechanisms in the brain that mediate a marked analgesia and produce long-lasting tolerance. Neither a comparable analgesia nor naloxone effects are seen in footshock-induced defensive reactions in rats (Rodgers & Deacon, 1981). It may be possible to relate the endorphin-mediated inhibition of distress vocalizations (see review by Panksepp, Herman, Vilberg, Bishop, & DeEskinazi, 1980) to the mechanisms for defeat behavior.

In female rats, vaginocervical stimulation also produces analgesia to painful stimuli (Komisaruk & Wallman, 1977). This effect is not dependent on gonadal

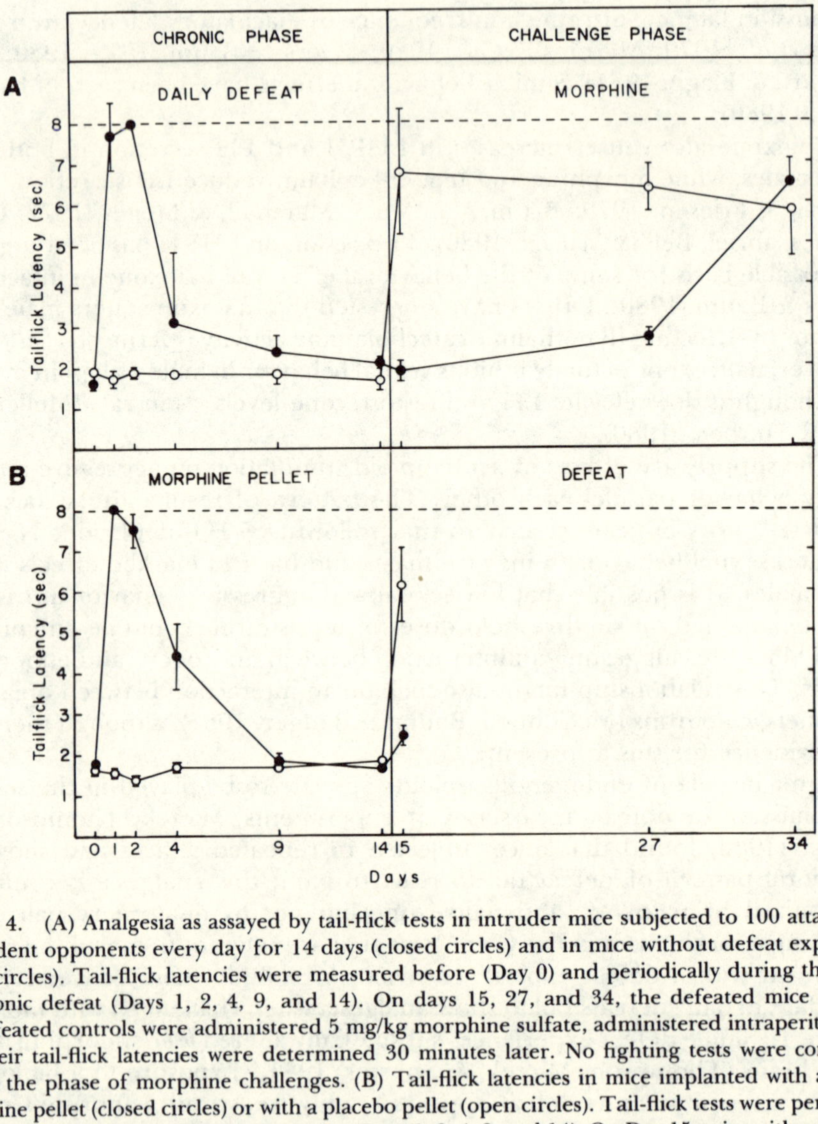

FIGURE 4. (A) Analgesia as assayed by tail-flick tests in intruder mice subjected to 100 attack bites by resident opponents every day for 14 days (closed circles) and in mice without defeat experience (open circles). Tail-flick latencies were measured before (Day 0) and periodically during the phase of chronic defeat (Days 1, 2, 4, 9, and 14). On days 15, 27, and 34, the defeated mice and the nondefeated controls were administered 5 mg/kg morphine sulfate, administered intraperitoneally, and their tail-flick latencies were determined 30 minutes later. No fighting tests were conducted during the phase of morphine challenges. (B) Tail-flick latencies in mice implanted with a 75-mg morphine pellet (closed circles) or with a placebo pellet (open circles). Tail-flick tests were performed before (Day 0) and after pellet implantation (Days 1, 2, 4, 9, and 14). On Day 15, mice with morphine and placebo pellets were exposed to 100 attack bites and subsequently tested for tail-flick latency. The vertical lines in each data point indicate ±1 S.E.M. The heat stimulus had an automatic cutoff at 8 seconds indicated by the dashed lines. (From "Opioid-like Analgesia in Defeated Mice" by K. A. Miczek, M. L. Thompson, and L. Shuster, *Science,* 1982, *215,* 1520–1522. Copyright 1982 by the American Association for the Advancement of Science. Reprinted by permission.)

hormones but may be relevant to the normal sexual behavior of female rats (Komisaruk, 1974). However, this analgesia is not readily blocked by naloxone. Naloxone does not alter vaginal-stimulation-produced analgesia to tailshock (Crowley, Rodriguez-Sierra, & Komisaruk, 1977), but it does reduce analgesia to heat stimuli (Hill & Ayliffe, 1981). These divergent observations may reflect the involvement of different opioid receptors or even nonopioid mechanisms.

Withdrawal from prolonged exposure to morphine engenders a syndrome of intense physiological and behavioral reactions. The appearance of certain aggressive and defensive behaviors during morphine withdrawal stands in contrast to the acute effects of the drug and appears to be specific to dependence on opiates (Gellert & Sparber, 1979; Gianutsos & Lal, 1978). The violent defensive posturing and biting in morphine-withdrawn rats can be further intensified by the administration of drugs with DA agonist properties, such as amphetamine, apomorphine, methylphenidate, and L-dopa. Lal (1975) has proposed that aggressive reactions during and after chronic exposure to opiates may result from the activation of supersensitive DA receptors. Indeed, dopaminergic nerve endings in the striatum are identified as sites of a large number of opiate-receptors (Pollard, Llorens, Schwartz, Gros & Dray, 1978); multiple opiate-receptors appear to be involved in the regulation of nigrostriatal DA function (Wood, Stotland, Richard, & Rackham, 1980). Further studies will undoubtedly elucidate the intriguing interaction between DA and opioid systems and delineate their role in the mechanisms mediating aggressive and defensive behavior. So far, the potential involvement of hormonal mechanisms in withdrawal aggression remains to be investigated.

ALCOHOL

In addition to drugs with behavioral effects linked to neurotransmitter systems, there are also drugs with less well understood mechanisms of action. This section will consider alcohol as a representative drug of abuse.

Most systematic studies on the effects of alcohol on aggression in animals have focused on the range of doses that produce a decrease in various kinds of aggressiveand defensive behavior (e.g., Crowley *et al.*, 1974; Cutler, 1976; Cutler, Mackintosh, & Chance, 1975; Ewart & Cutler, 1979; Irwin, Kinohi, Van Sloten, & Workman, 1971; Krsiak & Borgesova, 1973; Lagerspetz & Ekqvist, 1978; Yanai & Ginsburg, 1977). The evidence for a specific aggression-decreasing effect of alcohol is limited by the fact that the doses that reduce aggression are close to those that produce general sedation and ataxia.

There is suggestive evidence for an aggression-enhancing effect of alcohol from early single-dose studies in mice (Chance, Mackintosh, & Dixon, 1973), Siamese fighting fish (Raynes & Ryback, 1970; Raynes, Ryback, & Ingle, 1968), and macaque monkeys (Chamove & Harlow, 1970; Kamback, 1973). A systematic series of experiments in convict cichlid fish demonstrated the dose-dependent biphasic effect of alcohol on territorial aggression (Ellman, Herz, & Peeke, 1972;

Peeke & Figler, 1981; Peeke, Ellman, & Herz, 1973; Peeke, Peeke, Avis, & Ellman, 1975). Low doses of alcohol increased the frequency of attack bites and threats by resident male cichlids, whereas higher doses decreased these behaviors. A similar biphasic pattern of alcohol effects on attack behavior was also seen in mice and rats (Krsiak, 1976; Miczek & Barry, 1977; Miczek & O'Donnell, 1980), with low doses increasing aggression and higher doses decreasing it (Figure 5).

Surprisingly little is known about gonadal hormones as mediators of the behavioral effects of alcohol. Mendelson and Mello (1974) present data suggesting that alcohol effects on aggression in humans are correlated with the testosterone level in blood as assessed by radioimmunoassays. However, another laboratory, using a somewhat different population, did not find a significant correlation (Persky, O'Brien, Fine, Howard, Khan, & Beck, 1977). Hypotheses about a causative link among alcohol, testosterone, and aggression in humans need to be evaluated further.

A complex dose- and time-dependent action of alcohol on testosterone and LH appears to exist in rats (Cicero & Badger, 1977). Enhanced testosterone and LH levels were found with acute administration of low doses of alcohol, whereas higher doses decreased the level of both hormones. There was also indication that higher alcohol doses increase testosterone levels in the early phase of drug action. The decrease in testosterone after higher alcohol doses was confirmed in mice (Badr, Bartke, Dalterio, & Bulger, 1977). It is not clear in the rat studies how the changes in testosterone and LH are related to each other, and to what extent they reflect the peripheral or central actions of alcohol (Cicero, Newman, & Meyer, 1981).

Although the biphasic nature of these dose-dependent changes in testosterone is similar to the pattern of alcohol effects on aggression in rodents, it is

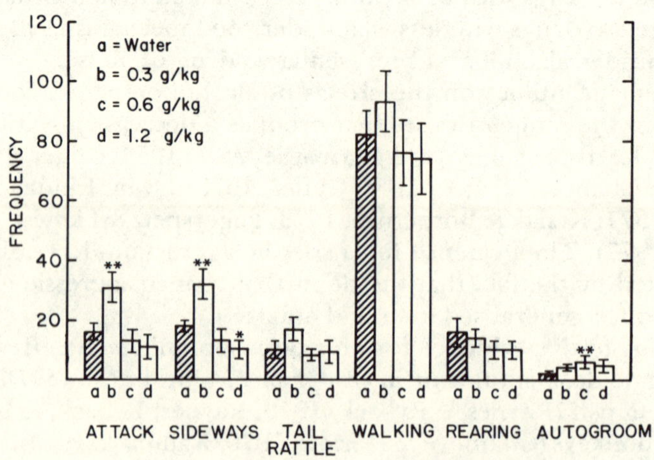

FIGURE 5. The effects of alcohol on the frequency of attack bites, sideways threats, tail-rattling, walking, rearing, and autogrooming. The resident mice were experienced attackers confronting group-housed opponents in a neutral test cage. (Data from Miczek & O'Donnell, 1980.)

unclear whether the magnitude of the changes in testosterone can account for the changes in aggression.

CONCLUSIONS

Within the last decade, research on neurotransmitter and neuroendocrine systems, including neuropeptides, yielded new possibilities for mechanisms mediating aggressive behavior. Drug and hormone treatments concurrently alter transmitter and endocrine systems, both in brain and peripheral organs. Various models for drug–hormone interactions can be entertained to account for the resulting changes in aggressive behavior. At present, the proposed models should be considered as heuristics awaiting experimental verification.

In recent years, drug–hormone interaction studies have been conducted on male and female sexual behavior. By contrast, comparable studies on aggression are lacking. One major reason for this difference in research progress stems from the complexity of aggressive behavior patterns, the variety of research paradigms for the study of aggression, and the problems in measuring the salient elements of aggressive behavior.

Inquiries into the neural mechanisms of aggressive behavior can benefit from the important conceptual and methodological pitfalls that are evident from the drug–hormone interaction studies on sexual behavior. For example, it is significant to compare hormone and drug action on animals that show aggressive behavior at a low level with the same action on those responding maximally. Moreover, experimental control over adrenal and gonadal hormone secretions appears to be an important prerequisite for pharmacological investigations.

Studies on drug–hormone interactions and male sexual behavior appear to offer the most promising similarities to those on male aggression. Particularly intriguing are the parallel patterns of results for both behaviors with manipulations of androgens, dopamine, opioids, and possibly serotonin. Whether the similar neurotransmitter and hormonal profiles for male aggressive and copulatory behavior imply identical mechanisms remains to be examined. Combined pharmacological and hormonal manipulations of animals that are concurrently tested for sexual and aggressive behavior will afford the opportunity to decipher potentially unique processes mediating either behavior pattern.

REFERENCES

Ahlenius, S., Engel, J., Eriksson, H., Modigh, K., & Södersten, P. Importance of central catecholamines in the mediation of lordosis behavior in ovariectomized rats treated with estrogen and inhibitors of monoamine synthesis. *Journal of Neural Transmission,* 1972, *33,* 247–255.

Ahlenius, S., Engel, J., Eriksson, H., Modigh, K., & Södersten, P. Involvement of monoamine in the mediation of lordosis behavior. In M. Sandler & G. L. Gessa (Eds.), *Sexual behavior: Pharmacology and biochemistry.* New York: Raven Press, 1975.

Allikmets, L. Kh. Cholinergic mechanisms in aggressive behaviour. *Medical Biology,* 1974, *52,* 19–30.

Anand, M., Gupta, G. B., & Bhargava, K. P. Modification of electroshock fighting by drugs known to interact with dopaminergic and noradrenergic neurons in normal and brain lesioned rats. *Journal of Pharmacy and Pharmacology,* 1977, *29,* 437–439.

Antelman, S. M., & Caggiula, A. R. Norepinephrine-dopamine interactions and behavior. *Science,* 1977, *195,* 646–653.

Arnauld, E., Dufy, B., Pestre, M., & Vincent, J. D. Effects of estrogens on the responses of caudate neurons to microiontophoretically applied dopamine. *Neuroscience Letters,* 1981, *21,* 325–331.

Badr, F. M., Bartke, A., Dalterio, S., & Bulger, W. Suppression of testosterone production by ethyl alcohol: Possible mode of action. *Steroids,* 1977, *30,* 647–655.

Baggio, G., & Ferrari, F. Role of brain dopaminergic mechanisms in rodent aggressive behavior: Influence of (+)N-n-propyl-norapomorphine on three experimental models. *Psychopharmacology,* 1980, *70,* 63–68.

Barr, G. A., Gibbons, J. L., & Bridger, W. H. Neuropharmacological regulation of mouse killing by rats. *Behavioral Biology,* 1976, *17,* 143–159.

Baum, M. J., & Starr, M. S. Inhibition of sexual behavior by dopamine antagonist or serotonin agonist drugs in castrated male rats given estradiol or dihydrotestosterone. *Pharmacology, Biochemistry and Behavior,* 1980, *13,* 57–67.

Beleslin, D. B., & Samardzic, R. Evidence of central cholinergic mechanisms in the appearance of affective aggressive behaviour: Dissociation of aggression from autonomic and motor phenomena. *Psychopharmacology,* 1979, *62,* 163–167.

Beleslin, D. B., Samardzic, R., & Stefanovic-Denic, K. 6-Hydroxydopamine and aggression in cats. *Pharmacology, Biochemistry, and Behavior,* 1981, *14* (Suppl. 1), 29–32.

Bell, R., & Brown, K. Lateral hypothalamus and shock-induced aggression—cholinergic mediation. *International Research Communication Systems (IRCS), Medical Science,* 1976, *4,* 83–84. (a)

Bell, R., & Brown, K. Failure of dopamine and spiroperidol, injected into the lateral hypothalamus, to affect shock-induced aggression in rats: Evidence against a mediating role for dopamine. *International Research Communication Systems (IRCS), Medical Science,* 1976, *4,* 527. (b)

Berntson, G. G., Beattie, M. S., & Walker, J. M. Effects of nicotinic and muscarinic compounds on biting attack in the cat. *Pharmacology, Biochemistry and Behavior,* 1976, *5,* 235–239.

Bignami, G. Pharmacological influences on mating behavior in the male rat. *Psychopharmacologia (Berl.),* 1966, *10,* 44–58.

Blank, M. S., Panerai, A. E., & Friesen, H. G. Opioid peptides modulate LH secretion during sexual maturation. *Science,* 1979, *203,* 1129–1131.

Boelkins, R. C. Effects of *para*chlorophenylalanine on the behavior of monkeys. In J. Barchas & E. Usdin (Eds.), *Serotonin and behavior,* New York: Academic Press, 1973.

Brain, P. F. What does individual housing mean to a mouse? *Life Sciences,* 1975, *16,* 187–200.

Brain, P. F., & Al-Maliki, S. Effects of lithium chloride injections on rank-related fighting, maternal aggression and locust-killing responses in naive and experienced "TO" strain mice. *Pharmacology, Biochemistry and Behavior,* 1979, *10,* 663–669.

Brodie, B. B., & Shore, P. A. A concept for a role of serotonin and norepinephrine as chemical mediators in the brain. *Annals of the New York Academy of Sciences,* 1957, *66,* 631–664.

Brunaud, M., & Siou, G. Action de substances psychotropes, chez le rat, sur un état d'aggressivité provoquée. In P. B. Bradley, P. Deniker, & C. Radouco-Thomas (Eds.), *Neuro-psychopharmacology.* Amsterdam: Elsevier, 1959.

Bruni, J. F., Van Vugt, D., Marshall, S., & Meites, J. Effects of naloxone, morphine and methionine enkephalin on serum prolactin, luteinizing hormone, follicle-stimulating hormone, thyroid-stimulating hormone, and growth hormone. *Life Sciences,* 1977, *21,* 461–466.

Butcher, L. L., & Dietrich, A. P. Effects on shock-elicited aggression in mice of preferentially protecting brain monoamines against the depleting action of reserpine. *Naunyn-Schmiedeberg's Archiv der experimentellen Pharmakologie,* 1973, *277,* 61–70.

Caggiula, A. R., Shaw, D. H., Antelman, S. M., & Edwards, J. Interactive effects of brain catecholamines and variations in sexual and non-sexual arousal on copulatory behavior of male rats. *Brain Research,* 1976, *111,* 321–336.

Caggiula, A. R., Herndon, J. G., Scanlon, R., Greenstone, D., Bradshaw, W., & Sharp, D. Dissociation of active from immobility components of sexual behavior in female rats by central 6-hydroxydopamine: Implications for CA involvement in sexual behavior and sensorimotor responsiveness. *Brain Research,* 1979, *172,* 505–520.

Carter, C. S., Bahr, J. M., & Ramirez, V. D. Monoamines, estrogen and female sexual behavior in the golden hamster. *Brain Research,* 1978, *144,* 109–121.

Chamove, A. S., & Harlow, H. F. Exaggeration of self-aggression following alcohol ingestion in rhesus monkeys. *Journal of Abnormal Psychology,* 1970, *75,* 207–209.

Chance, M. R. A., Mackintosh, J. H., & Dixon, A. K. The effects of ethyl alcohol on social encounters between mice. *Journal of Alcoholism,* 1973, *8,* 90–93.

Cicero, T. J., & Badger, T. M. Effects of alcohol on the hypothalamic-pituitary-gonadal axis in the male rat. *Journal of Pharmacology and Experimental Therapeutics,* 1977, *201,* 427–433.

Cicero, T. J., Meyer, E. R., Gabriel, S. M., Bell, R. D., & Wilcox, C. E. Morphine exerts testosteronelike effects in the hypothalamus of the castrated male rat. *Brain Research,* 1980, *202,* 151–164.

Cicero, T. J., Newman, K. S., & Meyer, E. R. Ethanol-induced inhibitions of testicular steroidogenesis in the male rat: Mechanisms of actions. *Life Sciences,* 1981, *28,* 871–877.

Clark, T. K. Male rat sexual behavior compared after 6-OHDA and electrolytic lesions in the dorsal NA bundle region of the midbrain. *Brain Research,* 1980, *202,* 429–443.

Clemens, L. G., & Dohanich, G. P. Inhibition of lordotic behavior in female rats following intracerebral infusion of anticholinergic agents. *Pharmacology, Biochemistry and Behavior,* 1980, *13,* 89–95.

Conner, R. L., Stolk, J. M., Barchas, J. D., Dement, W. C., & Levine, S. The effect of parachlorophenylalanine (PCPA) on shock-induced fighting behavior in rats. *Physiology and Behavior,* 1970, *5,* 1221–1224.

Cooper, J. R., Bloom, F. E., & Roth, R. H. *The biochemical basis of neuropharmacology.* New York: Oxford University Press, 1978.

Crowley, T. J., Stynes, A. J., Hydinger, M., & Kaufman, I. C. Ethanol, methamphetamine, pentobarbital, morphine, and monkey social behavior. *Archives of General Psychiatry,* 1974, *31,* 829–838.

Crowley, W. R., Feder, H. H., & Morin, L. P. Role of monoamines in sexual behavior of the female guinea pig. *Pharmacology, Biochemistry and Behavior,* 1976, *4,* 67–71.

Crowley, W. R., Rodriguez-Sierra, J. F., & Komisaruk, B. R. Analgesia induced by vaginal stimulation in rats is apparently independent of a morphine-sensitive process. *Psychopharmacology,* 1977, *54,* 223–225.

Crowley, W. R., O'Donohue, T. L., & Jacobowitz, D. M. Sex differences in catecholamine content in discrete brain nuclei of the rat: Effects of neonatal castration or testosterone treatment. *Acta Endocrinologica,* 1978, *89,* 20–28.

Cutler, M. G. Changes in the social behavior of laboratory mice during administration and on withdrawal from non-ataxic doses of ethyl alcohol. *Neuropharmacology,* 1976, *15,* 495–498.

Cutler, M. G., Mackintosh, J. H., & Chance, M. R. A. Effects of the environment on the behavioural response of mice to non-ataxic doses of ethyl alcohol. *Neuropharmacology,* 1975, *14,* 841–846.

DaVanzo, J. P., Daugherty, M., Ruckart, R., & Kang, L. Pharmacological and biochemical studies in isolation-induced fighting in mice. *Psychopharmacologia (Berl.),* 1966, *9,* 210–219.

Davis, G. A., & English, A. Monoamines and lordosis in the female guinea pig. *Hormones and Behavior,* 1977, *8,* 88–93.

Davis, G. A., & Kohl, R. The influence of α-receptors on lordosis in the female rat. *Pharmacology, Biochemistry and Behavior,* 1977, *6,* 47–53.

Davis, G. A., & Kohl, R. L. Biphasic effects of the antiserotonergic methysergide on lordosis in rats. *Pharmacology, Biochemistry and Behavior,* 1978, *9,* 487–491.

Delini-Stula, A., & Vassout, A. The effects of antidepressants on aggressiveness induced by social deprivation in mice. *Pharmacology, Biochemistry and Behavior,* 1981, *14* (Suppl. 1), 33–41.

Dlabac, A. Apomorphine-induced aggressivity in rats and its alterations. *Activitas Nervosa Superiora,* 1973, *15,* 133.

Dohanich, G. P., & Clemens, L. G. Brain areas implicated in cholinergic regulation of sexual behavior. *Hormones and Behavior,* 1981, *15,* 157–167.

Driscoll, P., & Baettig, K. Selective inhibition by nicotine of shock-induced fighting in the rat. *Pharmacology, Biochemistry and Behavior,* 1981, *14,* 175–179.

Dubinsky, B., Karpowicz, J. K., & Goldberg, M. E. Effects of tricyclic antidepressants on attack elicited by hypothalamic stimulation: Relation to brain biogenic amines. *Journal of Pharmacology and Experimental Therapeutics,* 1973, *187,* 550–557.

Eichelman, B. Role of biogenic amines in aggressive behavior. In M. Sandler (Ed.), *Psychopharmacology of aggression.* New York: Raven Press, 1979.

Eichelman, B., & Barchas, J. Facilitated shock-induced aggression following antidepressive medication in the rat. *Pharmacology, Biochemistry and Behavior,* 1975, *3,* 601–604.

Eichelman, B. S., Thoa, N. B., & Ng, K. Y. Facilitated aggression in the rat following 6-hydroxydopamine administration. *Physiology and Behavior,* 1972, *8,* 1–3.

Ellison, G. Monoamine neurotoxins: Selective and delayed effects on behavior in colonies of laboratory rats. *Brain Research,* 1976, *103,* 81–92.

Ellison, G. D., & Bresler, D. E. Tests of emotional behavior in rats following depletion of norepinephrine, of serotonin, or of both. *Psychopharmacologia (Berl.),* 1974, *34,* 275–288.

Ellison, G. D., & Bryan, K. S. Evidence that supersensitivity after recovery from 6-hydroxydopamine is an artifact of isolation housing. *Neuroscience Letters,* 1976, *3,* 229–232.

Ellman, G. L., Herz, M. J., & Peeke, H. V. S. Ethanol in a cichlid fish: Blood levels and aggressive behavior. *Proceedings of the Western Pharmacological Society,* 1972, *15,* 92–95.

Emley, G. S., & Hutchinson, R. R. Basis of behavioral influence of chlorpromazine. *Life Sciences,* 1972, *11,* 43–47.

Engel, J., Ahlenius, S., Almgren, O., Carlsson, A., Larsson, K., & Södersten, P. Effects of gonadectomy and hormone replacement on brain monoamine synthesis in male rats. *Pharmacology, Biochemistry and Behavior,* 1979, *10,* 149–154.

Eriksson, H., & Södersten, P. A failure to facilitate lordosis behavior in adrenalectomized and gonadectomized estrogen-primed rats with monoamine synthesis inhibitors. *Hormones and Behavior,* 1973, *4,* 89–97.

Everitt, B. J., & Fuxe, K. Dopamine and sexual behaviour in female rats. Effects of dopamine receptor agonists and sulpiride. *Neuroscience Letters,* 1977, *4,* 209–213.

Everitt, B. J., Fuxe, K., Hökfelt, T., & Jonsson, G. Role of monoamines in the control by hormones of sexual receptivity in the female rat. *Journal of Comparative and Physiological Psychology,* 1975, *89,* 556–572.

Ewart, F. G., & Cutler, M. G. Effects of ethyl alcohol on behaviour in nursing female mice. *Psychopharmacology,* 1979, *66,* 143–146.

Fanselow, M. S., Sigmundi, R. A., & Bolles, R. C. Naloxone pretreatment enhances shock-elicited aggression. *Physiological Psychology,* 1980, *8,* 369–371.

File, S. E., James, T. A., & Macleod, N. K. Depletion in amygdaloid 5-hydroxytryptamine concentration and changes in social and aggressive behaviour. *Journal of Neural Transmission,* 1981, *50,* 1–12.

Foreman, M. M., & Moss, R. L. Role of hypothalamic alpha and beta adrenergic receptors in the control of lordotic behavior in the ovariectomized-estrogen primed rat. *Pharmacology, Biochemistry and Behavior,* 1978, *9,* 235–241.

Gellert, V. F., & Sparber, S. B. Effects of morphine withdrawal on food competition hierarchies and fighting behavior in rats. *Psychopharmacology,* 1979, *60,* 165–172.

Gessa, G. L., Paglietti, E., & Pellegrini Quaranotti, B. Induction of copulatory behavior in sexually inactive rats by naloxone. *Science,* 1979, *204,* 203–205.

Gianutsos, G., & Lal, H. Narcotic analgesics and aggression. In L. Valzelli (Ed.), *Psychopharmacology of aggression. Modern problems of pharmacopsychiatry* (Vol. 13). Basel: Karger, 1978.

Gianutsos, G., Drawbaugh, R. B., Hynes, M. D., & Lal, H. Behavioral evidence for dopaminergic supersensitivity after chronic haloperidol. *Life Sciences,* 1974, *14,* 887–898.

Gleason, P. E., Michael, S. D., & Christian, J. J. Effects of gonadal steroids on agonistic behavior of female *Peromyscus leucopus. Hormones and Behavior,* 1979, *12,* 30–39.

Gorzalka, B. B., & Whalen, R. E. Inhibition not facilitation of sexual behavior by PCPA. *Pharmacology, Biochemistry and Behavior,* 1975, *3,* 511–513.

Goy, R. W., & McEwen, B. S. (Eds.). *Sexual differentiation of the brain.* Cambridge: MIT Press, 1980.

Hansult, C. D., Uphouse, L. L., Schlesinger, K., & Wilson, J. R. Induction of estrus in mice: Hypophyseal-adrenal effects. *Hormones and Behavior,* 1972, *3,* 113–121.

Heritage, A. S., Stumpf, W. E., Sar, M., & Grant, L. D. Brain-stem catecholamine neurons are target sites for sex steroid hormones. *Science,* 1980, *207,* 1377–1379.

Hill, R. B., & Ayliffe, S. J. The antinociceptive effect of vaginal stimulation in the rat is reduced by naloxone. *Pharmacology, Biochemistry and Behavior,* 1981, *14,* 631–632.

Hingtgen, J. N., & Hamm, H. D. Modification of tetrabenazine effects following pain-induced aggression in rats. *Life Sciences,* 1969, *8,* 1–7.

Hodge, G. K., & Butcher, L. L. 5-Hydroxytryptamine correlates of isolation-induced aggression in mice. *European Journal of Pharmacology,* 1974, *28,* 326–337.

Hodge, G. K., & Butcher, L. L. Catecholamine correlates of isolation-induced aggression in mice. *European Journal of Pharmacology,* 1975, *31,* 81–93.

Hoffmeister, F., & Wuttke, W. On the actions of psychotropic drugs on the attack- and aggressive-defensive behaviour of mice and cats. In S. Garattini & E. B. Sigg (Eds.), *Aggressive behaviour.* Amsterdam: Excerpta Medica Foundation, 1969.

Hökfelt, T., Johansson, O., Ljungdahl, A., Lundberg, J. M., & Schultzberg, M. Peptidergic neurons. *Nature,* 1980, *284,* 515–521.

Hole, K., Johnson G. E., & Berge, O. G. 5,7-Dihydroxytryptamine lesions of the ascending 5-hydroxytryptamine pathways: Habituation, motor activity and agonistic behavior. *Pharmacology, Biochemistry and Behavior,* 1977, *7,* 205–210.

Hughes, J., Smith, T. W., Kosterlitz, H. W., Fothergill, L. A., Morgan, B. A., & Morris, H. R. Identification of two related pentapeptides from the brain with potent opiate agonist activity. *Nature,* 1975, *258,* 577.

Hull, C. D., Buchwald, N. A., & Ling, G. M. Effects of direct cholinergic stimulation of forebrain structures. *Brain Research,* 1967, *6,* 22–35.

Hutchinson, R. R., & Emley, G. S. Effects of nicotine on avoidance, conditioned suppression and aggression response measures in animals and man. In W. L. Dunn (Ed.), *Smoking behavior: Motives and incentives.* Washington: V. H. Winston 1973.

Irving, S. M., Goy, R. W., Haning, R. V., & Davis, G. A. Prostaglandins, clonidine and sexual receptivity in the guinea pig. *Brain Research,* 1981, *204,* 65–77.

Irwin, S., Kinohi, R., Van Sloten, M., & Workman, M. P. Drug effects on distress-evoked behavior in mice: Methodology and drug class comparisons. *Psychopharmacologia (Berl.),* 1971, *20,* 172–185.

Johansson, G., Paakkonen, T., Ruusunen, S., Sandstrom, M., & Huttunen, M. Effect of 6-hydroxy-dopamine on agonistic behaviour elicited by stimulation of the hypothalamus in the cat. *Medical Biology,* 1974, *52,* 406–414.

Kamback, M. C. The effects of orbital and dorsolateral frontal cortical ablations on ethanol self-selection and emotional behaviors in monkeys (*Macaca nemestrina*), *Neuropsychologia,* 1973, *11,* 331–335.

Kantak, K. M., Hegstrand, L. R., & Eichelman, B. Dietary tryptophan modulation and aggressive behavior in mice. *Pharmacology, Biochemistry and Behavior,* 1980, *12,* 675–679.

Kantak, K. M., Hegstrand, L. R., Whitman, J., & Eichelman, B. Effects of dietary supplements and a tryptophan-free diet on aggressive behavior in rats. *Pharmacology, Biochemistry and Behavior,* 1980, *12,* 173–179.

Kantak, K. M., Hegstrand, L. R., & Eichelman, B. Facilitation of shock-induced fighting following intraventricular 5,7-dihydroxytryptamine and 6-hydroxydopa. *Psychopharmacology,* 1981, *74,* 157–160.

van de Kar, L., Levine, J., & van Orden, L. S., III. Serotonin in hypothalamic nuclei: Increased content after castration of male rats. *Neuroendocrinology,* 1978, *27,* 186–192.

Karczmar, A. G., & Scudder, C. L. Aggression and neurochemical changes in different strains and genera of mice. In S. Garattini & E. B. Sigg (Eds.), *Aggressive behaviour.* Amsterdam: Excerpta Medica Foundation, 1969.

Karczmar, A. G., Richardson, D. C., & Kindel, G. Neuropharmacological and related aspects of animal aggression. *Progress in Neuro-Psychopharmacology,* 1978, *2,* 611–631.

Kizer, J. S., & Youngblood, W. W. Neurotransmitter systems and central neuroendocrine regulation. In M. A. Lipton, A. DiMascio, & K. F. Killam (Eds.), *Psychopharmacology: A generation of progress.* New York: Raven Press, 1978.

Knapp, S., & Mandell, A. J. Short- and long-term lithium administration: Effects on the brain's serotonergic biosynthetic systems. *Science,* 1973, *180,* 645–647.

Komisaruk, B. R. Neural and hormonal interactions in the reproductive behavior of female rats. In W. Montagna & W. A. Sadler (Eds.), *Reproductive behaviour.* New York: Plenum Press, 1974.

Komisaruk, B. R., & Wallman, J. Antinociceptive effects of vaginal stimulation in rats: Neurophysiological and behavioral studies. *Brain Research,* 1977, *137,* 85–107.

Kraemer, G. W., Breese, G. R., Prange, A. J., Jr., Moran, E. C., Lewis, J. K., Kemnitz, J. W., Bushnell, P. J., Howard, J. L., & McKinney, W. T. Use of 6-hydroxydopamine to deplete brain catecholamines in the rhesus monkey: Effects on urinary catecholamine metabolites and behavior. *Psychopharmacology,* 1981, *73,* 1–11.

Krsiak, M. Effect of ethanol on aggression and timidity in mice. *Psychopharmacology,* 1976, *51,* 75–80.

Krsiak, M. Effects of drugs on behaviour of aggressive mice. *British Journal of Pharmacology,* 1979, *65,* 525–533.

Krsiak, M., & Borgesova, M. Effect of alcohol on behaviour of pairs of rats. *Psychopharmacologia (Berl.),* 1973, *32,* 201–209.

Krsiak, M., & Tomasikova, Z. Effects of scopolamine on agonistic behavior. *Activitas Nervosa Superior,* 1980, *22,* 201–303.

Krsiak, M., Sulcova, A., Tomasikova, Z., Dlohozkova, N., Kosar, E., & Masek, K. Drug effects on attack, defense and escape in mice. *Pharmacology, Biochemistry and Behavior,* 1981, *14* (Suppl. 1), 47–52.

Kruk, Z. L., & Pycock, L. J. *Neurotransmitters and drugs.* Baltimore: University Park Press, 1979.

Kulkarni, A. S., & Plotnikoff, N. P. Effects of central stimulants on aggressive behavior. In L. Valzelli (Ed.), *Psychopharmacology of aggression. Modern problems of Pharmacopsychiatry* (Vol. 13). Basel: Karger, 1978.

Lagerspetz, K. M. J., & Ekqvist, K. Failure to induce aggression in inhibited and in genetically nonaggressive mice through injections of ethyl alcohol. *Aggressive Behavior,* 1978, *4,* 105–113.

Lal, H. Narcotic dependence, narcotic action and dopamine receptors. *Life Sciences,* 1975, *17,* 483–496.

Le Douarec, J. C., & Neveu, C. Pharmacology and biochemistry of fenfluramine. In E. Costa & S. Garattini (Eds.), *Amphetamines and related compounds.* New York: Raven Press, 1970.

Lindstrom, L. H. Cholinergic mechanisms and sexual behavior in the female rat. In M. Sandler & G. L. Gessa (Eds.), *Sexual behavior: Pharmacology and biochemistry.* New York: Raven Press, 1975.

Luine, V., Park, D., Tong, J., Reis, D., & McEwen, B. S. Immunochemical demonstration of increased choline acetyltransferase concentration in rat preoptic area after estradiol administration. *Brain Research,* 1980, *191,* 273–277.

MacDonnell, M. F., Fessock, L., & Brown, S. H. Aggression and associated neural events in cats. Effects of P-chlorophenylalanine compared with alcohol. *Quarterly Journal of Studies on Alcohol,* 1971, *32,* 748–763.

Maj, J., Moglinicka, E., & Kordecka-Magiera, A. Effects of chronic administration of antidepressant drugs on aggressive behavior induced by clonidine in mice. *Pharmacology, Biochemistry and Behavior,* 1980, *13,* 153–154.

Malick, J. B., & Barnett, A. The role of serotonergic pathways in isolation-induced aggression in mice. *Pharmacology, Biochemistry and Behavior,* 1976, *5,* 55–61.

Malmnäs, C. O. Monoaminergic influence on testosterone-activated copulatory behavior in the castrated male rat. *Acta Physiologica Scandinavica,* 1973, *Suppl. 395,* 1–128.

Malmnäs, C. O. The significance of dopamine versus other catecholamines, for l-dopa induced facilitation of sexual behavior in the castrated male rat. *Pharmacology, Biochemistry and Behavior,* 1976, *4,* 521–526.

Malmnäs, C. O. Dopaminergic reversal of decline after castration of rat copulatory behaviour. *Journal of Endocrinology,* 1977, *73,* 187–188.

Malmnäs, C. O., & Meyerson, B. J. P-chlorophenylalanine and copulatory behavior in the male rat. *Nature,* 1971, *232,* 398–400.

Mann, M., Michael, S. D., & Svare, B. Ergot drugs suppress plasma prolactin and lactation but not aggression in parturient mice. *Hormones and Behavior,* 1980, *14,* 319–328.

Marini, J. L., & Sheard, M. H. The antiaggressive effect of lithium ion in man. *Acta Psychiatrica Scandinavica,* 1977, *55,* 269–286.

Marini, J. L., Walters, J. K., & Sheard, M. H. Effects of *d*-amphetamine and *l*-amphetamine on hypothalamically-elicited movement and attack in the cat. *Agressologie,* 1979, *20,* 155–160.

Marks, P. C., O'Brien, M., & Paxinos, G. 5,7-DHT-induced muricide: Inhibition as a result of preoperative exposure of rats to mice. *Brain Research,* 1977, *135,* 383–388.

McCarty, R. C., Whitesides, G. H., & Tomosky, T. K. Effects of *p*-chlorophenylalanine on the predatory behavior of *Onychomys torridus. Pharmacology, Biochemistry and Behavior,* 1976, *4,* 217–220.

McConnell, S. K., Baum, M. J., & Badger, T. M. Lack of correlation between naloxone-induced changes in sexual behavior and serum LH in male rats. *Hormones and Behavior,* 1981, *15,* 16–35.

McEwen, B. S., Davis, P. G., Parsons, B., & Pfaff, D. W. The brain as a target for steroid hormone action. *Annual Review of Neuroscience,* 1979, *2,* 65–112.

McIntosh, T. K., Vallano, M. L., & Barfield, R. J. Effects of morphine, β-endorphin and naloxone on catecholamine levels and sexual behavior in the male rat. *Pharmacology, Biochemistry and Behavior,* 1980, *13,* 435–441.

McKenzie, G. M. Apomorphine-induced aggression in the rat. *Brain Research,* 1971, *34,* 323–330.

Meller, R. E., Keverne, E. B., & Herbert, J. Behavioral and endocrine effects of naltrexone in male Talapoin monkeys. *Pharmacology, Biochemistry and Behavior,* 1980, *13,* 663–672.

Mendelson, J. H., & Mello, N. K. Alcohol, aggression and androgens. In S. H. Frazier (Ed.), *Aggression.* Baltimore: Williams & Wilkins, 1974.

Meyerson, B. J. Central nervous monoamines and hormone induced estrous behavior in the spayed rat. *Acta Physiologica Scandinavica,* 1964, *Suppl. 241,* 1–32.

Meyerson, B. J., & Lewander, T. Serotonin synthesis inhibition and estrous behavior in female rats. *Life Sciences,* 1970, *9,* 661–671.

Meyerson, B. J., & Terenius, L. β-endorphin and male sexual behavior. *European Journal of Pharmacology,* 1977, *42,* 191–192.

Meyerson, B. J., Palis, A., & Sietniks, A. Hormone-monoamine interactions and sexual behavior. In C. Beyer (Ed.), *Endocrine control of sexual behavior.* New York: Raven Press, 1979.

Michanek, A., & Meyerson, B. J. Copulatory behavior in the female rat after amphetamine and amphetamine derivatives. In M. Sandler & G. L. Gessa (Eds.), *Sexual behavior: Pharmacology and biochemistry.* New York: Raven Press, 1975.

Miczek, K. A. Intraspecies aggression in rats. Effects of *d*-amphetamine and chlordiazepoxide. *Psychopharmacolgia,* 1974, *39,* 275–301.

Miczek, K. A. Effects of L-DOPA, *d*-amphetamine and cocaine on intruder evoked aggression in rats and mice. *Progress in Neuro-Psychopharmacology,* 1977, *1,* 271–277.

Miczek, K. A. A new test for aggression in rats without aversive stimulation: Differential effects of *d*-amphetamine and cocaine. *Psychopharmacology,* 1979, *60,* 253–259.

Miczek, K. A. Differential antagonism of *d*-amphetamine effects on motor activity and agonistic behavior in mice. *Neuroscience Abstracts,* 1981, *7,* 343 (Abstract No. 113.12).

Miczek, K. A., & Barry, H., III. Pharmacology of sex and aggression. In S. D. Glick & J. Goldfarb (Eds.), *Behavioral pharmacology,* St. Louis: C. V. Mosby, 1976.

Miczek, K. A., & Barry, H., III. Effects of alcohol on attack and defensive-submissive reactions in rats. *Psychopharmacology,* 1977, *52,* 231–237.

Miczek, K. A., & Krsiak, M. Drug effects on agonistic behavior. In T. Thompson & P. B. Dews (Eds.), *Advances in behavioral pharmacology* (Vol. 2). New York: Academic Press, 1979.

Miczek, K. A., & Krsiak, M. Pharmacological analysis of attack and flight. In P. F. Brain & D. Benton (Eds.), *A multidisciplinary approach to aggression research.* Amsterdam: Elsevier, 1981.

Miczek, K. A., & O'Donnell, J. M. Intruder-evoked aggression in isolated and non-isolated mice: Effects of psychomotor stimulants and L-dopa. *Psychopharmacology,* 1978, *57,* 47–55.

Miczek, K. A., & O'Donnell, J. M. Alcohol and chlordiazepoxide increase suppressed aggression in mice. *Psychopharmacology,* 1980, *69,* 39–44.

Miczek, K. A., Altman, J. L., Appel, J. B., & Boggan, W. O. Para-chlorophenylalanine, serotonin and killing behavior. *Pharmacology, Biochemistry and Behavior,* 1975, *3,* 355–361.

Miczek, K. A., Woolley, J., Schlisserman, S., & Yoshimura, H. Analysis of amphetamine effects on agonistic and affiliative behavior in squirrel monkeys *(Saimiri sciureus). Pharmacology, Biochemistry and Behavior,* 1981, *14* (Suppl. 1), 103–107.

Miczek, K. A., Thompson, M. L., & Shuster, L. Opioid-like analgesia in defeated mice. *Science,* 1982, *215,* 1520–1522.

Morpurgo, C. Aggressive behavior induced by large doses of 2-(2,6-dichlorophenylamino)—imidazoline hydrochloride (ST 155) in mice. *European Journal of Pharmacology,* 1968, *3,* 374–377.

Mukherjee, B. P., & Pradhan, S. N. Effects of lithium on foot shock-induced aggressive behavior in rats. *Archives Internationales de Pharmacodynamie et de Therapie,* 1976, *222,* 125–131.

Myers, B. M., & Baum, M. J. Facilitation by opiate antagonists of sexual performance in the male rat. *Pharmacology, Biochemistry and Behavior,* 1979, *10,* 615–618.

Myers, B. M., & Baum, M. J. Facilitation of copulatory performance in male rats by naloxone: Effects of hypophysectomy, 17 α-estradiol, and luteinizing hormone releasing hormone. *Pharmacology, Biochemistry and Behavior,* 1980, *12,* 365–375.

Nagy, J., & Decsi, L. Simultaneous chemical stimulation of the hypothalamus and dorsal hippocampus in the waking cat. *Pharmacology, Biochemistry and Behavior,* 1974, *2,* 285–292.

Nakamura, K., & Thoenen, H. Increased irritability: A permanent behavior change induced in the rat by intraventricular administration of 6-hydroxydopamine. *Psychopharmacologia,* 1972, *24,* 359–372.

Noble, R. G. *Hamster sexual functioning: Modulation by opiate agonists, opiate antagonists, and exercise.* Paper presented at Eastern Conference on Reproductive Behavior, New York, 1980.

Nock, B., Blaustein, J. D., & Feder, H. H. Changes in noradrenergic transmission alter the concentration of cytoplasmic progestin receptors in hypothalamus. *Brain Research,* 1981, *207,* 371–396.

Ogren, S.-O., Holm, A. C., Renyi, A. L., & Ross, S. B. Anti-aggressive effect of zimelidine in isolated mice. *Acta Pharmacologica et Toxicologica,* 1980, *47,* 71.

Ozawa, H., Miyauchi, T., & Sugawara, K. Potentiating effect of lithium chloride on aggressive behaviour in mice by nialamide plus l-DOPA and by clonidine. *European Journal of Pharmacology,* 1975, *34,* 169–179.

Panksepp, J., Herman, B. H., Vilberg, T., Bishop, P., & DeEskinazi, F. G. Endogenous opioids and social behavior. *Neuroscience and Biobehavioral Reviews,* 1980, *4,* 473–487.

Paris, C. A., Resko, J. A., & Goy, R. W. A possible mechanism for the induction of lordosis by reserpine in spayed rats. *Biology of Reproduction,* 1971, *4,* 23–30.

Peeke, H. V. S., & Figler, M. H. Modulation of aggressive behavior in fish by alcohol and congeners. *Pharmacology, Biochemistry and Behavior,* 1981, *14* (Suppl. 1), 79–84.

Peeke, H. V. S., Ellman, G. E., & Herz, M. J. Dose dependent alcohol effects on the aggressive behavior of the convict cichlid *(Cichlasoma nigrofasciatum). Behavioral Biology,* 1973, *8,* 115–122.

Peeke, H. V. S., Peeke, S. C., Avis, H. H., & Ellman, G. Alcohol, habituation and the patterning of aggressive responses in a cichlid fish. *Pharmacology, Biochemistry and Behavior,* 1975, *3,* 1031–1036.

Persky, H., O'Brien, C. P., Fine, E., Howard, W. J., Khan, M. A., & Beck, R. W. The effect of alcohol and smoking on testosterone function and aggression in chronic alcoholics. *American Journal of Psychiatry,* 1977, *134,* 621–625.

Plotnik, R., Mollenauer, S., Gore, W., & Popov, A. Comparing the effects of scopolamine on operant and aggressive responses in squirrel monkeys. *Pharmacology, Biochemistry and Behavior,* 1975, *3,* 739–748.

Pollard, H., Llorens, C., Schwartz, J. C., Gros, C., & Dray, F. Localization of opiate receptors and enkephalins in the rat striatum in relationship with the nigrostriatal dopaminergic system: Lesion studies. *Brain Research,* 1978, *151,* 392–398.

Pöschlova, N., Masek, K., & Krsiak, M. Facilitated intermale aggression in the mouse after 6-hydroxydopamine administration. *Neuropharmacology,* 1976, *15,* 403–407.

Pöschlova, N., Masek, K., & Krsiak, M. Amphetamine-like effects of 5,6-dihydroxytryptamine on the social behavior in the mouse. *Neuropharmacology,* 1977, *16,* 317–321.

Poshivalov, V. Pharmaco-ethological analysis of social behaviour of isolated mice. *Pharmacology, Biochemistry and Behavior,* 1981, *14* (Suppl. 1), 53–59.

Powell, D. A., Milligan, W. L., & Walters, K. The effects of muscarinic cholinergic blockade upon shock-elicited aggression. *Pharmacology, Biochemistry and Behavior,* 1973, *1,* 389–394.

Puglisi-Allegra, S., & Oliverio, A. Naloxone potentiates shock-induced aggressive behavior in mice. *Pharmacology, Biochemistry and Behavior,* 1981, *15,* 513–514.

Rainbow, T. C., Degroff, V., Luine, V. N., & McEwen, B. S. Estradiol-17β increases the number of muscarinic receptors in hypothalamic nuclei. *Brain Research,* 1980, *198,* 239–243.

Raleigh, M. J., Brammer, G. L., Yuwiler, A., Flannery, J. W., McGuire, M. T., & Geller, E. Serotonergic influences on the social behavior of vervet monkeys *(Cercopithecus aethiops sabaeus). Experimental Neurology,* 1980, *68,* 332–334.

Raynes, A. E., & Ryback, R. S. Effect of alcohol and congeners on aggressive response in *Betta splendens. Quarterly Journal of Studies on Alcohol,* 1970, *5,* 130–135.

Raynes, A. E., Ryback, R., & Ingle, D. The effect of alcohol on aggression in *Betta splendens. Communications in Behavioral Biology,* 1968, *2,* 141–146.

Redmond, D. E., Maas, J. W., Kling, A., Graham, C. W., & Dkirmenjian, J. Social behavior of monkeys selectively depleted of monoamines. *Science,* 1971, *174,* 428–431.

Redmond, D. E., Hinrichs, R. L., Maas, J. W., & Kling, A. Behavior of free-ranging Macaques after intraventricular 6-hydroxydopamine. *Science,* 1973, *181,* 1256–1258.

Reis, D. J. Central neurotransmitters in aggression. In S. H. Frazier (Ed.), *Aggression.* Baltimore: Williams & Wilkins, 1974.

Reis, D. J., Moorhead, D. T., & Merlino, N. DOPA-induced excitement in the cat: Its relationship to brain norepinephrine concentrations. *Archives of Neurology,* 1970, *22,* 31–39.

Roche, K. E., & Leshner, A. I. ACTH and vasopressin treatments immediately after a defeat increase future submissiveness in male mice. *Science,* 1979, *204,* 1343–1344.

Rodgers, R. J. Effects of nicotine, mecamylamine, and hexamethonium on shock-induced fighting, pain reactivity, and locomotor behaviour in rats. *Psychopharmacology,* 1979, *66,* 93–98.

Rodgers, R. J., & Brown, K. Increased shock-induced attack in rats by physostigmine. *International Research Communication Systems (IRCS), Medical Science,* 1973, *5,* 38.

Rodgers, R. J., & Brown, K. Amygdaloid function in the central cholinergic mediation of shock-induced aggression in the rat. *Aggressive Behavior,* 1976, *2,* 131–152.

Rodgers, R. J., & Deacon, R. M. J. Foot-shock analgesia: Prevention by behavioural manipulation but not by naloxone. *Physiology and Behavior,* 1981, *26,* 183–187.

Rolinski, Z. Analysis of aggressiveness-stereotopy complex induced in mice by amphetamine or nialamide and L-DOPA. *Polish Journal of Pharmacology and Pharmacy,* 1973, *25,* 551–558.

Sachs, B. D., Valcourt, R. J., & Flagg, H. C. Copulatory behavior and sexual reflexes of male rats treated with naloxone. *Pharmacology, Biochemistry and Behavior,* 1981, *14,* 251–253.

Schlemmer, R. F., Jr., & Davis, J. M. Evidence for dopamine mediation of submissive gestures in the stumptail macaque monkey. *Pharmacology, Biochemistry and Behavior,* 1981, *14* (Suppl. 1), 95–102.

Senault, B. Comportement d'agressivité intraspecifique induit par l'apomorphine chez le rat. *Psychopharmacologia (Berl.),* 1970, *18,* 271–287.

Senault, B. Aggressive behaviour induced by apomorphine: Relations with some elements of the behavioural profile and with sensitivity to apomorphine. *Psychopharmacology,* 1976, *48,* 31–35.

Sheard, M. The role of drugs affecting catecholamines on shock-induced fighting in rats. In E. Usdin (Ed.), *Catecholamines: Basic and clinical frontiers* (Vol. 2). New York: Pergamon Press, 1979.

Sheard, M. H. Shock-induced fighting (SIF): Psychopharmacological studies. *Aggressive Behavior,* 1981, *7,* 41–49.

Sheard, M. H., & Davis, M. Shock elicited fighting in rats: Importance of intershock interval upon the effect of p-chlorophenylalanine (PCPA). *Brain Research,* 1976, *111,* 433–437.

Silverman, A. P. Behaviour of rats given a "smoking dose" of nicotine. *Animal Behaviour,* 1971, *19,* 67–74.

Singer, J. J. The effects of atropine upon the female and male sexual behavior of female rats. *Physiology and Behavior,* 1968, *3,* 377–378.

Södersten, P., & Ahlenius, S. Female lordosis behavior in estrogen-primed male rats treated with p-chlorophenylalanine or alpha-methyl-para-tyrosine. *Hormones and Behavior,* 1972, *3,* 181–189.

Södersten, P., Larsson, K., Ahlenius, S., & Engel, J. Sexual behavior in castrated male rats treated with monoamine synthesis inhibitors and testosterone. *Pharmacology, Biochemistry and Behavior,* 1976, *5,* 319–327.

Södersten, P., Berge, O. G., & Hole, K. Effects of p-chloroamphetamine and 5,7-dihydroxytryptamine on the sexual behavior of gonadectomized male and female rats. *Pharmacology, Biochemistry and Behavior,* 1978, *9,* 499–508.

Sofia, R. D. Effects of centrally active drugs on four models of experimentally-induced aggression in rodents. *Life Sciences,* 1969, *8,* 705–716.

Soulairac, M.-L., & Soulairac, A. Monoaminergic and cholinergic control of sexual behavior in the male rat. In M. Sandler & G. L. Gessa (Eds.), *Sexual behavior: Pharmacology and biochemistry.* New York: Raven Press, 1975.

Tedeschi, R. E., Tedeschi, D. H., Mucha, A., Cook, L., Mattis, P. A., & Fellows, E. J. Effects of various centrally acting drugs on fighting behavior of mice. *Journal of Pharmacology and Experimental Therapeutics,* 1959, *125,* 28–34.

Thoa, N. B., Eichelman, B., & Ng, L. K. Shock-induced aggression. Effects of 6-hydroxydopamine and other pharmacological agents. *Brain Research,* 1972, *43,* 467–475.

Thomas, G. S., Caccamise, D. J., & Clark, D. L. Aggression increase and water competition decrease in squirrel monkeys given physostigmine injections. *Pharmacology, Biochemistry and Behavior,* 1978, *8,* 633–639.

Thompson, M. L., Miczek, K. A., & Shuster, L. Changes in brain β-endorphin and tolerance to morphine analgesia after a single defeat in mice. *Neuroscience Abstracts,* 1981, 7, 881.

Thurmond, J. B., Lasley, S. M., Conkin, A. L., & Brown, J. W. Effects of dietary tyrosine, phenylalanine, and tryptophan on aggression in mice. *Pharmacology, Biochemistry and Behavior,* 1977, *6,* 475–478.

Thurmond, J. B., Lasley, S. M., Kramarcy, N. R., & Brown, J. W. Differential tolerance to dietary amino acid-induced changes in aggressive behavior and locomotor activity in mice. *Psychopharmacology,* 1979, *66,* 301–308.

Treiser, S. L., Cascio, C. S., O'Donohue, T. L., Thoa, N. B., Jacobowitz, D. M., & Kellar, K. J. Lithium increases serotonin release and decreases serotonin receptors in the hippocampus. *Science,* 1981, *213,* 1529–1531.

Valzelli, L. The "isolation syndrome" in mice. *Psychopharmacologia (Berl.),* 1973, *31,* 305–320.

Valzelli, L., Giacalone, E., & Garattini, S. Pharmacological control of aggressive behavior in mice. *European Journal of Pharmacology,* 1967, *2,* 144–146.

Van der Poel, A. M., & Remmelts, M. The effect of anti-cholinergics on the behaviour of the rat in a solitary and in a social situation. *Archives Internationales de Pharmacodynamie et de Therapie,* 1971, *189,* 394–396.

Vassout, A., & Delini-Stula, A. Effects of β-blocking agents (propranolol, oxprenolol) and diazepam on the aggressive behaviour in the rat. *Journal de Pharmacologie (Paris),* 1977, *8,* 5–14.

Vergnes, M., & Kempf, E. Tryptophan deprivation: Effects on mouse-killing and reactivity in the rat. *Pharmacology, Biochemistry and Behavior,* 1981, *14* (Suppl. 1), 19–23.

Waldbillig, R. J. Suppressive effects of intraperitoneal and intraventricular injections of nicotine on muricide and shock-induced attack on conspecifics. *Pharmacology, Biochemistry and Behavior,* 1980, *12,* 619–623.

Walters, J. K., Sheard, M. H., & Davis, M. Effects of N,N-dimethyltryptamine (DMT) and 5-methoxy-N,N-dimethyltryptamine (5-MeODMT) on shock elicited fighting in rats. *Pharmacology, Biochemistry and Behavior,* 1978, *9,* 87–90.

Weinstock, M., & Weiss, C. Antagonism by propranolol of isolation-induced aggression in mice: Correlation with 5-hydroxytryptamine receptor blockade. *Neuropharmacology,* 1980, *19,* 653–656.

Weischer, M.-L., & Opitz, K. Einfluss von Fenfluramin, Chlorphentermine und verwandten Verbindungen auf das Verhalten von aggressiven Mäusen. *Archives Internationales de Pharmacodynamie et de Therapie,* 1972, *195,* 252–259.

Welch, B. L., & Welch, A. S. Aggression and the biogenic amine neurohumors. In S. Garattini & E. B. Sigg (Eds.), *Aggressive behaviour.* Amsterdam: Excerpta Medica Foundation, 1969.

Whalen, R. E., & Luttge, W. G. P-Chlorophenylalanine methyl ester: An aphrodisiac? *Science,* 1970, *169,* 1000–1001.

Winslow, J. T., & Miczek, K. A. Habituation of aggression in mice: Pharmacological evidence for catecholaminergic and serotonergic mediation. *Psychopharmacology,* in press.

Wise, D. A., & Pryor, T. L. Effects of ergocornine and prolactin on aggression in the postpartum golden hamster. *Hormones and Behavior,* 1977, *8,* 30–39.

Wood, P. L., Stotland, M., Richard, J. W., & Rackham, A. Actions of *mu, kappa, sigma, delta* and agonist/antagonist opiates on striatal dopaminergic function. *Journal of Pharmacology and Experimental Therapeutics,* 1980, *215,* 697–703.

Yanai, J., & Ginsburg, B. E. Long term reduction of male agonistic behavior in mice following early exposure to ethanol. *Psychopharmacology,* 1977, *52,* 31–34.

Yen, C. I., Stanger, R. L., & Millman, N. Ataractic suppression of isolation-induced aggressive behavior. *Archives Internationales de Pharmacodynamie et de Therapie,* 1959, *123,* 179–185.

Zemlan, F. P., Ward, I. L., Crowley, W. R., & Margules, D. L. Activation of lordotic responding in female rats by suppression of serotonergic activity. *Science,* 1973, *179,* 1010–1011.

Zetler, G., & Otten, U. Aggressivität der Ratte nach kombinierter Behandlung mit Monoaminoxydase-Inhibitoren und anderen psychotropen Pharmaka, insbesondere Thymoleptica. *Naunyn-Schmiedeberg's Archiv der experimentellen Pathologie und Pharmakologie,* 1969, *264,* 32–54.

III

THE ENVIRONMENT, HORMONES, AND AGGRESSIVE BEHAVIOR

13

Conditions of Housing, Hormones, and Aggressive Behavior

Paul F. Brain and David Benton

INTRODUCTION

The relationships among environmental variables, the neuroendocrine systems, and aggression are examined, although the resulting correlations are often difficult to interpret. This chapter concentrates to a large extent on those studies that have used rodents and have studied intermale attack behavior. The importance of social structure as a variable is explored. Rather arbitrarily, the findings will be examined under the following headings: those studies in which the experimental variable is identified as "crowding" or housing density, those in which "isolation" or "social deprivation" is examined, and those in which the social organization of a group is looked at. The experimental designs can be remarkably similar in these approaches, but the interpretations of the results are often strikingly different.

STUDIES ON HOUSING DENSITY

The behavioral and endocrine correlates of housing density (numbers of animals per unit area) have been reviewed many times (Archer, 1970, 1979; Brain, 1971; Calhoun, 1962, 1973; Christian, 1963a,b). Housing density markedly alters intraspecific fighting in a vareity of species. Often, high densities initially stimulate attack but later suppress aggression. Social aggression in rats is said to be independent of cage size (Thor, 1976), but Veselovsky and Novacky (1976) reported that fighting in Golden hamsters was more likely to be found in groups of moderate density rather than in higher or lower concentrations.

PAUL F. BRAIN • Department of Zoology, University College of Swansea, Swansea SA2 8PP, Wales, United Kingdom. DAVID BENTON • Department of Psychology, University College of Swansea, Swansea SA2 8PP, Wales, United Kingdom.

Hull, Kastaniotis, L'Hommedieu, and Franz (1976) found increased marking and decreased fighting in densely housed male gerbils.

The endocrine correlates of housing density have been cited in many of the studies listed above. For example, increased density has been correlated with increased adrenocortical and reduced gonadal activity in many species; densely housed gerbils in the Hull *et al.* (1976) study had lighter ventral prostates and elevated plasma cortisol levels compared with less densely housed counterparts.

STUDIES ON ISOLATION

Behavioral Changes

The behavioral concomitants of isolation have been the subject of much research. Table I summarizes the influence of individual or group housing on a variety of characteristics. The term "individual housing" is a better label than isolation because the vast majority of studies separate the animals physically but do not exclude olfactory, auditory, and possibly visual communication with conspecifics. One study that has attempted the "total isolation" of male mice reported that fighting was suppressed in such animals in comparison with conventionally individually housed mice (Benton, Brain, & Goldsmith, 1979a).

Physical separation is often used to elevate social aggression in laboratory mice (reviewed in Brain, 1975). Indeed, this housing condition has a progressive effect on the intensity of this form of fighting behavior in subsequent encounters (Goldsmith, Brain, & Benton, 1976). A similar effect of isolation on aggression has been shown in other rodent species, including the guinea pig, *Cavia porcellus* (Fara & Catlett, 1971), deermouse, *Peromyscus* (Rosen & Hart, 1963), and mongolian gerbil, *Meriones unguiculatus* (Spencer, Gray, & Dalhouse, 1973) (although Hull, Chapin, & Kastaniotis, 1974, failed to find this in gerbils).

It is important to note that cohabitation with a female produces a more dramatic increase in social aggression in male mice than does a similar duration of individual housing (Brain, Benton, & Bolton, 1978). Much of the elevated motivation for fighting in these animals may be a consequence of the stimulated gonadal activity. Similar mechanisms have been suggested in the rat; Flannelly and Lore (1977) demonstrated that cohabitation with intact but not with ovariectomized females increased the social attack by male rats on unfamiliar male intruders, but it has been suggested that "in the gerbil at least, there is no simple relationship between basal levels of either cortisol or testosterone and aggressiveness" (Hull, Kastaniotis, L'Hommedieu, & Franz, 1976).

The results obtained when social aggression is compared in isolated and group-housed rats are equivocal and worthy of comment. Garattini, Giacalone, and Valzelli (1969) and Bernstein and Moyer (1970) both reported that this type of fighting was little modified by even seven weeks of individual housing. Although a number of reports (Bevan, Bloom, & Lewis, 1951; Seward, 1945; Sigg, Day, & Colombo, 1966; Welch & Welch, 1969; Yen, Day, & Sigg, 1963) have utilized isolation to generate social aggression, these papers do not report actual

TABLE I

Charactristics of Individual- and Group-Housing in Adult Male House Mice[a]

Characteristic	Individually housed	Group-housed	Reference
Intraspecific fighting	High	Low	Brain & Al-Maliki, 1978; Brain & Nowell, 1971a; Cairns, 1972; Charpentier, 1969; Levine, Diakow, & Barsel, 1965; Sigg, 1969; Valzelli, 1969, Valzelli & Garattini, 1968
Interspecific fighting	Equal	Equal	Al-Maliki & Brain, 1977
Reactivity in a novel arena	High	Low	Cairns, 1972; Svare & Leshner, 1973
Motor activity in response to novelty stress	High	Low	Brain, Haley, & Nowell, 1971
Activity/ambulation in novel arena	High	Low	Brain *et al.*, 1971; Essman, 1968; Priestnall, 1973
Activity/ambulation in novel arena	Low	High	Watson, Henry, & Haltmeyer, 1974
Turnover of brain catecholamines	Low	High	Garattini, Giacalone, & Valzelli, 1969; Modigh, 1973; Valzelli & Garattini, 1968; B. L. Welch & Welch, 1968
Adrenal weight	Low	High	Brain & Nowell, 1971a; Brain *et al.*, 1971; Christian, 1959; Welch & Welch, 1965
"Basal" glucocorticoid titer	Low	High	Brain & Nowell, 1971a
"Stress" glucocorticoid titer	High	Low	Henry, Ely, & Stephens, 1974; Sigg, 1969
"Stress" glucocorticoid titer	Low	High	Brain & Nowell, 1971a
Weight of sex accessory glands	High	Low	Brain & Nowell, 1970, 1971a; Christian, 1955a,b; Sigg, 1969; Svare & Leshner, 1973; Vanderbergh, 1960
Gonadal steroid secretion	High	Low	McKinney & Desjardins, 1973
Level of adrenal medullary epinephrine	Low	High	Welch & Welch, 1971
Level of adrenal medullary epinephrine	Equal	Equal	Anton, 1969; Anton, Schwartz, & Kramer, 1968; B. L. Welch & Welch, 1968
Coat condition	Good	Poor	Houssay, Epper, & Pazo, 1965
Response to Helminth parasites	Good	Poor	Brayton & Brain, 1974a
Immunological response to sheep red blood cells	Good	Poor	Brayton & Brain, 1974b

[a] Adapted from Brain and Benton (1977).

fighting. Recently, Brain, Benton, Howell, and Jones (1980) have demonstrated that following isolation for 95 days, Lister rats showed substantial levels of attack directed to intruder males (they differentiated among isolated intact, isolated castrated, and intact group-housed intruders). Powell and Buchanan (1978) also found that isolated rats attacked "intruders" more readily than their group-housed counterparts did.

Different testing situations may be devised to make fighting more likely. Barfield, Busch, and Wallen (1972) for example, suggested that rats were much more likely to fight in their home cages than in neutral arenas. The introduction of male rats into mixed-sex colonies consistently produces fighting behavior (Blanchard, Fukunaga, Blanchard, & Kelley, 1975; Flannelly & Lore, 1977) that may be more vigorous when the intruder has been isolated prior to its introduction (Brain *et al.*, 1980; Flannelly & Lore, 1975). Connor and Gregor (1973) recorded increased intermale fighting in individually maintained rats that were paired in a Skinner box with a single lever to press for a sugar solution reward. Zook and Adams (1975) found high incidences of social aggression in pairs of like-sexed laboratory rats in a situation involving competition for a food pellet.

Systematic studies with laboratory mice have shown the importance of the *duration of* and *age at* individual housing, in isolation–grouping comparisons. Goldsmith *et al.* (1976) studied the effects of different durations of isolation–grouping on social aggression. Male mice were differentially housed at 7 weeks of age and left for periods up to 41 weeks. After a little as 2 weeks, isolates showed significantly more attack than did grouped mice, although the most rapid increase in isolation-induced fighting occurred between 4 and 8 weeks. In a second experiment, the influence of age at isolation on fighting was investigated. The duration of differential housing was kept constant at one month and the age of commencement was varied; animals isolated earlier in life showed greater amounts of subsequent attack. It was suggested that younger mice are more able to take on the characteristics of dominant territory holders, their behavior being more plastic. Similarly, Brain and Evans (1975) found that older subjects from a well-established social dominance hierarchy may be less easy to convert to dominants by short periods of isolation. It must be noted that individually and group-housed rodents differ with respect to many other behaviors (see Table I). This means that interpretation of differences in fighting behavior is often complicated and that only full "ethological" analyses enable one to eliminate many of the alternative explanations to changes in "aggressiveness" (Benton, 1981).

Endocrine Factors

Adrenocortical Hormones

Much of the data supports the view that adrenocortical activities differ in isolated and group-housed mice. Brain and Benton (1977) have argued that, although several workers have talked about "isolation stress," the literature favors the view that isolation *per se* is no more stressful than group-housing. For example Benton, Goldsmith, Gamal-el-Din, Brain, and Hucklebridge (1978) examined adrenal activity in mice that had not been subjected to behavioral testing: they reported no difference in plasma corticosterone titer and a higher adrenal weight in grouped than in isolated mice. The studies that have reported higher corticosterone activity in isolates tend to have stressed these animals by prior behavioral testing and/or the method of blood sampling. However, isolated mice seem to show hypersensitivity to different types of stress, including handling, noise, and being involved in a fight. Isolates seem to habituate to novelty more slowly than do group-housed counterparts (Benton & Brain, 1979).

Much of the support for an "isolation stress syndrome" is derived from studies employing relative adrenal weights that are repeatedly found to be heavier in individually than in group-housed female mice. Unfortunately, estrogen titers (which are augmented in isolated females) may produce marked adrenocortical hypertrophy that is not related to increased steroidogenesis. Sex steroids have, indeed, profound influences on adrenocortical size and activity in a variety of rodent species (reviewed in Brain, 1979).

In rat studies, the methods of assessing adrenal function, the number of individuals in the grouped category, and the familiarity of members of the group vary greatly. These variables may also account for discrepancies between papers. A majority of studies suggested that individually housed rats have heavier adrenals than their group-housed counterparts (reviewed in Brain & Benton, 1979). Most of these investigations have, however, utilized female subjects from whom stressful fighting is rarely evident. In studies in which plasma levels of corticosterone are determined after nonstressful sampling (e.g., decapitation), equivocal results have been obtained. Some investigations suggested that there were few differences between these titers in isolated and grouped rats, whereas others suggested that isolates had augmented or reduced glucocorticoid levels compared with group-housed counterparts. In studies in which plasma samples were obtained after deliberately exposing the animal to a stressor, some indication of adrenocortical reactivity may be obtained. However, results obtained in isolation–grouping comparisons after this procedure are also not consistent. Again, much appears to depend on the period of isolation, the age of the subjects, their prior familiarity, etc.

The Goldsmith *et al.* (1976) study of the effects of increasing durations of individual housing on mice examined the effects on "basal" and "stressed" corticosteroid values. Samples for the former determinations were taken by the rapid decapitation of subjects that had not been subjected to ether vapor or to behavioral testing. Samples in the latter category were obtained by subjection to ether anesthesia before decapitation. When increasing durations of differential housing were examined for up to 40 weeks, no differences in "basal" corticosterone levels were found: Isolates did not differ from group-housed mice. When "stress" levels of glucocorticoids were examined, isolated mice had lower levels in the early phases of differential housing. However, after about 8 weeks of differential housing, the picture reversed: The isolates had higher levels of circulating hormone. There is an interesting similarity in the time-scale of the increase in fighting, change of ambulation, and stress corticosterone titer, with increasing durations of differential housing (Goldsmith, Brain, & Benton, 1978). Around the time an isolated mouse becomes increasingly aggressive, it becomes less active in the open-field and adrenal reactivity increases. It is possible that these changes have a common genesis although this question awaits further research.

Adrenomedullary Hormones

Differences in adrenomedullary hormones have also been recorded when individual- and group-housed mice have been compared. Welch (1965) found that isolated (aggressive compared with grouped counterparts) mice had a higher proportion of norepinephrine (NE) to total medullary catecholamines. Hucklebridge, Reid, Benton, and Brain (1976) found a similar result; both isolated and dominant male mice had a higher proportion of adrenal NE than did subordinates. Sigg *et al.* (1966) also found that isolated mice had greater adrenal NE content than grouped mice did. Other reports have failed to replicate these findings (Anton, 1969; Anton, Schwartz, & Kramer, 1968; Benton *et al.*, 1978; Welch & Welch, 1969).

Studies in this laboratory have further investigated adrenomedullary activity, by examining phenylethanolamine N-methyl transferase (PNMT) activity in differentially housed mice. PNMT activity was significantly higher in naive grouped mice than in isolates ($p < 0.01$). The lower medullary methylation may reflect the lower adrenocortical activity in isolates, since PNMT activity is stimulated by glucocorticoids (Wurtman & Alexrod, 1966). Ciaranello, Wooten, and Axelrod (1976) have demonstrated the enhancement of PNMT synthesis following splanchnic nerve stimulation. Thus the high PNMT activities in isolated mice may result from neural stimulation rather than adrenocortical changes.

Many of the inconsistencies in the adrenal medullary data disappear when one realizes that the relationship between isolation-induced aggression and high medullary NE exists only in mice that have been allowed to attack and defeat opponents. On the basis of studies involving aggressive and timid species, it has been suggested that the release of adrenal NE is associated with anger or aggression. Such situations can be regarded as challenging but predictable, allowing an active response. In contrast, the release of epinephrine has been linked to anxiety, or threatening situations in which responses have not been made (Schildkraut & Kety, 1967). Although it has been suggested that this is too simple a view (Frankenhaeuser, 1971), it is possible to explain the present results in these terms. Benton *et al.* (1978) explained the high levels of NE in dominant compared with untested isolated mice as a reflection of the lack of challenge associated with individual housing. One would predict that allowing isolated mice to fight would be associated with an increased production of norepinephrine, and this seems to be the case (Hucklebridge *et al.*, 1976). The reported high epinephrine turnover in grouped compared with isolated mice (B. L. Welch & Welch, 1968) is consistent with this view. Most grouped mice are subordinates and hence act passively in a threatening situation by showing submissive behavior.

Gonadal Hormones

Brain (1975) has reviewed the evidence that isolated mice are similar to dominant mice when compared on a number of physiological measures. For example, both in isolates and dominants there is a relatively high level of gonadal steroid secretion compared with grouped or subordinate mice (McKinney & Desjardins, 1973).

Some doubt concerning a simple relationship between sex steroid metabolism and social aggression in mice is provided by a study by Dessi-Fulgheri, Lucarini, and Lupo di Prisco (1976). They found no significant correlations among plasma concentrations of testosterone and estradiol, *in vitro* steroid metabolism of the gonads or the adrenals, and fighting in this species.

In rats, Dessi-Fulgheri, Lupo di Prisco, and Verdarelli (1976) established that males maintained in individual cages or separated by wire-mesh screens from males or females had lower plasma testosterone and estradiol titers than did males housed in groups of four with four females. These differences were evident at 60 and 180 days of age. Although the study was discussed in terms of "isolation" and deprivation, it could be argued that mating activity in the social groups accounts for the gonadal stimulation. Dessi-Fulgheri, Lupo di Prisco, and Verdarelli (1975) earlier demonstrated that long-term isolation of rats resulted in circulating testosterone levels higher than those in group-housed animals. No striking differences were evident in circulating 17β-estradiol levels, but there were many instances of altered steroid biosynthesis in isolated versus group-housed animals. There was a suggestion that isolation also increased "irritable aggression" in response to a plastic bar. One should also note the observation by Dessi-Fulgheri, Lucarini, and Lupo di Prisco (1975) that individual housing and deprivation of contact with females both result in altered activities of neurally located aromatizing enzymes compared with heterosexually housed males in this species.

Further support for the view that exposure of male mice to females alters their gonadal function is provided by Jean-Faucher, Berger, de Turkheim, Veyssiere, and Jean (1978) who recorded that males isolated since birth had impaired testicular activity compared with sexually active animals (from groups of three males and three females). Two interesting studies by Batty (1978a,b) established that exposure of male mice to estrus females causes a profound increase in plasma testosterone. Interestingly, although this phenomenon was evident in a number of strains, mice showing more vigorous mating behavior had *lower* elevations of circulating androgens than did those animals that showed less marked mating behavior. It was argued that perhaps strains that are sensitive to testosterone show more rapid negative feedback regulation of circulating hormones and greater behavioral responses to the hormone. This again emphasizes the difficulty of establishing simple relationships between hormones and behavior.

STUDIES OF SOCIAL STATUS

A frequent approach in studies that examine the influence of housing on aggression is to distinguish animals of different social status. A repeated finding is that both the behavior and physiology of dominant animals differs from that of subordinate animals. Some of the range of variables on which dominant and subordinate mice differ are listed in Table II. Although the term "dominance"

TABLE II

Characteristics of Dominant and Subordinate Mice from Small Groups[a]

Characteristic	Dominant	Subordinate	Reference
Intraspecific fighting	High	Low	Brain & Al-Maliki, 1978
Interspecific fighting	Equal	Equal	Al-Maliki & Brain, 1977
Reactivity in a novel arena	High	Low	Benton & Brain, 1979
Motor activity in response to novelty	High	Low	Benton & Brain, 1979
Turnover of brain catecholamines	Low	High	Welch & Welch, 1971
Adrenal weight	Low	High	Brain, 1972; Brain & Nowell, 1971b; Chapman, Desjardins, & Bronson, 1969; Louch & Higginbotham, 1967; B. L. Welch & Welch, 1968
"Basal" glucocorticoid titer	Equal	Equal	Bronson, 1973; Chapman *et al.*, 1969; Henry, Ely, & Stephens, 1974; Louch & Higginbotham, 1967
"Basal" glucocorticoid titer	Low	High	Bronson, 1973; Ely, Henry, & Ciaranello, 1974
"Stress" glucocorticoid titer	Low	High	Chapman *et al.*, 1969; Henry *et al.*, 1974; Louch & Higginbotham, 1967
Weight of sex accessory glands	High	Low	Brain, 1972; Brain & Nowell, 1971a; Bronson, 1973; Bronson, Stetson, & Stiff, 1973; Chapman *et al.*, 1969
Gonadal steroid secretion	High	Low	McKinney & Desjardins, 1973
Level of adrenal medullary epinephrine	Low	High	Welch & Welch, 1971
Percentage of adrenal medullary norepinephrine to total catecholamine	High	Low	Hyckelbridge, Reid, Benton, & Brain, 1976
Coat condition	Good	Poor	Brain, 1972
Reproductive success	High	Low	De Fries & McClearn, 1970

[a] Adapted from Brain and Benton (1977).

is widely used in biology, it has been defined in a variety of ways. For example, Van Kreveld (1970) suggests that "dominance is a priority of access to an approach situation or of leaving an avoidance situation" (p. 146). A different approach stresses overt fighting, so Klopfer (1974) states, "Dominance is inferred whenever one individual is able to chastise another with impunity" (p. 154). Unfortunately the diverse methods of assessing dominance do not always produce consistent results. Benton, Dalrymple-Alford, and Brain (1980) examined six measures of dominance in male mice and concluded that there was no evidence of an underlying dimension that could be labeled dominance. They caution against necessarily generalizing their results because the conditions used many be important in determining the findings. Care should be taken when extrapolating data when dominance has been measured in a particular way.

A particular problem in many studies results from keeping several male mice in a small laboratory cage. The work of Crowcroft (1966) and Mackintosh (1970) suggests that the male mouse may, when given enough space, defend a territory and drive off subordinate animals. Thus, group-housing in a laboratory cage may induce an unnatural social structure.

Behavioral Concomitants of Dominance-Subordination

Unremarkably, it has been reported that mice designated as dominant, following behavioral observation in groups, attack unfamiliar mice in neutral situations more vigorously than subordinates do (Benton & Brain, 1979; Brain & Al-Maliki, 1978). There is also evidence that a different type of encounter results when a dominant or a subordinate meets an unfamiliar mouse. Some recent work in our laboratory has analyzed the area of the body to which attack is directed. Dominants attacked the opponent's back more frequently than did subordinates and showed relatively high incidences of mounting. Subordinates generally attacked the flanks and showed little mounting. The finding in this study that subordinates fought unfamiliar mice, although to a lesser extent than dominants, illustrates the point that dominance-subordination is a relationship between particular animals.

There is evidence that in other respects, the behavior of dominants differs from that of subordinates, and also that dominants are similar to isolates. Isolates and dominants differ from subordinates in that animals from the last category are more active on the first day in an open-field but habituate relatively quickly. In contrast both isolates and dominants are relatively inactive initially (Benton & Brain, 1979). There are suggestions that the novelty of the test arena influences whether dominants show relatively fast or slow habituation. They habituate slowly in novel situations but show a quicker response in more familiar arenas. Using a seminatural situation, Ely and Henry (1974) reported that dominant males were more active over a 24-hour period than were subordinates. It seems that this activity difference may reflect the patrolling of boundaries by the dominants.

Similarly, differences in the activity of isolated and group-housed rats have been found. Open-field tests of short duration have shown that group-housed rats are more active than isolated rats (Ader & Friedman, 1964; Dalrymple-Alford & Benton, 1981; Moyer & Korn, 1965). However tests of longer duration have found isolates to be more active (Dalrymple-Alford & Benton, 1981; Domjan, Schorr, & Best, 1977; Einon & Morgan, 1977).

Endocrine Concomitants

Adrenocortical Hormones

The altered adrenocortical function in animals of differing social status has been reviewed by Brain (1975, 1979). An attempt will be made to illustrate only the basic findings. Dominant male mice have lower relative adrenal weights than do their subordinate partners (Brain, 1972; Brain & Al-Maliki, 1978; McKinney & Pasley, 1973). However, Buhl, Hasler, Tyler, Goldberg, and Banks (1978) found no evidence that social rank influenced this parameter in collared lemmings. Louch and Higginbotham (1967) also recorded that subordinates had elevated plasma corticosterone titers compared with dominant mice. Chapman, Desjardins, and Bronson (1969) were, however, able to show this differential only after superimposing stressors before plasma sampling. Brain and Benton's (1977) review suggested that "basal" levels of glucocorticoids do not differ in these two categories. It is also interesting to note that Angelucci and Valeri (1978) have demonstrated that specific corticosterone uptake in the hippocampal and septal region is lower in subordinate than in dominant mice.

Adrenomedullary Hormones

Attempts have been made to correlate social status and various measures of adrenomedullary activity in rats and mice. For example, dominant mice have lower levels of adrenal epinephrine (Welch, 1967) and higher proportions of norepinephrine (Hucklebridge *et al.*, 1976) than do subordinates. Levels of tyrosine hydroxylase and PNMT in the medulla are higher in subordinate mice than in dominants (Henry, Stephens, Axelrod, & Mueller, 1971). Ng, Marsden, Colburn, and Thoa (1973) also recorded that "socially withdrawn" mice from large colonies had elevated PNMT activities compared with territorial animals.

Studies in this laboratory have also investigated adrenomedullary correlates of dominance and subordination. The dominant mouse in a group was identified by the characteristic postures observed. The E and total catecholamine content of the adrenal glands were significantly greater in dominants than in subordinates. The high adrenal E content in dominants compared with subordinates may reflect the higher release of E from the adrenals of subordinates. For example Hucklebridge, Nowell, and Dilks (1973) reported that plasma E levels were elevated immediately after fighting in defeated subjects. In groups of four mice, we have further distinguished subordinates that display scarring, especially on the tail and rump, from subordinates with no scars. The scarred subordinates showed higher levels of PNMT activity ($p < 0.01$). Henry (1976) has suggested that dominance leads to the stimulation of the adrenomedullary system while subordination leads to stimulation of the adrenocortical system. Since PNMT activity is altered by both these systems (Bhagat & Horenstein, 1976), one would expect this enzyme activity to be elevated in both highly stressed subordinates and (to a lesser extent) in dominants. This view is supported by the present results.

There seems good evidence for the existence of two types of subordinate that differ in their endocrine profiles. The "avoiding" subordinate has a low PNMT activity, whereas repeatedly attacked counterparts have elevated enzyme. Leshner, Moyer, and Walker (1975) demonstrated that mice with either very low or very high levels of glucocorticoids have an increased tendency to avoid attack by fighter mice compared with mice with intermediate hormone levels. One must consider the possibility that subordinates experiencing low levels of attack do so as a result of low pituitary–adrenocortical activity (and PNMT activity). This is in contrast to the traditional view that the low adrenal activity in "avoiding" subordinates (reflected by low PNMT) is a consequence of the low level of attack to which they were subjected.

Gonadal Hormones

Although it is rarely possible to identify the gonadal factor that modulates aggression, there are numerous accounts suggesting that testicular androgen production is higher in dominant than in subordinate males. The typical approach is to identify dominants and subordinates (a possibly unreliable process, as emphasized by Benton *et al.*, 1980) and then to assess testicular weight, weights of sex accessories, testicular histology, fructose content of the seminal vesicles, circulating plasma androgen levels, plasma titer of "free" (i.e., unbound) androgens, blood production rate of androgenic secretion, or metabolic conversion by target tissues. It will be immediately appreciated that these methods differ in sophistication and that no one method gives a complete picture of normal functioning.

Most observations in this area have used rodents. There are, for example numerous studies (Brain, 1972; Brain & Benton,1977) suggesting that dominant mice have sex accessories heavier than those of their subordinates. The former category is also characterized by increased seminal vesicle fructose, elevated plasma androgen, and enhanced testosterone synthesis (McKinney & Desjardins, 1973). Indeed, physiological changes associated with repeated defeat may suppress androgen production in subordinates (Bronson, Stetson, & Stiff, 1973). Buhl *et al.* (1978) found that dominant male collared lemmings have higher plasma testosterone levels and heavier ventral prostates than do both active and passive subordinates. However, serum testosterone levels in dominant inbred mice are not always elevated compared with subordinates (Barkley & Goldman, 1977). In mice, gonadal functioning is generally elevated and adrenocortical activity lowered in dominants compared with subordinates; this reciprocal relationship has been repeatedly noted (Selye, 1971).

THE INVOLVEMENT OF THE DIFFERENT SENSORY MODALITIES IN HOUSING STUDIES

We have already noted that "isolation," as employed in the majority of rodent studies, is better termed "individual housing" because subjects can often hear, smell, and see conspecifics. One method of studying isolation is to look at different sensory modalities that may be deprived by such housing. A strategy employed in this laboratory involved deprivation of a particular sensory modality or the enhancement of cues perceived by it.

Olfactory Influences

Accumulating experimental evidence supports the unremarkable conclusion that in general, nocturnal, group-living mice rely heavily on olfactory communication. For example, sex attractants and aggression-modifying (depending on context) odors are produced by both male and female mice.

One may directly study olfactory influences on aggression by suppressing the sense of smell. A surprising variety of techniques have been used in diverse studies (reviewed by Alberts, 1974). They include bilateral bulbectomy, lateral olfactory tract lesions, surgical removal of the neuroepithelium, intranasal zinc sulfate treatment, local anesthesia of nasal mucosa, and nasal blockade with tracheostomy. Olfactory bulbectomy is, however, liable to induce peculiar behavioral and physiological consequences in animals in addition to producing an olfactory deficit (Edwards, 1974). Nasal blockade with tracheostomy has been used successfully with dogs but appears unsuitable for rodents. Bilateral occlusion of the nostrils produces suffocation or severe behavioral disruption due to abnormal mouth breathing in rats. In view of the above problems, intranasal zinc sulfate administration was the technique selected—it has been used successfully in mice (e.g., Edwards, 1974, but see Sieck & Baumbach, 1974, for a contrary view). We found that both the acute and chronic induction of zinc sulfate-induced anosmia to a large extent eliminates social aggression (Benton *et al.*, 1979a; Brain & Al-Maliki, 1978). A loss of olfactory capacity has also been found to reduce social aggression in feral rats (Alberts & Galef, 1973), laboratory rats (Luciano & Lore, 1975; Flannelly & Thor, 1976), Golden hamsters (Devor & Murphy, 1973), and gerbils (reviewed by Alberts, 1974). Hull, Hamilton, Engwall, and Rosselli (1974) found anosmic male gerbils did not show decreased marking and dominance scores when grouped, unlike intact animals. It has also been found that male Swiss-Webster albino mice made anosmic by zinc sulfate perfusion and tested against a castrate opponent showed a similar response; this applied whether they were individually housed for 24 days (Edwards, Thompson, & Burge, 1972) or trained as fighters (Edwards, 1974). The majority of these animals fought and showed the same frequency and duration of fighting as did control animals, although the formers' attack latencies were significantly extended. These results could be confounded by the fact that strange male urine is decidely more effective in eliciting attack if obtained from an intact than from a castrate male (Mugford & Nowell, 1970).

A second approach to this problem involves trying to enhance olfactory communication. The approach we have taken involves introducing soiled sawdust bedding into the cage in a petri dish with holes in the lid. Isolated males in the presence of bedding from groups of female mice displayed significantly less aggression than did those in the presence of the bedding from group-housed male mice (Benton *et al.*, 1979a).

These findings generally confirm the observations of Mugford (1973), who investigated the effects of odors on fighting in trained "fighter" albinos in encounters with castrated opponents. In contrast, Harmatz, Bodkins, & Kessler (1975) found that isolated DBA2 mice chronically subjected to soiled bedding from group-housed males (changed daily over a four-week period) showed *less* fighting when tested in their home cages against opponents from groups of 15 than did animals subjected to fresh sawdust or soiled bedding from other individually housed males. These discrepancies may be due to differences in their frequency of changing sawdust, size of the donor group, constitution of the group, site of testing, opportunity to ingest sawdust, etc.

Mugford (1973) claimed that fighting was stimulated to the same degree by recently deposited odors of fighting mice and home cage material from male mice accumulated over long periods. In contrast, Haug (1970) found that effectiveness was influenced by the stability of the group. Male donors of bedding in the present experiment had been regularly subjected to defeat by trained fighters or test males. It is possible that these manipulations changed the character of odor production.

Mugford (1973) suggests that there is an aggression-inhibiting "pheromone" in urine from estrus, diestrus, and spayed females. This factor could be responsible for the decreased tendency to fight in males raised on bedding from cages containing grouped females.

Tactile Influences

There is growing evidence that tactile information influences rodent aggression. Katz (1976) found that the removal of the mystacial vibrissae and anesthesia of the vibrissal pad reduces fighting in isolated male mice. We have also studied the effect of removing the vibrissae from mice experienced in fighting; this operation increased the tendency to attack. The results suggest that facial cues are important in triggering aggression in differentially housed rodents. Ghiselli and Thor (1975) compared shock-induced fighting in control, naive blind, anosmic, and devibrissated male rats. They found that both olfaction and touch were important in the control of this activity. A similar relationship may also apply to social aggression in mice.

Visual Influences

We have investigated the effects of visual contact on isolation-induced aggression, by using specially constructed cages. These cages had central partitions of different materials: perspex, so another mouse could be seen; stainless steel, so visual contact was impossible; and perforated zinc, which obscured vision but increased the passage of odors. We found that male mice able to see other animals through perspex screens were as aggressive as traditionally housed isolates, and those mice separated by zinc screens displayed nonsignificantly lower levels of fighting. Takemoto, Suzuki, and Miyama (1975) found that mice that had been chronically individually housed (24–28 weeks) were less prone to convulsive attack if they had visible neighbors. Thiessen, Zolman, and Rodgers (1962) compared C3H/Crgl/2 male mice separated by wire mesh in a hole-in-the-wall emergence test. They found that the latency of emergence of the mice separated by wire mesh fell between the long emergence latencies of the "isolated" and the short latencies of the grouped mice. Senault (1971) found that rats were more aggressive if they had been previously housed in opaque cages rather than wire boxes. The effect of physical "isolation" thus seems to vary from species to species.

Mice appear to suffer from "psychological stress" if housed in visual contact with a mouse that has previously defeated it. Bronson and Eleftheriou (1965b) found that, after short exposure, such mice had increased plasma corticosterone levels; a similar design by Hucklebridge and Nowell (1974), however, produced little or no evidence of an adrenomedullary response. A. S. Welch and Welch (1968) found that norepinephrine in the lower brain stem of "isolates" was reduced after being allowed to witness former "isolates" fighting for 75 minutes.

Discussion of the Involvement of Sensory Modalities in Housing Studies

The importance of olfactory, tactile, and visual cues on the development and expression of social aggression has been investigated, although the relevance of auditory cues has been little investigated. Auditory information could prove to be important, since ultrasounds have been reported to be employed in social behaviors in a variety of rodent species. Ultrasounds have been found to be implicated in maternal retrieving in rats (Bell, Nitschke, Gorry, & Sachman, 1971), in mating in rats (Barfield & Geyer, 1972), in mouse courtship (Whitney, Alpern, Dizinno, & Horowitz, 1974), and in fighting in rats (Sales, 1972). Prior auditory exposure has also been shown to influence susceptibility to audiogenic seizure in mice (Henry, 1967) and there are indications that short and long term "isolates" behave differently in a situation with a disturbing noise (Brain, Haley, & Nowell, 1971). However, there is no direct information on the effect of auditory deprivation on the development of "social aggression."

It seems that physical separation is the most important factor associated with individual-induced aggression. The mystacial vibrissae were found to have only a slight influence on the expression of isolation-induced fighting. Although data of Katz (1976) indicates that the innervation of the facial area is important in the mouse, it seems unlikley that these organs play a major role in the genesis of isolation-induced aggression.

Olfaction has been found to have significant influences on fighting: Chronically anosmic individually housed mice did not fight, and it was shown that the extent of development of isolation-induced fighting in intact mice could be influenced by odors in the home cage. Moreover, mice that were separated by perforated zinc partitions showed levels of fighting intermediate between those of pairs separated by opaque or transparent partitions (which are presumably less permeable to odoriferous cues than is perforated zinc). Consequently, odor cues may partially account for variations between different individual-housing and group-housing studies. Visual influences seem relatively unimportant: Mice that could see adjacent mice fought as much as those that could not.

GENERAL DISCUSSION

It seems likely that subjection to defeat causes both group-housed and subordinate mice to differ from their individually housed and dominant counterparts (see Table III).

A general conclusion is that workers in this area should pay attention to methodological factors; apparently similar experimental conditions can produce opposite effects. For example Goldsmith *et al.* (1976) found that short durations of isolation were associated with lower levels of stress glucocorticoids than in grouped mice, whereas after long periods, the opposite was true. However, "basal" level of corticosterone did not differ with housing. The duration of housing and method of hormone sampling was critical.

The life history of the animal is important; apparently minor differences in methodology change the results. Benton, Brain, and Goldsmith (1979b) examined one small difference in methodology, whether one starts with grouped animals and then isolates them, or *vice versa.* They reported that mice that had been continuously isolated or grouped had lower basal corticosterone levels than did mice rehoused 32 days prior to plasma sampling, but with this exception had received similar nonstressful treatment. Isolated mice that 32 days previously

TABLE III

Effects of Subjection to Defeat by 'Trained Fighters' Followed by Periods of Recovery on the Behavioral and Physiological Characteristics of Male Mice[a]

Characteristic	Change	Reference
Intraspecific fighting	Reduced	Bronson & Eleftheriou, 1965a
Levels of brain catecholamines	Increased	Eleftheriou & Church, 1968; Welch & Welch, 1971
Adrenocortical activity	Increased	Archer, 1970; Brain & Nowell, 1970; Bronson & Eleftheriou, 1965a,b
Weight of sex accessory glands	Decreased	McKinney & Pasley, 1973
Gonadal steroid secretion	Decreased	McKinney & Desjardins, 1973
Adrenal medullary catecholamine content	Increased	Goldberg & Welch, 1972; Welch & Welch, 1971

[a] Adapted from Brain and Benton (1977).

had been grouped had lower androgen-dependent organ weights than did continuously isolated mice. Again we see that the history of the animal is important.

The drawing of conclusions concerning hormones, housing, and aggression is extremely difficult. Workers should pay attention to the following variables:

(a) As soon as a generalization is made, it should be qualified by the duration of differential housing—a more extensive range of housing durations is strongly advocated. Attention should be paid to the fact that age is an accompanying variable in many experimental designs.

(b) Age at differential housing.

(c) Species, strain, and sex of the animal employed. There are a number of indications that these may influence the direction of the changes observed in individual and group-housing comparisons. The social organization assumed by small rodents (and the consequent relationships between behavior and endocrine function) may be drastically altered by the housing conditions employed—mice may evidence territoriality but only at densities of around 2 square meters per animal. Note that Parmigiani, Mainardi, and Pasquali (1981) have presented evidence that small rodents are very adaptable in their range of possible social structure.

(d) Source of experimental animals. It seems inadvisable to use purchased animals in this type of study. Superimposing the "stress" of transportation appears likely to confound the subtle effects of differential housing.

(e) Interactions of tests. One obvious source of variation is introduced by workers applying repeated behavioral–biological sampling techniques to their differentially housed animals. This makes unraveling the effects of housing condition very difficult.

(f) Constitution of the group for comparison with individual housing. It is likely that group size and/or housing density can influence the results of individual and group-housing comparisons. There are, however, further complications, since groups can be

1. Constituted from previously individually housed animals.
2. Formed by combining smaller groups.
3. Produced by splitting larger groups.
4. Reconstituted by taking individuals from groups of the same eventual size.
5. Varied by changing the original housing condition of the weanling animals.
6. Limited to a single sex or mixed sex.

Progress in understanding isolation influences on rodents seems dependent on future studies systematically examining these variables.

Our full understanding of this topic awaits additional study, paying attention to detail; any simple generalization is almost bound to be proved wrong.

Acknowledgments

We thank the various people who have collaborated in these studies. They include Dr. C. M. Evans, Dr. J. F. Goldsmith, Mr. G. Childs (M. Sc.), Dr. L. A.

Gamal-el-Din, Dr. F. H. Hucklebridge, Dr. S. Al-Maliki, Dr. S. Parmigiani, Mr. J. Dalrymple-Alford, and Miss S. Jones.

REFERENCES

Ader, R., & Friedman, S. B. Social factors affecting emotionality and resistance to disease in animals: IV. Differential housing, emotionality, and Walker 257 carcinosarcoma in the rat. *Psychological Reports,* 1964, *15,* 535–541.

Alberts, J. R. Producing and interpreting experimental olfactory deficits. *Physiology and Behavior,* 1974, *12,* 657–670.

Alberts, J. R., and Galef, B. G. Olfaction and movement: Stimuli mediating intra specific aggression in wild rats. *Journal of Comparative and Physiological Psychology,* 1973, *85,* 233–242.

Al-Maliki, S., & Brain, P. F. Influences of social rank and isolation on two forms of murine aggression. *I.R.C.S. Medical Science,* 1977, *5,* 290.

Angelucci, L., & Valeri, P. Influence of social environment and of drugs on 3H-corticosterone uptake in various brain areas. *Annals of 1st Super Sanita,* 1978, *14,* 49–58.

Anton, A. H. Effect of group size, sex and time on organ weights, catecholamines and behavior in mice. *Physiology and Behavior,* 1969, *4,* 483–487.

Anton, A. H., Schwartz, R. P., & Kramer, S. Catecholamines and behavior in isolated and group mice. *Journal of Psychiatric Research,* 1968, *6,* 211–220.

Archer, J. Effects of population density on behaviour in rodents. In J. H. Crook (Ed.), *Social behaviour in birds and mammals.* New York: Academic Press, 1970.

Archer, J. *Animals under stress.* Studies in Biology No. 108. London: Edward Arnold, 1979.

Barfield, R. J., & Geyer, L. A. Sexual behavior: Ultrasonic postejaculatory song of the male rat. *Science,* 1972, *176,* 1349–1350.

Barfield, R. J., Busch, D. E., & Wallen, K. Gonadal influences on agonistic behavior in the male domestic rat. *Hormones and Behavior,* 1972, *3,* 247–259.

Barkley, M. S., & Goldman, B. D. A quantitative study of serum testosterone, sex accessory organ weight growth and the development of intermale aggression in the mouse. *Hormones and Behavior,* 1977, *9,* 32–48.

Batty, J. Plasma levels of testosterone and male sexual behavior in strains of the house mouse *(Mus musculus). Animal Behaviour,* 1978, *26,* 339–348. (a)

Batty, J. Acute changes in plasma testosterone levels and their relation to measures of sexual behaviour in the male house mouse *(Mus musculus). Animal Behaviour,* 1978, *26,* 349–357. (b)

Bell, R. W., Nitschke, W., Gorry, T. H., & Sachman, T. A. The ontogeny of ultrasounds in *Rattus norvegicus. Developmental Psychobiology,* 1971, *4,* 181–191.

Benton, D. The measurement of aggression in the laboratory. In P. F. Brain & D. Benton (Eds.), *The biology of aggression.* The Netherlands: Noordhoff/Sijthoff, 1981.

Benton, D., & Brain, P. F. Behavioural comparisons of isolated, dominant and subordinate mice. *Behavioural Processes,* 1979, *4,* 211–219.

Benton, D., Goldsmith, J. F., Gamal-el-Din, L., Brain, P. F., & Hucklebridge, F. H. Adrenal activity in isolated mice and mice of different social status. *Physiology and Behavior,* 1978, *20,* 459–464.

Benton, D., Brain, P. F., & Goldsmith, J. F. The involvement of various sensory modalities and isolation-induced aggression in laboratory mice. *Aggressive Behavior,* 1979, *5,* 198–199. (a)

Benton, D., Brain, P. F., & Goldsmith, P. F. Effects of prior housing on endocrine responses to differential caging in male TO strain mice. *Physiological Psychology,* 1979, *7,* 89–92. (b)

Benton, D., Dalrymple-Alford, J. C., & Brain, P. F. Comparisons of measures of dominance in the laboratory mouse. *Animal Behaviour,* 1980, *28,* 1274–1279.

Bernstein, H., & Moyer, K. E. Aggressive behaviour in rats: Effects of isolation and olfactory bulb lesions. *Brain Research,* 1970, *20,* 75–84.

Bevan, W., Bloom, W. L., & Lewis, G. T. Levels of aggressiveness in normal and amino acid deficient albino rats. *Physiological Zoology,* 1951, *24,* 231.

Bhagat, B. D., & Horenstein, S. Modulation of adrenal medullary enzymes by stress. In E. Usdin, R. Kvetnansky, & J. Kopin (Eds.), *Catcholamines and stress.* New York: Pergamon Press, 1976.

Blanchard, R. J., Fukunaga, K. K., Blanchard, D. C., & Kelley, M. J. Conspecific aggression in the laboratory rat. *Journal of Comparative and Physiological Psychology,* 1975, *89,* 1204–1209.

Brain, P. F. The physiology of population limitation in rodents—a review. *Communications in Behavioral Biology,* 1971, *6,* 115–123.

Brain, P. F. Endocrine and behavioural differences between dominant and subordinate male house mice in pairs. *Psychonomic Science,* 1972, *28,* 260–262.

Brain, P. F. What does individual housing mean to a mouse? *Life Sciences,* 1975, *16,* 187–200.

Brain, P. F. Effects of the hormones of the pituitary-adrenal axis on behaviour. In K. Brown & S. J. Cooper (Eds.), *Chemical influences on behaviour.* New York: Academic Press, 1979.

Brain, P. F., & Al-Maliki, S. A comparison of 'Intermale Fighting in Standard Opponent' tests and attack directed towards locusts by 'TO' strain mice: Effects of simple experimental manipulation. *Animal Behaviour,* 1978, *26,* 723–737.

Brain, P. F., & Benton, D. What does individual housing mean to a research worker? *I.R.C.S. Journal of Medical Science,* 1977, *5,* 459–463.

Brain, P. F., & Benton, D. The interpretation of physiological correlates of differential housing in laboratory rats. *Life Sciences,* 1979, *24,* 99–116.

Brain, P. F., & Evans, C. M. Attempts to influence fighting and threat behaviors in adult isolated female CFW mice in "standard opponent" aggression tests using injected and subcutaneously implanted androgens. *Physiology and Behavior,* 1975, *14,* 551–556.

Brain, P. F., & Nowell, N. W. Some endocrine and behavioural changes in the development of the albino laboratory mouse. *Communications in Behavioral Biology,* 1969, *4,* 203–220.

Brain, P. F., & Nowell, N. W. Some observations on intermale aggression testing in albino mice. *Communications in Behavioral Biology,* 1970, *5,* 7–17.

Brain, P. F., & Nowell, N. W. Isolation versus grouping effects on adrenal and gonadal function in albino mice: 1. The male. *General and Comparative Endocrinology,* 1971, *16,* 149–154. (a)

Brain, P. F., & Nowell, N. W. The effects of differential grouping on endocrine function of mature male albino mice. *Physiology and Behaviour,* 1971, *5,* 907–910. (b)

Brain, P. F., Haley, P. G., & Nowell, N. W. Activity studies on long-term and recently isolated male albino mice using a number of different pieces of equipment. *Communications in Behavioral Biology,* 1971, *6,* 259–270.

Brain, P. F., Benton, D., & Bolton, J. C. A comparison of agonistic behavior in individually housed male mice and those cohabiting with females. *Aggressive Behavior,* 1978, *4,* 201–206.

Brain, P. F., Benton, D., Howell, P. A., & Jones, S. E. Resident rats aggression towards intruders. *Animal Learning and Behavior,* 1980, *8,* 331–335.

Brayton, A. R., & Brain, P. F. Effects of 'crowding' on endocrine function and retention of the digenean parasite *Microphallus pygmaeus* in male and female albino mice. *Journal of Helminthology,* 1974, *48,* 99–106. (b)

Brayton, A. R., & Brain, P. F. Studies on the effects of differential housing on some measures of disease resistance in male and female laboratory mice. *Journal of Endocrinology,* 1974, *61,* xlvii–xlix. (b)

Bronson, F. H. Establishment of social rank among grouped male mice: Relative effects on circulating FSH, LH and corticosterone. *Physiology and Behaviour,* 1973, *40,* 947–951.

Bronson, F. H., & Eleftheriou, B. E. Relative effects of fighting on bound and unbound corticosterone in mice. *Proceedings of the Society of Experimental Biological Medicine,* 1965, *118,* 146–149. (a)

Bronson, F. H., & Eleftheriou, B. E. Adrenal response to fighting in mice: Separation of physical and psychological causes. *Science,* 1965, *147,* 627–628. (b)

Bronson, F. H., Stetson, M. H., & Stiff, M. E., Serum FSH and LH in male mice following aggressive and non-aggressive interaction. *Physiology and Behavior,* 1973, *10,* 369–370.

Buhl, A. E., Hasler, J. F., Tyler, M. C., Goldberg, N., & Banks, E. M. The effects of social rank on reproductive indices in groups of male collared lemmings (Dicrostonyx groenlandicus). *Biology of Reproduction,* 1978, *18,* 317–324.

Cairns, R. B. Fighting and punishment from a developmental perspective. In J. K. Cole & D. D. Hensen (Eds.), *Nebraska Symposium on Motivation.* Lincoln: University of Nebraska Press, 1972.

Calhoun, J. B. Population density and social pathology. *Scientific American,* 1962, *206,* 139–148.

Calhoun, J. B. Death squared: The expolsive growth and demise of a mouse population. *Proceedings of the Royal Society of Medicine,* 1973, *66,* 80–88.

Chapman, V. M., Desjardins, C., & Bronson, F. H. Social rank in male mice and adrenocortical response to open field exposure. *Proceedings of the Society of Experimental Biology and Medicine,* 1969, *130,* 624–627.

Charpentier, J. Analysis and measurement of aggressive behaviour in mice. In S. Garattini & E. B. Sigg (Eds.), *Aggressive behaviour.* Amsterdam: Excerpta Medica Foundation, 1969.

Christian, J. J. Effects of population size on the weights of the reproductive organs of white mice. *American Journal of Physiology,* 1955, *181,* 477–488. (a)

Christian, J. J. Effect of population size on the adrenal glands and reproductive organs of male mice in populations of fixed size. *American Journal of Physiology,* 1955, *182,* 242–300. (b)

Christian, J. J. The role of endocrine and behavioural factors in the growth of mammalian populations. In A. Gorbman (Ed.), *Comparative endocrinology.* New York: Wiley, 1959.

Christian, J. J. The pathology of overpopulation. *Military Medicine,* 1963, *128,* 571–603. (a)

Christian, J. J. Endocrine adaptive mechanisms and the physiologic regulation of population growth. In W. V. Moyer & R. G. Van Gelder (Eds.), *Physiological mammalogy.* New York: Academic Press, 1963. (b)

Ciaranello, R. D., Wooten, G. F., & Axelrod, J. Regulation of rat adrenal dopamine β-hydroxylase 11: Receptor interaction in the regulation of enzyme synthesis and degradation. *Brain Research,* 1976, *113,* 349–362.

Connor, H. S., & Gregor, G. L. Crowding and isolation: Determinants of agonistic and food-seeking behavior in Norway rats. *Journal of Comparative and Physiological Psychology,* 1973, *84,* 593–597.

Crowcroft, P. *Mice all over.* London: G. T. Foulis, 1966.

Dalrymple-Alford, J. C., & Benton, D. Activity differences of individual and group-housed male and female rats. *Animal Learning and Behavior,* 1981, *9,* 50–55.

Dessi-Fulgheri, F., Lucarini, N., & Lupo di Prisco, C. Environmental influences on testosterone aromatization in the rat brain. *Proceedings of the 8th Conference of European Comparative Endocrinologists, Bangor,* 1975, 146.

Dessi-Fulgheri, F., Lupo di Prisco, C., & Verdarelli, P. Influence of long-term isolation on the production and metabolism of gonadal sex steroids in male and female rats. *Physiology and Behavior,* 1975, *14,* 495–499.

Dessi-Fulgheri, F., Lucarini, N., & Lupo di Prisco, C. Relationships between testosterone metabolism in the brain, other endocrine variables and intermale aggression in mice. *Aggressive Behavior,* 1976, *2,* 223–232.

Dessi-Fulgheri, F., Lupo di Prisco, C., & Verdarelli, P. Effects of two kinds of social deprivation on testosterone and estradiol-17β plasma levels in the male rat. *Experientia,* 1976, *32,* 114–115.

De Fries, J. C., & McClearn, G. E. Social dominance and Darwinian fitness in the laboratory mouse. *American Naturalist,* 1970, *104,* 408–410.

Devor, M., & Murphy, M. R. The effect of peripheral olfactory blockade on the social behavior of the male golden hamster. *Behavioral Biology,* 1973, 31–42.

Domjan, M., Schorr, R., & Best, M. Early environmental influences on conditioned and unconditioned ingestinal and locomotor behaviours. *Developmental Psychobiology,* 1977, *10,* 499–506.

Edwards, D. A. Non-sensory involvement of the olfactory bulbs in the mediation of social behaviour. *Behavioral Biology,* 1974, *11,* 287–302.

Edwards, D. A., Thompson, M. L., & Burge, K. G. Olfactory bulb removal in peripherally induced anosmia: Differential effects on the aggressive behaviour of male mice. *Behavioral Biology,* 1972, *7,* 823–824.

Einon, D. F., & Morgan, M. J. A critical period for social isolation in the rat. *Developmental Psychobiology,* 1977, *10,* 123–132.

Eleftheriou, B. E., & Church, R. L. Levels of hypothalamic luteinizing hormone-releasing factor after exposure to aggression (defeat) in C57BL/6J mice. *Journal of Endocrinology,* 1968, *42,* 347–348.

Ely, D. L., & Henry, J. P. Effects of prolonged social deprivation on murine behavior patterns, blood pressure, and adrenal weight. *Journal of Comparative and Physiological Psychology,* 1974, *87,* 733–740.

Ely, D. L., Henry, J. P., & Ciaranello, R. D. Long-term behavioral and biochemical differentiation of dominant and subrodinate mice in population cages. *Psychosomatic Medicine,* 1974, *36,* 463.

Essman, W. B. Differences in locomotor activity and brain-serotonin metabolism in differentially housed mice. *Journal of Comparative and Physiological Psychology,* 1968, *66,* 244–246.

Fara, J. W., & Catlett, R. H. Cardiac response and social behaviour in the guinea pig *(Cavia porcellus). Animal Behaviour,* 1971, *11,* 475–479.

Flannelly, K., & Lore, R. Dominance-subordination in cohabiting pairs of adult rats: Effects on aggressive behaviour. *Aggressive Behavior,* 1975, *1,* 331–340.

Flannelly, K., & Lore, R. The influence of females upon aggression in domesticated male rats *(Rattus norvegicus). Animal Behavior,* 1977, *25,* 654–659.

Flannelly, K. J., & Thor, D. H. Territorial behavior of laboratory rats under conditions of peripheral anosmia. *Animal Learning and Behaviour,* 1976, 337–340.

Frankenhaeuser, M. Behavior and circulating catecholamines. *Brain Research,* 1971, *31,* 241–262.

Garattini, S., Giacalone, E., & Valzelli, L. Biochemical changes during isolation-induced aggressiveness in mice. In S. Garattini & E. B. Sigg (Eds.), *Aggressive Behaviour.* Amsterdam: Excerpta Medica Foundation, 1969.

Ghiselli, W. B., & Thor, D. H. Visual, tactual and olfactory deprivation effects on irritable fighting behaviour of male hooded rats. *Physiological Psychology,* 1975, *3,* 47–50.

Goldberg, A. M., & Welch, B. L. Adaption of the adrenal medulla: Sustained increase in choline acetyl transferase by psycho-social stimulation. *Science,* 1972, *178,* 319–320.

Goldsmith, J. F., Brain, P. F., & Benton, D. Effects of age at differential housing and the duration of individual housing/grouping on intermale fighting behaviour and adrenocortical activity in TO strain mice. *Aggressive Behavior,* 1976, *2,* 207–223.

Goldsmith, J. F., Brain, P. F., & Benton, D. Effects of the duration of individual or group housing on behavioural and adrenocortical reactivity in male mice. *Physiology and Behavior,* 1978, *21,* 757–760.

Harmatz, P., Bodkins, R. C., & Kessler, S. Post isolation aggression and olfactory cues. *Behavioral Biology,* 1975, *13,* 219–224.

Haug, M. Mise en evidence de deux odeurs aux effets opposés de facilitation et d'inhibition des conduites agressivit chez la souris màle. *C. R. Academic Science (Paris),* 1970, *271,* 1567–1570.

Henry, J. P. Mechanisms of psychosomatic disease in animals. In C. A. Brandly, C. E. Cornelius, & W. I. B. Beveridge (Eds.), *Advances in veterinary science and comparative medicine.* New York: Academic Press, 1976.

Henry, J. P., Stephens, P. M., Axelrod, J., & Mueller, R. A. Effects of psychosocial stimulation on the enzymes involved in the biosynthesis and metabolism of noradrenaline and adrenaline. *Psychosomatic Medicine,* 1971, *33,* 227–237.

Henry, J. P., Ely, D. L., & Stephens, P. M. The role of psychosocial stimulation in the pathogenesis of hypertension. *Verhandlung Deutschen Gesellschaft für Innere Medizin,* 1974, *80,* 1724–1740.

Henry, K. R. Androgenic seizure suceptibility induced in C57B1/6J mice by prior auditory exposure. *Science,* 1967, *158,* 938–940.

Houssay, A. B., Epper, C. E., & Pazo, J. H. In A. C. Lyne & B. F. Short (Eds.), *Biology of skin and hair growth.* New York: Plenum Press, 1965.

Hucklebridge, F. H., & Nowell, N. W. Plasma catecholamines response to physical and psychological aspects of fighting in mice. *Physiology and Behavior,* 1974, *13,* 35–40.

Hucklebridge, F. H., Nowell, N. W., & Dilks, R. A. Plasma catecholamine response to fighting in the male albino mouse. *Behavioral Biology,* 1973, *8,* 785–800.

Hucklebridge, F. H., Reid, A., Benton, D., & Brain, P. F. A comparison of the levels of adrenal catecholamines and aggression in isolated, dominant and subordinate mice. *I.R.C.S. Medical Science,* 1976, *4,* 154.

Hull, E. M., Chapin, E., & Kastaniotis, C. Effects of crowding and intermittent isolation on gerbils *(Meriones unguiculatus). Physiology and Behavior,* 1974, *13,* 723–727.

Hull, E. M., Hamilton, K. L., Engwall, D. B., & Rosselli, L. Effects of olfactory bulbectomy and peripheral deafferentation on reactions to crowding in gerbils, *Meriones unguiculatus. Journal of Comparative and Physiological Psychology,* 1974, *86,* 247–254.

Hull, E. M., Kastaniotis, C. L'Hommedieu, G., & Franz, J. Environmental enrichment and crowding: Behavioral and hormonal effects. *Physiology and Behavior,* 1976, *17,* 735–741.

Jean-Faucher, C., Berger, M., de Turckheim, M., Veyssiere, G., & Jean, C. Effets de la restriction sociale et de l'absence d'experience sexuelle sur la maturation sexuelle de la souris màle. *Physiology and Behavior,* 1978, *21,* 491–496.

Katz, R. V. Role of the mystacial vibrissal in the control of isolation induced aggression in the mouse. *Behavioral Biology,* 1976, *17,* 399–402.

Klopfer, P. H. *An introduction to animal behavior: Ethology's first century.* Englewood Cliffs, N. J.: Prentice-Hall, 1974.

Leshner, A. I., Moyer, J. A., & Walker, W. A. Pituitary-adrenocortical activity and avoidance-of-attack in mice. *Physiology and Behavior,* 1975, *15,* 689–693.

Levine, L., Diakow, C. A., & Barsel, G. E. Interstrain fighting in male mice. *Animal Behaviour,* 1965, *13,* 52–58.

Louch, C. D., & Higginbotham, M. The relation between social rank and plasma corticosterone levels in mice. *General and Comparative Endocrinology,* 1967, *8,* 411–414.

Luciano, D., & Lore, R. Aggression and social experience in domesticated rats. *Journal of Comparative and Physiological Psychology,* 1975, *88,* 917–923.

Mackintosh, J. H. Territory formation by laboratory mice. *Animal Behaviour,* 1970, *18,* 177–183.

McKinney, T. D., & Desjardins, C. Intermale stimuli and testicular function in adult and immature house mice. *Biology of Reproduction,* 1973, *9,* 370–378.

McKinney, T. D., & Pasley, J. N. Effects of social rank and social disruption in adult male house mice. *General and Comparative Endocrinology,* 1973, *20,* 579–582.

Modigh, K. Effects of isolation and fighting in mice on the rate of synthesis of noradrenaline, dopamine and 5-hydroxy-tryptamine in the brain. *Psychopharmacologia,* 1973, *33,* 1–17.

Moyer, K. E., & Korn, J. H. Behavioral effects of isolation in the rat. *Psychonomic Science,* 1965, *3,* 503–504.

Mugford, R. A. Intermale fighting affected by home cage odors of male and female mice. *Journal of Comparative and Physiological Psychology,* 1973, *84,* 289–295.

Mugford, R. A., & Nowell, N. W. Pheromones and their effect on aggression in mice. *Nature,* 1970, *226,* 967–968.

Ng, L. K. Y., Marsden, H. M., Colburn, R. W., & Thoa, N. B. Population density and social pathology in mice: Differences in catecholamine metabolism associated with differences in behavior. *Brain Research,* 1973, *59,* 323–330.

Parmigiani, S., Mainardi, D., & Pasquali, A. A comparison of aggressiveness in dominant, subordinate and isolated house mice. In P. F. Brain & D. Benton (Eds.), *The biology of aggression.* Alphen aan den Rijn, Netherlands: Noordhoff/Sijthoff, 1981.

Powell, D. A., & Buchanan, S. L. A comparison of several aggression models in Long-Evans (Hooded) rats: (i) effects of sex and isolation. *Psychological Reports,* 1978, *43,* 703–720.

Priestnall, R. The effects of litter size and post weaning isolation or grouping on adult emotionality in C3H mice. *Developmental Psychobiology,* 1973, *6,* 217–224.

Rosen, F., & Hart, F. Effects of early social isolation upon adult timidity and dominance in the peromyscus. *Psychological Reports,* 1963, *13,* 47–50.

Sales, G. D. Ultrasounds and aggressive behaviour in rats and other small mammals. *Animal Behaviour,* 1972, *20,* 88–100.

Schildkraut, J. J., & Kety, S. S. Biogenic amines and emotion. *Science,* 1967, *156,* 21–30.

Selye, H. *Hormones and resistance.* Berlin: Springer-Verlag, 1971.

Senault, B. Influence of isolation on the aggressive behaviour induced by apomorphine in the rat. *Psychopharmacologia,* 1971, *20,* 389–394.

Seward, J. P. Aggressive behaviour in the rat: 1. General characteristics: Age and sex differences. *Journal of Comparative Psychology,* 1945, *38,* 175–179.

Sieck, M. H., & Baumbach, H. D. Differential effects of peripheral and central anosmia producing techniques on spontaneous behaviour patterns. *Physiology and Behavior,* 1974, *13,* 407–425.

Sigg, E. B. Relationship of aggressive behaviour to adrenal and gonadal function in male mice. In S. Garattini & E. B. Sigg (Eds.), *Aggressive Behaviour.* Amsterdam: Excerpta Medica Foundation, 1969.

Sigg, E. B., Day, C., & Colombo, C. Endocrine factors in isolation-induced aggressiveness in rodents. *Endocrinology,* 1966, *78,* 679–684.

Spencer, J., Gray, J., & Dalhouse, A. Social isolation in the gerbil: Its effect on exploratory or agonistic behaviour and adrenocortical activity. *Physiology and Behavior,* 1973, *10,* 231–237.

Svare, B. B., & Leshner, A. I. Behavioral correlates of intermale aggression and grouping in mice. *Journal of Comparative and Physiological Psychology,* 1973, *85,* 203–201.

Takemoto, T. I., Suzuki, T., & Miyama, J. Effects of isolation on mice in relation to age and sex. *Tohoku Journal of Experimental Medicine,* 1975, *117,* 153–165.

Thiessen, D. D., Zolman, J. P., & Rodgers, D. A. Relation between adrenal weight, brain cholinesterase activity, and hole-in-wall behavior of mice under different living conditions. *Journal of Comparative and Physiological Psychology,* 1962, *55,* 186–190.

Thor, D. H. Intraspecific aggression in rats and dimensions of enclosure. *Psychological Reports,* 1976, *38,* 1254.

Valzelli, L. Aggressive behavior induced by isolation. In S. Garattini & E. B. Sigg (Eds.), *Aggressive Behaviour.* Amsterdam: Excerpta Medica Foundation, 1969.

Valzelli, L., & Garattini, S. Behavioral changes and 5-HT turnover in animals. *Advances in Pharmacology,* 1968, *68,* 249–260.

Vandenbergh, J. G. Eosinophil response to aggressive behaviour in CFW albino mice. *Animal Behaviour,* 1960, *8,* 13–18.

van Kreveld, D. A. Selective review of dominance-subordination relations in animals. *Genetic Psychology Monograph,* 1970, *81,* 143–173.

Veselovsky, J., & Novacky, M. Effects of size of territory on social behaviour, conflict situations and level of free amino acids of brain, myocardium and muscle of back leg of Golden hamster. *Biologia,* 1976, *31,* 851–860.

Watson, F. M. C., Henry, J. P., & Haltmeyer, G. L. Effects of early experience on emotional and social reactivity in CBA mice. *Physiology and Behavior,* 1974, *13,* 9–14.

Welch, A. S., & Welch, B. L. Reduction of norepinephrine in the lower brainstem by psychological stimulus. *Proceedings of the National Academy of Science,* 1968, *60,* 478–481.

Welch, A. S., & Welch, B. L. Isolation, reactivity and aggression: Evidence for an involvement of brain catecholamines and serotonin. In B. E. Eleftheriou & J. P. Scott (Eds.), *The physiology of aggression and defeat.* New York: Plenum Press, 1971.

Welch, B. L. Psychophysiological response to the mean level of environmental stimulation: A theory of environmental integration. In *Medical aspects of stress in the military climate.* Washington, D.C.: U.S. Government Printing Office, 1965.

Welch, B. L. Aggression and defense: Neural mechanisms and social patterns. In C. D. Clemente & D. B. Lindsley (Eds.), *Brain function.* Los Angeles: University of California Press, 1967.

Welch, B. L., & Welch, A. S. Effect of grouping on the level of brain norepinephrine in white Swiss mice. *Life Science,* 165, *4,* 1011–1018.

Welch, B. L., & Welch, A. S. Greater lowering of brain and adrenal catecholamines in group-housed than in individually-housed mice administered DL-alpha-methyltyrosine. *Journal of Pharmacy and Pharmacology,* 1968, *20,* 244–246.

Welch, B. L., & Welch, A. S. Aggression and the biogenic amines. In S. Garattini & S. B. Sigg (Eds.), *Aggressive Behavior.* Amsterdam: Excerpta Medica Foundation, 1969.

Whitney, G., Alpern, M., Dizinno, G., & Horowitz, G. Female odors evoke untrasounds from male mice. *Animal Learning and Behavior,* 1974, *2,* 13–18.

Wurtman, R. J., & Axelrod, J. Control of enzymatic synthesis of adrenaline in the adrenal medulla by adrenal cortical steroids. *Journal of Biological Chemistry,* 1966, *241,* 2301–2305.

Yen, H. C. Y., Day, C. A., & Sigg, E. B. Influences of endocrine factors on development of fighting behavior in rodents. *Pharmacologist,* 1962, *4,* 1973.

Zook, J. M., & Adams, D. B. Competitive fighting in the rat. *Journal of Comparative and Physiological Psychology,* 1975, *88,* 418–423.

14

Pheromonal Influence on Aggressive Behavior

Ching-tse Lee and David W. Ingersoll

Two functionally related biochemical modulators of animal agonistic behavior are the hormonal and pheromonal systems. These two communicative systems can be viewed as regulators or, perhaps more accurately, as modulators of physiobehavioral systems, such as agonistic behavior.

Unfortunately, the concepts of hormones, pheromones, and agonistic behaviors lack clear or universally acceptable definitions. These terms are currently used as generic concepts with the following general properties. A hormone is a physiological regulator having the properties of being synthesized by a living cell of a gland and is directly secreted into the bloodstream (see Martin, 1976). A pheromone is a substance synthesized by a living cell and is expelled into the animal's external environment; the substance has the property of inducing a specific reaction in a conspecific, that is, a species-specific physiobehavioral function (see Kirschenblatt, 1962). At present, the terms "pheromones," "chemosignal," and "chemical stimuli" will be used interchangeably. "Agonistic behavior" is a term used to encompass all potential behavioral components of a fight situation, for instance, behavioral signals, attack, submissive postures, and escape behaviors (see Scott, 1966).

In general, aggression-modulating pheromones are dependent on endocrine activities for both their biochemical synthesis and physiobehavioral effects. For example, castrated male mice lack a behaviorally active aggression-promoting urinary chemosignal, and the agonistic behavior of these animals is not enhanced by exposure to the pheromone (Lee, 1976; Sawyer, 1980). Related to the issues of pheromone synthesis and its physiobehavioral effects are the respective topics of pheromonal chemical characterization and chemosensory detection. These topics have been primarily investigated in the house mouse, *Mus musculus,* and

---

CHING-TSE LEE • Department of Psychology, Brooklyn College of the City University of New York, Brooklyn, New York 11210. DAVID W. INGERSOLL • Department of Psychology, Fordham University, Rose Hill Campus, Bronx, New York 10458. This chapter was supported by funds provided by the Departments of Psychology of Fordham University and Brooklyn College.

more in terms of reproductive than agonistic behaviors (see Doty, 1976). Consequently, our discussion will focus on murine aggression-modulating chemocommunicative systems.

PHEROMONE MODULATION OF AGGRESSIVE BEHAVIOR

Most behavioral and physiological tests of chemosignal activity do not assess for particular chemical substances. Rather, these procedures permit only an inference of a chemical compound's communicative properties—that is, chemical identification is rarely obtained and independently tested. Consequently, pheromonal systems are typically referred to by their biological (e.g., pregnancy blocking) or behavioral (e.g., aggression-promoting) activity and not by the mediating chemical substance. From this consideration, it should be clear that a specific chemical substance with communicative properties may mediate several species' chemocommunicative systems.

Basically, two paradigms have been used to demonstrate that a particular behavioral system is modulated by a chemosignal. The first behavioral paradigm assays for the presence of chemosensory cues by assuming that the chemical stimulus is an odorant (see Johns, 1980, for a critique) and therefore is detected by the primary olfactory system (cranial nerve I). It is further assumed that if removal of the olfactory bulbs results in the elimination of the behavior in question, then the behavior is modulated by chemoreceptive stimuli. In the latter case, it is presumed that the olfactory bulbs possess only sensory functions. As will later be discussed (see section on pheromone detection and aggression), the olfactory bulbectomy procedure usually denervates most, if not all, of the five nasal chemoreceptive systems (see Wysocki, 1979). Consequently, a loss of pheromone detection subsequent to olfactory bulbectomy should not be interpreted only as olfactory mediation, but could also be interpreted as intranasal chemoreceptive mediation of the chemosignal.

The second approach attempts to identify the chemosignal source (e.g., urine) and then modulates levels of the chemosignal in the source or medium. The murine aggression-promoting pheromone found in adult male urine will be used to exemplify this approach. It has been observed that females are attacked less frequently by adult male fighters than are other adult males (Lee & Brake, 1971). Also, adult males who were "deodorized" subsequently received fewer attacks by trained fighter males. The sniffing behavior of these fighters is typically directed toward the uroanal area, suggesting urine and feces as potential sources of the pheromone.

Subsequent work showed that both attack and urogenital sniff behaviors decrease when a male fighter is exposed to adult castrated males (Lee & Brake, 1972; Mugford & Nowell, 1970). However, urine from adult intact males was shown to possess aggression-promoting properties when coated on the urogenital area of a castrated male (Mugford & Nowell, 1970). It was also shown that administration of testosterone injections to castrated males restores the urinary chemosignal. These studies jointly considered clearly indicate that the pheromone is found in urine and is androgen mediated.

In addition to the male aggression-promoting pheromone, there is evidence for a female urinary aggression-inhibiting chemosignal. Adult females are attacked less frequently than are adult males (Lee & Brake, 1971), and males coated with female urine receive fewer attacks than non-urine-coated males (Mugford & Nowell, 1970). The hormone mediation of this female pheromone has yet to be resolved. Mugford and Nowell (1971c) have suggested that estradiol injections produce aggression-promoting cues, whereas Ferrera (1975) has shown that such a treatment is responsible for producing the aggression-inhibiting chemosignal.

SOURCES OF PHEROMONES

The area of chemocommunication has been extensively reviewed and evaluated (see Aron, 1979; Beauchamp, Doty, Moulton, & Mugford, 1976; Bronson, 1971; Brown, 1979; Cheal, 1975; Doty, 1976; Shorey, 1976). Müller-Schwarze (1967) has noted that specialized scent glands are present in 15 of 19 mammalian orders. Social chemosensory stimuli can be released into the external environment by (1) the secretion of specialized scent glands (see Thiessen & Rice, 1976) as well as by (2) excretory by-products, for instance, feces (Leidahl & Moltz, 1977; Moltz & Leidahl, 1977) and urine (Lee, 1976; Mackintosh & Grant, 1966); and by (3) exocrine glands, for example, salivary secretions (Lee & Ingersoll, 1979).

It is of interest that some accessory sexual glands have been shown to contain pheromone substances. These glands have been shown to be androgen dependent. Gonadectomy reduces the weights, and subsequent testosterone treatment restores the weights of these glands (Jones, 1977). Consequently, much attention has been given to accessory sex glands as sources of pheromone production.

PREPUTIAL GLAND

The preputial gland has been shown to contain substances that affect aggression (Mugford & Nowell, 1971a), odor preference (Thody & Dijkstra, 1978), and territorial marking (Gawienowski, DeNicola, & Stacewicz-Sapuntzakis, 1976). The preputial gland may produce chemical substances that facilitate aggressive behaviors (Jones & Nowell, 1973a). McKinney and Christian (1970) have shown that preputialectomy will subsequently reduce intermale aggression in male mice. Also, Mugford and Nowell (1971b) reported a preputial-secreted pheromone in mice stressed with electric shocks.

The preputial glands of female mice have been shown to produce sex attractant(s) (Bronson & Caroom, 1971). A similar finding was reported in female rats by Thody and Dijkstra (1978). The "attractiveness" of this secretion may deter male aggression in a potential reproductive encounter, although experimental evidence is needed for such an assertion.

Coagulating Glands

Jones and Nowell (1973a,b) have conducted a series of studies investigating the role of the coagulating glands in chemocommunication. These authors have provided evidence that these glands contain a substance that can activate urinary chemosignals. When tested alone, secretions of the coagulating gland have no behavioral effects; when mixed with urine obtained from a male's bladder, however, the composite stimulus was found to be aversive to male conspecifics. Furthermore, the coagulating glands were shown to release urinary aggression-inhibiting cues, whereas secretions of the preputial gland and seminal vesicles were found not to alter the chemocommunicative properties of adult male mouse urine. The contribution of the prostate and cowpers glands to pheromone production is not clear at present.

Facial Cues

Glands associated with the ears, eyes, and mouth appear also to be related to chemocommunication. Landauer, Liu, and Goldberg (1980) have shown an attractive function of ear-gland secretions in hamsters. Thiessen and his colleagues have reported that the harderian gland, which is located near the eye tear glands of the Mongolian gerbil, secretes a substance that mediates social recognition (Thiessen, Clancy, & Goodwin, 1976) and thermoregulation (Thiessen & Kittrell, 1980; Thiessen, Graham, Perkins, & Macks, 1977). In hamsters, a sexual dimorphism of these glands is evident. Males possess heavier glands that contain two cell types, while females possess glands with only one cell type (Payne, McGadey, Moore, & Thompson, 1977). Secretions from these glands were shown to have attractive properties to male conspecifics. Harderian substances from castrated males lack this behavioral property (Payne, 1979). Furthermore, a homogenate of harderian substances from a female hamster has aggression-inhibiting properties, whereas a similar preparation from males has no effect on aggressive behaviors. Consequently, female hamsters possess harderian substance(s) that have attractive as well as aggression-inhibiting properties.

The salivary gland has also been implicated as a source of chemosignal cues. Signoret and du Mesnil du Buisson (1961) have reported a sex pheromone in the boar's saliva that induces a mating stance in the sow. The putative active ingredient of the pheromone is 3α-hydroxy-5α-androst-16-ene (Melrose, Reed, & Patterson, 1971). For the mouse, Yee (1977) did not find chemosignal activity of mouse saliva-gland homogenates in an agonistic situation. However, Lee and Ingersoll (1979) have demonstrated for the first time that saliva of male mice is aggression promoting and that this behavioral effect is androgen dependent. The saliva of females did not alter male aggressive behavior (Ingersoll & Lee, unpublished observations).

Cues from Other Sources

Thiessen and Rice (1976) have reviewed the use of specialized skin glands in pheromone production of various mammalian species. For the gerbil, the ventral scent gland is used for territorial marking (Nyby, Thiessen, & Wallace, 1970) and recognition (Thiessen & Dawber, 1972). Also, these animals use harderian secretions in individual recognition tasks (Thiessen *et al.,* 1976). The hamster possesses flank glands for territorial marking. Of interest is that these specialized glands are responsive to seven androgen levels. Castration reduces the weight and size of the gerbil's ventral gland and the hamster's flank glands. Subsequent testosterone treatment restores glandular weights.

HORMONE ACTION OF PHEROMONE ACTIVITY

Androgen

Testosterone can be metabolized into several compounds, two of which are of interest to the present discussion: dihydrotestosterone (DHT) and estradiol-17β (E). The aromatization hypothesis suggests that aggressive behavior is mediated not by the direct actions of testosterone on brain sites but by its estrogen metabolite (see Chapters 9 and 10 for more details on the aromatization hypothesis). Also, evidence is beginning to show that DHT and not E mediates the production of the male aggression-promoting pheromone (cf. Simon & Gandelman, 1978). Thus, two metabolites of testosterone may have different physiological roles in aggressive activities.

Conspecific-aggressive behavior is more frequently observed among males than females of a species. Consequently, the male gonadal steroids—androgens—have become a target for independent investigation of aggressive behavior and pheromone activity. These two androgen processes may have different physiological mechanisms. Aggressive behavior appears to be mediated by aromatization of testosterone to E in the brain (Bowden & Brain, 1978; Christie & Barfield, 1979), whereas aggression-pheromone output is dependent on the presence not of E but of DHT (Ingersoll & Sikes, in preparation), a putative nonaromatizable testosterone.

Behavior

Bowden and Brain (1976) have shown that intermale aggressive behavior can be maintained in castrated male mice by daily injections of E but not with moderate dosages of DHT. With respect to accessory sexual organ weights, the reverse effect was obtained: DHT produced heavier (and presumably more active) accessory sexual glands than did administration of E. The findings of this study clearly suggest different mechanisms for the production of aggressive behavior and aggression-promoting cues.

Pheromone

The production of pheromonal substances does not appear to be dependent on aromatization of testosterone. Duvall, Schmitt, Erpino, and Rabedeau (1978) have reported that urinary aggression-promoting cues can be obtained in castrated mice subsequent to DHT implantation. Moreover, the aggression-promoting properties of urine from these males is different from that of castrates receiving testosterone implantation.

Antiandrogen and Antiestrogen

Findings obtained from studies using the antiandrogen cyproterone acetate (CA) provide further support for the hypothesis of independent hormone mechanisms for mediating aggressive behavior and aggression promoting cues.

This drug has been reported to have progestational (El Etreby, 1979), antiandrogenic (Neumann, von Berswordt-Wallrobe, Elger, Steinback, Hahn, & Kramer, 1970; de la Torre, Noreń, Hedman, & Diczfalusy, 1979), antigonadotropic (Neumann & Schenck, 1976), and cortical-like activities (Neumann & Steinback, 1974), as well as prolactin-stimulating properties (Herbert, Schuppler, Poggel, Grünzel, & El Etreby, 1977). This compound has also been reported to antagonize the growth of androgen-dependent accessory sexual glands (Jones, 1977; Jones & Nowell, 1974; Neumann & Schenck, 1976; Urry, Dougherty, & Cokett, 1976).

Behavior

With respect to the effects of CA on aggressive behavior, the literature is somewhat confusing. In mice, it has been clearly shown that intermale fighting is androgen mediated (Lee & Brake, 1972; Svare & Leshner, 1973). Kurischko and Oettel (1977) showed that 3 mg CA injected subcutaneously into intact adult males daily for at least four days was sufficient to reduce attack behavior to an equivalent of castration levels (see also Oettel & Kurischko, 1978). However, this finding was not supported by Brain, Benton, Goldsmith, and Bowden (1976), Brain and Bowden (1976), and Ingersoll, Lee, and Caroff (in preparation). These investigators and Luttge (1979) clearly support the aromatization hypothesis of androgen-mediated aggressive behavior. In general, administration of MER25, an antiestrogen, produced an aggression deficit, whereas CA injections had no significant effect on this behavior.

Ingersoll, Lee, and Caroff (in preparation) used the urine donors discussed above to test for behavioral effects of CA and MER25 on urinary pheromone synthesis. Directly opposite findings were obtained: (1) CA inhibits pheromone production but has no effect on aggressive behavior, and (2) MER25 shows no effect on pheromone production but reduces aggressive behaviors. Thus, the biomechanisms of these two processes are different and, perhaps, independent.

Pheromone

Since the accessory sexual glands have been reported to either synthesize, store, enzymatically activate, and/or release pheromonal activity (see Lee, 1976), it may be that CA intervenes in these processes. Bloch and Wyss (1973) showed that males who received daily 1-mg injections of CA had a reduced pregnancy block effect (8/50) relative to control males (21/50) who received only oil injections. Similarly, Pandey and Dominic (1978) completely eliminated the estrous induction phenomenon by administering 0.5 mg CA into adult males for 21 days. Perhaps more germane is the finding by Jones and Nowell (1974) that injections of 2 mg CA every 48 hours for 36 days eliminate the aversive pheromone in male urine that is normally present in the coagulating gland (Jones & Nowell, 1973b).

Ingersoll *et al.* (in preparation) have shown that CA but not MER25 inhibits the presence of the aggression-promoting substance(s). In this study, intact male animals received 0.5-, 1-, or 2-mg injections of either MER25 or CA for 7 days prior to urine collection and for 14 additional days during urine collection. The pooled urine for each condition was behaviorally assayed for aggression-promoting properties. Only urine obtained from donors injected with MER25 was shown to be behaviorally active. Thus, CA inhibited the production of this pheromone in excreted urine. These studies jointly suggest that pheromone production is mediated by testosterone and not estrogen metabolites.

With regard to male reproductive behavior, the aromatization hypothesis appears to be supported (Davidson, 1980). Gonadectomy eliminates ejaculatory behavior (Champlin, Blight, & McGill, 1963). Both E and DHT are weak stimulants of reproductive behavior when administered separately but can synergistically restore ejaculatory behavior when concurrently injected into castrated males (Baum & Vreeberg, 1973; Beyer, de la Torre, Larsson, & Pérez-Palacios, 1975). Moreover, it has been reported that cyproterone does not block the behavioral actions of testosterone (Whalen & Edwards, 1969), whereas the antiestrogen MER25 blocks the actions of testosterone (Komisaruk & Beyer, 1972; Luttge, 1975). These findings clearly indicate the role of aromatization of testosterone to estradiol in mediating male reproduction behavior.

Progesterone

Apart from its normal function in female reproductive physiology, progesterone (P) is also antiandrogenic in the somatic organ weight and in masculinization effects of androgen in small rodents (Cagnoni, Fantini, Morace, & Ghetti, 1965; Junkmann & Neumann, 1976; Kincl & Maqueo, 1965). Progesterone's antagonism of androgen function on behavior has also been reported in various animal species (Morin, 1977).

With regard to aggression, progesterone is antiandrogenic (Erpino, 1975; Erpino & Chappelle, 1971; Luttge, 1972; Soares, Kalberer, & Erpino, 1978). Progesterone also antagonizes androgen-dependent olfactory preference of a conspecific in the treated animals. However, progesterone's effect on aggression chemosignal output appears to be inconsistent in the literature.

Lee and Griffo (1974) and Lee, Griffo, Braunstein, Mars, and Stein (1976) have reported that progesterone also inhibits androgen's effect on production of aggression-promoting urinary cues. The effect was observed in freshly collected urine but not in aged urine. These investigators have suggested that progesterone may produce aggression-inhibiting cues or "attractive" cues to override aggression-provoking signals produced by androgen. Both hypotheses seem to be plausible, as supported by our findings that urine of progesterone-treated castrates was aggression inhibiting as compared with the chemosignal properties of water (Lee, 1976).

Soares, Kalberer, and Erpino (1977) have presented results different from those of Lee and Griffo (1974) and Lee *et al.* (1976). They have shown that the simultaneous implantation of testosterone and progesterone into anosmic castrated male mice did not reduce the likelihood of their being attacked by a conspecific male, whereas anosmic castrates implanted with testosterone alone were more likely to attack. Duvall *et al.* (1978) have found that progesterone did not inhibit aggression-promoting urinary cues in androgen- and progesterone-treated animals. The discrepancy of these results may be due to differences in hormone administration or in the dosages and mouse strains used by Lee and Erpino's associates.

Lee and Griffo (1973) have studied the early progesterone effect on pheromone output. The results showed that neonatal progesterone treatment results in two kinds of females, one with constant estrus and one with constant diestrus. The urine of constant-estrous females was aggression-inhibiting. Another synthetic progesterone, medroxyprogesterone, did not appear to alter the treated female's urinary output (Griffo, 1976).

Estrogen

The role of estrogen in pheromone production may be viewed from two vantage points—as the naturally occurring female hormone or as one of the metabolites of androgen. These different views may partially explain the confusion about the role of estrogen in pheromone production. If estrogen, as a female hormone, plays a role in social communication related to aggression and reproduction, then it should be expected to produce sex attractants and/or aggression-inhibiting cues. Thus, male mice, rats, dogs, hamsters, monkeys, cats, foxes, etc. (see Doty, 1976; Müller-Schwarze & Silverstein, 1980) were found to be attracted to female conspecifics in estrus, implicating an estrogen-dependent mechanism in sexual attraction. As far as aggression is concerned, Ferrera (1975) found that estrogen, when compared with a control substance, produces aggression-inhibiting cues in treated ovariectomized female mice.

Since testosterone can be converted to estradiol, the so-called androgenic effect can be produced by the latter hormone. For example, early estrogen treatment has been reported to masculinize treated females to exhibit male aggressiveness (Gorski, 1980). Luttge (1979) and Clark and Nowell (1979) found that an antiestrogen, CI-628, inhibits testosterone-stimulated aggression in mice. These findings seem to support the aromatization hypothesis of androgen action mentioned earlier. Mugford and Nowell's finding (1971c), however, was incongruent with the above mentioned hypothesis. They found that estrogen injection of ovariectomized females produced an aggression-promoting pheromone. Further work is needed to clarify the difference between the findings of Mugford and Nowell (1971c) and those of Ferrera (1975).

Melanocyte-Stimulating Hormone

The alpha melanocyte-stimulating hormone (MSH), a pituitary hormone, has recently been suggested to modulate aggressive behavior as well as aggression-promoting chemosignals. MSH was found to reduce marking behavior in gerbils (Meddis, Thody, Bowden, & Shuster, 1977) and to inhibit murine aggression in a paired fighting situation (Nowell, Thody, & Woodley, 1980; Paterson, Rickerby, Simpson, & Vickers, 1980). It is postulated by Paterson *et al.* (1980) that the antiaggression effect of MSH was due to its inhibition of melatonin secretion in the pineal gland.

In addition to modulation of aggressive behavior, MSH was shown to be sebotrophic, namely, it affects sebum secretion in rodents (Thody & Shuster, 1975). Drugs that facilitate or inhibit MSH output also affect sebum output in a similar manner (Thody & Shuster, 1977). Thus, MSH may also modulate aggressive chemosignals. Nowell and Wouters (1975) have reported that MSH enhances the release of urinary aggression-promoting cues in treated mice. It was postulated that MSH produces its effect through a preputial gland–urinary axis; however, definite experimental evidence is lacking.

ACTIONS OF PHEROMONES ON HORMONES

Under the rubric of aggression, no work has been performed regarding the effects of pheromone detection on subsequent hormone activity. The potential of such research can be exemplified in the work by Bronson in puberty onset, Macrides on male testosterone levels, and Bronson on aggressive defeat.

Bronson and Desjardins (1974) examined the male puberty-accelerating pheromone in juvenile female mice (see also Vandenbergh, 1976; Wilson, Beamer, & Whitten, 1980). Typically, females exposed to male "odorants" will become sexually mature more rapidly than females not receiving male exposure. Bronson and Desjardins (1974) periodically measured plasma leutinizing hormone (LH), follicle stimulating hormone (FSH), and progesterone by radioimmunoassay in females either exposed or not exposed to adult male cues. The assays of plasma from these females clearly suggest an LH role in this phenomenon. Within three hours after male exposure, LH showed a temporary elevation in plasma concentration followed by a 15- to 20-fold increase in estradiol. No change in FSH serum concentration was observed. Consequently, these authors have proposed that male chemosignals may attain their effects through the actions of LH.

Macrides, Bartke, and Dalterio (1975) have shown that male mice will increase their plasma testosterone levels 30 to 60 minutes after odorant exposure to an unfamiliar female. This effect was not obtained if a female cagemate was removed and a male placed in the cage of the test male. Thus, it may be concluded that female chemosignals can have a hormonal effect on male hormone activity.

Bronson and Eleftheriou (1964, 1965a) have provided evidence that the experience of defeat can have adrenal-stimulating effects. In both studies, mice who repeatedly experienced defeat developed enlarged adrenal glands and increased levels of circulating corticosterones. Bronson and Eleftheriou (1965b) have also shown that this hormonal response to defeat is not dependent on continual physical contact with the dominant male. However, the communicative source of "continued threat," whether, odorant, visual, and/or auditory stimuli, is not clear. Demonstration of a chemosignal source would support our argument for a pheromone-modulating system of hormone activity in male mice. Such investigations in the area of aggression would aid in the understanding of the physiological effects of pheromones on hormone activity.

PARTIAL PURIFICATION AND CHARACTERIZATION OF A PHEROMONE

The chemical identification of pheromones is currently being performed on a variety of mammalian species (see Preti, Smith, & Beauchamp, 1977). For rodent species, work is being conducted on the hamster (Singer, Macrides, & Agosta, 1980), guinea pig (Berüter, Beauchamp, & Muetterties, 1973), and rabbit (Hesterman, Goodrich, & Mykytowycz, 1976). For the mouse, isolation procedures have been implemented to uncover the chemical structure of the aggression-promoting pheromone (Lee, 1976; Lee, Lukton, Bobotas, & Ingersoll, 1980), puberty-accelerating pheromone (Novotny, Jorgenson, Carmack, Wilson, Boyse, Yamazaki, Wilson, Beamer, & Whitten, 1980; Vandenbergh, Whitsett, & Lombardi, 1975), and pregnancy-blocking pheromone (Marchlewska-Koj, 1980). Of interest is the work performed on the aggression-modulating pheromone, predominantly by Lee and his colleagues. This work is presented below and confirms the findings previously reported by Lee (1976) with regard to its relative volatility.

Separation Techniques

For several years, Lee and colleagues (Ingersoll, Bobotas, Lee, & Lukton, 1982; Lee *et al.*, 1980) have been attempting to isolate and characterize the male mouse urinary aggression-promoting chemosignal. Although the pheromone has not been identified, the mode of its partial purification and the properties of this sample will be discussed.

Thin layer chromatography (TLC) has shown that the behaviorally active component of a dichloromethane male urine extract migrates (silica Gel G: CH_2Cl_2) with an R_F of 0.76 (Lee *et al.,* 1980). In addition, the remaining areas of the TLC plate were found not to contain a behaviorally active chemosignal—at least not detected by our behavioral assay. Subsequent examination of this zone by gas chromatography (GC) indicated 4 major and 10 minor peaks as monitored by flame ionization detection.

Analyses of a dichloromethane extract by high pressure liquid chromatography (HPLC) revealed three peaks, detection at 254 nm (Bobotas, Ingersoll, Lee & Lukton, 1979). Only one of these peaks was found to be behaviorally active. This active fraction was subsequently submitted to TLC for further separation and behavioral tests. It was found that only the zone corresponding to an $R_F = 0.76$ was behaviorally active. To confirm our finding, a TLC separation was performed using fresh unadulterated intact urine. Behavioral activity was found only in a spot at $R_F = 0.76$. Subsequently, this spot was eluted and applied to HPLC. A peak identical in elution volume with the previous behaviorally active HPLC peak was found. Thus, the chemosignals separated by HPLC and TLC appear to be identical. Currently, the TLC active zone has been further separated by eluting this spot for a second TLC analysis.

Chemosignal Properties

The chemosignal sample is relatively volatile. Urine will lose its aggression-promoting properties unless stored frozen (Lee & Griffo, 1974). A behaviorally active sample can be obtained by cryogenically trapping incubated acidified urine. Moreover, this sample has aggression-promoting properties similar to those of untreated intact-male urine (Lee *et al.,* 1980).

The chemosignal sample is lipophilic, since it is easily extracted with dichloromethane (Lee, 1976). A chemosignal sample obtained from TLC was shown to contain a lipid (iodine reagent-positive), a sterol ($SbCl_5$-reagent positive), and a phenolic compound (Folin reagent-positive). These properties, however, refer to a sample from a TLC spot and not a single compound. Recently, we have found that the chemosignal is present in bladder urine as a conjugate of glucuronide (Ingersoll *et al.,* 1982). This behaviorally inactive form of the chemosignal can be converted to the active-free chemosignal ·β-glucuronidase treatment.

In summary, the androgen dependence of the aggression-promoting chemosignal may be due to (1) testosterone as a precursor substrate or (2) the need of testosterone to activate pheromone synthesis and/or (3) to activate a transport system to make a lipid more soluble in urine. Evidence reviewed above suggests that all of these systems are employed in the synthesis and release of the aggression-promoting pheromone.

PHEROMONE DETECTION AND AGGRESSION

Rowe and Edwards (1971) have shown that removal of the olfactory bulbs in male mice subsequently eliminates attack behavior. In a later report, Edwards, Thompson, and Burge (1972) provided evidence to confirm this finding and also reported no significant change in aggressive behavior subsequent to peripheral deafferentation by intranasal zinc sulfate irrigation. These authors attributed the observed behavioral effects of bulbectomy to the elimination of nonsensory olfactory bulb functions (see Alberts, 1974; Edwards, Schlosberg, McMaster, & Harvey, 1977). The discrepancy between the results of these two deafferentation procedures can, however, be attributed to functional vomeronasal organs (of Jacobson) subsequent to peripheral but not central deafferentation (see Ingersoll, 1980; Meredith, 1980; Singh, Tucker, & Hofer, 1976; Winans & Powers, 1977). These investigators have provided both behavioral and histological evidence that the vomeronasal organ is not affected by intranasal zinc sulfate irrigation. Consequently, the vomeronasal organ may mediate the urinary aggression-promoting pheromone.

The murine vomeronasal organ (of Jacobson) is a cul-de-sac located anterior to the "septal window" (Keleman, 1953) of the nasal septum. In the mouse, it is approximately 4 mm from the external nare and approximately 4 mm in length. The receptor cells, found on the medial wall of the organ's lumen, project primary sensory fibers to only the accessory olfactory bulb. The latter structure directly innervates the medial and posteromedial sectors of the cortical medial amygdala. Tertiary fibers project via the stria terminalis to the medial preoptic area and to cells surrounding the ventromedial nucleus of the hypothalamus. These neurostructures have been shown to modulate reproductive functions (see Wysocki, 1979). Consequently, the vomeronasal organ is neuroanatomically associated with neural structures that modulate social behaviors. In turn, these social behaviors have been shown to be regulated by pheromonal activity. Thus, the function of the vomeronasal organ may be to modulate neurophysiological processes by the detection of chemocommunicative stimuli.

The primary olfactory system is also associated with limbic structures (Scalia & Winans, 1976). For example, the lateral hypothalamic area is innervated by fibers from the entorhinal and basolateral areas via the amygdalofugal pathway. Consequently, the olfactory system may also modulate pheromone-mediated behaviors.

The vomeronasal system has been shown by Ingersoll (1980, 1983) and others to mediate various mouse reproduction-related chemosignal systems such as (1) pregnancy blocking, (2) estrus synchrony, (3) estrus suppression (Reynolds & Keverne, 1979,) (4) puberty acceleration (Kaneko, Debski, Wilson, & Whitten, 1980), and (5) mating behavior. In addition, Ingersoll (1980, 1983) has provided evidence that the vomeronasal system also mediates murine aggression-promoting cues.

Using standard opponent-test and castrated males as stimulus animals, Ingersoll (1981) has shown that naive test males with lesioned vomeronasal organs but not olfactory peripheral deafferentation eliminate certain aggressive behaviors such as bites, chases, and attacks and significantly reduce urogenital sniff behaviors. Also, joint vomeronasal lesioning and olfactory destruction completely eliminate the above agonistic behaviors. However, for males who previously received fighter training experience, all deafferentation procedures (vomeronasal lesioning, olfactory mucosa deafferentation, and joint vomeronasal lesioning with olfactory mucosa deafferentation) significantly reduce but do not eliminate aggressive behavior. The joint deafferentation procedure produces the greatest behavioral decrement, and the vomeronasal lesioning and olfactory mucosa deafferentation produce similar behavioral deficits. These findings vis-a-vis the results of the nonfighter-trained males indicate that fighter training profoundly affects the role of each chemosensory system in mediating pheromone activity. The significant decrement in aggression subsequent to destruction of either the vomeronasal organ or olfactory mucosa indicates a mutual dependence between the two systems. Training experience may have modified the total stimulus complex to which the animal attends and/or the role of each sensory system in chemosensory detection.

It is not clear if the above fighter-training effect on chemoreception occurs under more natural situations. Ingersoll (1981) has shown that male mice can lose their social dominance status after vomeronasal lesioning but not after olfactory mucosa zinc sulfate treatment. In this experiment, four males were group-housed for five days to obtain a socially dominant male. The latter animals from different cages subsequently received the various intranasal deafferentation procedures. Eighty percent of the males who received vomeronasal lesioning and 95% of joint deafferented males lost their social dominance status. None of the males who received only olfactory musosa zinc sulfate treatment as well as none of the sham-treated males lost their social status. It may be concluded that the vomeronasal organ predominantly mediates the aggression-promoting pheromone. Perhaps fighter training creates a different sensory pheromone-detection system.

Ingersoll (1980) also investigated the role of the vomeronasal organ in a pheromone-discrimination task, with attack behavior used as a measure of discrimination performance. Males were tested daily in their home cages, with a submissive male intruder, either intact or castrated. Within a few days, castrated male intruders were attacked significantly less than intact male intruders, demonstrating sequential discrimination of the intruder's pheromones. Subsequent to deafferenting the test males, vomeronasal lesioning produced a greater deficit in attack behavior than olfactory deafferentation. Also, the latter treatment effect was not significantly different from sham male performance. Consequently, the vomeronasal organ has explicitly been shown to mediate chemosensory discrimination.

These studies performed in our laboratories clearly indicate vomeronasal sensory mediation of chemocommunicative substances. As with reproductive behavior (see Wysocki, 1979), the vomeronasal organ may mediate aggressive behavior by its influence on neuroendocrine systems. A pheromone–vomeronasal–hormone model, however, is in need of supportive evidence.

CONCLUDING REMARKS

Hormones and pheromones may be viewed as two different aspects of chemical communication. Consequently, specific definitions of these substances are clouded by the ambiguities of the concept of communication (Burghardt, 1970; Cullen, 1972). Barnett (1977) has suggested that the latter concept should not be defined but heuristically treated as a paradigm for the study of biological interaction—in other words, communication is viewed as a heuristic tool. Of interest is the relative heuristic value in analyzing hormones and pheromones under a communication paradigm, compared with other paradigms such as "systems analysis." Unfortunately, evidence is not available to evaluate such a query.

Ingersoll (1980) views both hormones and pheromones as modulators of physiobehavioral systems within the context of a social structure (see Mykytowycz, 1974). Consequently, he examined chemosignal activity in laboratory-established mouse colonies (i.e., four males and four females). In dominant males, destruction of selective intranasal chemoreceptive systems changes their social status by decreasing their reproductive and agonistic activities. This study showed that disruption of pheromone detection in dominant male mice modifies various social interactions and the social structure within the colony.

Of great complexity are the potentially interacting effects of hormones, pheromones, and physiobehavioral systems. Currently presented is evidence that gonadal and other hormones can modulate levels of the murine aggression-promoting pheromone. The converse statement cannot be asserted, since experimental evidence has not been provided. With regard to reproductive systems, hwoever, there is unambiguous evidence that pheromones can modulate endocrine activity (Bruce, 1959; Svare, Bartke, & Macrides, 1978; Vandenbergh, 1969; Van der Lee & Boot, 1956; Whitten & Bronson, 1970). Consequently, there may be aggression-modulating pheromone–hormone systems.

The behaviors that may affect hormonal and/or pheromonal systems are also of interest. With reference to agonistic behavior, Bronson and his colleagues (see Bronson, 1976) have shown that urine-marking behavior of male mice can be reduced by social subordination. Of concern is whether this experience also depresses pheromone production, perhaps through a neuroendocrine system. A requisite for such a study is chemical identification of the urinary aggression-promoting pheromone. Evidence of depressed pheromone production would be indicative of a behavior modifying a pheromonal system.

An interesting query derived from studies on the sensory mediation of pheromones is whether all chemosignals are mediated by the vomeronasal organ. This is not to suggest a sensory mediation dichotomy between the olfactory and vomeronasal systems. More likely, there are varying degrees of mediation of these two and other intranasal chemoreceptive systems. Consequently, it should be clear that a pheromone is not an olfactory cue unless the substance is shown to be exclusively mediated by the olfactory system.

REFERENCES

Alberts, J. R. Producing and interpreting experimental olfactory deficits. *Physiology and Behavior,* 1974, *12,* 657–670.

Aron, C. Mechanisms of control of the reproductive function by olfactory stimuli in female mammals. *Physiological Reviews,* 1979, *59,* 229–283.

Barnett, C. Aspects of chemical communication with special reference to fish. *Bioscience Communication,* 1977, *3,* 331–392.

Baum, M. J., & Vreeburg, J. T. M. Copulation in castrated male rats following combined treatment with estradiol and dihydrotestosterone. *Science,* 1973, *182,* 283–285.

Beauchamp, G. K., Doty, R. L., Moulton, D. G., & Mugford, R. A. The pheromone concept in mammalian chemical communication: A critique. In R. L. Doty (Ed.), *Mammalian olfaction, reproductive processes and behavior.* New York: Academic Press, 1976.

Berüter, J., Beauchamp, G. K., & Muetterties, E. L. Complexity of chemical communication in mammals. Urinary components mediating sex discrimination by male guinea pigs. *Biochemical and Biophysical Research Communication,* 1973, *53,* 264–271.

Beyer, C., de la Torre, L., Larsson, K., & Pérez-Palacios, G. Synergistic actions of estrogen and androgen on the sexual behavior of the castrated male rabbit. *Hormones and Behavior,* 1975, *6,* 301–306.

Block, S., & Wyss, H. I. An anti-androgen (cyproterone acetate) inhibits the pregnancy block in mice caused by the presence of strange males. *Journal of Endocrinology,* 1973, *59,* 365–366.

Bobotas, G., Ingersoll, D. W., Lee, C. T., & Lukton, A. *Purification and characterization of urinary chemosignals from Mus musculus.* Paper presented at the Northeast Regional Meeting of the Animal Behavior Society, Woods Hole, Massachusetts, 1979.

Bowden, N. J., & Brain, P. F. Maintenance of intermale fighting in castrated "agressive mice" by sex steroids: A dose response study. *Journal of Endocrinology,* 1976, *71,* 90–91.

Bowden, N. J., & Brain, P. F. Blockade of testosterone-maintained intermale fighting in albino laboratory mice by an aromatization inhibitor. *Physiology and Behavior,* 1978, *20,* 543–546.

Brain, P. F., & Bowden, N. J. Are androgens converted into estrogens before they exert their behavioral influences via the brain. *Neuroscience Letters,* 1976, *3,* 78–79.

Brain, P. F., Benton, D., Goldsmith, J. F., & Bowden, N. J. Influences of cyproterone acetate and ethamoxytriphetol on fighting behavior and sex accessory weights of castrated, testosterone-implanted or sham-implanted "aggressive mice." *Journal of Endocrinology,* 1976, *69,* 16.

Bronson, F. H. Rodent pheromones. *Biology of Reproduction,* 1971, *4,* 344–357.

Bronson, F. H. Urine marking in mice: Causes and effects. In R. L. Doty (Ed.), *Mammalian olfaction, reproductive processes and behavior.* New York: Academic Press, 1976.

Bronson, F. H., & Caroom, D. Preputial gland of the male mouse: Attractant function. *Journal of Reproduction and Fertility,* 1971, *25,* 279–282.

Bronson, F. H., & Desjardins, C. Circulating concentration of FSH, LH, estradiol, and progesterone associated with acute, male-induced puberty in female mice. *Endocrinology,* 1974, *94,* 1658–1668.

Bronson, F. H., & Eleftheriou, B. E. Chronic physiological effects of fighting in mice. *General and Comparative Endocrinology,* 1964, *4,* 9–14.

Bronson, F. H., & Eleftheriou, B. E. Relative effects of fighting on bound and unbound corticosterone in mice. *Proceedings of the Society of Experimental Biology and Medicine,* 1965, *118,* 146–149. (a)

Bronson, F. H., & Eleftheriou, B. E. Adrenal response to fighting in mice: Separation of physical and psychological causes. *Science,* 1965, *147,* 627–628. (b)

Brown, R. E. Mammalian social odors: A critical review. *Advances in the Study of Behavior,* 1979, *10,* 103–162.

Bruce, H. M. An exteroceptive block to pregnancy in the mouse. *Nature,* 1959, *184,* 105.

Burghardt, G. M. Defining communication. In W. J. Johnston, Jr., D. G. Moulton, & A. Turk (Eds.), *Advances in chemoreception* (Vol. 1): *Communication by chemical signals.* New York: Appleton-Century-Crofts, 1970.

Cagnoni, M., Fantini, F., Morace, G., & Ghetti, A. Failure of testosterone propionate to induce the early-androgen syndrome in rats previously injected with progesterone. *Journal of Endocrinology,* 1965, *33,* 527–528.

Champlin, A. K., Blight, W. C., & McGill, T. E. The effects of varying levels of testosterone on the sexual behavior of the male mouse. *Animal Behaviour,* 1963, *11,* 244-245.

Cheal, M. Social olfaction: A review of the ontogeny of olfactory influences on vertebrate behavior. *Behavioral Biology,* 1975, *15,* 1–25.

Christie, M. H., & Barfield, R. J. Effects of aromatizable androgens on aggressive behaviour among rats *(Rattus norvegicus). Journal of Endocrinology,* 1979, *83,* 17–26.

Clark, C. R., & Nowell, N. W. Effect of the antiandrogen cyproterone acetate on neural testosterone binding and on intermale aggressive behaviour in male mice. *Journal of Endocrinology,* 1979, *81,* 137–138.

Cullen, J. M. Some principles of animal communication. In R. A. Hinde (Ed.), *Non-verbal communication.* London: Cambridge University Press, 1972.

Davidson, J. M. Hormones and sexual behavior in the male. In D. T. Krieger & J. C. Hughes (Eds.), *Neuroendocrinology.* Sunderland, Mass.: Sinauer Associates, 1980.

Doty, R. (Ed.). *Mammalian olfaction, reproductive processes and behavior.* New York: Academic Press, 1976.

Duvall, D., Schmitt, C. V., Erpino, M. J., & Rabedeau, R. G. Androgen and concurrent androgen-progesterone maintenance of attack-eliciting characteristics in male mouse urine. *Behavioral Biology,* 1978, *22,* 343–353.

Edwards, D. A., Thompson, M. L., & Bruge, K. G. Olfactory bulb removal vs peripherally induced anosmia: Differential effects on the aggressive behavior of male mice. *Behavioral Biology,* 1972, 7, 823–828.

Edwards, D. A., Schlosberg, A. J., McMaster, S. E., & Harvey, J. A. Olfactory system damage and brain catecholamines in the rat. *Brain Research,* 1977, *121,* 121–130.

El Etreby, M. F. Effect of cyproterone acetate, levonorgestrel and progesterone on adrenal glands and reproductive organs in the beagle bitch. *Cell and Tissue Research,* 1979, *200,* 229–243.

Erpino, M. J. Androgen-induced aggression in neonatally androgenized female mice: Inhibition by progesterone. *Hormones and Behavior,* 175, *6,* 149–157.

Erpino, M. J., & Chappelle, T. C. Interactions between androgens and progesterone in mediation of aggression in the mouse. *Hormones and Behavior,* 1971, *2,* 265–272.

Ferrera, D. *The role of estrogen in the production of urinary aggression-inhibiting cues.* Paper presented at the meeting of the Eastern Psychological Association, New York, 1975.

Gawienowski, A. M., DeNicola, D. B., Stacewicz-Sapuntzakis, M. Androgen dependence of a marking pheromone in rat urine. *Hormones and Behavior,* 1976, 7, 401–405.

Gorski, R. A. Sexual differentiation of the brain. In D. T. Krieger & J. C. Hughes (Eds.), *Neuroendocrinology.* Sunderland, Mass.: Sinauer Associates, 1980.

Griffo, W. *Disruption of the processes of sex differentiation by progestational steroids.* Unpublished doctoral dissertation, Graduate School and University Center of the City University of New York, 1976.

Herbert, D. C., Schuppler, J., Poggel, A., Grünzel, P., & El Etreby, M. F. Effect of cyproterone acetate on prolactin secretion in the female rhesus monkey. *Cell and Tissue Research,* 1977, *183,* 51-60.

Hesterman, E. R., Goodrich, B. S., & Mykytowycz, R. Behavioral and cardiac responses of the rabbit, *Oryctolagus cuniculus* to chemical fractions from anal gland. *Journal of Chemical Ecology,* 1976, *2,* 25–37.

Ingersoll, D. W. *Role of the Jacobson's organ in murine priming and signaling chemocommunication systems.* Unpublished doctoral dissertation, Graduate School and University Center of the City University of New York, 1980.

Ingersoll, D. W. *Vomeronasal organ mediation of the murine aggression-promoting pheromone.* Manuscript submitted for publication, 1983.

Ingersoll, D. W. & Sikes, C. *Gonadal steroid modulation of both social behaviors and urinary chemocommunicative substances.* In preparation.

Ingersoll, D. W., Bobotas, G., Lee, C. T., & Lukton, A. β-Glucuronidase activation of latent aggression-promoting cues in mouse bladder urine. *Physiology and Behavior,* 1982, *29,* 789–793.

Ingersoll, D. W., Lee, C. T., & Caroff, A. *Cyproterone acetate and MER25 modulation of murine chemocommunication and agonistic behavior.* In preparation.

Johns, M. A. The role of the vomeronasal system in mammalian reproductive physiology. In D. Müller-Schwarze & R. M. Silverstein (Eds.), *Chemical signals vertebrates and aquatic invertebrates.* New York: Plenum Press, 1980.

Jones, R. B. Effects of testosterone, testosterone metabolites and anti-androgens on the function of the male accessory glands in the rabbit and rat. *Journal of Endocrinology,* 1977, *74,* 75–88.

Jones, R. B., & Nowell, N. W. Effects of preputial and coagulating gland secretions upon aggressive behaviour in male mice: A confirmation. *Journal of Endocrinology,* 1973, *59,* 203–204.(a)

Jones, R. B., & Nowell, N. W. The coagulating glands as a source of aversive and aggression-inhibiting pheromone(s) in the male albino mouse. *Physiology and Behavior,* 1973, *11,* 455–462.(b)

Jones, R. B., & Nowell, N. W. Effects of cyproterone acetate upon urinary aversive cues and accessory sex glands in male albino mice. *Journal of Endocrinology,* 1974, *62,* 167–168.

Junkmann, K., & Neumann, F. Mechanism of action of progesterone having an antimasculine effect on fetuses. *Acta Endocrinologica Supplementum,* 1964, *90,* 139–154.

Kaneko, N., Debski, E. A., Wilson, M. C., & Whitten, W. K. Puberty acceleration in mice: II. Evidence that the vomeronasal organ is a receptor for the primer pheromone in male mouse urine. *Biology of Reproduction,* 1980, *22,* 873–878.

Keleman, G. Nonexperimental nasal and paranasal pathology in hereditarily obese mice. *Archives of Otolaryngology,* 1953, *57,* 143–151.

Kincl, F. A., & Maqueo, M. Prevention by progesterone of steroid-induced sterility in neonatal male and female rats. *Endocrinology,* 1965, *77,* 859–862.

Kirschenblatt, J. Terminology of some biologically active substances and validity of the term "pheromones." *Nature,* 1962, *195,* 916–917.

Komisaruk, B. R., & Beyer, C. Differential antagonism, by MER-25, on behavioral and morphological effects of estradiol benzoate in rats. *Hormones and Behavior,* 1972, *3,* 63–70.

Kurischko, A., & Oettel, M. Androgen-dependent fighting behavior in male mice. *Endokrinologie,* 1977, *70,* 1–5.

Landauer, M. R., Liu, S., & Goldberg, N. Responses of male hamsters to the ear gland secretion of conspecifics. *Physiology and Behavior,* 1980, *24,* 1023–1026.

Lee, C. T. Agonistic behavior, sexual attraction, and olfaction in mice. In R. L. Doty (Ed.), *Mammalian olfaction, reproductive processes and behavior.* New York: Academic Press, 1976.

Lee, C. T., & Brake, S. C. Reactions of male fighters to male and female mice, untreated or deodorized. *Psychonomic Science,* 1971, *24,* 209–211.

Lee, C. T., & Brake, S. C. Reaction of male mouse fighters to male castrates treated with testosterone propionate or oil. *Psychonomic Science,* 1972, *27,* 287–288.

Lee, C. T., & Griffo, W. Early androgenization and aggression pheromone in inbred mice. *Hormones and Behavior,* 1973, *4,* 181–189.

Lee, C. T., & Griffo, W. The progesterone antagonism of androgen-dependent aggression-promoting pheromone in inbred mice *(Mus musculus). Journal of Comparative and Physiological Psychology,* 1974, *87,* 150–155.

Lee, C. T., & Ingersoll, D. W. Salivary cues in the mouse: A preliminary study. *Hormones and Behavior,* 1979, *12,* 20–29.

Lee, C. T., Griffo, W., Braunstein, A., Mars, H., & Stein, J. Progesterone antagonism of aggression-promoting olfactory signals: A time-dependent phenomenon. *Physiology and Behavior,* 1976, *17,* 319–323.

Lee, C. T., Lukton, A., Bobotas, G., & Ingersoll, D. W. Partial purification of male *mus musculus* urinary aggression-promoting chemosignal. *Aggressive Behavior,* 1980, *6,* 149–160.

Leidahl, L. C., & Moltz, H. Emission of the maternal pheromone in nulliparous and lactating females. *Physiology and Behavior,* 1977, *18,* 399–402.

Luttge, W. G. Activation and inhibition of isolation induced inter-male fighting behavior in castrated male CD-1 mice treated with steroidal hormones. *Hormones and Behavior,* 1972, *3,* 71–81.

Luttge, W. G. Effects of antiestrogens on testosterone stimulated male sexual behavior and peripheral target tissues in the castrate male rat. *Physiology and Behavior,* 1975, *14,* 839–846.

Luttge, W. G. Anti-estrogen inhibition of testosterone-stimulated aggression in mice. *Experientia,* 1979, *35,* 273.

Mackintosh, J. H., & Grant, E. C. The effect of olfactory stimuli on the agonistic behaviour of laboratory mice. *Zeitschrift für Tierpsychologie,* 1966, *23,* 584–587.

Macrides, F., Bartke, A., & Dalterio, S. Strange females increase plasma testosterone levels in male mice. *Science,* 1975, *189,* 1104–1106.

Marchlewska-Koj, A. Partial isolation of pregnancy block pheromone in mice. In D. Müller-Schwarze & R. M. Silverstein (Eds.), *Chemical signals: Vertebrates and aquatic invertebrates.* New York: Plenum Press, 1980.

Martin, C. R. *Endocrine physiology.* Baltimore: Williams & Wilkins, 1976.

McKinney, T. D., & Christian, J. J. Effect of preputialectomy on fighting behavior in mice. *Proceedings of the Society of Experimental Biology and Medicine,* 1970, *134,* 291–293.

Meddis, D., Thody, A. J., Bowden, P. E., & Shuster, S. α-melanocyte stimulating hormone and territorial marking in the mongolian gerbil. *Journal of Endocrinology,* 1977, *72,* 38–39.

Melrose, D. R., Reed, H. C. E., & Patterson, R. L. S. Androgen steroids associated with boar odour as an aid to the detection of oestrus in pig artificial insemination. *British Veterinary Journal,* 1971, *127,* 497–502.

Meredith, M. The vomeronasal organ and accessory olfactory system in the hamster. In D. Müller-Schwarze & R. M. Silverstein (Eds.) *Chemical signals: Vertebrates and aquatic invertebrates.* New York: Plenum Press, 1980.

Moltz, H., & Leidahl, L. C. Bile, prolactin and the maternal pheromone. *Science,* 1977, *196,* 81–83.

Morin, L. P. Progesterone: Inhibition of rodent sexual behavior. *Physiology and Behavior,* 1977, *18,* 701–715.

Mugford, R. A., & Nowell, N. W. Pheromones and their effect on aggression in mice. *Nature,* 1970, *226,* 967–968.

Mugford, R. A., & Nowell, N. W. The preputial glands as a source of aggression-promoting odors in mice. *Physiology and Behavior,* 1971, *6,* 247–249.(a)

Mugford, R. A., & Nowell, N. W. Shock-induced release of the preputial gland secretions that elicit fighting in mice. *Journal of Endocrinology,* 1971, *15,* 16–17.(b)

Mugford, R. A., & Nowell, N. W. Endocrine control over production and activity of the anti-aggression pheromone from female mice. *Journal of Endocrinology,* 1971, *49,* 225–232.(c)

Müller-Schwarze, D. Social odors in young male deer. *American Zoologist,* 1967, *7,* 807.

Müller-Schwarze, D., & Silverstein, R. M. (Eds.). *Chemical signals: Vertebrates and aquatic invertebrates.* New York: Plenum Press, 1980.

Mykytowycz, R. Odor in the spacing behavior of mammals. In M. C. Birch (Ed.), *Pheromones.* London: North-Holland, 1974.

Neumann, F., & Schenck, B. New antiandrogens and their mode of action. *Journal of Reproduction and Fertility Suppl. 24,* 1976, 129–145.

Neumann, F., & Steinback, H. Antiandrogens. In O. Eichler, A. Farah, H. Herken, & A. D. Welch (Eds.), *Handbook of experimental pharmacology.* New York: Springer-Verlag, 1974.

Neumann, F., von Berswordt-Wallrobe, R., Elger, W., Steinback, H., Hahn, J. D., & Kramer, M. Aspects of androgen-dependent events as studied by antiandrogens. *Recent Progress in Hormone Research,* 1970, *26,* 337–410.

Novotny, M., Jorgenson, J. W., Carmack, M., Wilson, S. R., Boyse, E. A., Yamazaki, K., Wilson, M., Beamer, W., & Whitten, W. K. Chemical studies of the primer mouse pheromones. In D. Müller-Schwarze & R. M. Silverstein (Eds.), *Chemical signals: Vertebrates and aquatic invertebrates.* New York: Plenum Press, 1980.

Nowell, N. W., Wouters, A. Release of aggression-promoting pheromone by male mice treated with α-melanocyte-stimulating hormone. *Journal of Endocrinology,* 1975, *65,* 36.

Nowell, N. W., Thody, A. J., & Woodley, R. α-melanocyte stimulating hormone and aggressive behavior in the male mouse. *Physiology and Behavior,* 1980, *24,* 5–9.

Nyby, J., Thiessen, D. D., & Wallace, P. Social inhibition of territorial marking in the mongolian gerbil *(Meriones unguiculatus). Psychonomic Science,* 1970, *21,* 310–312.

Oettel, V. M., & Kurischko, A. Physiologisch-endokrinologische studien zum aggressionsverhalten von Mäuseböcken. *Zeitschrift für Versuchstierk unde,* 1978, *20,* 186–193.

Pandey, S. D., & Dominic, C. J. Effect of cyproterone acetate on the production of the estrous-inducing pheromone in the wild mouse. *Indian Journal of Experimental Biology,* 1978, *16,* 887–889.

Paterson, A. T., Rickerby, J., Simpson, J., & Vickers, C. Possible interaction of melanocyte-stimulating hormone (MSH) and the pineal in the control of territorial aggression in mice. *Physiology and Behavior,* 1980, *24,* 843–848.

Payne, A. P. The attractiveness of Harderian gland smears to sexually naive and experienced male golden hamsters. *Animal Behaviour,* 1979, *27,* 897–904.

Payne, A. P., McGadey, J., Moore, M. R., & Thompson, G. Androgenic control of the Harderian gland in the male golden hamster. *Journal of Endocrinology,* 1977, *75,* 73–82.

Preti, G., Smith, A. B., III, & Beauchamp, G. K. Chemical and behavioral complexity in mammalian chemical communication systems: Guinea pigs *(cavia porcellus),* marmosets *(Saguinus fuscicollis)* and humans *(Homo sapiens).* In D. Müller-Schwarze & M. M. Mozell (Eds.), *Chemical signals in vertebrates.* New York: Plenum Press, 1977.

Reynolds, J., & Keverne, E. B. The accessory olfactory system and its role in the pheromonally mediated suppression of oestrus in grouped mice. *Journal of Reproduction and Fertility,* 1979, *57,* 31–35.

Rowe, F. A., & Edwards, D. A. Olfactory bulb removal: Influences on the aggressive behaviors of male mice. *Physiology and Behavior,* 1971, *7,* 889–892.

Sawyer, T. F. Androgen effects on responsiveness to aggression and stress-related odors of male mice. *Physiology and Behavior,* 1980, *25,* 183–187.

Scalia, F., & Winans, S. S. New perspectives on the morphology of the olfactory system: Olfaction and vomeronasal pathways in mammals. In R. L. Doty (Ed.), *Mammalian olfaction, reproductive processes and behavior.* New York: Academic Press, 1976.

Scott, J. P. Agonistic behavior of mice and rats: A review. *American Zoologist,* 1966, *6,* 683–701.

Shorey, H. H. *Animal communication by pheromones.* New York: Academic Press, 1976.

Signoret, J. P., & du Mesnil du Buisson, F. Etude du comportement de la truie en oestrus. *The proceedings of the Fourth International Congress of Animal Reproduction* (The Hague), 1961, 171–175.

Simon, N. G., & Gandelman, R. The estrogenic arousal of aggressive behavior in female mice. *Hormones and Behavior,* 1978, *10,* 118–127.

Singer, A. G., Macrides, F., & Agosta, W. C. Chemical studies of hamster reproductive pheromones. In D. Müller-Schwarze & R. M. Silverstein (Eds.), *Chemical signals vertebrates and aquatic invertebrates.* New York: Plenum Press, 1980.

Singh, P. J., Tucker, A. M., & Hofer, M. A. Effects of nasal $ZnSO_4$ irrigation and olfactory bulbectomy on rat pups. *Physiology and Behavior,* 1976, *17,* 373–382.

Soares, M. J., Kalberer, W. D., & Erpino, M. J. Progestogen effects on elicitation of aggressive behavior in male mice. *Journal of Endocrinology,* 1977, *73,* 507–510.

Soares, M. J., Kalberer, W. D., & Erpino, M. J. Progesterone: Effects on investigatory preferences, aggression, and olfaction in orchidectomized, testosterone-treated mice. *Behavioral Biology,* 1978, *23,* 260–266.

Svare, B. B., & Leshner, A. I. Behavioral correlates of intermale aggression and grouping in mice. *Journal of Comparative and Physiological Psychology,* 1973, *85,* 203–210.

Svare, B., Bartke, A., & Macrides, F. Juvenile male mice: An attempt to accelerate testis function by exposure to adult female stimuli. *Physiology and Behavior,* 1978, *21,* 1009–1013.

Thiessen, D. D., & Dawber, N. Territorial exclusion and reproductive isolation. *Psychonomic Science,* 1972, *28,* 159–160.

Thiessen, D. D., & Kittrell, E. M. W. The harderian gland and thermoregulation in the gerbil *(Meriones unguiculatus). Physiology and Behavior,* 1980, *24,* 417–424.

Thiessen, D. D., & Rice, M. Mammalian scent gland marking and social behavior. *Psychological Bulletin,* 1976, *83,* 505–539.

Thiessen, D. D., Clancy, A. N., & Goodwin, M. Harderian gland pheromone in the mongolian gerbil, *Meriones unguiculatus. Journal of Chemical Ecology,* 1976, *2,* 231–238.

Thiessen, D. D., Graham, M., Perkins, J., & Macks, S. Temperature regulation and social grooming in the mongolian gerbil *(Meriones unguiculatus). Behavioral Biology,* 1977, *19,* 279–288.

Thody, A. J., & Dijkstra, H. Effect of ovarian steroids on preputial gland odours in the female rat. *Journal of Endocrinology,* 1978, *77,* 397–403.

Thody, A. J., & Schuster, S. Control of sebaceous gland function in the rat by α-melanocyte-stimulating hormone. *Journal of Endocrinology,* 1975, *64,* 503–510.

Thody, A. J., & Schuster, S. Sebum secretion after drugs which affect melanocyte-stimulating hormone secretion. *Journal of Endocrinology,* 1977, *72,* 412–414.

de la Torre, B., Norén, S., Hedman, M., & Diczfalusy, E. Effect of cyproterone acetate (CPA) on gonadal and adrenal function in men. *Contraception,* 1979, *20,* 377–396.

Urry, R. L., Dougherty, K. A., & Cokett, A. T. K. Dose-related effect of testosterone dihydrotestosterone, and cyproterone acetate on male organ weights. *Surgical Forum,* 1976, *27,* 584–586.

Vandenbergh, J. G. Male odor accelerates female sexual maturation in mice. *Endocrinology,* 1969, *84,* 658–660.

Vandenbergh, J. G. Chromatographic separation of puberty acceleration pheromone from male mouse urine. *Biology of Reproduction,* 1976, *15,* 260–265.

Vandenbergh, J. G., Whitsett, J. M., & Lombardi, J. R. Partial isolation of a pheromone accelerating puberty in female mice. *Journal of Reproduction and Fertility,* 1975, *43,* 515–523.

Van der Lee, S., & Boot, L. M. Spontaneous pseudopregnancy in mice. *Acta Physiologica Pharmacologica, Neerlandica,* 1956, *4,* 442–443.

Whalen, R. G., & Edwards, D. A. Effects of the anti-androgen cyproterone acetate on mating behavior and seminal vesicle tissue in male rats. *Endocrinology,* 1969, *84,* 155–156.

Whitten, W. K., & Bronson, F. H. The role of pheromones in mammalian reproduction. In W. J. Johnston, Jr., D. G. Moulton, A. Turk (Eds.), *Advances in chemoreception* Vol. 1). New York: Appleton, 1970.

Wilson, M. C., Beamer, W. C., & Whitten, W. K. Puberty acceleration in mice: I. Dose-response effects and lack of critical time following exposure to male mouse urine. *Biology of Reproduction,* 1980, *22,* 864–872.

Winans, S. S., & Powers, J. B. Olfactory and vomeronasal deafferentation of male hamsters: Histological and behavioral analyses. *Brain Research,* 1977, *126,* 325–344.

Wysocki, C. J. Neurobehavioral evidence for the involvement of the vomeronasal system in mammalian reproduction. *Neuroscience and Biobehavioral Reviews,* 1979, *3,* 301–341.

Yee, A. L. *The source of the aggression-promoting pheromone in rodents.* Masters Thesis, Brooklyn College of the City University of New York, 1977.

15

The Hormonal Responses to Competition and Their Behavioral Significance

Alan I. Leshner

By far, the vast majority of studies concerned with the relationship between hormones and the behaviors that occur in competitive situations—the agonistic behaviors—have been directed toward explaining the ways in which hormones affect those behaviors. However, that relationship really is bidirectional. Not only do hormones affect aggressive and submissive responses, but experiencing agonistic interactions alters the individual's endocrine characteristics. Those hormonal responses to agonistic experiences are the topic of this discussion.

This chapter is organized into two parts. The first part provides a brief summary of what is known about the effects of agonistic experiences on circulating hormone levels, and the discussion focuses on those studies using rodents as subjects. The effects of agonistic experiences on endocrine function in other animals are discussed in other chapters of this book. The second part of this chapter then asks about the functions of those hormonal changes occurring after agonistic experiences. Specifically, the question is asked whether those hormonal changes are simply physiological responses to the experience, or whether these hormonal responses might play some important role in determining the individual's future agonistic behavior patterns. The discussions in both parts of this chapter will emphasize responses involving the pituitary-gonadal and pituitary-adrenocortical axes, since these have received the most attention. The reader interested in the responses of other endocrine subsystems might consult the recent review by Brain (1980).

THE HORMONAL RESPONSES TO AGONISTIC EXPERIENCES

Interest in the hormonal responses to competition seems to have emerged following the observation in the mid-1950s that, for rodents living in groups, as group size increases so does the average weight of the individual's adrenal glands.

ALAN I. LESHNER • National Science Foundation, Washington, D.C. 20550.

Furthermore, these early studies noted that within those groups, the subordinate animals have heavier adrenal glands than the dominant animals (e.g., Southwick & Bland, 1959). These findings, in combination with somewhat later observations that agonistic experiences also might alter gonadal function (e.g., Bronson & Marsden, 1973), led to a sporadic series of studies investigating those hormonal changes after agonistic experiences more intensely.

Since the picture one can derive from those many studies is by now fairly clear, it would serve no special function to repeat again here the detailed sequence that they took. The interested reader might consult the reviews of this literature by Brain (1980), by Leshner (1975, 1980), and by Schuurman (1980). Instead, I shall summarize here what is currently known.

First, there are marked stimulatory effects of agonistic experiences on pituitary-adrenocortical activity. Furthermore, although both victorious and defeated animals respond to competition with increased levels of corticosterone, and presumably of ACTH as well, defeated animals show a much more pronounced effect of competition than do the winners. Elevated corticosteroid levels can be observed within 10 minutes of the experience in both winners and losers, although the corticosteroid levels of the losers rise significantly above those of the winners. That difference between winners and losers lasts for at least 30 minutes, and the corticosterone levels of both competitors remain above control levels for at least one hour after the fight (e.g., Leshner, 1980). After repeated encounters, defeated animals will even respond to the threat of defeat in the form of exposure to a novel conspecific, with increases in corticosterone levels (Bronson & Eleftheriou, 1964). Thus, agonistic experiences lead to dramatic elevations in pituitary-adrenocortical activity levels, and the effect on defeated animals is greater than that on victorious animals.

The second endocrine subsystem receiving an appreciable amount of attention is the pituitary-gonadal axis. Although in the case of the pituitary-adrenal hormones the effects were established relatively clearly and early on, there initially was some controversy about the effects of competition on pituitary-gonadal function. Using rather indirect measures of pituitary-gonadal activity, some investigators found suggestions of no effect of victory or defeat on pituitary-gonadal hormone levels (e.g., Vale, Lee, & Ray, 1970; Vale, Vale, & Harley, 1971), whereas the findings of other studies suggested that pituitary-gonadal hormone levels were affected by competition (e.g., Bronson & Marsden, 1973; Bronson, Stetson, & Stiff, 1973). At about the time this disagreement had surfaced, more direct measures had just been perfected, and the controversy has since been resolved.

Beginning with the pituitary portion of that axis, Bronson and his colleagues (e.g., Bronson *et al.*, 1973) studied the effects of victory and defeat on circulating levels of luteinizing hormone (LH) and follicle stimulating hormone (FSH). They found that although both victory and defeat lead to decreases in the levels of these tropic hormones, defeat has a more dramatic and long-lasting effect than does victory. Therefore, competition has been shown to markedly alter gonadotropin levels, and, as was true in the case of the pituitary-adrenocortical hormones, defeat has a more dramatic effect than does victory.

Turning to direct measurement of gonadal hormone secretion, Bernstein and his colleagues (Rose, Holaday, & Bernstein, 1971; Rose, Gordon, & Bernstein, 1972) have conducted a long series of studies on Rhesus monkeys and have found defeat experiences to depress testosterone secretion markedly. Their work is described in more detail elsewhere in this book (see chapter by Bernstein). In rats, Schuurman (1980) has conducted a very elegant and extensive series of studies examining the testosterone responses to competition. In those studies, Schuurman found no effect of competition on the testosterone levels of winners but very marked depressive effects on those of the losers. Thus, although there initially was some controversy over the effects of competition on pituitary-gonadal function, those effects now seem much clearer. Defeat leads to a marked and prolonged depression of gonadotropin secretion and to a decrease in circulating testosterone levels. Victory also leads to a depression in gonadotropin levels, although less marked than that following defeat, but to no change in testosterone levels.

Putting together the data from studies focusing separately on competition's effects on pituitary-adrenocortical and pituitary-gonadal activity levels, the picture emerges of defeated animals showing very marked hormonal responses to agonistic experiences. Those responses take the form of heightened pituitary-adrenocortical activity and depressed pituitary-gonadal activity. Victorious animals also experience hormonal changes after competition, but these changes are less dramatic than those characteristic of defeated animals. Victorious animals show increases in pituitary-adrenal activity, but to a lesser degree than do defeated animals. And victorious animals show a depression, although less of one than defeated animals, in gonadotropin secretion and no change in testosterone levels. We turn now to the question of the function of those hormonal responses to competition.

PHYSIOLOGICAL SIGNIFICANCE OF THE HORMONAL RESPONSES TO COMPETITION

The first explanations of the hormonal responses to competition were physiological in nature. Perhaps the most obvious explanation is that these hormonal changes occur simply because the competing animals have been stressed. Furthermore, since everyone believes it is worse to lose than to win, defeated animals would be more stressed than would victorious animals. This kind of explanation follows logically from Selye's formulation of the hormonal components of the organism's stress-response/stress-coping mechanism (e.g., Selye, 1956). That now commonly held view of stress reactions suggests that when an individual is stressed, it responds with heightened secretion of the pituitary-adrenal hormones and depressed secretion of the pituitary-gonadal hormones. Further, it suggests that the stress response pattern is exactly the characteristic of animals just following competition. (Note: The reader interested in seeing in full the endocrine responses to stress might consult an early, but very comprehensive, series of studies by Mason, 1968, which detailed the endocrine pattern of monkeys following exposure to the stress of having to engage in avoidance behaviors).

That the hormonal responses to competition are merely a part of the organism's stress-response or stress-adaptation mechanisms really seemed rather simplistic, and some more elaborate explanations have been provided. Of particular note has been the suggestion of Christian and his colleagues (e.g., Christian, 1955; Christian & Davis, 1964) relating these hormonal responses to competition to the regulation of population size or numbers. Recall that the original observations of there being hormonal changes following competition were made in situations where the animals were living in groups. In fact, the purpose of those early studies was to examine the role of endocrine parameters in population regulation rather than in competition *per se*.

What Christian and his colleagues first observed was that as population size increases, so do the adrenal weights of the population members. In addition, these investigators observed that accessory reproductive organ sizes decrease as adrenal weights, and population size, increase. Therefore, Christian and Davis (1964) proposed a theory connecting these elements.

Their suggestion was quite sophisticated, although fairly straightforward. They suggested that the endocrine responses to increasing population number and pressure serve as a mechanism for controlling the population's size. The sequence posited is as follows: First, population size begins to increase. With that increase in population size is an increase in social pressure. This increase in social pressure takes the form of increased competition, as well as increases in other modes of social interaction. As competition increases, there is a resultant increase in pituitary-adrenocortical activity and a decrease in pituitary-gonadal activity. (It is not clear whether the altered pituitary-adrenal activity causes the change in pituitary-gonadal hormone levels or whether they are coincident. It is well documented (Leshner, 1978) that these endocrine subsystems can affect each other, but their interaction in this situation is not clear.) Those hormonal changes, then, particularly the changes in pituitary-gonadal activity, decrease reproductive efficiency such that the rate of population growth eventually decreases and the population pressure is eased. The pituitary-adrenocortical responses to increased population size and to increased competition provide another mechanism for regulating population growth, since increases in adrenal glucocorticoid secretion can have a suppressive effect on resistance to disease.

This explanation, then, takes a simple stress-response explanation one step further. It suggests that not only do the hormonal responses to competition help the individual adapt physiologically to the stresses imposed by competing, particularly by being defeated, but they also provide a feedback mechanism whereby the need for competition is reduced by a decrease in population size and pressure.

BEHAVIORAL SIGNIFICANCE OF THE HORMONAL RESPONSES TO COMPETITION

This basically physiological view was predominant until the mid-1970s, when it was suggested that the hormonal responses to competition might have behavioral significance as well. Elsewhere (Leshner, 1975), I have put forth the hypothesis that the hormonal responses to competition, particularly to defeat, not only provide a mechanism for physiological adaptation to the stresses imposed by that experience but also feed back and affect ongoing and future agonistic response patterns.

That hypothesis was based on the following logic, focusing on the case of defeat. First, the experience of defeat leads not only to hormonal change but to specific behavioral changes as well. Defeated animals become less aggressive and more submissive in the same and in near-future encounters. Second, as discussed above, defeat leads to increases in pituitary-adrenocortical activity and decreases in the gonadal secretion of the androgens. Third, as discussed in other chapters in this book, when naive animals are subjected to experimentally induced increases in pituitary-adrenocortical activity and decreases in androgen levels, they become less aggressive and more submissive than normal animals. Therefore, I suggested that the hormonal changes following defeat might cause or play a part in the behavioral changes following that experience. The sequence proposed is that defeat leads to altered pituitary-adrenocortical and gonadal hormone levels, which, then, lead to altered agonistic response patterns. This kind of hormone–behavior interaction has more recently (Leshner, 1979) been termed a "feedback effect of hormones on behavior," and this hypothesis might be called a "behavioral feedback hypothesis" of the functions of the hormonal responses to competition.

Methods for Testing the Behavioral Feedback Hypothesis

Having proposed that the hormonal responses to competition have behavioral significance in the form of feeding back and modifying ongoing or future behavior, a reasonable question to ask is how to test that hypothesis. Traditional behavioral endocrinological techniques of simply removing the gland and providing replacement therapy would not be appropriate here, because those manipulations would alter the test animal's hormonal state before entering the situation, as well as altering its hormonal responses to competition. I have argued elsewhere (Leshner, 1979) that there are two separable kinds of hormonal effects on behavior, one class reflecting effects exerted before entering the test situation (called baseline effects) and the second class being the feedback types of effects being discussed here. Therefore, it would be important to manipulate the animal's hormonal responses to competition without altering its hormonal state before entering the testing situation.

Two approaches to testing the feedback hypothesis seem appropriate. One would involve preventing, or diminishing, the normal hormonal responses to competition. The other would involve augmenting or exaggerating those responses. In the first case, after diminishing the hormonal responses to the encounter, one would expect to see decreased behavioral change consequent to the experience in the encounter. In the second case, one would expect the reverse: There should be greater than normal behavioral changes after the critical competitive experience.

Both approaches have been used to test whether the hormonal responses to competition are behaviorally effective, and the significance of both the pituitary-adrenocortical and pituitary-gonadal responses to agonistic experiences have been examined. The usual technique for diminishing postexperience hormonal responses has been to remove the relevant gland and replace the test hormone with a "behavioral replacement" dosage of that hormone. Using this technique results in animals entering the situation with normal hormonal states, but, since their hormone levels are being controlled experimentally, they cannot respond normally (endocrinologically) to the agonistic experience. The usual technique for augmenting the normal hormonal responses to competition (at least for those cases where the response is an increase in hormone levels) is simply to inject the animals with a high dosage of the test hormone immediately after the experience and observe later behavior patterns.

Behavioral Significance of the Pituitary-Gonadal Responses to Competition

The two endocrine subsystems being discussed in this chapter will be considered separately here, although I am aware that they normally do interact (Leshner, 1978). Since my colleagues and I have shown elsewhere (Leshner, Walker, Johnson, Kelling, Kreisler, & Svare, 1973) that pituitary-adrenal and androgenic influences on agonistic behavior are separable, I believe it reasonable to discuss them separately.

Two series of studies have examined the potential behavioral significance of there being altered pituitary-gonadal activity following competition, focusing only on changes in androgen levels. In the first set of studies to be published, Nock and Leshner (1976) examined the effects of castrating mice and treating them with a dosage of testosterone adequate to restore their initial aggressiveness to fully normal levels. They examined both (1) the pattern of agonistic responding of winners and losers with these controlled testosterone levels throughout a single encounter, and (2) the level of aggressiveness displayed by both winners and losers after many successive winning or losing experiences. They found no effect of preventing the normal androgen changes after competition on either ongoing or future agonistic responses. Therefore, they argued that, counter to the feedback hypothesis discussed above, there is no behavioral significance to the gonadal hormone responses to agonistic experiences.

In a series of studies published a year later, Maruniak, Desjardins, and Bronson (1977) employed more sophisticated methods for controlling testosterone levels. In their studies they substituted silastic implants of testosterone, which lead to steadily released and constant circulating testosterone levels, for the testosterone injection regimen used by Nock and Leshner, which would have resulted in greater, and perhaps quite marked, fluctuations in circulating testosterone levels. Using these more sophisticated techniques, Maruniak and his colleague found the same thing as did Nock and Leshner. Preventing the normal testosterone response to competition had no effect on ongoing or future agonistic responses.

These two sets of studies argue strongly against the hypothesis that at least the gonadal hormone responses to agonistic experiences have behavioral significance. In both cases, preventing the testosterone responses to competition had no effect on late agonistic response patterns. Therefore, if those hormonal responses have any significance, it is either relevant to some nonagonistic behavior, or it is purely physiological.

Behavioral Significance of the Pituitary-Adrenocortical Responses to Competition

Turning to the second set of hormonal responses to competition, those of the pituitary-adrenocortical axis, the situation is quite different. The findings to date strongly support the hypothesis that these hormonal responses to competition do have behavioral significance, at least for the case of defeated animals.

Using the same basic paradigm as they used when studying the androgen responses to competition, Nock and Leshner (1976) examined the effects of preventing the pituitary-adrenocortical responses to agonistic experience on both ongoing and future agonistic response patterns. In one study, they compared the agonistic response patterns of mice that had been hypophysectomized and treated with replacement dosages of both ACTH and testosterone, with the agonistic responses of intact, placebo-treated mice. The mice in the first condition, the controlled-hormone mice, fought longer during the initial encounter. In addition, although the victorious animals behaved basically the same as did victorious mice without controlled hormone levels, the losers in the encounters became nonaggressive and submissive later than did the normal defeated animals. This finding suggests that preventing the normal hormonal responses to defeat delays and decreases the behavioral responses that normally follow that experience, nonaggressiveness and submissiveness. Furthermore, with regard to future agonistic response patterns, the defeated mice with controlled corticosterone levels did not become significantly less aggressive after repeated defeat experiences, as did the intact defeated animals. Therefore, not only does controlling the hormonal responses to defeat affect ongoing agonistic response patterns in a feedback way, but this manipulation affects future responding as well.

In a second study, Nock and Leshner examined the effects of controlling the corticosterone responses to defeat, rather than multiple hormones at the same time. In this study, the controlled-hormone mice had been adrenalectomized and treated with a fixed behavioral replacement dosage of corticosterone. As in the first study of this series, victorious animals were not appreciably affected by this manipulation. However, defeated mice with controlled corticosterone levels became submissive later in the encounter than did intact defeated mice. This finding, again, supports the feedback hypothesis that the pituitary-adrenocortical responses to defeat feed back and affect ongoing and future agonistic response patterns.

My colleagues and I then turned to the second approach: We examined the effects on agonistic response patterns of experimentally increasing, rather than decreasing, the pituitary-adrenocortical responses to defeat experiences. We stopped studying the case of victorious animals, because the Nock and Leshner

(1976) study discussed above had shown no effect of preventing these responses on victorious animals. In addition, we had by then become interested primarily in the factors controlling submissiveness (e.g., Leshner, 1981), the principle one being the defeat experience. In these studies we used a two-test paradigm, in which the defeated animals were treated with the test hormone immediately after the initial defeat, and then tested a second time for submissiveness following some postdefeat interval.

In the first study using the "augmenting" approach, Roche and Leshner (1979) examined the effects of postdefeat injections with ACTH on the level of submissiveness exhibited either 24 hours, 48 hours, or seven days after that initial experience. Mice treated immediately after the initial defeat with this tropic hormone that leads to increases in both ACTH and corticosterone levels were more submissive than placebo-treated controls at both the 24- and 48-hour postdefeat intervals. They behaved normally at the seven-day testing interval. This finding supported the hypothesis that the pituitary-adrenal responses to defeat feed back and alter later agonistic responses.

In the interim, we had been studying the pituitary-adrenal control of submissive behavior directly (e.g., Leshner & Politch, 1979), and we had found corticosterone to be the most important hormone of this axis in the control of this behavior. Therefore, we conducted a series of studies testing the behavioral feedback hypothesis augmenting the corticosterone, but not ACTH, responses to defeat.

The basic finding from that series of studies (e.g., Leshner, 1980; Leshner, Korn, Mixon, Rosenthal, & Besser, 1980) was that corticosterone treatment has the same effect as did ACTH treatment in the Roche and Leshner (1979) study mentioned above. That is, postdefeat treatment with corticosterone increases later submissiveness.

Mediating Mechanisms for Behavioral Feedback

The findings discussed in the last section strongly support the hypothesis that the corticosterone responses to defeat experiences are behaviorally significant. That is, these postdefeat hormonal changes appear to lead to decreased levels of aggressiveness and increased levels of submissiveness in subsequent agonistic encounters. The next reasonable question seems to be: How do these hormones exert such effects?

On a physiological level, it is most likely that these hormonal effects are mediated centrally. Other chapters in this book discuss in detail the neural mediation of steroid hormone effects on agonistic behavior, and that issue will not be raised again here. What will be discussed here is the question of what might be called the "behavioral" mediation of these feedback effects. That is, what behavioral processes are being affected such that the corticosterone responses to defeat have these behavioral feedback effects? Do these hormones affect agonistic behaviors directly, or do they alter agonistic responding because they alter some other, related behavioral trait so that agonistic responding is incidently affected?

In an earlier chapter in this book, while discussing the mediation of pituitary-adrenocortical effects on aggressiveness, I suggested that these hormones affect competitive behaviors by altering the arousal level of the test animal. This increase in arousal level, then, would increase agonistic responding because the more aroused animal would perceive the incoming stimuli as more aggression-provoking or more submission-provoking. I believe that a somewhat similar explanation is appropriate here, focusing instead, however, on hormonal change itself as a stimulus or part of a stimulus complex.

One point that should be emphasized early on is that in studying feedback effects of hormones on behavior, one is in fact using a repeated testing paradigm. That is, first the test animal is subjected to an initial behavioral test, hormone levels change or are changed, and then behavioral reactions are assessed again. This is true whether one is examining feedback effects on ongoing behavioral reactions or on future behavioral response patterns. The use of this kind of repeated testing paradigm opens the way to a wider variety of interpretations than were possibly appropriate in the single-test paradigms discussed in other chapters.

For example, an obvious interpretation which could be applied to any case where a treatment after an initial experience affects behavior during a later experience, is that the treatment has affected memory processes directly. The reader may recall that these kinds of interpretations have been offered frequently when discussing hormonal effects on avoidance conditioning (e.g., Gold & McGaugh, 1977). However, it is unlikely that corticosterone affects memory processes directly (cf. Leshner, Merkle, & Mixon, 1981), and, therefore, that kind of interpretation seems inappropriate here.

A second reasonable mediational hypothesis might be that the corticosterone responses to defeat feed back and increase later submissiveness because they increase the test animal's fear of other, strange animals. This explanation has been offered before in interpreting pituitary-adrenal effects on aggressiveness (e.g., Brain, 1971; Leshner *et al.*, 1973). Applied in this situation, a fear-mediated interpretation would suggest that the animal is defeated, it experiences an increase in corticosterone levels, and this hormonal change leads to an increase in general fearfulness. The increased fear would then motivate the animal to react submissively in future situations, since submissive behaviors are in fact means of avoiding being attacked (Leshner & Politch, 1979).

A fear interpretation, however, seems inappropriate here, since there is little evidence to suggest that this steroid hormone increases fear in other situations. In fact, in standard fear-testing situations, such as avoidance responding in shock-mediated situations, increases in corticosterone levels decrease "fear" (e.g., Bohus, 1975; Weiss, McEwen, Silva, & Kalkut, 1969). Therefore, since increases in corticosterone levels do not increase fear in other situations, there is no reason to suspect they would be doing that in agonistic situations.

I suggest here a more phenomenological/experiential interpretation of corticosterone's behavioral feedback effects on agonistic responding. By definition, defeat experiences, as is true of all experiences, have certain stimulus characteristics associated with them. That is, there are stimulus elements that identify an experience's unique characteristics; this is, of course, true of the experience of defeat as well. Those stimulus characteristics are what is "remembered" in future similar situations. Since we know that corticosterone levels always rise after defeat experiences, and since we have shown that those corticosterone changes are behaviorally significant, it seems reasonable to suggest that the corticosterone changes following defeat are a part of the stimulus complex characterizing defeat experiences. That is not to suggest that corticosterone changes are unique to the experience of defeat. Corticosterone levels rise following all stressful experiences. But, since they always rise following defeat, and since the animal must be able in some way to "sense" their changes because those changes are behaviorally significant, it seems likely that corticosterone elevations are a normal part of the stimuli characterizing defeat.

I then argue that altering the corticosterone responses to defeat affects ongoing or future agonistic responses because the intensity of the normal defeat stimulus complex has been perturbed. Decreases in the "corticosterone stimuli" of defeat decrease the behavioral changes typical of recently defeated animals because defeat's total stimulus intensity has been decreased, and therefore its "meaningfulness." Conversely, increasing those corticosterone responses to defeat increases the intensity of the total defeat stimulus complex, and therefore increases later expressions of the normal behavioral responses to defeat—nonaggressiveness and submissiveness.

SUMMARY REMARKS

In this chapter, I have reviewed summarily what is known about the pituitary-gonadal and pituitary-adrenocortical responses to competitive or agonistic experiences. Basically, both victorious and defeated animals exhibit altered levels of the hormones of both of these endocrine axes, although defeated animals react much more dramatically than do victorious animals.

I then asked whether these hormonal responses to competition serve simply as a part of the normal physiological adaptive mechanisms called into operation following any stressful experience, or whether these responses might serve some unique functions in competitive situations. On a physiological level, it appears likely that the endocrine responses to competition, particularly in the case of defeated animals, can play a role in the regulation of population size or population number.

In terms of their behavioral significance, whether or not the hormonal responses to competition are important is rather complex. In the case of victorious animals, their hormonal responses to agonistic encounters do not seem to affect their later agonistic response patterns. But, in the case of defeated animals, the pituitary-adrenocortical responses to competition do seem to be behaviorally significant, although their pituitary-gonadal responses do not.

The final question posed concerned the mechanisms by which the pituitary-adrenocortical, particularly corticosterone, responses to defeat might alter future agonistic responding. A variety of explanations were discussed, and rejected, and it finally was suggested that the pituitary-adrenocortical responses to defeat affect future agonistic response patterns because these hormonal changes are a normal part of the stimulus complex associated with defeat. Any change in that stimulus complex, or in its intensity, would affect future related response patterns.

REFERENCES

Bohus, B. The hippocampus and the pituitary-adrenal system hormones. In R. L. Isaacson & K. H. Pribram (Eds.), *The hippocampus* (Vol. I). New York: Plenum Press, 1975.

Brain, P. F. Possible role of the pituitary-adrenocortical axis in aggressive behaviour. *Nature,* 1971, *233,* 489.

Brain, P. F. Adaptive aspects of hormonal correlates of attack and defense in laboratory mice: A study in ethobiology. *Progress in Brain Research,* 1980, *53,* 391–413.

Bronson, F. H., & Eleftheriou, B. E. Chronic physiological effects of fighting in mice. *General and Comparative Endocrinology,* 1964, *4,* 9–14.

Bronson, F. H., & Marsden, H. M. The preputial gland as an indicator of social dominance in male mice. *Behavioral Biology,* 1973, *9,* 625–628.

Bronson, F. H., Stetson, M. H., & Stiff, M. E. Serum FSH and LH in male mice following aggressive and nonaggressive interactions. *Physiology and Behavior,* 1973, *10,* 369–372.

Christian, J. J. Effect of population size on the adrenal glands and reproductive organs of male mice in populations of fixed size. *American Journal of Physiology,* 1955, *182,* 292–300.

Christian, J. J., & Davis, D. E. Endocrines, behavior and population. *Science,* 1964, *146,* 1550–1560.

Gold, P. E., & McGaugh, J. L. Hormones and memory. In L. H. Miller, C. A. Sandmann, & A. J. Kastin (Eds.), *Neuropeptide influences on the brain and behavior.* New York: Raven Press, 1977.

Leshner, A. I. A model of hormones and agonistic behavior. *Physiology and Behavior,* 1975, *15,* 225–235.

Leshner, A. I. *An introduction to behavioral endocrinology.* New York: Oxford University Press, 1978.

Leshner, A. I. Kinds of hormonal effects on behavior: A new view. *Neuroscience and Biobehavioral Reviews,* 1979, *3,* 69–73.

Leshner, A. I. The interaction of experience and neuroendocrine factors in determining behavioral adaptations to aggression. *Progress in Brain Research,* 1980, *53,* 427–438.

Leshner, A. I. The role of hormones in the control of submissiveness. In P. F. Brain & D. A. Benton (Eds.), *A multidisciplinary approach to aggression research.* Amsterdam: Elsevier, 1981.

Leshner, A. I., & Politch, J. A. Hormonal control of submissiveness in mice: Irrelevance of the androgens and relevance of the pituitary-adrenal hormones. *Physiology and Behavior,* 1979, *22,* 531–534.

Leshner, A. I., Walker, W. A., Johnson, A. E., Kelling, J. S., Kreisler, S. J., & Svare, B. B. Pituitary adrenocortical activity and intermale aggressiveness in isolated mice. *Physiology and Behavior,* 1973, *11,* 705–711.

Leshner, A. I., Korn, S. J., Mixon, J. F., Rosenthal, C., & Besser, A. K. Effects of corticosterone on submissiveness in mice: Some temporal and theoretical considerations. *Physiology and Behavior,* 1980, *24,* 283–288.

Leshner, A. I., Merkle, D. A., & Mixon, J. F. Pituitary–adrenocortical effects on learning and memory in social situations. In J. L. Martinez, R. A. Jensen, R. B. Messing, H. Rigter, & J. L. McGaugh (Eds.), *Endogenous peptides and learning and memory processes.* New York: Academic Press, 1981.

Maruniak, J. A., Desjardins, C., & Bronson, F. H. Dominant-subordinate relationships in castrated male mice bearing testosterone implants. *American Journal of Physiology,* 1977, *233,* E495–E499.

Mason, J. W. Organization of the multiple endocrine responses to avoidance in the monkey. *Psychosomatic Medicine,* 1968, *30,* 774–790.

Nock, B. L. & Leshner, A. I. Hormonal mediation of the effects of defeat on agonistic responding in mice. *Physiology and Behavior,* 1976, *17,* 111–119.

Roche, K. E., & Leshner, A. I. ACTH and vasopressin treatments immediately after a defeat increase future submissiveness in mice. *Sicence,* 1979, *204,* 1343–1344.

Rose, R. M., Holaday, J. W., & Bernstein, I. S. Plasma testosterone, dominance rank and aggressive behaviour. *Nature,* 1971, *231,* 366–368.

Rose, R. M., Gordon, T. P., & Bernstein, I. S. Plasma testosterone levels in the male Rhesus: Influences of sexual and social stimuli. *Science,* 1972, *178,* 643–645.

Schuurman, T. Hormonal correlates of agonistic behavior in adult male rats. *Progress in Brain Research,* 1980, *53,* 415–420.

Selye, H. *The stress of life.* New York: McGraw-Hill, 1956.

Southwick, C. H. & Bland, V. P. Effect of population density on adrenal glands and reproductive organs of CFW mice. *American Journal of Physiology,* 1959, *197,* 111–114.

Vale, J. R., Lee, C. T., & Ray, D. The effect of defeat upon the reproductive behavior and upon adrenal, testis and seminal vesicle weights of C57BL/6 mice. *Communications in Behavioral Biology,* 1970, *5,* 225–236.

Vale, J. R., Vale, C. A., & Harley, J. P. Interaction of genotype and population number with regard to aggressive behavior, social grooming and adrenal and gonadal weight in male mice. *Communications in Behavioral Biology,* 1971, *6,* 209–221.

Weiss, J. M., McEwen, B. S., Silva, M. T. A., & Kalkut, M. F. Pituitary-adrenal influences on fear responding. *Science,* 1969, *163,* 197–199.

IV

HORMONES AND AGGRESSIVE BEHAVIOR IN SELECTED VERTEBRATES

16

Hormones and Aggressive Behavior in Teleost Fishes

Trudy A. Villars

Teleost fishes form a highly diversified group of vertebrates that offers a rich source of material for comparative studies of neuroendocrine integration in the regulation of aggressive behavior. With some 20,000 species adapted to a wide range of ecological demands, they can be used to study the range of functioning and the adaptive capacity of neuroendocrine and behavioral systems. Hence, fishes are good subjects with which to study the relationship between the natural history of an animal and the behavioral-endocrine systems utilized by that animal.

Compared with mammals, there is little information available regarding the hormonal factors involved in regulation of fish aggressive behavior. The role of hormones in aggressive behavior of fishes has previously been reviewed only in the course of discussions of reproductive behavior (Fiedler, 1974; Liley, 1969).

In teleosts, the study of aggressive behavior encounters the same difficulties as do comparable studies in other species. "Aggressive behavior" refers to a heterogenous class of behaviors that inflict damage or threaten to inflict damage on a conspecific. The class of aggressive behaviors overlaps with a number of other conceptual classes including territorial behaviors, dominance hierarchies, and agonistic behaviors. The operational definitions of these concepts provide further difficulties, since researchers from a range of backgrounds, interests, and theoretical orientations have contributed to the field. The analysis of the function of aggressive behavior is made more difficult by the fact that function may vary from species to species and, within a single species, may vary with the social and environmental conditions. Finally, the problem is complicated by use of a diversity of experimental procedures—field research, laboratory/experimental work, isolation-induced aggression, etc.

The majority of research reviewed here has used the term "aggressive behavior" to refer either to the frequency of selected species-typical movements and postures, such as fin extensions (lateral display) and gill cover extension

TRUDY A. VILLARS • Department of Psychology, Wheaton College, Norton, Massachusetts 02766.

(frontal display), or to the frequency of explicitly damaging acts such as biting and butting. These studies measure the frequency of such behaviors in direct paired encounters or in pairings where only visual encounters are possible. Such operational definitions are not satisfactory in part because the behavior patterns might also be considered sexual or courtship behaviors when they occur in interactions between a male and female and in part because of the narrowness of their focus. However, they form the core of the available research.

This review begins with brief overviews of the systematics and endocrine system of teleosts, which provide the reader with an indication of the diversity of teleost species and the general similarity between teleost and mammalian endocrine systems. We then review the role of gonadal hormones in aggression of five major groups of teleosts and briefly comment on possible neural integration of behavioral and pituitary-gonadal interaction. The third section addresses two areas of exploration regarding the function of nonsteroidal hormones in behavior. Finally, there is a brief discussion focusing on the place of teleosts in the study of hormone–behavior interactions and the value of an ecological perspective.

SYSTEMATICS AND PHYLOGENY OF TELEOSTS

Teleost fishes comprise a diverse group of approximately 18,000 species, approximately double the number of avian and mammalian species and 96% of all fish species (Greenwood, Rosen, Weitzman, & Myers, 1966; Nelson, 1976). Greenwood *et al.* (1966) have proposed a systematic classification of teleosts based on known and hypothesized phylogenetic relationships. More recently, Nelson (1976) has provided an introduction to the systematics of fishes and reviewed some of the major differences of opinion regarding fish systematics. Greenwood *et al.* divide teleosts into three lines or divisions. Division I, which includes species such as eels and herrings, and Division II, which includes mormyrid and gymnarchid "electric fish," are smaller divisions, with only two and one superorders, respectively. The third and largest division contains five separate superorders: Protocanthopterygii, Ostariophysi, Paracanthopterygii, Atherinomorpha, and Acanthopterygii. Division I and Division II species have not been studied with regard to behavioral-endocrine interactions.

In Division III, the majority of attention has focused on sticklebacks and perciform species (Acanthopterygii), though some attention has been directed toward the behavioral endocrinology of salmonids (Protocanthopterygii), cyprinoids (Ostariophysi), and cyprinodontoids (Atherinomorpha). Acanthopterygii constitute the most diversified superorder of teleosts, with 12 orders and over 9,000 species. Within Acanthopterygii and among all teleosts, Perciformes form the most diversified order, with some 20 suborders and almost 7,000 species. Hormone–behavior studies have used only three suborders of Perciformes: the Percoidei, the Anabantoidei, and the Gobioidei. The Percoidei form the largest suborder, with over 70 families. A considerable amount of endocrine and behavioral research has been done with percoid species, namely *Lepomis* (family:

Centrarchidae), *Tilapia, Haplochromis, Aequidans,* and *Cymatogaster* (family: Cichlidae).

Several members of the Anabantoidei have been extensively studied. Commonly referred to as anabantids, this suborder contains a number of species that are quite popular with ethologists and psychologists as well as tropical fish hobbyists. The most familiar is *Betta splendens* (Siamese Fighting Fish), which has been selectively bred for aggressiveness and which is a popular choice for the introduction of fledgling zoology students to animal behavior (e.g., Southwick, 1968). *Betta* males cannot be easily housed together because of their aggressiveness; consequently, other anabantids such as the Paradise Fish *(Macropodus opercularis)* and the Blue Gourami *(Trichogaster trichopterus)* are frequently used in hormone–behavior studies.

In addition to the perciform fishes, one species of the Gasterosteiformes *(Gasterosteus aculeatus),* the three-spined stickleback, has been the subject of many of the studies reported in this review. *Gasterosteus* has also been the subject of much ethological research and competes with the goldfish as the "white rat" of fish behavioral research (cf. Wooton, 1976).

Among the Antherinomorpha, a number of families of Cyprinodontoidei have been subjects of endocrinological and/or behavioral, studies including Poecilidae (which includes the common guppy), Cyprinodontidae (which includes the genus *Fundulus*), Jennysidae (which includes mollies), and Oryziatidae (including the medaka). The Cyprinodontoidei are one of three suborders in the superorder Antherinomorpha, a group quite distant from the perciform teleosts.

Ostariophysi is another superorder that is quite distant from perciform teleosts. Among the ostariophysian teleosts are catfish, minnows, and carp, including the common goldfish *(Carassius auratus).* The goldfish is extensively used in physiological studies because of its size and the consequent ease in gathering adequate tissue for various physiological assays.

According to Greenwood *et al.* (1966), present-day salmonids, members of the superorder Protocanthopterygii, may be very similar to the primitive forms from which the Ostariophysi diverged. Salmonids include the genera *Oncorhynchus* and *Salmo.* All *Oncorhynchus* species and some *Salmo* species are migratory, hatching in freshwater streams, emigrating to the ocean or a large lake, and then returning to their homestream for spawning. The migratory species are generally known as salmon. The nonmigratory stream-dwelling members of the genus *Salmo* form the group referred to as trout.

On examination of teleost systematics, it is apparent that studies of hormone–behavior systems have focused on a limited number of orders and species of fishes. The selection of a species for study has been based primarily on practical rather than theoretical considerations. Subjects have been selected because of easy accessibility or simplicity in care and breeding. For example, sunfish and sticklebacks are widely available from local ponds or streams.

Examination of teleost systematics indicates the diversity of teleost fishes. This phylogenetic diversity greatly limits our ability to make general conclusions based on the experimental results obtained in any one species; any one species of fish might be less representative of all teleosts than would a single mammal be representative of all mammals. However, the advantage of this diversity is the opportunity it provides for the study of diversity in hormone–behavior systems and the opportunity for the observation of relationships between ecological factors and hormone–behavior systems.

TELEOST ENDOCRINE SYSTEM

A description of teleost endocrine systems is beyond the scope of this chapter. However, there are several recent sources available to the interested reader. A recent volume on evolution of vertebrate endocrine systems (Pang & Epple, 1980), provides an excellent collection of articles that include comparisons of nonmammalian with mammalian systems. Other reviews of nonmammalian and teleost systems provide overviews of pituitary gland (Holmes & Ball, 1974; Sage & Bern, 1971; Schreibman, Leatherland, & McKeown, 1973), evolution of pituitary hormones (Wallis, 1975), regulation of testicular and ovarian steroids (Callard, Callard, Lance, Bolaffi, & Rosset, 1978; Reinboth, 1972), hypothalamic–hypophysiotropic hormones (Crim, Dickoff, & Gorbman, 1978), neuroendocrinology (Kim, Stumpf, Sar, & Martinez-Vargas, 1978; Morrell & Pfaff, 1978; Peter, 1970, 1973, 1977), and endocrinology of reproduction (Donaldson, 1973; Fontaine, 1976).

The endocrine systems of teleosts are largely similar to those of other vertebrates. Sex steroids and pituitary regulation of their synthesis are essentially the same in teleosts and mammals, though teleosts show noteworthy unique characteristics (Lance & Callard, 1980). Some teleosts have been shown to produce testosterone metabolites, not seen in mammals, which may be more potent in fishes than are heterologous mammalian androgens (Idler, Horne, & Sangalang, 1971; Schmidt & Idler, 1962). Thus, where steroids produce a behavioral effect, the steroid metabolite that is actually effective may be unique in teleosts.

Another characteristic feature of teleost reproductive endocrinology is the ovarian production of a corticosteroid (DOC) and a debated role of ovarian DOC and interrenal steroids in the induction of ovulation. If those hormones, frequently associated with stress, do in fact play a role in ovulation, it would be interesting to speculate further that prespawning ("courtship") aggression might facilitate sexual readiness in females by stimulation of those hormones.

A third noteworthy aspect of teleost endocrinology is the continuing debate over whether teleosts have more than one gonadotropic hormone and the chemical nature of that (those) hormone(s) (see Fontaine, 1976, for a brief review). Protein hormones show greater phylogenetic variability than steroid hormones, introducing greater difficulties in the interpretation of experimental data when mammalian protein hormones are used to study teleost behavioral endocrinology. The use of mammalian protein hormones demands cross-validation to verify that the mammalian hormone produces homologous effects in teleosts. Recent advances in the understanding of teleost pituitary hormones should allow the use of teleost hormones, when necessary, in future research.

ROLE OF GONADAL HORMONES

Traditional wisdom and research in mammals suggest that the pituitary-gonadal axis is involved in aggressive behavior, especially when the behavior is associated with reproduction. The vast majority of the research with teleosts exploring the hormonal factors involved in aggression has focused on the pituitary-gonadal system, yet there is still only limited evidence that gonadal hormones are directly involved in aggressive behavior of the species studied. Androgens may directly affect the frequency of aggressive behaviors in some cichlids and anabantids. Alternatively, where gonadal hormones have been implicated in centrarchids and gasterosteiformes, it has been suggested that the effect of the gonadal hormones is mediated indirectly by an alteration in nestbuilding behavior. Finally, the gobies provide an interesting case in which androgen effects may indirectly affect aggression by influencing sex recognition.

Centrarchids

Reports are contradictory regarding the role of androgens and gonadotropins in aggressive behavior of *Lepomis* (Family: Centrarchidae). Studying *Lepomis gibbosus* and *L. megalotus,* Smith (1969) concluded that androgens and gonadotropic hormones had no direct effect on aggressive behaviors; rather, he suggested that androgens were involved in the establishment of spawning sites and nestbuilding, and that the increase in aggressive behaviors that occurs concurrently with reproduction is secondary to the crowding involved in establishment of nest sites. In contrast, Kramer (1971, 1973) concluded that androgens play a direct role in stimulating aggressive behaviors in *L. gibbosus,* and Avila (1976) proposed that gonadotropin and androgens synergize to maintain aggressive behavior in *L. macrochirus.* These differences in opinion probably arise from procedural differences, though they might reflect species differences.

Nestbuilding and sexual behavior are clearly affected by androgens in *Lepomis.* Methallibure, an antigonadotropic agent, suppresses nestbuilding and sexual behavior and arrests gametogenesis in *L. gibbosus* (Kramer, 1971, 1973). Methyl testosterone and testosterone proprionate stimulate sexual behavior and spawning in both male and female *L. macrochirus* (Avila, 1976). Castration blocks nestbuilding, a preliminary to sexual behavior, in both *L. megalotis* and *L. gibbosus,* and treatment with methyltestosterone reinstates nestbuilding (Smith, 1969).

Despite the apparent effect of androgens on components of reproductive behavior, the most complete evidence suggests little or no direct effect of androgens on aggressive behavior. Castration does not reduce the levels of aggressive behavior shown by *L. megalotus* and *L. gibbosus* whose territories were crowded artificially (Smith, 1969). Under natural conditions, *L. megalotis* and *L. gibbosus* move to shallow water in the spring to establish their nests. *L. megalotis* form colonies of nests with adjacent boundaries, while *L. gibbosus* space their nests a few meters apart. Castrates of both species do not show the normal preference for shallow water and do not build nests. When *L. megalotus* castrates are treated with testosterone, they move toward shallow water, build nests, and show a slight increase in aggressive behavior. Smith concludes that the aggressive behavior is secondary to the crowding following establishment of the nests. He does not report any comparable observation for *L. gibbosus,* but since they normally space their nests some distance apart, Smith's "crowding hypothesis" would predict that there would be no increase in aggressive behavior following treatment with androgen.

Aggressive behaviors and nesting increase in *L. gibbosus* and *L. megalotis* if the fish are subjected to long photoperiod and warm temperatures (Smith, 1970). Photoperiod but not temperature is necessary for nestbuilding behavior, and nestbuilding is stimulated by treatment with human chorionic gonadotropin (HCG). In contrast, aggressive behavior is stimulated by warm temperatures, regardless of the length of the photoperiod, and is unaffected by HCG. It appears that the mechanisms of environmental stimulation of nestbuilding involve pituitary-gonadal hormones, while the environmental influences on aggression do not. These results would seem also to suggest that changes in aggression can occur independently of changes in nesting. Smith suggests that warming temperatures might be involved in an increase in aggressive behavior that results in the disintegration of the schools formed over the winter. He does not speculate on any endocrine mechanism. Temperature-induced changes in aggression occur very quickly in wild populations (Breder, 1936), perhaps too quickly to be mediated by endocrine changes. Interestingly, the highest aggression Smith observed occurred in fish held on short photoperiod and in warm temperatures. Though the differences were slight, it may be that lengthening photoperiod may actually inhibit aggressive behavior.

A number of contradictory observations have been reported regarding the effects of androgens and antiandrogens on aggressive responses in direct paired encounters. Cyproterone acetate (CA), a reputed androgen receptor blocker, increases aggressive responses in male and female *L. macrochirus* (Avila, 1976), an effect that surprisingly lasts two weeks after a single CA injection. Reports are contradictory regarding the effects of CA on *L. gibbosus* aggression (Kramer, 1971; Kramer, Molenda, & Fiedler, 1969).

Methyl testosterone and testosterone propionate have been reported to have no effect on aggression of *L. macrochirus* (Avila, 1976) and to depress aggressive behavior of *L. gibbosus* (Kramer *et al.*, 1969). However, methyl testosterone administration to *L. gibbosus* pretreated with a gonadotropin inhibitor reportedly enhances aggressive behavior (Kramer, 1971, 1972, 1973).

The reported effects of gonadotropins on aggressive behavior are also contradictory. Treatment of *L. gibbosus* with the gonadotropin inhibitor methallibure had no effect on aggressive behavior, though it did affect nestbuilding and spawning responses (Kramer, 1973). However, ovine LH and perch GTH stimulate aggressive as well as nesting and sexual responses (Kramer *et al.*, 1969), and the administration of mammalian LH to methallibure-pretreated *L. gibbosus* slightly and briefly stimulated both sexual and aggressive responses (Kramer, 1971, 1973).

Variation in results might be due to species differences and/or differences in the reproductive/physiological state of the treated subjects. Species such as *L.*

megalotis, which nest in colonies, might show reduced aggression at the height of their reproductive cycle, or aggression might not be tied to their gonadal state. In contrast, species such as *L. gibbosus,* which nest in greater isolation, might show some connection between gonadal hormones and aggression. Some of the contradictions in the results obtained with *Lepomis* might be owed to small sample sizes and inadequate control groups. Differences in experimental procedures might also account for some variation in results. Both Avila and Kramer observed aggressive and sexual behavior in direct encounters in a small space, while Smith observed fish that had only visual access to each other or fish in a seminatural setting. If Smith is correct, then direct encounters in a crowded environment may not accurately reflect the normal function of gonadal hormones in aggression. Until further research is done, the most conservative interpretation of the data is that gonadal hormones do not directly regulate aggressive behavior of *Lepomis.*

Perhaps we should not expect that levels of aggression would be androgen dependent in *Lepomis,* since they show minimal sexual dimorphism in aggressive behavior. Greenberg (1947) and Hale (1956) note that mature and immature *Lepomis cyanellus* (green sunfish) of both sexes show aggressive behavior. *L. macrochirus* do not show sex differences in aggressive behavior when tested in paired encounters, either before they are reproductively ready or during the reproductive season (Chiszar, Ashe, Seixas, & Henderson, 1976; Henderson & Chiszar, 1977). Henderson and Chiszar suggest that nestbuilding and territorial behaviors are performed solely by males not because of physiological differences but because males enter the spawning grounds prior to the females and preempt available nest sites.

Chiszar *et al.* (1976) observed the intensity of aggressive behavior of *L. macrochirus* at varying stages of reproductive readiness and found that the length of isolation necessary to produce maximal isolation-induced aggression shortened during the reproductive season. In such experimental paradigms, reproductive state may enhance aggression nonspecifically by affecting the general reactivity of the fish (Davis, 1975).

Cichlids

In contrast to *Lepomis,* there is evidence that androgens stimulate aggressive behavior in cichlids. Androgen levels, as measured by the size of the androgen-dependent genital papilla, appear to be more effective than body size in predicting dominance in *Tilapia mariae* (Schwanck, 1980).

Citing Molenda (unpublished observations), Fiedler (1974) reports that CA reduces aggressive behavior in *Aequidans pulcher* and *Tilapia mossambica.* Testosterone apparently enhances aggression in *Aequidans* and depresses aggression in *Tilapia.* These conclusions must be regarded as tentative, since procedural details of Molenda's work are not available. Castration is reported to reduce aggressive behavior in *Haplochromis* and *Hemihaplochromis,* and androgen to reinstate aggressive behavior in castrates and enhance agonistic behavior in intact males and females (Fernald, 1976; Reinboth, 1972; Wapler-Leong & Reinboth, 1974). These effects could be mediated, in part, through androgen-induced changes in body markings. Sexually dimorphic and androgen-dependent body markings can influence the aggressive behavior of opponents (Heilegenberg & Kramer, 1972; Leong, 1969). The response of *Haplochromis* to conspecifics of differing intensity of body markings could be due to socially induced changes in androgen levels; however, androgens stimulate aggressive behavior even toward neutral nonconspecifics.

Investigations in Baerands' laboratory (Greenberg, Zijlstra, & Baerands, 1965; Metuzals, Ballintijn-de Vries, & Baerands, 1968; Polder, 1971) have dealt with pituitary and gonadal correlates of the behavioral patterns occurring during the two- to six-day spawning cycle of the cichlid *Aequidans portalegrensis*. In the male, the frequency of aggressive acts is highest two to three days prior to spawning. In the female, the frequency surges immediately after spawning. Histological data suggest that there is enhanced secretory activity preceding spawning in gonadotropic cells and in the neurohypophysis. Immediately after spawning there is evidence of activation of prolactin cells. These observations are consistent with the interpretation that changes in male but not female agonistic behavior are a consequence of pituitary-gonadal activity. Female agonistic behavior is correlated with pituitary prolactin activity.

The most convincing evidence of an androgen-induced effect on aggression was provided by Fernald (1976). Testosterone-treated male *Haplochromis* show increases in approach and attack responses to blinded nonconspecifics. No changes were seen in courtship or activity, and the aggressive behavior of control fish was unaffected by placebo injections. Fernald suggests that the motivational state of males might be amplified by interactions between androgens and body markings: A brightly marked conspecific might elicit increases in testosterone, enhancing both aggressive motivation and "territorial" markings, which in turn feed back and enhance the endocrine and motivational state of the initial intruder.

Both Baerand and colleagues and Fernald have speculated on the interplay between endocrine and behavioral states in the unfolding of the reproductive sequence. For example, Fernald suggests not only that androgens enhance "territorial" markings and aggressive motivation in *Haplochromis* but also that such changes in one fish stimulate androgen levels in its opponent, which in turn trigger changes that feed back to further amplify the behavior of the first fish. These positive feedback loops might serve to amplify the impact of hormones and of social stimuli.

Anabantids

Use of anabantids for the study of aggression is in part due to the conspicuous aggressivity of *Betta splendens*. The *Betta* and other anabantids share a very rich repetoire of social displays (Goldstein, 1974). The reinforcing effects of conspecific social stimuli have been extensively studied using anabantids. Presentation of a conspecific model, a mirror, or visual access to a conspecific is reinforcing in anabantids (Davis, Harris, & Shelby, 1974; Davis, Mitchell, & Dolson, 1976; Hogan, 1967, 1974; Thompson, 1963).

Anabantids exhibit clear sexual dimorphism in aggressive behavior. Davis *et al.* (1974) looked at levels of aggressive displays toward a mirror image and between pairs of *Macropodus opercularis*. Subjects were isolated for varying periods of time prior to exposure to the conspecific stimulus. Males show significantly higher levels of lateral display but not frontal display, attacks, or activity. Isolation-induced increases in the frequency of lateral display and frontal display occur in both sexes. In intrasexual encounters, males show higher levels of lateral display and quiver. Males display less to females than to males, while females show equal levels of agonistic display to male and female partners. Males find the mirror stimulus more attractive, as measured by the amount of time spent near the mirror. However, there are no sex differences in the rate of operant responding for visual access to a live conspecific (Davis *et al.*, 1976). Although the males show high levels of agonistic display toward a mirror and spend more time near a mirror than females, it appears that the attractiveness of the mirror and its ability to elicit agonistic display are independent (Davis *et al.*, 1974).

Macropodus also show sex differences in the levels of aggressive display shown in intrasexual groups of fish; groups of males show considerably more lateral displays than groups of females. These groups were held together for a number of days and the levels of aggressive behavior were stabilized in both male and female groups prior to comparison (Villars & Davis, in preparation).

Davis and Kassel (1975) found sex differences in the frequency of lateral displays shown by adult *Macropodus*; however, there were no sex differences in subadults. They studied the development of behavior and of the gonads in both sexes. Both males and females show increasing levels of aggressive display with sexual maturation; however, the rate of the behavioral enhancement is slightly greater in the males than in the females, resulting in the observed sex difference of adults.

Despite reliable sex differences in aggressive behavior, experiments have not clearly demonstrated an effect of gonadal hormones on agonistic behavior. Castration of *Macropodus* has no effect on spawning behavior, nestbuilding, lateral displays, or other sociosexual displays in males paired with females, though castration does reduce fertility (Villars, 1977; Villars & Davis, 1977). Castration does not affect responding in an operant situation where opportunity to display and to view a live conspecific is the reinforcer. However, the failure to obtain a behavioral effect of castration may be due to rapid testicular regeneration; *Macropodus* can regenerate significant quantities of testicular tissue within one or two weeks (Villars & Davis, 1977). When testes regeneration is inhibited by treatment with methallibure, an antigonadotropic agent, castrate males show impaired spawning and decreased frequency of lateral display to females. Replacement with testosterone enanthate reinstates sexual and aggressive behavior (Villars, 1977; Villars & Davis, 1977). Treatment of *Macropodus* with chorionic gonadotropin and methyl testosterone have been reported to enhance aggression and overall activity (Machemer, 1971).

Methallibure administration reduces the frequency of lateral displays but not frontal displays in a 10-minute trial with a conspecific (Davis *et al.*, 1976). Gonadosomatic index and fertility are decreased by methallibure, providing indirect evidence of an effect of methallibure on pituitary gonadotropin. However, a dose regime of methallibure that produces no histological evidence of gonadal atrophy is adequate to impair operant responding for conspecific visual stimulation in both males and females. Thus, it remains to be demonstrated whether any of these effects were mediated through gonadal steroids.

In castrate *Trichogaster trichopterus* there is no significant effect of gonadectomy on dominance established in paired encounters nor any significant difference in approach or lateral display. However, there is slightly less butting in pairs of castrated males than in pairs of intact males or in castrate-intact pairs. Parental behavior is also normal if castrates show spawning responses and are presented with fertile eggs. In contrast, nestbuilding is clearly impaired and spawning behavior reduced following castration (Johns & Liley, 1970). Methyl testosterone replacement restores nestbuilding and spawning. It appears that the *Trichogaster* either does not show testicular regeneration or that regeneration occurs more slowly than in *Macropodus.* Examination of castrates for testicular remnants more than three months after surgery reveals no apparent remnants in most castrates. However, Johns and Liley speculated that some spawning castrates may retain undetected sources of androgens, since they continue to show secondary sex characteristics known to be androgen dependent. *Betta splendens* also show no effect of castration on aggressive displays but, like *Macropodus,* show evidence of rapid testicular regeneration (Weiss & Coughlin, 1979).

Thus, in contrast to *Lepomis,* anabantids do show sexual dimorphism in aggressive behavior and a moderate effect of pituitary-gonadal hormones on aggressive behavior. Suppression of pituitary-gonadal hormones reduces some aggressive displays in pairs of *Macropodus,* and androgen treatment increases or reinstates levels of aggressive display of males. Further studies are needed to determine whether the reinforcing value of social-aggressive stimuli are affected by gonadal steroids. In contrast to *Macropodus,* aggressive behavior of *Trichogaster* does not appear to be influenced by gonadal steroids. It would be interesting to determine whether aggressive display of *Trichogaster* is less sexually dimorphic than in *Macropodus.* Since there are procedural differences among these studies, it would be valuable to observe sexual dimorphism of both species within the same experiment.

Gasterosteiformes

Although a great deal of work has been done on the stickleback *(Gasterosteus aculeatus),* the role of hormones in their aggressive behavior is far from clearly understood. The physiology, systematics, and behavior of *Gasterosteus* has been extensively reviewed in a recent monograph (Wooton, 1976). Wooton (1970) reported that males in a seminatural setting are more aggressive after building a nest and prior to spawning than are males prior to nestbuilding. Prior to nestbuilding, males do not distinguish between male and female intruders; however, after nestbuilding begins, males show more aggression toward male intruders than toward female intruders. The levels of aggressive behaviors displayed toward a conspecific across a glass partition are one quarter the levels of aggression shown during the spawning period (Baggerman, 1966). The levels of display then double in the week preceding onset of nestbuilding and double again during the nesting period. Aggression declines after actual spawning but shows a sharp increase following emergence of the young (Wooton, 1971).

It has been suggested that there are several distinct categories of aggressive behavior in mammals, each with a different physiological base (Moyer, 1968, 1976). In teleosts the function and physiology of aggression varies across the annual cycle of reproductive behavior. Aggression might function in the dispersal of schools during the transition from the schooling behavior of the winter fish to the establishment of territories and nests in the early spring (Hoar, 1962a). Aggression could acquire other functions during the territorial and nesting phase of reproduction. Wooton (1970) has suggested that during the nesting phase, aggressive behavior is cued by the presence of the nest. Some aggressive behavior may be appropriately labeled "parental" and may be cued by the presence of young in the nest, since there are increases following spawning and at the time the young are hatching (Wooton, 1971). The possible endocrine substrates of these various types of aggression are as yet undemonstrated.

The effects of experimental and physiological manipulations of gonadal hormones in *Gasterosteus* vary with the season and with the behavioral or physiological state of the subjects. Hoar (1962a,b) has reported that in prenesting males, the already low level of aggressive behavior of fish held on short photoperiod is unaffected by castration, while aggressive behavior of castrates held on long photoperiod increases in a manner qualitatively similar to increases shown by unoperated controls. In contrast, reproductive behavior and nestbuilding are decreased by castration regardless of photoperiod. Males castrated prior to the onset of nesting show no effect of castration on their aggressive behavior; their moderate levels of aggressive behavior continue at the same levels or increased to half the levels of intact fish that were building nests and establishing territories (Baggerman, 1966). In these same prenesting fish, the onset of nesting and courtship was prevented by castration. In contrast, castration of fish that had already begun nesting and were showing high levels of aggressive behaviors resulted in a decline in aggressive, nesting, and courtship behaviors (Baggerman, 1966; Wai & Hoar, 1963). The aggressive behavior of castrates previously nestbuilding was less than that of subjects castrated prior to onset of nesting. Levels of courtship and nestbuilding were completely suppressed in all subjects regardless of the timing of castration.

Two explanations have been proposed for the interaction of castration with photoperiod and behavioral condition of the subject. Hoar (1962a,b) and Baggerman (1966) suggest that prenesting aggression is stimulated by pituitary gonadotropin levels. Hoar (1962b) suggests that increasing the photoperiod stimulates pituitary gonadotropin levels, which in turn are instrumental in dispersing the winter schools. According to Hoar, this would explain why long-photoperiod castrates show increasing aggressive behaviors: Castration removes gonadal feedback inhibition and further stimulates gonadotropin already stimulated by long photoperiod, while short photoperiod inhibits gonadotropin levels regardless of gonadectomy. Baggerman (1966) extends these speculations by proposing that after nestbuilding, control of aggressive behaviors shifts from gonadotropins to gonadal hormones. She suggests that such a shift would explain why castration has no effect on aggressive behaviors in prenesting fish while it decreases aggressive behavior in nesting males. The gonadotropin hypothesis will be further discussed in a later section.

Wooton (1970) proposed an alternative explanation to Baggerman's results. He confirmed Baggerman's observations, finding that castration reduces the level of aggressive behaviors in postnesting but not prenesting males. However, unlike Baggerman, he found that prenesting males, both castrate and intact, show levels of aggression equal to the levels displayed by castrate postnesting males. Baggerman had reported that prenesting castrates show higher levels of aggression than males castrated after onset of nesting. Since Wooton found an absence of nesting in both castrate groups and in the prenesting group, and since those groups showed comparable levels of aggressive behaviors in his study, he suggests that during the reproductive cycle aggressive behaviors are stimulated by the presence of a nest and not by a hormonal state. Consistent with this interpretation is the observation that aggressive interactions decrease following nest disruption in sticklebacks (Symons, 1965). Also consistent with Wooton's proposal is Hoar's (1962a) comment that the aggressive behavior of intact long-photoperiod males with nests is "more intense" than that of non-nesting castrates on long photoperiod. Wooton suggests that the effects of castration on the aggressive behavior of *Gasterosteus* are secondary to the effects of castration on nesting behavior. This suggestion parallels a similar suggestion by Smith (1970) for the regulation of aggressive behavior in *Lepomis*.

Thus, Wooton's results and conclusions differ critically from Baggerman's. Liley (1969) suggests that differences between the researchers may be a function of the race of three-spined stickleback studied. Baggerman was working with the *trachurus* form, which is a migratory form spawning in freshwater and maturing in seawater. In contrast, Wooton was working with the *leirus* form, a nonmigratory, wholly freshwater form. With such dramatic differences in life history, it would not be surprising to find other differences in physiological or behavioral response to castration.

Photoperiod alters the effect of androgens on nestbuilding. Testosterone treatment—which is adequate in long-photoperiod fish to induce secondary sex characteristics, namely, kidney tubule development and "nuptual" coloration—does not affect nestbuilding in short-photoperiod fish (Baggerman, 1966; Hoar 1962a). Methyl testosterone induces nestbuilding in males held on a long photoperiod but does not induce adequately organized nestbuilding in gonadectomized females (Baggerman, 1968; Wai & Hoar, 1963).

Testosterone propionate has no effect on male nesting or on aggressive or courtship displays toward male or female stimulus fish (Rouse, Coppenger, &

Barnes, 1977). Treatment with CA, which presumably blocks androgen receptors (Schreck, 1973), decreases the frequency of charges and displacement fanning in the presence of another male and decreases zigzagging and charging directed toward a gravid female (Rouse *et al.,* 1977). Nest-tending and displacement fanning decrease in CA-treated fish. The effects of CA on nest-tending were not statistically significant; however, the number of subjects was small and a larger sample size might yield statistical significance.

Incompletely mature sticklebacks tested in the early spring show significant sex differences in agonistic behaviors. The males show significantly more lunges and bites, significantly less threat (displaced digging), and levels of approach and zigzagging equal to those of females (Huntingford, 1979). The behavior of fully reproductive individuals might show a different pattern, with even stronger sexual dimorphism.

BATHYGOBIUS

Bathygobius soporator is a species of seasonal breeders that spawns repeatedly over the course of the reproductive season. Both males and females show territorial and aggressive behavior year round (Tavolga, 1954). Castration of male *Bathygobius* during the reproductive season blocks aggressive behavior toward intruding males and nonreproductive females (Tavolga, 1955). Surprisingly, the castrates show increased courtship toward those intruders and the operation produces a slight (though statistically insignificant) increase in courtship directed toward reproductively ready females. A number of castrate males show testicular regeneration 40–50 days following surgery. Those regenerated males also show normal responses toward intruding males, while castrates that show no regeneration continue to respond toward intruding males with courtship.

Informal observations of groups of 30 individuals during early spring when the gonads of both sexes are regressed indicate that both sexes are territorial and display agonistic behavior. No courtship responses were ever seen in the early spring. During the reproductive season, *Bathygobius* intruders of both sexes, when enclosed in a glass tube, elicit courtship responses from resident males. Tavolga (1956, 1976) concluded that a pheromonal cue allows discrimination of the reproductive state of an intruder and that testicular hormones activate the male's ability to discriminate. However, it is difficult to reconcile that conclusion with the observation that males showing testicular regression during the spring show agonistic behavior and not courtship, the reverse of the effects of castration. *Bathygobius* is a very interesting and promising model for further behavior–endocrine study.

OTHER TELEOSTS

The species discussed in detail thus far represent only two orders of the 12 orders of Acanthopterygii. Representing Atherinomorpha, platyfish *(Xiphophorus maculatus)* castrates paired with females are less aggressive and less sexually active than sham-operated controls (Chizinsky, 1968). The role of hormones in female sexual behavior of other cyprinodonts has been studied (cf. Liley, 1972; Stacey, 1981) but little work is available on endocrine factors in male or female aggression.

The Ostariophysi is an extensive group whose behavior and physiology has been well studied. Representing the cyprinids, the endocrinology of the goldfish has been well described. The role of hormones in sexual behavior of females has recently been well examined (Stacey, 1981; Stacey & Liley, 1974). Recent work has also examined the effect of social stimuli on the endocrine condition of male goldfish (Kyle, Stacey, Billard, & Peter, 1979). However, goldfish reportedly do not show territorial or aggressive behavior.

The aggressive behavior of some species of siluriform catfish has been well documented, as has the role of chemosensory stimuli in individual recognition and maintenance of territory (Atema, 1969; Atema, Todd, & Bardach, 1969; Bardach & Villars, 1974). This would seem an appropriate group in which to explore further comparative aspects of behavior–endocrine mechanisms in aggressive and territorial behavior.

Little information is available regarding hormonal factors in the social behavior of adult salmonids. Jones and King (1952) report a decline in territorial and agonistic behavior of castrate male Atlantic salmon. Some possible extragonadal factors in juvenile aggression of salmon will be discussed in a later section.

Neuroendocrinology and Teleost Behavior

Recent extensions of autoradiographic techniques to teleosts has provided valuable information on the sites of steroid binding in teleost brain (Davis, Morrell, & Pfaff, 1977; Demski & Hornby, 1982; Kim *et al.*, 1978; Stumpf, Sar, Kim, Deefer, & Martinez-Vargas, 1980). Like other vertebrates, teleosts bind steroids in the tuberal hypothalamus, pituitary, and preoptic area. Binding sites have also been reported in ventral parts of the telencephalon. The absence of binding in the mesencephalon and areas homologous to the amygdala and hippocampus have been noted (Davis *et al.*, 1977; Stumpf *et al.*, 1980).

Advances in the understanding of CNS integration of endocrine physiology (e.g., Peter, 1973, 1977; Peter, Crim, Goos, & Crim, 1978) have laid a foundation for an increased understanding of the neuroendocrinology of behavior in teleosts. Thus far, few conclusions can be made. The hypothalamus has been implicated in feeding, courtship, and agonistic behavior (Demski, 1973; Demski & Knigge, 1971; Savage & Roberts, 1975) and in the regulation of gonadotropin and thyrotropin release (Peter & McKeown, 1975; Peter *et al.*, 1978). The preoptic area has been implicated in regulation of the spawning reflex (Macey, Pickford, & Peter, 1974). Lesions of the telencephalon affect aggression, nestbuilding, and reproduction (Atema, 1969; Davis & Kassel, in press; Davis, Kassel, & Martinez, in press). In goldfish, the steroid binding supracommissural region of the ventral telencephalon appears to play a central role in regulating or initiating sexual behavior, since lesions there (but not in preoptic or dorsal telencephalic regions) reduced initiation of spawning acts by males (Kyle & Peter, 1982). Regulation of teleost aggressive behavior has not been studied using such discrete lesions of steroid-binding CNS sites.

NONSTEROIDAL HORMONES

Most research on hormonal regulation of teleost aggressive behavior has focused on gonadal hormones. Research with mammals increasingly attends to the behavioral actions of extragonadal/nonsteroidal hormones. With notable exceptions (e.g., Diakow & Raimondi, 1981; Stacey, 1981), very little research on nonmammalian behavior has been directed toward such possibilities.

Recent work on regulation of female sexual behavior in goldfish (Stacey, 1981) has shown that prostaglandins play a critical role in that behavior. Treatment with prostaglandins stimulates female sexual activity (Stacey, 1976), and inhibition of prostaglandin synthesis blocks sexual activity (Stacey & Liley, 1974). The prostaglandins may operate directly on the central nervous system, since intracranial application is effective (Stacey & Peter, 1979). Treatment of the females with a variety of steroids is ineffective in initiating spawning (Stacey, 1976).

A pituitary factor may be necessary for the effect of prostaglandins; hypophysectomy blocks the effects of prostaglandins on sexual activity. Prostaglandins are released into the bloodstream during or immediately prior to ovulation. Stacey (1976, 1981) suggests that, in the goldfish, increasing levels of prostaglandins, following ovulation, directly or indirectly initiate sexual behavior.

In amphibia, Diakow and Raimondi (1981) have demonstrated a relationship between posterior pituitary hormones and female sexual receptivity. Female frogs will produce a distinctive vocalization if they are not sexually ready when mounted by a male. This vocalization is referred to as a "release call," since it interrupts sexual behavior. Arginine vasotocin (AVT) stimulates female receptivity as measured by the decrease in release calls (Diakow, 1978). As in goldfish, prostaglandins appear to facilitate frog receptivity; prostaglandin inhibition blocks the effects of AVT on release calls (Diakow & Nemiroff, 1980). Posterior pituitary hormones have also been suggested to synergize with androgens to stimulate male sexual behavior of newts (Moore & Zoeller, 1979).

These examples illustrate a role of nonsteroidal hormones in reproductive behavior. It is very likely that, as we come to understand more about hormone–behavior interactions in a variety of species, the number of examples will expand. In the regulation of fish aggressive behavior, there are two examples of situations in which nongonadal hormones may regulate aggressive behavior. There is a considerable body of evidence that indirectly implicates pituitary gonadotropins in the direct regulation of aggressive behavior. There also exists the interesting possibility that aggressive behavior of juvenile salmonids may be affected by hormonal shifts accompanying physiological changes in preparation for downstream emigration to the ocean.

Aggression and Downstream Emigration in Salmonids

Hoar (1976) has discussed the probable phylogenetic relationships between various species of salmonids, a group that includes species commonly referred to as trout and salmon. All species in the genus *Oncorynchus* and some in the genus *Salmo* are migratory, hatching and developing for some time in freshwater streams, then traveling downstream to the ocean or, in some cases, to estuaries or large lakes, where they form schools traveling, feeding, and developing to reproductive maturity. Prior to spawning, these migratory salmon return to the site of their hatching. The homeward migration prior to spawning and the olfactory basis of that migration have been well described and documented (Harden-Jones, 1968; Hasler, Scholz, & Horrall, 1978; Scholz, Horrall, Cooper, & Hasler, 1976).

The length of time juveniles remain in their homestream prior to downstream emigration varies with the species (Harden-Jones, 1968; Hoar, 1976); some species such as chum salmon *(Oncorhynchus gorbuscha)* may emigrate to the ocean within days or months of hatching, while some such as the coho *(O. kisutch)*, spend their first year or two in freshwater. The physiological and behavioral changes preparatory for the ocean-going (pelagic) stage of the salmon life cycle are referred to as the "smolt transformation" (smoltification), and juveniles in the process of emigration are generally referred to as "smolts" (Hoar, 1976).

Juvenile salmon that show a prolonged freshwater phase (parr stage) also show territorial and aggressive behavior (Dill, 1978; Kalleberg, 1958). A number of authors have suggested that this territorial behavior functions to provide adequate food supply for the juveniles (Dill, 1978; Kalleberg, 1958; Symons, 1968), though such a conclusion has been challenged by the research of others (Chapman, 1962; Keenleyside & Yamamoto, 1962). Food deprivation has been shown to increase aggressive behavior of other teleost species (Albrecht, 1966).

Aggressive mechanisms might also play a role in downstream emigration of juvenile salmonids. The causes of emigration of juvenile salmon remain unclear. In species that emigrate shortly after hatching, the mechanisms may be passive: As the juveniles emerge from the substrate they may simply be carried downstream. In species that reside in the freshwater stream for some time prior to emigration, other mechanisms need to be postulated. Thorpe and Morgan (1978) reviewed a number of hypothesized mechanisms in their discussion of the emigration of Atlantic salmon *(Salmo salar)*. Prior to emigration, presmolt salmon demonstrate site-attachment, territorial and aggressive behavior, and preference for freshwater—all potential mechanisms that could account for retention of the salmon in the stream. One potential mechanism of emigration might involve changes in the motivation or physical ability of the smolt to maintain a stationary position in the stream (Flagg & Smith, 1979; Thorpe & Morgan, 1978). Parr maintain a constant position in stream speeds faster than those in which smolts will maintain position (Flagg & Smith, 1979). Flagg and Smith suggest that the decline in maximum swim speed is secondary to osmotic stress in the smolts, which, because of the smoltification process, become increasingly less physically tolerant of the freshwater. However, their data do not show any difference between oxygen consumption of smolts and parr in freshwater, suggesting that the smolts are not more stressed. Thorpe and Morgan (1978) cite data by Wankowski that also shows reduced swim speeds in smolts. They suggest that the changes in swim tolerance are not simply due to fatigue, but rather that they involve a motivational change in which the fish are less inclined to attempt to maintain position in a rapidly flowing stream.

Changes in salinity preference might induce emigration once the juveniles are near the mouth of the river; smolts show increased preference for seawater compared with parr (Baggerman, 1963; Houston, 1957). Emigration could also result from changes in activity level, in preference for being in a downstream current, or in preference for light or horizontal location in the stream (Hoar, 1951).

A number of theories of emigration involve changes in territoriality or aggressive behavior. One speculation is that decreased territoriality and decreased site-attachment might result in emigration (Kalleberg, 1968; Keenleyside & Yamamoto, 1962). Both authors report a decline in territoriality of Atlantic salmon and trout prior to emigration; however, they do not provide detailed analysis of that change. An alternative theory might suggest that increases in aggressive behavior serve a function in the onset of oceangoing emigration. Chapman (1962) has found that fish that are not able to establish territories drift downstream. If smoltification is correlated with increases in territorial size and aggressive behavior, such a behavioral mechanism might combine with other physiological and behavioral mechanisms to induce the emigration of smolts into the ocean environment.

Interestingly, many of the physiological changes and behavioral changes occurring during smoltification involve increases in thyroid activity and increasing levels of thyroid hormones. Prior to parr–smolt transformation, there is an increasing activity in the thyroid (Hoar, 1939). There is accumulating evidence that the thyroid is involved in a variety of physiological changes accompanying smoltification (cf. reviews by Fontaine, 1975; Woodhead, 1975). Recent work by Dickhoff, Folmar, and Gorbman (1978) has demonstrated that plasma levels of thyroxine increase substantially immediately prior to smoltification in coho salmon.

Behavior changes have been related to thyroid activity. Preference for salinity has been shown to be related to thyroid activity in premigratory sticklebacks (Baggerman, 1966). Thyroid-stimulating hormone increases salinity preference in coho salmon (Baggerman, 1963). Activity, rheotaxis, and aggressive behavior decline in yearling Atlantic salmon treated with thyroxine (Godin, Dill, & Drury, 1974); triiodothyronine produces similar results. Thyroxine treatment reduces swim activity in cod (Woodhead, 1970). In mammals, thyrotropin-releasing hormone reduces isolation-induced aggression (Malick, 1976).

Increased territory size and dominance was found in two species of cichlids given a prolonged thyroxine treatment (Spiliotis, 1974); there were also hormone-related increases in activity. Thiourea, which blocks thyroxine production, reduces activity and results in abandonment of the territories. However, the effects of thiourea should be interpreted with caution since that drug can produce toxic side effects. Territory size is also increased in Pacific salmon treated with thyroxine (Hoar, Keenleyside, & Goodall, 1955).

The possible relationship between thyroxine and aggressive behavior in salmonids and the function of such changes in aggressive behavior are still speculative. However, it would be reasonable to suspect that thyroxine, which is involved in the physiological preparation for emigration into a salt-water environment, might also produce behavioral changes conducive to that environmental shift.

Gonadotropins

A number of researchers have speculated that gonadotropins are somehow directly involved in teleost aggressive behavior (Baggerman, 1966; Hoar, 1962a,b), and results of other studies could be interpreted in terms of a function of gonadotropin (e.g., Polder, 1971; Weiss & Coughlin, 1979). Gonadotropin regulation of behavior has been suggested to stimulate agonistic behavior of starlings (Davis, 1964; Mathewson, 1961) and quail (Cook & Butterfield, 1968). Starlings administered LH show an increase in their dominance position (Mathewson, 1961), while testosterone injections are followed by a decrease in dominance position (Davis, 1957). Davis (1964) discusses how gonadotropin might account for hitherto unexplained failures of the testosterone theory in birds. The evidence currently available to support a gonadotropin theory in fish is limited and largely circumstantial.

Some researchers (Baggerman, 1966; Hoar, 1962a,b; Tavolga, 1955) have turned to a gonadotropin theory to account for some of their results. Hoar observed that the effectiveness of castration in reducing agonistic behavior was dependent on the light cycle. Studying *Gasterosteus aculeatus,* he found that castration was ineffective in reducing aggression when fish were maintained on a long (16-hour) light cycle. He hypothesized that long daylight resulted in stimulation of the pituitary and that gonadotropin maintained the aggressive behavior of castrates on the long-day cycle.

Baggerman (1966) reported some data providing further indirect evidence for a gonadotropin theory. In stickleback, she noted that castration reduced agonistic behavior only after nestbuilding began. Castration had no effect on agonistic behavior in the prenesting period. Referring back to Hoar's work, she interpreted her data to mean that early, prenesting agonistic behavior was under the control of gonadotropins, which were rising in response to lengthening daylight; once androgen levels had risen and nestbuilding had begun, maintenance of agonistic behavior shifted to androgens.

In Tavolga's work with *Bathygobius* (Tavolga, 1955, 1956), castration reduces aggressive behavior while increasing sexual behavior. In contrast, hypophysectomy blocked both aggressive and sexual behavior. This might indicate a role of gonadotropin in stimulation of sexual behavior or inhibition of aggressive behavior, although the effects of hypophysectomy could be due to surgical trauma.

Mammalian hormones with LH-like activity (mLH, HCG) produce increases in aggressive behavior in some groups of *Gasterosteus* (Hoar, 1962a,b); however, hormones with follicle-stimulating activity do not seem to enhance aggressive behavior. Injections of salmon and cod pituitary homogenates produce increases in aggression of some groups of males. Carew (1968) found that males held on long photoperiod but treated with a gonadotropin inhibitor (methallibure) show decreased aggressive behavior. Since castration of long-photoperiod males does not reduce aggression, Carew's observation suggests a possible direct effect of gonadotropin, although further work is needed to rule out the nonspecific effect of methallibure.

The comparison of a gonadotropin theory with an androgen theory requires manipulation of gonadotropins. Such manipulations are complicated by the difficulty in hypophysectomizing fishes. Until recently, the only chemical inhibitors of gonadotropins were steroidal agents that produce many of or all the effects of gonadal hormones. They were of little use in dissecting out pituitary-gonadal function. The dithiocarbamoylhydrazine methallibure, mentioned earlier, is a nonsteroidal inhibitor of gonadotropins which has been shown to cause gonadal atrophy in both mammals and fish (for brief review see Davis *et al.*, 1976; Hyder, 1972). In goldfish it reduces plasma levels of gonadotropin (Bretan, Jalabert, & Billard, 1973); pituitary levels of gonadotropin were not reduced, suggesting an interference with releasing factors. Studies in rodents have suggested that methallibure acts at the level of the hypothalamus, not the pituitary (Malven, 1971). Methallibure might exert its effect through an interaction with catecholamines (Davis *et al.*, 1976), which may be involved in regulation of releasing factors (Brown & Fawke, 1972; Kalra & McCann, 1973; deWied & deJong, 1974).

In addition to methallibure's effects on gonadotropins, some impairment of thyroxine production may occur (Mackay, 1973; Tulloch, Crooks, & Brown, 1963). Apparently, no effect on ACTH or adrenal function is produced (Bielanski, 1973). Methallibure's effect on other pituitary factors is unclear, although it may inhibit growth hormone (Cargill-Thompson, 1963) and stimulate prolactin release (Dies & Vermouth, 1973).

Work by Villars and Davis (unpublished observations) explored the role of pituitary hormones in *Macropodus* aggression, using methallibure. Small groups of male *Macropodus* establish stable communities within two days and show levels of aggressive behaviors that are quite stable from day to day and are remarkably similar from group to group. Methallibure produces a significant, dose-related suppression in levels of aggressive behavior of treated groups. Control groups show slight increases in levels of aggressive interactions. Methallibure also reduces the operant rate for conspecific stimuli (Davis *et al.*, 1976) and the frequency of lateral display, but not frontal display, during a brief exposure to a restrained conspecific of the same sex. Methallibure alone has no effect on sexual or nestbuilding behavior (Davis *et al.*, 1976; Villars & Davis, 1977).

Castration has no effect on the levels of aggression in groups of male *Macropodus;* moreover, the effect of methallibure on aggression in castrates is no different from that in intact groups of males (Villars & Davis, in preparation). This might suggest that gonadal hormones are not involved in levels of group aggression or in the effects of methallibure on that aggression; however, potential testicular regeneration (Villars & Davis, 1977) confounds this conclusion.

Methallibure produces a decline in the levels of lateral display in males paired with females (Villars, 1977) but does not block spawning behavior unless it is given in conjunction with castration (Villars & Davis, 1977).

In a number of cases, inhibition of gonadotropin by pharmacological means reduces aggressive behavior while surgical castration does not, or treatment of subjects with gonadotropic hormones enhances aggression where androgen treatment does not. There are, however, a number of observations that caution against concluding that gonadotropins operate directly on mechanisms of aggressive behavior in fish. (1) Pharmacological procedures used to reduce gonadotropins might affect behavior nonspecifically (e.g., through mild toxicity) or through endocrine or physiological systems other than gonadotropin. (2) The inability of surgical castration to affect a behavior does not necessarily prove that the behavior in question is under direct gonadotropin control, as has been inferred by some authors, nor does it even conclusively demonstrate that testicular hormones are not involved in regulation of the behavior in question. Some fish species are capable of regenerating testicular tissue (e.g., Tavolga, 1954; Villars & Davis, 1977), and small remnants of testicular tissue may be adequate to maintain an apparently androgen-dependent behavior (Johns & Liley, 1970; Villars & Davis, 1977). (3) Some observations are inconsistent with a gonadotropic theory. For example, Rouse *et al.* (1977) report that CA, an androgen receptor blocker, decreases agonistic display of *Gasterosteus.* A gonadotropic theory would predict that CA would enhance or have no effect on aggression. (4) Comparison of results across studies should be done cautiously. Various researchers have used differing experimental and observational procedures, different species, and, in the same species, individuals of possibly differing physiological condition.

DISCUSSION

Despite the paucity of information on hormonal factors in the aggressive behavior of fishes, it is clear that the study of fishes can contribute to our understanding of the evolution and adaptation of hormone–behavior systems. Earlier discussion of teleost systematics illustrates the diversity of fish species. Fishes also show a great diversity of social systems (e.g., Breder & Rosen, 1966; Keenleyside, 1979; Perrone & Zaret, 1979). By using teleost models we can advance our understanding of the ways in which ecological pressures have shaped hormone–behavior systems.

Awareness of the relationship between behavior and ecology has been developing over the last century (Davis, 1974). Work on the ecology of aggressive behavior has focused on the relationship between population density and aggressive behavior; population density influences levels of aggressive behavior, and aggressive behavior affects population density through influences on reproductive capacities, mortality, and dispersion (King, 1973). The endocrine system mediates many of these interactions between behavior and ecosystem. Because of the nature of the behavior–ecology relations that have been studied, attention has focused on pituitary-adrenal and pituitary-gonadal systems as mediators.

However, aggressive behavior may interact with the ecosystem in other ways, and these behavior–environment interactions may be mediated through hormones other than those in the adrenal and gonadal systems. The speculated role of aggressive behavior in salmonid emigration that was mentioned earlier and the postulated role of thyroxine in that behavioral system provides one example. The role of prostaglandins in goldfish sexual behavior provides another example.

Cycles of reproductive physiology and behavior are tied to environmental cycles, presumably to optimize conditions for growth and/or rearing of the offspring and to maximize the probability that the young will reach reproductive maturity (e.g., Peter & Hontela, 1978; Scott, 1979; deVlaming, 1974). Endogenous cycles are synchronized (entrained) with environmental cycles by external signals (timing cues) (Scott, 1979). The most commonly studied timing cue is photoperiod, but other environmental factors might come to serve as well (Scott, 1979). Likewise, the endocrine system is a major mediator of the environmental control of reproductive cycles, with the majority of the research on mediation between environmental and behavioral/physiological cycles focusing on pituitary-gonadal systems. Aggressive behavior also occurs outside of the reproductive context (Moyer, 1976) and, that being the case, might involve physiological systems other than those involved in reproduction. Further comparative work should expand our knowledge of the diversity of environmental cues, mediating hormones, and behavioral systems. The study of the behavior and physiology of nonmammalian vertebrates has already resulted in a number of examples of hormone–behavior systems involving hormones not traditionally conceptualized as behavioral hormones. The type, or even the presence, of hormonal involvement in aggressive behavior must depend on the relationship of the aggressive behavior to the whole context of the animal's behavior. Thus, the behavior–endocrine systems utilized by an organism can and should be viewed in the context of the natural history or ecology of that animal. Teleost fishes are excellent models for studying the complex relationships between neuroendocrine, behavioral, and ecological systems.

Acknowledgments

I would like to thank Professor R. E. Davis for reading and commenting on preliminary drafts of this chapter. Also, thanks are due to Linda Guyton and Alice Peterson for their assistance in typing.

REFERENCES

Albrecht, H. Zur Stammesgeschichte einiger Bewegungsweisen bei Fischen, untersucht am Verhalten von *Haplochromis* (Pisces, Cichlidae). *Zeitschrift für Tierpsychologie,* 1966, *23*,270–302.

Atema, J. *The chemical senses in feeding and social behavior of the catfish, Ictalurus natalis.* Doctoral dissertation, University of Michigan, 1969.

Atema, J., Todd, J. H., & Bardach, J. E. Olfaction and behavioral sophistication in fish. In C. Pfaffman, (Ed.), *Third international symposium on olfaction and taste.* New York: Rockefeller University Press, 1969.

Avila, V. L. *Effects of 17 alpha-methyl testosterone, testosterone proprionate and cyproterone acetate on aggressive and reproductive behavior in bluegill sunfish (Lepomis macrochirus Rafinesque).* Paper presented at the meeting of the Animal Behavior Society, Boulder, Col., June 1976.

Baggerman, B. The effect of TSH and antithyroid substances on salinity preference and thyroid activity in juvenile pacific salmon. *Canadian Journal of Zoology,* 1963, *41,* 307–319.

Baggerman, B. On the endocrine control of reproductive behavior in the male three-spined stickleback *(Gasterosteus aculeatus L.). Symposia of the Society for Experimental Biology,* 1966, *20,* 427–456.

Bardach, J. E., & Villars T. A. Chemical senses of fishes. In P. T. Grant & A. M. Mackie (Eds.), *Chemoreception in marine organisms.* New York: Academic Press, 1974.

Bielanski, A. Effect of methallibure on the endocrine glands and the reproductive tract in gilts (hypophysis, thyroid, adrenal). *Bulletin de l'Académie Polonaise des Sciences, Série des Sciences Biologiques* (Warszawa), 1973, *21,* 415–420.

Breder, C. M., Jr. The reproductive habits of the North American sunfish (Centrarchidae). *Zoologica,* 1936, *21,* 1–48.

Breder, C. M., & Rosen, D. E. *Modes of reproduction in fishes.* New York: Natural History Press, 1966.

Bretan, B., Jalabert, B., & Billard, R. Pituitary and plasma gonadotrophin levels and spermatogenesis in the goldfish *Carassius auratus* after methallibure treatment. *Journal of Endocrinology,* 1973, *59,* 415–420.

Brown, P. S., & Fawke L. Effects of reserpine, p-chlorophenylalanine, alpha-methyl tyrosine, thymoxamine or methallibure on pituitary FSH in male rats. *Reproduction and Fertility,* 1972, *28,* 167–175.

Callard, I. P., Callard, G. V., Lance, V., Bolaffi, J. L., & Rosset, J. S. Testicular regulation in nonmammalian vertebrates. *Biology of Reproduction,* 1978, *18,* 16–43.

Carew, B. A. M. *Some effects of methallibure (I.C.I.33,828) on the stickleback Gasterosteus aculeatus L.* Masters Thesis, Department of Zoology, University of British Columbia, 1968. (Cited in Wooton, 1976.)

Cargill-Thompson, H. E. C. The effect of a derivative of dithiocarbamoylhydrazine on the tibiae of hypophysectomized rats. *Journal of Endocrinology,* 1963, *26,* 447–448.

Chapman, D. W. Aggressive behavior in juvenile coho salmon as a cause of emigration. *Journal of the Fisheries Research Board, Canada,* 1962, *19,* 1047–1080.

Chiszar, D., Ashe, F., Seixas, S., & Henderson, D. Social-aggressive behavior after various intervals of social isolation in bluegill sunfish (*Lepomis macrochirus* Rafinesque) in different states of reproductive readiness. *Behavioral Biology,* 1976, *16,* 475–487.

Chizinsky, W. Effects of castration upon sexual behavior in male platyfish, *Xiphophorus maculatus. Physiological Zoology,* 1968, *41,* 466–475.

Crim, J. W., Dickoff, W. W., & Gorbman, A. Comparative endocrinology of piscine hypothalamic hypophysiotropic peptides: Distribution and activity. *American Zoologist,* 1978, *18,* 411–424.

Crook, J. H., & Butterfield, P. A. Effects of testosterone proprionate and luteinizing hormone on agonistic and nest-building behavior of *Quelea quelea. Animal Behavior,* 1968, *16,* 370–384.

Davis, D. Aggressive behavior in castrated starlings. *Science,* 1957, *126,* 253–254.

Davis, D. The physiological analysis of aggressive behavior. In W. Etkin (Ed.), *Social behavior and organization among vertebrates.* Chicago: University of Chicago Press, 1964.

Davis, D. E. *Behavior as an ecological factor.* Stroudsburg, Pa.: Dowden, Hutchinson, & Ross, 1974.

Davis, R. E. Readiness to display in paradise fish *Macropodus opercularis,* L., (Belontiidae): The problem of general and specific effects of social isolation. *Behavioral Biology,* 1975, *15,* 419–433.

Davis, R. E., & Kassel, J. The ontogeny of agonistic behavior and the onset of sexual maturation in the paradise fish, *Macropodus opercularis* (Linnaeus). *Behavioral Biology,* 1975, *14,* 31–39.

Davis, R. E., & Kassel, J. The telencephalon and behavior. In R. E. Davis & R. G. Northcutt (Eds), *Fish neurobiology* (Vol. II). Ann Arbor, Mich.: University of Michigan Press, in press.

Davis, R. E., Harris, C., & Shelby, J. Sex differences in aggressivity and the effects of social isolation in the anabantoid fish, *Macropodus opercularis. Behavioral Biology,* 1974, *11,* 497–509.

Davis, R. E., Mitchell, M., & Dolson, L. The effects of methallibure on conspecific visual reinforcement, social display frequency, and spawning in the paradise fish, *Macropodus opercularis* (L.) Belontiidae. *Physiology and Behavior,* 1976, *17,* 47–52.

Davis, R. E., Morrell, J. I., & Pfaff, D. W. Autoradiographic localization of sex steroid-concentrating cells in the brain of the teleost *Macropodus opercularis* (Osteichthyes: Belontiidae). *General and Comparative Endocrinology,* 1977, *33,* 496–505.

Davis, R. E., Kassel, J., & Martinez, M. The telencephalon and reproductive behavior in the teleost *Macropodus opercularis.* In P. R. Laming (Ed.) *Brain mechanisms in fish and amphibians.* New York: Cambridge University Press, in press.

Demski, L. S. Feeding and aggressive behavior evoked by hypothalamic stimulation in a cichlid fish. *Comparative Biochemistry and Physiology,* 1973, *44,* 685–692.

Demski, L. S., & Hornby, P. J. Hormonal control of fish reproductive behavior: Brain-gonadal steroid interactions. *Canadian Journal of Fisheries and Aquatic Sciences,* 1982, *39,* 36–47.

Demski, L. S., & Knigge, K. M. The telencephalon and hypothalamus of the bluegill *(Lepomis macrochirus):* Evoked feeding, aggressive and reproductive behavior with representative frontal sections. *Journal of Comparative Neurology,* 1971, *143,* 1–16.

Diakow, C. A hormonal basis for breeding behavior of female frogs: Vasotocin inhibits the release call of *Rana pipiens. Science,* 1978, *199,* 1456–1457.

Diakow, C., & Nemiroff, A. Vasotocin, prostaglandin, and female reproductive behavior in the frog, *Rana pipiens. Hormones and Behavior,* 1981, *15,* 86–93.

Diakow, C., & Raimondi, D. Physiology of *Rana pipiens* reproductive behavior: A proposed mechanism for inhibition of the release call. *American Zoologist,* 1981, *21,* 295–304.

Dickhoff, W. W., Folmar, L. C., & Gorbman, A. Change in plasma thyroxine during smoltification of coho salmon, *Oncorhynchus kisutch. General and Comparative Endocrinology,* 1978, *36,* 229–232.

Dies, R., & Vermouth, N. Antiovulatory effect of ICI 33,828 without affecting prolactin release in normal rats. *Neuroendocrinology,* 1973, *13,* 351–356.

Dill, L. M. Aggressive distance in juvenile coho salmon *(Oncorhynchus kisutch). Canadian Journal of Zoology,* 1978, *56,* 1441–1446.

Donaldson, E. M. Reproductive endocrinology of fishes. *American Zoologist,* 1973, *13,* 909–927.

Fernald, R. D. The effect of testosterone on the behavior and coloration of adult male cichlid fish (*Haplochromis burtoni* Gunther). *Hormones and Research,* 1976, *7,* 172–178.

Fiedler, K. Hormonale Kontrolle des Verhaltens bei Fischen. *Fortschrift für Zoologie,* 1974, *22,* 268–309.

Flagg, T. A., & Smith, L. S. *Changes in swimming performance, critical fatigue levels, and metabolic rates during the Parr-smolt transformation of Toutle (accelerated underyearlings) coho salmon (Oncorhynchus kisutch).* Unpublished report prepared for National Marine Fisheries Service. Seattle: Fisheries Research Institute, College of Fisheries, University of Washington, 1979.

Fontaine, M. Physiological mechanisms in the migration of marine and amphihaline fish. *Advances in Marine Biology,* 1975, *13,* 241–355.

Fontaine, M. Hormones and the control of reproduction in aquaculture. *Journal of the Fisheries Research Board, Canada,* 1976, *33,* 922–939.

Godin, J. G., Dill, P. A., & Drury, D. E. Effects of thyroid hormones on behavior of yearling *Salmo salar. Journal of the Fisheries Research Board, Canada,* 1974, *31,* 1787–1790.

Greenberg, B. Hierarchies in *Lepomis cyanellus. Physiological Zoology,* 1947, *20,* 267–299.

Greenberg, B., Zijlstra, J. J., & Baerands, G. P. A quantitative description of the behavior changes during the reproductive cycle of the Cichlid fish *Aequidans portalegrensis* Hensel. *Koninklijke Nederlandse Akademie van Wetenschappen,* Amsterdam, 1965, *3,* 135–149.

Greenwood, P. H., Rosen, D. E. Weitzman, S. H., & Myers, G. S. Phyletic studies of teleostean fishes, with a provisional classification of living forms. *Bulletin of the American Museum of Natural History,* 1966, *131,* 339–456.

Hale, E. B. Effects of forebrain lesions on the aggressive behaviour of green sunfish *Lepomis cyanellus. Physiological Zoology,* 1956, *29,* 107–127.

Harden-Jones, F. R. *Fish migration.* London: Edward Arnold, 1968.

Hasler, A. D., Scholz, A. T., & Horrall, R. M. Olfactory imprinting and homing in salmon. *American Scientist,* 1978, *66,* 347–355.

Heiligenberg, W., & Kramer, U. Aggressiveness as a function of external stimulation. *Journal of Comparative Physiology,* 1972, *77,* 332–340.

Henderson, D. L., & Chiszar, D. A. Analysis of aggressive behavior in the bluegill sunfish *Lepomis macrochirus* Rafinesque: Effect of sex and size. *Animal Behavior,* 1977, *25,* 122–130.

Hoar, W. S. The thyroid gland of the Atlantic salmon. *Journal of Morphology,* 1939, *65,* 257–295.

Hoar, W. S. The behavior of chum, pink and coho salmon in relation to their seaward migration. *Journal of the Fisheries Research Board, Canada,* 1951, *8,* 241–263.

Hoar, W. S. Hormones and reproductive behaviour of the male three-spined stickleback *(Gasterosteus aculeatus). Animal Behavior,* 1962, *10,* 247–266.

Hoar, W. S. Reproductive behavior of fish. *General and Comparative Endocrinology,* 1962, *Suppl. 1,* 206–216.

Hoar, W. S. Smolt transformation: Evolution, behavior and physiology. *Journal of The Fisheries Research Board, Canada,* 1976, *33,* 1234–1252.

Hoar, W. S., Keenleyside, M. H. A., & Goodall, R. G. The effects of thyroxine and gonadal steroids on activity of salmon and goldfish. *Canadian Journal of Zoology,* 1955, *33,* 428–439.

Hogan, J. A. Fighting and reinforcement in the siamese fighting fish *(Betta splendens). Journal of Comparative Physiology and Psychology,* 1967, *64,* 356–359.

Hogan, J. A. On the choice between eating and aggressive display in the siamese fighting fish *(Betta splendens). Learning and Motivation,* 1974, *5,* 273–287.

Holmes, R. L., & J. N. Ball. *The pituitary gland: A comparative account.* New York: Cambridge University Press, 1974.

Houston, A. H. Responses of juvenile chum, pink, and coho salmon to sharp sea-water gradients. *Canadian Journal of Zoology,* 1957, *35,* 371–383.

Huntingford, F. A. Pre-breeding aggression in male and female three-spined sticklebacks *(Gasterosteus aculeatus). Aggressive Behaviour,* 1979, *5,* 51–58.

Hyder, M. Endocrine regulation of reproduction in *Tilipia. General and Comparative Endocrinology,* 1972, *3,* 729–740.

Idler, D. R., Horne, D. A., & Sangalang, G. B. Identification and quantification of the major androgens in testicular and peripheral plasma of Atlantic Salmon *(Salmo salar)* during sexual maturation. *General and Comparative Endocrinology,* 1971, *16,* 257–267.

Johns, L. S., & Liley, N. R. The effects of gonadectomy and testosterone treatment on the reproductive behavior of the male blue gourami *Trichogaster trichopterus. Canadian Journal of Zoology,* 1970, *48,* 977–987.

Jones, J. W., & King, G. M. The spawning of the male salmon parr (*Salmo salar Linn* juv.) *Journal of Zoology, Zoological Society, London,* 1952, *122,* 615–619.

Kalleberg, H. Observations in a stream tank of territoriality and competition in juvenile salmon and trout (*Salmo salar* L. and *S. trutta* L.) *Reports of the Institute of Freshwater Research,* 1958, *39,* 55–98.

Kalra, P. S., & McCann, S. M. Involvement of catecholamines in feedback mechanisms. *Progress in Brain Research,* 1973, *39,* 185–198.

Keenleyside, M. H. A. *Diversity and adaptation in fish behaviour.* New York: Springer-Verlag, 1979.

Keenleyside, M. H. A., & Yamamoto, F. T. Territorial behaviour of juvenile Atlantic salmon (*Salmo salar* L.). *Behaviour,* 1962, *19,* 139–169.

Kim, Y. S., Stumpf, W. E., Sar, M., & Martinez-Vargas, M. C. Estrogen and androgen target cells in the brain of fishes, reptiles and birds: Phylogeny and ontogeny. *American Zoologist,* 1978, *18,* 425–434.

King, J. A. The ecology of aggressive behavior. *Annual Review of Ecology and Systemtics* 1973, *4,* 117–138.

Kramer, B. Zur hormonalen steuerung von Verhaltenweisen der Fortpflanzung beim Sonnenbarsh *Lepomis gibbosus* (L.) (Centrarchidae, Teleostei). *Zeitschrift für Tierpsychologie,* 1971, *28,* 351–386.

Kramer, B. Behavioural effects of an antigonadotrophin, of sexual hormones and of psychopharmaka in the pumpkinseed sunfish, *Lepomis gibbosus* (Centrarchidae). *Experientia,* 1972, *28,* 1195–1196.

Kramer, B. Chemische Wirkstoffe im Nestbau-, sexual- und Kampfverhalten des Sonnenbarsches *Lepomis gibbosus* (L.) (Centrarchidae, Teleostei). *Zeitschrift für Tierpsychologie,* 1973, *32,* 353–373.

Kramer, B., Molenda, W., & Fiedler, K. Behavioral effect of the antiandrogen cyproterone acetate (Schering) in *Tilapia Mossambica* and *Lepomis gibbosus. General and Comparative Endocrinology,* 1969, *13,* 515.

Kyle, A. L., & Peter, R. E. Effects of forebrain lesions on spawning behaviour in the male goldfish. *Physiology and Behavior,* 1982, *28,* 1103–1109.

Kyle, A. L., Stacey, N. E., Billard, R., & Peter R. E. *Sexual stimuli rapidly increase milt and gonadotropin levels in goldfish. American Zoologist,* 1979, *19,* 854 (abst. 9).

Lance, V., & Callard, I. P. Phylogenetic trends in hormonal control of gonadal steroidogenesis. In P. K. T. Pang & A. Epple (Eds.), *Evolution of vertebrate endocrine systems.* Lubbock, Tex.: Texas Technical Press, 1980.

Leong, C. Y. The quantitative effect of releasers on the attack readiness of the fish *Haplochromis burtoni* (Cichlidae, Pisces). *Zeitschrift für vergleichende Physiologie,* 1969, *65,* 29–50.

Liley, N. R. Hormones and reproductive behavior in fishes. In W. S. Hoar & D. J. Randall (Eds.), *Fish physiology* (Vol. III). New York: Academic Press, 1969.

Liley, N. R. The effects of estrogens and other steroids on the sexual behavior of the female guppy, *Poecilia reticulata. General and Comparative Endocrinology,* 1972, *Suppl. 3,* 542–552.

Macey, N. J., Pickford, G. E., & Peter, R. E. Forebrain localization of the spawning reflex response to exogenous neurohypophysial hormones in the killifish, *Fundulus heteroclitus. Journal of Experimental Zoology,* 1974, *190,* 269–280.

Machemer, L. Synergistische Wirkung von Sauger-Prolacktin und Androgen beim Schaumnestbau des Paradiesfishes, *Macropodus opercularis* L. (Anabantidae). *Zeitschrift für Tierpsychologie,* 1971, *28,* 33–53.

Mackay, N. Effects of methallibure and thiourea on gametogenesis in the firetail gudgeon, *Hypseleotris galii. General and Comparative Endocrinology,* 1973, *20,* 221–235.

Malick, J. B. Antagonism of isolation induced aggression by thyrotropin-releasing hormone (TRH). *Pharmacology, Biochemistry and Behavior,* 1976, *5,* 665–669.

Malven, P. Hypothalamic sites of action for methallibure inhibition of gonadotropin secretion. *Journal of Animal Science,* 1971, *32,* 912–918.

Mathewson, S. Gonadotropic control of aggressive behavior in starlings. *Science,* 1961, *134,* 1522–1523.

Metuzals, J., Ballintijn-de Vries, G., & Baerands, G. P. The correlation of cytological changes in the adenohypophysis of the cichlid fish *Aequidens portalegrensis* (Hensel) with behaviour changes during the reproductive cycle. *Koninklijke Nederlandse Akademie van Wetenschappen,* Proc. C. 1968, *71,* 391–410.

Moore, F. L., & Zoeller, R. T. Endocrine control of amphibian sexual behavior: Evidence for a neurohormone-androgen interaction. *Hormones and Behavior,* 1979, *13,* 207–213.

Morrell, J. I., & Pfaff, D. W. A neuroendocrine approach to brain function: Localization of sex steroid concentrating cells in vertebrate brains. *American Zoologist,* 1978, *18,* 447–460.

Moyer, K. E. Kinds of aggression and their physiological basis. *Communications in Behavioral Biology,* 1968, *2,* 65–87.

Moyer, K. E. *The psychobiology of aggression.* New York: Harper & Row, 1976.

Nelson, J. S. *Fishes of the world.* New York: Wiley, 1976.

Pang, P. K. T., & Epple, A. *Evolution of vertebrate endocrine systems.* Lubbock, Tex.: Texas Technical Press, 1980.

Perrone, M., & Zaret, T. Parental care patterns of fish. *American Naturalist,* 1979, *113,* 351–361.

Peter, R. E. Hypothalamic control of thyroid gland activity and gonadal activity in the goldfish, *Carassius auratus. General and Comparative Endocrinology,* 1970, *14,* 334–356.

Peter, R. E. Neuroendocrinology of teleosts. *American Zoologist,* 1973, *13,* 743–755.

Peter, R. E. The preoptic nucleus in fishes: A comparative discussion of function-activity relationships. *American Zoologist,* 1977, *17,* 775–785.

Peter, R. E., & Hontela, A. Annual gonadal cycles in teleosts: Environmental factors and gonadotropin levels in blood. In I. Assenmacher & D. S. Farner, *Environmental endocrinology.* New York: Springer-Verlag, 1978.

Peter, R. E., & McKeown, B. A. Hypothalamic control of prolactin and thyrotropin secretion in teleosts, with special reference to recent studies on the goldfish. *General and Comparative Endocrinology,* 1975, *25,* 153–165.

Peter, R. E., Crim, L. W., Goos, H. J. Th., & Crim, J. W. Lesioning studies on the gravid female goldfish: Neuro-endocrine regulation of ovulation. *General and Comparative Endocrinology,* 1978, *35,* 391–401.

Polder, J. J. W. On gonads and reproductive behavior in the cichlid fish, *Aequidans portalegrensis* (Hensel). *Netherlands Journal of Zoology,* 1971, *21,* 265–365.

Rådesater, T., & Ferno, A. On the function of 'eye spots' in agonistic behavior in the fire-mouth cichlid *(Cichlasoma meedi). Behavioral Processes,* 1979, *4,* 5–13

Reinboth, R. Some remarks on secondary sex characters, sex and sexual behavior in teleosts. *General and Comparative Endocrinology,* 1972, *3,* 565–570.

Rouse, E. F., Coppenger, C. J., & Barnes, P. R. The effect of an androgen inhibitor on behavior and testicular morphology in the stickleback *Gasterosteus aculeatus. Hormones and Behavior,* 1977, *1,* 8–18.

Sage, M., & Bern, H. A. Cytophysiology of the teleost pituitary. *International Review of Cytology,* 1971, *31,* 339–376.

Savage, G. E., & Roberts, M. G. Behavioral effects of electrical stimulation of the hypothalamus of the goldfish *(Carassius auratus). Brain, Behavior and Evolution,* 1975, *12,* 42–56.

Schmidt, P. J., & Idler, D. R. Steroid hormones in the plasma of salmon at various stages of maturation. *General and Comparative Endocrinology,* 1962, *2,* 204–214.

Schreibman, M. P., Leatherland, J. F., & McKeown, B. A. Functional morphology of the teleost pituitary gland. *American Zoologist,* 1973, *13,* 719–742.

Scholz, A. T., Horrall, R. M., Cooper, J. C., & Hasler, A. D. Imprinting to chemical cues: The basis for home stream selection in salmon. *Science,* 1976, *192,* 1247–1249.

Schreck, C. B. Uptake of ^{3}H-testosterone and influence of antiandrogen in tissues of rainbow trout *(Salmo gairdneri). General and Comparative Endocrinology,* 1973, *21,* 60–68.

Schwanck, E. The effect of size and hormonal state on the establishment of dominance in young males of *Tilapia mariae* (Pisces, Cichlidae). *Behavioral Processes,* 1980, *5,* 45–53.

Scott, D. B. C. Environmental timing and the control of reproduction in teleost fish. In P. J. Miller (Ed.), *Fish phenology: Anabolic adaptiveness in teleosts.* London: Academic Press, 1979.

Smith, R. J. F. Control of prespawning behaviour of sunfish (*Lepomis megalotis* and *L. gibbosus*): I. Gonadal androgen. *Animal Behavior,* 1969, *17,* 279–285.

Smith, R. J. F. Control of prespawning behaviour of sunfish (*Lepomis gibbosus* and *Lepomis megalotis*): II. Environmental factors. *Animal Behavior,* 1970, *18,* 575–587.

Southwick, C. H. Display patterns in anabantid fish. In A. N. Stokes (Ed.), *Animal behavior in laboratory and field.* San Francisco: W. H. Freeman, 1968.

Spiliotis, P. H. The effect of thyroxine and thiourea on territorial behavior in cichlid fish. *Dissertation Abstracts International,* 1974, *35B,* 1030.

Stacey, N. E. Effects of indomethacin and prostaglandins on the spawning behaviour of female goldfish. *Prostaglandins,* 1976, *12,* 113–126.

Stacey, N. E. Hormonal regulation of female reproductive behavior in fish. *American Zoologist,* 1981, *21,* 305–316.

Stacey, N. E., & Liley, N. R. Regulation of spawning behaviour in the female goldfish. *Nature,* 1974, *247,* 71–72.

Stacey, N. E., & Peter, R. E. Central action of prostaglandins in spawning behavior of female goldfish. *Physiology and Behavior,* 1979, *22,* 1191–1196.

Stumpf, W. E., Sar, M., Kim, Y. S., Deefer, D. A., & Martinez-Vargas, M. C. Evolutionary aspects of estrogen-sensitive structures in the vertebrate forebrain. In P. K. T. Pang, & A. Epple (Eds.), *Evolution of vertebrate endocrine systems.* Lubbock, Tex.: Texas Technical Press, 1980.

Symons, P. E. K. Analysis of spine-raising in the male three-spined stickleback. *Behaviour,* 1965, *28,* 1–74.

Symons, P. E. K. Increase in aggression and in strength of the social hierarchy among juvenile Atlantic salmon deprived of food. *Journal of The Fisheries Research Board, Canada,* 1968, *25,* 2387–2401.

Tavolga, W. N. Reproductive behavior of the gobiid fish, *Bathygobius soporator. Bulletin of The American Museum of Natural History,* 1954, *104,* 429–459.

Tavolga, W. N. Effects of gonadectomy and hypophysectomy on prespawning behavior in males of tbe gobiid fish, *Bathygobius soporator. Physiological Zoology,* 1955, *28,* 218–233.

Tavolga, W. N. Visual, chemical and sound stimuli as cues in the sex discriminatory behavior of the gobiid fish *Bathygobius soporator. Zoologica,* 1956, *41,* 49–64.

Tavolga, W. N. Chemical stimuli in reproductive behavior in fish: Communication. *Experientia,* 1976, *32,* 1091–1093.

Thompson, T. Visual reinforcement in Siamese fighting fish. *Science,* 1963, *141,* 55–57.

Thorpe, J. E., & Morgan, R. I. G. Periodicity in Atlantic salmon *Salmo salar* L. smolt migration. *Journal of Fish Biology,* 1978, *12,* 541–548.

Tulloch, M. I., Crooks, J., & Brown, P. S. Inhibition of thyroid function by a dithiocarbamoylhydrazine. *Nature,* 1963, *199,* 288–289.

Villars, T. A. *Castration and reproductive behavior in the paradise fish, Macropodus opercularis.* Unpublished doctoral dissertation, University of Michigan, 1977.

Villars, T. A., & Davis, R. E. Castration and reproductive behavior in the paradise fish, *Macropodus opercularis* (L.) (Osteichthyes: Belontiidae). *Physiology and Behavior,* 1977, *19,* 371–375.

Villars, T. A., & Davis, R. E. Effects of methallibure and castration on levels of agonistic behavior in stabilized groups of male paradise fish, *Macropodus opercularis* (L.). In preparation.

deVlaming, V. L. Environmental and endocrine control of teleost reproduction. In C. B. Schreck (Ed.), *Control of sex in fishes.* Blacksburg, Va., Department of Fisheries and Wildlife Sciences, Virginia Polytechnic Institute and State University, 1974.

Wai, E. H., & Hoar, W. S. The secondary sex characters and reproductive behaviour of gonadectomised sticklebacks treated with methyl testosterone. *Canadian Journal of Zoology,* 1963, *41,* 611–628.

Wallis, M. The molecular evolution of pituitary hormones. *Biological Reviews,* 1975, *50,* 35–98.

Wapler-Leong, D. C. Y., & Reinboth, R. The influence of androgenic hormone on the behavior of *Haplichromis burtoni* (Cichlidae). *Fortschrift für Zoologie,* 1974, *22,* 334–339.

Weiss, C. S., & Coughlin, J. P. Maintained aggressive behavior in gonadectomized male siamese fighting fish. *Physiology and Behavior,* 1979, *23,* 173–177.

deWied, D., & deJong, W. Drug effects and hypothalamic-anterior pituitary function. *Annual Review of Pharmacology,* 1974, *14,* 389–412.

Woodhead, A. D. Endocrine physiology of fish migration. *Oceanography and Marine Biology Annual Review,* 1975, *13,* 287–382.

Woodhead, P. M. J. An effect of thyroxine upon the swimming of cod. *Journal of the Fisheries Research Board of Canada,* 1970, *27,* 2337–2338.

Wooton, R. J. Aggression in the early phases of the reproductive cycle of the male three-spined stickleback *(Gasterosteus aculeatus). Animal Behavior,* 1970, *18,* 740–746.

Wooton, R. J. Measures of the aggression of parental male three-spined sticklebacks. *Behaviour,* 1971, *40,* 228–262.

Wooton, R. J. *The biology of sticklebacks.* New York: Academic Press, 1976.

17

Hormonal Influences on Avian Aggressive Behavior

Cheryl F. Harding

INTRODUCTION

Aggressive behavior in birds involves a variety of complex visual and vocal displays, most of which appear to be under hormonal control. Singing is a good example of a complex vocal display used in aggressive interactions, and singing behavior is controlled by a series of recently evolved distinct neural nuclei, several of which are hormone sensitive. Visual displays are enhanced by various morphological adaptations such as striking plumage patterns; brightly colored bills, eye rings, and legs; or specialized wattles and combs. These, too, are often under hormonal control. Overall, much less is known about the relationship between hormones and aggressive behavior in birds than in mammals. In part, this may reflect the difficulty of manipulating the avian endocrine system. In general, removal of the avian gonads is much more difficult than gonadectomy in mammals, and other manipulations such as adrenalectomy or hypophysectomy are rarely attempted. The lack of data is certainly not due to any dearth of aggressive behavior among bird species. Bennett (1939), in summarizing her observations on the aggressive behavior of doves, noted that the high frequency of aggressive contacts and the violence of many encounters "tend to discredit the expression 'gentle as a dove' and to raise a question of the suitability of the dove as an emblem of peace" (p. 356). While studies of aggressive behavior are available on a wide variety of avian species, studies of hormone–behavior relationships have tended to focus on more common groups of birds such as pigeons, ducks, chickens, finches, and blackbirds. In comparison to the mammalian literature, there is a greater emphasis on field studies in avian behavioral endocrinology.

Aggressive behavior in birds has been studied in a variety of contexts, including dominance hierarchies in captive groups, dominance interactions in

CHERYL F. HARDING • Department of Psychology, Hunter College, New York, New York 10021, and Department of Ornithology, American Museum of Natural History, New York, New York 10024. My unpublished research was supported by CUNY PSC-BHE Awards #13011 and #13495.

feeding flocks in the field, territorial encounters, and interspecific interactions. While a wide variety of behaviors may be labeled "aggressive," it is important to distinguish among different forms of aggressive behavior and the circumstances in which they occur if one is to draw meaningful conclusions from the available data. For example, Bennett (1939), in her classic study of the social hierarchy in a flock of ring doves, pointed out that the relative dominance of individuals in the flock was affected by the time of day at which they were observed. An individual's ability to compete for various resources—such as food, nest material, or a place to roost—differed. Thus, the position of birds in the social dominance hierarchy varied over the day as individuals competed for the different resources.

In many species, courtship and aggressive displays are closely linked. In fact, one behavior, singing, may serve to repel males but at the same time attract females, and can thus be classified as either an aggressive or a courtship behavior depending on the social context in which it occurs or the observer's point of view. Or if one looks at aggressive behavior over the breeding cycle, one finds that as the frequency of one aggressive behavior declines, that of a different aggressive behavior increases. Thus, in redwinged blackbirds the frequency of territorial encounters between males declines over the breeding season, but the incidence of interspecific nest defense increases (Peek, 1971). The fact that different aggressive behavior patterns are often studied in different species makes it difficult to generalize across species.

Studies utilizing hormone implants in the brain (Barfield, 1971; Komisaruk, 1967) have shown that hormones may exert direct effects on aggressive behavior. But hormones may also modulate aggressive behavior indirectly by affecting morphological characters that play an important role in behavioral displays. Secondary sexual characteristics such as size, plumage pattern, bill, leg or skin color, and such other species-specific characters as combs, wattles, or spurs may be profoundly influenced by endocrine condition. Changes in such morphological characteristics should tend to amplify changes in behavior by signaling an individual's general hormonal status to interested individuals.

Although this review will focus on the relationship between hormones and aggressive behavior, it is important to remember that other factors can also exert a profound influence on aggressive behavior. One group (Ficken, Ficken, & Hailman, 1978) has found that a genetic mutation in the white-throated sparrow which results in white rather than tan median stripes on the head is associated with increased aggressive behavior. Observations of feeding flocks indicated that white morphs attacked other birds more frequently regardless of the color of the recipient or the size and composition of the flock. Both morphs, tan and white, are equally likely to be attacked. The mechanism causing increased aggressive behavior in the white morphs is unknown.

Brown (1963) established artificial feeding sites in a regional park and observed the dominance interactions of Steller jays that came to feed. He found that the most important determinant of dominance rank was distance from the nest, regardless of the time of year at which observations were made. At each feeding site, the dominant male in the hierarchy was the male whose nesting area included the area around the feeder. Males on contiguous nest areas were intermediate in rank, while males whose nest areas were separated from the feeding site by one or more intervening nest areas held the lowest ranks. A similar relationship between rank and distance from the nest area was found among females, which were, in general, subordinate to males. When birds shifted the position of their nests, their area of dominance shifted as well. Hormone levels also affected a bird's aggressivity, if not its dominance rank. Birds of both sexes were most aggressive during courtship, with levels of aggression peaking just before incubation, when circulating levels of gonadal hormones are presumed to be at their peak. Like most jays, these birds were quiet and inconspicuous during incubation, but levels of aggression increased when birds renested. A bird's area of dominance became larger during courtship and smaller during incubation. Thus, although the birds' aggressive behavior was presumably affected by their hormonal status, a more salient factor was where the observations took place in relation to the birds' nest area.

A variety of techniques have been used to examine the relationship between hormones and aggressive behavior (see Leshner, 1978, pp. 19–33, for a discussion of the various techniques and the advantages and disadvantages associated with each). Early studies of birds concentrated on documenting correlations between naturally occurring changes in endocrine function and aggressive behavior. Thus, workers noted that in birds, as in other vertebrates, males are generally more aggressive than females (Bennett, 1939; Collias 1950), though there are a few notable exceptions to this, including phalaropes and jacanas (Daly & Wilson, 1978; Höhn & Cheng, 1967). Reproduction in most nontropical species of birds is seasonal, and a great deal of work has been done documenting seasonal changes in aggressive behavior. In many studies, seasonal changes in behavior were correlated with changes in gonad size or activity, but with the advent of modern assay techniques for measuring circulating hormone levels, more recent studies have generally measured plasma hormone levels directly. All of these techniques are correlational, however, and conclusions one can draw from such data are limited. Causal relationships cannot be deduced from correlational data, and thus, if androgen levels are positively correlated with aggressive behavior, one does not know whether high androgen levels cause high levels of aggression, whether high levels of aggression cause high androgen levels, or whether these two variables are only indirectly related through a third variable that affects them both. In spite of this limitation, most field studies employ correlational techniques to study hormone–aggression relationships, and such studies will be reviewed first.

To establish a causal relationship between a hormone and a particular behavior pattern, one must manipulate the animal's hormone level and demonstrate that the behavior pattern varies accordingly. The most common manipulation is to administer exogenous hormones to see if administration of a particular hormone increases the quantity or quality of an animal's aggressive behavior. When intact birds are hormonally treated, the resulting data are difficult to interpret, since any change in behavior could be mediated indirectly through an effect of the administered hormone on endogenous hormone levels. Therefore, the method of choice is the ablation–replacement therapy experiment, although even this is not without its drawbacks (see Leshner, 1978). Berthold (1849) examined the effects of castration in roosters in what is generally acknowledged to be the first formal study of behavioral endocrinology. He found that this operation reduced the birds' aggressive and sexual tendencies. Reimplantation of one testis in the body cavity restored the males' behavior to normal, leading Berthold to conclude that some substance produced by the testis was involved in the control of these behaviors. Determining which of the many testicular hormones are behaviorally active was a task that had to wait until the production of purified steroids became commonplace in the 1940s. The question of which hormones actually cause changes in aggressive behavior is still a topic of active debate today, because recent studies have demonstrated that avian target tissues, like those of other vertebrates, possess enzymes capable of metabolizing endogenous or exogenous hormones to a variety of other metabolites.

A fair number of studies have looked for sites of hormone action in the avian brain. By and large, the results of these studies are similar to those on other vertebrate groups and will be discussed briefly. Data on the effects of early hormonal environment on the differentiation and development of later behavior patterns will also be briefly reviewed. The data on this topic are sparse and indicate that the sexual differentiation of behavior may proceed differently among different avian orders. The final topic addressed in this chapter is complex hormone–behavior interactions. Recent data indicate that engaging in social interactions results in rapid, dramatic changes in circulating hormone levels, and the implications of such changes for the modulation of aggressive behavior will be discussed.

SEASONAL CHANGES IN HORMONE LEVELS AND AGGRESSION

Most nontropical species of birds show marked seasonal changes in gonadal size and activity correlated with changes in aggressive and sexual behavior. Since males are usually more aggressive than females, most studies have focused on the relationship between aggressive behavior and endocrine status in the male. In general, increased aggressive behavior marks the onset of the breeding season, and levels of aggression increase further as birds compete for the resources needed for breeding, which are often in limited supply (Butterfield & Crook, 1968; Collias, 1950; Wingfield & Farner, 1978). In many species, changes in testis size from the nonbreeding to the breeding season are quite striking. For example, in their studies of the annual endocrine cycle in the white-crowned sparrow, Wingfield and Farner (1978) found that estimated testis weight increased from a low of approximately 5 mg in February to over 450 mg in early May (see Figure 1). In a study of the annual reproductive cycle in the starling, Temple (1974) found that changes in testis size were correlated with the number of interstitial cells in the testis, the primary source of androgen production. There was a 200-fold increase in interstitial cells during the breeding season, and sexual accessory tissues, dependent on androgen secretion, showed large

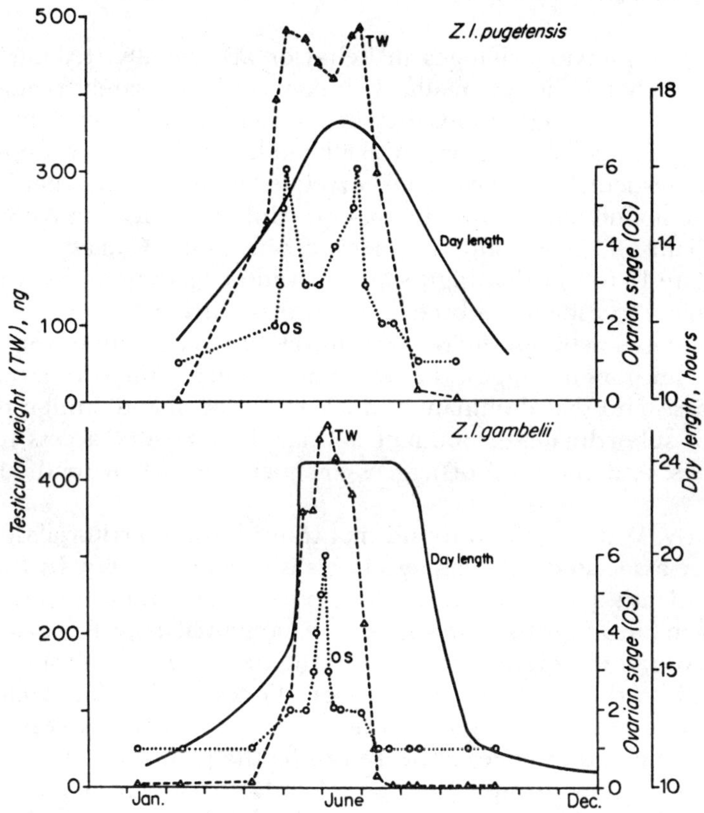

FIGURE 1. Annual cycle in testicular weight and ovarian stage in two taxa of the white-crowned sparrow *(Zonotrichia leucophrys)*, on Camano Island, Washington (*Z. l. pugetensis*), and in central Alaska *(Z. l. gambelii)*. Winter samples were collected either in central California or in Washington. In this figure and Figures 2 and 3 (all on the same scale), the data are presented as stages in the cycle and spaced according to the mean time required by a pair for each stage. Because these events are not synchronous among individuals of the population, the duration does not necessarily represent the time periods of such events as egg-laying and incubation for the population as a whole. The vernal migratory periods are in April and early May, with nesting for the first brood in early to mid-May in *Z. l. pugetensis* followed by nesting for a second brood in late June and early July, with incubation and feeding of the young during stages in between. In the single-brooded *Z. l. gambelii,* the nesting phase occurs in late May and June followed by the incubation and feeding of young extending to very early July. Photorefractoriness sets in by late June and early July, perhaps a little later in *Z. l. pugetensis,* and is closely followed by the onset of postnuptial molt. The autumnal migratory periods are in August, September, and October. Ovarian stages are: (1) ovary in winter condition, follicles less than 0.5 mm in diameter; (2) follicles 0.5–1.0 mm in diameter; (3) follicles 1–2 mm in diameter; (4) follicles 3–5 mm in diameter, about to ovulate; (5) egg in oviduct, recently ovulated follicle apparent. (From "Control of Seasonal Reproduction in Temperate-Zone Birds" by J. C. Wingfield and D. S. Farner, *Progress in Reproductive Biology,* P. O. Hubinot, Ed. Copyright 1980 by S. Karger AG, Basel. Reprinted by permission.)

increases in size, for instance, over a 475% increase in the size of the seminal sacs.

The accompanying changes in behavior are equally striking. Armstrong (1965) noted that although mallard drakes could be safely penned together during the nonbreeding season, they began to fight furiously with the onset of the breeding season in February, until all had been seriously wounded or killed. Increases in testicular size have been correlated with the transition from feeding flocks to the maintenance of individual breeding territories in Anna hummingbirds (Williamson, 1956) and ring-necked pheasants (Collias & Taber, 1951). The latter study found that high status in winter-feeding flocks was correlated with the ability to establish a breeding territory during the spring. Reversals in dominance status between the winter and the breeding season were rarely seen, leading the authors to suggest that dominance during the two periods shared a common basis and that dominant animals might be able to inhibit the breeding activities of subordinates. Dominant animals had greater access to food than subordinates, and the level of aggression increased when food supplies were scarce.

Similarly, Watson (1970) found that transitions in territorial defense in red grouse were associated with changes in testis size and activity. In June and July, the testes of adult males were fully regressed, and these males were found wandering in family groups. In August, the interstitial tissue began to regenerate, and males began to defend territories in the morning but still spent afternoons wandering with their families. By October, the testes of males hatched that year were comparable in size to those of adult males, and the increased aggressive behavior of the young males appeared to be the primary factor leading to the dissolution of family groups. Young males also began challenging older males in territorial disputes. The testes of nonterritorial males were consistently smaller and less active than those of territory holders during the period of November through April, and males that replaced territory owners, either through combat or through filling vacancies caused by male deaths, consistently had larger testes than males that remained nonterritorial. Watson also found that males became more aggressive and held larger territories when food was scarce or poor in quality.

A study of captive male weaver finches (Butterfield & Crook, 1968), indicated that both aggressive encounters over individual distance and aggressive interactions over nesting material were most frequent during the period of July to October when the birds were in full nuptial plumage. Hormonal activity in the study was monitored by examining the birds' plumage patterns, since nuptial plumage in this species is dependent on high levels of luteinizing hormone (LH), and plumage coloration can therefore be used as a bioassay for LH. Androgen levels would presumably also be high when LH levels are increased, since LH is one of the primary factors controlling androgen production by the testes. In November the birds began to molt to eclipse plumage, and the frequency of aggressive interactions decreased, with a small increase in the following March.

However, one must be cautious in inferring anything about the rate or level of hormone production from data on gonad size. Several studies (Kerlan & Jaffe,

1974; Kerlan, Jaffe, & Payne, 1974; Wingfield & Farner, 1978) present evidence that while testis size and androgen production correlate well when both are increasing rapidly during the photosensitive period at the beginning of the breeding season, androgen production falls off while the testes are still large. By the time the gonads reach their greatest size, androgen production may, in fact, be quite low. Wingfield and Farner found that plasma testosterone levels in white-crowned sparrows had fallen from their peak values at least one month before the testes began to regress (compare these data in Figures 1 and 2).

Actual measurements of plasma hormone levels over the reproductive cycle have consistently revealed a strong correlation between intermale aggression and high androgen concentrations. In their comprehensive study of the behavioral endocrinology of the white-crowned sparrow, Wingfield and Farner (1978; see Figure 2) found that territorial encounters were most frequent when testosterone levels were at their peak. At this time, when females were arriving and birds were pairing, males also sang frequently, averaging 400 songs/hour. Testosterone levels fell during incubation, as did the frequency of aggressive interactions and singing (5–10 songs/hour). Although the frequency of singing increased to 300

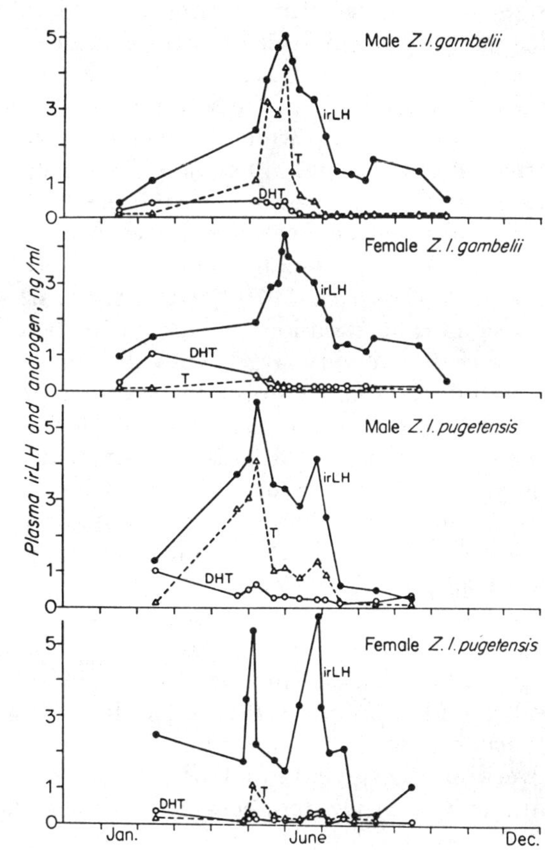

FIGURE 2. Annual cycle in plasma levels of immunoreactive luteinizing hormone (irLH), testosterone (T), and dihydrotestosterone (DHT) under natural conditions in the white-crowned sparrow, *Z. leucophrys*. See legend to Figure 1 for correlation of stage in cycle to time of year. (From "Control of Seasonal Reproduction in Temperate-Zone Birds" by J. C. Wingfield and D. S. Farner, *Progress in Reproductive Biology*, P. O. Hubinot, Ed. Copyright 1980 by S. Karger AG, Basel. Reprinted by permission.)

songs/hour during the period of courtship preceding the second brood, no significant increase in either testosterone levels or aggressive behavior was seen. Males of the double-brooded Puget Sound race maintained higher testosterone levels longer and remained on their territories longer than males of the single-brooded Gambel's race (Wingfield & Farner, 1980). However, in the latter race, losing the first clutch of eggs resulted in a surge of LH and testosterone in males as pairs renested, though it was somewhat lower than the one preceding the first clutch (Wingfield & Farner, 1979).

Data for starlings (Temple, 1974) are similar. The period of greatest intermale aggression coincided with the period of greatest androgen secretion. Males were found in large flocks only from July to October. As soon as androgen levels began to rise in late autumn, males began to remain near their breeding territories. During periods of extreme cold, males would roost together at night, but aggressive interactions in the roost were commonplace.

Putting together data on behavior (Peek, 1971) with data on hormone levels in redwinged blackbirds (Harding & Follett, unpublished observations; Kerlan & Jaffe, 1974) yields a similar story. Testosterone levels are highest early in the breeding season, as the females arrive and pairing begins. This is also the period of strongest territory defense and most frequent territory intrusions. Males sing significantly more often during this period. When one male was removed from his territory during this period of intense competition, Peek recorded 51 incidences of trespassing in the first hour after removal. Within three hours, another male had established himself on the territory. Following mating, intermale aggression declines, while the incidence of interspecific aggression increases as males attempt to defend females and their young from intruders. Alarm calls rather than song appear to be the predominant vocalization during this period when testosterone levels are decreasing.

Studies of plasma hormone levels and behavior in the domestic duck (Balthazart & Hendrick, 1976, 1979; Balthazart, Hendrick, & Deviche, 1977) found no significant variations in aggressive behavior over the year, but individual variations were very large. They did, however, find a large positive correlation between males' testosterone levels and aggressive behavior towards females. On days when a male's testosterone levels were high, he exhibited high levels of sexual behavior, but aggressive interactions with females were also more frequent. Other species in which high levels of LH and testosterone have been found during territory establishment and/or courtship include the snow goose, domestic pigeon, common eider, red grouse, and Western gull (See Wingfield & Farner, 1980, for review).

One study (Herzog & Boag, 1977) examined seasonal changes in the behavior of female spruce grouse. Tape recordings of a female attacking a decoy were played back to females in the field at various times of the year. Females called and approached the tape only during the period of March to May. Only females in breeding condition responded in this way; migratory females and overwintering juveniles left the area. Yearlings began to respond aggressively in May when they, too, became territorial. Stage of the breeding cycle was also important, since incubating females responded to the tape while females with chicks ignored it.

Studies on aggressive behavior in female birds also indicate the importance of androgens. The avian ovary has been shown to secrete fairly high levels of androgen, and in some species, androgen secretion has been implicated in the timing of ovulation (Meier & Ferrell, 1978). The incidence of territorial defense and/or singing in females has been associated with increased levels of androgen in robins, loggerhead shrikes, mockingbirds, phalaropes, white-crowned sparrows, and wren-tits (Collias, 1950; Höhn & Cheng, 1967; Wingfield & Farner, 1980; see Figures 2 and 3). In the Western gull, in which there is no major sexual dimorphism in aggressive displays or territorial defense, androgen levels in the female are equivalent to those in the male year round (Wingfield, Martin, Hunt, Hunt, & Farner, 1980; Wingfield, Newman, Hunt, & Farner, 1982), and the authors hypothesize that this may be the case in other monomorphic species.

The relationships of other hormones to aggressive behavior remain unclear. One correlational study on ring doves (Vowles & Harwood, 1966) indicates that prolactin may be involved in an increase in nest defense at the time the young hatch. However, the unusual stimuli used to elicit defense reactions in the behavioral tests (e.g., hand-held dove, model spider) make the generalizability of the data questionable. Meier (for review see Meier & Ferrell, 1978; Meier & Fivizzani, 1975) has suggested that corticosterone can stimulate or inhibit gonadal activity depending on the timing of the corticoid rhythm in relation to other

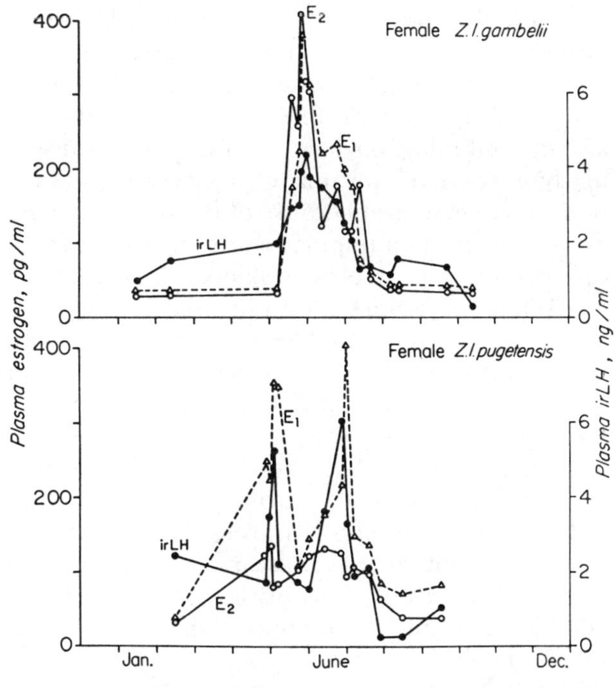

FIGURE 3. Annual cycle in plasma levels of immunoreactive luteinizing hormone (irLH), estrone (E1), and estradiol-17 (E2) under natural conditions in female white-crowned sparrows, *Z. leucophrys*. See legend to Figure 1 for correlation of stage in cycle and time of year. (From "Control of Seasonal Reproduction in Temperate-Zone Birds" by J. C. Wingfield and D. S. Farner, *Progress in Reproductive Biology*, P. O. Hubinot, Ed. Copyright 1980 by S. Karger AG, Basel. Reprinted by permission.)

endocrine and environmental variables. Any effects on aggression would be indirect. There have been no studies on the effects of the pituitary–adrenal axis on aggressive behavior in birds. Some studies have suggested that gonadotropins are involved in the control of aggressive behavior, and these studies will be discussed below, in the section on the LH controversy.

Thus, there appears to be a clear association between reproductive hormone secretion and aggressive behavior in all species studied. The annual fluctuations of reproductive hormones and aggressive behavior are positively correlated, and both reach their yearly maxima at the onset of the breeding season. Several studies suggest that gonadal androgens are the hormones that show the best correlation with aggressive behavior. Increases in hormone secretion with the onset of the breeding season are attributable in large part to changing environmental conditions, such as increased photoperiod (Wingfield & Farner, 1980), but social interactions also play a role. Social modulation of circulating hormone levels will be discussed in the section on complex hormone–behavior interactions.

EFFECTS OF HORMONAL MANIPULATIONS ON AGGRESSIVE BEHAVIOR

Hormonal Treatment of Intact Birds

With hormone treatments, as with other techniques of studying the relationship between hormones and aggression in birds, attention has focused for the most part on gonadal hormones, particularly androgens. Many of the pioneering studies of the effects of hormone treatment on aggression in various species of birds dealt with intact birds (Allee, Collias, & Lutherman, 1939; Bennett, 1940; Boss, 1943; Emlen & Lorenz, 1942). Some fairly recent studies have also utilized intact birds, at least in part because complete gonadectomy, particularly in some of the smaller species, is quite difficult. In other studies, intact birds have been used because of the field orientation of the researcher, for while gonadectomizing a population of birds is clearly impractical, the administration of long-lasting hormone implants to many birds is not.

While data from such studies may be suggestive, they are difficult to interpret because it is difficult to determine if a hormone treatment is affecting behavior directly or only indirectly through its effects on other hormones. Data from the LH controversy, which will be discussed below, are a good case in point. Several investigators have injected intact male birds with LH and then reported increases in aggressive behavior, which they attributed to the increased LH levels. However, LH is the primary hormone regulating androgen secretion by the testes, so the increased levels of aggressive behavior seen in these studies might just as easily be attributable to increased androgen secretion caused by the LH treatment. Treatment of intact birds with testosterone might modify their aggressive behavior by changing the output of other testicular hormones or by decreasing LH secretion.

Many studies have examined the hormonal control of secondary sexual characteristics that are often used in aggressive and courtship displays. Bill color is under the control of testosterone in both sexes in many species. In the house sparrow, the bill changes from ivory to black as testosterone levels increase (Haase, 1975; Keck, 1934). In herring gulls, adults acquire an ivory bill under the influence of androgen, and during the breeding season an orange spot develops on the lower mandible (Boss, 1943). The starling bill changes from black to bright yellow each spring as androgen levels increase (Witschi & Miller, 1938). Both male and female zebra finches have orange bills, but the orange color is darker and more intense in males. Castration causes the intensity of bill color in the male to fade to the characteristic female shade (Harding, unpublished observations). Plumage patterns of some species are under hormonal control, but the pattern of hormonal control varies greatly from one species to another. Domestic chickens and turkeys (Scott & Payne, 1934) will develop the male plumage pattern unless estrogen is present, but the adult plumage pattern in herring and black-headed gulls is controlled by androgen in both males and females (Boss 1943; Silver, O'Connell, & Saad, 1979). In ruffs, the male breeding plumage is controlled by androgen (Silver *et al.*, 1979), but in weaver finches it is controlled by LH (Butterfield & Crook, 1968). In line with the "role reversal" in phalaropes, female phalaropes have more colorful breeding plumage than males, and this plumage is thought to be induced by their high androgen levels (Höhn, 1970; Johns, 1964). Comb, wattle, and spur development are also androgen dependent (Collias, 1950). Body weight is affected by androgens, though once again, effects vary from species to species. In pigeons, as in most mammals, androgens cause an increase in body weight (Pietras & Wenzel, 1974), but in zebra finches, androgen treatment results in a decrease in body weight (Harding, unpublished observations). This is not an exhaustive list of the effects of hormone treatment on secondary sexual characteristics, but it should give you an idea of the variety of ways in which hormones may affect a bird's ability to display and thereby indirectly affect aggressive behavior.

Indeed, the importance of some of these hormone-dependent characteristics in aggressive interactions has been clearly demonstrated. For example, in red grouse the comb is an important social signal in eliciting aggressive behavior (Moss, Kolb, Marquiss, Watson, Treca, Watt, & Glennie, 1979). Aggressive birds raise their combs and submissive birds lower their combs out of sight underneath their feathers when they meet another bird. Comb size varies with age, season, dominance, and reproductive activity and, in studies of captive birds, was found to covary with a bird's aggressiveness. Comb size increases with age and is larger in the spring when reproductive and aggressive behaviors are most frequent. When captive males were moved from isolation to a cage containing hens, their comb size increased within 48 hours. This increase in comb size was associated with an increase in the bird's aggressiveness, and both changes were presumably caused by an increase in circulating androgen levels. Comb size was also positively correlated with dominance status in groups of males, with dominant males having larger combs than subordinate males. Comb size after grouping was usually a better indicator of dominance status than comb size before grouping, as the combs of dominant males increased further in size while those of subordinate males decreased further after grouping. The combs of intact males could be enlarged by treating the males with exogenous androgens, a treatment which also increased the males' aggressive behavior. Thus, comb size in this species is affected by a variety of environmental and social variables and is a good indicator of a bird's general aggressive status and dominance rank.

One of the earliest studies of the effects of androgen on aggressive behavior in birds was that of Allee and his collaborators (1939), who examined the effects of androgen treatment on the aggressive behavior of low-ranking hens. In all cases, testosterone propionate (TP) treatment increased the bird's aggressiveness, and in many cases it also increased the hen's status within her home flock. In other cases, although androgen-treated females were unable to better their status in their home flock, they were more successful in encounters with unfamiliar hens. The best predictor of success in an aggressive encounter was comb size, a character directly related to the amount of androgen present. Some hens were also observed to crow and court females. While these behaviors disappeared once hormone treatment was halted, higher social position, once achieved, was retained. Treating low-ranking female canaries with TP has also been shown to increase their standing in an established hierarchy (Armstrong, 1965).

Bennett (1940) treated intact ring doves of both sexes with TP and found that these androgen-treated birds became more aggressive and advanced in social status, maintaining their new status as long as hormone treatment was continued. These data led to the conclusion that aggressiveness was correlated with circulating testosterone levels, and that an increase in social rank was related to an increase in aggressive behavior.

Moss *et al.* (1979) attempted to reverse dominance ranks in captive red grouse by treating subordinates with TP. They found that TP treatment caused an immediate increase in the aggressive behavior of the implanted birds, but that it took a long time, sometimes almost three months before reversal of rank occurred. Ramenofsky (Ramenofsky & Gorbman, 1980; M. Ramenofsky, personal communication, 1980) also found evidence that hormonal manipulation cannot easily alter the effects of prior experience. She found that established dominance relationships in pairs of Japanese quail were rarely reversed by treating the subordinate bird with an implant of testosterone that resulted in a physiological increase in circulating testosterone levels of 2–4 ng/ml. The testosterone-treated subordinants increased their attack rates on the dominants, but only 18% of the birds managed to reverse the dominance order over the course of fourteen tests. Clearly, previous experience is a very important factor in determining dominance status.

Emlen and Lorenz (1942) examined the effects of implanting either TP or diethylstilbestrol (DES) in flock-living valley quail during the winter months. They found that TP-treated males became aggressive toward other birds in the flock within 48 hours. This increase in aggression did not increase the bird's status within the group, but these males paired earlier than usual as a result of the hormone treatment. TP-treated females did not become more aggressive, though they showed some male behaviors in pairing. DES had no noticeable effects on behavior.

Boss (1943) investigated the effects of TP and stilbestrol on the development of adult morphology and sexually dimorphic behaviors in herring gulls. Herring gulls do not normally develop adult plumage or behavior patterns until their third year, but TP treatment of immature birds caused them to develop adult behavior, bill color, and plumage patterns in that order. TP-treated males became extremely aggressive and dominated all other birds in the flock. Lowering the dominant male's TP dosage caused him to lose his rank to another TP-treated male. TP-treated females also became more aggressive, but they dominated only other females, not males. When hormone treatments were stopped, all characteristics and behavior reverted to their immature state. Stilbestrol was not found to have any significant effects on aggressive behavior or morphology. Combining stilbestrol with TP antagonized the effects of androgen on aggressive behavior. Similar studies conducted with chicks (see Collias, 1950) indicated that early androgen treatment resulted in development of secondary sexual characteristics, precocious crowing, aggressive behavior, and attempts to mate. Noble and Wurm (1940) treated month-old black-crowned night herons with TP. These chicks showed precocious aggressive behavior, setting up territories and engaging in adult aggressive displays.

A more recent study on laughing gulls (Terkel, Moore, & Beer, 1976) examined the effects of hormone treatment on the production of an adult vocalization used by both sexes, the long call. Long calls occur in several functional contexts including aggressive encounters, courtship, and parental care and provide a basis for individual recognition. The frequency of long calling typically increases at puberty and also at the beginning of each reproductive cycle. Both testosterone and estradiol increased the frequency of long calling in juveniles, but testosterone was more effective in both sexes. Males called more than females following both hormone treatments. Changes in eye ring and bill color were also noted in hormone-treated birds. Given the relative efficacies of androgen and estrogen, the authors concluded that vocal development in both sexes was controlled by androgen.

Searcy and Wingfield (1980) examined the effects of androgen levels on aggressive behavior in captive male redwings by giving them implants of either (1) TP, (2) flutamide, a potent antiandrogen, or (3) cholesterol, the control treatment. Males were housed in groups of three, one from each treatment group, matched for their previous level of aggressive behavior. Dominance rank was strongly related to hormone treatment, with the TP-treated males being dominant. TP treatment also resulted in a significant increase in attacks on subordinates but had no clear effect on song or display rates.

Two small field studies have indicated that androgen titers exert profound effects on aggressive behavior, body condition, social status, territory size, and reproductive success in natural populations. Watson (1970) found that the aggressiveness of male grouse could be increased out of season by TP treatment. Further, two males that had been unable to obtain territories during the normal period of territory defense (August–March) obtained territories from active males within nine days of TP implantation. Their physical condition improved markedly, and they were able to attract females to their territories. When territory owners received implants of TP, they consistently increased the sizes of their territories and usually attracted additional females. When one male was treated with estrogen, he became less aggressive, his mate deserted him, and he finally disappeared. One other male was given a very large implant of TP, 50 mg, and he became exceedingly aggressive. His courtship was so aggressive that his mate would run away when he approached her, and she finally deserted him.

J. C. Wingfield (personal communication, 1980) recently completed a small field study in which male white-crowned sparrows were given implants of testosterone during the breeding season. These implants profoundly affected the males' behavior. Testosterone-treated males continued to sing at high rates throughout the entire breeding cycle, adopting the typical drooped-wing advertising posture. Testosterone-treated males defended larger areas, remained highly aggressive during the later part of the breeding season, and, unlike normal males, would not allow neighboring males to forage on their territories. Three males were polygynous, having managed to attract additional females onto their territories. Apparently, none of these males managed to fledge any young, however. Males normally help to feed the young, particularly after fledging, but presumably the behavior of the testosterone-treated males was incompatible with the feeding of young.

While these data must be regarded as preliminary, given the small samples of treated males, they are tremendously exciting and illuminating in demonstrating the social repercussions of increased androgen levels. These studies also suggest that beyond a certain point, increased androgen levels and the resulting increase in aggressive behavior tend to lower a bird's reproductive success. This is probably an important factor in limiting the evolution of increased aggressiveness in males.

A study by Lumia (1972) underlines the importance of experience as well as hormonal status in determining a bird's aggressive behavior. Lumia took low-ranking pigeons from their home loft and either trained them to peck dominant birds, gave them TP, or gave them TP plus the training. TP alone had no effect on the bird's aggressive behavior in the home loft, while the conditioning did increase the bird's aggressiveness somewhat. But the combination of TP and training was by far the most effective treatment. TP treatment both potentiated the effects of the training and caused the birds to assume species-specific attack postures during their aggressive encounters, unlike the birds which had received the training alone. Thus, high testosterone levels appear to increase a bird's potential for acquiring and using aggressive behaviors.

A series of studies by Rohwer also illustrates how the effects of androgen treatment may be determined by other characteristics of the individual, in this case the bird's plumage pattern. Rohwer (1975) found that dominance in Harris sparrows in winter feeding flocks was positively correlated with the degree of black feathering on the throat, breast, and crown. Almost without exception, birds with more black feathers in these areas won encounters with birds with less black feathering; subordinates appeared actively to avoid such black-feathered birds. When the feathers of light-colored subordinates were dyed to resemble the plumage pattern of dominant birds, these birds were persecuted by the dominant birds in the flock (Rohwer, 1977). However, when subordinate males or females were dyed and given a large silastic implant of TP at the same time, their status improved markedly although they were still dominated by the original α and β birds in the flock (Rohwer & Rohwer, 1978). The status of birds that received the TP implants without having their feathers dyed did not improve. Birds were much more likely to dispute the attempts of the androgen-implanted birds to supplant them if the plumage of the implanted bird had not been altered. It is not known if the normal plumage pattern is hormone dependent or how it is controlled. It should be noted that the number of birds in this study was quite small, but the results appear to be consistent throughout.

A few studies have examined the effects of nongonadal hormones on aggression. One study (Murton, Thearle, & Lofts, 1969) found that both follicle-stimulating hormone and LH, as well as androgen, could increase aggressive behavior in male pigeons. Davis (1963) and Crook and Butterfield (1974) presented evidence for starlings and African weaver finches, respectively, that LH treatments increase aggressive behavior in intact males while testosterone treatment does not. Of course, treating intact males with gonadotropins is likely to increase levels of gonadal hormones as well, making the studies difficult to interpret. These studies will be discussed further in the section on the LH Controversy.

Estradiol benzoate treatment was reported to increase levels of aggressive behavior and dominance status in female pigeons (Murton *et al.*, 1969), while progesterone and prolactin were not noted to have any strong effects. Vowles and Harwood (1966) reported that prolactin and progesterone both increased defense of the nest in incubating ring doves of both sexes. The effect of progesterone appeared to be indirect, since it had a 12- to 18-hour latency. The authors felt that the increase in defense behavior was associated with the increased probability that the birds would stay on the nest when presented with novel stimuli. As mentioned previously, the results of this study should be interpreted cautiously, given the artificiality of the testing situations.

Thus, available data indicate that testosterone treatment of intact birds of both sexes often results in increased aggressiveness and is sometimes sufficient to increase an individual's status in a dominance hierarchy. The probability that androgen treatment will result in increased aggressive behavior and status is increased when the treated animals are tested with unfamiliar individuals, since hormone treatment alone is often insufficient to overcome the effects of previous aggressive interactions with a known opponent. In addition to affecting a bird's behavior, androgen treatment has been shown to affect various secondary sexual and morphological characteristics, such as bill color, plumage pattern, and body weight, which can have important implications on how an individual is perceived by other birds and on its ability to win an encounter.

The Effects of Castration and Replacement Therapy

As mentioned in the introduction, the first formal castration experiment was published by Berthold in 1849. Goodale (1913) castrated male leghorns, and demonstrated that they developed normal male plumage but their combs and wattles were small. These birds were not aggressive and showed no interest in mating. Ovariectomized females developed male plumage, but were extremely docile. However, if only the left ovary is removed, the right ovary in birds being rudimentary, the right ovary often develops into an ovotestis. An ovotestis, as its name implies, contains both ovarian and testicular tissue and shows a high level of androgen production. Females with an ovotestis show increased levels of aggressive behavior (Collias, 1950). Aggressive behavior in pigeons was also decreased by castration (Carpenter, 1933), with chasing being the behavior most affected. The decrease in aggressive behavior caused by castration in these species can be reversed with testosterone therapy (Collias, 1950).

Selinger and Bermant (1967) paired male Japanese quail and identified the dominant and the subordinate bird in each pair. When the dominant bird of the pair was castrated, its aggressive behavior declined rapidly. Within a few days after castration, none of these previously dominant males exhibited any aggressive behavior. Aggressive behavior initiated by the subordinates increased. When the castrated males were treated with testosterone, they once again became dominant. When intact subordinates received testosterone injections, they became dominant over their intact partners but only as long as the androgen injections were continued. In this study, levels of aggressive activity were easily manipulated by changing circulating androgen levels. Kuo (1960) noted that it was difficult to train castrated quail to become good fighters, and treatment of poor fighters with TP caused 63% of these to respond to training and become good fighters.

Hutchison (1975) found that the quality of courtship behavior in male ring doves was related to circulating androgen level. Following castration, the aggressive components of courtship, particularly bowing, disappeared within 1–3 days, while the nest-oriented components took 15–20 days to disappear, suggesting that relatively high androgen levels are required to elicit the aggressive displays. This hypothesis is supported by further experiments showing that to reinstate aggressive displays takes more testosterone than to reinstate nest-oriented displays. However, a seasonal factor also appears to be involved, since Hutchison found that TP implants given to sexually inactive intact males in the late fall elicited aggressive rather than nest-oriented displays. The androgen threshold to elicit aggressive behavior appears to remain low during testicular recrudescence until the testes are fully developed.

Arnold (1975) observed the behavior of male zebra finches when a castrated male was placed in a pair's cage. Zebra finches are highly gregarious and normally show low levels of aggressive behavior. Castration of the resident decreased the frequency of aggressive acts toward the intruder and increased the latency to attack. TP injections increased aggressive behavior in four out of six males, although this effect was not significant. Using a similar test situation but an intact intruder, studies in my lab have also found levels of aggressive behavior to be fairly low in both intact and castrated zebra finches. Castrates given testosterone implants that maintain circulating levels of testosterone in the high normal range for breeding males, 5–7 ng/ml, attack the intruder furiously. The behavior of the testosterone-treated males is so striking that it is impossible to run the experiment blind. When these males are observed in a different setting, a group aviary with limited nest sites, deficits in the aggressive behavior of the castrates become apparent; most castrates are unable to defend a nest site successfully. Other birds trespass frequently and often steal nest material, behavior that evokes quick attack when an intact male's nest is plundered.

Although there are few successful ablation–replacement therapy experiments in birds, once again the data indicate that gonadal androgens are the most important class of hormones involved in the activation of aggressive behavior in birds. Castration reduces aggressive behavior in males of all species examined, and replacement therapy with testosterone restores aggressive behavior to normal or above-normal levels.

The LH Controversy

There has been a long-standing controversy in the avian literature over whether aggressive behavior in passerine (i.e., song) birds is controlled by gonadotropins or gonadal hormones. This controversy originated with a report by Davis (1957) that neither castration nor testosterone replacement therapy had any effect on dominance hierarchies in captive starlings. Davis suggested that pituitary hormones might be involved instead. This report was followed by another (Mathewson, 1961) indicating that testosterone treatment of the two lowest-ranking starlings in a five-bird hierarchy had no effect on dominance ranks or the frequency of fighting, while LH treatment increased fighting and caused one reversal of rank. Further, when a castrated and an intact male were put in a competitive situation in which only one perch was available, the castrated male usually was dominant, and this was attributed to the high LH levels presumably resulting from the lack of negative androgenic feedback following castration. When the intact males were given LH, they then became dominant over the castrates. These data were expanded and discussed at length by Davis (1963).

Later workers studying aggressive behavior in African weaver finches (Butterfield & Crook, 1968; Crook & Butterfield, 1974) suggested that gonadal hormones control aggressive interactions over nesting material while LH modulates aggressive interactions over individual distance. Looking at the frequencies of the two different types of aggressive behavior over the year, Butterfield and Crook argued that the two vary independently, and therefore are probably controlled separately. They used exogenous hormone treatments to test their hypotheses. They found that injecting intact males with LH increased the frequency of aggressive interactions over individual distance and position on perches. TP treatment had no effect on such such behavior but increased the number of aggressive encounters over nesting material. The authors concluded that the hormonal control of aggressive behavior was context specific, with androgens controlling interactions over sexually relevant resources, while LH was involved in the control of resources not directly linked to reproduction.

A similar study in female weaver birds (Lazarus & Crook, 1973) found that LH increased the aggressiveness of ovariectomized females (these were not complete ovariectomies, since they reported ovarian regeneration in females that were not estrogen treated), although it had no effect on the behavior of intact females during the breeding season. The latter finding was attributed to the antagonistic effects of endogenous estrogens, estrogen levels naturally being high during the breeding season. In another experiment during the breeding season, ovariectomy itself was found to increase female aggressiveness, presumably acting through the dual mechanism of increasing LH levels while decreasing estrogen levels. Treating ovariectomized females with estrogen reversed this effect.

All of these studies are subject to criticism on methodological grounds. All are based on relatively small numbers of birds, and in the earlier studies, the same hormone-manipulated animals were retested with several animals, so that the reported data do not represent independent samples. In most of the work, LH was administered to intact males, so that any increase in LH was probably confounded by increased androgen levels as well. Hormone dosages were often not reported, and it seems likely that the short-term testosterone treatments used (2 shots over 3 days in Crook & Butterfield, 1974) would be insufficient to cause behavioral changes in most species for which testosterone is known to control aggressive behavior. When castration was attempted, it was often incomplete. Often, statistical tests were not used, and when they were, they were sometimes used in a rather arbitrary manner. It should also be noted that the exogenous LH administered in these studies was mammalian rather than avian. More recent studies on the biological effects of gonadotropin administration in birds have shown gonadotropins to have a high degree of species specificity. The use of mammalian LH in these studies may have caused anomalous effects. Thus, the case for gonadotropic control of aggression in passerines is not a strong one. While it has not been clearly refuted, this hypothesis does not gain support from any of the recent studies on the hormonal control of aggressive behavior in passerines.

Steroid Specificity in the Control of Aggressive Behavior

While gonadal steroids have clearly been implicated in the control of aggressive behavior in a wide variety of avian species, which hormone(s) may actually be causing these effects remains unknown. Most studies using exogenous steroids have used testosterone, the primary active androgen secreted by the testes. Testosterone can, however, be metabolized to a variety of different androgens and estrogens. Several metabolites of testosterone are found in significant quantities in plasma, including dihydrotestosterone (DHT), androstenedione, dehydroepiandrosterone, and androstanediol (Wingfield & Farner, 1980); and further metabolism may also take place in the various target tissues. Raith, Bauer, and Karl (1977) demonstrated that a wide variety of extrahepatic tissues, including both classical target tissues such as liver and kidney and such unexpected organs as heart and stomach, showed active metabolism of androgens in birds. The patterns of androgen metabolism found in avian tissues differed from those found in mammals, resulting in the production of respectable quantities of androgens such as androsterone and 3β,5β-androstanediol. In passerine birds, the source of plasma DHT appears to be extratesticular, since castrated males of several species given testosterone implants show the same 3:1 ratio of testosterone to DHT as intact males (Harding, unpublished observations; Luine, Nottlebohm, Harding, & McEwen, 1980). Effects on body weight in both pigeons and zebra finches appear to be mediated by 5α-reduced androgens, such as DHT (Harding, unpublished observations; Pietras & Wenzel, 1974). Many male secondary sexual characteristics, such as the comb and wattles of chickens, contain the necessary enzymes for 5α-reduction and appear to be maintained by 5α-DHT.

Various studies have indicated that a combination of androgens and estrogens is necessary for courtship behavior in several species of birds, and Adkins and Nock (1976) found that estradiol was both necessary and sufficient to elicit copulation in Japanese quail. Cheng and Lehrman (1975) found that TP but not estradiol benzoate reinstated hop-charging and other aggressive male behavior patterns in castrated ring doves.

In my laboratory, we have been examining the efficacy of several metabolites, including testosterone (T), androstenedione, α-DHT, androsterone, estradiol (E), and a combination of E and α-DHT, to elicit a variety of social behaviors in castrated male zebra finches compared to control groups of castrated or intact males given cholesterol (Harding, in press). Males whose hormone treatments provided estrogenic metabolites (i.e., E, T, androstenedione, and E + α-DHT) showed significantly more high-intensity aggressive behaviors, such as pecking and chasing, in individual tests than did intact or castrate controls. Males receiving E treatment were just as aggressive as males receiving T, and these were the two most aggressive groups. When tested in a different situation, in which one male from each hormone-treatment and control group, together with the females with which they had been paired, were released into an aviary with limited nest sites, males whose hormone treatments supplied estrogenic metabolites were significantly more likely to secure a preferred nest site. Thus, in these birds as in many species of mammals, estrogenic metabolites increase the frequency of high-intensity aggressive behaviors, and the effects of T and other aromatizable androgens on aggressive behavior appear to be mediated by their conversion to estrogenic metabolites.

NEURAL SITES OF HORMONE ACTION

Of course, the effects of hormones on an individual's aggressive behavior are presumably mediated through the brain. The process of identifying which neural tissues are involved in the hormonal activation of aggressive behavior involves the coordination of data from experiments using a wide variety of experimental techniques. Autoradiographic and hormone uptake studies identify which neural tissues are hormone sensitive. Stimulation and lesion studies are used to identify which neural areas are involved in the control of particular patterns of behavior. By combining data from these techniques, one should be able to predict which hormone-sensitive brain areas are involved in the control of aggressive behavior, and then test whether hormone implants in these brain areas do actually elicit aggression in animals deprived of that hormone (e.g., castrated males if the effects of testosterone are being examined).

There have been a number of studies of the uptake of tritiated steroids in the avian brain (chaffinch: Zigmond, Nottebohm, & Pfaff, 1973; chick: Kim, Stumpf, Sar, & Martinez-Vargas, 1978; Meyer, 1973; chicken: Barfield, Ronay, & Pfaff, 1978; Wood-Gush, Langley, Leitch, Gentle, & Gilbert, 1977; ring dove: Kim *et al.*, 1978; Stern, 1972; zebra finch: Arnold, Nottebohm, & Pfaff, 1976). In brief, the results of these studies indicate that the distribution of radioactively labeled steroids in the avian brain is quite similar to the pattern seen in the mammalian brain following an injection of tritiated steroid. Steroid-concentrating cells are found in specific, limited areas of the telencephalon, diencephalon, mesencephalon, and medulla, with the heaviest uptake occurring in the medial preoptic area, the anterior and infundibular hypothalamus, the lateral septum, and the nucleus taeniae of the limbic archistriatum, which appears to be analagous to the medial amygdaloid nucleus of mammals. The distributions of estrogen and androgen-concentrating neurons in these areas appear quite similar, though there are some small differences. The similarity between the patterns of neural uptake seen in all vertebrate species examined suggests that the steroid-binding ability of hypothalamic, preoptic, and limbic structures may have existed since the early stages of vertebrate evolution.

In contrast to this evolutionarily stable pattern of uptake, neural uptake of steroids is found in a second group of brain structures in song birds, structures related to the production of song. Six discrete areas are known to be involved in the control of vocalization in songbirds (Arnold *et al.*, 1976; see Figure 4). Three of these structures, the hyperstriatum ventrale pars caudale (HVc), the nucleus robustus archistriatalus (RA), and the nucleus hypoglossi pars tracheosyringealis (nXIIts), are part of a direct efferent system linking the telencephalon to muscles of the syrinx, the avian vocal organ. The three other areas, area X of the lobus parolfactorius, the nucleus intercollicularis of the mesencephalon (ICo), and the magnocellular nucleus of the anterior neostriatum (MAN), receive projections from HVc or RA. Four of these areas, MAN, HVc, ICo, and nXIIts, show heavy labeling following an injection of tritiated testosterone, which is not surprising since singing is hormone dependent. Heavy labeling of ICo is also found in the chicken, a nonsongbird (Barfield *et al.*, 1978). However, the uptake of steroids in MAN and HVc is unusual in that they are evolutionarily recent structures, and comparably recent structures in mammals do not show steroid uptake (Arnold *et al.*, 1976). MAN and HVc appear to be present only in oscine songbirds (i.e., the suborder of songbirds with the most highly specialized vocal apparatus); there is no evidence of these structures in chickens, pigeons, or a nonoscine songbird species—the king bird (Arnold *et al.*, 1976). Another difference between hormone uptake in these and that in other neural structures is that the song-control areas concentrate only androgens (Arnold & Saltiel, 1979). There is no evidence of estrogen uptake in these areas, while patterns of ^{3}H-T and ^{3}H-DHT uptake are comparable. The syrinx itself, including both the muscle and the structural elements, contains high concentrations of high-affinity androgen-binding proteins (Lieberburg & Nottebohm, 1979),

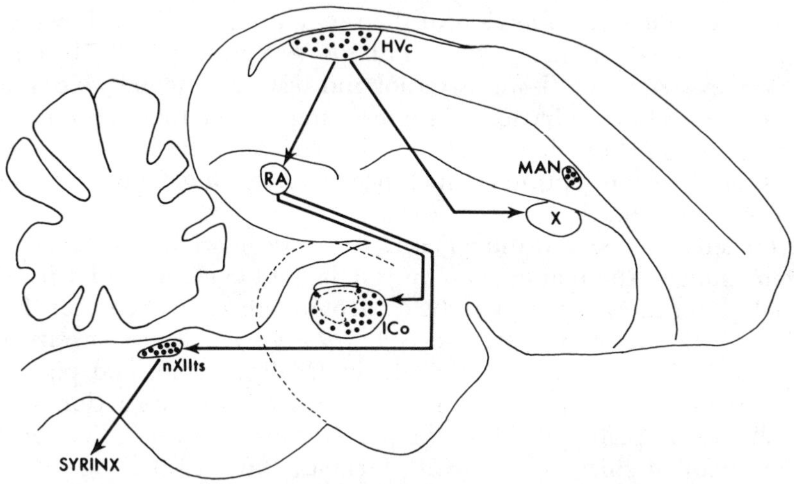

FIGURE 4. Parasagittal view summarizing anatomical and autoradiographic evidence on neural control of song. Lesions of HVc and RA disrupt singing behavior in the canary, and these nuclei project to other nuclei (X, ICo, nXIIts). Black dots indicate the presence of labeled cells in a subset of these nuclei in the zebra finch brain but do not represent actual numbers of cells detected. Labeled cells in other areas of the brain are not depicted. See text for explanation of abbreviations. (From "Hormone Concentrating Cells in Vocal Control and Other Areas of the Brain of the Zebra Finch" by A. P. Arnold, F. Nottlebohm, and D. W. Pfaff, *Journal of Comparative Neurology*, 1976, *165*, 487–512. Copyright 1976 by *Journal of Comparative Neurology*. Reprinted by permission.

indicating that androgenic control of singing behavior extends from the highest neural centers to the avian vocal organ.

Having briefly outlined the areas that concentrate sex steroids in the avian brain, the obvious question is which of these areas are involved in the control of aggressive behavior. Data on this point are sparse but suggestive. Certainly the song-control areas of the passerine brain should be included, given the use of song in aggressive displays in these species. Other hormone-sensitive areas that are probably involved in the control of aggressive behavior include the preoptic area, lateral septum, archistriatum, and paleostriatum. Preoptic implants of TP have been shown to elicit aggressive as well as sexual behavior in castrated ring doves (Barfield, 1971; Hutchison, 1970, 1971, 1975), although such implants are not effective in chickens (Barfield, 1969; Phillips & Barfield, 1977).

Lesions of the tractus infundibuli have been found to depress attack behavior in chicks (Mitchell & Andrew, 1976). The ability of preoptic-area implants to elicit aggression depends on both the androgen used and the rate of release (Hutchison, 1975). Hutchison found that implants of DHT were ineffective while equivalent implants of TP were successful in eliciting aggressive courtship in doves. Furthermore, the effectiveness of TP implants in this area was determined by their rate of diffusion. Low-diffusion implants reinstated the nest-oriented components of courtship without affecting aggressive behavior, while high-diffusion implants increased the aggressive components. Data on the avian septal area are rather limited. Stimulation of the septal area has been shown to cause some agonistic behavior in chickens (Phillips & Youngren, 1971; Putkonen, 1966). Given evidence that both estradiol and testosterone are able to activate high levels of aggressive behavior, it is interesting to note that the septum is one of the few brain areas in which there is evidence for differential uptake of sex steroids; labeling of the septum is much heavier following ^{3}H-testosterone than ^{3}H-estradiol treatment.

The nucleus taeniae and the adjacent area of the archistriatum are considered homologous to the mammalian amygdala, and evidence in birds, like that in mammals, implicates this area in the control of aggressive behavior. Electrical stimulation of the archistriatum induces aggressive attack in chickens (Phillips & Youngren, 1971; Putkonen, 1966), ducks (Maley, 1969), and pigeons (Akerman, 1966). Lesions of this area reduce aggressive responses in ducks (Phillips, 1964). However, implants of TP in the archistriatum had no effect on the aggressive behavior of chickens or doves (Barfield, 1969, 1971).

Thus, a number of androgen-sensitive neural areas in the avian brain appear to be involved in the production of aggressive behavior. The androgen-sensitive song-control areas in the passerine—HVc, MAN, ICo, and nXIIts—are involved in the control of singing behavior, one of the most common aggressive displays in territorial interactions in songbirds. There is also evidence linking the preoptic area, lateral septum, archistriatum, and paleostriatum to the control of attack behavior. However, the precise control of aggressive behavior remains to be elucidated. So far a direct effect of androgen on these neural areas to elicit aggressive behavior has been demonstrated only in the preoptic area of the ring dove, although electrical stimulation has been found to be effective in all areas listed above.

SEXUAL DIFFERENTIATION OF BEHAVIOR

Sexual differentiation of avian behavior has been closely examined in only two species, Japanese quail and zebra finches. Although aggressive behavior was not examined directly in any of these studies, the process of sexual differentiation in birds deserves mention because the underlying hormonal mechanism differs from that in mammals, and it also differs between these two species. For a discussion of the available data on other species, see Adkins (1978).

Adkins (1975) injected Japanese quail eggs with either estradiol benzoate or TP during incubation, and the quail were tested for sexual behavior in adulthood. Birds were treated with TP or estradiol benzoate prior to being tested for male or female behavior, respectively, to ensure that any deficits in behavior were not caused by a deficiency in gonadal hormones caused by the early steroid treatment. The female quail showed perfectly normal feminine sexual behavior, while the males showed good feminine sexual behavior but tremendous deficits in male behavior. Thus, early hormone treatment at about days 9–11 of incubation demasculinized males and had no effect on females, indicating that in quail, unlike mammals, the male is the neutral sex. Further study indicated that antiestrogen treatment during days 9–11 of incubation masculinized females, while having no effect on males (Adkins, 1976).

The process of sexual differentiation in the zebra finch is quite different. Gurney and Konishi (1980) investigated sexual differentiation of the song-control nuclei and singing behavior in zebra finches. Sexual differentiation in this species may be altered by hormone treatment after hatching, and in this species the female is the neutral sex. Song development in males was not affected by early hormone treatment, but females that received silastic implants of estradiol on the day of hatching and either testosterone or DHT in adulthood were masculinized. They showed typical male singing and courtship behavior, behaviors never seen in normal female zebra finches. DHT implantation on the day of hatching had no effect on the females' behavior when they were given further androgen in adulthood, indicating the importance of female hormones in male differentiation in the finch. In this species, females have never been known to sing, and three song-control nuclei—HVc, ICo, and RA—are much larger in the male than the female. Examination of the neural song-control nuclei showed that in females, both RA and HVc increased in size following early estradiol or DHT treatment, with estradiol exerting a more profound effect. In contrast, ICo was affected only by DHT treatment. All three nuclei were further enlarged when early estradiol, but not DHT, treatment was followed by DHT or testosterone treatment in adulthood, with DHT exerting a larger effect than testosterone. Such effects are not seen if female zebra finches are given these hormone treatments only in adulthood. Sexual differentiation of the zebra finch brain appears to be a two-stage process, although the authors did not determine whether the effects of adult androgen treatment caused a permanent increase in nuclear size that would persist after androgen treatment was withdrawn.

Thus, masculinization in zebra finches appears to result from early estrogen exposure followed by exposure to androgen as an adult, and is more similar to the pattern of differentiation seen in mammals then to that of quail. Although none of these studies examined the effects of early hormone treatment on aggressive behavior, there is no reason to assume that the sexual differentiation of aggressive behavior would differ from that of other social behaviors.

COMPLEX HORMONE–BEHAVIOR INTERACTIONS

EFFECTS OF MALE–FEMALE INTERACTIONS ON HORMONE LEVELS AND BEHAVIOR

Although seasonal factors such as increasing photoperiod play an important role in increasing gonadal activity and hormone secretion in most species prior to breeding (Wingfield & Farner, 1980), data from several sources indicate that part of the increase in testosterone levels seen during the breeding period is attributable to the male's interactions with the female (ring doves: Feder, Storey, Goodwin, Reboulleau, & Silver, 1977; pigeons: Haase, Paulke, & Sharp, 1976;

zebra finches: Pröve, 1978; white-crowned sparrows: Wingfield & Farner, 1980). Social interactions may be the major influence on androgen secretion in species such as the ring dove and zebra finch, which do not show strong seasonality. Wingfield and Farner reported that plasma testosterone levels were always higher in male white-crowned sparrows in the field than in those kept in aviaries, and under both conditions males paired with females had higher testosterone levels than unpaired males. Feder and his collaborators found that testosterone levels in male ring doves began to rise within four hours of being paired with a female, continued to rise until the third day of courtship, declined during incubation, and rose again when the birds began to court for a second time. Males' initial behavior toward both females and other males was aggressive, but only pairing with females caused a significant increase in androgen levels. Interestingly, aggressive behavior patterns, such as bow-coo, chase, and cackle, occur within seconds of introducing another bird even though testosterone levels are quite low. The same patterns also occur if a strange bird is introduced during incubation, when androgen levels are also low, indicating that these behaviors probably have a low hormonal threshold for stimulation. However, castrated males do not show these behavior patterns even though their baseline androgen levels appear similar to those of isolated intact males.

Another study (Moss *et al.*, 1979) found that when captive male red grouse were moved from isolation to group housing with several hens, the size of their combs increased within 48 hours, presumably because of increased androgen secretion. Pairing with a hen also increased the male's aggressive behavior.

The importance of female endocrine status and behavior in modulating the male's endocrine status was conclusively demonstrated in a series of laboratory and field experiments on the white-crowned sparrow by Moore (1980, in press; Moore & Wingfield, 1980). In the laboratory, photostimulated males exposed to estradiol-treated, receptive females had elevated levels of LH and testosterone compared with control photostimulated males. The testes of the males paired with such females remained enlarged for a longer period before regressing. In a second experiment, free-living females were implanted with long-lasting estradiol capsules and released. Surprisingly, the estradiol implants did not affect the females' breeding cycles. Implanted females laid normal clutches, incubated them, and successfully fledged their offspring in four of six nests observed. However, implanted females continued to solicit and copulate through at least the nestling phase. Normally these behaviors disappear when the females begin to incubate. Plasma testosterone and dihydrotestosterone levels in males paired with the implanted females did not decline in the usual fashion at the beginning of incubation. Their plasma androgen levels during incubation were not significantly different from their levels during courtship and were significantly higher than those of control males during incubation. Moore argues that these increased androgen levels in the male caused by female behavior are more important in activating aggressive than sexual behavior. Although male white-crowns with basal testosterone levels have been observed to mount females, only males with high androgen levels maintain exclusive territories. Such exclusive territories are usually maintained until incubation begins; males then allow neighboring birds to forage quietly, though not display, on their territories. Moore argues that such exclusive territories function both to help assure male paternity and to provide females with the necessary resources for egg production at a time when their energy needs are quite high and resources are quite low. (See section on hormonal treatment of intact birds for a discussion of the effects of testosterone implants in the male white-crown on aggression and territorial defense.)

Another study (Hutchison & Lovari, 1976) demonstrated that the level of aggressive behavior that the male directs toward the female can adversely affect her endocrine condition and reproductive state. One group of female ring doves was paired with males that courted them very aggressively, with twice as much chasing and six times as much bowing as that shown by another group of males on the initial test after pairing. Aggressive males showed only rudimentary nest-building. The females paired with aggressive males had longer latencies to nest-solicit and sit, and they laid their clutches significantly later than the other group. The testes of the aggressive males were less well developed than those in the control group and their plasma testosterone levels were also lower (Hutchison & Katangole, 1975). This seeming inconsistency, higher levels of aggression associated with lower plasma testosterone levels, appears to be related to seasonal differences in neural responsiveness to androgen stimulation. When both groups of males received hypothalamic implants of testosterone, the group of aggressive males whose testes were still recrudescing showed more aggression and no nest-soliciting, unlike the control group whose testes were fully developed. (See section on the effects of castration and replacement therapy for additional discussion.)

Hormone Changes Elicited by Aggressive Interactions

Although studies that measure an individual's plasma hormone levels either during or immediately following an aggressive encounter are fairly common in the mammalian literature, this paradigm has rarely been used in avian research. Ramenofsky (Ramenofsky & Gorbman, 1980; M. Ramenofsky, personal communication, 1980) examined the relationship between endogenous hormone levels and aggression in Japanese quail. Male quail were paired for brief encounters in a neutral area following a round-robin design, and blood samples were obtained before or 24 hours after an encounter. Levels of LH, testosterone, DHT, androstenedione, and corticosterone were determined. Only plasma testosterone levels appeared to be related to aggressive behavior. Within each pair of birds, the dominant bird had a significantly higher plasma testosterone concentration 24 hours after the encounter. The mean difference in plasma testosterone levels between winners and losers decreased over tests. Ramenofsky also divided the birds into three classes on the basis of their behavior during the encounters. (1) Winners consistently attacked and won their encounters. (2) Intermediates did not attack but displayed aggressive postures. These individuals were subordinate to attackers but dominant to losers. (3) Losers did not attack and displayed less frequently than intermediates. Using this classification system, there was a significant correlation between endogenous testosterone levels and fighting competence, with winners having the highest testosterone levels and losers the lowest. Ramenofsky concluded that testosterone affects an animal's ability to fight but does not necessarily determine the outcome of a given fight. Testosterone's effects on fighting competence were greatest when dominance status was first being established, and over tests the relative importance of plasma testosterone levels seemed to decrease, while the influence of other factors, such as experience, appeared to increase. This conclusion is supported by the results of other studies indicating that dominance hierarchies, once established, are often difficult to manipulate by altering the circulating hormone levels of the individuals involved (see discussion of various studies in section on hormonal treatment of intact birds). This suggests that in naive animals, behavior may be more open to influence by circulating hormone levels, while in socially experienced individuals, the reverse may be true: Behavior may exert a stronger influence on plasma hormone levels.

Harding and Follett (1979) examined hormone changes triggered by aggressive interactions in a natural population of redwinged blackbirds. In this study, plasma hormone levels of male redwings caught on their territory at the height of an aggressive encounter with an intruder were compared with those of males caught while foraging. There was no difference in the plasma corticoid levels of the aggressive and foraging males. This finding was particularly interesting in view of the many laboratory studies that have found increased corticoid levels in both dominant and subordinate individuals following aggressive interactions. It could represent a species difference in the way animals respond to aggressive interactions, or it could reflect the difference between the freedom to initiate and pace an aggressive encounter in contrast to confinement with a conspecific in a small space.

Aggressive males had significantly lower LH levels than foraging males, which is consistent with data from mammals showing that LH concentrations are lower in both dominant and subordinate individuals following an aggressive interaction. Fighting males also had significantly different DHT levels compared with foraging birds, but this was a difference in variability rather than mean hormone level. DHT concentrations in aggressive males were more variable than those in foraging males. The same trend was evident in both the testosterone and total androgen (testosterone + DHT) data, but it was significant only for the total androgen data. There were significant positive correlations between levels of LH and testosterone and LH and DHT in aggressive males but not foraging males (see Figure 5), suggesting that aggressive interactions might alter neuroendocrine control mechanisms.

In an attempt to explain the greater variability in the DHT levels of the aggressive males, these males were divided into two groups, those whose DHT levels fell above the median level of the foraging males and those whose DHT levels fell below it (Harding, 1981). There were several significant differences in the behavior of these two groups of males as they attempted to defend their territories from the intruder. Males with high androgen levels performed significantly more bill-ups, a common aggressive display in blackbirds, both at a distance and within close range of the intruder. Although the singing rate of the two groups did not differ, high- and low-androgen males differed in when

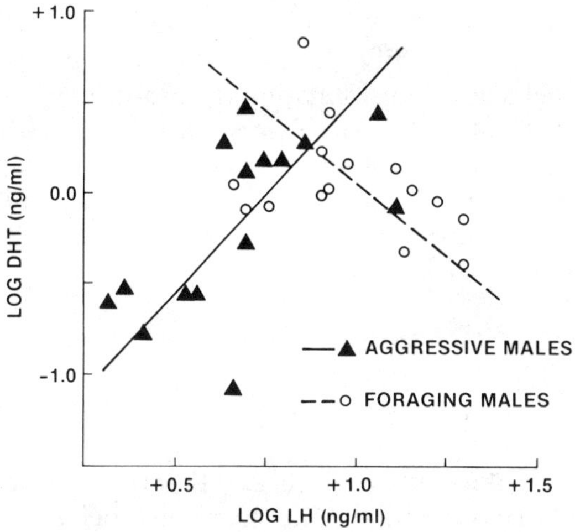

FIGURE 5. The relation between concentrations of LH and DHT in an individual differs in aggressive and foraging males. There is a significant positive correlation between LH and DHT in aggressive blackbirds but not in foraging blackbirds. (From "Hormone changes triggered by aggression in a natural population of blackbirds" by C. F. Harding and B. K. Follett, *Science*, 1979, *203*, 918–920. Copyright 1979 by the American Association for the Advancement of Science. Reprinted by permission.)

they sang the most. Low-androgen males sang more during silent intervals when the intruding redwing did not appear to be singing, while high-androgen males were more likely to sing in response to advertising song playback by the experimenter. Low-androgen males sidled up to the intruder indirectly and uttered more alarm calls than high-androgen males. Thus, low-androgen males appeared "less confident" in their defense of their territory, although they, like the high-androgen males, attempted to drive off the intruder.

Why DHT levels reacted more strongly to behavioral stimulation than did testosterone is unknown. In most studies of androgen–behavior interactions, DHT assays are omitted, although circulating hormone with the mobility of DHT has been found in several other species (Luine *et al.*, 1980; Wingfield & Farner, 1975). Exogenous DHT does not appear to alter levels of aggressive behavior in castrated zebra finches (Harding, unpublished observations).

This study demonstrated that under natural conditions, an individual's plasma hormone concentrations fluctuate, reflecting the ongoing social interaction. Furthermore, individuals showing different patterns of behavior show different patterns of circulating hormones. Males showing the most aggressive displays during a territorial interaction have higher androgen levels than those that appear more cautious and/or alarmed. It is important to determine whether changing hormone levels during social interactions are merely a consequence of the ongoing stimulation or whether they are in fact modulating the individual's behavior. This is a difficult question to investigate, but a very important one. There are some data for mammals indicating that changing hormone levels during a social interaction can indeed affect both an individual's ongoing behavior and its response to similar situations in the future (see Nock & Leshner, 1976; Roche & Leshner, 1979). These data also suggest a way in which initially small differences between individuals can be amplified with repeated social interactions. Small differences in experience may cause slight differences in hormonal response between individuals when they are first exposed to a particular social situation. These small differences in response to the initial exposure may further modify the individual's responsivity, leading to larger differences in behavior and circulating hormone levels between individuals on later exposures to similar situations. Thus, the rapidity with which the endocrine system responds to social stimulation suggests that circulating hormone levels may play an important role in modulating an individual's minute-to-minute behavior during aggressive interactions, as well as functioning in the classical sense as long-term modulators of behavior.

SUMMARY

All evidence indicates that the androgen, testosterone, is the most important hormone influencing aggressive behavior in birds. Testosterone appears to activate aggressive behavior in both males and females of practically every species that has been studied. Yet testosterone's efficacy may stem from its ability to be converted to a variety of other hormones. There is evidence implicating both androgenic and estrogenic metabolites of testosterone in modulating avian aggressive behavior. Besides direct effects on behavior, androgens can modify an animal's status and behavior indirectly by changing morphological characteristics, such as plumage, body weight, and bill color. In many species, changes in androgen level and aggressive behavior are seasonal, resulting in increased levels of plasma testosterone and aggressive behavior at the onset of the breeding season. Hormone levels, and ultimately behavior, may be further modified by social interactions,. Interactions with members of the opposite sex and pair formation have been shown to be important determinants of increased androgen secretion in many species.

There are, however, important gaps in our knowledge of the hormonal influences on aggressive behavior in birds. Nothing is known about the adrenal glands and their ability to secrete androgens or other hormones that might modulate aggressive behavior. Although it now appears unlikely that LH plays a major role in regulating aggressive behavior in passerines, additional well-controlled studies are needed. Very few studies have focused on aggressive behavior in the female, and much remains to be done. Although the sex-steroid sensitive areas of the avian brain have been well mapped, only one study has succeeded in eliciting aggression through implanting androgen in the brain. Clearly, a great deal of work will have to be done before we know precisely which neural areas are involved in the control of aggressive behavior. The results of studies on avian sexual differentiation make it obvious that more data are needed if we are to understand why the female is the neutral sex in some species and the male in others. Finally, additional effort must be made to understand complex hormone–behavior interactions. One of the important tasks before us is determining whether the rapid changes in hormone levels that occur during social interactions are capable of modulating the behavior of the individual. The next 10 years should be a particularly exciting time in avian behavioral endocrinology as researchers try to answer these questions.

ACKNOWLEDGMENTS

I would like to thank Kevin Sheridan, Michael Walters, and especially Kathy Woods for their assistance in locating some of the references. I would also like to thank Alan Leshner, Fred Stollnitz, and John Wingfield for their critical comments on an earlier draft of this manuscript.

REFERENCES

Adkins, E. K. Hormonal basis of sexual differentiation in Japanese quail. *Journal of Comparative and Physiological Psychology,* 1975, *89,* 61–71.

Adkins, E. K. Embryonic exposure to an antiestrogen masculinized behavior of female quail. *Physiology and Behavior,* 1976, *17,* 357–359.

Adkins, E. K. Sex steroids and the differentiation of avian reproductive behavior. *American Zoologist,* 1978, *18,* 501–509.

Adkins, E. K., & Nock, B. L. The effects of the antiestrogen CI-628 on sexual behavior activated by androgen or estrogen in quail. *Hormones and Behavior,* 1976, 7, 417–429.

Akerman, B. Behavioral effects of electrical stimulation in the forebrain of the pigeon: II. Protective behaviour. *Behaviour,* 1966, *26,* 339–350.

Allee, W. C., Collias, N.E., & Lutherman, C. Z. Modification of the social order in flocks of hens by the injection of testosterone propionate. *Physiological Zoology,* 1939, *12,* 413–439.

Armstrong, E. A. *Bird display and behavior.* New York: Dover Publications, 1965.

Arnold, A. P. The effects of castration and androgen replacement on song, courtship, and aggression in zebra finches, *Poephila guttata. Journal of Experimental Zoology,* 1975, *191,* 309–326.

Arnold, A. P., & Saltiel, A. Sexual difference in pattern of hormone accumulation in the brain of a song bird. *Science,* 1979, *205,* 702–705.

Arnold, A. P., Nottebohm, F., & Pfaff, D. W. Hormone concentrating cells in vocal control and other areas of the brain of the zebra finch *(Poephila guttata). Journal of Comparative Neurology,* 1976, *165,* 487–512.

Balthazart, J., & Hendrick, J. Annual variation in reproductive behavior, testosterone, and plasma FSH levels in the Rouen duck, *Anas platyrhynchos. General and Comparative Endocrinology,* 1976, *28,* 171–183.

Balthazart, J., & Hendrick, J. C. Relationships between the daily variations of social behavior and of plasma FSH, LH and testosterone levels in the domestic duck *(Anas platyrhynchos). Behavioral Processes,* 1979, *4,* 107–128.

Balthazart, J., Hendrick, J. C., & Deviche, P. Diurnal variations of plasma gonadotrophins in male domestic ducks during the sexual cycle. *General and Comparative Endocrinology,* 1977, *32,* 376–389.

Barfield, R. J. Activation of copulatory behavior by androgen implants in the preoptic area of the male fowl. *Hormones and Behavior,* 1969, *1,* 37–52.

Barfield, R. J. Activation of sexual and aggressive behavior by androgen implanted into the male ring dove brain. *Endocrinology,* 1971, *89,* 1470–1476.

Barfield, R. J., Ronay, G., & Pfaff, D. W. Autoradiographic localization of androgen-concentrating cells in the brain of the male domestic fowl. *Neuroendocrinology,* 1978, *26,* 297–311.

Bennett, M. A. The social hierarchy in ring doves. *Ecology,* 1939, *20*(3), 337–356.

Bennett, M. A. The social hierarchy in ring doves: II. The effect of treatment with testosterone propionate. *Ecology,* 1940, *21,* 148–165.

Berthold, A. A. Transplantation der Hoden. *Archiven für Anatomie und Physiologie,* 1849, *16,* 42–46.

Boss, W. R. Hormonal determination of adult characters and sex behavior in herring gulls *(Larus argentatus)*. *Journal of Experimental Zoology,* 1943, *94,* 181–203.

Brown, J. L. Aggressiveness, dominance and social organization in the Steller Jay. *Condor,* 1963, *65,* 460–484.

Butterfield, P. A., & Crook, J. H. The annual cycle of nest building and agonistic behaviour in captive *Quelea quelea* with reference to endocrine factors. *Animal Behavior,* 1968, *16,* 308–317.

Carpenter, C. R. Psychobiological studies of social behavior in aves: I. The effect of complete and incomplete gonadectomy on the primary sexual activity of the male pigeon. *Journal of Comparative Psychology,* 1933, *16,* 25–97.

Cheng, M. F., & Lehrman, D. Gonadal hormone specificity in the sexual behavior of ring doves. *Psychoneuroendocrinology,* 1975, *1,* 95–102.

Collias, N. E. Hormones and behavior with special reference to birds and the mechanisms of hormone action. In E. S. Gordon (Ed.), *Steroid hormones.* Madison, Wisc.: University of Wisconsin Press, 1950.

Collias, N. E., & Taber, R. D. A field study of some grouping and dominance relations in ring-necked pheasants. *Condor,* 1951, *53,* 265–275.

Crook, J. H., & Butterfield, P. A. Gender role in the social system of Quelea. In J. H. Crook (Ed.), *Social behaviour in birds and mammals.* New York: Academic Press, 1974.

Daly, M., & Wilson, M. *Sex, evolution, and behavior.* North Sciutate, Mass.: Duxberry Press, 1978.

Davis, D. E. Aggressive behavior in castrated starlings. *Science,* 1957, *126,* 253.

Davis, D. E. The hormonal control of aggressive behavior. *Proceedings XIII International Ornithological Congress* (Baton Rouge), 1963, *2,* 994–1003.

Emlen, J. T., Jr., & Lorenz, F. W. Pairing responses of free-living valley quail to sex-hormone pellet implants. *Auk,* 1942, *59,* 369–378.

Feder, H. H., Storey, A., Goodwin, D., Reboulleau, C., & Silver, R. Testosterone and "5 α-dihydrotestosterone" levels in peripheral plasma of male and female ring doves *(Streptopelia risoria)* during the reproductive cycle. *Biology of Reproduction,* 1977, *16,* 666–677.

Ficken, R. W., Ficken, M. S., & Hailman, J. P. Differential aggression in genetically different morphs of the white-throated sparrow *(Zonotrichia albicollis)*. *Zeitschrift für Tierpsychologie,* 1978, *46,* 43–57.

Goodale, H. D. Castration in relation to the secondary sexual characters of brown leghorns. *American Naturalist,* 1913, *47,* 159–169.

Gurney, M. E., & Konishi, M. Hormone-induced sexual differentiation of brain and behavior of zebra finches. *Science,* 1980, *208,* 1380–1383.

Haase, E. The effects of testosterone propionate on secondary sexual characters and testes of house sparrows, *Passer domesticus. General and Comparative Endocrinology,* 1975, *26,* 248–252.

Haase, E., Paulke, E., & Sharp, P. J. Effects of seasonal and social factors on testicular activity and hormone levels in domestic pigeons. *Journal of Experimental Zoology,* 1976, *197,* 81–88.

Harding, C. F. Social modulation of circulating hormone levels in the male. *American Zoologist,* 1981, *21,* 223–231.

Harding, C. F. Hormonal specificity and activation of social behavior in the male zebra finch. In Balthazart, J. (Ed.), *Hormones and behaviour in higher vertebrates.* Basel: Springer Verlag, in press.

Harding, C. F., & Follett, B. K. Hormone changes triggered by aggression in a natural population of blackbirds. *Science,* 1979, *203,* 918–920.

Herzog, P. W., & Boag, D. A. Seasonal changes in aggressive behavior of female spruce grouse. *Canadian Journal of Zoology,* 1977, *55,* 1735–1739.

Höhn, E. O. Gonadal hormone concentrations in northern phalaropes in relation to nuptial plumage. *Canadian Journal of Zoology,* 1970, *48,* 400–401.

Höhn, E. O., & Cheng, S. C. Gonadal steroids in Wilson's phalaropes and certain other birds in relation to nuptial plumage and sex behaviour of phalaropes. *General and Comparative Endocrinology,* 1967, *8,* 1–11.

Hutchison, J. B. Influence of gonadal hormones on the hypothalamic integration of courtship behavior in the Barbary dove. *Journal of Reproduction and Fertility* (Supplement), 1970, *11,* 15–41.

Hutchison, J. B. Effects of hypothalamic implants of gonadal steroids on courtship behavior in Barbary doves *(Streptopelia risoria). Journal of Endocrinology,* 1971, *50,* 97–113.

Hutchison, J. B. Target cells for gonadal steroids in the brain: Studies on steroid-sensitive mechanisms of behaviour. In P. Wright, P. G. Caryl, & D. M. Vowles (Eds.), *Neural and endocrine aspects of behaviour in birds.* Amsterdam: Elsevier, 1975.

Hutchison, J. B., & Katongole, C. B. Plasma testosterone in courting and incubating male Barbary doves *(Streptopelia risoria). Journal of Endocrinology,* 1975, *65,* 275–276.

Hutchison, J. B., & Lovari, S. Effects of male aggressiveness on behavioural transitions in the reproductive cycle of the Barbary dove. *Behaviour,* 1976, *59,* 296–317.

Johns, J. E. Testosterone-induced nuptial feathers in phalaropes. *Condor,* 1964, *66,* 449–455.

Keck, W. N. The control of secondary sexual characteristics in the English sparrow. *Journal of Experimental Zoology,* 1934, *67,* 315–367

Kerlan, J. T., & Jaffe, R. B. Plasma testosterone levels during the testicular cycle of the redwinged blackbird *(Agelaius phoeniceus). General and Comparative Endocrinology,* 1974, *22,* 428–432.

Kerlan, J. T., Jaffe, R. B., & Payne, A. H. Sex-steroid formation in gonadal tissue homogenates during the testicular cycle of the redwinged blackbird *(Agelaius phoeniceus). General and Comparative Endocrinology,* 1974, *24,* 352–363.

Kim, Y. S., Stumpf, W. E., Sar, M., & Martinez-Vargas, M. C. Estrogen and androgen target cells in the brain of fishes, reptiles and birds: Phylogeny and ontogeny. *American Zoologist,* 1978, *18,* 425–433.

Komisaruk, B. R. Effects of local brain implants of progesterone on reproductive behavior in ring doves. *Journal of Comparative and Physiological Psychology,* 1967, *64,* 2, 219–224.

Kuo, Z. Y. Studies on the basic factors in animal fighting: III. Hormonal factors affecting fighting in quails. *Journal of Genetic Psychology,* 1960, *96,* 217–223.

Lazarus, J., & Crook, J. H. The effects of luteinizing hormone, oestrogen and ovariectomy on the agonistic behaviour of female *Quelea quelea. Animal Behavior,* 1973, *21,* 49–60.

Leshner, A. I. *An introduction to behavioral endocrinology.* New York: Oxford University Press, 1978.

Lieberburg, I., & Nottebohm, F. High-affinity androgen binding proteins in syringeal tissues of songbirds. *General and Comparative Endocrinology,* 1979, *37,* 286–293.

Luine, V., Nottebohm, F., Harding, C., & McEwen, B. S. Androgen affects cholinergic enzymes in syringeal motor neurons and muscle. *Brain Research,* 1980, *192,* 89–107.

Lumia, A. R. The relationships among testosterone, conditioned aggression, and dominance in male pigeons. *Hormones and Behavior,* 1972, *3,* 277–286.

Maley, M. J. Electrical stimulation of agonistic behavior in the mallard. *Behaviour,* 1969, *34,* 138–160.

Mathewson, S. Gonadotropic hormones affect aggressive behaviour in starlings. *Science,* 1961, *134,* 1522–1523.

Meier, A. H., & Ferrell, B. R. Avian endocrinology. In A. H. Brush (Ed.), *Chemical zoology* (Vol. 10): *Aves.* New York: Academic Press, 1978.

Meier, A. H., & Fivizzani, A. J. Changes in the daily rhythm of plasma corticosterone concentration related to seasonal conditions in the white-throated sparrow, *Zonotrichia albicollis. Proceedings of the Society for Experimental Biology and Medicine,* 1975, *150,* 356–362.

Meyer, C. C. Testosterone concentration in the male chick brain: An autoradiographic study. *Science,* 1973, *180,* 1381–1383.

Mitchell, J., & Andrew, R. J. A copulatory system in the hypothalamus and tegmentum of the domestic chick. *Journal of Comparative and Physiological Psychology,* 1976, *90*(7), 643–652.

Moore, M. C. *The role of behavioral interactions within members of a pair in the endocrine synchronization of reproduction.* Paper presented at the American Ornithologist's Union meetings, Denver, August 11–15, 1980.

Moore, M. C. Female behavior controls testosterone levels of free-living white-crowned sparrows. *Journal of Zoology,* in press.

Moore, M. C., & Wingfield, J. C. Hormone-behavior interactions during reproduction in white-crowned sparrow. *American Zoologist,* 1980, *20,* 788.

Moss, R., Kolb, H. H., Marquiss, M., Watson, A., Treca, B., Watt, D., & Glennie, W. Aggressiveness and dominance in captive red grouse. *Aggressive Behaviour,* 1979, *5,* 59–84.

Murton, R. K., Thearle, R. J. P., & Lofts, B. The endocrine basis of breeding behaviour in the feral pigeon (*Columba livia*): I. Effects of exogenous hormones on the pre-incubation behaviour of intact males. *Animal Behavior,* 1969, *17,* 286–306.

Noble, G. K., & Wurm, M. The effect of testosterone propionate on the black-crowned night heron. *Endocrinology,* 1940, *26,* 837–850.

Nock, B. L., & Leshner, A. I. Hormonal mediation of the effects of defeat on agonistic responding in mice. *Physiology and Behavior,* 1976, *17,* 111–119.

Peek, F. W. Seasonal change in the breeding behavior of the male red-winged blackbird. *The Wilson Bulletin,* 1971, *83,* 383–395.

Phillips, R. E. 'Wildness' in the mallard duck: Effects of brain lesions and stimulation on 'escape behavior' and reproduction. *Journal of Comparative Neurology,* 1964, *122,* 139–155.

Phillips, R. E., & Barfield, R. J. Effects of testosterone implants in midbrain vocal areas of capons. *Brain Research,* 1977, *122,* 378–381.

Phillips, R. E., & Youngren, O. M. Brain stimulation and species typical behavior: Activities evoked by electrical stimulation of the brains of chickens. *Animal Behaviour,* 1971, *19,* 757–779.

Pietras, R. J., & Wenzel, B. M. Effects of androgens on body weight, feeding, and courtship behavior in the pigeon. *Hormones and Behavior,* 1974, *5,* 289–302.

Pröve, E. Quantitative Untersuchungen zu Wechselbeziehungen zwischen Balzaktivität und Testosterontitern bei männlichen Zebrafinken *(Taeniopygia guttata castanotis). Zeitschrift für Tierpsychologie,* 1978, *48,* 47–67.

Putkonen, P. T. S. Attack elicited by forebrain and hypothalamic stimulation in the chicken. *Experientia,* 1966, *22,* 405–406.

Raith, L., Bauer, H. J., & Karl, H. J. The extrahepatic metabolism of androgens in birds and mammals. *Acta Endocrinologica,* 1977, *84,* 74–75.

Ramenofsky, M., & Gorbman, A. Plasma hormones and aggressive behavior in adult male Japanese quail. *American Zoologist,* 1980, *20,* 791. (Abstract)

Roche, K. E., & Leshner, A. I. ACTH and vasopressin treatments immediately after a defeat increase future submissiveness in male mice. *Science,* 1979, *204,* 1343–1344.

Rohwer, S. The social significance of avian winter plumage variability. *Evolution,* 1975, *29,* 593–610.

Rohwer, S. Status signalling in Harris Sparrows: Some experiments in deception. *Behaviour,* 1977, *61,* 107–129.

Rohwer, S., & Rohwer, F. C. Status signalling in Harris sparrows: Experimental deceptions achieved. *Animal Behaviour,* 1978, *26,* 1012–1022.

Scott, H. M., & Payne, L. F. The effect of gonadectomy on the secondary sexual characters of the bronze turkey *(M. gallopavo). Journal of Experimental Zoology,* 1934, *69,* 123–136.

Searcy, W. A., & Wingfield, J. C. The effects of androgen and antiandrogen on dominance and aggressiveness in male red-winged blackbirds. *Hormones and Behavior,* 1980, *14,* 126–135.

Selinger, H. E., & Bermant, G. Hormonal control of aggressive behavior in Japanese quail *(Coturnix coturnix japonica). Behavior,* 1967, *28,* 255–268.

Silver, R., O'Connell, M., & Saad, R. Effect of androgens on the behavior of birds. In C. Beyer (Ed.), *Endocrine control of sexual behavior.* New York: Raven Press, 1979.

Stern, J. M. Androgen accumulation in hypothalamus and anterior pituitary of male ring doves: Influence of steroid hormones. *General and Comparative Endocrinology,* 1972, *18,* 439–449.

Temple, S. A. Plasma testosterone titers during the annual reproductive cycle of starlings *(Sturnus vulgaris). General and Comparative Endocrinology,* 1974, *22,* 470–479.

Terkel, A. D., Moore, C. L., & Beer, C. G. The effects of testosterone and estrogen on the rate of long-calling vocalization in juvenile laughing gulls, *Larus atricilla. Hormones and Behavior,* 1976, *7,* 49–58.

Vowles, D. M., & Harwood, D. The effect of exogenous hormones on aggressive and defensive behaviour in the ring dove *(Streptopelia risoria). Journal of Endocrinology,* 1966, *36,* 35–51.

Watson, A. Territorial and reproductive behaviour of red grouse. *Journal of Reproduction and Fertility,* 1970, *Suppl. II,* 3–14.

Williamson, F. S. L. The molt and testis cycles of the Anna Hummingbird. *Condor,* 1956, *58,* 342–366.

Wingfield, J. C., & Farner, D. S. The determination of five steroids in avian plasma by radioimmunoassay and competitive protein-binding. *Steroids,* 1975, *26*(3), 311–327.

Wingfield, J. C., & Farner, D. S. The endocrinology of a natural breeding population of the white-crowned sparrow *(Zonotrichia leucophrys pugetensis)*. *Physiological Zoology*, 1978, *51*, 188–205.

Wingfield, J. C., & Farner, D. S. Some endocrine correlates of renesting after loss of clutch or brood in the white-crowned sparrow, *Zonotrichia leucophrys gambelii*. *General and Comparative Endocrinology*, 1979, *38*, 322–331.

Wingfield, J. C., & Farner, D. S. Control of seasonal reproduction in temperate-zone birds. In P. O. Hubinot (Ed.), *Progress in reproductive biology*. (Vol. 5) Basel: Karger, 1980.

Wingfield, J. C., Martin, A., Hunt, M. W., Hunt, G. L., Jr., & Farner, D. S. Origin of homosexual pairing of female Western gulls on Santa Barbara Island. In D. M. Power (Ed.), *The California islands*. Santa Barbara, Cal.: Santa Barbara Museum of Natural History, 1980.

Wingfield, J. C., Newman, A. L., Hunt, G. L., Jr., & Farner, D. S. Endocrine aspects of female-female pairing in the Western gull *(Larus occidentalis wymani)*. *Animal Behavior*, 1982, *30*, 9–22.

Witschi, E., & Miller, R. A. Ambisexuality in the female starling. *Journal of Experimental Zoology*, 1938, *79*, 475–487.

Wood-Gush, D. G. M., Langley, G. A. S., Leitch, A. F., Gentle, M. J., & Gilbert, A. B. An autoradiographic study of sex steroids in the chicken telencephalon. *General and Comparative Endocrinology*, 1977, *31*, 161–168.

Zigmond, R. E., Nottebohm, F., & Pfaff, D. W. Androgen-concentrating cells in the midbrain of a songbird. *Science*, 1973, *179*, 1005–1007.

18

Physiological Ethology of Aggression in Amphibians and Reptiles

Neil Greenberg and David Crews

INTRODUCTION

Amphibians and reptiles have always excited the curiosity and interest of researchers, partly because of their status as transitional forms in the evolution of terrestrial life, and partly because, although representatives of our ancestors, they possess remarkably alien forms and habits. They have infiltrated the heart of many religious and cultural traditions (e.g., Frazer, 1935; Morris & Morris, 1965), and in many cases the thoughts of man and the habits of reptiles and amphibians have become intertwined in a way that challenges objectivity.

However fascinated scientists might be by these forms, the selection of amphibians or reptiles as subjects of study is typically guided by the research tradition into which training has guided them. While this tradition has made amphibians and reptiles attractive to workers in the principally descriptive sciences, they have been largely ignored by students in the experimental disciplines: Their study did not seem to be cost-effective when applying research findings to man, "the measure of all things."

This chapter will survey the diversity of agonistic behavior in amphibians and reptiles, with emphasis on its physiological bases where information is available. The behavioral patterns of principal interest are those characteristic of competitive aggression, including defensive and submissive reactions. There is considerable overlap of motor patterns and possibly motivation associated with reproduction and various forms of aggression (discussed by Fentress, 1973, 1978). We will, therefore, discuss reproductive behavior and other forms of aggression when they contribute to our understanding of competitive aggression.

NEIL GREENBERG • Department of Zoology, University of Tennessee, Knoxville, Tennessee 37916. DAVID CREWS • Department of Zoology, University of Texas, Austin, Texas 78712. Unpublished research was supported in part by NINCDS 15305, NICHHD 2709, and NIMH RSDA MH00135 to D. C.; and by a Schweppe Research and Education Fund Award and NIH BRSG RR07088 to N. G. through the University of Tennessee.

In reviewing this material, we will show how far this field has come from stultifying anthropocentric views and prejudices and will indicate both the advantages and disadvantages of utilizing amphibians or reptiles as subjects in experimental research. We will also try to share our optimism about the utility of amphibians and reptiles in research when approached from the perspective of physiological ethology—the study of the causation of behavior spontaneously expressed in a naturalistic context.

Key references to central nervous system influences on aggressive behavior will be provided to encourage a broad perspective on the problem of endocrine control. That hormones affect behavior by means of their action on the central nervous system is the central dogma of behavioral endocrinology. There is considerable documentation of feedback and regulatory axes involving the hypothalamus and hypophysis (see Krieger & Hughes, 1980). Hormones organize specific aspects of the developing central nervous system as well as activating integrated behavioral patterns in the adult (e.g., Beach, 1974; Gorski, 1979). The importance of recent "hormonal history" on behavior, including adjustments in baseline levels of hormonal output or responsiveness, has been emphasized by Leshner (1979). This diversity of hormonal effects on complex brain functions as well as neuronal mechanisms has been summarized by Verniko-Danellis (1972).

Amphibians and Reptiles

The three principal orders of contemporary amphibians are the Anura (tailless forms—frogs and toads), Caudata or urodeles (possessing tails—salamanders), and Apoda (legless forms—caecilians). A fourth order, Trachystomata (sirens), is completely aquatic, contains only three species, and is often included within the Caudata. Anurans, in part because of their hopping locomotion, are among the most specialized amphibians (Romer, 1968, p. 168). Caudate amphibians are somewhat more generalized than Anurans, and the Apoda consists of tropical burrowing worm-like forms. While often terrestrial in many respects, amphibians are still bound to an aquatic environment for reproduction.

Unlike amphibians, reptiles have scales. A more fundamental distinction, however, between amphibians and reptiles lies in their reproductive anatomy and physiology: All living reptiles have a well-developed amnion (Carroll, 1969), and the cleidoic egg is generally regarded as a dividing line between amphibians and reptiles.

Reptiles are grouped into four orders: Rhyncocephalia (one species—the tuatara), Chelonia (Testudinata—turtles), Crocodilia (alligators and crocodiles), and Squamata (suborders: Serpentes or Ophidia—snakes; Sauria or Lacertilia—lizards; and Amphisbaenia—ringed lizards). The single Rhyncocephalian has survived almost unchanged from the early Mesozoic. Chelonians are grouped into two suborders distinguished mainly by their ways of retracting their heads under their shells. Crocodilians are conservative archosaurs, the only survivors of the group that included pterosaurs and dinosaurs. The squamates are the most successful modern reptiles and have undergone extensive adaptive radiation.

Standard texts that provide basic information on amphibians and reptiles are those of Porter (1972) and Goin, Goin, and Zug (1978). The landmark three-volume series on amphibian physiology edited by Moore (1964) and Lofts (1974, 1976) contains important summaries of endocrinology (Gorbman, 1964), neurobiology (Jorgensen, 1974; Oksche & Jeck, 1976), and reproductive and courtship patterns (Salthe & Mecham, 1974). Amphibian endocrinology was also reviewed by Hanke (1974). Reproductive biology of amphibians was extensively dealt with in a more recently published symposium (Taylor & Guttman, 1977). An enduring classic on amphibian biology is Noble's (1931) monograph. Bellairs (1970) and Grzimek, Hediger, Klemmer, Kuhn, and Wermuth (1975) provide a somewhat more detailed account of reptilia but, for a state-of-the-art summary of specialized aspects in reptile biology relevant to our present discussion, readers must refer to the magisterial series edited by Gans and others, including volumes containing articles on endocrine morphology (Gans & Parsons, 1970), physiology (Gans & Dawson, 1976a), ecology and behavior (Gans & Tinkle, 1977), and neurology (Gans, Northcutt, & Ulinski, 1979). Some of the recently published symposia of relevance to our topic deal with turtle biology (Harless & Morlock, 1979), lizard ecology and behavior (Huey, Schoener, & Pianka, 1983; Milstead, 1967), neurology and behavior in lizards (Greenberg & MacLean, 1978), and reptile social behavior (Greenberg & Crews, 1977).

Comparative information on the neurology of reptiles and amphibians is still best sought in the classic work of Ariens Kappers, Huber, and Crosby (1936) and in the major new summary by Kuhlenbeck (1975, 1977). Functional and experimental studies were included in the comparative review by Pearson and Pearson (1976). Ebbesson's (1980) edited collection of papers emphasizes the telencephalon, and frog neurobiology is dealt with by Llinas and Precht (1976).

Evolutionary Considerations

Modern orders of amphibians have a fossil record that begins in the Mesozoic, but there is little if any evidence indicating their earlier history or their links to the Paleozoic Labyrinthodontia, the group that arose from fish ancestors and gave rise to reptiles.

The oldest undoubted reptiles have been dated to the middle of the upper Paleozoic (Pennsylvanian) at about the same time that the reptilian ancestors of mammals appeared. The four orders of living reptiles, however, cannot be traced further back than the early Mesozoic, the time of the greatest reptilian differentiation.

It is easy to argue for the study of amphibians and reptiles as representatives of the groups that gave rise to our more immediate relatives (e.g., MacLean, 1978). It is also fair to say that much of the value in studying these groups arises from their diversity and radiation into new habitats, which presented unique challenges to be overcome by varying proportions of physiological and behavioral adaptations. For example, while Northcutt (1978) samples variation in the brains of reptile taxa in order to identify common patterns of adaptation, Gans and Dawson (1976b) seek to encourage a systematically comparative approach to reptilian ecological physiology that should lead to a particular appreciation for the diversity of adaptive strategies. These perspectives have the conceptual advantage of recognizing fundamental commonalities while leading us to seek the positive attributes of variation among taxa. We cannot merely regard amphibians and reptiles as evolutionary steps, burdened with primitive imperfections, on the way to more "advanced" mammals and birds.

Utility as Models in Behavioral Physiology

Animal models are employed whenever we perceive that a phenomenon of interest is in some way made more accessible to study in a surrogate animal. Such "species of convenience" are utilized extensively in teaching as well as research. It may be fair to say, however, that their historical use as generalized models based on commonalities with other forms is now being transcended by their utility as "natural experiments." That is, these species often present us with specific attributes that are not available in more conventional research species. These attributes, in combination with each other and the habitat, allow us to test specific hypotheses about physiological and behavioral adaptations that may prove to be generalizable to larger animal groupings.

A crucial challenge that must be met in any investigation into the physiology of behavior is to identify the behavioral pattern of interest and to bring it under replicable stimulus control. To understand behavioral causes and consequences, we must build on a base of extensive and detailed intraspecific behavioral data complemented by anatomical, developmental, and physiological knowledge (Beach, 1978). The ecological correlates that bear on both stimulus control and ultimate causation must also be taken into consideration (Greenberg, 1978). All these factors affect the balance between the generalizability of a behavioral pattern and its replicability under appropriate study conditions.

Necessity of an Ethological Perspective

A significant virtue of amphibians and reptiles that commends their use in experimental investigations of the physiology of behavior is their stereotyped behavior. Once ecological constraints on their behavior are understood and accommodated, these classes are easily adapted to the laboratory habitat. Behavioral patterns are under the control of stimulus configurations that are often unique to the ecological niche of a species (Morris, 1964; Seligman, 1970). Investigators must become familiar with field studies and natural units of behavior before a laboratory paradigm can be developed and experimentally induced changes in behavior can be reliably interpreted. Examples of often ignored constraints on behavior that can affect behavior in surprising ways include microclimate (e.g., humidity and temperature), observer effects (e.g., tonic immobility), housing conditions (adequate space and shelter), and feeding regimen (e.g., random or scheduled) (reviewed by Greenberg, 1978; and see Burghardt, 1977).

A general strategy of introducing an unfamiliar species to the laboratory for behavioral experiments is to duplicate as much of the natural environment as may be necessary to elicit the behavioral pattern of interest. Then, by a process of elimination one can reduce the complexity (and inconvenience) of the habitat until only the relevant cues necessary to maintain the behavioral pattern are retained (Greenberg, 1978).

AGGRESSIVE BEHAVIOR AND PHYSIOLOGY IN AMPHIBIANS

Research into the physiological causation of aggressive behavior in amphibia is, to date, almost nonexistent. With several notable exceptions (discussed below), much of the information we have on amphibian aggression consists of descriptive accounts that emphasize ecological correlates or develop phylogenetic inferences. Nevertheless, recent reviews of adaptive amphibian autecology (Salthe & Mecham, 1974), anuran social behavior (Wells, 1977), and comparative sexual behavior of salamanders (Arnold, 1976, 1977) indicate that sufficient descriptive information may now be available to provide a cohesive base on which experimentalists can build.

Reproductive and courtship patterns were summarized by Salthe and Mecham (1974), who considered the proximal cues of behavioral relevance as well as the ecology of amphibian reproduction. Such reviews of the social behavior of a taxon, that document behavioral diversity without losing sight of commonalities and associate behavioral patterns with specific ecological adaptations, are of special value in developing research models.

Examples of Amphibian Aggression

Anura

In a recent review of the social behavior of anurans, Wells (1977) developed the thesis that male–male competition is largely dependent on reproductive patterns. He identified a continuum of basic reproductive patterns, from prolonged (year-round) to explosive (single-night) breeding. Of particular interest is Wells' discussion of amplexus, vocalizations, and aggressive interactions associated with territoriality.

Amplexus is the male's grasp of a receptive female's body just behind the front legs (axillary amplexus) or just before her hind legs (inguinal amplexus). Vocalizations are a prominent feature of anuran social behavior and consist of four general types of long-range calls: a species-typical mating or "advertisement" call emitted by breeding males that appears to announce territorial status and aid species discrimination and orientation to a breeding site; a territorial call that appears to maintain spacing of males at the breeding site; an encounter call, evoked by territorial interactions; and a reciprocation call, emitted by females of some species in response to advertisement calls of conspecific males (Littlejohn, 1977; Wells, 1977). A short-range release call is produced by another male or an unreceptive female when amplexus is attempted. These categories, however, merely reflect terms of convenience; the precise function of a vocalization is often unclear, making the use of such terms as "territorial" or "mating" potentially misleading.

Territorial aggression is generally associated with defense of calling stations or oviposition sites (Salthe & Mecham, 1974; Wells, 1977). In many cases, agonistic encounters are associated with visual threat signals. Aggressive territoriality has been documented for green frogs *Rana clamitaus* (Wells, 1978) and male bullfrogs (Wiewandt, 1969). Emlen (1968) noted that a calling, apparently territorial, male assumes an elevated posture by inflating his lungs and thus revealing a brilliant yellow gular area. Sometimes males approaching each other will pause to tilt their heads back, further displaying the gular sac (Ryan, 1980). If approached persistently by another male, distinctive "hiccup" vocalizations escalate to approach behavior and finally to pushing or clasping, kicking contests. The "territorial" and "mating" vocalizations of bullfrogs have been illustrated by Capranica (1968); Wiewandt (1969) used playbacks of recordings to investigate bullfrog territoriality. The treefrogs *Hyla versicolor* use both mating calls and encounter calls to defend their perches (Fellers, 1979).

In the South African clawed frog *(Xenopus laevis)*, aggressive tendencies are weight correlated and, in increasing intensity, consist of approach, push, and biting (Haubrich, 1961). These behavioral patterns were very similar to eating behavior, and in fact, were facilitated by the presence of food.

Other forms of anuran aggression have been documented in several species of dendrobatid frogs. In this group, most males are territorial, but females of *Colostethus trinitatus* and *C. collaris* are more prominently marked than are males and show more marked territorial behavior. Some resident females threaten by means of a high-head posture that reveals their slowly pulsating bright yellow throats. The threat may be repeated after an advance and finally escalate to the resident's jumping on the intruder's back or, less commonly, a chin-hitting or pushing contest (Durant & Dole, 1975). Other interactions may include a stiff-legged posturing that is often followed by a "wrestling" bout (Sexton, 1960; Test, 1954; Wells, 1980).

Even in those species of dendrobatids in which male aggression is less associated with territorial defense of a specific site, encounters are frequently vigorous and may serve to displace potential competitors. Males will leap toward each other, one individual sometimes slapping another's snout with his chin or landing directly on his back (Crump, 1972; Durant & Dole, 1975; Goodman, 1971). In *Dendrobates galindoi,* a male will interrupt his mating call to threaten another male vocally and then posturally in a sequence that may culminate in a forelimb wrestling bout (Duellman, 1966). Sometimes, approach alone is sufficient to chase away an intruder; in one case, an intruder retreated after a resident's approach and performance of push-ups (Durant & Dole, 1975). One description of combat included forearm ventrum-to-ventrum grappling, with both inguinal and axillary amplexus-like grasps (Goodman, 1971).

An apparently status-related color dimorphism capable of rapid (1–10 minutes) change was recently observed in male *Colostethus trinitatis* (Wells, 1980). In this species, foraging males or those carrying tadpoles are always light brown, while calling males are always black. Aggressive interactions are almost exclusively limited to black males; when fights end, one or both males turn brown.

Interference by a male in the reproductive activities of other males is well known in anurans. In most cases, interference occurs during crowded breeding aggregations (Davies & Halliday, 1979; Wells, 1977, 1979), but clearly directed male–male aggression in such situations, while less common, is documented for two Mexican leaf frogs, *Phylomedusa* (Pyburn, 1970) and *Leptodactylus pentadactylus* (Rivero & Esteves, 1969). In these frogs, single males will attack males clasping females in amplexus and attempt to dislodge them (summarized by Salthe & Mecham, 1974). Size is an advantage in such "displacement" interactions in the toad *Bufo bufo* (Davies & Halliday, 1979). In this species, males in amplexus may ride females for days until eggs are deposited, but they may be dislodged by a larger male (Davies & Halliday, 1977). There is evidence that the pitch of the male's call may be used by potentially aggressive males to assess the body size of their adversary: A deep croak will inhibit a challenger (Davies & Halliday, 1978).

Parental care of eggs or brood is widely distributed among frog families (reviewed by McDiarmid, 1978; Ridley, 1978; Wells, 1981), and in several species, the eggs or young are apparently defended or are the focus of a form of territorial defense. Wells (1981) presents arguments suggesting that, at least in some frogs, male parental care has evolved from territorial behavior.

Predatory aggression in anurans has elicited considerable interest, partly because several species are especially appropriate for neurobehavioral experimentation. Predatory patterns of hunting and feeding were examined in detail by Eibl-Eibesfeldt (1951), with particular attention to effective stimuli for eliciting key responses. Visual cues were originally believed to be the most significant stimulus in feeding, but various observations (e.g., Alexander, 1969; Eibl-Eibesfeldt, 1951) led Sternthal (1974) to demonstrate that while vision is a sufficient cue, it may not be absolutely necessary. For example, the leopard frog *(Rana pipiens)* can learn olfactory discriminations. Prey size, activity, and other characteristics have also been found to be important stimuli (Ewert, 1974; Ingle, 1968, 1973).

Feeding hierarchies have been observed in captive populations of several anuran species, but it is not clear that they are related to social behavior in nature, where populations are much less dense. In the cannibalistic South African clawed frog *Xenopus laevis,* predatory and intermale aggression utilize the same behavioral patterns (Haubrich, 1961). A remarkably broad range of stimuli can elicit predatory behavior in toads (Bragg, 1957). This raises the question as to whether social aggressiveness is merely misdirected predation (Tracy, 1973). This is probable in some cases but not all. Territorial aggression increases in *Leptodactylus* in the presence of food (Brattstrom & Yarnell, 1968), and intermale aggression with occasional "challenge" posturing stimulated by the presentation of food to captive *Rana* can actually interfere with feeding (Emerson & Fenton, 1971).

Caudata

There is remarkably little known about aggression in salamanders, but this may be due more to their secretive habits than to an intrinsic placidity. In an extensive study of the behavioral ecology of the genus *Desmognathus,* Organ (1961) observed that captive males become highly active during the breeding season and, on encountering another individual, initiate courtship. The response of the encountered individual appears to be a method of sex discrimination: While females will either flee or remain passive, fights generally occur when one male encounters or begins to court another. In some species, male–male aggression involves little more than a nipping pursuit of potential competitors, but in other species, males show tenacious biting.

Arnold (1976) studied three species of salamander with behaviorally dissimilar courtship patterns to determine the selective pressures that influence reproductive patterns. The behavior described generally includes patterns that have often been characterized as aggressive in other taxa: male nudging of females in *Plethodon jordani* and *Ambystoma maculatum,* and "forceful transportation" (shoving) in *Ambystoma tigrinum.* Arnold also described sexual competition among males forced into close proximity by their reproductive ecology. The interactions that attend courtship under these crowded conditions will result in a variety of forms of sexual interference, including interruption of male–female interactions, deliberately covering a male's spermatophore, or interfering with spermatophore deposition. Male *Ambystoma tigrinum,* as noted, will also try to transport a female away from a competitor by pushing.

Genuinely aggressive male–male interactions (biting) have been observed in several species of *Plethodon* (Hutchison, 1959) as well as in other salamander genera. Thurow (1975) described aggression in several species of captive *Plethodon* and noted the occurrence of interspecific fighting as well as intraspecific aggression with the formation of social hierarchies. In *Plethodon jordani,* intruding males will flee if bitten but will also act like females, inducing the resident to "waste" a spermatophore in homosexual courtship. This latter problem may require segregation of male–female pairs in the laboratory if sufficient data on heterosexual interactions is to be collected (e.g., Arnold, 1976). The green salamander *Aneides aeneus* was studied in the laboratory by Cupp (1980), who was able to demonstrate male territoriality and the importance of residency to the outcome of interactions.

There are several examples of paternal defense of eggs or brood in a few families of salamanders that have external fertilization (summarized by Ridley, 1978). Parental aggression in salamanders was described in a brief note by Thorn (1962). During the 25- to 30-day incubation period, a captive male *Hynobius nebulosis* tended to remain near the eggs it had fertilized. Whether resting near the eggs, foraging for food, or fed to satiety, it would furiously attack and bite a pencil or an earthworm that approached or touched the egg bags. If eggs were removed overnight and reintroduced at another site, the salamander would adapt and defend them at the new site.

Physiology of Amphibian Aggression

Against the growing background of amphibian life-history data, only the neurobiology of predatory aggression and the endocrinology of reproductive behavior have received notable attention.

Predatory Aggression

The neurobiology of predatory aggression in amphibians has been the subject of considerable interest and research, particularly since the fundamental work on mechanisms of visual–neural integration of the 1950s that employed the frog as a model (Lettvin, Maturana, McCullough, & Pitts, 1959). By building on basic ethological work such as that described above and exploiting species-specific characteristics of predatory behavior, researchers such as Ewert (1970, 1974) and Ingle (1975, 1976) have made significant contributions to our understanding of the manner in which external information is conveyed to the central nervous system, its subsequent integration, and the consequent selection of an appropriate adaptive response. Research on the stimulus parameters of visually guided prey-catching (including size, velocity and orientation) is progressing for urodeles as well as anurans (references in Luthardt & Roth, 1979). While motivational state may, of course, be a significant element in such studies, hormonal influences as such have not been investigated.

Reproductive Behavior

Endocrine influences on amphibian behavior have been investigated only with respect to reproductive patterns of males (e.g., Obert, 1977), although an interest in female behavior is emerging (Diakow, 1978). In addition to the behavioral and physiological overlap of reproductive behavior and aggression, the intriguing complexities of hormonal control of sexual behavior in amphibians may presage the situation in aggression research.

Researchers encouraged by the ample data about androgen influences on male amphibian reproductive behavior in other taxa (review in Beyer, 1979; Crews & Silver, 1983; Hutchison, 1978) will be impressed immediately by the complexities indicated in the literature on amphibians. While reproductive behavior can be influenced in some amphibian males by exogenously administered androgen (Kelley & Pfaff, 1978 on *Xenopus*), other species have presented considerable difficulties (Schmidt, 1966, on *Rana* and *Hyala;* Palka & Gorbman, 1973, on *Rana*). In some instances, the mode of administration of hormones can also be an important factor. Androgen administered subcutaneously will not stimulate sexual behavior in *Rana pipiens* (Palka & Gorbman, 1973) but will rapidly elicit calling if implanted directly into the preoptic area (Wada & Gorb-

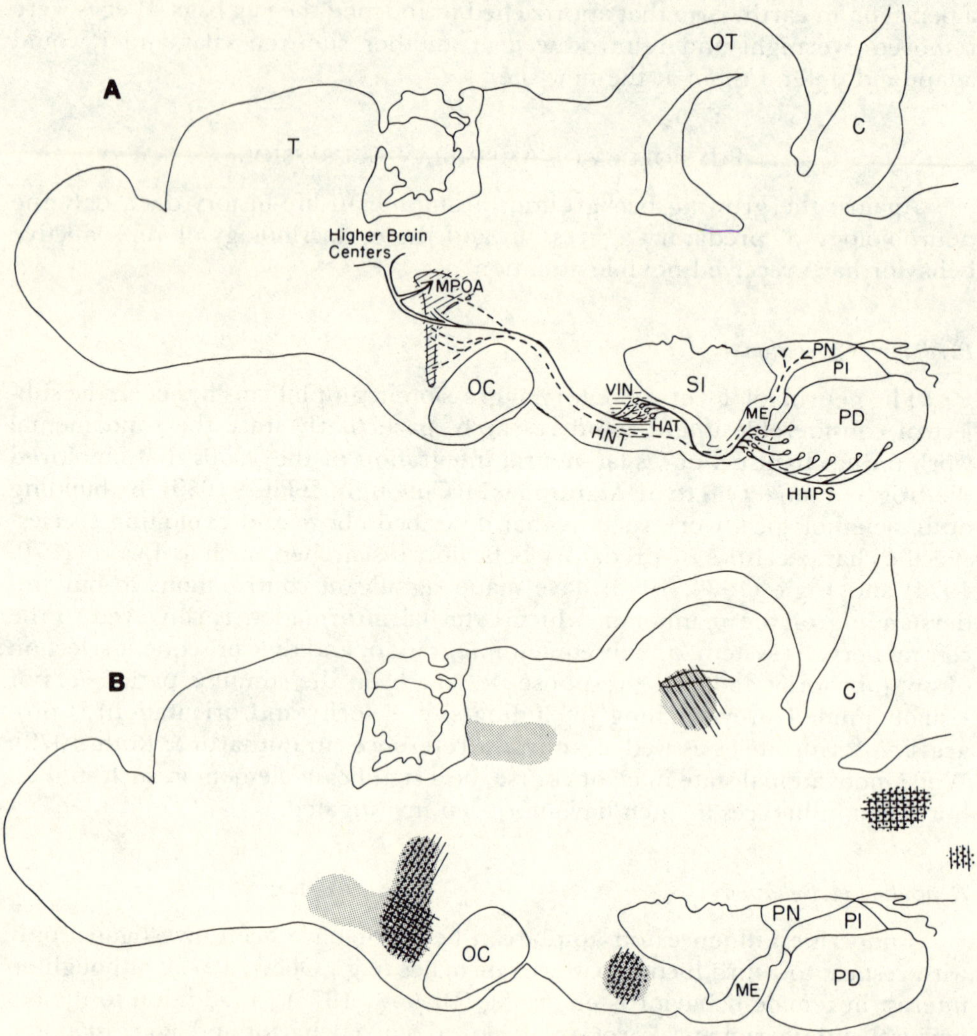

FIGURE 1. (A) A parasagittal view of the neuroendocrine-regulating regions in the amphibian brain. Experiments indicate that the magnocellular preoptic nucleus (MPOA) is involved in ovulation (////) and the production of pituitary gonadotropin-releasing hormone, which is transported to the ventral infundibular nucleus (VIN) (☰). Other abbreviations: C, cerebellum; HAT, hypothalamo-adenohypophyseal tract; HHPS, hypothalamo-hypophyseal portal system; HNT, hypothalamo-neurohypophyseal tract; ME, median eminence; OC, optic chiasm; OT, optic tectum; PD, pars distalis; PI, pars intermedia; PN, pars nervosa, SI, saccus infundibuli; T, telencephalon. (B) A parasagittal view of the steroid-concentrating regions in the amphibian brain and their relationship to neural areas known to be involved in the control of social behavior. Experiments indicate that the trigeminoisthmic tegmentum, medial preoptic area, and "arcuate" nucleus are involved in calling behavior (☰, release call; ////, mating call); the ventral preoptic area and torus semicircularis are involved in orientation behavior (\\\\), and the clasping reflex is situated in deep medullary areas (||||||). Shaded areas indicate regions that have been reported to concentrate testosterone (▩) and estradiol (░). (Adapted from Crews & Silver, 1983.)

man, 1977). In the rough-skinned newt *Taricha granulosa* androgens are necessary but not sufficient to initiate reproductive behavior (Moore, 1978); subsequent experiments have indicated that a neurohypophyseal peptide (vasopressin) is required (Morre & Zoeller, 1979). Neurohormones have also been implicated in the inhibition of release calls emitted by nonreceptive female *Rana pipiens* (Diakow, 1978).

Lesion and stimulation techniques were used by Schmidt (1966, 1973) to identify a magnocellular area in the ventral preoptic nucleus as important in mating calling of leopard frogs *(Rana pipiens)* and tree frogs *(Hyla versicolor)*. The regions of the hypoglossal and vagal motor nuclei were found to be necessary for release calling. Warning calls were occasionally elicited by electrical stimulation in the region of the main Sensory Nucleus V. However, testosterone will stimulate male frog calls and clasps typical of reproductive behavior only if implanted in the preoptic area (Wada & Gorbman, 1977); this area is a major site of testosterone concentration (Kelley, Morrell, & Pfaff, 1975). Hypoglossal and vagal motor nuclei and the principal nucleus of Cranial Nerve V also concentrate androgens (Kelley, 1980), as does a mesencephalic site (torus semicircularis) that is stimulated by sound. The neural control of amplexus in *Rana* and *Bufo* was investigated by successive transections of the central nervous system by Hutchison (1967), who postulated a self-regulatory hindbrain system with minor but distinct midbrain and forebrain contributions (see Figure 1A). The neuroanatomical and endocrine mechanisms of sexual behavior have been extensively reviewed in the context of other taxa (Crews & Silver, 1983; Kelley & Pfaff, 1978). Hormone-concentrating sites in the brain of *Xenopus* (Figure 1B) were identified for testosterone (Kelley *et al.*, 1975) and estradiol (Morrell, Crews, & Pfaff, 1975).

AGGRESSIVE BEHAVIOR AND PHYSIOLOGY IN REPTILES

Although more is known about the physiological bases of aggressive behavior in reptiles than in amphibians, that knowledge is largely limited to a few species of lizards. Our knowledge of lizards partly reflects the traditional interest in reptiles as representatives of the first truly terrestrial vertebrates and the facts that lizards are the least alien reptile in form and habits, and their behavior is relatively amenable to observation. Although the numbers of species on which we have information is limited, the forms of aggressive behavior that have been documented are quite varied.

Examples of Reptilian Aggression

We will present several examples of aggression from each group of reptiles to indicate their behavioral diversity, but we will concentrate on the lizards, about which we have the most physiological, as well as behavioral, information. An attempt has been made to inventory and interpret the stereotyped behavioral patterns of aggression exhibited by all classes of reptiles (Carpenter & Ferguson, 1977).

Rhyncocephalia

The tuatara *Sphenodon punctatus* is the only living representative of this ancient order. It is a solitary, nocturnal, lizard-like animal found only on the barrier islands off the coast of New Zealand. Because of its limited range and rarity, few opportunities have existed to study this important group. The behavioral patterns believed to be associated with aggression are trunk inflation, gaping, dorsal crest erection, and vicious biting (reviewed in Carpenter & Ferguson, 1977).

Chelonia

Although the diversity of aggressive behavior in turtles is constrained by their morphology, both terrestrial and aquatic species utilize visual as well as chemical or tactile signals. Chelonian social behavior and its stimulus control was reviewed by Harless (1979). Terrestrial species are also known to exploit their shells as weapons during contact aggression.

The few reports of aggression in aquatic or semiaquatic turtles were briefly reviewed by Bury and Wolfheim (1973). While most species are both wary and difficult to observe, aggression in the Pacific pond turtle *Clemmys marmorata* is amenable to study when they emerge from the water to bask. In this species, both sexes will direct gaping threats at other animals of all ages, often as a prelude to episodes of biting, pushing, or ramming.

There is more known about aggression in terrestrial turtles (reviewed by Carpenter & Ferguson, 1977). Tactile and chemical signals are relatively more important in social communication than visual or auditory cues in the display behavior of land tortoises (Auffenberg, 1977). It has only recently become clear that many turtles are sensitive to airborne sounds, although in a limited range. Nevertheless, several species are known to roar, bellow, grunt, moan, or hiss during interactions, including a few male combat situations (Auffenberg, 1977).

Visual signals utilized by tortoises include ritualized vertical or horizontal head movements (Eglis, 1962); however, distinctive variations in color associated with seasonal breeding have been noted (Auffenberg, 1964a). *Geochelone carbonaria* utilize head color as a discriminative cue in sex recognition; artificial heads painted black with contrasting light-colored markings similar to those of a living animal will elicit challenge behavior from resident males, while all-black heads will not (Auffenberg, 1965).

Tactile signals are also important in the aggressive interactions of male tortoises. Most contact aggression is directed toward displacing the adversary. When biting occurs, it is directed toward the head or limbs and may serve to unbalance an antagonist, allowing flipping or more effective pushing. Males in combat often ram one another with epiplastral projections on the anterior portion of the shell. Xiphiplastral ramming or hooking, utilizing a thickened posterior portion of the shell, is seen only during courtship. In the tortoise *Geochelone travancorica,* shell-ramming contributes to sex-recognition, and a male may ram a female for days or weeks before she will respond with a sexually receptive immobility. This is regarded by Auffenberg (1964b) as a tactile cue that might be important in synchronizing reproductive behavior.

The possible significance of chemical cues is indicated by the observation that some tortoises deposit their feces in specific areas. The function of this behavior is to signal territorial boundaries and status (Auffenberg, 1977). Male tortoises possess mental glands beneath the chin. These glands show cyclic secretory activity that coincides with reproduction, but the role of their secretions in aggression is not known. The tortoise *G. travancorica* will ram and mount a block of wood smeared with the cloacal exudation of sexually active females (Auffenberg, 1964b).

The literature on combat in the genus *Gopherus* was summarized by Weaver (1970). Male tortoises establish dominance in ramming-pushing-flipping contests. Status relationships are then indicated by posture: Dominants elevate their bodies on all four limbs and carry their heads characteristically high, while subordinates hold their heads low. Nonaggressive individuals either escape or withdraw their necks and appendages. Weaver also noted that specific elements in combat and courtship behavior of *G. berlandieri* were the same. Fritts (1979) emphasized the significance of body elevation in determining dominance among giant Galapagos tortoises. During periods of intense competition for food, water, or shade, animals with longer appendages and elevated carapaces will become dominant over more massive but lower individuals.

Crocodilia

Although behavioral research on crocodilians presents formidable difficulties, several investigators have been able to provide some basic information on intraspecific aggression in three species: *Alligator mississippiensis, Crocodylus acutus,* and *C. niloticus* (reviewed in Garrick & Lang, 1977). In the species observed, dominance combat precedes the establishment of mating territories. Territorial defense is similar in all three species and utilizes distinctive postures, chases, lunges, real or mock biting, "roaring" vocalizations, and geysering of water from the nares, often preceding a "head-slap" of the water. Although subdominant and other males will be excluded from a territory, females are tolerated and may move freely among territories. Dominant males may interrupt the courtship of these peripheral subdominants. Female *A. mississippiensis* display submissive behavior to males and occasionally to other females, but in one observation, a female alligator was seen defending an area in which her future nest was to be built (Garrick, Lang, & Herzog, 1978). Maternal defense of the nest site and hatchlings is well known (Pooley & Gans, 1976).

Serpentes

The image of a cobra with a raised head and an expanded hood is the most familiar example of reptile aggression. Recent detailed observations by Carpenter and his colleagues (e.g., Carpenter, 1977; Carpenter, Murphy, & Mitchell,

1978) have revealed a diversity of aggressive behaviors not previously suspected in this relatively secretive suborder. Their studies demonstrate that the communicative channels available to snakes include the visual, chemosensory, tactile, and acoustical, although visual displays and physical contact dominate aggressive interactions.

Combat rituals have been documented for three families of snakes: the colubrids, elapids, and viperids (Bogert & Roth, 1966). All groups utilize body contact extensively as apparent tests of strength. Elapids such as cobras *(Naja)* tend to entwine, while colubrids such as corn snakes *(Elaphe)* align themselves in more parallel fashion and press against each other with head and neck. Vipers, including the crotaline snakes such as rattlesnakes *(Crotalus),* raise their heads, necks, and anterior trunks high during contact and sway back and forth until one male pushes the other down. A visual signal that often precedes such contests is raised body-swaying, which may function to elicit a reciprocal response in an adversary. Courtship and aggressive patterns are fairly distinct in crotalines (Lowe & Norris, 1950). Although aggressive behavior associated with territorial defense is known in a few species such as cobras (Smith, 1939), the relationship is uncertain for most snakes. Several species (e.g., *Natrix, Thamnophis,* and *Storeria*) have been extensively studied and have never been observed fighting (see Noble, 1937).

Snakes have radiated into many habitat types including arboreal, fossorial, and aquatic. A habitat-related adaptation of combat has been documented in the arboreal Madagascan boa *Sanzinia madagascariensis* (Carpenter *et al.,* 1978), which employs its pelvic spurs, a vestigial structure of boids. These spurs are proportionally larger in males than in females and can be erected and retracted from a slight pocket in the body well. The use of spurs in male–male combat is facilitated by a powerful purchase on limbs with the anterior body. The combatants wrestle with the pelvic area in an attempt to dislodge an adversary. The use of spurs is also a well-known feature of boid courtship, where they are used for scratching, stroking, or vibrating against a female.

Sauria

The predominantly diurnal terrestrial habits of lizards have led to their being the most studied reptile group. Lizards have radiated into all terrestrial niches except the coldest, and they display a comparable diversity of morphological, physiological, and behavioral adaptations. For example, body sizes vary from the massive Komodo dragon *Varanus komodoensis* of southeast Asia (3 m/165 kg) to the minute gekkonid *Sphaerodactylus sp.* of the tropical Americas (4 cm/2 g).

Lizard research has led the revolution in thinking about physiological traits once thought to be "primitive" or "inferior." These new findings reveal that lizards, like other reptiles, possess traits that make them more adaptive in certain ecological contexts than their mammalian or avian counterparts (Gans & Dawson, 1976b; Greenberg, 1980; Regal, 1978).

Lizards are frequently divided into two major groups, reflecting both phyletic history and adaptive specializations: Autarchoglossia and Ascalabota (Camp, 1923). A crucial difference between these groups is the possession by the former of the ventral superficial rectus muscle, a trait that profoundly affects locomotory abilities and associated characteristics. Autarchoglossid lizards are basically ground-dwelling, with forked tongues and well-developed chemosensory abilities, while ascalabotid lizards have fleshy tongues and reduced chemosensory abilities but excellent vision (Evans, 1961). The tendency of most ascalabotids to be diurnal and utilize conspicuous visual displays has led to their being much favored in research, although considerable basic information is being acquired about many autarchoglossids.

Autarchoglossia. The Autarchoglossid division includes teids, lacertids, varanids, and skinks. The genus *Varanus* provides some of the most dramatic examples of aggression among the autarchoglossids. Threat often consists of actions that increase apparent size, such as rearing up on the hindlimbs, dorsoventral trunk compression, or throat inflation. Tail "rattling" or lashing is also employed. Rearing-up often grades into bipedal approach and "wrestling." During wrestling, the ventral surfaces are adpressed, with contact being maintained by tight mutual clasping as the contestants twist, arch, and eventually tumble back to the substrate (Carpenter, Gillingham, Murphy, & Mitchell, 1976; Murphy & Mitchell, 1974).

Aggressive behavior of the Komodo dragon *V. komodoensis* was reported by Auffenberg (1978, 1981) who regarded it mainly as a way of defending carrion/prey from potential competitors; "appeasement" displays were common in females or subadults.

Much basic behavioral ecology is known for the teid lizard genus *Cnemidophorus* (e.g., Hardy, 1962). In this group, males are aggressive and will lunge at and may bite other males encountered during the breeding season. Interactions often involve open-jaw approach with sagittal-plane profile changes and tail-vibration (Hardy, 1962). It is not clear if there are defended territories in nature (Carpenter, 1960); in captivity, *C. sexlineatus* will form a linear hierarchy. It is interesting that lunging attacks have also been observed in species of all-female parthenogenetic *Cnemidophorus* that have been studied in the field and laboratory (Crews, Gustafson, & Tokarz, 1983; Mitchell, unpublished observations).

Male five-lined skinks *Eumeces fasciatus* are known to be aggressive in the breeding season. They are not territorial but will vigorously attack other males encountered while foraging. In this species, aggression appears to be released by a sexually dimorphic, seasonally appearing bright red head. Fighting is *not* preceded by stereotyped displays and often results in trauma (Fitch, 1954).

Ascalabota. Ascalabotid lizards include the iguanids, agamids, and gekkonids. Most iguanids and agamids are diurnal and have excellent vision and elaborate visual displays, often employing distinctive colors as signals. Many visual displays make use of special morphological structures involving the hyoid apparatus. These structures can often be suddenly erected, as in the case of neck frills or dewlaps, to startle an enemy as well as to alter an animal's profile and greatly increase its apparent size. Gular or trunk changes, often involving inflation with air or with the aid of special skeletal or muscular mechanisms, will also increase apparent size.

With one notable exception (see below) auditory "display" is insignificant in this group. Olfactory or chemical cues are also relatively unimportant in social communication; however, there is an interesting report of chemosensory control of display behavior in the iguanid *Sceloporus occidentalis*. In this species, males display more often when presented with surfaces "labeled" with the droppings of conspecifics than when presented with unlabeled surfaces (Duvall, 1980). Many iguanid species have enlarged proctodeal or femoral glands that are seasonally active (Cole, 1966), and reproductively active males often scrape their hindlegs or vents against surfaces in their territories.

Gekkonid lizards are both nocturnal and diurnal and utilize both acoustic and visual displays in varying combinations. The acoustic displays—squeaks, chirps, growls, and barks—seem to parallel functionally the specific visual displays of iguanid lizards in several circumstances (reviewed by Marcellini, 1977, 1978).

Display Behavior. Many agonistic signals utilize stereotyped body movements such as tail displays, neckbends (up or down), and foreleg waves or "circumduction." Bobbing movements of the head and body movements are the most ubiquitous and specialized of these displays and have generated considerable interest. It is fair to say that behavioral herpetology took great strides following Carpenter's (1961, 1962) attempts to delineate objective categories for social behavior and his development of the "display-action-pattern" (time/motion) technique for analyzing bobbing patterns.

Several patterns of bobbing displays have been distinguished, but their forms cannot be reliably interpreted as conveying similar messages in different species. Each species must be studied separately, and the contexts and consequences of display behavior must be analyzed before function can be attributed to a specific display form. Further, some species have extremely variable displays, with little consistency in form between individuals and even changes in pattern within individuals (Greenberg & Jenssen, 1982).

The *Anolis* lizards are characterized by an extensible throat pouch known as a dewlap. The dewlap is sexually dimorphic in size (Crews & Greenberg, 1981) and is used in species recognition (reviewed in Crews & Williams, 1977). The dewlap is usually either monochromatic (e.g., red, white, or brown) or biocolored, with a spot of one color centrally placed; species with three colors exist such as *A. tropidonotus,* in which the dewlap has a central black spot, surrounded by orange with a yellow border. During aggressive encounters, the dewlap is often extended in association with a bobbing display or an elevation of the forepart of the body.

The displays of the green anole *Anolis carolinensis* are fairly well defined (see Table I). Male aggressive behavior often appears as a continuum of displays recognizable by a specific motor pattern or posture (Figure 2) that are emitted in a fairly predictable sequence (Figure 3). If an unfamiliar *A. carolinensis* enters the territory of a mature, active male, the resident will immediately exhibit the *assertion display.* This display is given while the animal is stationary and is char-

TABLE I
Behavioral Units in Anolis carolinensis Displays

Behavioral pattern	Behavioral units
Subordination display (SD)	1.[a] *Head nods* (HN)—Cadenced vertical movements of the head effected by neck musculature.
Assertion display (AD)	1.[a] *Bobbing* (BB)—Vertical movements of the body in a characteristic up-and-down pattern effected by *pushups* (PU) with or without coordinated *head nods* (HN). 2. *Dewlap* (DW)—Unfolding of dewlap effected by erection of the retrobasal hyoid process in coordination with *bobbing* (BB).
Threat display (TD)	1. *Bobbing* (BB) 2. *Dewlap* (DW) 3.[a] *Extended throat* (ET)—Lowering of basal hyoid component causing expansion of sagittal profile of throat. 4. *Crests* (CC)—erection of nuchal and dorsal crests along midline.
Challenge display (CD)	1. *Bobbing* (BB) 2. *Dewlap* (DW), as above but more variable. 3. *Extended throat* (ET) 4.[a] *Sagittal expansion* (SE)—Lateral compression of thorax causing enlargement of sagittal profile of the torso. 5. *Crests* (CC) 6. *Eyespot* (ES)—Darkening of patch of skin adjacent to the posterior margin of the orbit.
Courtship display (MD)	1. *Bobbing* (BB) 2. *Dewlap* (DW) 3.[a] *Rapid nod* (RN)—Rapid vertical movements of the head effected by neck musculature.

[a] Diagnostic criterion for display type. (From Greenberg, 1982.)

acterized by species-typical bobbing movements (Greenberg & Jenssen, unpublished data) in concert with the display of a red dewlap extended by abduction of the retrobasal hyoid element. The assertion display can also be expressed spontaneously in nonsocial contexts, where it is associated with elevated behavioral arousal (Greenberg, 1982). If the eliciting stimulus is a potential adversary, the resident may perform the *threat display*. The threat display is similar in form to the assertion display but includes an engorged or extended throat. This is distinct from dewlap extension because the profile change is effected by a partial abduction of a basal hyoid element; the red of the dewlap is not seen. The

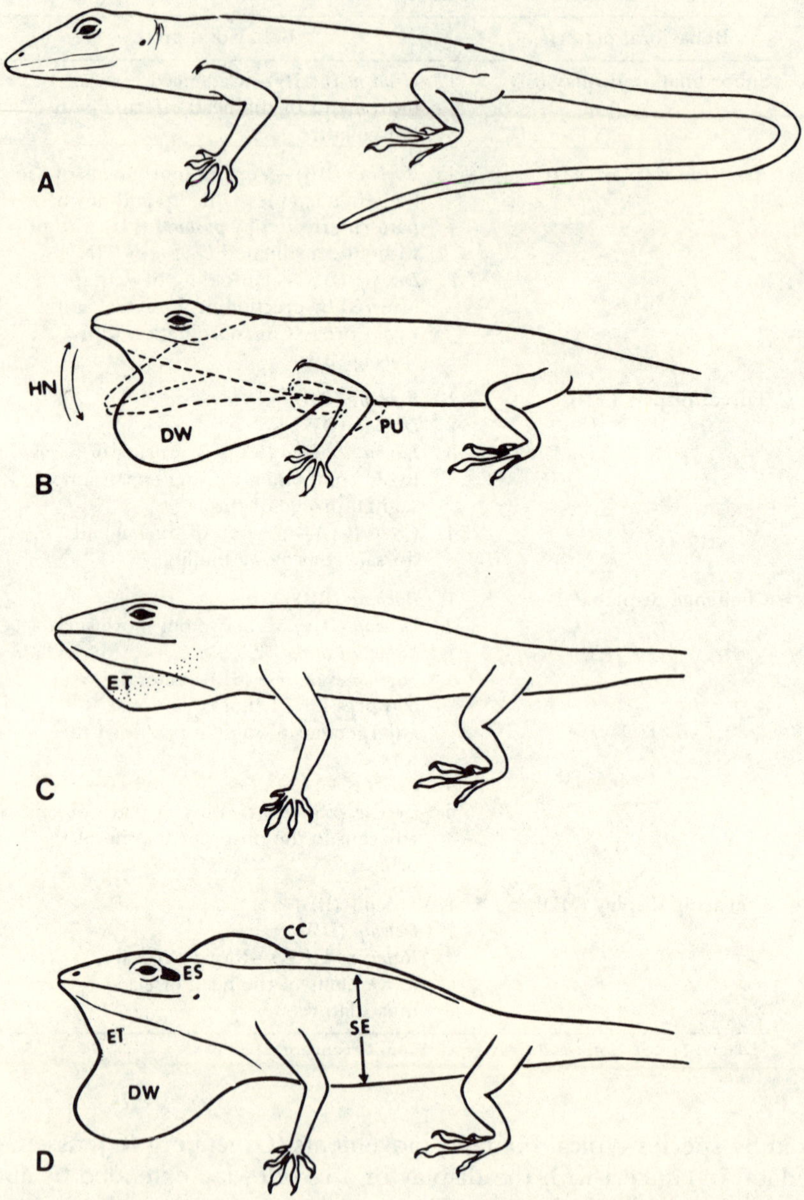

FIGURE 2. Characteristic appearance of *Anolis carolinensis:* (A) At rest. (B) Coordinated head nods (HN), pushups (PU), and dewlap extension (DW) of the assertion display. (C) Extended throat (ET) of the threat display. (D) Postural modifiers of the assertion display that communicate "challenge," including eyespot (ES), crest erection (CC), and sagittal expansion (SE) of the profile. (Modified from Greenberg, 1977a.)

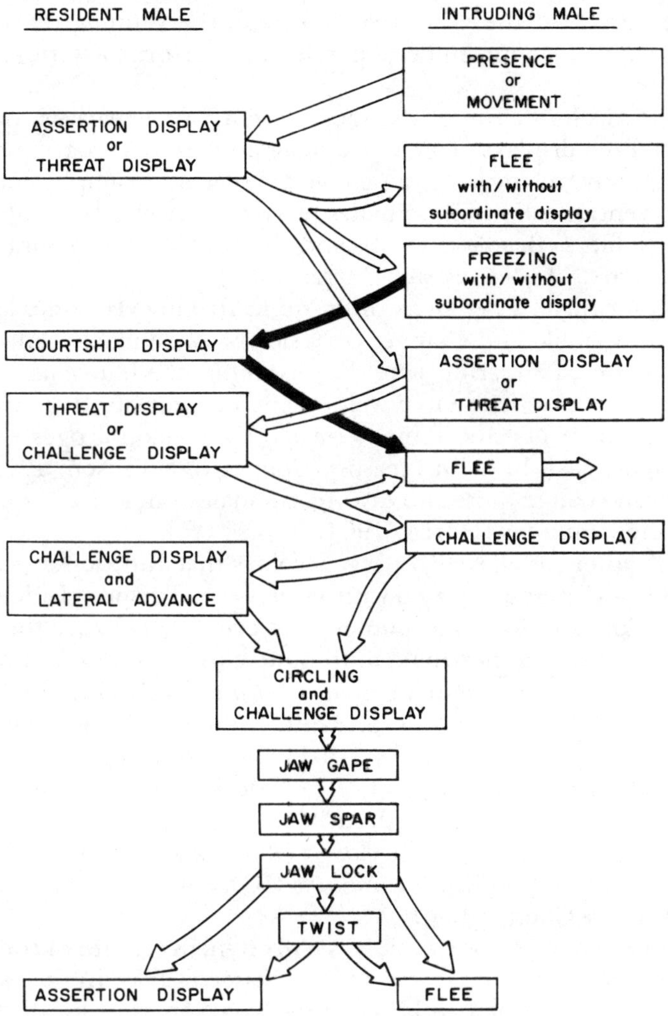

FIGURE 3. Characteristic sequence of behavioral patterns utilized by *Anolis carolinensis* territorial resident males and intruders. Few encounters progress to the final stages. Solid arrows indicate an uncommon sequence.

response of the intruder will determine the next act. For example, if the intruding animal is a male who does not give the *subordination display* (head nodding without push-ups or dewlap) or who reacts aggressively to the resident, the resident male will challenge the intruder. The *challenge display* is identified by an extreme lateral compression of the body (enlarging the sagittal profile) in concert with the stereotyped bobbing movement. During an aggressive bout, the males will circle one another, each pausing occasionally to display. As the encounter progresses, a crest will form above the back and neck. Color changes may occur, including the formation of a black spot immediately behind the eye. It is common for the males to lock jaws and twist, each male trying to throw the other off the perch. At the end of such a bout, the winning (dominant) male will typically climb to a prominent perch and perform a series of assertion displays.

Body Color. Color is one of the most potent visual releasers of aggression in lizards. Ritualized displays in iguanid and agamid lizards are frequently related to emphasizing, advertising, or hiding color. Excellent examples may be drawn from the blue ventral coloration in males of several species of *Sceloporus* (Greenberg, 1977a; Noble & Bradley, 1933), and the brilliant red or orange head of male rainbow lizards *Agama agama* (Harris, 1964).

Sexual dimorphism in color is often quite striking. In some lizards, color appears relatively stable and changes only slowly, presumably reflecting an adjustment in the tonic endocrine state. For example, the blue color of *Sceloporus* undergoes androgen-dependent development in intensity during maturation (Kimball & Erpino, 1971), and female *Crotaphytus* develop progesterone-dependent orange spots each breeding season (Cooper & Ferguson, 1972). In other species, color can change quite rapidly. *Agama* males can lose or adopt their red head within a few minutes (Harris, 1964).

Lizards of other families also possess colors that function as social signals, but they are not generally associated with a stereotyped display. Indeed, relatively secretive male skinks *Eumeces fasciatus* are not even likely to see the aggression-releasing red head of another male until quite close to it (Fitch, 1954).

The body color of a dominant male *Agama* or *Sceloporus* also functions to advertise status. While *Agama* males can rapidly adopt the inconspicuous coloring of a female or juvenile to denote subordination to an aggressive challenger, *Sceloporus* adopts a body-down posture to hide its blue ventrum (Greenberg, 1977a). Association of body color with social status has also been noted for several other lizard species including *Anolis cuvieri* (Rand & Andrews, 1975), *A. agassizi* (Rand, Gorman, & Rand, 1975), *A. carolinensis* (Greenberg, 1983), and *Crotaphytus collaris* (Ferguson & Chen, 1973).

The ability to change color rapidly is a prominent feature of the Fiji banded iguana *Brachylophus fasciatus* (Greenberg & Jenssen, 1983). In this large iguanid lizard, the male's alternating bands of color will darken dramatically (from green to black) during aggressive interactions, often in concert with body profile (apparent size) changes.

In *Anolis carolinensis,* colors alternate between green and brown during agonistic interactions, but the colors have no obvious significance as a social signal. Color state does, however, indicate something about the underlying motivational state of an animal: The average color state (over 10 or more days) appears to be related to the probability of color change during a social encounter (Greenberg, 1983). Animals with a pretest color score of "green" respond to antagonists with a color change significantly less often than do animals scored as "green/brown."

Body colors of *A. carolinensis,* while rapidly reversible in agonistic situations, remain relatively stable in hierarchical situations. Under these conditions, dominant animals are generally green while subordinates are typically brown (Greenberg, 1983). In one series of observations, 30 isolated adult males were scored as to body color and three behavioral traits associated with behavioral dominance: posture, perching site selection, and activity levels. When pairs matched along these dimensions were placed in neutral habitats, aggressive interactions occurred within 30 minutes and differences between body color appeared or increased. Over a period of several days, the color of one individual remained relatively stable (green) while the other became darker (more brown). In concert with this, the darker individual was often less active, usually adopted a lower body posture, and invariably selected lower perching sites than the lighter lizard (Figure 4).

When a dominant-subordinate pair of male *A. carolinensis* is broken up, the subordinate animal's color often becomes lighter again. In certain cases, however, color change is *not* reversible and may reflect pathological processes that attend status change. In such animals, color becomes progressively darker and the animals eventually die. Similarly in *Agama,* the red head of a once-dominant male reduced in status may reappear and become quite bright along with signs of a progressive and fatal behavioral depression (Greenberg, unpublished observations).

Female lizards also make use of color, often in connection with display behavior, to signal reproductive condition. During most of the year, male and female *Sceloporus virgatus* have blue throats, but during the breeding season, females develop an orange throat color that surrounds or replaces the blue color. The orange becomes most intense after ovulation and gradually fades after

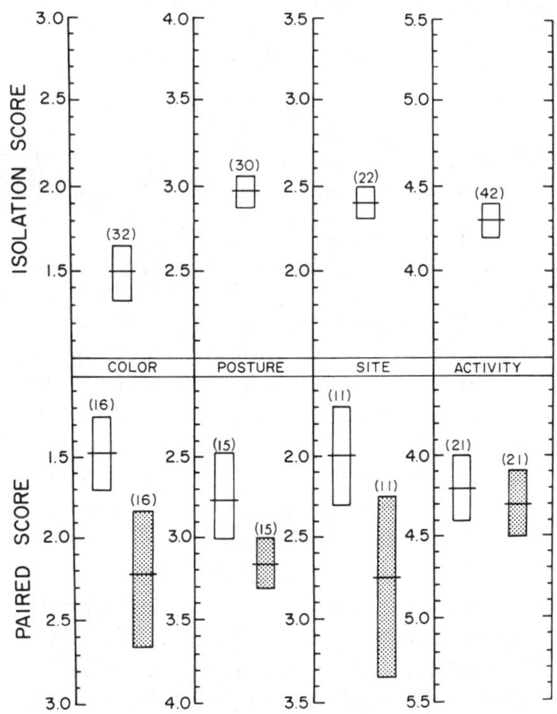

FIGURE 4. Changes in body color and behaviors associated with social dominance in cohabiting adult male *Anolis carolinensis.* Isolation scores (above) and paired condition scores (below) are of lizards matched for body color, posture elevation, perch site, or activity. Increasing scores reflect darker color, lower body posture, lower site-selection, and reduced activity. Bars represent two standard errors of the mean; shaded bars represent subordinates. (Adapted from Greenberg, 1983.)

oviposition (Vinegar, 1972). *S. virgatus* also utilizes a distinctively female display known as "sidlehopping," which appears to inhibit the attention of courting males (Carpenter, 1967; reviewed in Crews & Silver, 1983). Captive female *S. virgatus* will display the sidlehopping pattern to introduced females or males when oviductal eggs are present. "Threat" displays (involving sagittal profile changes and dewlap extension) are exhibited to intruding males or females before the oviductal stage and after oviposition (Vinegar, 1972). Female *Crotaphytus collaris* also develop orange spots on the lateral trunk and back when gravid. These spots occur in conjunction with the female "rejection" display (sidlehopping) and are shown to courting males, effectively inhibiting the male's advances (Cooper & Ferguson, 1972). Other species that develop breeding-season coloration are *Holbrookia propinqua, Cophosaurus texanus,* and *Callisaurus draconoides* (Clarke, 1965). These species utilize distinctive postures to display the colors. The female will also be aggressive toward males when gravid, although the form of the aggression in these species is similar to male behavior, unlike the distinctive sidlehopping of female *S. virgatus.*

Status Relationships. Hierarchical status relationships are readily elicited in many species of lizards when grouped in captivity. Some reptile species, when maintained in relatively small habitats, establish clear dominant–subordinate relationships that are upset when the animals are moved to larger habitats (Harris, 1964, with *Agama;* Mayhew, 1963, with *Amphibolorus;* Greenberg, unpublished observations on *Dipsosaurus*). In these new habitats, the subordinate is presumably no longer subject to the continual inhibitory effect of the presence of some attribute of the dominant (Greenberg, 1978). Individual distance or the presence of visual barriers between potentially antagonistic individuals is probably a significant consideration in this phenomenon. Many species that are territorial in nature readily form dominance hierarchies under special conditions such as a sudden local food abundance (Evans, 1951a, with *Ctenosaura;* Fitch, 1940, with *Sceloporus;* Norris, 1953, with *Dipsosaurus*), indicating that the behavior pattern is adaptive and firmly fixed in the animal's biology.

Physiology of Reptilian Aggression

There is more known about the physiological bases of aggressive behavior in reptiles than in amphibians. Even so, reptile behavioral endocrinology is still in its infancy when compared with our understanding of the relationships between hormones and aggression in birds and mammals (as evidenced by this volume).

Aggression is not only reciprocally related to an animal's internal state but also is a factor in the external environment of other animals, affecting their internal state as well, and thus the probability of their expression of a specific behavioral pattern. This has been effectively demonstrated for *Anolis carolinensis,* in which male aggressive behavior inhibits environmentally induced (seasonal) ovarian growth (Crews, 1974, 1975, 1981).

Virtually all of our information about the hormonal control of agonistic behavior in reptiles is limited to lizards. In other reptile groups, the association of aggression with seasonal reproductive and territorial behavior implicates involvement of the sex steroid hormones, but there is as yet no significant experimental confirmation of this. For example, in the tortoise *Gopherus berlandieri,* males are more combative during the breeding season (Weaver, 1970). However, injection of exogenous androgens will not elicit aggression even though the development of secondary sexual characteristics are stimulated. Most of our knowledge of the behavioral endocrinology of turtles was contributed by Evans (e.g., 1940, 1951b, 1952), who emphasized hormonal influences on secondary sex characteristics. Hormonal influences on the development of secondary sexual characteristics has been reviewed by Forbes (1961). It is important to point out that we cannot extrapolate from mammalian studies, as evidenced by recent work on the garter snake *Thamnophis.* In this species courtship behavior appears to be independent of gonadal hormones (Camazine, Garstka, Tokarz, & Crews, 1980).

Gonadal Hormones

The hormonal condition of male lizards is significantly related to their performance of displays in an appropriate context. Castration of sexually active male *Anolis carolinensis* abolishes courtship behavior although challenge behavior may be retained (Crews, 1974; Crews, Traina, Wetzel, & Muller, 1978; Mason & Adkins, 1976). The role of androgens in the control of male aggressive behavior is still unclear. Several authors have reported that castration of *A. carolinensis* results in a decline in the level of aggressiveness although this reduction is not significant (Crews *et al.,* 1978; Mason & Adkins, 1976). In some experiments, androgen replacement therapy using testosterone propionate (TP) or free testosterone has no effect on male challenge behavior (Crews *et al.,* 1978; Mason & Adkins, 1976), but in other studies, both TP and dihydrotestosterone propionate are potent stimulators of male aggression (Adkins & Schlesinger, 1979; Huang, 1979). One possible reason for this discrepancy is a difference in the procedure for dealing with observer effects during behavior observations.

Challenge behavior is also performed by juvenile males in a relatively disorganized fashion and is directed toward both males and females. After the administration of androgen, however, juveniles will respond appropriately to the characteristic female subordinate display and shift from challenge to courtship behavior (Evans, 1957).

Winter-collected animals with regressed gonads will also fail to respond appropriately to social signals (Crews, unpublished observations). These males will not exhibit either defensive or subordinate behavior when placed in a territory with a male housed under an environmental regimen that will maintain reproductive activity. Winter males will also ignore receptive females (Crews & Greenberg, 1981), as well as challenging dominants (Greenberg, unpublished observations).

In *A. carolinensis,* the physiological consequences of status are also indicated by observations of reduced testis weight and loss of fat bodies in subordinates, although the animals observed were also part of a highly unstable hierarchical organization (Tsui, unpublished observations cited in Pearson, Tsui, & Licht, 1976). When hierarchies were not disturbed, however, subordinates did not have a lower testosterone level than dominants.

Female aggression is also related to reproductive condition. An increase in aggression in female *A. carolinensis* following ovariectomy has been reported by Evans (1936). Subsequent experiments by Mason and Adkins (1976), however, indicated that female aggression was independent of gonadal hormones. The hormonal involvement in the control of female aggression is still open to question, since Adkins has reported recently that dihydrotestosterone propionate stimulates female aggressiveness (Adkins & Schlesinger, 1979).

As mentioned earlier, several inguanid species develop special coloration and will direct aggressive displays to other lizards when gravid (Clarke, 1965). The orange spots that develop on the lateral trunk and back of female *Crotaphytus collaris* when gravid can be induced by exogenous testosterone or progesterone, the latter hormone being more effective (Cooper & Ferguson, 1972), especially when the animals are estrogen primed (Cooper & Ferguson, 1973). *Sceloporus virgatus* will shift from the characteristically female "sidlehopping" rejection display shown to other lizards when gravid to more male-like aggressive displays after oviposition (Vinegar, 1972). Interestingly, reproductively inactive or post-ovulatory females in several species of parthenogenetic *Cnemidophorus* will court and attempt copulation with preovulatory females in a manner much like their male congeners (Crews & Fitzgerald, 1980).

Vigorus nest-site competition has been observed in marine iguanas *Amblyrhynchus* during the egg-laying period. The females also adopt a male-like coloration during this time (Eibl-Eibesfeldt, 1966). Nest-site competition also occurs in *Iguana iguana* (Rand, 1968; Rand & Rand, 1976). After nesting, iguanas will suddenly (after less than 30 minutes) cease to defend their nest-sites (Rand, 1968). This rapid change in behavior recalls the situation in the frog *Hyla,* in which a rapid loss of orientation to a male call follows oviposition (R.R. Capranica, personal communication) or the sudden change in behavior in goldfish after spawning (Stacey, 1981). The mechanism responsible is known to be under prostaglandin control in the goldfish and under neurohormonal control in the frog (Diakow, 1978).

Adrenal Hormones

The association of aggressiveness, status, and body color implicates adrenal hormones as a significant element in the endocrinology of reptile agonistic behavior. The acute stress that may attend agonistic interactions and the possible chronic stress that may be a factor in long-term status or hierarchical relationships involves both adrenal chromaffin hormones (epinephrine and norepinephrine) and interrenal (homologous to mammalian adrenocortical) hormones, most notably, corticosterone (reviewed in detail by Gabe, 1970).

The relationship of symptoms of stress to the formation of social hierarchies was discussed by Rowell (1974), whose research on mammals suggested that subordinates play a major role in maintaining a hierarchy. Subordinate status is presumably reinforced and maintained by dominant animals through agonistic interactions or by conspicuous status signs to which the subordinate is continually exposed.

The involvement of the adrenal gland in behaviorally induced stress is well known (Christian & Davis, 1964) and has recently been implicated in social stress or status relationships of lizards. Callard, Chan, and Callard (1973) observed enlarged adrenals in dense colonies of *Dipsosaurus dorsalis,* and Brackin (1978) found that adrenal volume was inversely related to rank in captive *Cnemidophorus sexlineatus.*

The hormones involved in acute and chronic stress responses are also implicated in body color changes (Figure 5). As mentioned earlier, body color in some species is an important signal in their status hierarchies and may be affected by hormones related to agonistic behavior. In other species, the signal function of color change is not obvious; however, its relationship to hormone flux may be more clear. For example, the endocrinology of color change in *A. carolinensis* is relatively well understood (Hadley & Goldman, 1969; Hadley & Bagnara,

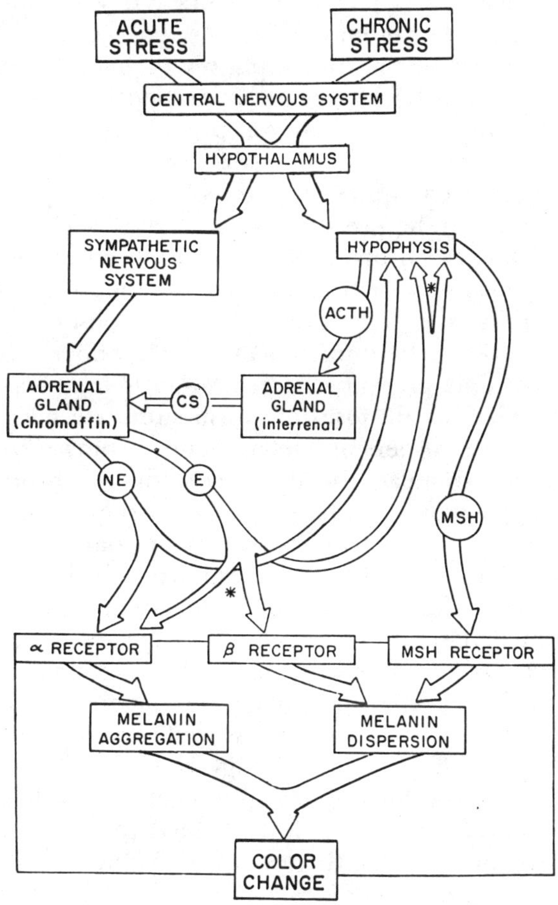

FIGURE 5. Putative relationships between chronic and acute stress and body color in *Anolis carolinensis.* Asterisk: Biphasic effect of epinephrine (E), causing darkening of body color (at low levels) or lightening (at high levels). Corticosterone (CS) stimulates additional synthesis of E by facilitating the norepinephrine (NE) to E methylating enzyme PNMT. ACTH, adrenal corticotrophic hormone; MSH, melanocyte stimulating hormone.

1975), and this species thus appears to be an excellent model for investigating adrenal influences on agonistic interactions and related color change. In contrast to most reptile species described, *A. carolinensis* has no sympathetic innervation of the chromatophores that regulate skin color (Kleinholtz, 1938a,b), indicating that only blood-borne agents affect dispersion or aggregation of chromatophores. Briefly, these agents are melanocyte stimulating hormone (MSH, controlled by visual or dermal photoreceptor response to background or illumination level), epinephrine (E, aggregates or disperses melanin depending on concentration), and norepinephrine (NE, usually aggregates but sometimes disperses melanin) (Hadley & Goldman, 1969).

Release of MSH from the pars intermedia of the lizard pituitary is not subject to direct innervation but is probably under inhibitory control of circulating catecholamines (Hadley & Bagnara, 1975). However, low levels of epinephrine may facilitate rather than inhibit MSH release (Hadley & Hruby, 1977), in a manner similar to its biphasic effect on the chromatophores (see below). Serotonin is believed to have a tonic stimulatory effect on MSH release and is presumed to reach to the pars intermedia by way of the neurointermediate plexus (Levitin, 1980). Dopamine, like NE, will inhibit MSH, probably by inhibiting the effect of serotonin.

In most chromatophores of *A. carolinensis,* color is due to the combined effects of hormones on alpha and beta adrenegic receptors, but while all melanophores have beta receptors (responsible for darkening), a few do not have alpha receptors (responsible for lightening). While E can affect alphas or betas, NE affects alphas only; thus, a darkening of skin with beta-only cells positively indicates the presence of E (Hadley & Goldman, 1969). Most of such skin with beta-only chromatophores is localized in small patches, whose stimulation by E can cause a spotted appearance. One such patch has been identified in a postorbital area (the "eyespot"), and is observed in animals under acute or chronic social or environmental stress (Greenberg, 1977b; Hadley & Goldman, 1969).

Our present understanding is that, when light and temperature are controlled, small doses of E will stimulate melanophore beta receptors and MSH release, dispersing melanin and turning skin color brown. NE or large doses of E will stimulate the alpha receptors on chromatophores and induce aggregation of melanin, resulting in a green skin color (see Figure 5), but if E is present, beta-only skin patches will remain dark.

Since body color under stress is, in this species, indicative of the ratio of chromaffin E to NE, it is interesting to note that stress can increase the rate of methylation of NE to E. This increase is due to a facilitory effect of adrenal corticosteroid secretion on the methylating enzyme phenylethanolamine-N-methyltransferase (PNMT) (Wurtman & Axelrod, 1965). This effect is known to be present in reptiles as well as in mammals (Wurtman, Axelrod, & Tramezzani, 1967). Thus, although the adrenal gland (adrenal "medulla") is probably under the control of hypothalamic sites that control the secretion of the E and NE (Folkow & von Euler, 1954) (possibly in concert with "fight or flight" tendencies), under the long-term influence of corticosteroids, an altered ratio of these catecholamines, biased toward E, may result. Body color may then reflect this by a direct effect of catecholamines on chromatophores (Hadley & Goldman, 1969) and through their effect on MSH secretion (Hadley & Hruby, 1977). Aggressiveness in other taxa is known to be increased or reduced by low or high glucocorticoid level, respectively (Leshner, 1978). Recently, Dr. Thomas Chen and I have been able to demonstrate that socially subordinate *Anolis carolinensis* have significantly elevated levels of serum corticosterone (Greenberg, 1983).

The effect of E on MSH secretion is of interest considering the known behavioral effects of MSH on resistance to extinction of conditioned active avoidance, passive avoidance, and appetitive responses (reviewed by LaHoste, Olson, Kastin, & Olson, 1980), effects that are possibly the result of enhanced attention to selected environmental cues (Kastin, Miller, Nockton, Sandman, Schally, & Stratton, 1973, with rats). After both physical and nonphysical stress, MSH can be detected in the blood of mammals (Sandman, Kastin, Schally, Kendall, & Miller, 1973); and in mice, MSH reduces territorial aggression (Patterson, Rickerby, Simpson, & Vickers, 1978). The putative fear response to predation (tonic immobility) will be reduced in lizards treated with MSH (Stratton & Kastin, 1976), although the administration of serotonin, believed to stimulate MSH release (Levitin, 1980), will produce little change in tonic immobility (Hennig, Dunlap, & Harston, 1979).

Central Neural Control of Aggression

Both androgens and estrogens are localized in areas of the central nervous system of *A. carolinensis* that are important to the expression of aggressive behavior (Morrell, Crews, Ballin, Morgentalin, & Pfaff, 1979). These areas include the septum, amygdala, preoptic area, anterior hypothalamus, ventromedial hypothalamus, central mesencephalic gray, and torus semicircularis (Figure 6).

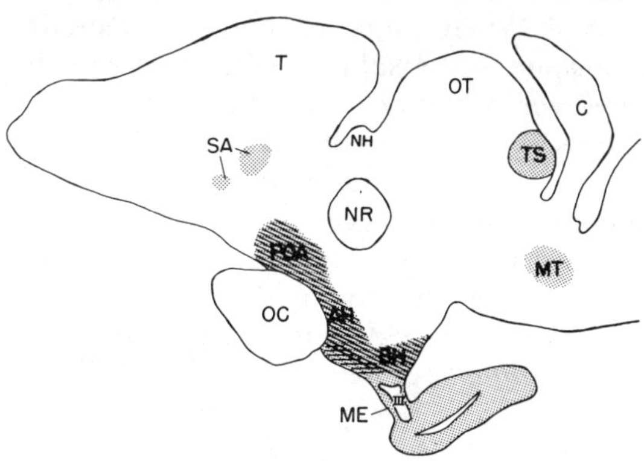

FIGURE 6. A parasagittal view of the neuroendocrine-regulating regions in the brain of the lizard *Anolis carolinensis* and their relationship to neural areas known to be involved in the control of reproductive behavior and pituitary function. Stippled areas indicate steroid-concentrating sites. Lesion studies as well as studies involving intracranial implantation of steroid hormones indicate that the anterior basal hypothalamus-median eminence region (horizontal lines) is involved in the feedback control of gonadotropin secretion and that the preoptic area and anterior and basal hypothalamus (diagonal lines) are involved in the regulation of reproductive behavior. Abbreviations: AH, anterior hypothalamus; BH, basal hypothalamus; C, cerebellum; ME, median eminence; MT, midbrain tegmentum; NH, nucleus habenularis; NR, nucleus rotundus; OC, optic chiasm; OT, optic tectum; POA, preoptic area; SA, strioamygdaloid area (not in plane of section); T, telencephalon; TS, torus semicircularis; III, third ventricle.

Electrical stimulation of some of these structures in other iguanid species elicits components of agonistic responses. For example, stimulation of several midbrain areas of the collared lizard *Crotaphytus collaris* results in gular extension (Sugerman & Demski, 1978). Stimulation of sites near the torus semicircularis (homologous in part to the mammalian inferior colliculus) will elicit vocalizations in the tokay gecko *Gekko gecko* (Kennedy, 1975) and dewlap extension in the green iguana *Iguana iguana* (Distel, 1978). In the crocodilian *Caiman sclerops,* stimulation of the amygdala will result in flight responses, while diencephalic stimulation elicits attack (Keating, Korman, & Horel, 1970).

Lesion studies of the preoptic area and basal hypothalamus of *A. carolinensis* have shown that the integrity of these structure is essential to the expression of courtship and aggressive behavior (Farragher & Crews, 1979; Wheeler & Crews, 1978). In the fence lizard *Sceloporus occidentalis,* bilateral lesions of the amygdala will alter the responses of dominant males and abolish aggressive behavior (Tarr, 1977). Similar findings were made after amygdala area lesions in *Caiman sclerops,* and decreased attack and flight responses to handling could be correlated with the extent of damage to the ventromedial nucleus (Keating *et al.,* 1970).

Brain areas that do not concentrate sex steroids but are implicated in aggressive behavior may be sites of complex integration. The paleostriatum of *A. carolinensis* (homologous in part to the mammalian basal ganglia) probably possesses "multimodal" neurons that may function to facilitate species-typical behavior organized at lower neural levels (Northcutt, 1978). The importance of this area in integrating display responses of dominant males to the sight of potential competitors was tested (Greenberg, 1977b; Greenberg, MacLean, & Ferguson, 1979). To do this, we took advantage of the predominantly unilateral optic projections in this species (Butler & Northcutt, 1971) and placed radiofrequency lesions in this forebrain site in one hemisphere only. We then utilized removable eyepatches in order to present stimuli to the lesioned or to the unlesioned hemisphere, thus allowing the animal to function as its own postoperative control. When a stimulus male was seen only with the eye leading to the normal hemisphere, the subject behaved normally and challenged potential intruders. However, when the stimulus was presented to the lesioned side, the subject showed a significant deficit in the challenge display, while assertion displays and activity levels were left intact. Interestingly, the experimental animal acted as though it did not recognize that the intruder was a conspecific, an effect sometimes termed "social agnosia."

Although specific sites associated with adrenal hormone concentration have not been identified in the reptile brain, it is likely that they exist in a pattern similar to that seen in other taxa (McEwen, 1973).

ENVOI

Physiological ethology is viewed as an effective approach to the study of the causation of behavior because it directly engages the problem of environmental constraints on the expression of behavior. Thus, it includes stimulus control, ultimate form, and synchronization of organisms with each other and the habitat. This approach is strategically important in the study of behavioral physiology in amphibians and reptiles because of their many habitat-related stereotyped behavioral patterns.

The physiological ethology of amphibian aggression is unexplored, although many species are known to engage in vigorous combat or competition. Various aspects of the neurobiology of predatory aggression have, however, been developed with considerable sophistication, largely because of their value as models for visual–neural integration in other taxa. The budding interest in the endocrinology of amphibian aggression may gather some impetus from the more developed field of amphibian reproductive endocrinology. This area has only recently been experimentally investigated, possibly because it is fraught with apparent difficulties that, in the light of new findings, constitute fascinating challenges.

Reptilian aggressive behavior (particularly of the diurnal, terrestrial forms) has been the subject of considerably more research during the past 10 years. There is a relatively well-developed data base in basic anatomy and physiology, which, with new ethological and experimental techniques, provides a reasonable prospect for this field to mature into a discipline that will be influential in many other areas of theoretical and applied behavior and physiology.

Recent work on the relative roles of gonadal and adrenal hormones on aggressiveness and submissiveness indicates that these attributes of agonistic behavior are not simply opposite ends of a continuum. For example, when androgen levels of mice are manipulated (by castration), there is no significant influence on the probability of expressing submissive behavior, but corticosterone is crucial (Leshner & Politch, 1979). In addition, ACTH, generally associated with the release of adrenal corticosterone, seems to have an extra-adrenal suppressive effect on aggressiveness (Leshner, Walker, Johnson, Kelling, Kreisler, & Svare, 1973). In the green anole lizard *A. carolinensis,* castration abolishes sexual behavior but has little effect on challenge behavior if the male is returned to his home cage (Crews, 1974; see also Mason & Adkins, 1976); however, if he is transferred to an unfamiliar territory, he will not act aggressively to male intruders. Interestingly, displacement from home cage or confrontation with novel stimuli has been associated with increased corticosterone production in mammals (Barrett & Stockham, 1963). Preliminary studies on *A. carolinensis* indicate that a comparable situation may exist in reptiles (Greenberg, Chen, & Green, 1982).

As usual, what begins as a curiosity or an apparent anomaly in behavior promises to provide additional insights and future breakthroughs. Such phenomena as change of body color, synchronization of autonomic and somatic elements in displays, changes in female aggressiveness associated with color changes or oviposition, masculine behavior in parthenogenetic females, and the association of stress with aggression, status, and color are just a few phenomena observed in reptiles that may provide such breakthroughs.

ACKNOWLEDGMENTS

We are grateful for the critical readings and useful suggestions of A. C. Echternacht, G. Burghardt, K. Wells, and A. B. Greenberg.

REFERENCES

Adkins, E. K., & Schlesinger, L. Androgens and the social behavior of male and female lizards *(Anolis carolinesis). Hormones and Behavior,* 1979, *13,* 139–152.

Alexander, T. R. Observations on the feeding behavior of *Bufo marinus* (Linné). *Herpetologica,* 1964, *20,* 255–259.

Ariens Kappers, C. U., Huber, G. C., & Crosby, E. C. *The comparative anatomy of the nervous system of vertebrates, including man* (reprinted in 3 vols.). New York: Hafner, 1936.

Arnold, S. J. Sexual behavior, sexual interference and sexual defense in the salamanders *Ambystoma maculatum, Ambystoma tigrinum* and *Plethodon jordani. Zeitschrift für Tierpsychologie,* 1976, *42,* 247–300.

Arnold, S. J. The evolution of courtship behavior in new world salamanders with some comments on old world salamandrids. In D. H. Taylor & S. I. Guttman (Eds.), *The reproductive biology of amphibians.* New York: Plenum Press, 1977.

Auffenberg, W. A first record of breeding color changes in a tortoise. *Journal of Bombay Natural History Society,* 1964, *61,* 191–192. (a)

Auffenberg, W. Notes on the courtship of the land tortoise *Geochelone travancorica* (Boulenger). *Journal of Bombay Natural History Society,* 1964, *61,* 247–253. (b)

Auffenberg, W. Sex and species discrimination in two sympatric South American tortoises. *Copeia,* 1965, *1965,* 335–342.

Auffenberg, W. Display behavior in tortoises. *American Zoologist,* 1977, *17,* 241–250.

Auffenberg, W. Social and feeding behavior in *Varanus komodoensis.* In N. Greenberg & P. D. MacLean (Eds.), *Behavior and neurology of lizards.* Rockville, Md.: NIMH, 1978.

Auffenberg, W. *The behavioral ecology of the Komodo monitor.* Gainesville, Fla.: University Presses of Florida, 1981.

Barrett, A. M., & Stockham, M. A. The effect of housing conditions and simple experimental procedures upon the corticosterone level in the plasma of rats. *Journal of Endocrinology,* 1963, *26,* 97–105.

Beach, F. A. Behavioral endocrinology and the study of reproduction. *Biology of Reproduction,* 1974, *10,* 2–18.

Beach, F. A. Animal models and psychological inference. In H. A. Katchadourian (Ed.), *Human sexuality: A comparative and developmental perspective.* Berkeley, Ca.: University of California Press, 1978.

Bellairs, A. *The life of reptiles* (2 vols.). New York: Universe Books, 1970.

Beyer, C. (Ed.). *Endocrine control of sexual behavior.* New York: Raven Press, 1979.

Bogert, C. M., & Roth, V. D. Ritualistic combat of male gopher snakes. *Pituophis melanoleucus affinis* (Reptilia, Colubridae). *American Museum Novitates,* 1966, *2245,* 1–27.

Brackin, M. F. The relation of weight to physiological state in *Cnemidophorus sexlineatus* dominance hierarchies. *Herpetologica,* 1978, *34*(2), 185–191.

Bragg, A. N. Some factors in the feeding of toads. *Herpetologica,* 1957, *13,* 189–191.

Brattstrom, B. H., & Yarnell, R. M. Aggressive behavior in two species of Leptodactylid frogs. *Herpetologica,* 1968, *24,* 222–228.

Burghardt, G. M. Learning processes in reptiles. In C. Gans & D. W. Tinkle (Eds.), *Biology of the reptiles* (Vol. 7). New York: Academic Press, 1977.

Bury, B. R., & Wolfheim, J. H. Aggression in free-living pond turtles *(Clemmys marmorata). BioScience,* 1973, *23,* 659–662.

Butler, A. B., & Northcutt, R. G. Ascending tectal efferent projections in the lizard *Iguana iguana*. *Brain Research,* 1971, *35,* 597–601.

Callard, I. P., Chan, S. W. C., & Callard, G. V. Hypothalamic-pituitary-adrenal relationships in reptiles. In A. Brodish & E. S. Redgate (Eds.), *Brain-pituitary-adrenal-relationships.* Basel: Karger, 1973.

Camazine, B., Garstka, W., Tokarz, R., & Crews, D. The effects of castration and androgen replacement therapy on male courtship behavior in the red-sided garter snake *(Thamnophis sirtalis parietalis). Hormones and Behavior,* 1980, *14,* 358–372.

Camp, C. L. Classification of the lizards. *Bulletin of the American Museum of Natural History,* 1923, *48,* 289–481.

Capranica, R. R. The vocal repertoire of the bullfrog *(Rana catesbeiana). Behavior,* 1968, *31,* 302–325.

Carpenter, C. C. Aggressive behavior and social dominance in the sixlined racerunner *(Cnemidophorus sexlineatus). Animal Behavior,* 1960, *8*(1–2), 61–66.

Carpenter, C. C. Patterns of social behavior of Merriam's canyon lizard (*Sceloporus m. merriami* Iguanidae). *SW Naturalist,* 1961, *6*(3–4), 138–148.

Carpenter, C. C. A comparison of the patterns of display of *Urosaurus, Uta,* and *Streptosaurus. Herpetologica,* 1962, *18*(3), 145–152.

Carpenter, C. C. Aggression and social structure in iguanid lizards. In W. W. Milstead (Ed.), *Lizard ecology: A symposium.* Columbia, Mo.: University of Missouri Press, 1967.

Carpenter, C. C. Communication and displays of snakes. *American Zoologist,* 1977, *17,* 217–223.

Carpenter, C. C., & Ferguson, G. W. Variation and evolution of stereotyped behavior in reptiles. In C. Gans & D. W. Tinkle (Eds.), *Biology of the reptilia* (Vol. 7): *Ecology and behavior A.* New York: Academic Press, 1977.

Carpenter, C. C., Gillingham, J. C., Murphy, J. B., & Mitchell, L. A. A further analysis of the combat ritual of the pygmy mulga monitor, *Varanus gilleni* (Reptilia: Varanidae). *Herpetologica,* 1976, *32*(1), 35–40.

Carpenter, C. C., Murphy, J. B., & Mitchell, L. A. Combat bouts with spur use in the Madagascan boa *(Sanzinia madagascariensis). Herpetologica,* 1978, *32*(2), 207–212.

Carroll, R. L. The origin of reptiles. In C. Gans, A. Bellairs, & T. Parsons (Eds.), *Biology of the reptilia* (Vol. 1). New York: Academic Press, 1969.

Christian, J. J., & Davis, D. E. Endocrines, behavior, and population. *Science,* 1964, *146,* 1550–1560.

Clarke, R. F. An ethological study of the iguanid lizard genera *Callisaurus, Cophosaurus,* and *Holbrookia. Emporia State Research Studies,* 1965, *13*(4), 1–66.

Cole, C. J. Femoral glands in lizards: A review. *Herpetologica,* 1966, *22,* 199–206.

Cooper, W. E., Jr., & Ferguson, G. W. Relative effectiveness of progesterone and testosterone as inductors of orange spotting in female collared lizards. *Herpetologica,* 1972, *28*(1), 64–65.

Cooper, W. E., & Ferguson, G. W. Estrogenic priming of color change induced by progesterone in the collared lizard, *Crotaphytus collaris. Herpetologica,* 1973, *29,* 107–110.

Crews, D. Effects of group stability, male-male aggression, and male courtship behavior on environmentally-induced ovarian recrudescence in the lizard, *Anolis carolinensis. Journal of Zoology (London),* 1974, *172,* 419–441.

Crews, D. Psychobiology of reptilian reproduction. *Science,* 1975, *189,* 1059–1065.

Crews, D. Interrelationships between ecological, behavioral and neurendocrine processes in the reproductive cycle of *Anolis carolinensis* and other reptiles. In J. S. Rosenblatt, R. A. Hinde, C. G. Beer, & M. Busnel (Eds.), *Advances in the study of behavior* (Vol. 11). New York: Academic Press, 1981.

Crews, D., & Fitzgerald, K. T. "Sexual" behavior in parthenogenetic lizards *(Cnemidophorus). Proceedings of the National Academy of Science,* 1980, *77*(1), 499–502.

Crews, D., & Greenberg, N. Function and causation of social signals of lizards. *American Zoologist,* 1981, *21*(1), 273–294.

Crews, D., & Silver, R. Reproductive physiology-behavior interactions in non-mammalian vertebrates. In R. W. Coy & D. W. Pfaff (Eds.), *Handbook of behavioral neurobiology.* Book in preparation, 1983.

Crews, D., & Williams, E. E. Hormones, reproductive behavior, and speciation. *American Zoologist,* 1977, *17,* 271–286.

Crews, D., Traina, V., Wetzel, F. T., & Muller, C. Hormone control of male reproductive behavior in the lizard, *Anolis carolinensis:* Role of testosterone, dihydrotestosterone and estradiol. *Endocrinology,* 1978, *103,* 1814–1821.

Crews, D., Gustafson, J. E., & Tokarz, R. R. Psychobiology of parthenogenesis in reptiles. In R. Huey, T. Schoener, & E. Pianka (Eds.), *Lizard ecology.* Cambridge: Harvard University Press, 1983.

Crump, M. L. Territoriality and mating behavior in *Dendrobates granuliferus. Herpetologica,* 1972, *28,* 195–198.

Cupp, P. V., Jr. Territoriality in the green salamander, *Aneides aeneus. Copeia,* 1980, *1980,* 463–468.

Davies, N. B., & Halliday, T. R. Optimal mate selection in the toad *Bufo bufo. Nature,* 1977, *269,* 56–58.

Davies, N. B., & Halliday, T. R. Deep croaks and fighting assessment in toads *Bufo bufo Nature,* 1978, *274,* 683–685.

Davies, N. B., & Halliday, T. R. Competitive mate searching in male common toads, *Bufo bufo. Animal Behavior,* 1979, *27,* 1253–1267.

Diakow, C. Hormonal basis for breeding behavior in female frogs: Vasotocin inhibits the release call of *Rana pipiens. Science,* 1978, *199,* 1456–1457.

Distel, H. Behavior and electrical brain stimulation in the green iguana, *Iguana iguana* L.: II. Stimulation effects. *Experimental Brain Research,* 1978, *31,* 353–367.

Duellman, W. E. Aggressive behavior in dendrobatid frogs. *Herpetologica,* 1966, *22,* 217–221.

Durant, P., & Dole, J. W. Aggressive behavior in *Colostethus* (= *Prostherapis*) *collaris* (Anura: Dendrobatidae). *Herpetologica,* 1975, *31,* 23–26.

Duvall, D. Western fence lizard *(Sceloporus occidentalis)* chemical signals: I. Conspecific discriminations and release of a species-typical visual display. *Journal of Experimental Zoology,* 1980, *210*(2), 321–325.

Ebbesson, S. O. E. (Ed.), *Comparative neurology of the telencephalon.* New York: Plenum Press, 1980.

Eglis, A. Tortoise behavior: A taxonomic adjunct. *Herpetologica,* 1962, *18,* 1–8.

Eibl-Eibesfeldt, I. Nahrungserwerb und beuteschema der erdkrote (*Bufo bufo* L.). *Behavior.* 1951, *4*(1), 1–35.

Eibl-Eibesfeldt, I. Das verteidigen der eiablageplatze bei der hood-meerechse *(Amblyrhynchus cristatus venustissimus). Zeitschrift für Tierpsychologie,* 1966, *23,* 627–631.

Emlen, S. T. Territoriality in the bullfrog, *Rana catesbeiana. Copeia,* 1968, *2,* 240–243.

Emerson, F. H., & Fenton, K. R. Agonistic behavior in some captive frogs *(Rana)* from Nevada. *Herpetological Review,* 1971, *3*(2), 39.

Evans, L. T. Territorial behavior of normal and castrated females of *Anolis carolinensis. Journal of Genetic Psychology,* 1936, *49,* 49–60.

Evans, L. T. Effects of testosterone propionate upon social dominance in young turtles, *Chrysemys picta. Biological Bulletin,* 1940, *79,* 371.

Evans, L. T. A field study of the social behavior of the black lizard *Ctenosauria pectinata. American Museum Novitates #1493,* 1951. (a)

Evans, L. T. Endocrines and secondary sex characters of the box tortoise, *Terrapene carolina. Anatomical Record,* 1951, *111,* 139. (b)

Evans, L. T. Endocrine relationships in turtles; III. Some effects of male hormones in turtles. *Herpetologica,* 1952, *8,* 11–14.

Evans, L. T. The effect of hormones upon juvenile lizards. *Anatomical Record,* 1957, *128*(3), 545. (Abstract)

Evans, L. T. Structure as related to behavior in the organization of populations in reptiles. In W. F. Blair (Ed.), *Vertebrate speciation.* Austin: University of Texas Press, 1961.

Ewert, J. P. Neural mechanisms of prey-catching and avoidance behavior in the toad (*Bufo bufo* L.). *Brain, Behavior and Evolution,* 1970, *3,* 36–56.

Ewert, J. P. The neural basis of visually guided behavior. *Scientific American,* 1974, *230*(3), 34–42.

Farragher, K., & Crews, D. Role of basal hypothalamus in the regulation of male reproductive behavior in the lizard, *Anolis carolinensis:* Lesion studies. *Hormones and Behavior,* 1979, *13,* 185–206.

Fellers, G. M. Aggression, territoriality, and mating behavior in North American treefrogs. *Animal Behavior,* 1979, *27,* 107–119.

Fentress, J. C. Specific and nonspecific factors in the causation of behavior. In P. P. G. Bateson & P. H. Klopfer (Eds.), *Perspectives in ethology*. New York: Plenum Press, 1973.

Fentress, J. C. Conflict and context in sexual behavior. In J. B. Hutchison (Ed.), *Biological determinants of sexual behavior*. New York: Wiley, 1978.

Ferguson, G. W., & Chen, C. L. Steroid hormones, color change and ovarian cycling in free-living collared lizards *Crotaphytus collaris*. *American Zoologist*, 1973, *13*, 119.

Fitch, H. S. A field study of the growth and behavior of the fence lizard. *University of California Publication in Zoology*, 1940, *44*(2), 151–172.

Fitch, H. S. Life history and ecology of the five-lined skink, *Eumeces fasciatus*. *University of Kansas Publication of the Museum of Natural History*, 1954, *8*(1), 1–156.

Folkow, B., & von Euler, U. S. Selective activation of noradrenaline and adrenaline producing cells in the cat's adrenal gland by hypothalamic stimulation. *Circulation Research*, 1954, *2*, 191–195.

Forbes, T. H. Endocrinology of reproduction in cold-blooded vertebrates. In W. C. Young (Ed.), *Sex and internal secretions*. Baltimore: Williams & Wilkins, 1961.

Frazer, J. G. *The golden bough*. New York: Macmillan, 1935.

Fritts, T. H. Carapace shape, aggression and Galapagos ecology—a movie of giant tortoise behavior. *Proceedings of the Joint Annual Meeting of the Herpetologists League and the Society for the Study of Amphibians and Reptiles*, 1979, 30. (Abstract)

Gabe, M. The adrenal. In C. Gans & T. S. Parsons (Eds.), *Biology of reptilia* (Vol. 3): *Morphology C*. New York: Academic Press, 1970.

Gans, C., & Dawson, W. R. (Eds.). *Biology of the reptilia* (Vol. 5): *Physiology A*. New York: Academic Press, 1976. (a)

Gans, C., & Dawson, W. R. Reptilian physiology: An overview. In C. Gans & W. R. Dawson (Eds.), *Biology of the reptilia* (Vol. 5): *Physiology A*. New York: Academic Press, 1976. (b)

Gans, C., & Parsons, T. (Eds.). *Biology of the reptilia* (Vol. 2): *Morphology B*. New York: Academic Press, 1970.

Gans, C., & Tinkle, D. W. (Eds.). *Biology of the reptilia* (Vol. 7): *Ecology and behavior A*. New York: Academic Press, 1977.

Gans, C., Northcutt, R. G., & Ulinski, P. (Eds.). *Biology of the reptilia* (Vol. 10): *Neurology B*. New York: Academic Press, 1979.

Garrick, L. D., & Lang, J. W. Social signals and behaviors of adult alligators and crocodiles. *American Zoologist*, 1977, *17*, 225–239.

Garrick, L. D., Lang, J. W., & Herzog, H. A., Jr. Social signals of adult American alligators. *Bulletin of the American Museum of Natural History*, 1978, *160*, 153–192.

Goin, C. J., Goin, O., & Zug, G. *Introduction to herpetology* (3rd ed.). San Francisco: W. H. Freeman, 1978.

Goodman, D. E. Territorial behavior in a neotropical frog, *Dendrobates granuliferus*. *Copeia*, 1971, *2*, 365–370.

Gorbman, A. Endocrinology of the amphibia. In J. A. Moore (Ed.), *Physiology of the Amphibia* (Vol. 1). New York: Academic Press, 1964.

Gorski, R. A. The neuroendocrinology of reproduction: An overview. *Biology of Reproduction*, 1979, *20*(1), 111–127.

Greenberg, N. An ethogram of the blue spiny lizard, *Sceloporus cyanogenys* (Reptilia, Lacertilia, Iguanidae). *Journal of Herpetology*, 1977, *11*(2), 177–195. (a)

Greenberg, N. A neuroethological study of display behavior in the lizard *Anolis carolinensis* (Reptilia, Lacertilia, Iguanidae). *American Zoologist*, 1977, *17*, 191–201. (b)

Greenberg, N. Ethological considerations in the experimental study of lizard behavior. In N. Greenberg & P. D. MacLean (Eds.), *Behavior and neurology of lizards*. Rockville, Md.: NIMH, 1978.

Greenberg, N. Physiological and behavioral thermoregulation in living reptiles. In R. D. K. Thomas & E. C. Olson (Eds.), *A cold look at the warm-blooded dinosaurs* (AAAS Selected Symposium #28). Washington: AAAS, 1980.

Greenberg, N. *Display behavior in the lizard, Anolis carolinensis*. Working paper No. 1, Laboratory of Physiological Ethology, Department of Zoology, University of Tennessee, Knoxville, Tennessee, 1982.

Greenberg, N. Central and autonomic aspects of aggression and social status in the lizard, *Anolis carolinensis*. In J. P. Ewert, R. R. Capranica, & D. J. Ingle (Eds.), *Advances in vertebrate neuroethology*. New York: Plenum Press, 1983.

Greenberg, N., & Crews, D. (Organizers). Social behavior in reptiles: A symposium. *American Zoologist,* 1977, *17,* 151–186.

Greenberg, N., & Jenssen, T. A. Displays of the banded iguana, *Brachylophus fasciatus*. In G. Burghardt & A. S. Rand (Eds.), *Iguanas of the world.* Park Ridge, N. J.: Noyes Publications, 1982.

Greenberg, N., & MacLean, P. D. (Eds.). *The behavior and neurology of lizards.* DHEW Publication no. (ADM) 77–491. Rockville, Md.: NIMH, 1978.

Greenberg, N., MacLean, P. D., & Ferguson, L. Role of the paleostriatum in species-typical display behavior of the lizard *(Anolis carolinensis). Brain Research,* 1979, *172,* 229–241.

Greenberg, N., Chen, T., & Green, H. *Corticosterone response to psychological stress in the lizard, Anolis carolinensis.* Working Paper No. 12, Laboratory of Physiological Ethology Department of Zoology, University of Tennessee, Knoxville, Tenn., 1982.

Grzimek, B., Hediger, H., Klemmer, K., Kuhn, O., & Wermuth, H. (Eds.). *Animal life encyclopedia* (Vol. 6): *Reptiles.* New York: Van Nostrand Reinhold, 1975.

Hadley, M. E., & Bagnara, J. T. Regulation of release and mechanisms of action of MSH. *American Zoologist,* 1975, *15*(Suppl. 1), 81–104.

Hadley, M. E., & Goldman, J. M. Physiological color changes in reptiles. *American Zoologist,* 1969, *9*(2), 489–504.

Hadley, M. E., & Hruby, V. J. Neurohypophysial peptides and the regulation of melanophore stimulating hormone (MSH) regulation. *American Zoologist,* 1977, *17,* 809–821.

Hanke, W. Endocrinology of amphibia. In M. Florkin & B. T. Scheer (Eds.), *Chemical Zoology* (Vol. IX). New York: Academic Press, 123–159.

Hardy, D. F. Ecology and behavior of the six-lined racerunner *Cnemidophorus sexlineatus. University of Kansas Science Bulletin,* 1962, *43*(1), 3–73.

Harless, M. Social behavior. In M. Harless & H. Morlock (Eds.), *Turtles: Perspectives and research.* New York: Wiley, 1979.

Harless, M., & Morlock, H. (Eds.). *Turtles: Perspectives and research.* New York: Wiley, 1979.

Harris, V. A. The life of the rainbow lizard. *Hutchinson Tropical Monographs.* London: Hutchinson, 1964.

Haubrich, R. Hierarchical behavior in the South African clawed frog *Xenopus laevis* Daudin. *Animal Behavior,* 1961, *9*(1–2), 71–76.

Hennig, C. W., Dunlap, W. P., & Harston, C. T. Tonic immobility and skin color in anoles: effects of serotonic precursors and metabolites. *Physiology and Behavior,* 1979, *22,* 1079–1088.

Huang, J. C. Effects of castration and sex steroid replacement therapy on male to male aggression and male courtship behavior in the lizard *Anolis carolinensis.* Master's thesis, Michigan State University, 1979.

Huey, R., Schoener, T., & Pianka, E. (Eds.). *Lizard ecology.* Cambridge, Mass.: Harvard University Press, 1983.

Hutchison, J. B. A study of the neural control of sexual clasping behavior in *Rana angolensis* Bocage and *Bufo* regulars. Reuss with a consideration of self-regulatory hindbrain systems in the anura. *Behavior,* 1967, *28,* 1–57.

Hutchison, J. B. (Ed.). *Biological determinants of sexual behavior.* New York: Wiley, 1978.

Hutchison, V. C. Aggressive behavior in *Plethodon jordani. Copeia,* 1959, *1959,* 72–73.

Ingle, D. Visual releasers of prey-catching behavior in frogs and toads. *Brain, Behavior and Evolution,* 1968, *1,* 500–518.

Ingle, D. Size preferences for prey-catching in frogs: Relationships to motivational state. *Behavioral Biology,* 1973, *(9),* 485–491.

Ingle, D. Focal attention in the frog: Behavioral and physiological correlates. *Science,* 1975, *188,* 1033–1035.

Ingle, D. *Behavioral correlates of central visual function in anurans.* In R. Llinas & W. Precht (Eds.), *Frog neurobiology.* New York: Springer-Verlag, 1976.

Jorgenson, C. B. Integrative functions of the brain. In B. Lofts (Ed.), *Physiology of the amphibia* (Vol. 2), New York: Academic Press, 1974.

Kastin, A. J., Miller, H., Nockton, R., Sandman, G. A., Schally, A. V., & Stratton, L. O. Behavioral aspects of melanocyte-stimulating hormone (MSH). *Progress in Brain Research, 39,* 1973, 461–470.

Keating, E. G., Korman, L. A., & Horel, J. A. The behavioral effects of stimulating and ablating the reptilian amygdala. *Physiology and Behavior,* 1970, *5,* 55–59.

Kelley, D. B. Auditory and vocal nuclei in the frog brain concentrate sex hormones. *Science,* 1980, *207,* 553–555.

Kelley, D. B., & Pfaff, D. W. Generalizations from comparative studies on neuroanatomical and endocrine mechanisms of sexual behavior. In J. B. Hutchison (Ed.), *Biological determinants of sexual behavior.* New York: Wiley, 1978.

Kelley, D. B., B., Morrell, J. L., & Pfaff, D. W. Autoradiographic localization of hormone-concentrating cells in the brain of an amphibian, *Xenopus laevis:* I. Testosterone. *Journal of Comparative Neurology,* 1975, *164,* 47–62.

Kennedy, M. C. Vocalization elicited in a lizard by electrical stimulation of the midbrain. *Brain Research,* 1975, *91,* 321–325.

Kimball, F. A., & Erpino, M. J. Hormonal control of pigmentary sexual dimorphism in *Sceloporus occidentalis. General Comparative Endocrinology,* 1971, *16,* 375–384.

Kleinholz, L. H. Studies in reptilian color changes. II. The pituitary and adrenal glands in the regulation of the melanophores of *Anolis carolinensis. Journal of Experimental Biology,* 1938, *15,* 474–491. (a)

Kleinholz, L. H. Studies in reptilian color changes: III. Control of the light phase and behavior of isolated skin. *Journal of Experimental Biology,* 1938, *15,* 492–499. (b)

Krieger, D. T., & Hughes, J. C. (Eds.). *Neuroendocrinology.* Sunderland, Mass.: Sinauer, 1980.

Kuhlenbeck, H. *The central nervous system of vertebrates* (Vol. 4). Basel: Karger, 1975.

Kuhlenbeck, H. *The central nervous system of vertebrates* (Vol. 5, Part 1). Basel: Karger, 1977.

LaHoste, G. J., Olson, G. A., Kastin, A. J., & Olson, R. D. Behavioral effects of melanocyte stimulating hormone. *Neuroscience and Biobehavior Review,* 1980, *4,* 9–16.

Leshner, A. I. *An introduction to behavioral endocrinology.* New York: Oxford University Press, 1978.

Leshner, A. I. Kinds of hormonal effects on behavior: A new view. *Neuroscience and Biobehavioral Review,* 1979, *3,* 69–73.

Leshner, A. I., & Politch, J. A. Hormonal control of submissiveness in mice: Irrelevance of the androgens and relevance of the pituitary-adrenal hormones. *Physiology and Behavior,* 1979, *22,* 531–534.

Leshner, A. I., Walker, W. A., Johnson, A. E., Kelling, J. S., Kreisler, S. J., & Svare, B. B. Pituitary adrenocortical activity and intermale aggressiveness in isolated mice. *Physiology and Behavior,* 1973, *11,* 705–711.

Lettvin, J. W., Maturana, J. W., McCullough, H. R., & Pitts, W. H. What the frog's eye tells the frog's brain. *Proceedings of the Institute of Radio Engineers,* 1959, *47,* 1940–1951.

Levitin, H. P. Monoaminergic control of MSH release in the lizard, *Anolis carolinensis. General and Comparative Endocrinology,* 1980, *41,* 279–286.

Littlejohn, M. J. Long-range acoustic communication in anurans: An integrated and evolutionary approach. In D. H. Taylor & S. I. Guttman (Eds.), *The reproductive biology of amphibians.* New York: Plenum Press, 1977.

Llinas, R., & Precht, W. (Eds.). *Frog neurobiology.* New York: Springer-Verlag, 1976.

Lofts, B. (Ed.). *Physiology of the amphibia* (vols. 2 & 3). New York: Academic Press, 1974, 1976.

Lowe, C. H., Jr., & Norris, K. S. Aggressive behavior in male sidewinders, *Crotalus cerastes,* with a discussion of aggressive behavior and territoriality in snakes. *Natural History Miscellanea,* 1950, *66,* 1–13.

Luthardt, G., & Roth, G. The relationship between stimulus orientation and stimulus movement pattern in the prey catching behavior of *Salamandra salamandra. Copeia,* 1979, *1979*(3), 442–447.

MacLean, P. D. Why do brain research on lizards? In N. Greenberg & P. D. MacLean (Eds.), *The behavior and neurology of lizards.* Rockville, Md.: NIMH, 1978.

Marcellini, D. L. The function of a vocal display of the lizard *Hemidactylus Frenatus* (Sauria: Gekkonidae). *Animal behavior,* 1977, *25,* 414–417.

Marcellini, D. L. The acoustic behavior of lizards. In N. Greenberg & P. D. MacLean (Eds.). *The behavior and neurology of lizards.* Rockville, Md.: NIMH, 1978.

Mason, P., & Adkins, E. K. Hormones and social behavior in the lizard *Anolis carolinensis*. *Hormones and Behavior,* 1976, *7,* 75–86.

Mayhew, W. W. Observations on captive *Amphibolorus pictus* an Australian Agamid lizard. *Herpetologica,* 1963, *19*(2), 81–88.

McDiarmid, R. W. Evolution of parental care in frogs. In G. M. Burghardt & M. Beckoff (Eds.), *The development of behavior: Comparative and evolutionary aspects.* New York: Garland Press, 1978.

McEwen, B. S. Glucocorticoid binding sites in rat brain: Subcellular and anatomical localizations. *Progress in Brain Research,* 1973, *39,* 87–96.

Milstead, W. W. (Ed.). *Lizard ecology: A symposium.* Columbia, Mo.: University of Missouri Press, 1967.

Moore, F. L. Differential effects of testosterone plus dihydrotestosterone on male courtship of castrated newts, *Taricha granulosa. Hormones and Behavior,* 1978, *11,* 202–208.

Moore, F. L., & Zoeller, R. T. Endocrine control of amphibian sexual behavior: Evidence for a neurohormone-androgen interaction. *Hormones and Behavior,* 1979, *13,* 207–213.

Moore, J. A. (Ed.). *Physiology of the amphibia* (vol. 1). New York: Academic Press, 1964.

Morrell, J. L., Kelley, D. B., & Pfaff, D. W. Autoradiographic localization of hormone concentrating cells in the brain of an amphibian, *Xenopus laevis:* II. Estradiol. *Journal of Comparative Neurology,* 1975, *164,* 63–78.

Morrell, J. L., Crews, D., Ballin, A., Morgentalin, A., & Pfaff, D. W. ^{3}H-estradiol, ^{3}H-testosterone, and ^{3}H-dihydrotestosterone localization in the brain of the lizard, *Anolis carolinensis. Journal of Comparative Neurology,* 1979, *188,* 201–224.

Morris, D. The response of animals to a restricted environment. *Symposium at the Zoological Society (London),* 1964, *13,* 99–118.

Morris, R., & Morris, D. *Men and snakes.* New York: McGraw-Hill, 1965.

Murphy, J. B., & Mitchell, L. A. Ritualized combat behavior of the pygmy mulga monitor lizard, *Varanus gillenti* (Sauria: Varanidae). *Herpetologica,* 1974, *30,* 90–97.

Noble, G. K. *Biology of the amphibian.* New York: McGraw-Hill, 1931.

Noble, G. K. The sense organs involved in the courtship of *Storeria, Thamnophis,* and other snakes. *Bulletin of the American Museum of Natural History,* 1937, *73,* 673–725.

Noble, G. K., & Bradley, H. T. The mating behavior of lizards: Its bearing on the theory of sexual selection. *Annals of the New York Academy of Science,* 1933, *35,* 25–100.

Norris, K. S. The ecology of the desert iguana, *Dipsosaurus dorsalis. Ecology,* 1953, *34*(2), 265–287.

Northcutt, R. G. Forebrain and midbrain organization in lizards and its possible evolutionary significance. In N. Greenberg & P. D. MacLean (Eds.), *Behavior and neurology of lizards.* Rockville, Md.: NIMH, 1978.

Obert, H. J. Hormonal influences on calling and reproductive behavior in anurans. In D. H. Taylor & S. J. Guttman (Eds.)., *The reproductive physiology of amphibians,* New York: Plenum Press, 1977.

Oksche, A., & Jeck, M. The nervous system. In B. Lofts (Ed.), *The physiology of the amphibia* (Vol. 3). New York: Academic Press, 1976.

Organ, J. A. Studies of the local distribution, life history, and population dynamics of the salamander genus *Desmognathus* in Virginia. *Ecological Monographs,* 1961, *31,* 189–200.

Palka, V. S., & Gorbman, A. Pituitary and testicular influenced sexual behavior in male frogs, *Rana pipiens. General and Comparative Endocrinology,* 1973, *21,* 148–151.

Patterson, A. T., Rickerby, J., Simpson, J., & Vickers, C. Melanocyte-stimulating hormone and the pineal in the control of territorial aggression. *Journal of Physiology,* 1978, *285,* 45.

Pearson, A. K., Tsui, H. W., & Licht, P. Effect of temperature on spermatogenesis, on the production and action of androgens and on the ultrastructure of gonadotropic cells in the lizard *Anolis carolinensis. Journal of Experimental Zoology,* 1976, *195,* 291–304.

Pearson, R., & Pearson, L. *The vertebrate brain.* New York: Academic Press, 1976.

Pooley, A. C., & Gans, C. The Nile crocodile. *Scientific American,* 1976, *234*(4), 114–124.

Porter, K. R. *Herpetology.* Philadelphia: W. S. Saunders, 1972.

Pyburn, W. F. Breeding behavior of the leaf-frogs *Phyllomedusa callidrias* and *Phyllomedusa dacnicolor* in Mexico. *Copeia,* 1970, *1970,* 209–218.

Rand, A. S. A nesting aggregation of iguanas. *Copeia,* 1968, *1968*(3), 552–561.

Rand. A. S., & Andrews, R. Adult color dimorphism and juvenile pattern in *Anolis cuvieri. Journal of Herpetology,* 1975, *9*(2), 257–260.

Rand, A. S., Gorman, G. C., & Rand, W. M. Natural history, behavior and ecology of *Anolis agassizi*. *Smithsonian Contributions to Zoology*, 1975, *176*, 27–38.

Rand, W. M., & Rand, A. S. Agonistic behavior in nesting iguanas: A stochastic analysis of dispute settlement dominated by the minimization of energy cost. *Zeitschrift für Tierpsychologie*, 1976, *40*, 279–299.

Regal, P. J. Behavioral differences between reptiles and mammals: An analysis of activity and mental capabilities. In N. Greenberg & P. D. MacLean (Eds.), *Behavior and neurology of lizards*. Rockville, Md.: NIMH, 1978.

Ridley, M. Paternal care. *Animal Behavior*, 1978, *26*, 904–932.

Rivero, J. A., & Esteves, A. E. Observations on the agonistic and breeding behavior of *Leptodactylus pentadactylus* and other amphibian species in Venezuela. *Breviora*, 1969, *No. 321*, 1–14.

Romer, A. S. *The procession of life*. New York: World Publishing, 1968.

Rowell, T. E. The concept of social dominance. *Behavioral Biology*, 1974, *11*, 131–154.

Ryan, M. J. The reproductive behavior of the bullfrog *(Rana catesbeiana)*. *Copeia*, 1980, *1980*(1), 108–114.

Salthe, S. N., & Mecham, J. S. Reproductive and courtship patterns. In B. Lofts (Ed.), *Physiology of the amphibia* (Vol. 2), New York: Academic Press, 1974.

Sandman, C. A., Kastin, A. J., Schally, A. V., Kendall, J. W., & Miller, L. H. Neuroendocrine responses to physical and psychological stress. *Journal of Comparative Physiology and Psychology*, 1973, *84*, 386–390.

Schmidt, R. S. Central mechanisms of frog calling. *Behavior*, 1966, *26*(3–4), 251–285.

Schmidt, R. S. Central mechanisms of frog calling. *American Zoologist*, 1973, *13*, 1169–1177.

Seligman, M. E. P. On the generality of the laws of learning. *Psychology Review*, 1970, *77*, 406–418.

Sexton, O. J. Some aspects of the behavior and of the territory of a dendrobatid frog, *Prostherapis trinitatis*. *Ecology*, 1960, *41*, 107–115.

Smith, M. A. *The fauna of British India, Ceylon and Burma* (Vol. 3): *Reptilia and Amphibia*. London: Taylor & Francis, 1939.

Stacey, N. E. Hormonal regulation of female reproductive behavior in fish. *American Zoologist*, 1981, *21*(1), 305–316.

Sternthal, D. E. Olfactory and visual cues in the feeding behavior of the leopard frog *(Rana pipiens)*. *Zeitschrift für Tierpsychologie*, 1974, *34*, 239–249.

Stratton, L. O., & Kastin, A. J. Melanocyte stimulating hormone and MSH/ACTH reduce tonic immobility in the lizard. *Physiology and Behavior*, 1976, *16*, 771–774.

Sugerman, R. A., & Demski, L. S. Agonistic behavior elicited by electrical stimulation of the brain in western collared lizards, *Crotaphytus collaris*. *Brain, Behavior and Evolution*, 1978, *15*, 446–469.

Tarr, R. S. Role of the amygdala in the intraspecies aggressive behavior of the iguanid lizard, *Sceloporus occidentalis*. *Physiology and Behavior*, 1977, *18*, 1153–1158.

Taylor, D. H., & Guttman, S. I. (Eds.). *The reproductive biology of amphibians*. New York: Plenum Press, 1977.

Test, F. H. Social aggressiveness in an amphibian. *Science*, 1954, *120*, 140–141.

Thorn, R. Protection of the brood by the male of the salamander *Hynobius nebulosis*. *Copeia*, 1962, *1962*(3), 639–640.

Thurow, G. Aggression and competition in Eastern *Plethodon* (Amphibia, Urodela, Plethodontidae). *Journal of Herpetology*, 1975, *10*(4), 277–291.

Tracy, C. R. Observations on social behavior in immature California toads *(Bufo boreas)* during feeding. *Copeia*, 1973, *1973*(2), 342–345.

Verniko-Danellis, J. Effects of hormones on the central nervous system. In S. Levine (Ed.), *Hormones and behavior*. New York: Academic Press, 1972.

Vinegar, M. B. The function of breeding coloration in the lizard *Sceloporus virgatus*. *Copeia*, 1972, *1972*(4), 660–664.

Wada, M., & Gorbman, A. Relation of mode of administration of testosterone to evocation of male sex behavior in frogs. *Hormones and Behavior*, 1977, *8*, 310–319.

Weaver, W. G., Jr. Courtship and combat behavior in *Gopherus berlandieri*. *Bulletin Florida State MUS, Biological Sciences*, 1970, *15*(1), 1–43.

Wells, K. D. The social behavior of anuran amphibians. *Animal Behavior,* 1977, *25,* 666–693.
Wells, K. D. Territoriality in the green frog *(Rana clamitans):* Vocalizations and agonistic behavior. *Animal Behavior,* 1978, *26,* 1051–1063.
Wells, K. D. Reproductive behavior and male mating success in a neotropical toad, *Bufo typhonius. Biotropica,* 1979, *11*(4), 301–307.
Wells, K. D. Social behavior and communication of a dendrobatid frog *(Colostethus trinitatis). Herpetologica,* 1980, *36*(2), 189–199.
Wells, K. D. Parental behavior of male and female frogs. In R. D. Alexander & D. W. Tinkle (Eds.). *Natural selection and social behavior: Recent research and new theory.* New York: Chiron Press, 1981.
Wheeler, J. M., & Crews, D. The role of the anterior hypothalamus-preoptic area in the regulation of male reproductive behavior in the lizard, *Anolis carolinensis:* Lesion studies. *Hormones and Behavior,* 1978, *11,* 42–60.
Wiewandt, T. A. Vocalization, aggressive behavior, and territoriality in the bullfrog, *Rana catesbeiana. Copeia,* 1969, *1969*(2), 276–285.
Wurtman, R. J., & Axelrod, J. Adrenaline synthesis: Control by the pituitary gland and adrenal glucocorticoids. *Science,* 1965, *150,* 1464–1465.
Wurtman, R. J., Axelrod, J., & Tramezzani, J. Distribution of the adrenaline-forming enzyme in the adrenal gland of a snake, *Xenodon merremi. Nature,* 1967, *215,* 879–880.

19

Hormonal Influences on Aggressive Behavior in Ungulates

Marie-France Bouissou

Castration of domestic animals with the aim of controlling sexual and aggressive behavior dates from the earliest times of the human race. Thus, before systematic research, man had intuitively established the relationship between these behaviors and the presence of sexual organs.

The fact that endocrine and especially gonadal secretions are involved in the control of agonistic behavior is well established in several species. In most mammals, the expression of agonistic behavior is sexually dimorphic, with males being generally more aggressive than females (Collias, 1944). In rodents, a species in which extensive work has been done to describe the relationship between hormones and aggressiveness (cf. Bronson & Desjardins, 1971; Conner & Levine, 1969; Hart, 1974; Sigg, 1969), the aggressiveness of males appears at puberty, is reduced by castration, and is restored by androgen replacement. On the other hand, androgen treatment of the female is without effect unless she has been previously "masculinized" by androgen exposure during the neonatal period.

The role of androgens with respect to aggressive behavior varies considerably according to species and according to developmental and environmental factors (e.g., Dixon, 1980). Nevertheless, a great number of studies have demonstrated that androgens influence aggressive motivation in most species; some forms, such as intermale aggression, seem to be more easily influenced than others. Hormones can exert their effects both during adult life and during neonatal life. In some species, both actions may involve conversion of androgens to estrogenic metabolites. Other hormones of the pituitary-adrenal axis (adrenocorticotropic hormone) and the pituitary-gonadal axis (estrogen, progesterone, prolactin, luteinizing hormone) also modulate the behavior (for review, see Brain, 1979a,b).

Among ungulates, males of some domesticated species (e.g., bulls and stal-

MARIE-FRANCE BOUISSOU • Station de Physiologie de la Reproduction, I.N.R.A., 37380 Nouzilly, France.

lions) are frequently castrated in order to reduce their aggressiveness and to facilitate handling and utilization for work. Surprisingly, the relationship between hormones and aggressive behavior in this zoological group has until recently received little attention. In this review, we shall (1) survey the differences in agonistic and related behaviors between males and females, (2) examine the relationships between spontaneous variation in hormonal levels and concomitant variations in agonistic behavior, and (3) review the experimental findings involving manipulation of hormonal levels and the resulting effects on behavior.

However, it is useful first to recall that aggressive behavior is not a unitary concept and may be difficult to define. Moyer (1968) attempted to classify the various types of aggressive behavior according to the situations in which they were observed. It is thought that different forms of aggression arise from different motivations and have different physiological correlates. The present review will confine interest to agonistic behaviors among individuals of the same species in a social context. The social rank of animals within hierarchically structured groups will also be treated, but it is beyond the scope of this discussion to debate the concept of social dominance and its validity or usefulness (for further discussion, cf. Richards, 1974; Rowell, 1974; Syme, 1974). In this regard, let us emphasize that aggressiveness and dominance are two separate concepts that may vary independently.

Before considering the relationship between hormones and aggressive behavior, it would be useful to describe briefly the social context in which this aggressiveness occurs. With the exception of some species of African antelope (in particular, the Cephalophinae, Neotraginae, and Reducinae) and the moose *(Alces alces),* most ungulates are gregarious. It is difficult, however, to classify the social structures of the various species, because they may vary considerably within any single species, particularly under varying environmental conditions. For example, in the Grant gazella *(Gazella granti),* typically, males are territorial in bushy areas while mixed groups are observed in open areas.

One method of classification would be to distinguish those species in which males and females always belong to the same functional unit from those species in which males and females are segregated except during the breeding season. In the first case, a herd composed of males and females is a structural characteristic of most bovine species (domestic bovines, buffalo, and bison), certain Tragelaphinae (e.g., the eland), and elephants. However, groups of males and solitary males are known to join the herd periodically, especially during the rut. In these species, the existence of a social hierarchy between adult males is a characteristic phenomenon. The social structure of the Equidae (horses and zebras, with the exception of *Dolichohippus grevyi grevyi*) is characterized by permanent harems and stable family groups composed of an adult male and one or several mares and their progeny. There also exist some male bachelor groups. These species are not territorial. In a great number of cases, however, males and females segregate themselves except during the breeding season. Males typically form same-sex groups with a hierarchical organization, but these groups are generally of a less stable composition than are female groups. Male members are known to disperse during the rut either to rejoin females (e.g., wild ovines) or to migrate to a "rutting area" where they may form a temporary harem (e.g., red deer).

Many species of ungulates are territorial. In some species, breeding males defend their territory all the time (e.g., the kob, *Adenota kob thomasi*), while males of other species limit their territorial behavior to the breeding season (e.g., the roe deer, *Capreolus capreolus*). This brief survey of ungulate social structure demonstrates the diversity of social context in which aggressive behavior occurs.

SEX DIFFERENCES IN AGONISTIC AND RELATED BEHAVIORS

In ungulates, as in most mammals, aggressive behavior is sexually dimorphic, in that males are generally more aggressive than females. Fights between males of certain species, especially camels and rams, are commonly organized by man for entertainment in many Eastern countries. There is one notable exception to the predominant use of males in this activity: In parts of Europe (Switzerland and France), cow fights are organized each year when animals are turned out to pasture after the winter period.

In some species, there are sex differences in motor patterns observed during agonistic encounters. For example, while the stag tends to attack frontally (at least during that time when his antlers have lost their velvet), the hind uses her forefeet. During the rut, the stag also shows a stiff-legged approach threat and uses a lateral display not seen in the hind (Lincoln, 1970, 1971). In the horse, the method of fighting differs considerably between stallions and mares: the former rear and often use their forefeet, whereas the females kick with their hindfeet.

Territorial possession and its defense are nearly always the prerogative of the male, but in some species of antelope (of the Madoquinae family) females also participate. Still, marking is mostly performed by males, either by rubbing against vegetation and thus depositing odorous substances (as in the Cervidae) or by depositing feces on certain parts of the territory (as observed in stallions).

Even in young animals, sex differences in agonistic displays are apparent. In primates (Goy & Phoenix, 1971; Phoenix, Goy & Resko, 1968) and in dogs (Bekoff, 1974), males exhibit more mounts, rough-and-tumble play, and threats than their female peers. Such sex differences have also been observed in cattle, sheep, and horses. In the young zebus *(Bos indicus)*, males are more often involved in bouts of pushing and shoving while females are more apt to strike blows (Reinhardt, Mutiso, & Reinhardt, 1978). In young lambs, it is the male that is more frequently involved in reciprocal and one-way butting (Sachs & Harris, 1978). In fact, aggressive play behavior is rarely observed in the female lamb after the fifteenth week, while in males it is preferentially directed toward other males and is apparent until the time of weaning. In this species, the development of this behavior does not seem to depend on the secretion of gonadotropic hormones (LH, FSH) or testosterone (J. P. Signoret, personal communication, 1980). The dimorphism in aggressive play is further illustrated in ponies (Tyler,

1972). Play-fighting between foals is primarily between males. Of the total recorded play bouts, 74% were observed between two males and only 3% between two females (Schoen, Banks, & Curtis, 1976).

VARIATION IN AGONISTIC AND RELATED BEHAVIORS AS THEY RELATE TO VARIATIONS IN HORMONAL LEVELS UNDER NATURAL CONDITIONS

Physiologic Generalities

The endocrinology of reproduction is, of course, well known in domestic mammals (reviewed by Levasseur & Thibault, 1980a,b). In all species studied, it is known that the secretion of gonadotropins induces important transition events during fetal life and again at puberty. The secretion of gonadotropin-releasing hormone (GnRH) and the gonadotropins begins during early fetal life without respect to the length of gestation or the timing of gonad formation. Secretion thereafter regresses, sometimes even before parturition, and later becomes active again during early puberty, until it recedes when adult levels of gonadotropins and steroids are reached.

From the moment of differentiation of the testes (i.e., the fetal period), the response of the interstitial tissue of the testes to endogenous or exogenous gonadotropins is an increase in the secretion of testosterone. However, the response of the ovaries in the form of estradiol secretion occurs only when antrum follicles are present. Thus, substantial amounts of androgen are found in the male fetus, and in some cases, these levels are sufficient to masculinize a female twin.

The prepuberal onset of gonadotropic function is characterized by a dramatic increase in the frequency and magnitude of secretion. The increase in the secretion of gonadotropins leads to a stepwise increase in the secretion of steroids by the gonads. Thus, the level of gonadotropin secretion is the determining factor in the onset of puberty. In spite of variations due to breed, climate, or nutritional or social conditions, it has been established that the age of puberty onset is 6–7 months in sheep, goats, and pigs; 12 months in cattle; and 15–18 months in horses. However, puberty onset appears to be more closely related to body weight than to chronological age. In those species with seasonal reproduction, puberty is reached by the young only during the breeding season. However, in almost all wild species, the "sociological" puberty of males (i.e., the time when they begin to mate and assume other male-typical roles within the group) lags well behind the time of physiological puberty.

A great number of mammals, including most species of ungulates, have a well-defined season of sexual activity that depends mainly on the duration of daylight, at least in temperate zones. The beginning of the sexually active period coincides with increased levels of gonadotropins, which in turn lead to increasing amounts of circulating steroids. Alternatively, an annual period of sexual quiescence can also be observed, involving testicular regression with cessation of spermatogenesis and a considerable reduction in circulating androgens. Seasonal variation in blood levels of testosterone have been reported in several species, both domestic and wild, including the ram (Attal, 1970; Katongole, Naftolin, & Short, 1974; Rouger, 1974), the billy goat (Leidl, Hoffmann, & Karg, 1970; Rouger, 1974), the bull (Katongole, Naftolin, & Short, 1971; Rouger, 1974), the stag (Lincoln, Youngson, & Short, 1970), and the roe deer (Sempere, Garreau, & Boissin, 1980; Short & Mann, 1966) (see Figure 1). Even in those ungulates that are sexually active throughout the year, there is often seasonal variation in intensity. For example, although the ram is capable of breeding the whole year, the weight of the testes and levels of gonadotropins and testosterone are at a minimum during the anestrous period of the female. In other species, such as the bull, testosterone levels are nearly constant yearlong, but secretion of LH undergoes wide 24-hour fluctuations, leading to variable testosterone secretion. In this regard, it should be noted that besides testosterone, the major secretory product of the testes, androstenedione is also produced by the testes of the prepuberal bull and, too, considerable amounts of estrogens in the stallion and boar.

Reproductive cycles of female ungulates may occur throughout the year (e.g., the cow and sow) or for a limited time only (e.g., the ewe, goat, most mares, and females of many wild species). And although the same pituitary and ovarian hormones are involved in the control of cycles, the way in which they are con-

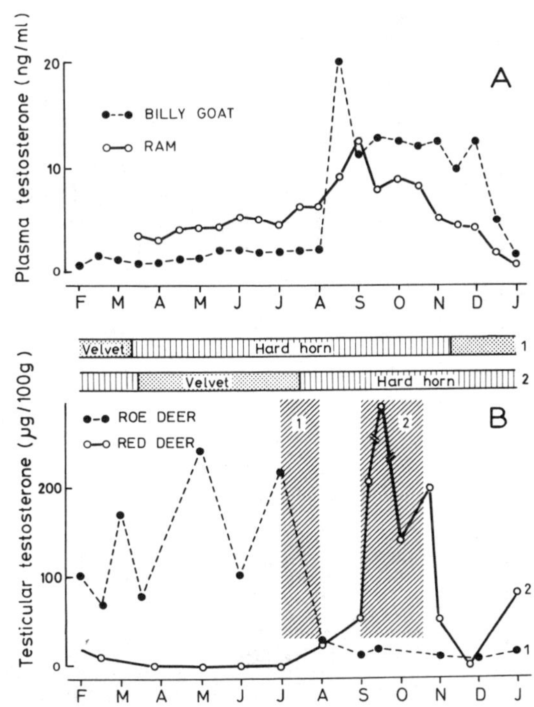

FIGURE 1. Seasonal variation of (A) plasma testosterone in two domestic ungulates and (B) testicular testosterone in two wild ungulates. Shaded area represents rut. (Adapted from Rouger, 1974, for billy goat and ram; Short & Mann, 1966, for roe deer; Lincoln, Youngson, & Short, 1970, for red deer.)

trolled and the effects they produce vary from one species to another (Figure 2). As an example, let us outline the events that occur prior to estrus in the ewe.

1. The corpus luteum of the previous cycle regresses.
2. The drop in the level of progesterone leads to secretion of estrogens (and androgens) by the graafian follicle, and these levels peak after 24 hours.
3. The increase in estrogen levels brings about a release of LH 12 hours later. LH release can occur only by estrogen induction in the absence of progesterone.
4. Ovulation occurs and eventually behavioral estrus occurs, while the secretion of estrogens diminishes.
5. Ovulation takes place about 24 hours following the LH surge.
6. The cells of the granulosa begin to secrete progesterone.

This sequence of events is identical in the cow and sow, but the timing of particular events is very different in these two species.

The participation of other steroid hormones (especially androgens) in the reproductive behavior of female ungulates is less well known. Interestingly, the

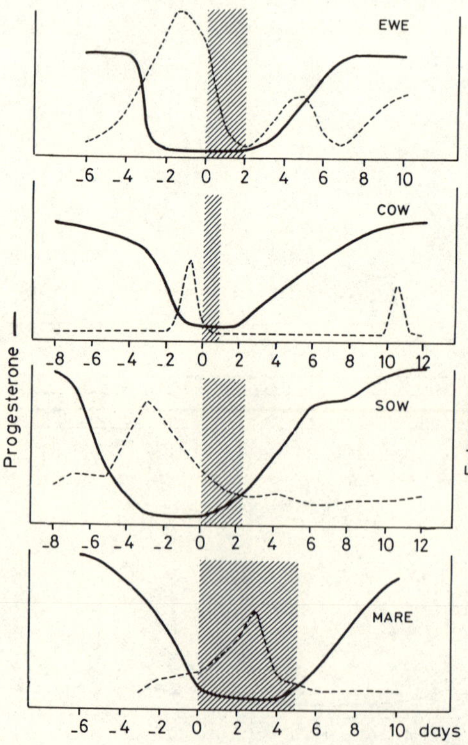

FIGURE 2. Diagrammatic representation of hormone changes during estrous cycle in domestic ungulates. Shaded area represents sexual receptivity.

main steroid secreted by the ovary of the ewe at estrus is not 17β-estradiol but androstenedione. This androgen may play an important (but as yet unknown) role in the induction of certain components of behavioral estrus.

The interaction between males and females considerably influences various aspects of the reproductive physiology of the female, including puberty onset, the length of estrus, synchronization of cycles, the timing of ovulation, and the seasonal reappearance of cycles at the end of anoestrus. In a reciprocal fashion, where social interaction produces sexual arousal in the male, there is a release of LH, which stimulates testosterone secretion.

Seasonal Variations in Agonistic and Territorial Behavior in Males: Relationship to Androgen Levels

Seasonal Variations

In animals with a seasonal reproduction, the "rut" refers to both the season during which reproduction takes place and the behavior of males during that period. All wild ungulates in the Northern hemisphere have a period of rut. Species such as the red deer, reindeer, elk, wapiti, musk-oxen, and bison all provide representative aspects of the different types of behavior that can be observed during that period (Fraser, 1968). Changes in the behavior of the male are dramatic indeed, culminating in an increase in the frequency and intensity of fighting in almost all species studied. This is typified in the American bison *(Bison bison)* (Lott, 1974; McHugh, 1958) and also in the musk-oxen *(Ovibos moschatus)* (Oeming, 1965); in the latter species, it has been estimated that as many as 5% to 10% of adult bulls die each year from wounds received during rutting (Wilkinson & Shank, 1977).

In Soay sheep, the rams begin to disperse as the rutting period approaches in late September and early October. At this time, the rams test the ewes for estrus more frequently and this coincides with an increase in the frequency of aggressive interactions (Grubb, 1974; Grubb & Jewell, 1973). This is also observed in wild American sheep and goats *(Ovis canadensis* and *Oreamnos americanus*) (Geist, 1964, 1968, 1971) as well as in most Cervidae. Stags are known to become highly aggressive even in the absence of females (Lincoln *et al.,* 1970; Lincoln, Guinness, & Short, 1972), and their agonistic behavior then includes lateral threat postures not observed in bachelor herds. Clutton-Brock, Albon, Gibson, & Guinness (1979) estimate that 23% of males studied on the island of Rhum are wounded during fighting and 6% are permanently injured each year. In reindeer, *(Rangifer tarandus* and *R. tarandus groenlandicus*), aggressiveness of males increases not only toward other males but is also directed against the whole herd (Espmark, 1964a). This irritability leads to outbreaks of fighting especially between males of the same size (Bergerud, 1974). In the Cervidae of the genera *Odocoileus,* some aggressive interactions involving the use of antlers are observed throughout the year, but serious fighting is confined to the breeding season (Brown, 1974; Michael, 1968). An increase in aggressiveness during the period of rut is also observed in the deer *(Dama dama)* (Alvarez, Braza, & Norzagaray,

1975a,b). In addition, some domestic species that reproduce seasonally exhibit an intensification of agonistic behavior during that period (e.g., ram, billy goat).

One exception to the usual seasonal variation in male aggressiveness is the moose *(Alces alces),* in which aggressiveness is exhibited all year long, especially between two females or between a female and young male. Another exception is the Ethiopian ibex *(Capra wallie),* in which there is a negative correlation between the frequency of rutting behavior and the incidence of fighting; but this species is atypical in a second respect, in that breeding can occur at any time during the year (Nievergelt, 1974).

In territorial species, defense of a territory and associated marking activity are generally limited to the reproductive period. For example, male pronghorns *(Antilocapra americana)* are territorial only from late March to early October (Kitchen & Bromley, 1974). Likewise, the roe deer defends its territory only during the rut although it uses the same territory as a home range during the rest of the year (Bramley, 1970). In many antelopes there is seasonal variation in the intensity of territorial behavior, with peak periods of territorial activity corresponding to peaks in sexual activity (Jarman, 1979; Montfort-Braham, 1975; Robbel & Child, 1975).

Relationships to Hormonal Levels

Thus, it has been well established that males of almost all domestic or wild species of ungulates exhibit variation in their aggressiveness and their territorial behavior that is closely linked to their sexual activity. Sexual activity itself reflects variation in the level of circulating androgens. In the past, only indirect measures of hormonal levels were available, such as the weight of the testes or the bulbourethral glands, the diameter of the seminiferous tubules, the fructose content of the seminal vesicles, or some other secondary sexual characteristic such as antlers in the Cervidae. It is only in recent years that reliable and easy methods of plasma hormone determinations have permitted accurate study of the relationships between hormonal levels and behavior.

The relationship between aggressiveness and androgen level has been demonstrated in the ram of the Soay breed. In these males, a peak of testicular activity (testosterone secretion) occured at the end of four months of short photoperiod days, followed by an increased incidence of aggressive behavior one month later (Lincoln & Davidson, 1977) (Figure 3). It has been well established that sexual arousal, either by copulation or simple proximity to a female, causes a transitory rise in LH and testosterone in the bull (Katongole *et al.,* 1971). In the adult ram, the proximity to estrous females also affects circulating levels of testosterone, and testosterone elevation is accompanied by an increase in aggressiveness (Illius, Haynes, & Lamming, 1976). In white-tailed deer as well, the presence or absence of estrous females influences the patterns of testosterone secretion (Bubenik, Bubenik, Brown, & Wilson, 1977).

The North American wild horse breeds seasonally from late March to mid-July; peak breeding activity occurs in early May at the same time as peak an-

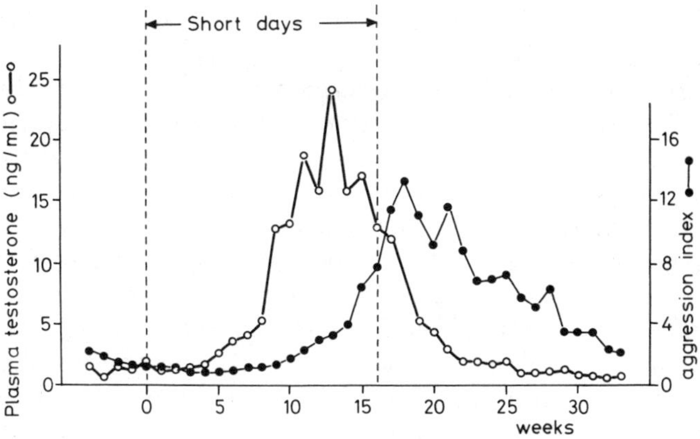

FIGURE 3. Plasma testosterone levels and aggression score for Soay rams housed under artificial lighting conditions. (Adapted from Lincoln & Davidson, 1977.)

drogen/testosterone concentration in stallions; an increased aggressiveness between males coincides with the higher concentrations (Kirkpatrick, Wiesner, Kenney, Ganjam, & Turner, 1977).

Among wild species, two remarkable examples of testosterone-dependent aggression are illustrated in the Asiatic elephant and the one-humped camel. A phenomenon known as "musth" in the captive Asiatic elephant *(Elephas maximus)* is observed only in adult males and is characterized by periods of aggressive and masturbatory behavior accompanied by abundant secretory activity from the temporal glands. Periods of musth can last from several weeks to several months. During this time, males become very aggressive toward conspecifics, toward man, and even toward inanimate objects. In fact they are too dangerous for domestic use and are kept chained up. Periods of musth are distributed over the whole year, with peak incidence in the months of January–April and August–November. Almost all males have one musth per year and some have two. The relationship between musth and sexual activity has not been well established. In fact, semen production, estrus, and mating seem to be temporally independent of this phenomenon since reproduction is known to occur all year long.

The relationship between testosterone titers and musth onset and decline has been established. Between periods of musth, testosterone levels are low (less than 0.2–1.4 ng/ml). As the temporal glands begin to grow, testosterone levels increase (4.3–13.7 ng/ml), and when the animals are in full musth, levels of testosterone are extremely high, sometimes reaching 30–65 ng/ml. These values greatly exceed maximal values reported for other species (e.g., bulls, rams, and stallions), whereas testosterone levels of the Asiatic elephant are comparable to these other species outside the period of musth. Thus, it has been established that the onset of musth is associated with and probably induced by increased secretion of testosterone. In addition, engaging in copulation may induce musth onset, since males used in breeding exhibit musth two to five months earlier than nonbreeding males (Jainudeen Katongole, & Short, 1972; Scheurmann & Jainudeen, 1972).

An analogous phenomenon was described recently in the one-humped camel *(Camelus dromedarius)* by Yagil and Etzion (1980). During the breeding season (January through April), males of this species become so aggressive that they cannot be used for domestic purposes. A brownish, fetid fluid is secreted by glands located on the dorsal part of the neck just under the ears. And, similar to the findings for Asiatic elephants, blood levels of androgens increase considerably during this period, ranging about 2 ng/ml from September to November, rising to 17 ng/ml in December, and attaining a level of 35 ng/ml in February—more than a tenfold increase from basal levels. In this species, the relationship between testosterone secretion and the onset of breeding is clear cut: Androgen levels increase in the male only at the time of the estrous period of the females.

The relationship between hormonal levels and territorial behaviors such as marking has also been well established in some ungulates. In the roe deer, androgenic activity (as measured by testes and seminal vesicle weights, the mean diameter of the seminiferous tubules, and the fructose level of seminal vesicles) increases in April and reaches a maximum in July. This coincides with peak periods of territorial activity (Bramley, 1970). Male roe deer mark their territory by rubbing their forehead against trees or other vegetation in the vicinity. The frequency of marking closely parallels the increasing levels of circulating testosterone as the rut approaches (Figure 4) (Sempere *et al.*, 1980). The growth and secretory activity of the cutaneous glands used in marking is under androgen control, as demonstrated by their timely development with the approach of the rut season and with gonadal secretory activity (Moy, 1970). Castration and androgen-replacement studies also support this view (Strauss & Ebling, 1970).

Comparisons made between territorial and nonterritorial impala males have revealed that testes weights, seminal vesicle weight, and mean diameters of the seminiferous tubules do not differ in these two groups. However, mean weights of the bulbourethral glands and levels of circulating testosterone are significantly higher (almost double) in territorial males, that is, those having access to females. Bramley and Neaves (1972) have interpreted this finding to mean that increased testosterone titers are a consequence rather than a cause of their territorial status, since it has been repeatedly shown that sexual activity and female proximity both increase levels of testosterone. It follows, then, that the separation of male ungulates into bachelors and those having access to receptive females (through territorial or harem ownership) may be reinforced through testosterone levels associated with age and social milieu. If one postulates that proximity to females and copulation increase testosterone levels, which in turn increase aggressiveness, then dominant males will be more able to maintain their social position than submissive ones (Illius *et al.*, 1976).

While the relationship between androgen levels and agonistic behavior is rather well documented, the relationship between hormonal levels and social rank have rarely been directly investigated. In one such study, Davant, Han, and Moody (1974) showed that there is no correlation between dominance and testosterone levels in the ram. On the other hand, when two deer *(Odocoileus*

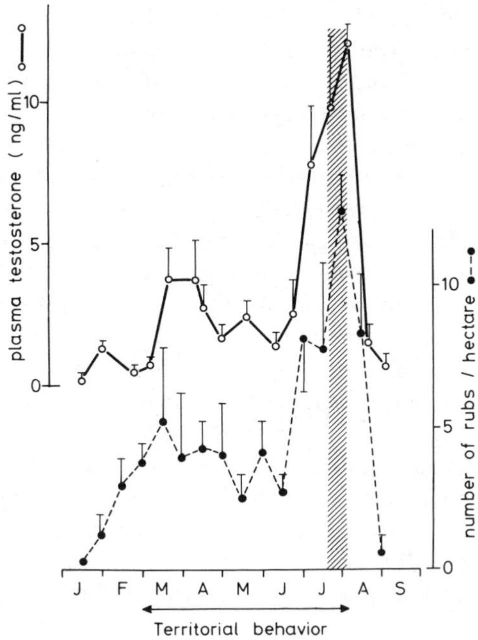

FIGURE 4. Plasma testosterone levels and territorial behavior as indicated by rubs in the roe deer. Shaded area represents rut. (Adapted from Sempere, Garreau, & Boissin, 1980.)

virginianus) are placed in same enclosure, testosterone is higher in the dominant male than in the submissive male. It is somewhat surprising that parallel studies have rarely been carried out in animals such as domestic bull, for which hormonal data are well known.

VARIATION IN AGONISTIC BEHAVIOR OF FEMALES ACCORDING TO THEIR PHYSIOLOGIC STATUS

While much is still lacking in our knowledge of male aggressiveness, even less is known about aggressive behavior and its physiologic correlates in female ungulates. Of the wild species, an increase in aggressiveness can be observed in the female Ethiopian ibex *(Capra wallie)* from December to May, corresponding to the postparturition period (Nievergelt, 1974). In moose, agonistic behavior is commonly seen among females especially during the rutting season (Geist, 1963). Female moose search for the wallows of males and compete to obtain those places (Altmann, 1959). In reindeer, the female is dominant over males during the winter, often chasing them from craters dug in the snow to search for food. This social dominance may be due to hormonal changes associated with gestation or may be related to females retaining their antlers longer than males (Espmark, 1964b). Female one-humped camels *(Camelus dromedarius)* lose their usual placidity and may even become dangerous during the estrous period, lasting from January to April (Yagil & Etzion, 1980).

Among domestic animals, an increase in aggressiveness is commonly observed during the estrous period of the ewe (Banks, 1964; Hulet, Alexander, &

Hafez, 1975). Mares in estrus seem to be more aggressive among themselves, at least in the absence of the stallion (Archambeaud, 1972). From anecdotal reports, it is also known that some mares are more difficult to handle when they are in heat. In the cow, onset of estrus affects the frequency of both fights and threats, nearly doubling their incidence during that period. In contrast, the initiator/victim ratio attained its highest value the day before estrus and its lowest value the day after (Hurnik, King, & Robertson, 1975a,b). In this study, however, the behavioral measures cited as agonistic responses were not clearly described. For instance, under "agonistic interactions," the authors have included all types of head contact (64% of the total agonistic responses), probably including mock-fighting, an activity normally considered a component of the sexual behavior of bovines rather than an agonistic response. Furthermore, it is not clear whether this study distinguished between aggressive and agonistic interactions (including spontaneous withdrawal and retreat). Obviously, estrous females (particularly the subordinate ones) who seek contact with other females without heed to social rank are more frequently involved in agonistic interactions just before and during the estrous period. When strange heifers are herded together into all-estrous and all-nonestrous groups, neither the number of aggressive interactions nor the number of spontaneous withdrawals differs during the two hours following gathering of the groups. Nevertheless, the number of fights is higher among estrous females. Moreover, the frequency of agonistic interactions among females (whether or not they are in estrus) largely exceeds that recorded in groups of males put together under the same conditions (Bouissou, 1975).

In the cow, we have found that the social hierarchy is not modified by the advent of estrus. Asa, Goldfoot, and Ginther (1979) have described similar observations for the mare.

Hormonal changes that accompany gestation and lactation may account for changes in the level of aggressiveness. This possibility was advanced earlier by Espmark (1964) to explain the behavioral changes observed in female reindeer during the winter months. In pregnant cows, the frequency of aggressive responses among females and of those directed toward foreign calves, as well as the social relationship between pregnant and nonpregnant females, remains unchanged (Bouissou, unpublished observations).

In all species, the female may be highly aggressive toward both conspecifics and members of other species (e.g., possible predators), and it is thought that this aggressiveness serves to defend their young. Very little is known about the physiological basis of this kind of aggression. In rodents, it has been shown that aggressiveness, as measured under a variety of test situations (frog-killing or shock-induced fighting) differs dramatically between lactating and nonlactating females (Endroczi, Lissak, & Telegdy, 1958; Thoman *et al.*, 1970). No comparable studies have been conducted in ungulates. In the cow, the level of aggressiveness among females does not seem to change after calving (Bouissou, unpublished observations). However, in the mare and the female zebra, changes in hierarchical structure may occur following parturition and during postpartum estrus. At these times, the dam becomes more aggressive toward other females and even toward the stallion in the protection of her young (Klingel, 1972; Sereni, 1977).

CONSEQUENCES OF AGONISTIC INTERACTIONS ON HORMONAL LEVELS

A great number of studies of various species, especially rodents and primates, demonstrate the consequences of aggressive behavior and social rank on the activity of the pituitary–gonadal and pituitary–adrenal axes (reviewed by Bronson & Desjardins, 1971; Conner, 1972; Dixon, 1980). In pigs and cattle, the social environment can affect adrenocortical activity. For example, adrenal gland weights of subordinate animals are significantly higher than those of dominant animals among groups of cattle (Bouissou & Signoret, 1971). Interestingly, there is no correlation between the level of circulating corticoids and social rank in cattle, and corticoid levels are not appreciably modified by experimentally induced rank changes. However, variations in the patterns of activity differ markedly between "aggressive" and "docile" animals (Arave, Mickelson, Lamb, Sevejda, & Canfield, 1976). In pigs, if two animals belonging to different social groups confront each other in a neutral environment, aggressive interactions are observed and fighting may ensue. In that case, activation of the adrenal cortex is reflected by alterations in circulating corticosteroids, with a greater increase in subordinate animals than in dominant animals (Dantzer & Mormède, 1979) (Figure 5).

Exposure to an unknown animal produces an increase in corticosteroid levels

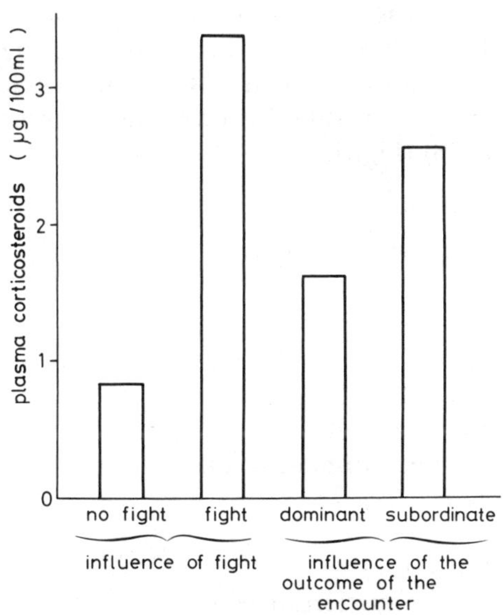

FIGURE 5. Influence of incidence and outcome of agonistic encounter on plasma corticosteroid levels in young pigs. (Adapted from Dantzer & Mormède, 1979.)

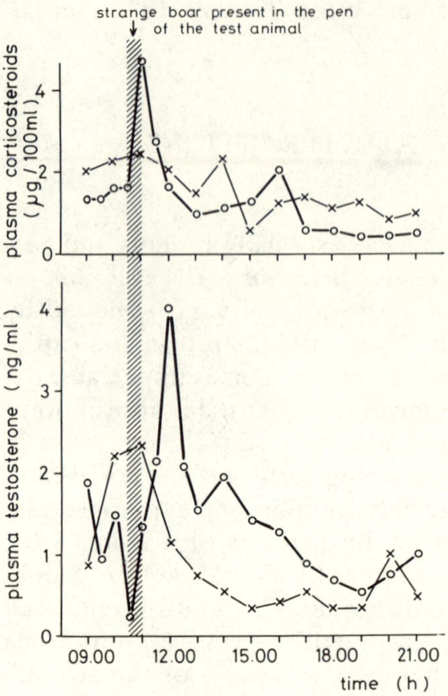

FIGURE 6. Plasma corticosteroids and plasma testosterone in a boar when exposed to a strange boar. Circles, test day; crosses, control day. (Adapted from Liptrap & Raeside, 1978.)

that is immediately followed by an increase in testosterone levels (events similarly reported for copulatory activity). Suppression of corticosteroid production by the enzymatic inhibitor, metyrapone, brings about a corresponding reduction in testosterone levels, suggesting that the increase in testosterone titers following the stress associated with aggression is mediated, in part, by the adrenal cortex. This hypothesis is well supported by the finding that injection of ACTH or cortisol in the boar, ACTH in the stallion, and dexamethasone in the bull all produce a transient increase in blood levels of testosterone (Liptrap & Raeside, 1978) (Figure 6).

EFFECTS OF EXPERIMENTAL MANIPULATION OF HORMONAL LEVELS ON AGONISTIC BEHAVIOR

To date, only the sex steroids have been studied. The classic methods used to manipulate hormonal levels include (1) castration or ovariectomy and (2) hormone replacement therapy by implant placement or injection of castrated or intact animals at different developmental periods.

CASTRATION AND OVARIECTOMY

Reduction of levels of sex steroids is most commonly obtained by removal of the organs that serve as the major source of these hormones—the gonads.

Castration of domestic animals (e.g., oxen and horses) with the aim of facilitating their handling and utilization for labor has been practiced since antiquity. Surprisingly, only a few systematic studies have been carried out until recently to quantify those changes in aggressiveness or social rank following removal of the gonads. It has been established that castrated males can occupy high ranks, as in the case of cattle as well as the horse (Tyler, 1972) and reindeer (Espmark, 1964b).

Androgen effects on aggression have been studied by comparing the frequency of aggressive interactions among males of a herd composed of equal numbers of bulls and steers. Bulls were sterilized by a surgically induced cryptorchidism or by surgical sectioning of the epididymis, and their behavior was followed longitudinally between 9 and 18 months of age. It was found that bulls are socially more active than steers and initiate more aggressive encounters (Hinch, 1978, 1980; Bass, Peterson, Payne, & Jarnet, 1977). And, although the overall level of aggressiveness is higher in bulls than in steers, this difference is not readily apparent until 14 months of age. Between 15 and 18 months, the frequency and intensity of agonistic interactions (especially "head pushing") increases considerably in the bull, and this may contribute to the increased incidence of wounding seen (Hinch, 1980) . In addition, the social distance maintained among bulls is greater than that among steers; in bulls, distancing is observed early on and increased by the presence of females. Finally, bulls tend to dominate steers (Hinch, Thwaites, Lynch, & Pearson, 1982).

The European red deer *(Cervus elaphus)* provides a good example of the endocrine control of social behavior (Lincoln, 1971; Lincoln *et al.,* 1970, 1972), but the picture is complicated by the partial mediation of hormonal influences on social behavior through a secondary sexual characteristic—antler growth. Antler growth is under androgenic control: A young male castrated before puberty will never develop pedicles or antlers, whereas antler growth can be induced in females by the administration of androgens. At the beginning of their growth, antlers are covered by a highly vascular tissue, called "velvet." This initial growth period lasts about four months and ends in August. The velvet is lost with a further increase in testosterone titers, so that antlers are completely free of velvet at the time of rutting, a condition known as "hard-horns." A decrease in testosterone levels is responsible for the casting of antlers the following spring. Importantly, the aggressive behavior of the stag changes according to the status of his antlers: When a stag is in "hard-horns," he uses them as a principal means of attack and defense; when his antlers are casted or in velvet, he uses his forefeet in fighting.

Antler growth is known to influence social structure. During the time when the antlers are in velvet, the hierarchy of male groups is less stable. In addition, the loss of velvet or the loss of the antlers themselves is not a synchronized event among males—old males lose their antlers and velvets earlier than younger males. Thus, during spring and fall transition periods, frequent hierarchical changes can be observed.

Behavioral changes occur abruptly at the casting of antlers or velvets while hormonal change proceeds progressively. Alternately, there is considerable variation in testosterone levels and little behavior change during the period of hard-horns. Taken together, these findings suggest that changes in aggressive behavior and social rank depend directly on antler development and only indirectly on androgen levels. By manipulating hormone levels (by castration or androgen administration) or by nonhormonally altering the appearance of the antlers (by amputation or addition of tines), it has been shown that it is the antlers that mainly govern social interactions. That is, behavior changes when some aspect of the antlers changes, and manipulation of hormonal levels is without effect unless the hormonal change is accompanied by a change in antler aspect.

The effect of castration of adult stags depends on the condition of the antlers at the time of surgical intervention. If antlers are already cleaned (free of velvet), they are cast some weeks after the operation and grow again but do so in an abnormal manner and remain in velvet for the rest of the animal's life. If the antlers are in velvet at the time of castration, they are not cast and remain in velvet. The castrated stags studied by Lincoln *et al.* (1970, 1972) showed a change in their agonistic behavior only when their antlers were cast. These males immediately reverted to the characteristic "velvet behavior," fighting with their forefeet, and they were relegated to the bottom of the hierarchy. They regained a few places in the social order only when other stags of the group came into velvet, but they never regained the position they had held prior to castration.

Information on the maintenance of aggressive behavior following castration is scarce. In one report, a roe deer castrated in January remained aggressive until March but lost its territory in April (Bramley, 1970). Even more scarce is information about the effects of ovariectomy on female aggression. Within a group of cows, ovariectomy had no effect on the frequency of aggressive interactions or on existing dominance relationships (Bouissou, unpublished observations).

Hormonal Treatment of Adults

Exogenous hormones have been administered to intact or castrated animals—by injection, subcutaneous implant, or, more rarely, intracerebral implant—in order to study their effects on sexual and aggressive behavior. The study of the agonistic behavior of ungulates has been made using two groups of steroids: the androgens—testosterone, the nonaromatizable androgen dihydrotesterone (DHT), and 19-hydroxytestosterone (19HT)—and the estrogens—estradiol and estrone.

Males

The effects of steroid treatment are rather variable. According to Lincoln and Davidson (1977), implantation of 200 mg testosterone to intact rams during their period of sexual inactivity does not change their aggression score. On the other hand, a daily intramuscular injection of 15 mg testosterone in rams that had been castrated as adults restored their aggressive behavior and libido within four weeks. Administration of 30 mg testosterone induced extreme aggressiveness without increasing libido. The daily administration of 5 mg testosterone, 15 mg DHT, or 100 μg estradiol had only a slight effect if given alone. However, using a combined therapy of testosterone or DHT with estradiol, a considerable increase in aggressiveness was observed, thus indicating a synergistic action of these hormones (Mattner, 1976). However, almost all rams castrated on the day after birth exhibited aggression in response to adult treatment with estrogen, 19HT, DHT, or testosterone (Parrott, 1978).

A territorial roe buck implanted with 100 mg testosterone in November (a time after the antlers had been cast) did not exhibit altered aggressiveness, and the size of its claimed home range did not change. In fact, the following year he even lost his territory. A young nonterritorial male did not exhibit any aggressiveness and did not acquire any territory after implantation of 1 g testosterone in April (Bramley, 1970). Androgen treatment was also without effect in the kob, but neither the dose nor the experimental conditions were indicated (Leuthold, 1966).

Implantation of 2 g testosterone induced increased aggressiveness in both castrated and intact stags for the duration of two months following treatment. The males also underwent a change in their social rank, but in the case of intact stags, this seemed to be related to a change in the status of their antlers. Testosterone implantation in December induced rutting behavior in castrated stags, whereas the same treatment given in April and June had no effect until the next breeding season, when both intact and castrated males exhibited rutting. This suggests that testosterone plays a permissive role in the control of sexual behavior, but is only effective during certain seasons. Conversely, androgens have a direct effect on aggressive behavior, apparently producing dose-dependent changes. Furthermore, androgens affect the mode of aggression (Lincoln *et al.*, 1972).

The administration of estrogen (estradiol-17β) produces effects that vary according to whether the males are intact or castrate: Castrated males exhibited a short period of rut one to two months following estrogen implantation, with increased aggressiveness toward other males, and these effects were not related to the season. Estrogen-treated males lost their velvet and remained in hard-horns the rest of their lives, providing them with a temporary advantage in agonistic encounters over untreated males that had cast their antlers or were in velvet. Treated animals eventually recovered their initial rank when other males lost their velvet. Intact males responded to the implant with continuous roaring until their deaths more than two years later (Fletcher, 1978).

Females

The effects of androgen treatment of ovariectomized and intact females have been studied in Cervidae and domestic ungulates (cow, mare, ewe, and goat). The administration of large doses of testosterone (4.5 g between February and March, then 8 g from September to January) to intact or ovariectomized hinds induced pedicle growth and increased aggressiveness. This treatment induced the hinds to fight in a manner similar to stags in hard-horns, with head-to-head combat. When left untreated, hinds normally use their forefeet (Jaczewski & Krzywinska, 1975). Similar findings have been reported following the administration of estrogen (dose range 0.5–1 g). The behavioral changes produced appear to depend more on the dose administered than on the particular steriod used (Jaczewski, Galka, & Krzywinska, 1979). In addition, testosterone treatment is accompanied by increased rank within the social hierarchy. One of the females treated by Fletcher (1978) climbed to the top of the social hierarchy of females within 70 days following implantation of 1 g testosterone. This female even dominated two males and maintained dominance for seven months after implantation. Two other treated females gained four and two places respectively. In all cases, the hinds showed threat and fight postures typical of rutting stags. Fletcher (1974) noted that these hierarchical changes were not accompanied by an increase in the number or intensity of agonistic encounters. Rather, the modifications in social rank seemed to result from the testosterone-treated hinds' refusal to submit to the threats of dominant animals.

Similar results have been obtained in the cow and mare (Figure 7). In cattle, the administration of androgen or estrogen to ovariectomized or intact heifers led to a complete restructuring of the hierarchy. In some cases, the absolute

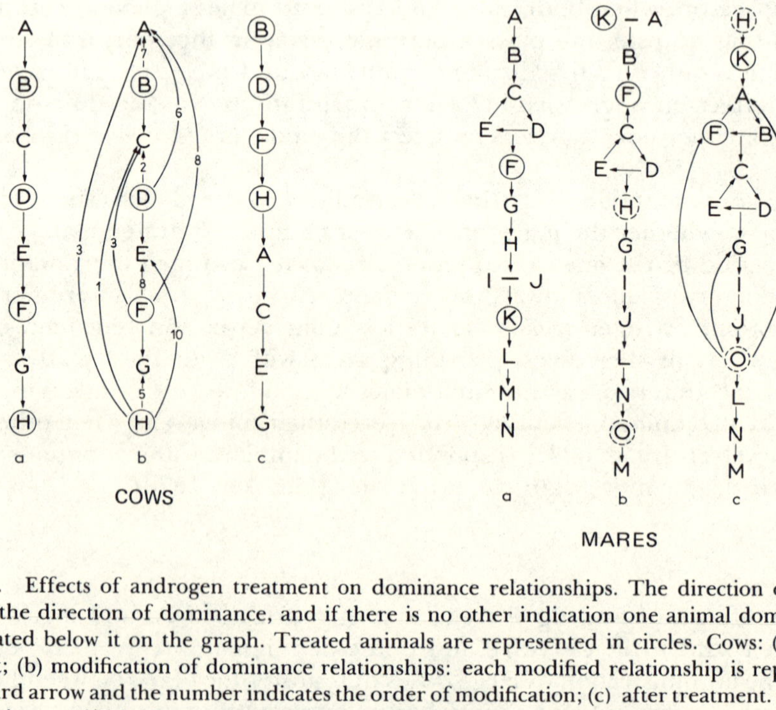

FIGURE 7. Effects of androgen treatment on dominance relationships. The direction of arrows indicates the direction of dominance, and if there is no other indication one animal dominates all those located below it on the graph. Treated animals are represented in circles. Cows: (a) before treatment; (b) modification of dominance relationships: each modified relationship is represented by a upward arrow and the number indicates the order of modification; (c) after treatment. (Adapted from Bouissou, 1978.) Mares: (a) before treatment; (b) after treatment of K and F, but before treatment of H and O; (c) after treatment of H and O. I–J and K–A, unknown relationships. (Adapted from Cougouille-Gauffreteau, 1982.)

frequency of agonistic interactions was increased by steroid treatment, but this increased incidence was observed among treated animals as well as untreated controls within the group. This generalized increase in aggressiveness may have resulted from the disturbance caused by females in permanent behavioral estrus following androgen or estrogen treatment. As a matter of fact, when compared with controls, androgen-treated females appeared to be less aggressive toward untreated strange animals introduced into their group. In treated animals, a reduction in submissive behavior toward hierarchical superiors was noticed. That is, treated animals reacted far less frequently to aggression by taking flight, the normal submissive response (Bouissou, 1978). When one group of females was treated with androgen for one month and then exposed to an equal number of untreated animals, the proportion of attacks received by treated heifers to which they did not respond submissively ranged around 65% while the proportion in control animals was 3.4% (Bouissou, unpublished observations).

In the mare, as in the cow, androgen treatment did not increase the number of aggressive interactions (bunts or threats). However, treated females became more efficient at winning in their aggressive encounters with time, while untreated females became less effective. Treated animals gained rank in the hierarchy and sometimes climbed to the top. A change in the mode of aggression was also observed—treated mares reared on their hind legs and used their forefeet to attack, like stallions. In addition, androgenized mares began to assume some of the social roles of the stallion, exhibiting "herding behavior," characteristic of a male gathering females into a harem (Cougouille-Gauffreteau, Jussiaux, & Trillaud, 1981).

Both the incidence and intensity of aggressive interactions in pairs of goats treated with 180–900 mg of testosterone propionate increased, and dose-dependent effects were evident. An increase in "assertive" behavior was also observed in the subordinate animal during direct competition (Barker & Bosu, 1980).

Pregnant ewes have been induced to exhibit increased aggressiveness if testosterone is implanted before day 70 of gestation (Clarke, Scaramuzzi, & Short, 1976). Ovariectomized ewes treated with androgen were more aggressive toward other females and the ram, and they frequently assumed male-typical roles in the herd (C. Fabre, personal communication, 1980).

Among free-martin heifers, the administration of estradiol or testosterone between 50 and 55 weeks of age, or the administration of a combined regime of estradiol or testosterone plus testosterone-oenanthate between 68 and 73 weeks of age, produced an increase in agonistic behavior. Animals treated with estradiol or testosterone were more frequently involved in aggressive interactions and tended to be dominant. On the other hand, DHT had no effect (Greene, Mogil, & Foote, 1978).

In females, the results of hormonal treatment are more clear cut than in males, since all studies show an effect of sex steroids (androgen and estrogen) on social behavior. In some instances, steroids directly modify aggression. However, it seems that the behavioral effects on rank and social role are not due directly to changes in the aggressiveness of treated animals, but rather due to a reduction in the fear experienced toward their superiors or even a certain indifference toward the attacks received from peers. Although the results so far obtained seem to support this hypothesis, the basis for the behavioral changes observed has not been established. In rodents, it has been reported that "fear of being attacked" does not seem to depend on the level of circulating androgens (Leshner, 1975).

Finally, it has been observed that cows, like mares, retain their new hierarchical positions (and the herding behavior seen in the mare) long after administered steroids reach undetectable levels in the blood. This emphasizes the importance of experience and so-called social inertia in considering steroidal effects on social behavior (Bouissou, 1978; Cougouille-Gauffreteau *et al.*, 1981).

Hormonal Treatment during the Fetal or Neonatal Period

The presence of androgen during the critical period of sexual differentiation produces the masculine pattern of gonadotropin secretion and male sexual behavior (Jost, 1965). When androgens are not present during the critical period, central structures develop into a female pattern. Differentiation is a species-specific process in which the critical period may extend to early neonatal life (as in most rodents) or may occur during fetal life (as in sheep).

The effects of hormonal treatment during the perinatal period on aggressive behavior are well documented in rodents (for review, see Brain, 1979a; Bronson & Desjardins, 1971; Leshner, 1975). In male mice, castration on the day of birth prevents the appearance of aggressiveness in adulthood, even following androgen replacement therapy in adulthood. Neonatal treatment of castrated males and females with androgen or estrogen renders them responsive to the aggression-promoting properties of androgen in adulthood, whereas untreated controls are unresponsive. Similarly, in primates, it has been shown that female rhesus monkeys that had been masculinized by prenatal androgen treatment showed anatomical modifications as well as increased aggressiveness in early and adult life (reviewed by Dixon, 1980). Thus, it appears that steroids organize central structures involved in the control of aggressive behavior during the perinatal period.

Masculinization of some aspects of the behavior of females by androgen treatment *in utero* has also been shown in sheep. Implantation of 1 g testosterone into ewes between days 30 and 120 of gestation caused total or partial masculinization of the external genitalia and the micturation posture of female progeny (Clarke *et al.,* 1976). As adults, these females exhibited masculine patterns of sexual behavior as well as a higher frequency of attacking. This was especially evident among those females exposed to androgen between days 50 and 120 of intrauterine life. During prenatal life, testosterone affects sexual differentiation of the genitalia between days 40 and 50, whereas behavior seems to be affected later (Clarke, 1977a).

In species that carry multiple fetuses to term (e.g., pigs and some breeds of sheep), the question may be raised as to whether the behavior of females can be modified by androgens secreted by the testes of male fetuses. It has already been shown in mice that the position of female fetuses relative to male fetuses *in utero* affects not only their morphology but also their behavior (Gandelman, Vom Saal, & Reinisch, 1977; Vom Saal & Bronson, 1978). This question has received little attention in ungulates, although it has been established that fetal male sheep secrete large amounts of androgen early in development (Attal, 1969). Along these lines, Sachs and Harris (1978) did not find any behavioral differences in young female lambs, comparing twin sisters of a male with twin sisters of a female.

The most commonly cited example of masculinization of female fetuses is the phenomenon known as the "free-martin" effect in cattle. As many as 92% of those females having a twin brother exhibit various degrees of masculinization and are usually sterile. Many studies have been devoted to the physiology of these animals (Marcum, 1974), whereas only a few investigations have been made of their behavior and of these, most have neglected comparisons with normal heifers. Preliminary results seem to show that free-martin heifers do not differ from control heifers with respect to their agonistic behavior and they do not attain a higher position in the social hierarchy (Bouissou & Gaudioso, unpublished observation).

With respect to the male, Bass *et al.* (1977) have attempted to alter the aggressiveness of bulls by neonatal estrogen treatment. At the age of four and six months, intact bulls were compared with castrated animals and to males that had received 6 mg estradiol benzoate on the day of birth, followed by 3 mg/day for 20 days. At either age, there was no difference in circulating testosterone in intact bulls and estrogen-treated bulls. Apart from a slight "quietening" effect at four months, there was no difference in levels of aggression. However, both intact and estrogen-treated bulls were much more socially active than castrated animals.

CONCLUSION

In concluding this review, we must admit that our knowledge of the influence of hormones on the aggressive behavior of ungulates is somewhat fragmentary, despite the fact that the endocrine systems of various domestic species are well-known. The sex steroids have been studied almost exclusively and even in those studies, methodological differences and the low number of investigations could account for discrepancies in the results obtained. As a matter of fact each study almost represents a unique case: The species, the particular breed, the steroids used, and their dose and route of administration vary as widely as do the types of responses and their measurement. Thus, it is difficult to survey this problem and arrive at generalizable principles.

It may be concluded, however, that aggressive behavior of the male depends on the presence of androgens. Similarly, in all species studied, androgen treatment of females produced changes in social rank with or without changes in aggressiveness. In some cases, social rank changed without any apparent change in aggressiveness, and the hypothesis of fear-reduction awaits experimental validation.

In some species, estrogens are also capable of increasing and restoring aggression in intact and castrated males and may alter the aggressiveness or social rank of females. The nonaromatizable androgen DHT is of little or no effect. It has been suggested that estrogen is the active metabolite in the central nervous system and is ultimately responsible for adult sexual behavior (Naftolin, Ryan, & Petro, 1971; Ryan, Naftolin, Reddy, Flores, & Petro, 1972). The hypothesis of testosterone aromatization has also been applied to intermale aggression in rodents (Brain, 1979a) and may explain some of the findings in ungulates.

As in many other species, androgens may act during fetal life in ungulates to direct the differentiation of those structures responsible for aggressive behavior. Until now, very few studies have been conducted on this subject. Finally, almost no data are available on the role of pituitary - adrenal hormones in aggression and the consequences of aggressive behavior for hormonal levels.

Ungulates, especially the domestic species of this group, provide a unique opportunity for studying hormonal influences on behavior. First, their physiology is well known and their physical size would facilitate various experimental interventions. Second, ungulates provide a diversity in social structures not seen among rodents commonly used in experimental work. Many species live permanently in groups with a social hierarchy. The social roles of males and females differ considerably, and some species have sexually dimorphic patterns of aggression. Therefore, ungulates provide an interesting and useful model for studying aggression and its relationship to social dominance as well as the physiologic bases for both.

REFERENCES

Altmann, M. Group dynamics in Wyoming moose during the rutting season. *Journal of Mammalogy,* 1959, *40,* 420–424.

Alvarez, F., Braza, F., & Norzagaray, A. Estructura social gel gamo (*Dama dama,* Mammalia, Cervidae) en Doñana. *Ardeola,* 1975, *21,* 1119–1142. (a)

Alvarez, F., Braza F., & Norzagaray, A. Etograma cuantificado del gamo *(Dama dama)* en libertad, Doñana. *Acta Vertebrata,* 1975, *2,* 93–142. (b)

Arave, C. W., Mickelsen, C. H., Lamb, R. C., Sevejda, A. J., Canfield, R. V. Effects of dominance rank changes age and body weight on plasma corticoids of mature dairy cattle. *Journal of Dairy Science,* 1976, *60,* 244–248.

Archambeaud, B. Comportement sexuel et social des chevaux: Mémoire de fin d'études. Unpublished report, 1972, ENSA, Dijon, France.

Asa, C. S., Goldfoot, D. A., & Ginther, O. J. Sociosexual behavior and the ovulatory cycle of ponies *(Equus caballus)* observed in harem groups. *Hormones and Behavior,* 1979, *13,* 49–65.

Attal, J. Levels of testosterone, androstenedione, oestrone and oestradiol 17β in the testis of foetal sheep. *Endocrinology,* 1969, *85,* 280–289.

Attal, J. Mesure des oestrogènes et des androgènes testiculaires et plasmatiques dans l'espèce ovine par des microméthodes de chromatographie en phase gazeuse: influence de l'âge, de la saison et du cycle diurne. Unpublished doctoral thesis, Université de Paris, 1970.

Banks, E. M. Some aspects of sexual behavior in domestic sheep, *Ovis aries*. *Behaviour,* 1964, *23,* 249–279.

Barker, C. A. V., & Bosu, W. T. K. Studies on experimentally androgenized female goats. *9th Congress on Animal Reproduction and Artificial Insemination* 1980, *2,* 593–599.

Bass, J. J., Peterson, A. J., Payne, E., & Jarnet, M. P. The effect of neonatal estrogen treatment on plasma hormone levels and behavior in pre and post-pubertal bulls. *Theriogenology,* 1977, *8,* 59–71.

Bekoff, M. Social play and play soliciting by infant canids. *American Zoologist,* 1974, *14,* 323–340.

Bergerud, A. T. Rutting behavior of Newfoundland caribou. In V. Geist & F. Walther (Eds.), *The behavior of ungulates and its relation to management* (Vol. 1). Morges, Switzerland: IUCN Publications (No. 24), 1974.

Bouissou, M.-F. Interactions des motivations sexuelles et agressives chez les Bovins domestiques. *Comptes Rendus de l'Académie des Sciences (Paris),* 1975, *280*(D), 121–124.

Bouissou, M.-F. Effect of injections of testosterone propionate on dominance relationships in a group of cows. *Hormones and Behavior,* 1978, *11,* 388–400.

Bouissou, M.-F., & Signoret, J. P. Cattle behaviour under modern management techniques. *The Farm Building Association Journal,* 1971, pp. 25–28.

Brain, P. F. Effects of the hormones of the pituitary-gonadal axis on behavior. In R. Brown & S. J. Cooper (Eds.), *Chemical influences on behavior.* New York: Academic Press, 1979.(a)

Brain, P. F. Effects of the hormones of the pituitary-adrenal axis on behavior. In R. Brown & S. J. Cooper (Eds.), *Chemical influences on behavior.* New York: Academic Press, 1979. (b)

Bramley, P. S. Territoriality and reproductive behavior of roe deer. *Journal of Reproduction and Fertility,* 1970, *11,* (Suppl.) 43–70.

Bramley, P. S. & Neaves, W. B. The relationship between social status and reproductive activity in male impala *(Aepyceros melampus). Journal of Reproduction and Fertility,* 1972, *31,* 77–81.

Bronson, F. H., & Desjardins, C. Steroid hormones and aggressive behavior in mammals. In B. Eleftheriou & J. P. Brain (Eds.), *The physiology of aggression and defeat.* New York: Plenum Press, 1971.

Brown, B. A. Social organization in male groups of white-tailed deer. In V. Geist & F. Walther (Eds.), *The behavior of ungulates and its relation to management* (Vol. 1). Morges, Switzerland: IUCN Publication (No. 24), 1974.

Bubenik, G. A., Bubenik, A. B., Brown, G. M., & Wilson, D. A. Sexual stimulation and variations of plasma testosterone in normal, antiandrogen and antiestrogen treated white-tailed deer *(Odocoileus virginianus)* during the annual cycle. *XIII Congress of Game Biologists,* 1977, 377–386. (a)

Clarke, I. J. The sexual behavior of prenatally androgenized ewes observed in the field. *Journal of Reproduction and Fertility,* 1977, *49,* 311–315. (a)

Clarke, I. J. The sexual and aggressive behavior of normal and masculinized ewes chronically treated with oestradiol 17 B(E_2) or testosterone (T). *Theriogenology,* 1977, *8,* 152. (b)

Clarke, I. J., Scaramuzzi, R. J., & Short, R. V. Effects of testosterone implants in pregnant ewes on their female offspring. *Journal of Embryology and Experimental Morphology,* 1976, *36,* 87–99.

Clutton-Brock, T. H., Albon, S. D., Gibson, R. M., & Guinness, F. E. The logical stag: Adaptive aspects of fighting in red deer *(Cervus elaphus L.). Animal Behavior,* 1979, *27,* 211–225.

Collias, N. E. Aggressive behavior among vertebrate animals. *Physiological Zoology,* 1944, *17,* 83–123.

Conner, R. L. Hormones, biogenic amines, and aggression. In S. Levine (Ed.), *Hormones and behavior.* New York: Academic Press, 1972.

Conner, R. L., & Levine, S. Hormonal influences on aggressive behavior. In S. Garrattini & E. B. Sigg (Eds.), *Aggressive behavior.* Amsterdam: Excerpta Medica, 1969.

Cougouille-Gauffreteau, B. *Etude longitudinale de l'évolution de la structure sociale d'un groupe de juments: Effets de l'androgénisation de certaines d'entre elles sur le comportement social et sexuel.* Unpublished doctoral thesis, Université d'Aix et de Marseille II, 1982.

Cougouille-Gauffreteau, B., Jussiaux, M., & Trillaud, C. Etude, après arrêt du traitement, des conséquences de l'injection d'hormones mâles à des juments, sur leur comportement social et sur leur position hiérarchique. *Comptes Rendus de l'Académie des Sciences* (Paris), 1981, *292,* 1073–1076.

Dantzer, R., & Mormède, P. Le stress en élevage intensif. In *Actualités scientifiques de l'INRA.* Paris: Masson, 1979.

Davant, J. H., Han, D. K., & Moody, E. L. Dominance in rams in relation to serum testosterone. *Journal of Animal Science,* 1974, *38,* 1333–1334.

Dixon, A. F. Androgens and aggressive behavior in primates. A review. *Aggressive Behavior,* 1980, *6,* 37–67.

Endroczi, E., Lissak, L., & Telegdy, G. Influence of sexual and adrenocortical hormones on the maternal aggressivity. *Acta Physiologica Academiae Scientiarum Hungaricae,* 1958, *14,* 353–357.

Espmark, Y. Rutting behavior in reindeer (*Rangifer tarandus* L.). *Animal Behaviour,* 1964, *12,* 159–163. (a)

Espmark, Y. Studies in dominance-subordination relationship in a group of semi-domestic reindeer (*Rangifer tarandus* L.). *Animal Behaviour,* 1964, *12,* 420–425. (b)

Fletcher, T. J. *The environmental and hormonal control of reproduction in male and female red deer (Cervus elaphus, L).* Unpublished doctoral thesis, University of Cambridge, 1974.

Fletcher, T. J. The induction of male sexual behavior in red deer *(Cervus elaphus)* by the administration of testosterone to hinds and estradiol 17β to stags. *Hormones and Behavior,* 1978, *11,* 74–88.

Fraser, A. F. *Reproductive behaviour in ungulates.* New York: Academic Press, 1968.

Gandelman, R., von Saal, F. S., & Reinisch, J. M. Contiguity to male foetuses affects morphology and behaviour of female mice. *Nature,* 1977, *266,* 722–724.

Geist, V. On the behaviour of the North American moose in British Columbia (*Alces alces andersoni,* Petersen 1950). *Behaviour,* 1963, *20,* 377–416.

Geist, V. On the rutting behaviour of the mountain goat. *Journal of Mammalogy,* 1964, *45,* 551–568.

Geist, V. On the inter-relation of external appearance, social behaviour and social structure of mountain sheep. *Zeitschrift für Tierpsychologie,* 1968, *25,* 199–215.

Geist, V. Behavior during the rutting season. In G. B. Schaller (Ed.), *Wildlife behavior and ecology series: Mountain sheep.* Chicago: University of Chicago Press, 1971.

Goy, R. W., & Phoenix, C. H. The effects of testosterone propionate administered before birth of the development of behavior in genetic female rhesus monkeys. In C. Sawyer & R. Gorki (Eds.), *Steroids hormones and brain function.* Berkeley: University of California Press, 1971.

Greene, W. A., Mogil, L., & Foote, R. H. Behavioral characteristics of Freemartins administered estradiol, estrone, testosterone, and dihydrotestosterone. *Hormones and Behavior,* 1978, *10,* 71–84.

Grubb, P. The rut and behaviour of Soay rams. In P. A. Jewell, C. Milner, & J. Morton-Boyd (Eds.), *Island survivors: The ecology of Soay sheep of St. Kilda.* London: Athlone Press, 1974.

Grubb, P., & Jewell, P. A. The rut and occurrence of oestrus in the Soay sheep on St. Kilda. *Journal of Reproduction and Fertility,* 1973, *19,* (Suppl.) 491–502.

Hart, B. L. Gonadal androgen and socio-sexual behavior of male mammals: A comparative analysis. *Psychological Bulletin,* 1974, *81,* 383–400.

Hinch, G. N. Social behaviour of young partially-castrated bulls and steers related to their management. *Proceedings of the Australian Society of Animal Production,* 1978, *12,* 265.

Hinch, G. N. Handling strategies for entire and castrated male cattle. In M. Wodzicka-Tomaszewska, T. N. Edey, & J. J. Lynch (Eds.), *Reviews in rural sciences (no. 4): Behaviour in relation to reproduction, management and welfare of farm animals.* Australia: University of New England, 1980.

Hinch, G. N., Thwaites, C. J., Lynch, J. J., & Pearson, A. J. Spatial relationships within a herd of young sterile bulls and steers. *Applied Animal Ethology,* 1982, *8,* 27–44.

Hulet, C. V., Alexander, G., & Hafez, E. S. E. The behaviour of sheep. In E. S. E. Hafez (Ed.), *The behaviour of domestic animals.* London: Baillière, Tindall, 1975.

Hurnik, J. F., King, G. J., & Robertson, H. A. Behavioral symptoms of estrus in post-partum Holstein cows. *Canadian Journal of Animal Science,* 1975, *55,* 786. (a)

Hurnik, J. F., King, G. J., & Robertson, H. A. Estrous and related behaviour in post-partum Holstein cows. *Applied Animal Ethology,* 1975, *2,* 55–68. (b)

Illius, A. W., Haynes, N. B., & Lamming, G. E. Effects of ewe proximity on peripheral plasma testosterone levels and behaviour in the ram. *Journal of Reproduction and Fertility,* 1976, *48,* 25–32.

Jaczewski, Z., & Krzywinska, K. The effect of testosterone on the behaviour of castrated females of red deer (*Cervus elaphus* L.) *Prace i Materialy Zooteckniczne,* 1975, *8,* 37–44.

Jaczewski, Z., Galka, B., & Krzywinski, A. Effect of sex hormones on the behavior of red deer. *Acta Physiologica Polonica,* 1979, *30,* 81–82.

Jainudeen, M. R., Katongole, C. B., & Short, R. V. Plasma testosterone levels in relation to musth and sexual activity in the male Asiatic elephant *(Elephas maximus). Journal of Reproduction and Fertility.* 1972, *29,* 99–103.

Jainudeen, M. R., McKay, G. M., & Eisenberg, J. F. Observations on musth in the domesticated Asiatic elephant *(Elephas maximus). Mammalia,* 1972, *36,* 247–261.

Jarman, M. V. Impala social behaviour, territory, hierarchy, mating and the use of space. *Advances in Ethology,* 1979, *21,* 1–92.

Jost, A. Gonadal hormones in the sex differentiation of the mammalian foetus. In R. L. De Haan & H. Ursprung (Eds.), *Organogenesis.* New York: Holt, Rinehart & Winston, 1965.

Katongole, C. B., Naftolin, F., & Short, R. V. Relationship between blood levels of luteinizing hormone and testosterone in bulls, and the effects of sexual stimulation. *Journal of Endocrinology,* 1971, *50,* 457–466.

Katongole, C. B., Naftolin, F., & Short, R. V. Seasonal variations in blood luteinizing hormone and testosterone levels in rams. *Journal of Endocrinology,* 1974, *60,* 101–106.

Kirkpatrick, J. F., Wiesner, L., Kenney, R. M., Ganjam, V. K., & Turner, J. W. Seasonal variation in plasma androgens and testosterone in the North American wild horse. *Journal of Endocrinology,* 1977, *72,* 237–238.

Kitchen, D., & Bromley, P. T. Agonistic behaviour of territorial pronghorn bucks. In V. Geist & F. Walther (Eds.), *The behaviour of ungulates and its relation to management* (Vol. 1). Morges, Switzerland: IUCN Publications (No. 24), 1974.

Klingel, H. Das Verhalten der Pferde (Equidae). *Handbuch der Zoologie VIII,* 1972, *10,* 1–66.

Leidl, W., Hoffman, D. & Karg, H. Endokrine Regulation und Jahreszeitlicher Rythms der Fortpflanzung beim Zeigenbock. *Zentralblatt für Veterinarmedecine,* 1970, *17,* 623–633.

Leshner, A. I. A model of hormones and agonistic behavior. *Physiology and Behavior,* 1975, *15,* 225–235.

Leuthold, W. Variations in territorial behavior of Uganda kob (*Adenota kob* Thomasi). *Behaviour,* 1966, *27,* 214–257.

Levasseur, M. C., & Thibault, C. De la puberté a la sénescence: La fécondité chez l'homme et les autres mammifères. *Actualités Scientifiques de l'INRA.* Paris: Masson, 1980. (a)

Levasseur, M. C., & Thibault, C. Reproductive life cycles. In E. S. E. Hafez (Ed.), *Reproduction in farm animals* (4th ed.). Philadelphia: Lea & Febiger, 1980. (b)

Lincoln, G. A. The seasonal reproductive changes in the red deer stag *(Cervus elaphus). Journal of Zoology,* 1971, *163,* 105–123.

Lincoln, G. A. The role of antlers in the behaviour of red deer. *Journal of Experimental Zoology,* 1972, *182,* 233–245.

Lincoln, G. A., & Davidson, W. The relationship between sexual and aggressive behaviour and pituitary and testicular activity during the seasonal sexual cycle of rams, and the influence of photoperiod. *Journal of Reproduction and Fertility,* 1977, *49,* 267–276.

Lincoln, G. A., Youngson, R. W., & Short, R. V. The social and sexual behaviour of the red deer stag. *Journal of Reproduction and Fertility,* 1970, *11,* (Suppl.) 71–103.

Lincoln, G. A., Guinness, F., & Short, R. V. The way in which testosterone controls the social and sexual behavior of the red deer stag *(Cervus elephus). Hormones and Behavior,* 1972, *3,* 375–396.

Liptrap, R. M., & Raeside, J. I. A relationship between plasma concentrations of testosterone and corticosteroids during sexual and aggressive behaviour in the boar. *Journal of Endocrinology,* 1978, *76,* 75–85.

Lott, D. F. Sexual and aggressive behaviour of American bison, *Bison bison.* In V. Geist & F. Walther (Eds.), *The behaviour of ungulates and its relations to management* (Vol. 1). Morges, Switzerland: IUCN Publications (No. 24), 1974.

Marcum, J. B. The Freemartin syndrome. *Animal Breeding Abstracts,* 1974, *42,* 227–242.

Mattner, P. E. Effect of androgens and oestradiol on libido and aggressiveness in rams castrated as adults. *Theriogenology,* 1976, *6,* 612.

McHugh, T. Social behavior of the American buffalo *(Bison bison bison). Zoologica,* 1958, *43,* 40.

Michael, E. D. Aggressive behavior of white-tailed deer. *The Southwestern Naturalist,* 1968, *13,* 411–420.

Monfort-Braham, N. Variations dans la structure sociale du topi (*Damaliscus korrigum* Ogilby), au Parc National de l'Akagera, Rwanda. *Zeitschrift für Tierpsychologie,* 1975, *39,* 332–364.

Moy, R. F. Histology of the subauricular and rump glands of the pronghorn (*Antilocapra americana* Ord.). *American Journal of Anatomy,* 1970, *129,* 65–88.

Moyer, K. E. Kinds of aggression and their physiological basis. *Communications in Behavioral Biology,* 1968, *2,* 65–87.

Naftolin, F., Ryan, K. J., & Petro, Z. Aromatization of androstenedione by the diencephalon. *Journal of Clinical Endocrinology and Metabolism,* 1971, *33,* 368–370.

Nievergelt, B. A comparison of rutting behaviour and grouping in the Ethiopian and Alpine ibex. In V. Geist & F. Walther (Eds.), *The behaviour of ungulates and its relation to management* (Vol. 1). Morges, Switzerland: IUCN Publications (No. 24), 1974.

Oeming, A. A herd of musk-oxen, Ovibos moschatus, in captivity. *International Zoo Yearbook,* 1965, 58–65.

Parrott, R. F. Courtship and copulation in prepuberally castrated male sheep (Wethers) treated with 17β-estradiol, aromatizable androgens, or dihydrotestosterone. *Hormones and Behavior,* 1978, *11,* 20–27.

Phoenix, C. H., Goy, R. W., & Resko, J. A. Psychosexual differentiation as a function of androgenic stimulation. In M. Diamond (Ed.), *Perspectives in reproduction and sexual behavior.* Bloomington, Ind.: Indiana University Press, 1968.

Reinhardt, V., Mutiso, F. M., & Reinhardt, A. Social behavior and social relationships between female and male prepubertal bovine cattle *(Bos indicus). Applied Animal Ethology,* 1978, *4,* 43–54.

Richards, S. M. The concept of dominance and methods of assessment. *Animal Behaviour,* 1974, *22,* 914–930.

Robbel, H., & Child, G. Notes on territorial behaviour in lechwe. *Mammalia,* 1975, *39,* 707–709.

Rouger, Y. *Etudes des interactions de l'environnement et des hormones sexuelles dans la régulation du comportement sexuel des Bovidae.* Unpublished doctoral thesis, Université de Rennes, France, 1974.

Rowell, T. E. The concept of social dominance. *Behavioral Biology,* 1974, *11,* 131–154.

Ryan, K. J., Naftolin, F., Reddy, V., Flores, F., & Petro, Z. Estrogen formation in the brain. *American Journal of Obstetrics and Gynecology,* 1972, *114,* 454–460.

Sachs, B. D., & Harris, V. S. Sex differences and developmental changes in selected juvenile activities (play) of domestic lambs. *Animal Behaviour,* 1978, *26,* 678–684.

Scheurmann, E., & Jainudeen, M. R. "Musth" beim Asiatischen Elefanten *(Elephas maximus). Zoologische Garten Leipzig,* 1972, *42,* 131–142.

Schoen, A. M. S., Banks, E. M., & Curtis, S. E. Behavior of young Shetland and Welsh ponies *(Equus caballus). Biology of Behaviour,* 1976, *1,* 192–216.

Sempéré, A., Garreau, J. J., & Boissin, J. Variations saisonnières de l'activité de marquage territorial et de la testostéronémie chez le Chevreuil male adulte (*Capreolus, capreolus,* L.). *Comptes Rendus de l'Academie des Sciences (Paris),* 1980, *290*(D), 803–806.

Sereni, J. L. *Recherches sur l'évolution de la structure sociale et du comportement alimentaire d'un groupe de chevaux camarguais.* Unpublished doctoral thesis, Université d'Aix et de Marseille, France, 1977.

Short, R. V., & Mann, T. The sexual cycle of a seasonally breeding mammal, the roebuck *(Capreolus capreolus). Journal of Reproduction and Fertility,* 1966, *12,* 337–351.

Sigg, E. B. Relationship of aggressive behaviour to adrenal and gonadal function in male mice. In S. Garrattini & E. B. Sigg (Eds.), *Aggressive behaviour.* Amsterdam: Excerpta Medica, 1969.

Strauss, J. S., & Ebling, F. J. Control and function of skin glands in mammals. *Memoirs of the Society for Endocrinology,* 1970, *18,* 341–371.

Syme, G. J. Competitive orders as measures of social dominance. *Animal Behaviour,* 1974, *22,* 931–940.

Thoman, E. B., Conner, R. L., & Levine, S. Lactation suppresses adrenal corticosteroid activity and aggressiveness in rats. *Journal of Comparative and Physiological Psychology,* 1970, *70,* 364–369.

Tyler, S. J. The behaviour and social organization of the New Forest ponies. *Animal Behaviour Monographs,* 1972, *5,* 87–196.

Vom Saal, F. S., & Bronson, F. H. *In utero* proximity of female mouse fetuses to males: Effects on reproductive performances during later life. *Biology of Reproduction,* 1978, *19,* 842–853.

Wilkinson, P. F., & Shank, C. C. Rutting fight mortality among musk-oxen on Banks Island, Northwest Territories, Canada. *Animal Behaviour,* 1977, *24,* 756–758.

Yagil, R., & Etzion, Z. Hormonal and behavioural patterns in the male camel *(Camelus dromedarius). Journal of Reproduction and Fertility,* 1980, *58,* 61–65.

20

The Interaction of Hormones, Behavior, and Social Context in Nonhuman Primates

Irwin S. Bernstein, Thomas P. Gordon, and Robert M. Rose

INTRODUCTION

Hormones are often regarded as one independent variable responsible for the behavior (the dependent variable) of an individual. Hormones are seen as the physiological mechanism underlying behavior and are treated as a "cause" of certain behavioral patterns—"behavioral patterns," because hormonal levels are not viewed as stimuli that release specific fixed action patterns but as the regulators of motivational states that increase or decrease the probabilities of specific classes of response to environmental stimuli.

Frank Beach (1965), however, indicated that the relationship between hormones and behavior was more complicated than a simple cause-and-effect sequence: A feedback loop was involved such that behavior also influences hormones. The life experiences of an individual may profoundly influence levels of circulating hormones, and these experiences may be entirely cognitive. Changes in circulating levels of hormones may then, in turn, alter the probability that the individual will engage in the very types of interaction that were responsible for the previous experiences that altered hormonal levels.

Hormones may therefore be seen as one of the regulators (Wilson, 1975) that modify an animal's activities and adapt them to particular situations within some time frame. Since the best prediction of a future outcome is based on the history of outcomes of similar past situations, the alteration of hormones as a function of experience, which then alters the propensity to engage in the same kind of behavior in the future, may be seen as the adaptive significance of the feedback loop system.

IRWIN S. BERNSTEIN • Department of Psychology, University of Georgia, Athens, Georgia 30602, and Yerkes Regional Primate Research Center, Emory University, Atlanta, Georgia 30322. THOMAS P. GORDON • Yerkes Regional Primate Research Center, Emory University, Atlanta, Georgia 30322. ROBERT M. ROSE • University of Texas Medical School Branch at Galveston, Galveston, Texas 77550. The research reported in this paper was supported by ADAMHA grants MH 20483 and MH 13864, and in part by NIH grant RR 00165.

We have directed our research to the study of gonadal hormones in nonhuman primates, primarily the rhesus monkey, *Macaca mulatta,* in order to identify the variables that influence circulating levels of hormone and to study the consequences of such variability for behavior. Our approach has included analyses of ontogenetic (e.g., puberty), cyclic (e.g., seasonal and circadian), and behavioral influences on hormonal levels, and the correlation of hormonal levels with behavioral measures, both with regard to spontaneous and manipulated (e.g., human chorionic gonadotropin-stimulated) hormonal levels. We have tried to identify the specific causal agents of hormonal variation by systematic manipulation of social experience, and we have searched for interaction effects among the identified potent variables. Virtually all of this work has been done within the context of social groups, for although more precise control of the experience of individually caged animals can be attained, it is within the context of the social group that the hormonal–behavioral system of primates has evolved, and where we could best hope to appreciate the functional significance of the mechanisms. In so doing, we recognized that social context is, in itself, an extremely potent variable with regard to behavior, and, as later demonstrated, with regard to circulating levels of gonadal hormones as well.

CIRCADIAN AND SEASONAL TESTOSTERONE CHANGES

Mature male rhesus monkeys exhibit both diurnal and seasonal changes in circulating levels of plasma testosterone, although there are variations in plasma testosterone mean levels in males living under different housing conditions, (Michael & Bonsall, 1977; Perachio, Alexander, Marr, & Collins, 1977; Phoenix, Dixson, & Resko, 1977). In our laboratory, males living in social groups outdoors had breeding-season mean plasma testosterone levels of 12.0 ng/ml at 10:00 A.M., (Gordon, Rose, & Bernstein, 1976). When sampled every 3 hours for 27 hours, the mean level in eight males showed a peak at about 4:00 A.M., a rapid decline at daybreak, and a more or less stable lower level from about 10:00 A.M. to 4:00 P.M.; after sunset, levels began to rise again (Rose, Gordon, & Bernstein, 1978). The pattern of change in testosterone levels was not a perfect correlate of cortisol levels, but cortisol levels were also at the highest around daybreak. This pattern is similar to that reported in other studies with rhesus monkeys (Goodman, Hotchkiss, Karsch, & Knobil, 1974; Michael, Wilson, & Plant, 1973; Michael, Setchell, & Plant, 1974; Perachio *et al.,* 1977) and humans (Nieschlag, 1974).

To determine whether variability in the level of hormone assayed might be accounted for in part by capture and handling, we obtained samples by venipuncture of four awake males five times within 90 minutes, and then 24 hours later. These samples revealed remarkable stability, with virtually no changes in hormonal levels as a consequence of repeated handling and venipuncture (Rose, Gordon, & Bernstein, 1978). We attribute this, at least in part, to the fact that all subjects are trained to the routine and appear to be fully habituated to our handling techniques.

The rhesus monkey is a seasonal breeder, and, in our colony, births and copulatory behavior have been restricted to discrete portions of the year (Bernstein & Gordon, 1977). Indirect evidence from other sources (Coimbra-Filho & Maia, 1977; Conaway & Koford, 1965; Koford, 1965; Sade, 1964; Zamboni, Conaway, & van Pelt, 1974) had suggested that there might be a hormonal seasonal change correlating with the breeding season, and we verified such a change over a three-year study with males living in multiple groups being sampled not less than once per month (Gordon *et al.*, 1976). The samples were all collected within 1 hour of 1000 hours in order to control for the circadian rhythm. Seasonal variation was found to be about the same magnitude as the circadian variation. All of the males living in heterosexual groups showed more or less synchronous rises from a mean of about 2.0 ng/ml in the birth season to about 6.0 ng/ml by late summer (a transition level), finally rising to 12.0 ng/ml by the breeding season. (Hereafter all testosterone values should be read as ng/ml.) Subsequent studies revealed that annual changes were produced by a seasonal upward displacement of the entire circadian curve (Rose, Gordon, & Bernstein, 1978). Similar seasonal patterns have also been reported in other laboratories (Michael & Bonsall, 1977; Michael & Zumpe, 1978; Plant, Zumpe, Sauls, & Michael, 1974) and for other species (Coe, Mendoza, Davidson, Smith, Dallman, & Levine, 1978).

These results emphasized the necessity of sampling male androgen levels at a fixed time of day and in accordance with expected seasonal levels. Since our measure had already proven reliable, we next assessed the degree of interindividual variability, which proved so great that we found it advisable to design our studies so that each subject served as his own control. The influence of social experience on gonadal hormones, however, proved so potent in some cases that even differences in group means between treated and control groups were readily demonstrated.

One factor contributing to individual variation was ontogeny. We found that puberty was not a simple all-or-none phenomenon and that mean seasonal levels for males might increase over a period of two or three years during puberty (Rose, Bernstein, Gordon, & Lindsey, 1978). The same period of prolonged changes following puberty have also been identified in chimpanzees *(Pan troglodytes)* (Martin, Swenson, & Collins, 1977) and in pigtail *(Macaca nemestrina)* and stumptail macaque males *(M. arctoides),* although these are not seasonal breeders and show no seasonal androgen rhythms (Bernstein, Rose, Gordon, & Grady, 1979). Snipes, Forest, and Migeon (1969) have previously reviewed the correlations between age and testosterone concentrations in primates. Our studies of adult males were conducted with males at least six years of age, as determined by birth records or dental age on arrival (Gavan & Hutchinson, 1973; Hurme & van Wagenen, 1961).

INFLUENCE OF FEMALES ON MALE TESTOSTERONE LEVELS

In noting the seasonal change in rhesus male androgen levels, we observed that rises to a transitional level occurred before the first copulations were seen, but that the first copulations coincided with the first seasonal maxima. In males living apart from females, those denied visual access did not exceed transitional hormonal levels and showed no change in male–male sexual behavior with season, whereas all-male groups living in visual contact with females showed full seasonal rises and a pattern of male–male copulation that mimicked the pattern seen in heterosexual pairs (Gordon & Bernstein, 1973; Gordon, Bernstein, & Rose, 1978). It was therefore suggested that male behavioral and hormonal responses during the breeding season were dependent on some sensory access to cycling females but that direct physical contact was not necessary. Considering the poor olfactory apparatus of the male rhesus, it seemed unlikely that chemical cues were being transmitted across the 7 m of open space between the groups, especially since an all-male group separated from the same heterosexual group by a 10-cm-wide hollow sheet metal fence did not respond with seasonal hormonal or behavioral changes.

In monitoring female gonadal hormones (estradiol and progesterone, to be specific) and copulatory behavior, we found that most group-living females engaged in copulatory behavior only during the follicular portion of their ovulatory cycles (the luteal phase being distinguished by progesterone peaking). Moreover, females living in the same group did not show synchronous cycling. Some females in fact did not show their first ovulatory cycle of the season until several months after the first females (Gordon, 1981). Females usually conceived (10 of 11) on their first ovulatory cycle when living with intact males, and only when prevented from becoming pregnant (we vasectomized the males) did they show multiple ovulatory cycles throughout the season. Under these conditions, all the females nonetheless ceased ovulatory cycling at a date that would be described as late for the end of the breeding season. Only then did the males in their group show testosterone declines typical of males at the end of the breeding season (Gordon, 1981).

The synchrony of male androgen elevations with the first ovulatory cycle of any female, suggested by these data and the data of Vandenbergh and Drickamer (1974), led us to suspect that females exerted strong influences over male gonadal hormones. In order to determine if any female or only an estrous female would influence a male, we introduced rhesus males to cycling rhesus females during the breeding season and to lactating rhesus females during the birth season. In both seasons, males showed significant testosterone rises on being introduced to females (Bernstein, Rose, & Gordon, 1977; Gordon *et al.*, 1978). Although there was some interaction with season (the males introduced to lactating females in the birth season showed significantly lower elevations), the major influence on the males seemed to be introduction to females per se, independent of their hormonal state or the season. We stess "introduction" because males living with females all year round show much lower androgen levels during the birth season than the introduced males, and the introduced males did not sustain the elevated androgen levels beyond the first week of introduction. The lactating females did not respond sexually to the males and, in fact, the males were generally avoided by these females.

The existence of receptivity, or sexual access, may be hypothesized to be a requirement for sustaining male androgen elevations, but we found that receptive extraspecific females that did copulate with the males not only failed to sustain higher male androgen levels, but failed to stimulate hormonal rises in or out of the rhesus monkey breeding season (Bernstein & Gordon, 1979). Moreover, rhesus males showed sexual responses to receptive pigtail females only during the rhesus monkey breeding season. Pigtail males also failed to respond hormonally to rhesus female sexual partners but, although pigtails are nonseasonal breeders, they did show androgen rises when introduced to pigtail females (Bernstein, Gordon, Rose, & Peterson, 1978). The extraspecific copulatory behavior witnessed in these studies included fully fertile matings resulting in the birth of two hybrids.

Male sexual behavior is therefore not dependent on current androgen levels, for males showed sexual behavior with extraspecific females despite unchanged androgen levels and showed copulatory behavior with receptive conspecific females within minutes of introduction. Changes in male androgen levels followed introductions to conspecific females and were independent of sexual behavior, although sexual activity may or may not have also influenced hormonal levels (Phoenix *et al.,* 1977).

The strong influence of females on male androgen levels was also demonstrated in both caged and group-living males. Testosterone levels in a caged male declined significantly after the removal of his female partner, (from 12.8 to 5.4 in 24 hours). When a single female was introduced to a large group of males, the three monitored males all showed androgen increases within 24 hours, despite the female's lack of receptivity and some male fighting following her introduction. The alpha male went from 3.5 to 10.0, the beta from 3.6 to 4.1, and a third male from 6.9 to 10.0.

When a group of eight females was introduced to an all-male group living in view of females during the breeding season, the additional access to females proved significant only for the alpha male, who went from 8.1 prior to introduction to 10.9 during the time they were present, falling to 5.7 following their removal. Values for two other males showed no significant rise as a function of female presence (7.0, 8.2, and 8.6 for the first male and 13.9, 8.6, and 6.9 for the second male during the respective phases of baseline, female introduction, and removal). The confounding of female visual access and introduction, and social events was further demonstrated when eight rhesus males were introduced to a large group of females in the first part of the breeding season. The males were attacked by the females, and serious fighting broke out among the males. Mean male testosterone levels during baseline, introduction, and followup in this study were 6.3, 4.2, and 5.1, respectively, showing no elevations as a function of introduction.

DEFEAT AND DEPRESSION

The failure of the eight males in the last-mentioned study to show testosterone changes when introduced to females who attacked and defeated them suggested that social defeat was a variable that could be as powerful as sexual stimulation in influencing circulating levels of male androgens. Female talapoin monkeys *(Miopithecus talapoin)* had also been suggested to show hormonal response to stimulation by the opposite sex (Rowell, 1977). Females living in isolation showed sex skin changes in coloration and tumescence within 15 minutes of being caged with a male. Baboon females *(Papio sp.)*, who also showed variation in their menstrual cycles as a function of isolation or exposure to males, were reported to show changes in their reproductive cycles when under social attack or harassment (Rowell, 1970). Dunbar and Dunbar (1977) also suggested that stress influenced the reproductive cycles of *Theropithecus gelada* females.

We began a series of systematic studies of the influence of social defeat on circulating levels of testosterone in male rhesus using brief introductions to groups and monitoring testosterone levels. In the first study, five males were individually introduced to an unfamiliar heterosexual group. As soon as a male crouched passively and ceased resistance to group attack, a state that could be identified within 15 minutes to 2 hours after introduction, he was removed. Hormonal assays revealed profound depressions in circulating levels of testosterone in all of these males (to 10%–15% of baseline) within 24 hours. These males did not show recovery when kept in single cages for followup observation for the next several weeks (Rose, Gordon, & Bernstein, 1972). This hormonal depression did not preclude the males from responding to females immediately on introduction to them, and 24 hours after introductions to females, these males had hormonal levels typical of males (themselves included) from a more typical baseline condition on introduction to females. It is not known if the males with depressed testosterone levels were fertile, but certainly they were not as profoundly influenced by social defeat as were the male tree shrews *(Tupaia glis)* described by von Holst (1977), which became sterile after losing a single fight.

Males living in a heterosexual group attacked and defeated by another group in a merger of two groups also showed significant hormonal declines, despite the constant presence of their original group and the presence of the females from the new group. Once again, hormonal recovery did not take place for several weeks (Rose, Bernstein, & Gordon, 1975). These studies were replicated with the same results in the nonseasonal pigtail monkey (Bernstein, Rose, Gordon, & Grady, 1979). The rapidity of hormonal change following defeat is suggested by measures taken on a rhesus male 150 minutes after introduction, which showed a decline from 14.0 to 10.3.

Further efforts to study the depression of testosterone levels in males introduced to unfamiliar social situations were conducted in the nonbreeding season and included both individual and group-to-group introductions. Although the general behavioral patterns were similar in and out of the breeding season, the much lower nonseasonal baseline levels made detection of significant individual declines difficult to detect. It was only in comparing the directionality of changes over males and conditions that we could demonstrate that social defeat produced a depression of male androgen levels in every season of the year.

Three introductions conducted at the end of the breeding season produced baseline and postintroduction measures of 8.7–4.6, 8.1–6.7, and 9.2–2.4. Three other males caged prior to introductions also showed the same directional changes (6.1–3.1, 8.6–3.2, and 2.4–1.2). Further, the first two males showed some recovery (to 4.5 and 4.7) following removal, despite the general seasonal downward tendency, and the third male showed no further change, with a measure of 0.9.

Hormonal declines were also produced when males were introduced to all-male groups and were defeated (5.0–2.0, 8.2–3.5, and 8.4–0.8 in three cases). In controlling for the effects of simple removal from a familiar social group as opposed to defeat by the new group, seven males were used in a yoked design with seven males introduced to a host group. The control males did not show any consistent response to caging alone, while the introduced males that had scored as showing behavioral depression showed lower hormonal levels following introduction, and four of the five for which recovery measures were available showed returns to baseline in the followup period. These data are summarized in Table I. Thus, while we could not demonstrate significant hormonal changes in each of the 16 males subjected to social defeat, it is significant that all 16 showed lower levels following social defeat.

In examining the interaction among female stimulation, season, and social defeat, we had introduced a group of six males living isolated from females into a heterosexual group at the end of the breeding season. The six males were attacked and defeated by the host group, and we found that although these males began with hormonal levels that were quite low for males at the end of the breeding season (but consistent with their having been isolated from females), whatever stimulatory effect the females had was more than offset by social defeat. Every one of the six males showed a lower hormonal value following introduction than his levels during baseline. Although the net changes were small, six out of six is significant (using a sign test) at $p \leq 0.02$. Data are summarized in Table II.

TABLE I
Testosterone Levels (ng/ml), Social Defeat, and Recovery[a]

Subjects	Baseline	Introduction and defeat	Followup
1	3.6	0.5	2.7
2	6.6	2.7	8.6
3	3.6	3.0	1.3
4	4.1	3.0	5.3
5	12.0	7.5	13.4
6	3.3	1.1	no data
6 Experimentals			
Mean	5.5	3.0	6.3
Standard deviation	3.4	2.5	4.9
7 Controls			
Mean	6.9	6.7	6.3
Standard deviation	4.6	5.3	4.3

[a] Testosterone levels of six experimental subjects and seven controls during baseline measurements, following the introduction of the experimental subjects to a new group and during recovery.

TABLE II
Testosterone Levels (ng/ml) of Males Defeated by a Heterosexual Group[a]

	1	2	3	4	5	6
Mean baseline	4.3	1.8	3.3	5.2	3.1	6.0
24 hours after introduction	3.3	1.7	1.6	3.1	1.6	3.8
Followup	2.0	1.4	1.4	2.1	2.1	3.3

[a] Six males living apart from females all show decreases in circulating levels of testosterone following introduction to and defeat by a heterosexual group.

It may immediately be noted, however, that five of the six males showed yet lower values in the followup samples collected during the birth season. The possibility therefore exists that we did not see hormonal depression as a consequence of defeat but rather had essentially produced no hormonal change by virtue of the manipulation and were only recording normal seasonal hormonal declines.

We therefore designed another study wherein we introduced groups of males to a heterosexual group from which they suffered social defeat. The first group of eight males was defeated toward the end of the breeding season, and the second group of five males was defeated a week later. The first group of eight was immediately returned to its original housing after defeat to see if return to a familiar social setting would produce some recovery of hormonal levels despite the general seasonal trend. The second group was housed in individual cages following defeat and was not reconstituted until two weeks later.

The results (summarized in Table III) show that every one of the total of 13 males in the two defeated groups showed a lower level of circulating levels of testosterone following defeat and that, of the eight males immediately reformed into their original group, seven showed a mean value during the followup period higher than that obtained immediately following social defeat and approximating their preintroduction baselines. The five males placed in individual cages for recovery did not show uniform recovery; two showed caged levels unchanged from those measured after social defeat and three showed some recovery. When they were reorganized into their original group, two weeks later, three of the five showed hormonal declines again, and in the followup period, now past the breeding season, three males showed levels higher than those recorded after social defeat, one equal, and one actually lower.

We therefore concluded that social defeat is a potent stimulus in depressing male hormonal levels, even when baseline levels are both seasonally depressed and measured at a low point on the circadian curve. Moreover, simple exposure to females does not counteract such hormonal depression when simultaneous with the experience of social defeat. Return to the original social situation did result in rapid recovery, even in the face of the ongoing seasonal declines, but introduction into a "recovery" cage was less effective in promoting recovery. Social depression was also shown not to interfere with normal seasonal trends, despite its effectiveness in producing a major short-term perturbation, and, while

TABLE III

Testosterone Levels (ng/ml) of Males in Groups Defeated on Introduction to a Heterosexual Group[a]

	Group I								Group II				
	1	2	3	4	5	6	7	8	1	2	3	4	5
Baseline	6.8	6.8	9.3	7.0	5.8	8.7	5.1	11.4	5.3	8.0	8.8	10.9	3.3
After defeat	5.8	5.4(−)	7.3	3.1	3.1	2.8	2.2	4.4	2.0	3.5	4.8	5.5	2.5
Followup in group	7.6	5.4(+)	8.4	6.9	3.3	5.1	5.2	3.6	—	—	—	—	—
Followup in cage	—	—	—	—	—	—	—	—	3.5	3.4	4.0	12.3	6.4
Reformation	—	—	—	—	—	—	—	—	5.5	6.7	2.9	4.5	3.0
Followup	—	—	—	—	—	—	—	—	3.7	5.0	2.9	10.8	2.5

[a] All males in both groups showed testosterone declines when their group was defeated by a heterosexual group, but, while all but one of the males in the group returned after the defeat showed some hormonal recovery, males in the second group caged for recovery showed more variable responses to such caging.

stimulation by introduction to females following social defeat had previously been demonstrated to produce male androgen levels approximating those of control males introduced to females, female stimulation at the time of defeat was ineffective in buffering the males from hormonal depression. Recovery from defeat is therefore seen not as a spontaneous event, but rather as a consequence of subsequent social experience.

The complicated nature of response to introduction was further demonstrated by taking a group of six males, forming a social group of three, and then adding the other three singly. All three males introduced singly showed immediate hormonal declines following their introduction and initial defeat (12.8–3.3, 13.0–5.5, and 5.6–2.2). Next, we introduced a group of 11 other males individually into the group, but did not remove any previously introduced male. The first male introduced was entering a new group and showed a hormonal decline (3.5–1.2) despite his having been the lowest-ranking member of his former group. Each subsequent introduction involved a male entering a group in which the senior members (and consistently top-ranking animals; Bernstein & Gordon, 1980) were new but an increasing proportion of other members were familiar partners from their previous group. Under these circumstances, only six of the next nine introductions resulted in a followup hormonal measure lower than baseline. (The eleventh male was agonadal.) We therefore concluded that prior social experience with the animals present in a new social context might influence the course of introductions and their hormonal consequents, much as had been demonstrated by introducing pigtail males into groups containing familiar or unfamiliar males—in other words, attack usually followed introduction to unfamiliar males, with hormonal depressions following, whereas animals submitting to familiar males in a new social context did not get mobbed by the group and did not show androgen declines (Bernstein, Rose, Gordon, & Grady, 1979).

Finally, we had the opportunity to monitor hormonal changes in males undergoing spontaneous defeat in an established group, and this convinced us that it was attack-and-defeat that was crucial and not introduction to a new social situation and context. A male attacked the alpha male in an all-male rhesus group and, after a period of severe fighting, lost in his bid to bypass the alpha male. His hormonal baseline had been collected for another study and his mean level prior to the fight was 11.9. Twenty-four hours after his unsuccessful attack on the alpha male, his circulating testosterone level was down to 1.8 (15% of his previous level). The alpha male, on the other hand, had been badly injured and had to be removed from the group for treatment, but had successfully fought off the challenge. His baseline of 7.0 was followed by a level of 9.6 24 hours after his successful defense and also after his removal from the group. He gradually recovered physically, and his hormonal mean approximated his preremoval level at 6.7 prior to his return to the group. To our surprise, he was attacked by the group *en masse* on reintroduction and was soundly defeated. We immediately removed him from the group and 24 hours later his hormonal levels had fallen to 1.9, or 28% of his baseline level.

AGONISTIC ENCOUNTERS NOT RESULTING IN DEFEAT

The hormonal response of the last-mentioned male following successful defense of his position was not a significant elevation but was suggestive since we had already reported transitory but significant androgen rises in some males following their defeat of a rival group on group merger (Bernstein, Rose, & Gordon, 1974). In several cases for which we had failed to obtain behavioral depression, we checked the hormonal levels of the undefeated males and noted transitory rises as well. In both types of cases (an alpha male successfully defeating a rival group, and a single male successfully defending himself on introduction to a new group), the transitory rises were approximately three times baseline levels. In another unreported case of successful defense on introduction, we saw a similar threefold increase from 9.5 to 30.3, dropping back to 16.2 over the next few days. In an additional unreported case, when we had subjected the same male to repeated introduction to different subsets of animals in an unsuccessful attempt to produce social defeat, his hormonal baseline at 6.3 was essentially unchanged 24 hours later at 7.0, but this was in the face of a rapid seasonal decline, for his followup measures over the next few days declined to 3.9. In a third case, an unsuccessful effort to depress a male in a group introduction resulted in a rise in that male from a baseline of 2.0 to 5.1 and a subsequent return to 2.3.

It thus appeared possible that being on the winning side of an agonistic encounter might result in hormonal elevation, and we looked at the hormonal values of host group males before and after introductions in which the subjects were defeated. We looked at four high-ranking males in one host group and found essentially random changes in their hormonal levels following defeat of an introduced subject (8.4–7.3, 9.4–9.8, 8.7–3.9, and 7.4–12.9). Repetition with a second subject followed, and the two males showing some rise in the first introduction showed similar minimal rises (8.2–9.2 and 8.4–11.4).

These data were considered insufficient to warrant any conclusions, so we designed a series of studies in which we would monitor preselected males in groups before and after each manipulation in which their group served as host in a series of single male introductions. In the first study, we formed a group of three males, all of whom were monitored before and after the introduction of each of three subjects. Each of the introduced subjects was expected to show some lowering of circulating levels of testosterone following his own introduction, but he was then monitored to see if there was any response to the next introduction, during which the new animal would now be a group member "winning" over the newly introduced animal.

None of the three original males showed a hormonal rise following the introduction of the first subject, but all three of them, as well as the first-introduced subject, showed a rise on the introduction of the second subject. Responses to the third subject were mixed, although the second-introduced subject did show a small rise that he later sustained, suggesting perhaps recovery from depressed levels after his own introduction rather than a response to winning.

TABLE IV
Testosterone Levels (ng/ml) in Resident Males Prior to and Following Introduction of New Males[a]

	Baseline	First introduction	Baseline	Second introduction	Baseline	Third introduction	Followup
Original male							
1	7.5	2.6	5.4	10.2	7.1	5.9	6.6
2	4.7	3.0	3.1	5.4	3.3	6.7	3.5
3	3.2	1.9	1.5	11.3	3.7	1.1	1.4
New male							
1	—	—	4.4	7.6	3.9	1.4	1.5
2	—	—	—	—	2.8	4.8	5.4

[a] Resident males do not show consistent hormonal response to the introduction of a new male into the group.

The rises following the introduction of the second male and the tenfold increase in one of the subjects could not be explained by differences in the behavioral data. Hormonal data are summarized in Table IV. This study was conducted early in the breeding season but no females were in view and baseline levels approximate the transitional levels previously described.

In an attempt to clarify these findings, we conducted a second study, with males in view of females but at the close of the breeding season. The three original males were followed during the first five introductions, but the sudden traumatic death of one of these males (not during an introduction) limits his data to the first four introductions. The first three males introduced were also followed during the introductions of the remaining males, but, beginning with the fourth introduction, the next 10 subjects were followed only for the first introduction following their own. This was to test to see if hormonal rises in these males would reflect a true response to having been on the winning side (if all the other males showed rises as well) or might reflect hormonal recoveries following their own introduction. In fact the subjects showed no consistent hormonal responses to new introductions nor to introductions of the next male following their own introduction. The only pattern suggested was the general

TABLE V
Male Testosterone Levels (ng/ml) Prior to and Following Successive Introductions to a Group[a]

	Base	1st	Base	2nd	Base	3rd	Base	4th	Base	5th	Followup
Original male											
1	30.7	13.6	8.2	26.2	22.6	22.7	8.7	8.9	—	—	—
2	13.4	13.1	14.6	17.8	8.8	7.5	10.5	14.5	3.9	3.6	2.2
3	3.0	5.3	9.4	6.3	7.0	4.6	3.6	1.6	1.8	2.6	4.5
Introduced male											
1	—	—	2.8	1.5	3.1	3.7	7.9	2.5	3.5	1.5	1.5
2	—	—	—	—	6.3	12.0	8.3	2.9	1.1	0.8	0.6
3	—	—	—	—	—	—	2.7	2.2	1.4	1.8	3.7

[a] Neither original resident males nor newly introduced males show consistent hormonal response to the introduction of additional males.

normal seasonal decline in male hormonal levels as the study progressed into the nonbreeding season. Data are summarized in Table V for the first six males and in Table VI for the 10 males followed only for the single introduction following their own. Several individual males did show suggestive hormonal changes, but no group phenomenon could be detected and the explanation for these individual changes must lie in the individual rather than group pattern of response and in the individual's experience in each encounter.

In reexamining our previously reported data on the hormonal rise of an alpha male during a group merger in which his group won, we saw a clear pattern, with a stable baseline that varied between 3.2 and 4.0 and a single measure one day following group formation at 12.6. Within 72 hours, his circulating levels were back to 4.0, and the next five measures again varied only between 3.2 and 4.0. The second-ranking adult male in his group had supported him but had never individually achieved any success over any of the new group members. His baseline had varied between 5.0 and 12.6; his testosterone level was at 12.0 following the group introduction and maintained its variability between 4.4 and 10.8 over the next followup measures. This difference between the two males suggests that actual individual experience in the encounter, rather than being operationally defined as having been on the winning side, was reflected in hormonal changes following introductions. This was further confirmed when we monitored three males in a group of six before and after they had been in a "victorious" group merger. None of the three males selected showed any hormonal response, even though one of them was the alpha male. Their "before," "after," and "followup" measures were, respectively, 7.2–4.3–4.5, 1.9–2.1–2.0, and 1.6–2.2–2.0.

We next hypothesized that hormonal response might occur when a socially significant event took place, such as the spontaneous overthrow of an alpha male. We fortunately had an adequate baseline of measures prior to three such events, two in which the alpha was killed and a third in which he was not killed but was permanently removed from the group because we feared for his life. The androgen levels of the males in the three groups do not show any consistent hormonal response to the alpha overthrow. Changes in both directions can be suggested depending on which individual male is under focus, and this once

TABLE VI

Testosterone Levels (ng/ml) of Newly Introduced Males and Their Responses Following the Introduction of the Next Subject[a]

	Subjects									
	1	2	3	4	5	6	7	8	9	10
Baseline following own introduction	6.4	3.9	1.2	0.8	5.1	1.8	0.6	2.0	8.1	1.0
Level after next introduction	3.4	4.1	4.3	0.4	2.7	5.7	0.7	1.9	1.2	2.7
Followup	1.4	7.0	2.6	0.7	4.5	1.7	0.8	1.8	1.2	1.5

[a] Males newly introduced to a group do not necessarily show testosterone recoveries when a new male is introduced to their group.

TABLE VII
Testosterone Responses (ng/ml) to Alpha Male Overthrows[a]

	Prior measure	Alpha male is overthrown	Followup measure
Group I			
1	9.8	7.9	9.2
2	7.3	3.1	2.8
3	2.9	7.5	3.1
4	2.8	4.7	2.9
5	1.8	2.9	4.2
6	4.4	3.1	3.6
7	2.5	17.8	10.1
Group II			
1	14.6	13.9	8.6
2	7.8	4.0	4.3
3	8.4	3.8	2.9
4	15.5	9.1	11.0
5	8.5	11.8	9.3
6	11.3	6.3	7.5
7	14.7	11.4	15.8
Group III			
1	2.9	1.1	1.1
2	1.6	1.5	2.0
3	2.2	1.4	1.4
4	14.5	5.5	2.4
5	2.5	6.4	0.5

[a] The overthrow of an alpha male produces differential testosterone level changes among the remaining group members.

again suggests that all group members do not respond to the overthrow of an alpha male in the same fashion. Much as only some males may participate in the actual challenge, the consequences of group reorganization following the overthrow will vary from individual to individual. The data for the males in the three groups are summarized in Table VII. Differences in the absolute levels among the three groups reflect seasonal factors at the time of the incident. All three groups were all-male groups with differential visual access to females; the group II overthrow occurred during a breeding season while overthrows in groups I and III took place at the close and after the close of the breeding season. Based on these results, we concluded that merely being in a group that defeats an individual, or overthrows an alpha male, is not sufficient social change to influence individual male hormonal levels. When hormonal levels are altered following successful defense or aggression, such an elevation must be attributed to the individual's experience and assessment of his experiences.

DOMINANCE STATUS

In an early paper (Rose, Bernstein, & Holaday, 1971), we had reported a small significant positive correlation between circulating levels of plasma testosterone and agonistic rank in an all-male group of rhesus monkeys. In analyzing the data, we had reported that most of the correlation could be attributed to the tendency for males in the upper quartile to have high testosterone levels and the balance of the subjects in the 34-male group to show a more or less random distribution of androgen level with respect to rank. The original correlation had been found within six months of group formation and at the end of the breeding season. Reexamination of the same group six months later failed to reveal any significant correlation of rank with testosterone. Eaton and Resko (1974) suggested that the original correlation may have been due to the correlation between attaining high rank in a newly formed group and successful agonistic encounters. In the data previously presented on agonistic encounters, it would seem that successful encounters produce a transitory rise unlikely to influence a randomly timed sample for androgen assay, but that the more prolonged depression of hormonal levels following defeat might be reflected. It does not seem reasonable that all lower-ranking males were depressed, but if some of them were, then the upper quartile of males might indeed have a mean level higher than the group mean. Eaton and Resko (1974) showed no significant correlation between male Japanese macaque ranks in the breeding season and androgen levels assayed once prior to the season in each of two successive years.

Inasmuch as numerical rank position results from the investigator's ability to count the number of individuals that any subject submits to, or is submitted to by, this is an easily defined rating but one of dubious social significance to a monkey. Each monkey in a group indeed knows which individuals it must submit to and which will submit to it, but the differential numerical significance of ranks 16 and 26 is probably lost on the subjects. A high-ranking male, say number 6 in the group, may be repeatedly harassed by two or three yet higher-ranking individuals, while the animal ranking 26 may seldom be subjected to any aggressive communication and may even be tolerated by the highest-ranking animals, to whom he poses no challenge. Numerical rank-ordering is a human attribute and is not significant to a monkey. On the other hand, being alpha means an individual submits to nobody, and this may be significant to a monkey.

In accordance with this analysis, we obtained androgen baseline measures on males prior to group formations and then followed these males during group formations while we independently measured dominance ranks, using the method of scoring directionality in unidirectional encounters (Bernstein, 1968; Rowell, 1966; Sade, 1967). Bidirectional confrontations were scored as unresolved, giving each subject an equal score.

We formed five small all-male groups. The first group consisted of four males, two of which fought without resolution on day one but established stable rankings of first and last on day two, the two nonparticipating males both asserting themselves over the loser. The losing male had had the highest baseline testosterone levels but fell to about 15% of this level following his defeat. The winning male had had the second highest baseline, but his levels after group formation steadily rose so that in the two weeks following his victory, even the highest levels of the other males were less than his lowest recorded level. Data are summarized in Table VIII.

The next three groups were formed with three males each. Differences in

TABLE VIII
Testosterone Levels (ng/ml) of Four Males During a Group Formation[a]

Final group rank	Baseline range	1st day	3rd day	Followup range
Group I				
1	5.4–5.9	6.4	8.5	7.5–12.6
2	4.4–4.4	7.7	5.1	6.2–7.4
3	3.0–4.2	—	9.5	3.2–5.4
4	9.1–13.0	8.0	1.6	1.4–6.6

[a] The winning male of the four did not have the highest initial testosterone levels, and testosterone levels did not correlate with male rank until after the dominance system stabilized.

visual access to females and seasons are reflected in the different baseline levels shown in Table IX, and baseline measures did not consistently correlate with rank attained. Despite these perturbations, the male who became alpha ended up with the highest circulating level of androgen in each case. In each of the four groups, then, the male who became alpha had the highest testosterone level on becoming alpha and maintained the highest level in the group during the following periods.

It is true that the group III baseline levels were predictive of final rank outcomes and that the baseline assays of group IV were not available, but in neither groups I nor II was there any correlation between baseline testosterone level and rank after group formation. A fifth group was formed with six males, and once again there was no correlation between baseline androgen measures and dominance rank attained, although it is to be noted that, while the alpha

TABLE IX
Testosterone Levels (ng/ml) of Males in Group Formations[a]

Male ranks	Baseline	Formation	Followup
Group II			
1	3.0	31.5	30.0
2	5.4	11.8	15.0
3	3.6	1.5	4.4
Group III			
1	12.8	19.6	17.1
2	7.6	2.1	2.7
3	4.7	3.0	2.1
Group IV			
1	—	7.2	7.5
2	—	2.0	4.7
3	—	Low, off scale	3.2

[a] In all three groups, the testosterone levels of the males matched their dominance ranks following group formation.

male in this group showed the largest androgen rise following group formation, the number-four male exceeded his circulating levels on day one and, in the following period, the lowest-ranking male bypassed both of these two. These data were collected during the breeding season with no females in view, and are summarized in Table X. The probability of the alpha male having the highest androgen level is, of course, related to the number of males in the group. The probability of the alpha male having the highest levels in the first four groups is $0.25 \times 0.33 \times 0.33 \times 0.33 = 0.009$. If we compute the probability of the alpha male in all five groups having an androgen level in the top third of the group, the p value is 0.004.

Inasmuch as the alpha male seemed to show a prompt response on becoming alpha, which was not always seen to result in his sustaining the highest circulating androgen levels in long-term groups, it is possible that a recent history of successful agonistic encounters, even the minor ones of supplanting and avoidance, may account for the high levels of alpha males immediately following group formation. In an attempt to separate out the attainment of high rank from the successful experiences in initial agonistic encounters during group formation, we designed a series of studies to test the effect of becoming alpha without going through any agonistic process. We achieved this by simply capturing and removing the alpha, or alpha and beta males, while monitoring the hormonal levels of the remaining group males, including the one who would become alpha by default. In using well-established groups, we also predicted that the temporary removal of an alpha male would not destroy the relationship of that male with the group. If the alpha male could reclaim his old position as number one on return to the group, then we would predict that the temporary alpha male might also show hormonal response to the loss of his alpha position in returning to his previous rank position. We predicted that the reclaiming of original positions would progress without any serious confrontations and that the loss of rank would be free from the experience of social defeat. Further, in accordance with

TABLE X
Testosterone Levels (ng/ml) in Males on Group Formation[a]

Male rank	Baseline	Day	Followup
Group V			
1	3.6	7.0	7.3
2	7.7	2.0	6.8
3	4.7	6.3	3.9
4	6.5	7.5	8.5
5	5.7	1.7	5.9
6	4.3	2.8	8.8

[a] Although there is no correlation between baseline testosterone levels and rank attained on group formation, the male attaining alpha status did not attain the highest circulating level of testosterone in the group despite doubling his circulating level over baseline on day one.

our data that rank has significance only within a social context (Bernstein & Gordon, 1980), we also expected hormonal decline in the original alpha male on his removal from the group and a rise following his return. In this case there could be no social defeat on removal and "loss" of rank.

This experiment was conducted using five different groups. In group II, the levels of the alpha and beta males on removal and return were reflective of their fight prior to removal, and therefore were not used in assessing the effect of removal and return on their androgen levels. No data following the return of the alpha and beta males are reported because severe fighting followed this return. Samples for the removed alpha and beta males in group III were not available, and in groups IV and V, the alpha and beta were removed and returned multiple times, but only their first removal data were used inasmuch as the removed alpha and beta males were subjected to other experimental manipulations during subsequent removals. The data are summarized in Table XI.

In group I, the alpha male shows a decline on removal from his group and a rise on return. In group V, the alpha and beta males were both removed and androgen levels in both males fell on removal and rose 24 hours after return. In the second repetition, androgen levels were assayed based on a sample taken 1 hour after the removal of the alpha and beta males, and once again both males were down.

In groups I, II, and III, the male who became alpha showed a rise in circulating levels of androgen 24 hours later. It is to be noted, however, that although the control males showed variable changes or stability when the alpha male was removed, the male becoming alpha neither had the highest baseline level prior to the removal of the original alpha male nor necessarily had the highest circulating level following the removal of the alpha male. In group II, the first male to claim the alpha position by default showed the expected hormonal rise, but he yielded the position without observed conflict by day 2 and his testosterone levels promptly fell. The male who took over had shown no change at 24 hours; after taking over, he had the highest levels in the group, but in the followup period his levels were second highest.

In group IV, the male who became alpha by default failed to show a hormonal rise on the removal of the alpha; he and the beta male showed lower levels 24 hours after the removal of the alpha and higher levels after the return of the original alpha and beta.

In group V, the alpha and beta males were removed repeatedly, four times in one year and then again a year later. On every removal, the male who assumed the alpha position showed a rise in testosterone, as revealed in both 24-hour and 1-hour measurements, although the 24-hour measurements were always higher. The male assuming the beta position showed the same pattern on the four manipulations conducted during the first year, but the replacement male used in the second year failed to show a hormonal rise on assuming the beta male slot during the removal of the alpha and beta males.

In summary, then, in the three cases of the alpha male being removed and in the two removals of the beta male, these males showed lower levels of circulating androgen 24 hours after removal and elevations on return. In eight of

TABLE XI
Testosterone Levels (ng/ml) in Males Assuming and Losing Alpha and Beta Ranks by Default[a]

	Baseline	After one hour	Removal (Day 1)	Followup	Return	Followup
Group I						
Original alpha	7.6	—	4.2	4.6	10.8	7.8
Becomes alpha	10.7	—	8.6	13.7	9.0	10.0
Control male	7.0	—	4.0	9.0	10.4	9.5
Group II						
Becomes alpha day 1	2.7	—	9.2	1.8	—	1.1
Becomes alpha day 2	1.1	—	1.2	9.3	—	3.7
Control male 1	1.0	—	11.5	0.5	—	1.1
Control male 2	8.9	—	2.5	2.1	—	6.9
Control male 3	1.5	—	1.2	1.0	—	1.1
Group III						
Becomes alpha	8.1	—	12.2	14.6	—	—
Control male 1	10.2	—	8.9	7.8	—	—
Control male 2	6.2	—	12.2	8.4	—	—
Control male 3	15.8	—	—	15.5	—	—
Control male 4	5.8	—	12.3	8.5	—	—
Group IV						
Becomes alpha	19.5	—	12.1	—	16.7	14.9
Becomes beta	8.9	—	7.8	—	16.8	9.4
	23.6	—	7.9	—	17.5	—
Group V						
Original alpha	14.5	7.6	—	—	—	—
Original beta	22.4	—	5.3	—	10.7	—
	10.0	8.7	—	—	—	—
Becomes alpha	12.2	—	20.8	—	10.3	—
Becomes beta	7.8	—	8.7	—	—	—
Repetition						
Becomes alpha	12.1	10.8	20.9	—	9.4	—
Becomes beta	5.4	7.2	15.7	—	—	—
Becomes alpha	6.6	10.5	10.7	—	Repeat	—
Repetition						
Becomes beta	4.8	9.7	15.0	—	5.3	—
Becomes alpha	14.7	17.4	—	—	12.1	—
Becomes alpha	10.3	—	25.4	—	14.7	—
Becomes beta	5.7	—	17.2	—	5.4	—
Becomes beta	7.6	—	10.4	—	5.7	—
Repetition						
Becomes alpha	8.8	—	18.1	—	9.1	8.4
One yr. later						
Becomes beta	24.9	—	18.1	—	19.6	8.8

[a] On their removal from their groups, the alpha and beta males show hormonal declines. Eight of the nine males assuming the alpha position and seven of the nine males assuming the beta position showed hormonal rises, only to come down again on return of the original alpha and beta. Males assuming higher ranks did not necessarily have high testosterone levels while the alpha was present and control males do not show consistent hormonal rises on the removal or return of the original alpha.

the nine cases where a male assumed the alpha position by default, and in seven of the nine cases where a male assumed the beta position by default, circulating levels of testosterone rose, only to come back down again on the return of the original alpha and beta. Thus, it appears that becoming alpha (and, to some extent, beta as well) stimulates an elevation in hormonal levels, while losing these positions produces a decline. This occurs even in the absence of overt agonistic confrontations. The mere removal of the alpha (and beta) males, on the other hand, does not necessarily produce a rise in androgen levels in the remaining group members, many of whom actually showed declines in circulating androgen levels.

When an alpha male is forcibly overthrown or killed, group members experience (1) his removal, (2) the agonistic experiences associated with his overthrow, and (3) possible rank improvements, especially for the male claiming the alpha position. We have earlier reported the inconsistent responses of group members to such an event, but we pooled our data on spontaneous overthrows of the alpha male from several species groups and added one more rhesus monkey case.

In the rhesus monkey group, the alpha male was killed in a spontaneous overthrow and the male who took over showed an extraordinarily high level (in the group) of circulating androgen the day before the alpha male was killed. The actual death of the former alpha did not produce a rise in any of the males, as can be seen in Table XII.

The alpha male in a pigtail monkey group had maintained a mean monthly circulating level of 19.6 of testosterone over a two-year period. When he was bypassed by another male, his level fell to 5.5. The new alpha had maintained an average of 8.6 for the same two-year period, but his mean had been only 4.6 in the year prior to his successful takeover and then it rose to 8.4. Other males in the group showed no consistent change (2.6–7.4, 11.1–15.7, 9.4–17.1, 1.7–3.4, 5.7–10.2, and 7.1–6.4). The former alpha male had maintained the highest mean androgen level in the group for two years and, in 15 of 17 assays, had the highest absolute level, but on losing his position he was eighth in a field of 10. The new alpha male, however, maintained a circulating level midway through the group (5 out of 10).

TABLE XII
Male Testosterone Levels (ng/ml) and Ranks on the Death of an Alpha[a]

New male ranks	Base	Death of alpha	Followup
1	14.5	5.5	2.4
2	2.2	1.4	1.4
3	1.6	1.5	2.0
4	2.5	6.4	0.5
5	2.9	1.1	1.1

[a] The death of the alpha male did not produce uniform hormonal responses among the males in the group.

In a group of stumptail males, the alpha male had also maintained the highest mean level of circulating androgen of the five males in the group, but due to fluctuations in month-to-month samples, his levels were absolutely highest in only 6 of the 17 months. Single samples are therefore considered poor indicators of a male's hormonal standing in a group.

In a group of sooty mangabeys, there was no suggestion of a correlation between male rank and hormonal level, and the alpha male had a mean circulating level of only 4.3, as compared with the 11.7 seen in the beta male as an average monthly measure. The gamma (2.6) and delta males (3.8) were lower than both of the top two males but on one month, nevertheless, bypassed both the top-ranking males.

Being an alpha male thus does not correlate with either hormonal levels prior to group formation, nor with hormonal levels in a long-term stable group, but the transition into a high-ranking (or lower-ranking) position may influence circulating levels of androgen in males. The acquisition or loss of the alpha position is therefore seen as an independent variable that, along with outcomes of agonistic encounters, female stimulation, season, circadian rhythm, and other factors, influences circulating levels of androgen in male monkeys.

It is important to note that in the case of status changes and the outcome of agonistic encounters, it is the cognitive evaluation of the situation that appears to be the key variable. Agonistic experiences are not influential because of the injuries inflicted or received, and merely participating in the expression of aggression has no influence on the circulating levels of testosterone in male rhesus monkeys. In a study designed to demonstrate just this point, a group containing six adult males was routinely captured and blood samples were obtained from the males. Two infants were caught out of the group during the handling, and when all the animals were back outside, an investigator held first one and then the other infant up to the outside fence and allowed the group to respond to the struggling infants. Most of the males were aroused enough to charge the fence, reach through, and try to attack the investigator's gloved hands. The investigator did not respond to the threats and charges of group members and, after 40 minutes, disappeared indoors. A team (not including this person) repeated the routine group capture and collected a second blood sample. The infants were reshuffled into the group during the release. The males therefore had differentially expressed aggression in responding to the infant removals, but there had been no consequence of their aggression in that they neither won (defeated the investigator or retrieved the infants) nor lost (suffered any injuries or indicated submission). None of the males showed any hormonal response to this manipulation.

HCG STIMULATION OF ANDROGEN PRODUCTION

The influence of hormones on behavior has been demonstrated in multiple species, and androgens have been shown to influence profoundly the behavior of animals tested in individual and pair test situations (Rose, Bernstein, Gordon, & Catlin, 1974). In our laboratory, however, we have tested the influence of increased circulating levels of androgens on males living in the context of a social group. In three separate studies, we administered HCG to adult rhesus males living in social groups and in every case observed significant rises in levels of circulating androgens. This procedure was used in lieu of testosterone propionate pellet implants because our subjects were all intact males and we felt it better to raise androgen levels by stimulation of the gonads rather than run the risk of negative feedback from the pituitary, in response to injected testosterone, acting to alter normal gonadal function and the output of other gonadal products. The effect of HCG on the intact male was greater than we anticipated, and circulating levels of testosterone rose initially to levels as much as 10 times those seen in any of our normal unstimulated males at the height of the breeding season. These levels were achieved in both the rhesus monkey birth and breeding seasons, and the data are reported in Gordon (1981; Gordon, Rose, Grady, & Bernstein, 1979). In the first study we looked for behavioral changes in two treated males living in a group of eight. Our measures failed to note any changes in status or aggressive behavior, but we did note that the two subjects carried fresh wounds. This curious state led us to hypothesize that the influence of increased testosterone on behavior was perhaps more subtle than anticipated and we refined our behavioral measurements.

In the next study, we stimulated the top two and bottom two males in a group of seven males isolated from females during the breeding season. None had previously shown seasonal rises nor did the unstimulated males show seasonal rises in the presence of the treated males, suggesting again the dependence of males on females for seasonal cues. The social structure of the group remained stable—that is, there were no rank changes observed. Moreover, within the context of the social group, the effect of hormonal elevation appeared to be to enhance existing behavioral patterns—for example, a male with a high baseline rate of aggressive behavior showed significant increases in aggression, a male with a high baseline rate of sexual behavior showed significant increases in sexual responses, etc., but the lowest-ranking male showed no sign of behavioral change, possibly indicating that the omega male has the least latitude in behavioral change before social mechanisms would come into play to restore the normal social pattern of the group.

In a third study, HCG was administered to all seven males resident in a heterosexual group during the birth season. The males showed significant weight changes, increased yawning and masturbation, and decreased play and agonistic behavior, but there were no changes in female responsiveness to the males. The females did not respond to hormonally primed males with any sign of receptivity, and there was no induced "breeding season" as had been suggested to occur when females had been hormonally manipulated in the nonbreeding season (Vandenbergh & Drickamer, 1974). This once again indicates that the male rhesus monkey's sexual seasonality is controlled by female stimuli and that male behavioral patterns are strongly influenced by social context rather than solely by hormonal states.

In accordance with the premises that hormonal states reflect recent experience and that they modify the probability of behavioral responses, functioning to adapt the individual to current circumstances, it might be expected that within the social context in which the system evolved, artificial perturbations of hormonal states may be overridden by more direct environmental influences on behavior. The influence of social context on behavior has been demonstrated in a number of studies, including the dramatic deaths of subordinate *Tupaia glis* merely forced to stay in visual contact with dominant individuals (von Holst, 1977). Social rank in a small group seems to be a sufficiently good predictor of social experience that it strongly influences sexual behavior and may supercede cycle stage as a predictor of copulation (Goldfoot, 1971; Perachio, Alexander, & Marr, 1973). The administration of estradiol benzoate (EB) to a female powerfully influences behavior in pair-tested animals, but is significantly less determinant in small groups. Keverne, Leonard, Scruton, and Young (1978) found that rank and partner preferences influenced the visual monitoring of both males and females receiving EB. Of course, social experience may influence hormones other than the one being artificially manipulated, and this may account for the lower influence of a particular manipulated hormonal state on behavior. Bowman, Dilley, and Keverne (1978) demonstrated that administered doses of estrogen to talapoin females triggered a luteinizing hormone surge only in the high-ranking females. It therefore seems clear that social context not only influences hormonal states but also may powerfully interact with the effects of hormones on behavior.

The logical alternative to studies of the influence of above-normal hormonal levels on behavior is the study of the influence of below-normal hormonal levels on behavior. Whereas HCG stimulation or hormonal implants can be used to increase circulating levels of androgens, castration or the use of chemical blocking agents can be used to decrease circulating levels of gonadal hormones. Few investigators have studied the influence of androgens on aggression in primates, although there is a report of observed changes in social behavior and social status in 10 castrated male rhesus monkeys living in a free-ranging colony (Wilson, 1969; Wilson & Vessey, 1968). Behavioral changes among the males were inconsistent, and several maintained high social positions and continued to defeat larger intact males in fights.

More systematic studies of the effects of castration and hormonal replacement therapies on sexual behavior have been conducted using caged animals tested under paired conditions. These studies also reveal inconsistent responses to castration, with some males showing continued sexual performance for years after castration and with sex differences for the species persisting even in neonatal castrates (Goy, Phoenix, & Resko, 1965; Michael & Wilson, 1974; Phoenix, 1973, 1974, 1978). The variability of the effect of castration on behavior in minimal social contexts suggests a strong psychogenic effect, possibly influenced by ontogenetic factors (i.e., age and experience at the time of castration). It would not be easy to extrapolate these findings to animals living in social contexts, but there is the possibility that an even greater dissociation between hormonal levels and behavior might be observed when all the mechanisms in a social context operate to continue to produce normal behavior. A pilot observation of a single agonadal adult male rhesus monkey living in a social group (Bernstein, Gordon,

& Peterson, 1979) as well as the residual seasonal influences in two castrated males implanted with testosterone proprionate (Michael & Wilson, 1975) are suggestive. Studies of hormonal influences on behavior using primates living in social groups have only just begun.

CONCLUSIONS

We conclude that there is a significant correlation between circulating androgen levels and behavior in male monkeys. The recent social experience of a male influences his circulating level of androgen, and this in turn may influence his propensity to engage in the same types of behavior that have contributed to his recent social experiences. We see social experience as regulating the hormonal condition of animals such that their subsequent behavior is more likely to be appropriate to their recent past history of social experiences. Inasmuch as the past is a reasonably good predictor of the future in stable social groups, hormones contribute to the maintenance of social behavior appropriate to an individual's position in a group. If, on the other hand, hormones are manipulated independently of an individual's social experience, they may directly influence the individual's tendencies to engage in particular types of behavior. Within a social context, however, social mechanisms operate to overcome the influence of inappropriate hormonal levels, and the individual may maintain stable social activity levels in an apparent dissociation with its hormonal condition.

Under normal circumstances, social mechanisms and hormonal levels would reinforce the same kinds of social behavior. In socially living animals, with the ability to modify behavior as well developed as that seen in monkeys, social mechanisms may be powerful indeed. An agonadal male may still follow seasonal activity patterns activated not by his seasonal hormonal changes but by the social facilitation of his group associates (Bernstein, Gordon, & Peterson, 1979). Likewise, males treated with HCG may sustain normal levels of social activity befitting their social positions in a group, using only socially acceptable outlets for any additional social activity induced by higher circulating levels of testosterone. Any departure from the normal behavioral pattern of an individual may provoke group aggression to modify the individual's behavior back to the previously established and accepted pattern. As K. R. L. Hall pointed out (1964), the most potent stimulus for aggression in a social group is a perceived violation of the social code. In animals with elaborate development of the cerebral cortex, there may well be less dependence on hormonally mediated regulation of behavior. Not only may an individual act in a manner not predictable based on hormonal levels alone, but other individuals may exert social pressures to modify an individual's behavior regardless of the underlying hormonal substrate. Under most natural circumstances, however, hormonal regulation of behavior will reinforce rather than conflict with social and cortical control of behavior. With the elaboration of the cerebral cortex in the primate, hormonal influences on behavior are not lost, but may be superceded when unusual circumstances result in hormonally induced behavioral propensities inappropriate for the particular circumstance in which an animal finds itself.

REFERENCES

Beach, F. A. *Retrospect and prospect in sex and behavior.* New York: Wiley, 1965.

Bernstein, I. S. Social status of two hybrids in a wild troop of *Macaca irus. Folia Primatologia,* 1968, *8,* 121–131.

Bernstein, I. S., & Gordon, T. P. Behavioral research in breeding colonies of Old World monkeys. *Laboratory Animal Science,* 1977, *27,* 532–540.

Bernstein, I. S., & Gordon, T. P. Inter and intraspecific sexual behavior in two species of macaque: A possible behavioral barrier to gene flow. *Behavioral Processes,* 1979, *4,* 265–272.

Bernstein, I. S., & Gordon, T. P. The social component of dominance relationships in rhesus monkeys *(Macaca mulatta). Animal Behaviour,* 1980, *28,* 1033–1039.

Bernstein, I. S., Rose, R. M., & Gordon, T. P. Behavioral and environmental events influencing primate testosterone levels. *Journal of Human Evolution,* 1974, *3,* 517–525.

Bernstein, I. S., Rose, R. M., & Gordon, T. P. Behavioral and hormonal responses of male rhesus monkeys introduced to females in breeding and non-breeding seasons. *Animal Behaviour,* 1977, *25,* 609–614.

Bernstein, I. S., Gordon, T. P., Rose, R. M., & Peterson, M. S. Influences of sexual and social stimuli upon circulation levels of testosterone in male pigtail macaques. *Behavioral Biology,* 1978, *24,* 400–404.

Bernstein, I. S., Rose, R. M., Gordon, T. P., & Grady, C. L. Agonistic rank, aggression, social context, and testosterone in male pigtail monkeys. *Aggressive Behavior,* 1979, *5,* 329–339.

Bernstein, I. S., Gordon, T. P., & Peterson, M. Role behavior of an agonadal male rhesus monkey in a heterosexual group. *Folia Primatologia,* 1979, *32,* 263–267.

Bowman, L. A., Dilley, S. R., & Keverne, E. B. Suppression of oestrogen-induced LH surges by social subordination in talapoin monkeys. *Nature (London),* 1978, *275,* 56–58.

Coe, C. L., Mendoza, S. P., Davidson, J. M., Smith, E. R., Dallman, M. F., & Levine, S. Hormonal response to stress in squirrel monkey *(Saimiri sciureus). Neuroendocrinology,* 1978, *26,* 367–377.

Coimbra-Filho, A. F., & Maia, A. de A. Phases of the reproductive process in *Macaca mulatta* (Zimmerman, 1780) at Ilha do Pinherio, Rio de Janeiro, Brazil (Cercopithecidae, Primates). *Review of Brazilian Biology,* 1977, *37,* 71–78.

Conaway, C. H., & Koford, C. B. Estrous cycles and mating behavior in a free-ranging band of rhesus monkeys. *Journal of Mammalogy,* 1965, *45,* 577–587.

Dunbar, R. I. M., & Dunbar, E. P. Dominance and reproductive success among female gelada baboons. *Nature (London),* 1977, *266,* 351–352.

Eaton, G. G., & Resko, J. A. Plasma testosterone and male dominance in a Japanese macaque *(Macaca fuscata)* troop compared with repeated measures of testosterone in laboratory males. *Hormones and Behavior,* 1974, *5,* 251–259.

Gavan, J. A., & Hutchinson, T. C. The problem of age estimation: A study using rhesus monkeys *(Macaca mulatta). American Journal of Physical Anthropology,* 1973, *38,* 69–81.

Goldfoot, D. A. Hormonal and social determinants of sexual behavior in the pigtail monkey *(Macaca nemestrina).* In G. B. A. Stoelinga & J. J. van der Werff ten Bosch (Eds.), *Normal and abnormal development of brain and behavior.* Leiden: University of Leiden Press, 1971.

Goodman, R. L., Hotchkiss, J., Karsch, F. J., & Knobil, E. Diurnal variations in serum testosterone concentrations in the adult male rhesus monkey. *Biology of Reproduction,* 1974, *11,* 624–630.

Gordon, T. P. Mating patterns in the rhesus monkey: Social, seasonal and hormonal variables. *American Zoologist,* 1981, *21,* 185–195.

Gordon, T. P., & Bernstein, I. S. Seasonal variation in sexual behavior of all-male rhesus troops. *American Journal of Physical Anthropology,* 1973, *38,* 221–225.

Gordon, T. P., Rose, R. M., & Bernstein, I. S. Seasonal rhythm in plasma testosterone levels in the rhesus monkey *(Macaca mulatta):* A three year study. *Hormones and Behavior,* 1976, *7,* 229–243.

Gordon, T. P., Bernstein, I. S., & Rose, R. M. Social and seasonal influences on testosterone secretion in the male rhesus monkey. *Physiology and Behavior,* 1978, *21,* 623–627.

Gordon, T. P., Rose, R. M., Grady, C. L., & Bernstein, I. S. Effects of increased testosterone secretion on the behavior of adult male rhesus living in a social group. *Folia Primatologica,* 1979, *32,* 149–160.

Goy, R. W., Phoenix, C. H., & Resko, J. A. Relationship of testosterone level to sexual behavior in juvenile rhesus monkeys. *American Zoologist,* 1965, *5,* 713.

Hall, K. R. L. Aggression in monkey and ape societies. In J. D. Carthy & F. J. Ebling (Eds.), *The natural history of aggression.* London: Academic Press, 1964.

von Holst, D. Social stress in tree shrews: Problems, results and goals. *Journal of Comparative and Physiological Psychology,* 1977, *120,* 71–86.

Hurme, V. O., & van Wagenen, G. Basic data on the emergence of permanent teeth in the rhesus monkey *(Macaca mulatta). Proceedings of the American Philosophical Society,* 1961, *105,* 105–140.

Keverne, E. B., Leonard, R. A., Scruton, D. M., & Young, S. K. Visual monitoring in social groups of talapoin monkeys *(Miopithecus talapoin). Animal Behaviour,* 1978, *26,* 933–944.

Koford, C. B. Population dynamics of rhesus monkeys on Cayo Santiago. In I. DeVore (Ed.), *Primate behavior: Field studies of monkeys and apes.* New York: Holt, Rinehart & Winston, 1965.

Martin, D. E., Swenson, R. B., & Collins, D. C. Correlation of serum testosterone levels with age in male chimpanzees. *Steroids,* 1977, *29,* 471–481.

Michael, R. P., & Bonsall, R. W. A three-year study of an annual rhythm in plasma androgen levels in male rhesus monkeys *(Macaca mulatta)* in a constant laboratory environment. *Journal of Reproduction and Fertility,* 1977, *49,* 129–131.

Michael, R. P., & Wilson, M. I. Effects of castration and hormone replacement in fully adult male rhesus monkeys *(Macaca mulatta). Endocrinology,* 1974, *95,* 150–159.

Michael, R. P., & Wilson, M. I. Mating seasonality in castrated male rhesus monkeys. *Journal of Reproduction and Fertility,* 1975, *43,* 325–328.

Michael, R. P., & Zumpe, D. Annual cycles of aggression and plasma testosterone in captive male rhesus monkeys. *Psychoneuroendocrinology,* 1978, *3,* 217–220.

Michael, R. P., Wilson, M., & Plant, T. M. Sexual behaviour of male primates and the role of testosterone. In R. P. Michael & J. H. Crook (Eds.), *Comparative ecology and behaviour of primates.* London: Academic Press, 1973.

Michael, R. P., Setchell, K. D. R., & Plant, T. M. Diurnal changes in plasma testosterone and studies on plasma corticosteroids in non-anaesthetized male rhesus monkeys *(Macaca mulatta). Journal of Endocrinology,* 1974, *63,* 325–335.

Nieschlag, E. Circadian rhythm of plasma testosterone. In J. Aschoff, F. Ceresa, & F. Halberg (Eds.), *Chronobiological aspects of endocrinology.* Stuttgart: Schattauer, 1974.

Perachio, A. A., Alexander, M., & Marr, L. D. Hormonal and social factors affecting evoked sexual behavior in rhesus monkeys *American Journal of Physical Anthropology,* 1973, *38,* 227–232.

Perachio, A. A., Alexander, M., Marr, L. D., & Collins, D. C. Diurnal variations of serum testosterone levels in intact and gonadectomized male and female rhesus monkeys. *Steroids,* 1977, *29,* 21–33.

Phoenix, C. H. The role of testosterone in the sexual behavior of laboratory male rhesus. In C. H. Phoenix (Ed.), *Symposia of the 4th International Congress of Primatology* (Vol. 2): *Primate reproductive behavior.* Basel: Karger, 1973.

Phoenix, C. H. The role of androgens in the sexual behavior of adult male rhesus monkeys. *Advances in Behavioral Biology,* 1974, *2,* 249–258.

Phoenix, C. H. Steroids and sexual behavior in castrated male rhesus monkeys. *Hormones and Behavior,* 1978, *10,* 1–9.

Phoenix, C. H., Dixson, A. F., & Resko, J. A. Effects of ejaculation on levels of testosterone, cortisol and luteinizing hormone in peripheral plasma of rhesus monkeys. *Journal of Comparative and Physiological Psychology,* 1977, *91,* 120–127.

Plant, T. M., Zumpe, D., Sauls, M., & Michael, R. P. An annual rhythm in the plasma testosterone of adult male rhesus monkeys maintained in laboratory. *Journal of Endocrinology,* 1974, *62,* 403–404.

Rose, R. M., Bernstein, I. S., & Holaday, J. W. Plasma testosterone, dominance rank and aggressive behavior in a group of male rhesus monkeys. *Nature,* 1971, *231,* 366–368.

Rose, R. M., Gordon, T. P., & Bernstein, I. S. Plasma testosterone levels in the male rhesus: Influences of sexual and social stimuli. *Science,* 1972, *178,* 643–645.

Rose, R. M., Bernstein, I. S., Gordon, T. P., & Catlin, S. F. Androgens and aggression. A review. In R. L. Holloway (Ed.), *Primate aggression, territoriality, and xenophobia.* New York: Academic Press, 1974.

Rose, R. M., Bernstein, I. S., & Gordon, T. P. Consequences of social conflict on plasma testosterone levels in rhesus monkeys. *Psychosomatic Medicine,* 1975, *37,* 50–61.

Rose, R. M., Gordon, T. P., & Bernstein, I. S. Diurnal variation in plasma testosterone and cortisol in rhesus monkeys living in social groups. *Journal of Endocrinology,* 1978, *76,* 67–74.

Rose, R. M., Bernstein, I. S., Gordon, T. P., & Lindsey, J. G. Changes in testosterone and behavior during adolescence in the male rhesus monkeys. *Psychosomatic Medicine,* 1978, *40,* 60–70.

Rowell, T. E. Hierarchy in the organization of a captive baboon group. *Animal Behaviour,* 1966, *14,* 430–443.

Rowell, T. E. Baboon menstrual cycles affected by social environment. *Journal of Reproduction and Fertility,* 1970, *21,* 133–141.

Rowell, T. E. Reproductive cycles of the talapoin monkey *(Miopithecus talapoin). Folia Primatologica,* 1977, *28,* 188–202.

Sade, D. S. Seasonal cycle in size of testes of free-ranging *Macaca mulatta. Folia Primatologica,* 1964, *2,* 171–180.

Sade, D. S. Determinants of dominance in a group of free-ranging rhesus monkeys. In S. A. Altmann (Ed.), *Social communication among primates.* Chicago: University of Chicago Press, 1967.

Snipes, C. A., Forest, M. G., & Migeon, C. J. Plasma androgen concentrations in several species of Old and New World monkeys. *Endocrinology,* 1969, *85*(5), 941–945.

Vandenbergh, J. G., & Drickamer, L. C. Reproductive coordination among free-ranging rhesus monkeys. *Physiology and Behavior,* 1974, *13,* 373–376.

Wilson, A. P. Behavior of agonadal free-ranging rhesus monkeys. *American Journal of Physical Anthropology,* 1969, *31,* 261.

Wilson, A. P., & Vessey, S. H. Behavior of free-ranging castrated rhesus monkeys. *Folia Primatologia,* 1968, *9,* 1–14.

Wilson, E. O. *Sociobiology: The new synthesis.* Cambridge, Mass.: Belknap Press of Harvard University Press, 1975.

Zamboni, L., Conaway, C. H., & van Pelt, L. Seasonal changes in production of semen in free-ranging rhesus monkeys. *Biology of Reproduction,* 1974, *11,* 251–267.

21

Hormones, Aggression, and Dominance in Humans

Allan Mazur

INTRODUCTION

The most important form of human aggression is organized large-scale warfare, found in no other species. Hormones will probably not explain war, which is best understood in terms of bureaucracies, complex organizations, and the politics of nations. Rather, it is at the level of face-to-face interaction, which occurs among all social vertebrates, where physiological mechanisms are likely to be influential in human as well as animal affairs.

The ethical and practical problems of hormonal studies of human aggression are formidable. Not much empirical work has been done, and much of what has been done is methodologically weak. The number of review articles on the topic may now exceed the number of relevant, original research reports (e. g., Bell, 1978; Leshner, 1978; Mazur, 1976; Moyer, 1974; Rada, Kellner, & Winslow, 1976). Lacking good data from human subjects, there is a heavy reliance on extrapolation from animal studies, a hazardous enterprise, given the importance of cultural influences on human behavior. Nonetheless, the extrapolations and the sketchy human data are sufficiently intriguing to justify the present strong interest in this field.

It is important to distinguish aggressive behavior from dominance behavior. Individuals will be said to act *aggressively* if their apparent intent is to inflict physical injury on other members of the species. This refers to face-to-face actions, such as punching, biting, shoving, or clubbing, but does not include verbal abuse or distal or symbolic actions, such as aerial bombing or voting for capital punishment. Individuals will be said to act *dominantly* if their apparent intent is to achieve or maintain a status advantage (i.e., power, prestige, or valued prerogatives) over other members of the species. Dominance behavior may be aggressive or intimidating but it need not be, and among adults of the higher primates, particularly man and apes, it usually takes nonaggressive forms, including vocalization and physical gesturing (Mazur, 1973).

ALLAN MAZUR • Department of Sociology, Syracuse University, Syracuse, New York 13210.

Most relevant animal research has looked at intermale aggression among species that achieve and maintain status aggressively, such as rodents. Most of these results may be regarded as linking hormones with dominance behavior as well as with aggressive behavior, since the two are intertwined. In apes and man, in which dominating and aggressive behaviors are often separated, hormones may be related to one but not the other. This review will suggest that apparent inconsistencies in the available data are explicable if one assumes that in humans, testosterone affects dominance behavior but not aggressive behavior.

The topics that follow are not all-inclusive but have been selected because they have received the most research attention or seemed most promising. The first of these is the early development of sex differences, producing males who are more aggressive than females. This is followed by an examination of testosterone and adult intermale aggression, and then of testosterone and intermale dominance. Beside testosterone, the major hormones of the adrenal cortex are known to be related to dominance/aggression among animals, and these are discussed next. Finally, there are indications that hormone changes during the menstrual cycle may affect irritable aggression in women.

DEVELOPMENT OF SEX DIFFERENCES IN AGGRESSION

Among humans, males are more aggressive than females (Maccoby & Jacklin, 1974). This may reflect a male tendency toward higher activity and more rough-and-tumble play, and perhaps boys are no more aggressive than girls of comparable activity level, a possibility which has not been tested. In any case, an overall sex difference in aggressiveness is found consistently across cultures. It is clear that boys are often trained to be more aggressive than girls (or girls are trained to be more passive). However, observed differences in aggression cannot be explained simply as the result of sex-role training because, in a study of six diverse cultures, sex differences in 3- to 6-year-old children were as large as those in 7- to 10-year-olds, who had had more role-learning opportunity (D'Andrade, 1966).

A hormonal explanation for the development of sex differences in aggression, which is well supported by studies of rodents and monkeys, seems attractive for humans as well. Genetically male fetuses develop testes, which produce androgens, the male sex hormones, including testosterone. These influence the development of secondary sexual characteristics. A genetically male fetus that is somehow deprived of androgen will have female-appearing genitals, whereas a genetically female fetus exposed to androgen will develop male-appearing genitals. The principle of differentiation, somewhat simplified, appears to be: Add androgen and obtain a male; add nothing and obtain a female (Money & Ehrhardt, 1972). This principle appears applicable to sexual patterns of behavior as well as to physical characteristics. If pregnant rhesus monkeys are given testosterone, their female infant offspring, presumably exposed to the testosterone as fetuses, show threat behavior and rough-and-tumble play that greatly exceed those shown by normal females and are more typical of infant males

(Goy, 1970). As adults, such females are more aggressive than normal females (Eaton, Goy, & Phoenix, 1973).

Obviously, humans cannot be experimentally manipulated in this way, but there have been unfortunate instances in which human female fetuses have received more androgen than normal, with effects on subsequent development. Genetically female fetuses may receive excess androgen from their own abnormally functioning adrenocortical glands, a condition known as adrenogenital syndrome (AGS), which is treatable by permanent cortisone therapy. Also, some normal female fetuses have been androgenized as a result of drugs given to their mothers to prevent miscarriage; these children require no hormone therapy. Females exposed to excess androgen *in utero,* for whatever reason, are born with varying degrees of masculinized genitals, from a slightly enlarged clitoris to a normal-looking penis and empty scrotum.

Money and his associates studied about 42 such girls and concluded that, compared with normal girls, they are tomboyish in tastes and behavior, often preferring boys as playmates, being highly active in rough outdoor play, and showing more interest in boys' than girls' games. The relevant question here is whether these girls are more aggressive than normal, and the answer is not clear-cut. Aggression was measured by asking the girls and their mothers who usually instigated fights. In one series of 17 AGS girls, the subjects were nearly twice as likely as control girls to initiate fighting, but this difference was not significant, given the small sample size (Ehrhardt & Baker, 1974). In an earlier series of 25 androgenized girls (both AGS and drug-induced), the comparison of their aggression with that of control girls was simply reported as having "no significant difference," nor were the androgenized girls perceived to be particularly dominant (Money & Ehrhardt, 1972, p. 106; also see Money & Schwartz, 1976).

A number of methodological issues have been raised about the Money studies (Quadagno, Briscoe, & Quadagno, 1977). The measure of aggression is very crude, and girls and their mothers may have underreported aggressiveness to conform to desired sex stereotypes. While the girls with masculinized genitals did have corrective surgery at an average age of about two years, the psychic effects of the genital abnormality on mother and child are unknown. Finally, in most experimental studies, testosterone was used to produce masculinization and aggression in female animals, whereas the masculinization of human fetuses seems to have involved other androgens. These issues make the Money and Ehrhardt studies difficult to interpret.

The androgen hypothesis of sex differences in human aggression is promising but certainly not well tested. Future research might profitably focus on androgen effects during adolescence, which seem more accessible to study in human subjects than do *in utero* effects. Boys experience rapidly rising testosterone levels approximately between the ages of 12 and 20 (Brotherton, 1976, p. 301; Frasier, Gafford, & Horton, 1969). The beginning of this period is a time of increased dominating behavior, fighting, and intermale competition, similar to the increased aggression and dominance behavior of male apes during puberty (Hamburg, 1971). If the timing of a boy's rise in testosterone and the development of his secondary sexual characteristics allow good predictions of subsequent aggressive or dominating behaviors, than a causal role for androgen in producing these behaviors would be increasingly plausible.

Before concluding this discussion of sex differences in aggression, it is worthwhile to comment on the reputed hyperaggressivity of XYY males, which has received much popular attention. It is well known that a normal person has two sex chromosomes: XY if male, XX if female. There are, however, abnormal individuals who have more than two or only one sex chromosome. Thus, some males have an "extra" Y chromosome, and it has been conjectured that if the presence of one Y chromosome in normal males is associated with high aggressiveness (relative to females), than an abnormal male with two Ys might be extra-aggressive, perhaps because of a heightened testosterone level. Reports that, compared with their incidence in the general population, relatively large numbers of XYY males are found in maximum security mental institutions support this hypothesis. It is well documented by now that XYY men are more likely than XY men to have criminal arrests, but when the crimes are examined in detail, they are not particularly violent ones, nor do fragmentary endocrine studies indicate abnormally high testosterone levels (Borgaonkar & Shah, 1974; Witkin, Mednick, Schulsinger, Bakkestrom, Christiansen, Goodenough, Hirschhorn, Lundsteen, Owen, Philip, Rubin, & Stocking, 1976). XYY men are tall and tend to have low measured intelligence, suggesting that their propensity for criminal acts, or more particularly for being arrested and sentenced for such acts, may be explained by these factors rather than by increased aggressiveness. Furthermore, it appears that men with abnormal XXY genotypes are also overrepresented in the same kinds of mental/penal institutions as XYY men, and this outcome cannot be explained as the result of an "extra Y" chromosome (Mazur & Robertson, 1972; Witkin *et al.*, 1976). At present, there seems little reason to believe that XYY males are hyperaggressive because of an androgen effect produced by the double Y; however, the issue has not been settled definitively.

TESTOSTERONE AND AGGRESSION IN MEN

Castration is a traditional method for gentling a bull, perhaps because it removes the animal's major source of testosterone. In America and Europe during this century, hundreds of deviate men, usually sex offenders of low intelligence, have been castrated. Reports of the "success" of this treatment have been cited in support of a testosterone effect on aggression in humans. However, these data are of questionable value, usually being anecdotal, lacking both proper control groups and direct measures of aggression, and sometimes showing distinct biases. For example, Hawke (1951) reports on his castrates:

> Many of these individuals so treated were vicious homesexuals [*sic*] and very brutal in attacks on small children. They were very unstable and would create a disturbance at every opportunity. After castration, they became stabilized, and those who cannot be paroled are good useful citizens in the institution.

> In our experimental work, we have administered Testosterone, the male hormone [to some castrates]. . . . In a number of cases, after we had treated them for a period of two or three weeks, the floor supervisor would call me up and ask if I would not be willing to stop administering Testosterone to certain individuals who had reverted to all of their anti-social tendencies, were attacking small children, starting fights, breaking windows and destroying furniture. We would stop the administration of Testosterone in these individuals, and within a few days they would be restabilized and cause no further ward disturbances. We have felt that this proves the male hormone is the exciting factor in these cases. (p. 222)

Castrated sex criminals, often with records of violent crime, appear to have low recidivism rates (LeMaire, 1956). This may be due to a hormone effect or, alternatively, to the effectiveness of castration as a symbolic deterrent, the parolee wondering what more would be cut off should he be caught again. Bremer (1959) reports that castration does not reduce aggressiveness *per se* but does reduce libido and might thereby remove the motivation for repeated sex offenses. While these castration studies suggest a causal link between androgen and aggression, they are surely not compelling. In this regard they may be clumped with the occasional anecdotal reports of aggressive individuals who became placid after receiving estrogen-like compounds or antiandrogen (Bell, 1978; Sands, 1954) and of the quiet individuals who became assertive after receiving testosterone therapeutically (Howard & Vest, 1939; Pratt, 1942).

With the development of radioimmunoassays for testosterone, several investigators attempted to correlate naturally occurring variation in testosterone with aggressiveness, as in many of the animal studies reported in this volume. While this is a relatively simple matter with laboratory animals, it is difficult and unethical to encourage or even allow high aggression in human subjects. Thus, when Persky, Smith, and Basu (1971) tried this approach with men as subjects, they measured aggression by administering paper-and-pencil tests, which are not reliably related to actual aggressive acts (Buss, Fischer, & Simmons, 1968; Ehrenkranz, Bliss, & Sheard, 1974; Kreuz & Rose, 1972). Their report of a high correlation between testosterone production rate and "aggression" has been widely cited, so it is worth examining this result in some detail. Their basic data display is a matrix of correlations among seven personality scales and three testosterone measures (plasma testosterone level, testosterone production rate, and metabolic clearance rate), producing 21 correlations. The correlations are calculated for two separate subject groups, one consisting of young men (n = 18) and the other of older men (n = 15), so there are a total of 42 testosterone–personality correlations. Three of these 42 correlations are significant at the 0.05 level or better, which is about what one would expect by chance, given the null hypothesis. Two of the significant correlations relate score on the Buss-Durkee Hostility Inventory to plasma testosterone level and to testosterone production rate, but only for the younger men; for the older men, one of these correlations is negative (but nonsignificant) and the other is near zero. (The third significant correlation relates testosterone production rate to an anxiety scale, among the younger subjects, and is not relevant here.)

The authors of this study consider these results to be evidence of a testosterone–aggression link in humans, which is a reasonable interpretation and one that has been widely accepted. However, one can equally well consider the few significant correlations to be chance outcomes, particularly since they are not repeated in the older-subject group or with other personality scales that purport to measure aggression and hostility. Note too that since plasma testosterone level and testosterone production rate were highly correlated to each other, their respective correlations with the Buss-Durkee scale are redundant.

Several subsequent attempts at other laboratories to relate testosterone levels to pencil-and-paper measures of aggression or hostility, including the Buss-Durkee Inventory and other scales, have failed to replicate significant correlations (Brown & Davis, 1975; Doering, Brodie, Kraemer, Moos, Becher, & Mechanic, 1975; Dotson, Robertson, & Tuchfeld, 1974; Ehrenkranz *et al.*, 1974; Kreuz & Rose, 1972; Meyer-Bahlburg, Boon, Sharma, & Edwards, 1973; Monti, Brown, & Corriveau, 1977; Rada *et al.*, 1976). It now seems clear that plasma testosterone level is not correlated in any substantial way with aggression as measured on common personality scales, and the Persky *et al.* (1971) result was most likely a chance outcome.

Kreuz and Rose (1972) tried to relate testosterone to observable aggressive behavior rather than limiting themselves to pencil-and-paper scales of aggressiveness. Among their subjects, 21 young male prisoners with records of violent crime, plasma testosterone was not related to frequency of fighting while in prison, to frequency of verbal aggression while in prison, or to scores on the Buss-Durkee Hostility Inventory. However, the 10 men with histories of more violent and aggressive offenses *during adolescence* had significantly higher testosterone levels than the men without such adolescent histories.

Kreuz and Rose suggest that the men who are currently high in testosterone were also relatively high as adolescents or entered maturity earlier than the others. Their high testosterone back in the teen-age years may have made these men relatively violent at that time. Unfortunately, there is little information on the stability of testosterone levels over years-long periods, but Doering *et al.* (1975) report large fluctuations in individuals' testosterone levels over a period of two months, so longer-term stability seems doubtful. In view of the tenuous causal link between present testosterone concentration and violent activity years earlier, one would like to see a replication of this result before accepting it as valid.

Ehrenkranz *et al.* (1974) measured testosterone levels in three groups of prison convicts (n = 12/group). One group was made up of "socially dominant" *but unaggressive* men who were in prison for nonviolent crimes. "They were recognized by prison staff and other inmates as socially dominant and had asserted themselves into prestigious jobs and positions in inmate hierarchies" (p. 470). A second group was composed of men who were "chronically aggressive"; all were in prison for violent crimes, and they continued to show aggressive and threatening behavior while in prison. The third group consisted of convicts who were neither dominant nor aggressive. The socially dominant but unaggressive group and the chronically aggressive group did not have significantly different mean testosterone levels, but both had significantly higher levels than the nondominant unaggressive group. Unfortunately, the interpretation of these results is clouded because the relative dominance of the aggressive men is not reported. It seems unlikely that they were a submissive lot, and perhaps they were more dominant than the "socially dominant" group. Thus, the aggressive group may have had high testosterone because of their aggressiveness or because of their dominance. In any case, it is clear that the "socially dominant" group had high testosterone without aggression.

In sum, the available data on testosterone and aggression in men provide a weak and inconsistent picture. The weight of evidence is against a relationship between testosterone and pencil-and-paper measures of aggression. Of two attempts to relate testosterone to currently occurring observable aggressive behavior in men, one gave a negative and one a positive result. One reported relationship between present testosterone level and long-past violent behavior is problematic and must be regarded as dubious unless a successful replication is reported. Taken together, there is not much evidence from human subjects favoring the hypothesized link between circulating testosterone level and aggression in postadolescent males. It should also be noted that if the link does exist, then a high testosterone level could be an effect of aggressiveness rather than its cause, or testosterone and aggressiveness could be correlated spuriously through some third (unknown) variable.

TESTOSTERONE AND DOMINANCE IN MEN

The distinction between aggressive behavior and dominating behavior is particularly important for humans, who often assert their dominance without any intent to cause injury. It may be the case that testosterone is related *primarily* to dominance behavior and not to aggression except in situations in which dominance happens to be asserted aggressively. Ehrenkranz *et al.* (1974) showed that socially dominant but *unaggressive* prisoners had relatively high plasma testosterone, not significantly different from the testosterone levels of aggressive prisoners (who may have been dominant too). Nearly all primate studies that have been interpreted as linking testosterone to aggression (Dixson, 1980) may as easily be interpreted as linking testosterone with dominating behavior (Mazur, 1976).

If there is a link between testosterone and dominance, the primate studies suggest that it works in two directions. First, changes in testosterone may influence changes in dominating behavior, with heightened testosterone facilitating attempts to achieve or maintain high status and decreased testosterone inhibiting such attempts. Various manipulations of testosterone in monkeys and apes (through castration or hormone injection) have had mixed results but have given some support for a causal link from hormones to behavior. Testosterone given therapeutically is sometimes said to increase a human subject's assertiveness (Howard & Vest, 1939; Pratt, 1942), but such anecdotal claims ought to be regarded cautiously. At present there are no firm data on the effect of testosterone on dominance in humans.

Causation may work in the opposite direction, with changes in dominating behavior causing changes in testosterone level, as has been demonstrated in monkeys. Successful attempts to achieve or maintain status seem to increase plasma testosterone, while unsuccessful attempts or defeats seem to produce a decline in testosterone (Rose, Bernstein, & Gordon, 1975). In order to test for this effect in humans, one might place men in a competitive situation in which some will win status and others lose it; testosterone in the winners should rise and in the losers should fall. In a small study of this sort, young male subjects competed in doubles tennis matches of which each winner received a prize of $100. In two matches with decisive triumphs, the four losers showed drops in testosterone one hour after the matches, while three of the four winners showed rises in testosterone. In a third match, in which the victory was by the slimmest of margins and the players did not feel that there had been a clear triumph, winners and losers all showed testosterone drops (Mazur & Lamb, 1980). Elias (1981) has recently reported a similar finding among college wrestlers, with winners of competitive matches showing greater increases in testosterone than losers.

Testosterone has been measured in a few other situations in which men have undergone status changes. Five male seniors in medical school showed relative rises in testosterone the day after they received their M.D. degrees (Mazur & Lamb, 1980). Twenty-seven army recruits during basic training (Rose, Bourne, Poe, Mougey, Collins, & Mason, 1969) and 18 young men in officer candidate school (Kreuz, Rose, & Jennings, 1972) showed very low testosterone levels compared with control populations, perhaps because of their degraded status in these training situations, though other stress factors might also account for these low levels. (Unfortunately, testosterone was not measured in their instructors, who would be expected to show high levels.) Perhaps these results will encourage more empirical work in this promising but barely explored area.

ADRENOCORTICAL ACTIVITY AND DOMINANCE

Several monkey studies show correlations between status rank and adrenocortical activity (measured in a variety of ways), but the direction of the relationship is not consistent. In some cases, dominant rhesus macaques showed less adrenocortical activity than subordinates (Chamove & Bowman, 1978; Golub, Sassenrath, & Goo, 1979; Keverne, 1979; Sassenrath, 1970), which is the relationship usually reported in rodents. In other studies using other monkey species, the dominant animal was relatively high in adrenocortical activity (Coe, Mendoza, & Levine, 1979; Hayama, 1966; Leshner & Candland, 1972). The different species of experimental animals might account for this difference. It will be difficult to extrapolate to humans until a clearer understanding is obtained of the nonhuman primates. However, a preliminary picture has begun to emerge that may clarify these inconsistencies.

Since adrenocortical activity is a manifestation of stress, one would expect the greatest activity in the individuals facing the greatest stressors. Stressors may originate within the primate group, as when there are challenges for status or disputes over a valued commodity, or they may originate outside the group, as from a predator or an experimenter trying to capture a group member for testing. If the source of stress originates within the group, it is usually a dominant individual who applies the stress, being able to turn it on or off at will. He will experience less stress than his subordinate targets and, therefore, show less adrenocortical activity than the subordinates. However, if the stress originates outside of the group and is not under the control of the dominant animals, then one would expect greater stress (and greater adrenocortical activity) in whoever meets the stressor most directly. It is the leaders of primate groups who confront external stressors such as predators or conspecific enemies; the subordinates are shielded (Mazur, 1973). Thus, stressors that originate outside of the group are likely to produce relatively high adrenocortical activity in the high-ranked members, while stressors originating inside the group are likely to produce high activity in the low-ranked members. Existing studies have not controlled for the differential exposure of high- and low-ranked members to stressors, which may account for the apparent inconsistency in results. Monkeys in newly formed groups or housed in crowded cages experience high internal conflict, whereas in established groups with roomy environs, the dominant animals allow the subordinates a more placid existence.

Two studies of adrenocortical activity in human groups used military teams under externally produced stress. Marchbanks (1958) studied a bomber crew on a long flight: The pilot showed the highest adrenocortical activity. Bourne, Rose, and Mason (1968) examined an army team anticipating an enemy attack: Adrenocortical activity was higher in the 2 officers than in the 10 enlisted men. No studies of human groups with pre-existing status hierarchies have been noted in which the stressors are primarily of internal origin.

THE MENSTRUAL CYCLE

According to folklore and common experience, women's moods are often correlated with their menstrual cycles. In particular, the days before menstruation and the first days of blood flow are regarded as negative mood days, marked by irritability, depression, and lethargy. Correlations between mood and menstrual cycle have been reported in a number of studies, some methodologically poor and with sexist biases (Sherif, 1980), but some without these weaknesses. Rossi and Rossi's (1980) study of 82 college women over a 40-day period, which is methodologically strong and without apparent sexist bias, found relatively positive moods at the middle (ovulatory) phase of the menstrual cycle and relatively negative moods in the week preceding menstruation and in the first day or two of bleeding. Women taking birth control pills, who therefore do not experience the normal monthly fluctuations in estrogen and progesterone, did not report these mood fluctuations, a fact that the Rossis regard as indirect evidence that the mood cycle is caused by hormone changes.

One might expect that mood cycles would be reflected in corresponding monthly cycles of misbehavior, mishap, or morbidity, and a number of studies, particularly by Dalton (1964, 1968), document such effects. The most often-reported of these effects is a significantly increased likelihood that women will attempt suicide and succeed in the few days before or after the beginning of menstruation (see citations in Dalton, 1964; also Mandell & Mandell, 1967; Tonks, Rack, & Rose, 1968), although Birtchnell and Floyd (1974) did not find this effect.

Of particular concern here is the claim that women, in their irritable phase, are prone to angry outbursts of aggression (Dalton, 1964; Moyer, 1974). Cooke (1945) reported that 84% of violent crimes committed by Parisian women occurred on menstrual or premenstrual days, but this often-cited number is based on hearsay evidence of doubtful veracity. A first-hand report of women on a prison farm claimed that, of 42 inmates guilty of violent crimes whose menstrual dates could be recalled, 62% committed their crimes during the premenstrual week and 17% while menstruating, giving a total figure very close to Cooke's 84% (Morton, Additon, Addison, Hunt, & Sullivan, 1953). However, the method of fixing the date of the crime relative to the date of menstruation, apparently based solely on the inmates' memory of these events months earlier, seems inadequate to the task, and the result may not be credible.

Dalton (1964) studied a girls' school in which older students, aged 16–18, served as prefects who were permitted to punish the younger girls for misbehavior. The 11 prefects punished significantly more than would be expected by chance during the first four days of their own periods, but they punished less than expected during four days of the premenstruum. The form of punishment is not reported, but presumably it was not an infliction of physical injury and thus was not really aggression as defined in this review.

Hands, Herbert, and Tennent (1974) studied 23 women who had been housed at least six months in a secure hospital ward for patients with dangerous, violent, or criminal propensities. If patients became particularly disturbed, and especially if they started "acting out," then it was the practice to confine them to their rooms. Records were kept of these confinements and also of menstrual dates, and a comparison of these data showed significantly more confinements in the week before menstruation than would be expected by chance (also see statistical corrections by James, 1974). Some caution is obviously required in interpreting this result since confinement may not be an adequate proxy measure for aggressive behavior.

In the best study to date on aggression and menstruation, Ellis and Austin (1971) followed 45 prison inmates through three complete menstrual cycles. During this time, prison officials recorded for the study 174 aggressive acts, about one-third being physical attacks and the rest verbal abuse. Both physical and verbal attacks were concentrated at greater-than-expected levels in the premenstrual and menstrual days, and these were also the days of highest feelings of irritability, according to inmates' self-reports.

A number of criticisms have been raised against the whole body of menstrual studies (Laws & Schwartz, 1977; Parlee, 1973), and the most interesting will follow.

First, the particular phase of the cycle that is associated with mishaps and morbidity shifts from study to study and from behavior to behavior, sometimes being the premenstruum, sometimes menstruation itself, and occasionally ovulation. Timing is confounded by the wide variation in cycle length from woman to woman, and for some women, from one cycle to the next. However, if one assumes that the "negative phase" of the cycle is the week prior to menstruation and the first day or two of bleeding—consistent with Rossi and Rossi's (1980) careful mood study—then most reported periods of high mishap misbehavior and morbidity fall within this phase (making allowances for small sample fluctuations).

Second, most menstrual studies have focused on unusual populations such as prisoners or patients in psychiatric wards, and it is not clear that effects observed in such settings are generalizable to the larger population.

Third, since it is known that stressors can alter the timing of menstruation, causing it to be early, late, or missed altogether (Dalton, 1968), the possibility remains that mishaps are the cause of menstrual onset rather than its effect. For example, a woman in midcycle, involved in an auto accident, might start to menstruate because of the stress experience. An investigator timing the accident from the woman's first subsequent period would falsely conclude that the accident had occurred during the premenstruum. One should note, however, that such effects could obscure a real premenstrual effect as easily as exaggerate it. If the accident had occurred in the premenstruum, and the resulting stress had delayed or eliminated the imminent onset of bleeding, then the investigator would erroneously place this mishap outside of the premenstruum.

Finally, it must be emphasized that even if behaviors and moods are firmly correlated with particular phases of the menstrual cycle, it is still not obvious that hormonal changes are the causes of these behaviors and moods. The strong social expectations that girls develop about the menstrual period and its dysfunctions, even before they reach menarche (Clarke & Ruble, 1978), may serve as self-fulfilling prophecies.

In sum, there is a substantial body of evidence indicating that women experience relatively negative moods during the week preceding and the first days of menstruation. There does appear to be an increase in morbidity and misbehavior at that time, including suicide and aggression against others. It is plausible that this aggression is the result of hormone-induced irritability. But valid objections can be raised against the data obtained so far, and even accepting the cyclicity of aggression, its causal link to hormonal change has not been demonstrated.

CONCLUSIONS

Hormonal influences on human aggression have not been firmly established and must be regarded as speculative at this time. However, there are promising leads that may eventually show hormonal effects. Boys are physically more aggressive than girls in diverse cultures, and since this difference is often as strong among younger as older children, it seems unlikely to be wholly due to sex-specific training. Androgens are known to be important in sex differentiation at the fetal and adolescent stages, so it seems plausible that similar hormonal effects will be implicated in the sex difference in aggression.

Among postadolescent males, level of plasma testosterone may be related to aggressiveness, but studies of humans to date are weak and inconsistent. It seems as plausible to link testosterone to dominance behavior (which is often unaggressive in humans) as to aggression per se. Furthermore, if such links exist, they may be reciprocal ones, with dominance or aggressive behavior affecting hormone levels as well as hormones affecting dominance or aggression.

Adrenocortical activity appears to be related to status ranking in monkey troops, but empirical studies are inconsistent and the precise nature of the relationship is unclear, so it is difficult to extrapolate to humans. Hypothetically, position in a group's status hierarchy affects one's exposure to stressors, and this in turn should affect one's adrenocortical activity. These responses have barely been studied in humans.

Irritable individuals of either sex may be particularly prone to outbursts of angry aggression. In females, however, the menstrual cycle is correlated to mood, with irritability being relatively high in the premenstrual and early menstrual phases. It appears that some women are more likely to be aggressive during these high-irritability phases, though it is not clear that this is an explicitly hormonal effect, since other explanations are feasible.

ACKNOWLEDGMENTS

I appreciate the critical comments and advice of Robert Brush, Lee Ellis, and Jill Quadagno.

REFERENCES

Bell, R. Hormone influences on human aggression. *Irish Journal of Medical Science,* 1978, *147*(Suppl. 1), 5–9.

Birtchnell, J., & Floyd, S. Attempted suicide and the menstrual cycle—a negative conclusion. *Journal of Psychosomatic Research,* 1974, *18,* 361–369.

Borgaonkar, D., & Shah, S. The XYY chromosome male—or syndrome? In A. Steinberg & A. Bearn (Eds.), *Progress in medical genetics* (Vol. 10). New York: Grune & Stratton, 1974.

Bourne, P., Rose, R., & Mason, J. 17-OHCS levels in combat. *Archives of General Psychiatry,* 1968, *19,* 135–140.

Bremer, J. *Asexualization.* New York: Macmillan, 1959.

Brotherton, J. *Sex hormone pharmacology.* London: Academic Press, 1976.

Brown, W., & Davis, G. Serum testosterone and irritability in man. *Psychosomatic Medicine,* 1975, *37,* 87.

Buss, A., Fischer, H., & Simmons, A. Aggression and hostility in psychiatric patients. *Journal of Consulting and Clinical Psychology.* 1968, *32,* 21.

Chamove, A., & Bowman, R. Rhesus plasma cortisol response at four dominance positions. *Aggressive Behavior,* 1978, *4,* 43–55.

Clarke, A., & Ruble, D. Young adolescents' beliefs concerning menstruation. *Child Development,* 1978, *49,* 231–234.

Coe, C., Mendoza, S., & Levine S. Social status constrains the stress response in the squirrel monkey. *Physiology and Behavior,* 1979, *23,* 633–638.

Cooke, W. Presidential address: Differential psychology of American women. *American Journal of Obstetrics and Gynecology,* 1945, *65,* 457.

Dalton, K. *The premenstrual syndrome.* Springfield, Ill.: Charles C Thomas, 1964.

Dalton, K. Menstruation and examinations. *Lancet,* 1968, *2,* 1386–1388.

D'Andrade, R. Sex differences and cultural institutions. In E. Maccoby (Ed.), *The development of sex differences.* Stanford, Ca.: Stanford University Press, 1966.

Dixson, A. Androgens and aggressive behavior in primates: A review. *Aggressive Behavior,* 1980, *6,* 37–67.

Doering, C., Brodie, J., Kraemer, H., Moos, R., Becker, H., & Mechanic, D. Negative affect and plasma testosterone: A longitudinal human study. *Psychosomatic Medicine,* 1975, *37,* 484–491.

Dotson, L., Robertson, L., & Tuchfeld, B. *Some correlations among alcohol, cigarettes, hormones, and hostility.* (Mimeo.) Washington: Insurance Institute for Highway Safety, 1974.

Eaton, G., Goy, R., & Phoenix, C. Effects of testosterone treatment in adulthood on sexual behavior of female pseudohermophrodite rhesus monkeys. *Nature,* 1973, *242,* 119–120.

Ehrenkranz, J., Bliss, E., & Sheard, M. Plasma testosterone: Correlation with aggressive behavior and social dominance in man. *Psychosomatic Medicine,* 1974, *36,* 469–475.

Ehrhardt, A., & Baker, S. Fetal androgens, human central nervous system differentiation, and behavior sex differences. In R. Friedman, R. Richart, & R. Vande Wiele (Eds.), *Sex differences in behavior.* New York: Wiley, 1974.

Elias, M. Serum cortisol, testosterone and testosterone binding globulin responses to competitive fighting in human males. *Aggressive Behavior,* 1981, *7,* 215–224.

Ellis, D., & Austin, P. Menstruation and aggressive behavior in a correctional center for women. *The Journal of Criminal Law and Police Science,* 1971, *62,* 388–395.

Frasier, J., Gafford, F., & Horton, R. Plasma androgens in childhood and adolescence. *Journal of Clinical Endocrinology and Metabolism,* 1969, *29,* 1404–1408.

Golub, M., Sassenrath, E., & Goo, G. Plasma cortisol levels and dominance in peer groups of rhesus monkey weanlings. *Hormones and Behavior,* 1979, *12,* 50–59.

Goy, R. Early hormonal influences on the development of sexual and sex-related behavior. In F. Schmitt, G. Quarton, T. Melnechuck, & G. Adelman (Eds.), *The neurosciences: Second study program.* New York: Rockefeller University Press, 1970.

Hamburg, D. Psychobiological studies of aggressive behavior. *Nature,* 1971, *230,* 19–23.

Hands, J., Herbert, V., & Tennent, G. Menstruation and behavior in a special hospital. *Medicine, Science, and the Law,* 1974, *14,* 32–35.

Hawke, C. Castration and sex crimes. *American Journal of Mental Deficiency,* 1951, *55,* 220–226.

Hayama, S. Correlation between adrenal gland weight and dominance rank in caged crab-eating monkeys *(Macaca irus). Primates,* 1966, *7,* 21–26.

Howard, J., & Vest, S. Clinical experiments with male sex hormones. *American Journal of Medical Science,* 1939, *198,* 823.

James, W. Letter. *Medicine, Science, and the Law,* 1974, *14,* 290.

Keverne, E. Sexual and aggressive behavior in social groups of talapoin monkeys. *Ciba Foundation Symposium,* 1979, *62* (new series), 271–286.

Kreuz, L., & Rose, R. Assessment of aggressive behavior and plasma testosterone in a young criminal population. *Psychosomatic Medicine,* 1972, *34,* 321–332.

Kreuz, L., Rose, R., & Jennings, J. Suppression of plasma testosterone levels and psychological stress. *Archives of General Psychiatry,* 1972, *26,* 479–482.

Laws, J., & Schwartz, P. *Sexual scripts.* Hinsdale, Ill.: Dryden Press, 1977.

LeMaire, E. Danish experiences regarding the castration of sexual offenders. *Journal of Criminal Law, Criminology, and Police Science,* 1956, *47,* 294–310.

Leshner, A. *An introduction to behavioral endocrinology.* New York: Oxford University Press, 1978.

Leshner, A., & Candland, D. Endocrine effects of grouping and dominance rank in squirrel monkeys. *Physiological Behaviour,* 1972, *8,* 441–445.

Maccoby, E., & Jacklin, C. *The psychology of sex differences.* Stanford, Ca.: Stanford University Press, 1974.

Mandell, A., & Mandell, M. Suicide and the menstrual cycle. *Journal of the American Medical Association,* 1967, *200,* 792–793.

Marchbanks, W. Effect of flying stress on urinary 17-hydroxycorticosteroid levels. *Aviation Medicine,* 1958, *29,* 676–682.

Mazur, A. Cross-species comparison of status in established small groups. *American Sociological Review,* 1973, *38,* 513–530.

Mazur, A. Effects of testosterone on status in primate groups. *Folia Primatologica,* 1976, *26,* 214–226.

Mazur, A., & Lamb, T. Testosterone, status, and mood in human males. *Hormones and Behavior,* 1980, *14,* 236–246.

Mazur, A., & Robertson, L. *Biology and social behavior.* New York: Free Press, 1972.

Meyer-Bahlburg, H., Boon, D., Sharma, M., & Edwards, J. Aggressiveness and testosterone measures in man. *Psychosomatic Medicine,* 1973, *35,* 453.

Money, J., & Ehrhardt, A. *Man and woman, boy and girl.* Baltimore: Johns Hopkins University Press, 1972.

Money, J., & Schwartz, M. Fetal androgens in the early treated adrenogenital syndrome of 46XX hermaphroditism: Influence on assertive and aggressive types of behavior. *Aggressive Behavior,* 1976, *2,* 19–30.

Monti, P., Brown, W., & Corriveau, D. Testosterone and components of aggressive and sexual behavior in man. *American Journal of Psychiatry,* 1977, *134,* 692–694.

Morton, J., Additon, H., Addison, R., Hunt, C., & Sullivan, J. A clinical study of premenstrual tension. *American Journal of Obstetrics and Gynecology,* 1953, *65,* 1182–1191.

Moyer, K. Sex differences in aggression. In R. Friedman, R. Richart, & R. Vande Wiele (Eds.), *Sex differences in behavior.* New York: Wiley, 1974.

Parlee, M. The premenstrual syndrome. *Psychological Bulletin,* 1973, *80,* 454–465.

Persky, H., Smith, K., & Basu, G. Relation of psychologic measures of aggression and hostility to testosterone production in man. *Psychosomatic Medicine,* 1971, *33,* 265–277.

Pratt, J. A personal note on methyl testosterone in hypogonadism. *Journal of Clinical Endocrinology and Metabolism,* 1942, *2,* 490–464.

Quadagno, D., Briscoe, R., & Quadagno, J. Effect of perinatal gonadal hormones on selected nonsexual behavior patterns: A critical assessment of the nonhuman and human literature. *Psychological Bulletin,* 1977, *84,* 62–80.

Rada, R., Kellner, R., & Winslow, W. Plasma testosterone and aggressive behavior. *Psychosomatics,* 1976, *17,* 138–142.

Rose, R., Bourne, P., Poe, R., Mougey, E., Collins, D., & Mason, J. Androgen response to stress: II. Excretion of testosterone, epitestosterone, androsterone, and etiocholanolone during basic combat training and under threat of attack. *Psychosomatic Medicine,* 1969, *31,* 418–436.

Rose, R., Bernstein, I., & Gordon, T. Consequences of social conflict on plasma testosterone levels in rhesus monkeys. *Psychosomatic Medicine,* 1974, *37,* 50–61.

Rossi, A., & Rossi, P. Body time and social time: Mood patterns by menstrual cycle phase and day of week. In J. Parsons (Ed.), *The psychobiology of sex differences and sex roles.* New York: McGraw-Hill, 1980.

Sands, D. Further studies on endocrine treatment in adolescence and early adult life. *Journal of Mental Science,* 1954, *100,* 211–219.

Sassenrath, E. Increased adrenal responsiveness related to social stress in rhesus monkeys. *Hormones and Behavior,* 1970, *1,* 283–298.

Sherif, C. A social psychological perspective on the menstrual cycle. In J. Parsons (Ed.), *The psychobiology of sex differences and sex roles.* New York: McGraw-Hill, 1980.

Tonks, C., Rack, P., & Rose, M. Attempted suicide and the menstrual cycle. *Journal of Psychosomatic Research,* 1968, *11,* 319–23.

Witkin, H., Mednick, S., Schulsinger, F., Bakkestrom, E., Christiansen, K., Goodenough, D., Hirschhorn, K., Lundsteen, C., Owen, D., Philip, J., Rubin, D., & Stocking, M. Criminality in XYY and XXY men. *Science,* 1976, *193,* 547–555.

22

Hormonal Influences on Invertebrate Aggressive Behavior

Michael D. Breed and William J. Bell

INTRODUCTION

Aggressive behavior has been described in nearly all of the invertebrate groups. In many cases the presence or absence of aggression, or the intensity of aggressive expression, is known to vary with time or with changing environmental conditions; such variation is suggestive of hormonal influences, but direct hormonal control of aggression has been demonstrated in only a few cases. The purposes of this chapter are to review critically the literature concerning hormonal influences on aggression in some of the invertebrates and to suggest areas for possible future studies.

Before proceeding, some general introduction is necessary. The invertebrates are a phyletically diverse assemblage, with many different sorts of endocrine systems. Although aggression is known in "higher" invertebrates, such as molluscs and asteroids, we have chosen to restrict our discussion to arthropods, thereby reducing the diversity of endocrine systems with which we will deal. Most arthropod endocrine systems are well understood and are quite different from vertebrate endocrine systems, thus providing an interesting contrast.

The functions of hormones in arthropods differ from those in vertebrates on two principal points. First, expansion of body components is a cyclic phenomenon, requiring the hormonally controlled periodic shedding of the old exoskeleton and formation of a new exoskeleton. Second, efforts to demonstrate gonadally derived hormones that exert an influence on behavioral expression have generally failed in invertebrates (but see Beach, 1979). Thus, all individuals in a population will undergo marked cycles at least through part of their lifespan, and their vulnerability to attack will vary with their stage in the life cycle. Despite the lack of evidence concerning gonadally produced hormones that might be

MICHAEL D. BREED • Department of Environmental, Population, and Organismic Biology, University of Colorado, Boulder, Colorado 80309. WILLIAM J. BELL • Department of Entomology, University of Kansas, Lawrence, Kansas 66045.

responsible for sexually related behavior, there is ample evidence for the endocrine regulation of reproduction in arthropods and for behavioral phenomena mediated by reproductive hormones.

The hormonal control of behavior in insects has been reviewed by Truman and Riddiford (1974) and for all invertebrates by Truman and Riddiford (1977). Much of the available literature on hormonal influences on insect behavior comes from the work of Truman (e.g., Truman & Riddiford, 1970; Truman & Sokolove, 1972) on eclosion behavior in moths and the work of Barth (1961, 1962, 1965, 1968), Loher (1962, 1966), and Loher and Huber (1964, 1966), reviewed by Barth and Lester (1973), on the hormonal control of sexual behavior in Orthoptera.

In the following sections, the emphasis is on cases in which one of three events suggests the possible involvement of a hormone in the control of aggression: (1) cyclic changes in aggression correlated with the molt cycle; (2) changes in aggression associated with reproductive cycles, mating, or the onset of reproductive maturity; or (3) differential aggression observed among members of a social hierarchy, correlated with reproductive capacity.

THE CRUSTACEAN ENDOCRINE SYSTEM

Since the behavioral changes that we discuss in Crustacea are associated with the molt cycle, only those hormones and glands that have a role in the molt cycle will be discussed. Crustacea, like other arthropods, molt periodically through their lives in order to create room within the exoskeleton for the growth of body tissues. In contrast to the insects, most Crustacea continue to molt after reaching reproductive maturity. The molt cycle involves several stages; in the proecdysis, the calcium in the integument is absorbed into the blood, softening the integument so that it may then be shed at ecdysis. Postecdysis animals swallow water to expand the integument, and then calcium is deposited, giving it its shell-like characteristic. Animals then enter an intermolt period, in which the state of the integument remains constant. Some authors have divided the proecdysial period into stages, D_0 to D_4, corresponding to diminishing cuticular thickness (see review in Tombes, 1970).

A set of neurosecretory cells on the eyestalk, the X-organ, apparently synthesizes a factor that inhibits the onset of ecdysis. These cells empty their product in the sinus gland, which then releases the product into the hemocoel. Removal of the sinus gland does not affect the molt cycle, but removal of the X-organ releases precocial ecdysis. Implantation of the X-organ in an animal without the organ or implantation of supernumary X-organs has the effect of lengthening the intermolt period. Another gland, the Y-organ, synthesizes and secretes the molting hormone, ecydsone (a steroid). The product of the Y-organ is necessary for normal molting, and molting can be stimulated by the application of ecdysone to an animal. The X-organ also controls diuresis and protein metabolism and consequently has a role in the completion of the normal molt (reviewed in Tombes, 1970).

In addition to those hormones involved in molting, there are a pair of glands, called the androgenic glands or the vas deferens glands, associated with the male reproductive organs. These glands are necessary for the development of the male reproductive system and can masculinize a female if implanted. In species that have protandric hermaphroditism, these glands degenerate or become inactive when the individual switches to being female.

BEHAVIORAL CORRELATES OF ENDOCRINE CHANGES IN THE CRUSTACEA

Although several authors have commented on variation in the aggressiveness of Crustacea through the molt cycle (Cobb & Tamm, 1975), only one study has quantified the aggressive responses in a way sufficiently detailed to allow possible analyses of hormonal influences on aggressiveness (Cobb & Tamm, 1975; Tamm & Cobb, 1978). Working on the American lobster, *Homarus americanus,* Tamm and Cobb (1978) found that the proportion of aggression observed increased as the animal approached molt state D_2 and then declined in the final days before the molt. This is correlated with an increase in serum ecdysone titers during D_1 and D_2 (Tamm & Cobb, 1978). Apparently, the increased aggressiveness serves to space individuals away from an animal that is about to become vulnerable in a molt (Tamm & Cobb, 1978). Interestingly, Cobb and Tamm (1974) found that lobsters may also delay the molt if kept under crowded social conditions. This sort of mechanism may serve to prevent physical damage during the vulnerable molt state, if less crowded conditions can be found.

R. L. Caldwell (personal communication, 1980) conducted a preliminary study in which ecdysterone injected into dominant stomatopods caused reversals of rank in some cases, while no reversals took place in control individuals. He cautions that these data are the result of a small sample and require further exploration. R. L. Caldwell (personal communication, 1980) has recently followed the specific changes in behavior that occur in association with the normal cycle. As the molt approaches, the cavity in which an individual lives is modified to provide additional protection from conspecifics during the vulnerable molting period. During the molting process, the animals change tactics and use exaggerated threats rather than direct aggressive acts. These changes in behavior occur quite predictably in the molt cycle and most likely have an endocrinological basis.

THE INSECTAN ENDOCRINE SYSTEM

The insect endocrine system consists of sets of neurosecretory cells located in the brain, which are connected via axons to a neurohemal organ, the corpus cardiaca, and then to the corpora allata (reviewed in Engelmann, 1970). The corpora allata are of particular interest in the context of behavioral studies, because they secrete juvenile hormone (JH). JH is a sesquiterpenoid compound, with an ester at one end (Menn & Beroza, 1972). In this section, we focus on JH production and control, since JH is the hormone that has been most often associated with behavioral changes in adult insects. Another hormone of interest is ecdysone; the presence of ecdysone has been demonstrated in the ovaries of the mosquito (Hagedorn, Fallon, & Laufer, 1973), and this hormone may have feedback effects on other aspects of the neuroendocrine system. In the housefly, Adams, Hintz, and Pominis (1968) demonstrated that "oostatic hormone" was produced by the ovaries and had a negative feedback effect on the corpora allata.

As the name implies, JH has functions in insect growth and development. JH titer, or JH titer relative to the titer of the molting hormone, ecdysone, apparently determines the degree of morphological juvenile characteristics expressed at each molt. At the imaginal molt, JH is absent, and the resulting insect is a morphological adult. Insects, as opposed to the Crustacea discussed above, undergo a fixed number of immature molt stages (instars) and do not molt subsequent to the imaginal molt.

The principal known function of JH in adult insects is in the control of female reproductive processes. Female insects undergoing vitellogenesis (yolk formation and deposition in the eggs) have large, actively secreting corpora allata and high blood JH titers. JH has a dual role in most insect species, stimulating vitellogenin synthesis by the fat body and uptake of these yolk precursors by vitellogenic follicles (reviewed in Engelmann, 1979). The role of ecdysone in reproduction is less clear, having been implicated in the stimulation of vitellogenin synthesis in mosquitoes (Hagedorn *et al.*, 1973) and inhibiting JH secretion in cockroaches (Stay, Friedel, Tobe, & Mundall, 1980).

The mechanisms for the control of endogenous JH titers are not completely understood. What is known makes a complicated picture that contains some seemingly contradictory elements. Corpora allata activity is, at least in some species, controlled by the neurosecretory cells, and JH titer is correlated with corpora allata activity (reviewed in Engelmann, 1970). However, highly active JH-specific esterases are found in the blood, and the synthesis of such esterases provides another possible pathway for the regulation of JH titers (Sandburg, Kramer, Kezdy, & Law, 1975). Finally, JH-binding proteins are also present. These proteins protect JH from general metabolic activity and, again, may be partially responsible for blood titer regulation (Kramer, Dunn, Peterson, & Law, 1976).

Homeostatic feedback systems that control the level of JH production are known to occur in insects. For example, in oviparous cockroaches, JH is secreted in fed and mated females, thus stimulating vitellogenesis (reviewed in Engelmann, 1970). Feedback from the uterus or mature eggs during pregnancy inhibits JH production and prevents development of the next batch of oocytes. At parturition, JH is again released and the next vitellogenic cycle begins (Engelmann, 1957, 1959; Roth, 1964; Scharrer, & von Harnack, 1958). Recently Stay *et al.* (1980) showed *in vitro* that ecdysterone, secreted from mature eggs in the cockroach, *Diploptera punctata*, inhibits JH synthesis and release. That JH synthesis is also modulated continuously by existing JH levels in the blood was suggested by increase in JH output when one corpus allatum (CA) was removed and by decrease when an additional CA was implanted; these results were confirmed *in vitro* (Tobe & Stay, 1979). It would seem, therefore, that viable models of JH homeostatic mechanisms are available and that what is needed is testing of these models in a variety of insect species. Currently, there are seemingly contradictory results in experiments concerning the control of the CA by the brain. For example, transection of the nerve ending leading from neurosecretory centers in the brain to the CA leads to viable results. In the Colorado potato beetle, the CA remain active after transection (Schooneveld, Kramer, Privee, & van Huis, 1979); in the locust, *Schistocerca* sp., the same operation leads to inactivity of the CA (Strong, 1965); and in a number of other insects, it leads to an increase in CA activity (Baehr, 1973; Fraser & Pipa, 1977; Roth & Stay, 1959). However, all of these discrepancies could be explained by the theory that the CA is inhibited electrically by one population of neurons from the brain and stimulated by another group of neurosecretory neurons (Scharrer, 1958). Variable results from nerve transection would be entirely expected if the experimenter is not aware of the on/off state of electrical and neurosecretory information flowing from the brain to the CA.

The roles of the esterases and of the binding proteins in regulation of blood JH titers is even less clear. Sparks and Hammock (1979) found that peaks of juvenile hormone esterase activity occurred at those stages of larval development when blood titers of JH were lowest. Further, for at least the second peak of esterase activity, the induction of the esterase seemed to be dependent on the presence of JH; thus, JH is implicated in regulating its own metabolism, at least during this larval instar. It may be that at this critical stage of development it is important to remove all of the JH from the tissue, so that cessation of synthesis of JH by the corpora allata is not a sufficient regulatory measure.

CORRELATIONS BETWEEN ENDOCRINES AND AGGRESSION IN INSECTS—GENERAL

Aggression is known to occur in several contexts in insects. Male Orthoptera such as crickets (Alexander, 1961) and cockroaches (Bell & Sams, 1973) form territories or dominance hierarchies. Female cockroaches are also sometimes aggressive, particularly when they are carrying egg cases. Other female Orthoptera become aggressive after mating and repel the advances of males. The most thoroughly studied systems of social dominance in insects occur in the eusocial Hymenoptera (bees, wasps, and ants that live in groups with a reproductive division of labor). In primitively eusocial bees and wasps, there is little or no morphological differentiation between castes, and social dominance is often enforced by direct aggressive behavior. Individuals in these colonies are all females. Consequently, in investigations of Hymenoptera there is an interest in hormones that relate to female reproduction.

Before discussing specific cases in detail, some comments about the relationships between our knowledge of insect endocrinology and the appearance of aggression in insects are relevant. Males of many insect species are known to fight and to form either hierarchies or territories. However, our knowledge of the endocrinology of the adult male insect is limited and studies on aggression have not focused on temporal variations in aggressiveness that might be due to hormonal factors. As pointed out above, male and female insect gonads do not usually produce hormones that have an impact on sex-related behavioral characteristics, although Beach (1979) found that biting behavior in the mosquito is inhibited by ecdysone, which can be secreted by the ovaries (Hagedorn, O'Connor, Fuchs, Sage, Schlaeger, & Bohm, 1975). In this way insects differ markedly from most vertebrate models that deal with the role of hormones in determining sex-specific behavioral characteristics. It has been shown in the cockroach, *Leucophea maderae* that gonadectomy does not change the expression of female sexual behavior, although such behavior is under the influence of juvenile hormone (Barth & Lester, 1973).

The effects of gonadectomy on agonistic behavior in male insects have been investigated primarily in cockroaches. Ziegler (1972) found that removing the testes had no influence on aggressiveness or territorial behavior in *Gromphadorhina brunneri.* Gonadectomy does not affect male sexual behavior in *Lymantria* (Oudemanns, 1899), *Gryllus* (Regen, 1910), or *Leucophea* (Engelmann, 1960). Schal (unpublished manuscript) implanted additional testes into males of *Nauphoeta cinerea* and paired these animals with sham-operated controls. Of 11 pairs, the sham-operated control established dominance over the testis-implanted male in all cases; in 3 pairs, the testis-implanted male was older than the sham, and although age always confers an advantage over younger males, the older males in this case were subordinate to the shams. Thus, testes do not confer advantage in establishing dominance and may even suggest a disadvantage. When last-instar nymphs were gonadectomized and then paired with controls as adults, they were dominant over younger males and submissive to older males; that is, the operation did not affect their agonistic behavior.

Ziegler (1972) mentions that removal of the corpora allata in male *Gromphadorhina brunneri* does not affect agonistic behavior, but no information about the age of participants is given. Temporal aspects of male agonistic behavior were thoroughly investigated in a cockroach, *Nauphoeta cinerea,* by C. Schal (unpublished observations). Factors such as body size and weight were not important in establishing dominant–subordinate relationships, whereas age was the most important factor in inexperienced males. Until about seven days after adult ecdysis, males either lose encounters or elicit sexual responses, such as wing-raising or pseudocopulatory attempts from older males. After about seven days, males gain the ability to win encounters and to establish dominance over other males, and experience becomes the most important predictor of eventual rank. The possibility that hormones were involved in the process of gaining dominance potential was therefore investigated. C. Schal (unpublished observations) applied JH III (50 μg) topically to young (2–7 days) male *Nauphoeta cinerea* and then paired them with older males. Normally, older males wing-raise when approached by males less than seven days of age. Responses to JH-treated males were more aggressive than to controls, 50% of encounters being initiated by an immediate aggressive lunge. These observations suggested that perhaps juvenile hormone alters the odors or behavior of male *Nauphoeta* in some way that is recognized as "male" by older males. Precocene treatments were therefore attempted to determine whether males lacking JH would continue to be perceived by older males as "nonmales." (Precocene is known to cause corpora allata degeneration in some insects, Unnithan, Nair, & Bowers, 1977.) These experiments were inconclusive in that older males responded to five precocene-treated males with wing-raising and to another five with aggressive lunging. When JH III was applied to those males that elicited wing-raising, older males altered their responses to lunging. Thus the stimuli (presumably pheromonal) that release intermale aggression in this species may be under pheromonal control.

Because JH has been studied primarily in females, the following sections are limited to a consideration of aggression by females. Since females do not often fight over access to mates or over territories that are held in conjunction with mate selection, this factor, which is a principal cause for aggression among animals in general, is not considered in detail below.

FEMALE AGGRESSIVENESS AND SEXUAL RECEPTIVITY IN THE ORTHOPTERA

In many orthopteran insect species, the female undergoes a period or periods of sexual receptivity. If a female has several periods of sexual receptivity, these are separated by spans of time in which oocyte development occurs, the eggs are fertilized, and the eggs are either oviposited or, in some cases, tranferred to a brood pouch to allow embryonic development. Females are often aggressive toward approaching males during periods of nonreceptivity.

Most of the work on female aggressiveness in Orthoptera was done by Loher (1966) and Loher and Huber (1966). Working with a grasshopper, *Gomphocerus rufus,* they characterized the female as going through distinct stages. For the first stage, *primary defense,* females aggressively maintain interindividual distances by kicking away approaching animals; this starts at the imaginal molt. After six to eight days, the oocytes have begun to develop and the female enters a period of sexual receptivity. Within two hours of mating, the female displays *secondary defense,* which, like primary defense, involves kicking away approaching animals. Secondary defense is initiated by neural feedback from the duct of the seminal receptacle, which is stimulated by the volume of the spermatophore inserted by the male (Loher & Huber, 1966).

Primary defense is considerably more complicated in its control. Removal of the CA immediately before or after the imaginal molt causes continuous primary defense. The major impact of CA removal is to eliminate JH from the animal's system. Thus, primary defense seems to be controlled by the combination of the imaginal molt and the absence of JH. If sexually mature females are allatectomized, receptivity declines and disappears within six days of the operation. Application of a JH analog to allatectomized individuals causes onset of sexual receptivity. Similarly, Breed, Hinkle, and Bell (1975) found in a cockroach, *Blattella germanica,* that pregnant females were more aggressive than females at other stages in the reproductive cycle. Unfortunately, no data are available concerning the endocrinology of the reproductive cycle in this species. Müller (1965) found in another grasshopper that allatectomy did not inhibit the onset of sexual receptivity. Thus, there may be considerable variation among species in the mechanism of control.

HORMONAL CONTROL OF DOMINANCE IN SOCIAL HYMENOPTERA

In some social Hymenoptera, the reproductive division of labor between the queen and the workers in a colony is based on behavioral dominance of the workers by the queen. Since reproduction is controlled by JH, several workers have hypothesized a controlling role for JH in dominance behavior. Correlational evidence supports this hypothesis, although nothing is known about the causal factors involved in this relationship. In this section, we review the data showing a relationship between reproductive status and dominance behavior and consider the work done showing the relationship between JH and dominance behavior.

Social Hymenoptera may be divided into two general categories. Primitively eusocial species are those in which the morphological distinction between castes is weak or absent; it is in the primitively eusocial species that reproductive division of labor is most often based on dominance hierarchies. In the highly eusocial species, there is a marked morphological differentiation between the reproductive and worker castes; control of reproduction may be accomplished by pheromones or by nutritional deprivation of workers during larval development (Engels, 1974). The most important point is that in the primitively eusocial species, all individuals are capable of full reproduction, while in the highly social species not all individuals have that capability.

The best-studied primitively eusocial species are in three groups. Wasps of the genus *Polistes,* bees of the genus *Lasioglossum,* and bees of the genus *Bombus* all have dominance hierarchies among the individuals in a colony, and the hierarchy is correlated with reproductive activity. The honeybee, *Apis mellifera,* is highly eusocial, and, under ordinary conditions, caste and reproductive behavior are controlled by pheromones and nutrition. Under certain circumstances, however, reproductive dominance is established through aggressive behavior in the honeybee. The known and hypothetical effects of JH on the expression of behavior dominance in each of these species is discussed below.

POLISTES WASPS

Pardi (1948) first demonstrated the presence of a behavioral dominance hierarchy in *Polistes gallicus.* Since then, several workers (Bohm, 1972; Gamboa, Heacock, & Wiltjer, 1978; West, 1967) have studied aspects of caste and behavioral dominance in a number of *Polistes* species; there is a great deal of consistency within the genus in the nature of the dominance hierarchy and in the behavioral acts used to establish dominance. Pardi (1948) first noted that the aggressive behavior in the dominance hierarchy was correlated with ovarian size. This correlation led Barth, Sroka, Kessler, and Hearn (1975) to hypothesize that JH is involved in the control of dominance behavior in *Polistes*. To test this, they maintained *Polistes* colonies in their lab and scored each individual in each colony for dominance. They then treated subdominant individuals with 10 μg JH, applied topically in an oil vehicle. They found that the treated individuals scored significantly higher than untreated individuals and sham-treated controls when observations of dominance were made subsequent to the treatment. Further, the treated individuals developed larger ovaries than the control groups (Bohm, 1972, also found that JH stimulated ovarian development). These results strongly implicate a function for JH at some point in the control mechanism of social dominance in *Polistes*. Two criticisms can be leveled concerning the way in which Barth *et al.* (1975) conducted this experiment, although neither negates the result obtained. First, the nutritional requirements of the individuals, particularly for protein, may not have been appropriately met (R. H. Barth, personal communication). Second, topical applications of hormone may have allowed some to rub off onto other colony members; the general rise in aggression observed subsequent to treatment may have been a result of disruption of the hierarchy caused in part by hormone transferred in this way.

Recently, Röseler, Röseler, and Strambi (1980) took a different approach to the same problem. CA activity in individual wasps was judged in two ways: The volume of the CA was measured, and the rate of JH synthesis was estimated by measuring the rate of uptake of a JH precursor, methionine, radiochemically. They found that synthetic activity of CA of dominant wasps was significantly higher than that of subordinate wasps, and they concluded that changes in endocrine activity are a primary result of the behavior that maintains the social hierarchy.

Lasioglossum Bees

These bees are ground-nesting members of the family Halictidae. Colonies are founded by single queens in the spring, and the daughters of the founding queen become the workers. In midsummer, colonies may attain sizes of up to 25 individuals, although a typical colony is smaller, containing 6–10 individuals. Within a colony, a gradation in ovarian size exists, from the queen, with very large ovaries containing oocytes that can be laid, to workers with no detectable vitellogenesis (Michener & Wille, 1961). In the most well-studied species in the genus, *Lasioglossum zephyrum,* a complex series of dominance behavioral acts enforces this hierarchy of ovarian development. The queen taps (nudges) the workers with her head and backs away from workers, inducing them to follow. Breed and Gamboa (1977) found that backing queens in some cases actually led workers to work-sites in the nest. These ritualized forms of dominance seem to take the place of more overt aggression in this species. However, Breed (1977) found that if a queen was removed from a colony, a worker would begin to behave in a queen-like manner within hours. If the old queen was then returned to the colony, in some cases direct aggression, including biting and stinging, occurred between the old and replacement queens. Although experiments by Bell (unpublished) attempting to manipulate the behavior of workers using JH treatments have failed in this species, Bell (1974) found that exogenously applied JH stimulated yolk deposition in workers. It is particularly interesting in this species that the behavioral change observed in the workers after queen removal occurs in a span of time that is too short for detectable changes in ovarian state but that is long enough to allow endocrine changes.

Bumblebees

Bumblebee hierarchies have not been studied in as quantitative a manner as have hierarchies in *Polistes* and *Lasioglossum.* Free (1955) found that in a colony of bumblebees, the queen is behaviorally dominant and two or three other bees may be aggressive, in a linear hierarchy, with the rest of the workers showing no aggression. Aggressiveness is, again, related to ovarian development. Röseler (1977) found that ovarian development could be stimulated in workers by treatment with JH, but no extensive work has been done on the relationship between such treatments and behavioral changes in the bumblebee.

The Honeybee

The situation in the honeybee is considerably more complicated. In European races of *Apis mellifera,* the bee most commonly kept for honey production, the queen controls workers reproduction by secreting a pheromone, 9-oxo-2-decenoic acid that inhibits worker ovarian development. Normally, workers have a period of increase of CA activity through the first few days of life and then steady CA activity through the remainder of their lives (Gast, 1967; Imboden, Wille, Gerig, & Luscher, 1977). Some workers show signs of vitellogenic activity but usually do not develop eggs large enough to lay. Associated with this changed hormonal state is a change in the workers' behavior patterns; early in life they are nurse bees, feeding the brood, and later in life they become field bees, foraging for nectar and pollen.

If for some reason a colony loses its queen and no new queen can be reared, a number of interesting things happen. Within 15 days, some workers develop ovaries large enough to lay eggs (Sakagami, 1954; Velthuis, 1970; Velthuis, Verheijen, & Gottenbos, 1965). Other workers become aggressive toward the laying workers, attacking them by mauling, biting, and sometimes stinging. These aggressive workers have intermediate levels of vitellogenesis between the laying workers and normal workers; this is thought to correspond to intermediate levels of CA activity and blood JH titer (Velthuis, 1976). Treatment of newly emerged workers with JH results in inhibition of CA development and a lower probability that the worker will be aggressive (Breed, 1982). Lecomte (1961) presents data that suggest a possible relationship between worker ovarian development and aggressiveness of worker bees defending the hive, although his results are not statistically conclusive.

Hormones and Social Dominance in Hymenoptera: An Overview

The association among JH, vitellogenesis, and aggression or dominance behavior in the eusocial Hymenoptera seems to be ubiquitous. In all species studied, this relationship exists, at least under special circumstances. Unfortunately, little is known concerning the causal basis for this correlation. Surgical removal of the CA has been accomplished in the honeybee (Rutz, Gerig, Wille, & Luscher, 1976), and Röseler *et al.* (1980) ovariectomized worker bumblebees. Such surgery usually allows continued physiological experimentation on the individual but may have a serious effect on postsurgical expression of social behavior. Neither of the authors mentioned discusses the behavioral effects of surgery. Nonetheless, this remains a promising avenue for attempting to discover whether the effect of JH on behavior is a direct one or is mediated by other hormones, afferent feedback (i.e., from the ovaries), or neurosecretory products that remain within the brain.

CONCLUSIONS

Arthropods provide a number of interesting areas for further work in the hormonal control of behavior. Detailed investigation of the relationship between behavior and the crustacean molt cycle will probably yield much information concerning endocrine control of cyclic behavioral phenomena. Similarly, the role of juvenile hormone in regulation of dominance behavior among individuals in colonies of social Hymenoptera opens up several promising avenues of research.

Our lack of knowledge concerning blood hormone titers in invertebrates, relative to vertebrates, is the major factor that must be overcome in order to advance our knowledge of hormonal correlates of invertebrate behavior. The key to increasing our understanding of these systems at this point is the precise quantification of behavioral correlates of endocrine changes, and concomitantly, the precise establishment of relationships between blood hormonal levels and behavioral expression. Most insects have blood volumes that are small relative to the amount of blood required to perform bioassays for juvenile hormone, but in the future, radioimmunoassays may be available that would allow for determinations of JH titers in individual animals. Rankin and Riddiford (1978) have provided excellent information relating juvenile hormone titers in *Oncopeltus fasciatus* (milkweed bug) to migratory behavior, although their data relied on blood samples pooled from 25 animals. Development of techniques for assays of individual animals will open the way for major increases in our understanding of these systems.

REFERENCES

Adams, T. S., Hintz, A. M., & Pomonis, J. G. Oostatic hormone production in houseflies, *Musca domestica,* with developing ovaries.. *Journal of Insect Physiology,* 1968, *14,* 983–993.

Alexander, R. D. Aggressiveness, territoriality, and sexual behaviour in field crickets. *Behaviour,* 1961, *17,* 130–223.

Baehr, J. C. Controle neuroendocrine du functionement du corpus allatum chez *Rhodnius prolixus. Journal of Insect Physiology,* 1973, *19,* 1041–1055.

Barth, R. H. Hormonal control of sex attractant production in the Cuban cockroach. *Science,* 1961, *133,* 1598–1599.

Barth, R. H. The endocrine control of mating behavior in the cockroach *Byrsotria fumigata* (Guerin). *General and Comparative Endocrinology,* 1962, *2,* 53–69.

Barth, R. H. Insect mating behavior: Endocrine control of a chemical communication system. *Science,* 1965, *149,* 882–883.

Barth, R. H. The comparative physiology of reproductive processes in cockroaches: I. Mating behavior and its endocrine control. *Advances in Reproductive Physiology,* 1968, *3,* 167–207.

Barth, R. H., & Lester, L. J. Neurohormonal control of sexual behavior in insects. *Annual Review of Entomology,* 1973, *18,* 445–472.

Barth, R. H., Lester, L. H., Sroka, P., Kessler, T., & Hearn, R. Juvenile hormone promotes dominance behavior and ovarian development in social wasps *(Polistes annularis). Experientia,* 1975, *31,* 691–692.

Beach, R. Mosquitoes: Biting behavior inhibited by ecdysone. *Science,* 1979, *205,* 829–831.

Bell, W. J. Factors controlling the initiation of vitellogenesis in a primitively social bee, *Lasioglossum zephyrum. Insectes Sociaux,* 1974, *20,* 253–260.

Bell, W. J., & Sams, G. R. Aggressiveness in the cockroach *Periplaneta americana. Behavioral Biology,* 1973, *9,* 581–593.

Bohm, M. K. Effects of environment and Juvenile Hormone on ovaries of the wasp, *Polistes metricus. Journal of Insect Physiology,* 1972, *18,* 1875–1883.

Breed, M. D. Interaction among individuals and queen replacement in a primitively eusocial bee. *Proceedings 8th International Congress Union for the Study of Social Insects,* 1977, 228–231.

Breed, M. D. Juvenile hormone and aggression in the honey bee. In M. D. Breed, C. D. Michener, & H. E. Evans (Eds.), *The biology of social insects.* Boulder, Colo.: Westview Press, 1982.

Breed, M. D., & Gamboa, G. J. Behavioral control of workers by queens in primitively eusocial bees. *Science,* 1977, *195,* 694–696.

Breed, M. D., Hinkle, C. M., & Bell, W. J. Agonistic behavior in the German cockroach, *Blattella germanica. Zeitschrift für Tierpsychologie,* 1975, *29,* 24–32.

Cobb, J. S., & Tamm, G. R. Social conditions increase intermolt period in juvenile lobsters, *Homarus americanus. Journal of the Fisheries Research Board, Canada,* 1974, *32,* 1941–1943.

Cobb, J. S., & Tamm, G. R. Dominance status and molt order in lobsters *(Homarus americanus). Marine Behavior and Physiology,* 1975, *3,* 119–124.

Engelmann, F. Die Steuerung der Ovarfunktion bei der ovoviviparen Schabe *Leuchophaea maderae* (Fabr.). *Journal of Insect Physiology,* 1957, *1,* 257–278.

Engelmann, F. The control of reproduction in *Diploptera punctata* (Blattaria). *Biological Bulletin (Woods Hole),* 1959, *116,* 406–419.

Engelmann, F. Hormonal control of mating behavior in an insect. *Experientia,* 1960, *41,* 69–70.

Engelmann, F. *The physiology of insect reproduction.* Oxford: Pergamon Press, 1970.

Engelmann, F. Insect vitellogenin: Identification, biosynthesis and role in vitellogenesis. *Advances in Insect Physiology,* 1979, *14,* 49–108.

Engels, W. Occurrence and significance of vitellogenins in female castes of social Hymenoptera. *American Zoologist,* 1974, *14,* 1229–1237.

Fraser, J., & Pipa, R. Corpus allatum regulation during the metamorphosis of *Periplaneta americana:* Axon pathways. *Journal of Insect Physiology,* 1977, *23,* 976–984.

Free, J. B. The behavior of egg-laying workers of bumblebee colonies. *British Journal of Animal Behaviour,* 1955, *3,* 147–153.

Gamboa, G. J., Heacock, B. D., & Wiltjer, S. J. Division of labor and subordinate longevity in foundress associations of the paper wasp, *Polistes metricus. Journal of the Kansas Entomological Society,* 1978, *51,* 343–352.

Gast, R. Untersuchungen uber den einfluss der Koniginnensubstanz auf die entwicklung der Endokrinen Drusen bei der Arbeiterin der Honigbiene *(Apis mellifera). Insectes Sociaux,* 1967, *14,* 1–12.

Hagedorn, H. H., Fallon, A. M., & Laufer, H. Vitellogenin synthesis by the fat body of the mosquito *Aedes aegypti:* Evidence for transcriptional control. *Developmental Biology,* 1973, *31,* 285–294.

Hagedorn, H. H., O'Connor, J. D., Fuchs, M. S., Sage, B., Schlaeger, D. A., & Bohm, M. K. The ovary as a source of α-ecdysone in an adult mosquito. *Proceedings of the National Academy of Science USA,* 1975, *72,* 3255–3259.

Imboden, H., Wille, H., Gerig, L., & Luscher, M. Die Vitellogeninsynthese bei der Bienen-Arbeiterin *(Apis mellifera)* und ihre Abhangigkeit von Juvenilhormon (JH). *Revues Suisse de Zoologie,* 1977, *83,* 928–933.

Kramer, K. J., Dunn, P. E., Peterson, R. C., & Law, L. H. Interaction of juvenile hormone with binding proteins in insect hemolymph. In L. G. Gilbert (Ed.), *The juvenile hormones.* New York: Plenum Press, 1976.

Lecomte, J. Le comportement agressif des ouvrieres d'*Apis mellifica.* Doctoral thesis, Université de Paris, 1961.

Loher, W. Die Kontrolle des Weibchengesanges von *Gomphocerus rufus* L. (Acridinae) durch die Corpora allata. *Naturwissenschaften,* 1962, *49,* 406.

Loher, W. Die steurung sexueller verhalten-sweisen und der oocytentwicklung bei *Gomphocerus rufus* L. *Zeitschrift für vergleichende Physiologie,* 1966, *53,* 277–316.

Loher, W., & Huber, F. Experimentellen Untersuchungen am Sexualverhalten des Weibchens der Heuschrecke *Gomphocerus rufus L. Journal of Insect Physiology,* 1964, *10,* 13–36.

Loher, W., & Huber, F. Nervous and endocrine control of sexual behavior in a grasshopper (*Gromphocerus rufus* L., Acrididae). *Symposia of the Society for Experimental Biology,* 1966, *20,* 381–400.

Menn, J. J., & Beroza, M. (Eds.). *Insect juvenile hormones.* New York: Academic Press, 1972.

Michener, C. D., & Wille, A. The bionomics of a primitively social bee, *Lasioglossum inconspicuum. University of Kansas Science Bulletin,* 1961, *42,* 1123–1202.

Müller, H. P. Zur der steuerung des Paarangsuerhaltens und dur Eirefung bei der Feldheuschreche *Euthystiva brachyptera* Osck. unter besonderer verüchsichfiyung der Rolle der Corpora allata. *Zeitschrift fur Vergleichende Physiologie,* 1965, *50,* 447–497.

Oudemanns, J. T. Falter aus castrirten Raupen, wie sie aussehen und wie sie sich benehmen. *Zoologie Jahrbuch,* (Syst.), 1899, *12,* 71–88.

Pardi, L. Dominance order in *Polistes* wasps. *Physiological Zoology,* 1948, *21,* 1–13.

Rankin, M. A., & Riddiford, L. M. Significance of haemolymph Juvenile Hormone titer changes in timing of migration and reproduction in adult *Oncopeltus fasciatus. Journal of Insect Physiology,* 1978, *24,* 31–38.

Regen, J. Kastration und ihre Folgeerscheinungen bei *Gryllus campestris* L. *Zoologie Anzheit,* 1910, *35,* 427–432.

Röseler, P.-F. Juvenile hormone control of oogenesis in bumblebee workers, *Bombus terrestris. Journal of Insect Physiology,* 1977, *23,* 985–992.

Röseler, P.-F., Röseler, I., & Strambi, A. The activity of corpora allata in dominant and subordinated females of the wasp *Polistes gallicas. Insectes Sociaux,* 1980, *27,* 97–108.

Roth, L. M. Control of reproduction in female cockroaches with special reference to *Nauphoeta cinerea:* II. Gestation and postparturition. *Psyche,* 1964, *71,* 198–244.

Roth, L. M., & Stay, B. Control of reproduction in cockroaches. *Science,* 1959, *130,* 271–272.

Rutz, W., Gerig, L., Wille, H., & Luscher, M. The function of juvenile hormone in adult worker honeybee *Apis mellifera. Journal of Insect Physiology,* 1976, *22,* 1485–1491.

Sakagami, S. F. Occurrence of an aggressive behavior in queenless hives, with considerations on the social organization of honeybees. *Insectes Sociaux,* 1954, *1,* 331–343.

Sandberg, L. L., Kramer, K. J., Kezdy, J. F., & Law, J. H. Juvenile hormone—specific esterases in the haemolymph of the tobacco hornworm *Manduca sexta. Journal of Insect Physiology,* 1975, *21,* 873–887.

Scharrer, B. Neuroendocrine mechanisms in insects. *Second International Symposium on Neurosecretion.* Berlin: Springer-Verlag, 1958.

Scharrer, B., & von Harnack, M. Histophysiological studies on the corpus allatum of *Leucophaea maderae:* I. Normal life cycle in male and female adults. *Biological Bulletin (Woods Hole),* 1958, *115,* 508–520.

Schooneveld, A., Kramer, S. J., Privee, H., & van Huis, H. Evidence of controlled corpus allatum activity in the adult Colorado potato beetle. *Journal of Insect Physiology,* 1979, *25,* 449–453.

Sparks, T. C., & Hammock, B. D. Induction and regulation of juvenile hormone esterases during the last larval instar of the cabbage looper, *Trichoplusia ni. Journal of Insect Physiology,* 1979, *25,* 551–560.

Stay, B., Friedel, T., Tobe, S. S., & Mundall, E. C. Feedback control of juvenile hormone synthesis in cockroaches: Possible role for ecdysterone. *Science,* 1980, *207,* 898–900.

Strong, L. The relationships between the brain, corpora allata, and oocyte growth in the Central American locust, *Schistocerca sp.:* II. The innervation of the corpora allata, the internal neurosecretory complex and oocyte growth. *Journal of Insect Physiology,* 1965, *11,* 271–280.

Tamm, G. R., & Cobb, J. S. Behavior and the crustacean molt cycle: Changes in aggression of *Homarus americanus. Science,* 1978, *200,* 79–81.

Tobe, S. S., & Stay, B. Modulation of juvenile hormone synthesis by an analogue in the cockroach. *Nature,* 1979, *281,* 481–482.

Tombes, A. S. *An introduction to invertebrate endocrinology.* New York: Academic Press, 1970.

Truman, J. W., & Riddiford, L. M. Neuroendocrine control of ecdysis in silkmoths. *Science,* 1970, *167,* 1624–1626.

Truman, J. W., & Riddiford, L. M. Hormonal mechanisms underlying insect behavior. *Advances in Insect Physiology,* 1974, *10,* 297–352.

Truman, J. W., & Riddiford, L. M. Invertebrate systems for the study of hormonal effects on behavior. *Vitamins and Hormones,* 1977, *35,* 283–315.

Truman, J. W., & Sokolove, P. G. Silkmoth eclosion: Hormonal triggering of a centrally programmed pattern of behavior. *Science,* 1972, *175,* 1491–1493.

Unnithan, G. C., Nair, K. K., & Bowers, W. S. Precocene-induced degeneration of the corpus allatum of adult females of the bug, *Oncopeltus fasciatus. Journal of Insect Physiology,* 1977, *23,* 1081–1094.

Velthuis, H. H. W. Ovarian development in *Apis mellifera* worker bees. *Entomology, Experimental and Applied,* 1970, *13,* 377–394.

Velthuis, H. H. W. Egglaying, aggression, and dominance in bees. *Proceedings of the XV International Congress of Entomology,* 1976, 436–449.

Velthuis, H. H. W., Verheijen, F. J., & Gottenbos, A. J. Laying worker bee: Similarities to the queen. *Nature (London),* 1965, *207,* 1314.

West, M. J. Foundress associations in polistine wasps: Dominance hierarchies and the evolution of social behavior. *Science,* 1967, *157,* 1584–1585.

Ziegler, R. Sexual and territorialverhalten der Schabe *Gromphadorhina brunneri* (Butler). *Zeitschrift für Tierpsychologie,* 1972, *31,* 531–541.

Index